Cell–Cell Junctions

A subject collection from *Cold Spring Harbor Perspectives in Biology*

ALSO FROM COLD SPRING HARBOR LABORATORY PRESS

OTHER SUBJECT COLLECTIONS FROM COLD SPRING HARBOR PERSPECTIVES IN BIOLOGY

Generation and Interpretation of Morphogen Gradients
Symmetry Breaking in Biology

RELATED LABORATORY MANUALS

Basic Methods in Microscopy: Protocols and Concepts from Cells: A Laboratory Manual
Live Cell Imaging: A Laboratory Manual, Second Edition

OTHER RELATED TITLES

Gastrulation: From Cells to Embryo
The TGF-β Family

Cell–Cell Junctions

A subject collection from *Cold Spring Harbor Perspectives in Biology*

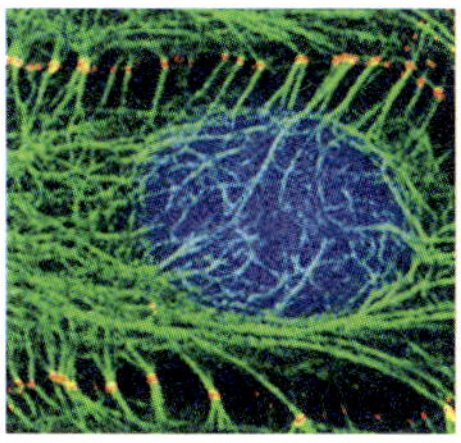

EDITED BY

W. James Nelson
Stanford University

Elaine Fuchs
The Rockefeller University

COLD SPRING HARBOR LABORATORY PRESS
Cold Spring Harbor, New York • www.cshlpress.com

Cell–Cell Junctions

A Subject Collection from *Cold Spring Harbor Perspectives in Biology*
Articles online at www.cshperspectives.org

Publisher	John Inglis
Acquisition Editor	Richard Sever
Director of Development, Marketing, & Sales	Jan Argentine
Project Coordinator	Inez Sialiano
Permissions Coordinator	Carol Brown
Production Editor	Kaaren Kockenmeister
Production Manager	Denise Weiss
Book Marketing Manager	Ingrid Benirschke
Sales Account Managers	Jane Carter and Elizabeth Powers
Cover Designer	Denise Weiss

Front cover artwork: Triple-color fluorescence micrograph showing a monolayer cell culture of human keratinocytes that are connected to each other by cell-to-cell bridges containing desmosomes. These desmosomes, visualized by antibodies to the protein desmoplakin (red), provide anchorage to the cytoplasmic bundles of intermediate filaments of the keratin type (green). The chromatin of the nuclei of these cells is stained with a specific blue dye reagent. For details, see Figure 2 of the article by W. Franke in this volume.

Library of Congress Cataloging-in-Publication Data

Cell–Cell Junctions/edited by W. James Nelson, Elaine Fuchs.
p. cm.
Includes bibliographical references and index.
ISBN 978-0-87969-884-3 (hardcover: alk. paper)
1. Cell junctions. 2. Cell membranes. 3. Cell. I. Nelson, W. J. (W. James).
II. Fuchs, Elaine. III. Title.

QH603.C4C455 2010
571.6–dc22

2009033079

10 9 8 7 6 5 4 3 2 1

All World Wide Web addresses are accurate to the best of our knowledge at the time of printing.

All Cold Spring Harbor Laboratory Press publications may be ordered directly from Cold Spring Harbor Laboratory Press, 500 Sunnyside Blvd., Woodbury, New York 11797–2924. Phone: 1-800-843-4388 in Continental U.S. and Canada. All other locations: (516) 422-4100. FAX: (516) 422-4097. E-mail: cshpress@cshl.edu. For a complete catalog of all Cold Spring Harbor Laboratory Press publications, visit our website at http://www.cshlpress.com/.

Contents

Preface

CELL–CELL JUNCTIONS EMERGED OVER 600 MILLION YEARS AGO, when metazoan life began and concomitant with the formation of a new basic structure—a multicellular layer surrounding an inner space that separated the inside of the organism from the outside. During subsequent evolution, this basic organization has been conserved in different types of organs and tissues of what are now modern day mammals, including humans. However, with increasingly more elaborate tissue structures, demands on intercellular organization intensified, presenting a number of structural and functional challenges that led to the increased diversity of cell–cell junctions that are highlighted in the ensuing collection of papers.

One hurdle was how to maintain the structural and functional integrity of the epithelium while simultaneously allowing for the dynamic flux in cells responsible for constantly rejuvenating the tissue through homeostasis. This problem was solved by the formation of specialized cell–cell junctions (desmosomes, adherens junctions) and intercellular communication (gap junctions) throughout the tissue. These junctions not only glue cells together, but also do it in such a way that permits junctional remodeling in response to various intrinsic as well as environmental cues. This volume starts with an essay on how each of these cell–cell junctions was discovered (Section I). This is followed by descriptions of the structural and molecular organization of these junctions, the proteins involved in extracellular contacts, and the intracellular links that integrate cell–cell adhesion to the cytoskeleton and signaling pathways (Sections II and III).

As multicellular organisms evolved, cells within complex tissue structures were confronted with the need to migrate, change shape, and reorganize during normal morphogenesis and homeostasis, as well as in response to injury and wound repair. Section IV focuses on the dynamics of cell–cell junctions and how their organization and functions are modified to control changes in cell–cell interactions.

The ability of cells to adhere to one another also endowed tissues with an ability to control the surrounding ionic environment and also provide a barrier to regulate the flow of proteins and other materials through or around the multicellular layer. Such properties were achieved in part by formation of tight junctions, and several contributions in Sections II and III provide the latest insights into the structure and functions of these specialized molecular barriers. Another opportunity was the ability to coordinate signals from the outside to regulate cell growth, differentiation, and organization. Section V includes several chapters on a variety of extracellular signals that impinge on different junctional complexes, and the cytoplasmic signaling pathways that emanate from the junctional complexes. Of particular interest are how these pathways regulate local events such as cytoskeleton dynamics, and also how they communicate with the nucleus to regulate gene transcription.

While the epithelium has been used as a paradigm for many of the contributions here that delineate the characteristic junction organization, others highlight how intercellular junctions have diversified to suit the particular needs of other tissues. In this regard, Section VI focuses on several specialized intercellular junctions, or synapses, that form between neurons and also between immune cells. These synaptic junctions contain non-junctional proteins that regulate functional outputs from cell–cell contacts. The role of cell–cell junctions in organizing non-junctional proteins is also reiterated by other cytoskeletal proteins that polarize cells in the apical-basal and anterior-posterior (planar cell polarity) orientation.

The importance of intercellular junctions in the structural and functional organization of multicellular tissues is also underscored by the finding that they are the target of pathogens seeking to gain entry into individual cells and tissues (Section VII), and of genetic defects that lead to significant tissue and organ defects in metabolic diseases and cancers (Section VII).

While not exhaustive in the full scope of functions of cell–cell junctions, these Chapters provide examples of the remarkable structural and functional diversity of cell–cell junctions, and glimpses into future directions for research in this extraordinarily broad and exciting field.

We are grateful to the authors who took time to contribute to this collection and, at CSHL Press, to Dr. Richard Sever who initiated the idea of this book on Cell–Cell Junctions, to Ms. Inez Sialiano who helped tirelessly in the management of the manuscripts, and to Denise Weiss and Kaaren Kockenmeister in Production.

Elaine Fuchs
The Rockefeller University

W. James Nelson
Stanford Universtiy

Discovering the Molecular Components of Intercellular Junctions—A Historical View

Werner W. Franke

Helmholtz Group for Cell Biology, German Cancer Research Center, Im Neuenheimer Feld 280, D-69120 Heidelberg, Germany

Correspondence: w.franke@dkfz.de

The organization of metazoa is based on the formation of tissues and on tissue-typical functions and these in turn are based on cell–cell connecting structures. In vertebrates, four major forms of cell junctions have been classified and the molecular composition of which has been elucidated in the past three decades: Desmosomes, which connect epithelial and some other cell types, and the almost ubiquitous adherens junctions are based on closely *cis*-packed glycoproteins, cadherins, which are associated head-to-head with those of the hemi-junction domain of an adjacent cell, whereas their cytoplasmic regions assemble sizable plaques of special proteins anchoring cytoskeletal filaments. In contrast, the tight junctions (TJs) and gap junctions (GJs) are formed by tetraspan proteins (claudins and occludins, or connexins) arranged head-to-head as TJ seal bands or as paracrystalline connexin channels, allowing intercellular exchange of small molecules. The by and large parallel discoveries of the junction protein families are reported.

In the year of the bicenturial jubilee of Charles Darwin (born 1809) and his 1859 publication of the concept of natural selection as the decisive driving force of evolution, it is perhaps appropriate to begin this review with the notion that the four major kinds of cell–cell junctions are among the oldest and most important structures contributing to the formation and functional diversification of multilayered metazoan organisms. From mere associations of individual cells, whether protozoan or parazoan, it was the cooperation of the molecular ensembles of these junctions to provide the basis for eumetazoan life. Note, however, that some protocadherin glycoproteins and *armadillo*-type proteins already occur in certain nonmetazoa (e.g., King et al. 2003; Nichols et al. 2006; for refs. Halbleib and Nelson 2006). In particular, diverse cell–cell junction molecules were—and are—needed to assemble and organize metazoan architecture, notably that of the Bilateria, to allow the formation of the epithelial layers of ectoderm and endoderm, mesoderm-derived tissues, and the segregation of diverse kinds of interstitial cells and the organs derived therefrom. So, it is not so surprising that major kinds of cell junctional structures already exist in the lowest divisions of eumetazoa (e.g., Hobmayer et al. 1996, 2000).

 doi: 10.1101/cshperspect.a003061
Cite this article as *Cold Spring Harb Perspect Biol* 2009;1:a003061

In general, metazoan animals possess three intercellular junction systems of the adhering type formed by characteristic transmembrane molecules and proteins that assemble into specific submembranous plaques (Table 1). A fundamentally different type, gap junctions, serves primarily as a system of assemblies of intercellular channels but also contributes to cell–cell adhesion. In summarizing the history of the discoveries of the molecular ensembles forming these junctions, it is striking that remarkably different arts of science and tribes of cell biologists have contributed to the progress in this field. The intercellular junctional systems are:

(1) *Desmosomes* (*maculae adherentes*) are by far the most abundant junctions in stratified epithelia and have been studied with special impetus by researchers analyzing the cytoskeleton and tissue architecture, and by dermatologists.

(2) *Adherens junctions,* including the *zonulae* and *fasciae adherentes*, have first attracted

Table 1. Constitutive molecular components of the major types of symmetrical (homotypic) junctions

	Occurrence	Associated filaments	Transmembrane proteins and glycoproteins	Specific plaque proteins
Desmosomes *Maculae adherentes*	Epithelial cells, various types of cardiomyocytes, meningothelial cells, dendritic reticulum cells of the thymus and lymph follicles	Intermediate-sized filaments (keratins, vimentin, desmin)	Desmogleins 1–4[a] desmocollins 1–3[a]	*Plakoglobin*[b] desmoplakin I/II *plakophilins 1–3*[a]
Adherens junctions *Zonulae adherentes* *Fasciae adherentes* *Puncta adhaerentia*	Epithelial cells, endothelial cells, various types of cardiomyocytes, mesenchymal and neural cells	Microfilaments (actin)	Typical cadherins[a] (e.g., E-cadherin, N-cadherin, P-cadherin, VE-cadherin, cadherin-11)	α- and *β-Catenin, plakoglobin, protein p120, protein ARVCF, protein p0071, neurojungin (δ-catenin)*[a], (*plakophilin-2*[c]) proteins ZO-1, ZO-2, ZO-3
Tight junctions *Zonulae occludentes* *Fasciae occludentes* *Puncta adhaerentia*	Epithelial cells, endothelial cells	–[d]	Occludin, claudins 1–24[a] tricellulin(s)[e] proteins of the JAMA-group, CAR[a], ESAM[a]	Proteins ZO-1, ZO-2, ZO-3, cingulin
Gap junctions (nexus)	All kinds of tissue-forming cells	–	Connexins 1–21[a]	Proteins ZO-1, ZO-2, ZO-3

Only established, i.e., repeatedly confirmed, constituent structural molecules are mentioned here; some further regulatory or peripherally associated molecules are not listed here but are discussed in the text.

[a]One or combinations of a few representatives, with cell type and cell layer specificities.

[b]Proteins of the so-called *armadillo* family are in italics.

[c]Only in specific proliferatively active cells (Rickelt et al. 2009).

[d]Actin microfilaments are seen near some tight junctions but their specific association is not clear.

[e]There are at least two mRNA splice products but only one protein has so far been localized.

the special interest of developmental biologists because of their importance in mammalian embryogenesis.

(3) *Tight junctions* (*zonulae occludentes*), and in particular the transmembrane molecules involved, had long been sought as this structure was important in controlling the paracellular transport of molecules and particles. Finally, the centrally important tetraspan proteins, desired by a generation of physiologists and membranologists, have been found as the result of careful cell particle fractionation work and immunoelectron microscopy.

(4) *Gap junctions* (*nexus*) had always fascinated electron microscopists and crystallographers because of their esthetic paracrystalline substructural order, as well as physiologists studying lateral direct molecule exchange from cell to cell. Here, the stability of the structure itself helped in its isolation and reconstitution in vitro.

Thus, at about the same time in the late 1970s, when the ultrastructural organization and specificities of the diverse kinds of these junctions had been well determined (for reviews, see Farquhar and Palade 1963; Staehelin 1974), the race by cell biologists to elucidate the molecular compositions of these junctions began. Although this analytical research period is not quite over and a few new junction diamonds may still just lie around the corner, the prime interest of the community of cell biological researchers has already moved on to the next research arena, studying the mechanisms of junction formation and the functions of the junctions. Over the last two decades, these major structural elements of our bodies have also become objects of intense medical research, already with some startling results.

In the following, the by and large parallel searches for constituent molecules of the four major categories of cell–cell junctions in vertebrate cells is described. However, only constituents localized and confirmed by different groups and with various methods are mentioned as generally accepted components. This, of course, does not exclude the presence of others for which the available evidence does not yet seem sufficient. Furthermore, certain special types of junctions that do not fit one of the four major categories will be reviewed elsewhere (Franke et al. 2009).

HARD-CORE ANCHORS OF CYTOSKELETAL ELEMENTS: THE DESMOSOMES

The desmosomes represent a category of intercellular junctions that typically reveal a highly distinctive subarchitecture. They are abundant in epithelial and meningothelial tissues and cell cultures, as well as in certain types and stages of cardiomyocyte differentiation, but are also known to connect reticulum cells of the thymus and lymphatic follicles. Because of their relatively good preservation after diverse electron microscopic fixation procedures and their abundance in certain tissues (e.g., in some regions of the *stratum spinosum* of the epidermis and other stratified squamous epithelia, they occupy 50% or even more of the total cell surface), the ultrastructural principles of desmosome architecture were more or less already clear in the 1970s (see, e.g., Farquhar and Palade 1963; Campbell and Campbell 1971; Staehelin, 1974; for cardiomyocytes see Fawcett and McNutt, 1969).

The appearance of desmosomes in electron micrographs of ultrathin sections impresses observers by the parallel alignment of two trilaminar plasma membrane domains, an organized 20–30-nm interface (the "*desmogloea*"), which often reveals a dense midline structure (the *glutinamentum*), and periodical cross-striations between this midline and the plasma membranes (Fig. 1A,B). On their cytoplasmic side, the membranes of the desmosomal halves are coated by an electron-dense, 15–25 nm thick plaque, which, in turn, is coated by somewhat less dense material onto which in many, but not in all cells, bundles of intermediate-sized filaments (IFs) insert. Freeze-fractures further reveal the membrane interior of desmosomes as relatively loosely arranged arrays of clustered transmembrane complexes (Fig. 1C) (Leloup et al. 1979).

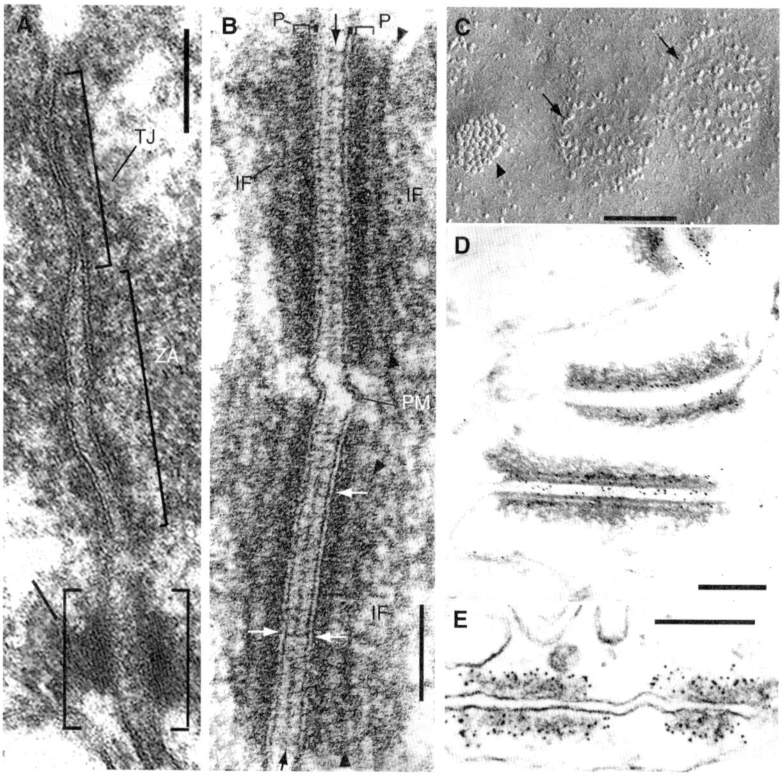

Figure 1. Electron microscopic aspects of intercellular junctions, in particular desmosomes. (*A*) Characteristic subapical trias of tight junction (TJ), *zonula adhaerens* (ZA), and desmosome (D) of murine intestinal epithelial cells in a longitudinal section. (*B*) Higher magnification of two adjacent desmosomes in fetal (20 wk) human foot-sole epidermis, showing desmosomal substructures (black arrows: midline structure; white arrows: trilaminar "unit membrane" structure of the plasma membrane domain proper, also pointing to electron dense cross-bridge structures; arrowheads: secondary dense layer of the plaque; P: electron-dense primary plaques; IF; anchoring structures of intermediate-sized filaments). (*C*) Freeze-fracture image of spinous layer of murine epidermis, showing the intramembranous fracture plane of the plasma membrane with two desmosomes (arrows) and a gap junction (arrowhead). Note the typical paracrystalline packing of the connexin substructures in the gap junction in comparison with the loosely and irregularly arranged transmembrane elements of the desmosomes. (*D*) Immunoelectron microscopy of an ultrathin section through an epithelium, showing the immunogold decoration (5-nm particles) of desmoplakin at—or near—the desmosomal plaque structures. (*E*) Immunoelectron microscopy of an ultrathin cross-section through the *zonula adhaerens* of plasma membranes connecting two endothelial cells of a cardiac capillary, showing the specific 5-nm immunogold decoration of the junctional plaque with plakoglobin antibodies, thus demonstrating that plakoglobin can also occur in the plaques of nondesmosomal adhering junctions. Bars denote 0.1 μm (*A*, *B*) and 0.5 μm (*C*–*E*). For details, see Cowin et al. (1985a), Franke et al. (1987b), Kapprell et al. (1987), and Schlüter et al. (2007).

The basis for biochemical analyses and the molecular identification of components was laid by a good choice of appropriate tissues, for example bovine muzzle epidermis, for the enrichment of isolated desmosomal structures. The results were markedly dependent on the specific fractionation method used, with more or less preserved plaque structures and "contaminations" by other elements, in particular IF bundle residues (e.g., Skerrow and Matoltsy 1974a,b; Drochmans et al. 1978; Colaco and Evans 1981; Franke et al. 1981; Gorbsky and Steinberg 1981; Mueller and Franke 1983; for review see Skerrow 1986). Although gel electrophoretic analyses of proteins from such fractions allowed some educated guesses about molecular components of desmosomes, it took the direct demonstration of specific proteins by immunofluorescence and immunoelectron microscopy to provide definitive evidence of desmosomal protein localizations. This was first presented at the Cold Spring Harbor Symposium on the "Organization of the Cytoplasm" in the summer of 1981 for the major desmosome-specific plaque protein, desmoplakin, using antibodies that identified desmoplakins I and II and some proteolytic breakdown products thereof (cf. Franke et al. 1982b; see also Franke et al. 1981, 1982a,b) (for an example, see Fig. 1D). Later, K. Green and her group determined that desmoplakins I and II are two splice mRNA forms derived from the same gene and that this large protein presents a very special organization with a near–amino-terminal plakin domain, a central coiled–coil region, and a so-called plakin-repeat domain in the carboxy-terminal portion (Green et al. 1988, 1990; Virata et al. 1992; Godsel et al. 2004), so to say the prototype of what later was termed the "plakin family" of proteins (Ruhrberg and Watt 1997).

The identification of the desmoplakins was soon followed by the localization of epidermal-type desmosomal proteins and glycoproteins in various cell types and species, although not in all cells that were known to contain true desmosomes (cf. Cohen et al. 1983; Cowin and Garrod 1983; Mueller and Franke 1983; Cowin et al. 1984a,b, 1985a,b; Giudice et al. 1984; Gorbsky et al. 1985). As an explanation for cell-type differences, the problem of cell-type specific proteins of complex gene families arose, an experience this research field had just come across with the bewildering complexity of IF proteins.

Two additonal constitutive proteins of desmosomal plaques were identified based on peptide mapping and immunoblotting: They were originally counted as desmosomal bands five- and six-proteins (Franke et al. 1983b) and later named plakoglobin and plakophilin-1 (see also Table 1). Plakoglobin was a particularly challenging component as it turned out also to be a major protein of the plaques of diverse morphotypes of adherens junctions (AJs), including the *zonulae adherentes* of endothelia (e.g., Fig. 1E) (Cowin et al. 1986; Franke et al. 1987a,b). Plakoglobin was also a very attractive molecule for cDNA- and gene-cloning and soon after the human amino acid sequence had been published (Franke et al. 1989, 1992; Fouquet et al. 1992), Peifer and Wieschaus (1990) recognized that plakoglobin was a close homolog to the *armadillo* protein encoded by the *Drosophila melanogaster* segment polarity gene (see also McCrea et al. 1991). This finding resulted in an avalanche of publications leading to the definition of a large and important multigene family, the *armadillo* (*arm*) proteins of adhering junctions in general, i.e., AJs and desmosomes. This family includes a number of components (e.g., plakoglobin, β-catenin, proteins p120, p0071, ARVCF, neurojungin, and the plakophilins) that serve more than one function and occur in junction-bound forms as well as in special cytoplasmic and nuclear forms (Peifer et al. 1992, 1994; for reviews, see Hatzfeld 1999; Schmidt and Koch 2008).

In parallel, several laboratories had begun to clone and identify the genes encoding the polypeptide chains of the two prominent desmosomal glycoproteins, desmoglein (Dsg) and desmocollin (Dsc), again starting with RNA isolated from bovine muzzle or related stratified epithelia. Both the Dsg and Dsc molecules revealed amino acid sequences and carbohydrate branch sites that were markedly homologous to those of the "classic" cadherins such as E- and N-cadherin, which had just been identified as the major transmembrane AJ molecules (Koch et al. 1990, 1991a,b; Goodwin et al. 1990; Holton et al. 1990; Amagai et al. 1991; Collins et al. 1991; Mechanic et al. 1991; Nilles et al. 1991; Parker et al. 1991; Wheeler et al. 1991a,b; King et al. 1993a,b; Theis et al. 1993). Remarkably, in less than 3 years, the cDNAs of all six known desmosomal cadherins were cloned and sequenced. Again, however, these amino acid sequences showed that Dsg and Dsc from different cell types differed from each other. Furthermore, the cytoplasmic carboxy-terminal portion of Dsg contained several distinct sequence subdomains (Koch et al. 1990; reviews: Franke et al. 1992; Koch and Franke 1994; Garrod et al. 2002; Godsel et al. 2004) and not only spans the plasma membrane but also extends through the cytoplasmic plaque, as shown by the accessibility to antibodies directly injected into living cultured cells (Schmelz et al. 1986a,b).

Both Dsg and Dsc appeared in cell type- and cell layer-specific subforms (Parrish et al. 1990; Buxton and Magee 1992; Koch et al. 1992; Buxton et al. 1993; King et al. 1993a,b, 1995, 1996, 1997; Koch and Franke 1994). Only one of each group, Dsg2 and Dsc2, was found in all proliferative cells such as one-layered ("simple") epithelia, basal cell layers of stratified epithelia, meningothelia, cardiomyocytes, and reticulum cells of lymphatic tissues (Schäfer et al. 1994, 1996; Nuber et al. 1995, 1996). By contrast, highly differentiated upper cell layers of several stratified epithelia contained Dsg1 and Dsc1 as predominant desmosomal cadherins, often together with Dsg3 and Dsc3 (Arnemann et al. 1993; King et al. 1993b, 1997; Theis et al. 1993; Legan et al. 1994; Nuber et al. 1996). Dsg4 was a very late addition, synthesized only in the uppermost living cell strata of the epidermis (Kljuic et al. 2003; Bazzi et al. 2006; reviews: Godsel et al. 2004; Schmidt and Koch 2008).

It is striking in desmosomes that both Dsg and Dsc almost always appear together as a pair, probably forming isostoichiometric oligomers that are stabilized by plakoglobin (see, e.g., Troyanovsky et al. 1993, 1994a,b, 1996; Chitaev et al. 1996; Witcher et al. 1996; Chitaev and Troyanovsky 1997; Marcozzi et al. 1998; review: Troyanovsky 2005). Therefore, it came as a surprise when it was recently noted that at least in some nonepithelial cell types (e.g., melanocytes and melanoma cells growing in culture, and melanoma cells in situ), a desmosomal glycoprotein, Dsg2, can occur as an abundant surface component out of any junctional complex (Schmitt et al. 2007; Rickelt et al. 2008).

The last elucidated subfamily of desmosomal components were the plakophilins, apparently located deep in the plaque (see, e.g., Mertens et al. 1996; for differential electron microscopy of desmosomal components, see North et al. 1999; Stokes 2007). Plakophilin was originally numbered as "band 6 protein" in bovine muzzle desmosomes and had been identified as a protein restricted to stratified and complex epithelia (Kapprell et al. 1988). However, it still took several years until the amino acid sequence of this protein, then named plakophilin-1, had been determined as another *arm*-protein (Schäfer et al. 1993; Hatzfeld et al. 1994; Schmidt et al. 1994). This protein was soon followed by plakophilin-2, the most widespread member of this subgroup of *arm*-proteins and the only one present in the desmosomes of simple epithelial cells as well as in nonepithelial desmosomes (Mertens et al. 1996, 1999). Finally, plakophilin-3, which coexists with plakophilin-2 in many desmosomes, has also been sequenced and localized (Bonné et al. 1999, 2003; Schmidt et al. 1999). Similar to plakoglobin and β-catenin, plakophilins are frequently found in distinct particles in the nucleus and the cytoplasm, and such particles may include essential nuclear components such as the complexes containing RNA polymerase III (e.g., Mertens et al. 1996, 2001; Schmidt et al. 1997; Bonné et al. 1999).

With remarkable speed, the cell biological insights obtained and the reagents generated for desmosomes were introduced into developmental biology and the medical sciences. Desmosomal molecules could be shown in very different normal and malignantly transformed cell types, either in layers of tightly associated cells or in arrays resembling spinous layer-type growth forms, in which the cells are connected by special bridges containing desmosomes (Fig. 2A,B) as well as AJs of the *puncta adherentia* type (Vasioukhin et al. 2000). Moreover, the use of subtype-specific antibodies against members of the plakophilin and the cadherin glycoprotein families allowed the determination of not only the cell type but also a certain state of epithelial differentiation in normal tissues and in tumors (Koch et al. 1992; King et al. 1995, 1996, 1997; Nuber et al. 1995, 1996; Schäfer et al. 1996). Consequently, very soon after their first publications in scientific journals, antibodies against

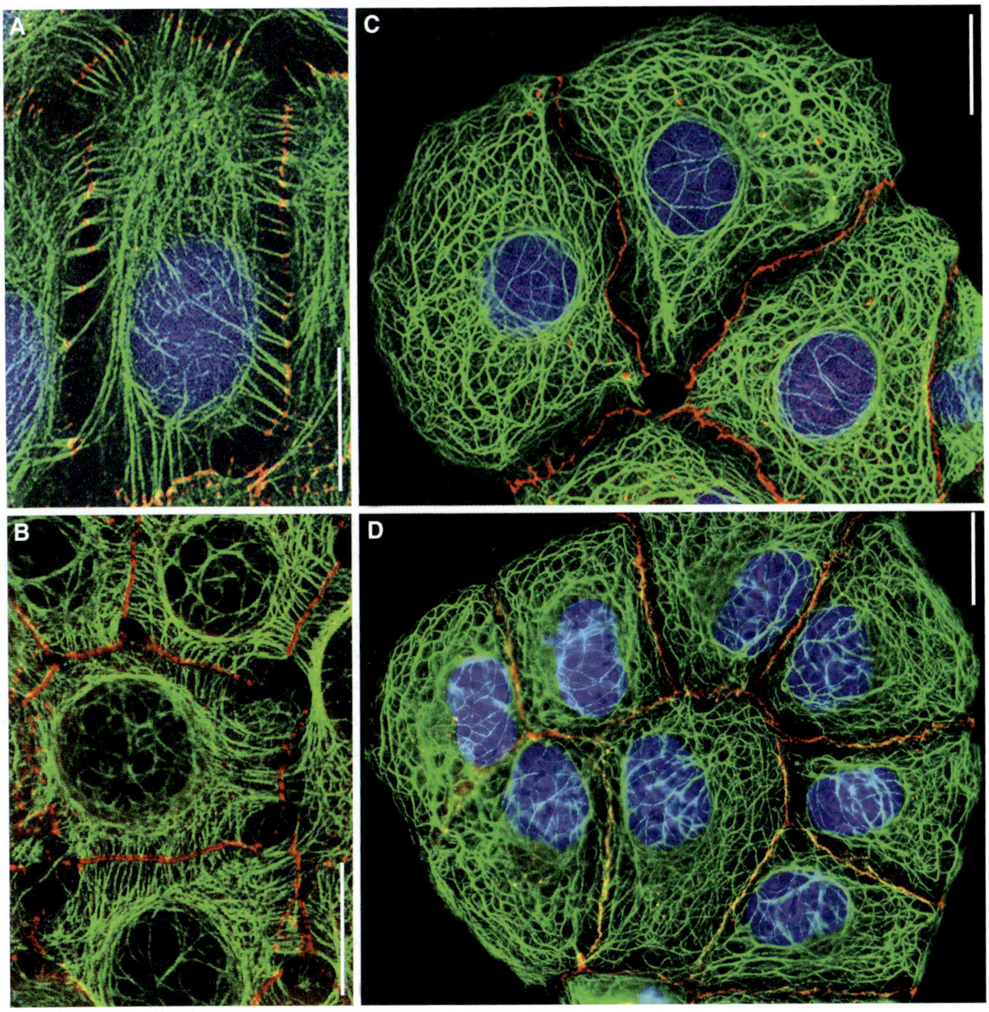

Figure 2. Double-label immunofluorescence microscopy of monolayer cultures of epithelial cells derived from human multilayered (*A*, keratinocytes of line HaCaT; *B*, squamous cell carcinoma-derived line A-431) or one-layered (*C*, *D*: liver carcinoma cells of line PLC) tissues. Two forms of attachment of bundles of keratin IFs (green, mouse monoclonal antibody mAb lu-5) to the plaques of desmosomes (*A*, *B*: red, desmoplakin, guinea pig antibodies) represent a continuous transcellular cytoskeletal system (the chromatin in the nuclei of *A*, *C*, and *D* is stained blue with DAPI reagent): The cells in *A* are connected by cell-to-cell bridges with near-centrally located desmosomes, whereas in *B* the bodies of the cells are directly and tightly associated with each other via numerous, closely spaced desmosomes. In contrast, no specific anchorage of keratin IF bundles is seen at cell junctions of the *zonula* or *punctum adherens* type, which are seen by immunoreaction for β-catenin (*C*, red, guinea pig antibodies) or protein p0071 (*D*, red, murine mAb). Bars: 20 μm.

cell-type-specific desmosomal proteins were added to the immunodiagnostic armamentarium of pathologists (e.g., Franke et al. 1983a; Schwechheimer et al. 1984; Denk et al. 1985b; Parrish et al. 1986, 1987; Vilela et al. 1987; reviews: Moll et al. 1986; Garrod et al. 1996; Kottke et al. 2006).

Specific probes for desmosomal molecules and their genes soon allowed valuable and pioneering contributions to the elucidation of, for example, autoantibody-caused diseases such as those of the *pemphigus* kind (Korman et al. 1989; Amagai et al. 1991; reviews: Amagai 1999; Kottke et al. 2006; Holthöfer et al. 2007;

Schmidt and Koch 2008; Waschke 2008; see also Delva et al. 2009). Notably, studies of the molecular biology of the genes encoding desmosomal proteins, in particular gene abrogation experiments in mice (e.g., Bierkamp et al. 1996; Ruiz et al. 1996; Gallicano et al. 1998, 2001; Grossmann et al. 2004; Zhou et al. 2004), paved the way for the identification of the causes of a large proportion of human hereditary cardiomyopathies, most strikingly the majority of the forms of arrhythmogenic cardiomyopathies, which often lead to so-called "sudden death" (Gerull et al. 2004; Franke et al. 2009; and Delva et al. 2009).

The definition of the molecular ensembles characteristic of various types of desmosomes and the reagents to distinguish desmosomal and desmosome precursor assemblies from components of AJs (Geiger et al. 1983) also provided the basis for studies of molecular mechanisms involved in the intracellular assembly of desmosomes, their IF anchorage, their exocytotic exposure on the cell surface, and their endocytotic resumption, notably on depletions of Ca^{2+} ions in the surrounding media (e.g., Kartenbeck et al. 1982, 1991; Watt et al. 1984; Duden and Franke 1988; Pasdar and Nelson 1988a,b; Stappenbeck et al. 1993; Kowalczyk et al. 1994, 1997, 1998; Demlehner et al. 1995; Bornslaeger et al. 1996; Palka and Green 1997; Smith and Fuchs 1998; Hofmann et al. 2000; Chen et al. 2002; Koeser et al. 2003; Goossens et al. 2007; see also Green et al. 2009). Desmosomes had changed from stable anchor structures to highly dynamic cell elements and had finally become subjects of general physiological research.

Aside from the structure-bound states of desmosomal proteins in junction plaques and in nuclear or cytoplasmic particles, it had also become clear that the dynamic basis of all the reactions, the diffusible "regulation-ready" form, must exist somewhere in the cell. This "free" soluble form, however, is one of the most under-researched entities. The major soluble state of only plakoglobin, a 7S dimer, has been determined so far (Kapprell et al. 1987; see also Fouquet et al. 1992).

The molecular definition of true desmosomal components had also allowed the observations that in certain cell types these junctions do not anchor bundles of keratin IFs but rather vimentin IFs (Kartenbeck et al. 1984; Schwechheimer et al. 1984; Moll et al. 1986). Desmin-rich IFs and actin-containing myofilaments have also been shown to insert at the desmosomes and *areae compositae* of cardiomyocytes (Table 1) (Franke et al. 1982a, 2006; Kartenbeck et al. 1983). Note also that it was shown early on that the formation and maintenance of desmosomes can occur independently of anchorage to IFs (Denk et al. 1985a), indicating that desmosomes have functions other than as anchors and organizers of IF bundles (Green and Gaudry 2000).

THE DECADE OF THE ARMADILLOS AND THE MOLECULAR DIVERSITY OF ADHERENS JUNCTIONS

The discovery of the molecular ensembles forming the plaque-coated AJs has a history very different from that of the research on desmosomes. Here, seminal findings had been made many decades back and almost exclusively in the field of developmental biology. At the turn of the previous century, embryologists and zoologists recognized the importance of certain cell-surface entities for specific cell–cell adhesion and cell sorting in animal embryology as well as in tissue-like cell assemblies in culture dishes, from sponges and polyps to amphibian and chicken embryos, and, ultimately, mammalian organ development and regeneration (reviews to be recommended here: Wilson 1907; Holtfreter 1939; Moscona 1962; Okada 1996).

In the late 1970s, several research groups attempted to identify the molecules and the molecular principles governing cell sorting and adhesion mechanisms essential in embryogenesis. Steinberg (1958) recognized the general importance of Ca^{2+} ions in these processes. Following the phenomenon of Ca^{2+}-mediated cell–cell assemblies to higher order structures, Takeichi's group in Kyoto identified a glycoprotein of around 150 kDa mol. wt. (Takeichi, 1977), which a few years later was described—somewhat down-sized to 124 kDa—as a central player in cell–cell adhesion events, in particular in "morula compaction" of early embryogenesis. Consequently, this molecule was named "cadherin" for its Ca^{2+} dependence and cell adhesion effect (Yoshida-Noro et al. 1984; see also Yoshida-Noro and Takeichi 1982).

The Kyoto group was not alone in their search for the cell surface molecules involved in cell–cell interaction and homophilic sorting in early embryogenesis. Using different experimental approaches, the groups of G. Edelman in New York and F. Jacob in Paris had also embarked on the elucidation of these embryonal cell-adhesion principles and had independently arrived at the same kind of major interaction molecule, which they named uvomorulin, a molecule and concept that was then followed further by R. Kemler's group (Kemler et al. 1977; Hyafil et al. 1980, 1981; Peyrieras et al. 1983; Vestweber and Kemler 1984, 1985; Boller et al. 1985; Ringwald et al. 1987), or L-CAM (Edelman et al. 1983; Gallin et al. 1983, 1987; Cunningham et al. 1984; see also Thiery et al. 1984). Other groups, using slightly different experimental approaches, called this glycoprotein A-CAM (Volk and Geiger 1984; Geiger et al. 1985a,b) or "glycoprotein Cell-CAM 120/80" (Damsky et al. 1983, 1985). In parallel studies using cell culture approaches, W. Birchmeier and his collaborators (Imhof et al. 1983; Behrens et al. 1985) had also identified this cell-surface glycoprotein, which turned out to be E-cadherin, and the "120/80 kDa" polypeptides were soon identified as a mixture of intact and truncated E-cadherin molecules. So, at a small expert meeting entitled "The Cell in Contact" on the campus of Rockefeller University in 1985, the representatives of these groups recognized that in the previous years they all had worked on the same molecule, the first AJ-specific component identified (for reviews, see Cunningham 1985; Damsky et al. 1985; Edelman 1985; Geiger et al. 1985a; Takeichi et al. 1985; Thiery et al. 1985).

As with the identification of desmosomal proteins, the identification of *a* cadherin was not the end of a research track but the beginning of a global outburst of hectic and cell biology-complicating work, as it started simultaneously two avalanches of cell type-specific proteins, the "cadherins" and the "*armadillos*." Luckily, effective cDNA cloning methods were already at hand so that in only a few years the major cell type-specific cadherins were sequenced and could be localized by effective antibodies: E-cadherin (Gallin et al. 1983; Schuh et al. 1986; Bussemakers et al. 1993), N-cadherin (Hatta and Takeichi 1986; Hatta et al. 1987, 1988; Nose and Takeichi 1986), P-cadherin (Nose and Takeichi 1986; Nose et al. 1987), and the vascular endothelium-characteristic VE-cadherin (Lampugnani et al. 1995; Dejana 1996; Dejana et al. 2000). Finally, more than 50 members of a large "superfamily" in mammals have been identified, including so-called "classic" cadherins, "type II cadherins," and "protocadherins" (reviews: Takeichi 1988, 1990; Gumbiner 1996; Nollet et al. 2000; Angst et al. 2001; Wheelock and Johnson 2003; Perez and Nelson 2004; Halbleib and Nelson 2006), not to mention the bewildering numbers of cadherins and their splice variants in some other parts of the animal kingdom (for arthropod cadherins, see, e.g., Hsu et al. 2009). Each of these newly discovered cadherins immediately produced "daughter avalanches" of publications dealing with the molecular interactions of the specific cadherins with each other and with AJ plaque proteins, the regulation of assembly or disassembly of AJs, and the functions of different cadherins during development and in the mature tissue, beginning with the study of Matsunaga et al. (1988, see also Meng and Takeichi 2009). In addition, each of the identified cadherins immediately produced questions as to its molecular complex state (for evidence of Ca^{2+}-dependent *cis*-homodimers, see, e.g., the E-cadherin study by Takeda et al. 1999).

During the early period of cadherin discoveries, a number of researchers noted the biological specificity of the cadherin ensemble of a given cell type but also that there are examples of drastic changes of the type of cadherin produced when cells change their character, for example during wound healing (e.g., Hinz et al. 2004). A very conspicuous change often occurs when epithelial cells turn into mesenchymal cells during development, and when cells transform to malignancy and develop to carcinomas. Increased malignancy often appeared to be associated with the down-regulation of E-cadherin and an up-regulation or even induction of N-cadherin (e.g., Behrens et al. 1989, 1993; Frixen et al. 1991; Mareel et al. 1991, 1994; Vleminckx et al. 1991; Birchmeier et al. 1993, 1995). In addition, an increasing number of cadherin mutations and mRNA splicing errors have been reported that result in, or at least contribute to, carcinogenesis (e.g., Kanai et al. 1994). These principles were in a short period of time confirmed many times and have opened a totally new avenue of cancer research (for recent reviews, see Brabletz 2004; Strumane et al. 2004; Wheelock et al. 2008; see also Berx and van Roy 2009).

In parallel, we all had to watch with care and concern the other avalanche of reports on discoveries of proteins located in the cytoplasmic AJ plaque and directly or indirectly complexed with the cadherins. Soon after the discovery of the prototypic *arm*-protein, β-catenin, in 1989, this name could then only be used in the plural form (Tables 1 and 2): Six *arm*-proteins plus the actin-binder and -regulator, α-catenin, were identified in a short time, including some that are restricted to certain cell types and diseases (Ozawa et al. 1989, 1990a,b; Peifer and Wieschaus 1990; Nagafuchi et al. 1991; McCrea and Gumbiner 1991; McCrea et al. 1991; Peifer et al. 1992). Indeed, some of them were so specific that diverse AJ subforms could readily be distinguished not only from desmosomes but also from each other (e.g., Figs. 2C,D and 3).

Perhaps the greatest surprise and an additional attraction for researchers were observations that several of the *arm*-proteins were detected and seemed to function in very distant and different places in the cell: as structure-integrated rather insoluble components of junctional plaques, and as components of diffusible, relatively small regulatory complexes in the nucleus, involved in diverse processes including gene transcription and other nuclear functions. Thus, it was almost to be expected that an ever increasing number of cell biology groups embarked on research projects trying to unravel the molecular interactions of these proteins, both in the AJ plaques and in nuclei. This led to the founding of several new fields of regulatory cell biology, perhaps best illustrated by the affluence of reports on *arm*-proteins involved in regulatory pathways, including the Wnt-pathway (Choi and Weis 2004; Brabletz 2004; Drees et al. 2005; Lewis-Tuffin and Anastasiadis 2008; see also Heuberger and Birchmeier 2009, and Cadigan and Peifer 2009). Finally, it should be noted that sizeable portions of *arm*-proteins generally occur in small diffusible forms in the cytoplasm, such as in the maternal pools of the amphibian ooplasm (e.g., Fouquet et al. 1992).

Some puzzling complications, however, were brought to the field of "cadherinology" by discoveries of a series of special, carboxy-terminally truncated cadherins, i.e., molecules with a significantly longer extracellular portion, but only a very short, e.g., 20 amino acids, cytoplasmic portion. These cadherins included mammalian LI-cadherin found in the polar simple epithelia of liver and/or intestine (Berndorff et al. 1994; Kreft et al. 1997), and Ksp-cadherin of certain kidney epithelia (Thomson et al. 1995). An even further truncated type of cadherin was found in which a normal extracellular element comprising five repeating elements lacked a membrane-anchor but was attached to the plasma membrane by a glycosylphospatidyl inositol residue (for T-cadherin, see Ranscht and Dours-Zimmermann 1991; Tanihara et al. 1994). However, these cadherins are not integrated into an AJ structure but widely scattered over the plasma membrane.

In the mid 1990s, just when the cell biological community thought that it finally knew the composition of AJs, again a disturbing call "halt—not so fast" was heard. The group of Y. Takai had identified yet another cell surface membrane complex, in this case with major transmembrane molecules of the immunoglobulin-like category: The nectins represent a family of four major and several minor (splice-) isoforms that can bind to

Table 2. The *armadillo*-report avalanche in the 1990s: Some results on *arm*-repeat and associated actin-interacting proteins in plaques of AJs[a]

Arm-proteins	References	
Plakoglobin	Cowin et al. 1986	Knudsen and Wheelock 1992
	Franke et al. 1987a–c, 1989	Troyanovsky et al. 1996
	McCrea et al. 1991	Wahl et al. 1996
	Fouquet et al. 1992	Zhurinsky et al. 2000
β-Catenin	Ozawa et al. 1989, 1990a,b	Hülsken et al. 1994
	Peifer and Wieschaus 1990	Näthke et al. 1994
	Nagafuchi et al. 1991	Stappert and Kemler 1994
	McCrea and Gumbiner 1991	Huber et al. 1997
	Ozawa and Kemler 1992	Zhurinsky et al. 2000
Protein p120	Reynolds et al. 1992, 1994, 1996	Anastasiadis et al. 2000
	Aghib and McCrea 1995	Golenhofen and Drenckhahn 2000
	Shibamoto et al. 1995	Iyer et al. 2004
	Aho et al. 1999	Setzer et al. 2004
	Ohkubo and Ozawa 1999	Hatzfeld 2005
	Anastasiadis and Reynolds 2000	Ferreri and Vincent 2008
Protein ARVCF	Sirotkin et al. 1997	Paulson et al. 2000
	Mariner et al. 1999, 2000	Waibler et al. 2001
	Borrmann et al. 2000	Kausalya et al. 2004
	Kaufmann et al. 2000	Walter et al. 2008
Protein p0071	Hatzfeld and Nachtsheim 1996	Deguchi et al. 2000
	Hatzfeld 1999, 2005	Hofmann et al. 2008, 2009
Neurojungin (δ-Catenin)	Paffenholz and Franke 1997	Paffenholz et al. 1999
	Zhou et al. 1997	Ho et al. 2000
	Lu et al. 1999	Kosik et al. 2005
Representatives of actin-interacting proteins associated with AJ plaques		
α-Catenins (E-, N-, T-)	Ozawa et al. 1989	Aberle et al. 1996
	Herrenknecht et al. 1991	Pai et al. 1996
	Nagafuchi et al. 1991, 1994	Koslov et al. 1997
	Nagafuchi and Tsukita 1994	Nieset et al. 1997
	Stappert and Kemler 1994	Watabe-Uchida et al. 1998
	Rimm et al. 1995	Weiss et al. 1998
	Yonemura et al. 1995	Janssens et al. 2001
MAGUK Family:	Itoh et al. 1991, 1993, 1999	Fanning et al. 1998, 2002
ZO-1, ZO-2, and ZO-3	Horwarth et al. 1992	Wittchen et al. 1999
	Yonemura et al. 1995	Kausalya et al. 2004
ERM-Family[b]	Bretscher 1986	Sato et al. 1991, 1992
	Gould et al. 1989	Gary and Bretscher 1993
	Tsukita et al. 1989, 1994, 1997a,b	Hirao et al. 1996
	Funayama et al. 1991	
IQGAP Complexes[c]	Brill et al. 1996	Izumi et al. 2004
	Hart et al. 1996	Noritake et al. 2004
	Kuroda et al. 1996, 1998	Lehtonen et al. 2005
	McCallum et al. 1996	Lui et al. 2005

[a]Including some recent review articles.
[b]Homologous proteins: ezrin, radixin, moesin, and merlin.
[c]IQ-domain GTPase-activating proteins.

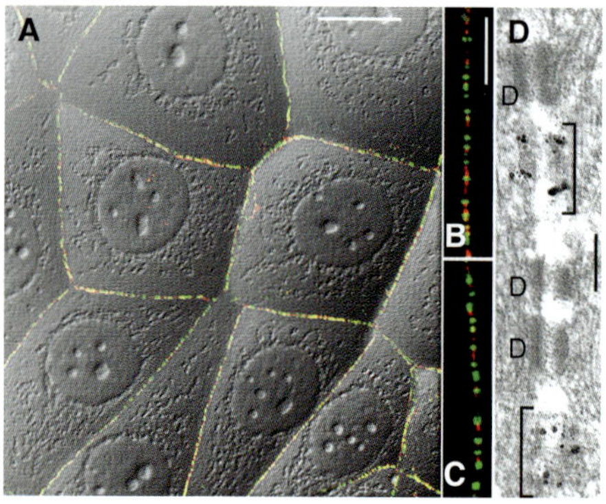

Figure 3. High-resolution double-label immunofluorescence (*A–C*) and immunoelectron (*D*) microscopy of monolayer cell cultures of human breast-carcinoma cells of line MCF-7, as seen after reactions with antibodies to the desmosomal plaque component, desmoplakin (green, guinea pig antibodies), or to the adherens plaque protein, p0071 (red, as in Fig. 2B), as seen on the background of differential interference contrast (DIC) (*A*), allowing for the most part to distinguish the small *puncta adhaerentia* from the similarly small desmosomes, located side-by-side. (*B, C*) Higher magnification micrographs of cells as shown in *A*, presenting details of cell–cell junctions along close membrane contact regions, allowing to distinguish in many places the alternating pairs of symmetrical plaques of desmosomes (green) and *puncta adhaerentia* (red). (*D*) Equivalent comparison at the electron microscopic level, showing an ultrathin section through junctions after immunogold reaction of *puncta adhaerentia* with antibodies to protein p0071 (brackets, immunogold granules enhanced by secondary silver reaction), in comparison with the negative small desmosomes (*D*). For details, see Hofmann et al. (2008, 2009). Bars: 20 µm (*A*), 5 µm (*B, C*), and 0.1 µm (*D*).

each other in *trans*-interactions in a Ca^{2+}-independent mechanism. They are bound to a submembranous plaque-type protein, afadin, which also occurs in several splice variants and in turn is associated with some specific actin-binding proteins, one of which has been named ponsin. This complex was observed close to, and often colocalizing with, AJ structures (Mandai et al. 1997; Takahashi et al. 1999; Satoh-Horikawa et al. 2000; for reviews see Takai and Nakanishi 2003; Irie et al. 2004; Takai et al. 2008a,b). Afadin apparently can also interact with α-catenin, thus providing the possibility of an indirect actin microfilament bridge connection between nectin and cadherin complexes (e.g., Tachibana et al. 2000; see also Takai et al. 2008a). So, why have these "Takai group" molecules still not been "officially" added to the AJ ensemble? One remaining problem is that although they colocalize, even at the electron microscopic level, with some of the typical AJ structures, for example with the *zonula adhaerens* of polar epithelia, they have not been seen to occur in other AJs such as those of the nearby *puncta adhaerentia* along the lateral surfaces of the same cells (e.g., Mandai et al. 1997; Takai et al. 2008a,b). Another problem is that components of these two strikingly close structural complexes have not been coimmunoprecipitated or cross-linked by short-distance chemical cross-linkers. However, recent studies indicate that the nectin–afadin transmembrane elements play an important pioneering and dynamic role in AJ assembly because the nectin complex provides the first cell–cell contact structures between two cells whereupon the cadherin-containing AJ elements follow (Takai et al. 2003, 2008a,b; see also Meng and Takeichi 2009).

Finally, a special "border guard" role has been ascribed to the groups of relatively small *puncta adhaerentia*-type AJs, which in certain brain regions surround neural synapses and seem to be important in the formation and maintained organization of the spatial junction relationship and the proper functioning of these specific synapses (Uchida et al. 1996; see Giagtzoglou et al. 2009).

THE SECRETS OF THE KISSES: THE MEMBRANE MOLECULES FORMING TIGHT JUNCTIONS

The electron microscopic appearance of the subapical TJs of simple epithelial cells (*zonulae occludentes*) was relatively well-known since the mid 1970s. Transmission electron microscopy of cross sections showed one or a few rather small direct contact sites of the two plasma membranes, commonly called "kisses."

Planar freeze-fractures through the membrane interior revealed, after metal-shadowing, a linear, or ornamentally woven in some tissues, branched relief of one or several ridges (Farquhar and Palade 1963; Goodenough and Revel 1970; Friend and Gilula 1972; Claude and Goodenough 1973; Staehelin 1974; Montesano et al. 1975; Hull and Staehelin 1976; Diamond 1977; reviews: Schneeberger and Lynch 1992; Mitic and Anderson 1998; Förster 2008; see also Goodenough and Paul 2009). Early on, it was also obvious that this structure must be most important in the establishment, maintenance, and functions of epithelia and endothelia, and hence in metazoan life in general. Most convincingly, the barrier role of TJ structures for paracellular diffusion, transport processes, and particle translocations had been directly shown by electrophysiological experiments and by electron microscopy using heavy metal-labeled tracers (see also Anderson and Van Itallie 2009). It had also been suggested that a number of physiological phenomena and diseases, from tissue swellings to liquid losses, were caused by partial or transient interruptions of such a barrier (e.g., Claude 1978; Madara and Dharmsathaphorn 1985; Simons 1990; Matter and Balda 1999; Van Itallie and Anderson 2006). Naturally, this field of research was closely watched by pharmacologists and drug developers awaiting the emergence of principles or molecules that would allow restoration of an interrupted barrier, or to induce a controlled transient permeability of the blood-brain-barrier for drugs of other components.

Molecular candidates and insights, however, arrived only dropwise. The first relevant discovery was made by Stevenson et al. (1986) who isolated and localized a component of the relatively thin plaque on the cytoplasmic face of TJs, therefore named protein ZO-1 ("*zonula occludens*"-1; identical to the ca. 220-kDa protein described by Itoh et al. 1991, 1993), which was later followed by the identification of ZO-2 (Gumbiner et al. 1991) and ZO-3 (Haskins et al. 1998). These proteins were founding members of the so-called MAGUK (membrane-associated guanylate kinase) family of proteins (for homologous proteins in invertebrates, see the finding in *Drosophila* by Willott et al. 1993), which soon turned out not to be exclusive for TJs but to be almost as regularly found in AJ- and GJ-associated plaques (Tables 1 and 2) (for reviews, see Anderson 1996; Balda and Matter 2000). These proteins can also occur in the nucleus (Gottardi et al. 1996) and in several other locations and functions, from transcriptional factors to regulators of the behavior of tumor cells (see the anthology edited by Cereijido and Anderson 2001, and the reviews of González-Mariscal et al. 2003; Gottardi and Niessen 2008).

Soon thereafter, Citi et al. (1988, 1989, 1991) discovered and localized another major TJ plaque protein of ~Mr 140 kDa, which they named cingulin (from the Latin *cingulum*, i.e., stringlike girdle), which is clearly TJ-specific, albeit seemingly confined to epithelial TJs (e.g., Cordenonsi et al. 1999; Citi 2001; Fanning 2001). Altogether, about a dozen TJ plaque proteins have now been described, but so far their detailed molecular interactions and locations have only partly been elucidated. Although for some of them only molecular complexes with classic TJ components have been reported, others, again, show the phenomenon of dual localization outside of TJs (reviews: Citi 2001; Fanning 2001).

But what was the "core of the barriers," the "real seal," the "nature of the kisses," the "hydrophobic transmembrane stuff?" Here, a first answer came at Christmas 1993 as a present, an unexpected major breakthrough in TJ membranology, a fruit of both "good old" cell fractionation work combined with biologically conceived strategic experimentation. "The Tsukitas," Sachiko and Shoichiro, already known for the high-quality standard of their plasma membrane fractions from liver tissue (Tsukita and Tsukita 1989) and profiting from the enthusiasm and courage of a young graduate student, M. Furuse, had decided to take the classic route by homogenizing the most suitable tissue and enrich the desired structure by the most suitable techniques. The biological source was cleverly chosen—the bile canalicular front fraction from chicken liver, which is naturally enriched in all major kinds of junctions. Their first "eureka" candidate was a tetraspan transmembrane protein, which they could immunolocalize to the TJ structure, the *zonula occludens*, of liver and other simple epithelia, and so they named it "occludin" (Furuse et al. 1993; for the appropriate welcome ceremony, see Gumbiner 1993). Particularly satisfying was the result that on transfection of occludin-encoding cDNA expression constructs into TJ-possessing cells, the newly synthesized occludin integrated into pre-existing TJ structures, and when injected into TJ-lacking cells, the gene product was able to form de novo TJ-typical intramembranous "ridges" as well as cytoplasmic stacks of lamellae closely resembling typical TJ membrane structures (Furuse et al. 1994, 1996; Saitou et al. 1997). So, in a wide range of mammals, occludin and its various phosphorylation forms appeared to be the basis of TJ membranes (Fig. 4A) (Ando-Akatsuka et al. 1996; Sahakibara et al. 1997).

Then, suddenly came a deep disappointment—not an uncommon one in modern biology, but in this case fortunately only a short one. Deletion of both alleles of the occludin gene had resulted in occludin-deficient mice, which, however, not only lived but formed typical TJ structures in the typical tissues, combined with plaque proteins of the ZO-1–3 protein family and TJ-typical functions (Saitou et al. 1998; only later, a series of occludin$^{-/-}$ mice with defects in several tissues and functions were reported: Saitou et al. 2000). Consequently, in their 1998 paper, Saitou et al. had to conclude: "These findings indicate that there are as yet unidentified TJ integral membrane protein(s) that can form strand structures, recruit ZO-1, and function as a barrier without occludin." And so, without much mourning, they went back to square one, the chicken liver bile canalicular membrane fraction, and noticed two further bands of smaller ($\sim$22 kDa) polypeptides that had amino acid sequences also indicative of molecules traversing the TJ membrane four times but without any occludin homology. Nevertheless, the new, very small, kind of protein immunolocalized to TJs, incorporated into TJs of living cells, and formed de novo intramembranous, TJ-typical ridge structures on cDNA transfection, even in TJ-lacking cells. Consequently, our Japanese friends had to expand their Latin vocabulary and named these new TJ proteins "claudins" nos. 1 and 2 (Furuse et al. 1998a,b; Tsukita and Furuse 1998).

But TJ life did not become much easier, and the ups and downs of TJ protein research emotions were not over. Japanese scientists are usually not regarded as hectic, but with the TJ proteins everything was new and different. Shoichiro (Tsukita) had just informed me and a few others about the first claudin amino acid sequence, but then came the stop-fax "hold it" and then the phone call "Sorry: Something strange has happened, these sequences are already in the computer." Very quickly, however, his good mood returned with the solution: Two claudins were identical to the receptors of a food poisoning enterotoxin of the bacillus *Clostridium perfringens*, a known cause of gastrointestinal disease (Sonoda et al. 1999). Thus, another medical research path had just been opened: Claudins were recognized as membrane-bound receptors of certain toxins produced by prokaryotic pathogens (for a special review, see, e.g., Hecht 2001).

As we have already experienced with desmosomal molecules, and with cadherin and *arm*-protein constituents of the AJs, the claudins likewise did not come only as a pair of genes. To the contrary, the Tsukita group had stumbled into another large, novel gene family, the claudins, with very complex and functionally important, cell-type-specific combinations of a few selected representatives (e.g., Furuse et al. 1999; Morita et al. 1999a-c; Tsukita and Furuse 2000). And at the end of that year, coincidentally also the end of the millenium, already 15 different claudins had been discovered and soon thereafter a total number of 24 claudins were found to be encoded in the human genome (Tsukita et al. 2001; Tsukita and Furuse 2002). Some of the claudins came with a list of specific medical research questions, the first example being the Mg^{++}-reabsorption

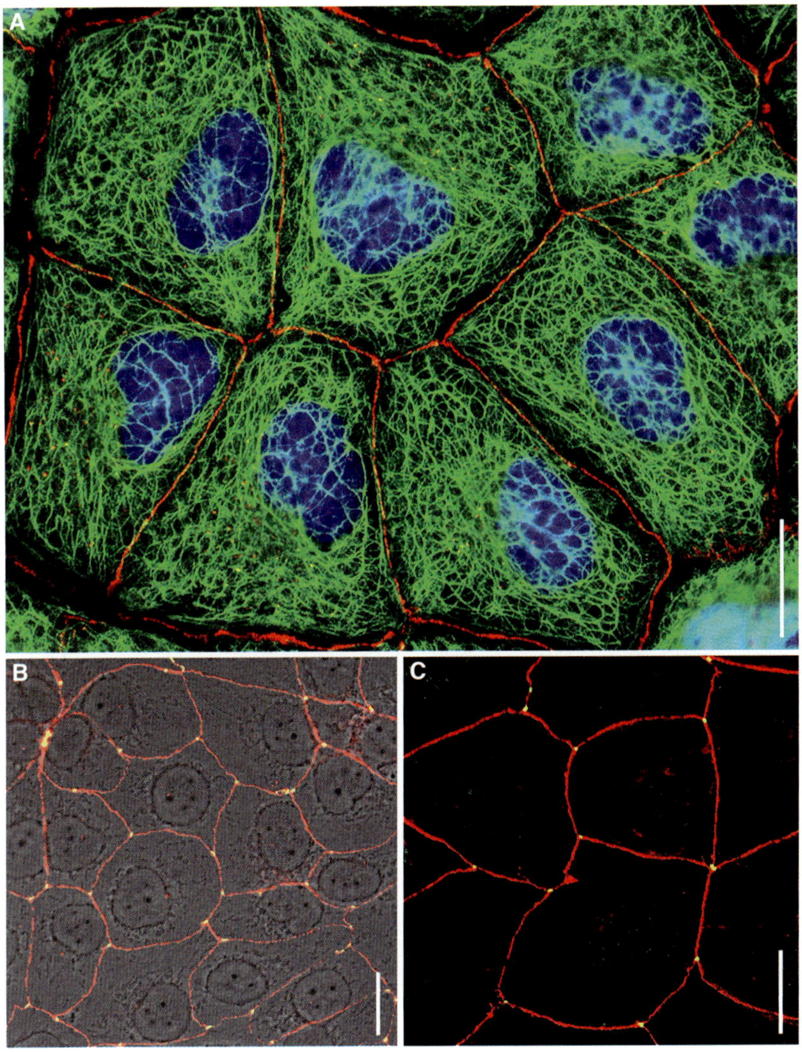

Figure 4. Double-label immunofluorescence microscopy showing the tight junction (TJ) systems of monolayer cell cultures of human hepatocellular carcinoma-derived cells of line PLC (*A*) and of cultures of human keratinocytes of line HaCaT beginning to form the first suprabasal cell layer (*B*, *C*), using antibodies to keratin (*A*, green, mAb lu-5), in comparison with the TJ protein, occludin (*A*, red), or comparing the localization of the protein tricellulin (*B*, green, rat antibodies) with the TJ plaque protein, ZO-1 (*B*, red, rabbit antibodies), or the transmembrane TJ protein, occludin (*C*, red, rabbit antibodies). Note that these cells are totally interconnected by the TJ system (*zonula occludens*), which is completed at the tricellular corners by the occludin-related, but very specifically located protein, tricellulin, presenting complete colocalization with protein ZO-1 and occludin at the tricellular corners. For details, see also Schlüter et al. 2007. Bars: 20 μm.

problem protein, just called "paracellin," now reclassified as claudin-16 (Simon et al. 1990; for a special review see Choate et al. 2001). Even higher claudin gene complexities are found in some other vertebrate species, with a—so far—record number of a total of 56 claudin genes in the—yes, no joke—Japanese puffer fish, *Fugu rubripes* (Loh et al. 2004; see there for further references). Thus, the smallest vertebrate genome known presents the highest claudin gene number known, an ironical paradox of Mother Nature. And while these sentences are written, a new claudin message comes in, indicating further bewildering complexity: Just one claudin, claudin-10, can occur in six spliced isoforms with different tissue and intracellular localizations and functions (Günzel et al. 2009). Evolution has indeed been merciless to cell biologists!

So the search for claudin functions in the different tissues and for claudin-connected diseases began. But halt once more! When the cell biological world thought that the TJ ensemble was finally complete, it turned out that not all essential facts of TJ structures had been carefully considered. Although complete TJs encircle individual cells of a simple, one-layered epithelium or an endothelium, thus forming a tight-looking barrier (Fig. 4A), in some epithelia, small intercellular "gaps" remain at places where three cells meet each other, a "three corner state" problem, representing gaps through which sizable molecules or even particles might translocate. Again, the Tsukita laboratory came through, identifying, as a very late addition to cell junction biology at the end of the year 2005, the protein "tricellulin," just in time to let the late Shoichiro be sure about the publication (Ikenouchi et al. 2005). This protein, related by partial sequence homology to occludin, assembles exactly at those "three corner" sites to make the epithelial tight junction system really tight (Fig. 4B,C; Table 1).

In the meantime, systematic function-oriented studies had tried to elucidate the biological roles as well as the potential importance of specific claudins by gene mutation and deletion experiments. However, it was important to recognize that certain claudins occur in very specific cell types: For example, claudin-11, formerly known as "oligodendrocyte-specific protein" (OSP), was identified in cells as different as Sertoli cells of testis and myelin sheath-forming oligodendrocytes, biological distances and differences difficult to bridge by simple biological hypotheses (e.g., Morita et al. 1999b). The first systematic study of specific claudin gene deletion deficiencies brought both surprises and clarity. Claudin-1-deficient mice died on the first day after birth with wrinkled, dried-out skin, caused by massive transepidermal water loss (Furuse et al. 2002), a finding difficult to explain on the basis of the then prevailing textbook dogma that there are no TJ structures in the upper living layer (*stratum granulosum*) of the epidermis and other stratified epithelia. However, it was a lucky coincidence that in the summer of 2002, typical TJ structures, positive for claudins and occludin, had also been noted and localized in these uppermost layers of the epidermis and other squamous stratified epithelia (Fig. 5A–C) (Brandner et al. 2002; Furuse et al. 2002; Langbein et al. 2002; Schlüter et al. 2004; see also Morita et al. 1998, 2002; Pummi et al. 2001). These findings have changed the view of how the body surface is sealed from the outside world. Recently, tricellulin was also identified in suprabasal cell layers of keratinocyte cultures (Fig. 5B,C) (Schlüter et al. 2007).

Here, the TJ story could end. There is, however, a remaining fundamental cell biological problem. TJ-type structures, appearing as "kisses" or other close membrane–membrane contacts, can also be seen in interdesmosomal regions of the spinous layer of the epidermis and the equivalent layers of several other stratified epithelia. They appear as small "lamellated" or "sandwich" junctions or even as "stud junctions" (*puncta occludentia*), exactly in places intensely immunopositive for occludin or certain claudins (Fig. 5B,C) (e.g., Brandner et al. 2002; Langbein et al. 2002, 2003; Schlüter et al. 2007). Could it be that there still exists a so far ignored world of TJ-related junctions? Could it be that discoveries of TJ molecule assemblies are not only history?

GAP JUNCTIONS: THE BEAUTY AND COMPLEXITY OF INTERCELLULAR CHANNELS

In the "good old," although rather controversial days of molecular discoveries, cell biological discussions of possible functions of a given protein or structure started with educated guesses but often ended with less-educated, semitired jokelike comments. In the case of GJs,

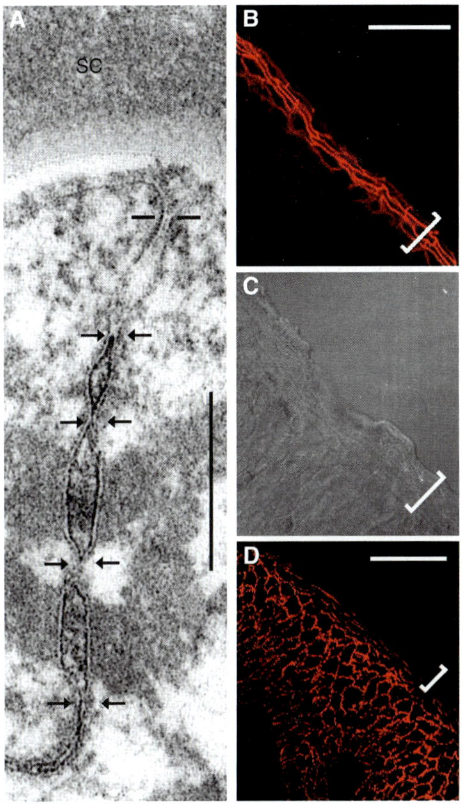

Figure 5. Electron (*A*) and immunofluorescence (*B–D*) microscopy, showing the tight junction (TJ) system in the granular layer (*stratum granulosum*) of stratified epithelia. (*A*) The uppermost living layer of human epidermis (the first cornified layer, *stratum corneum*, is labeled *SC*) contains a subapical system of TJ structures (horizontal bars and pairs of arrows denote distinct membrane contact and fusion points: "kisses"). Note that the four lowermost TJs (arrows) are interspersed between desmosomes. (*B–D*) Near-vertical cryostat section through bovine gingiva, showing the TJ system in the uppermost living layers by immunostaining with antibodies to occludin (*B*, bracket; *C*, corresponding phase contrast image), whereas the cell–cell junction structures positive for the TJ protein, claudin-1, are not limited to the granular layer but extend basally into a number of *stratum spinosum* cell layers as well, indicating that there exist further, not yet fully characterized, partly punctate structures containing claudin-1 but not occludin (for methods used, see also Brandner et al. 2002; Schlüter et al. 2007). Bars: 0.2 μm (*A*), 50 μm (*B–D*).

intercellular communication and exchange was one of the favorite discussion subjects early on, but non-GJ colleagues said that they were there "only to give pleasure to electron microscopists." Indeed, these near-ubiquitous tissue structures seemed to have been precipitated out of an electron microscopist's dreams as they comprised variously sized and shaped, densely packed, planar paracrystalline assemblies of short vertical hollow cylinders ("channels") made up of symmetrically head-on-head contacting hemichannels. All of these structural details were well-demonstrable in the electron microscope, either in transverse or horizontal ultrathin sections, by freeze fractures (see, e.g., Fig. 1C) or in negatively stained isolated GJs. Thus, it was well known already in the 1960s that the hemichannel interior could be filled, at least partly, with heavy metal molecules but also that the small extracellular spaces in between the channels, the "gaps," were also penetrable by such agents added from the outside. Moreover, in cell fractionation experiments, GJs were among the more stress-, detergent-, and even alkaline pH-resistant structures (e.g., Dewey and Barr 1964; Goodenough 1974; Henderson et al. 1979; Hertzberg and Gilula 1979; Hertzberg 1984; Stauffer et al. 1993). As a result of these near-optimal conditions, the ultrastructural organization of GJs was essentially known at the end of the 1970s (e.g., Goodenough and Revel 1970, 1971; McNutt and Weinstein 1970; Caspar et al. 1977; Unwin and Zampighi 1980; for reviews see, e.g., Loewenstein 1981; Gilula 1985, 1987, 1990; Revel et al. 1985, 1987; Unwin 1987; Goodenough 1990; Sosinsky 2000; see also Goodenough and Paul 2009). There was indeed no shortage of structural GJ models when the molecular phase began.

The major GJ proteins identified in fractions enriched in isolated junction structures from liver, heart, and eye lens were soon analyzed and characterized, including protein chemistry and reconstitution experiments, and cDNA cloning and sequencing (e.g., Goodenough 1974; Henderson et al. 1979; Hertzberg and Gilula 1979; Hertzberg et al. 1982; Hertzberg and Skibbens 1984; Kumar and Gilula 1986; Paul 1986; for review, see Goodenough 1990). A major protein constitutent was obvious, but it was also recognized early that the molecules representing this major GJ protein in different tissues were related but not identical. "Connexins" was chosen as the family name (Goodenough 1974) and, after some discussion on the molecular relatedness alphabetical nomenclature of the Westcoast and the "connexin plus apparent molecular weight" (e.g., Cx26, etc.) proposal from the Eastcoast, the "Cx plus M_r value" version found more friends. One inherent problem, however, was also noted from the beginning that the M_r value even of the ortholog protein could vary between species. Another problem was the same as encountered with the other cell–cell junctions. Yes, you are right: Connexins are another large junction protein family showing cell-type specificity.

Definitive evidence that a certain protein is indeed part of a GJ channel structure came when antibodies specific for a given type of connexin(s) had been generated. This enabled immunoblot testing for specific reactions with a connexin, or a few related ones, from GJ-enriched fractions, and the identification and exact GJ localization using immunoelectron and immunofluorescence microscopy with isolated gap junction membranes or, most convincing, GJ structures in situ (Willecke et al. 1982; Janssen-Timmen et al. 1983; Dermietzel et al. 1984; Paul 1985; Beyer et al. 1989; Fromaget et al. 1990, 1992, 1993; Gilula 1990; Yeager 1993). As with the transmembrane proteins of the other three major categories of junctions, antibodies that bound to certain epitopes located in the cell–cell bridge domains of connexins were also able to interfere with, or even disrupt, cell–cell coupling (e.g., Warner et al. 1984; Gilula 1990; Hertzberg et al. 1985).

In parallel, an increasing number of connexins had been characterized by cloning and sequencing, using cDNA clones from a wide variety of tissues (Kumar and Gilula 1986; Nicholson et al. 1985; Heynkes et al. 1986; Paul 1986; Beyer et al. 1987, 1990; Traub et al. 1989; Zhang and Nicholson 1989; Kanter et al. 1991; reviews: Beyer et al. 1993; Willecke et al. 1993; Yeager 1993). These studies led to an extremely complex cell-type-specific catalog of mammalian connexins. Twenty-one human Cx gene products have been identified so far, and so one has to recognize the great potential of Cx-type combinations to form GJs (homomeric, heteromeric, homotypic, or heterotypic), and further different properties that may result from phosphorylations and other modifications (reviews: Bruzzone et al. 1996; Goodenough et al. 1996; Kumar and Gilula 1996; Sosinsky 2000; see, e.g., the anthology edited by Peracchia 2000; for a special review integrating recent results and insights, see Goodenough and Paul 2009).

In view of the fundamentally different ultrastructural organization of the GJs, compared to the other three junctional complexes, and the obvious lack of homology between the connexins and the cadherins or claudins, it was finally some surprise, at least to this author, that connexin cytoplasmic domains also specifically associated with certain AJ-typical plaque proteins such as MAGUK proteins of the ZO-1 to ZO-3 group (Giepmans and Molenaar 1998; Toyofuku et al. 1998, 2001; Giepmans et al. 2001; Kausalya et al. 2001; Laing et al. 2001a,b; Nielsen et al. 2001, 2002, 2003; Barker et al. 2002; Jin et al. 2004; Singh et al. 2005; Talhouk et al. 2008). That this interaction reflects direct binding of the MAGUK protein to the connexin partner has perhaps most convincingly been shown using pure ZO-1 protein synthesized by translation in vitro (e.g., Laing et al. 2001a,b). So it is justified to hypothesize that MAGUK proteins are also important components involved in regulatory mechanisms of GJ formation and functions, and that the intracellular GJ membrane surface is a "hot spot" of interactions with diverse cytoplasmic proteins (Table 1) (review: Dbouk et al. 2009). The list of GJ-associated proteins and of interactions of GJs with other cell structures is still growing, although a distinct electron-dense plaque structure has not been demonstrated (for examples and references, see recent reviews: Park et al. 2007; Koval 2008; Prochnow and Dermietzel 2008).

The history of the elucidation of the molecular composition of GJs has been for some time under some strange clouds and question marks. For example, a radically alternative concept of GJ structures in various invertebrates and vertebrates was based on the existence of GJ-like structures assembled from very small (16–27 kDa) polypeptides, later named "ductins" (for references, see Finbow et al. 1985; Finbow and Pitts 1993; Finbow 1997; Ashrafi et al. 2000). Very recently, now such "GJ-mimicking" structures and the ductins could be definitively ascribed to proteins of the vacuolar ATPase proton pump and to be absent from GJs (M. Finbow, personal communication). Another long-standing question mark had already been answered at the turn of the millenium: The so-called major intrinsic polypeptides (MPs or MIPs) of the eye lens, one of the best studied structures in cell biology, were finally identified as true connexins (e.g., MP70 is Cx50; here, the reader is referred to "translations" of MP into Cx language as, e.g., in Graw et al. 2001).

In view of the important cell–cell comunication and exchange functions of GJs, it was expected that studies on the functions of specific GJs and GJ proteins would also bring valuable information about the functions of individual proteins, not only from gene deletion or mutation experiments but also from the growing list of human disorders and diseases that can be clearly related to mutant or disregulated connexins (for examples, see recent reviews: Meşe et al. 2007; Koval 2008). Thus, with connexins as with the constitutive molecules of the other three categories of cell–cell junctions, cell biologists can finally also conclude with satisfaction that basic cell research here has not only revealed the molecular organizations and functions of junction components, but has also made an unexpectedly high number of contributions to medical research and practice.

SOME CONCLUDING REMARKS

Although on a first glance the molecular ensembles and the morphology of the four homotypic cell–cell junctions differ in many aspects, one can also note a series of similarities in structural organization:

(1) All four forms of symmetrical junctions are clusters of densely packed, specific transmembrane molecules.

(2) The transmembrane molecules are either once-spanning glycoproteins (cadherins) or tetraspan proteins.

(3) The cytoplasmically projecting domains of the transmembrane molecules are intimately associated with proteins that form layers or plaques, which widely vary in thickness and density and in certain cases can connect the junction with the cytoskeleton.

(4) The transmembrane molecules of a hemijunction are connected head-to-head with their counterparts of a hemi-junction of an adjacent cell.

(5) The positions are often not at random but topogenically integrated into the specific cell and tissue architecture.

(6) The types, the sizes, and the topological positions of these junctions are under developmental and functional control. Consequently, disorders often result in disease.

REFERENCES

Aberle H, Schwartz H, Kemler R. 1996. Cadherin-catenin complex: Protein interactions and their implications for cadherin function. *J Cell Biochem* **61:** 514–523.

Aghib DF, McCrea PD. 1995. The E-cadherin complex contains the src substrate p120. *Exp Cell Res* **218:** 359–369.

Aho S, Rothenberger K, Uitto J. 1999. Human p120ctn catenin: Tissue-specific expression of isoforms and molecular interactions with BP180/type XVII collagen. *J Cell Biochem* **73:** 390–399.

Amagai M. 1999. Autoimmunity against desmosomal cadherins in pemphigus. *J Dermatol Sci* **20:** 92–102.

Amagai M, Klaus-Kovtun V, Stanley JR. 1991. Autoantibodies against a novel epithelial cadherin in *pemphigus vulgaris*, a disease of cell adhesion. *Cell* **67:** 869–977.

Anastasiadis PZ, Reynolds AB. 2000. The 120 catenin family: Complex roles in adhesion, signaling and cancer. *J Cell Sci* **113:** 1319–1334.

Anastasiadis PZ, Moon SY, Thoreson MA, Mariner DJ, Crawford HC, Zheng Y, Reynolds AB. 2000. Inhibition of RhoA by p120 catenin. *Nat Cell Biol* **2:** 637–644.

Anderson JM. 1996. MAGUK magic. *Curr Biol* **6:** 326–329.

Anderson JM, Van Itallie CM. 2009. Physiology and function of the tight junction. *Cold Spring Harb Perspect Biol* **1:** a002584.

Ando-Akatsuka Y, Saitou M, Hirase T, Kishi M, Sakakibara A, Itoh M, Yonemura S, Furuse M, Tsujita Sh. 1996. Interspecies diversity of the occludin sequence: cDNA cloning of human, mouse, dog, and rat-kangaroo homologues. *J Cell Biol* **133:** 43–47.

Angst BD, Marcozzi C, Magee AI. 2001. The cadherin superfamily: Diversity in form and function. *J Cell Sci* **114:** 629–641.

Arnemann J, Sullivan KH, Magee AI, King IA, Buxton RS. 1993. Stratification-related expression of isoforms of the desmosomal cadherins in human epidermis. *J Cell Sci* **104:** 741–750.

Ashrafi GH, Pitts JD, Faccini AM, McLean P, O'Brien V, Finbow ME, Campo MS. 2000. Binding of bovine papillomavirus type 4 E8 to ductin (16K proteolipid), down-regulation of gap junction intercellular communication and full cell transformation are independent events. *J Gen Virol* **81:** 689–694.

Balda MS, Matter K. 2000. The tight junction protein ZO-1 and an interacting transcription factor regulate ErbB-2 expression. *EMBO J* **19:** 2024–2033.

Barker RJ, Price RL, Gourdie RG. 2002. Increased association of ZO-1 with connexin43 during remodeling of cardiac gap junctions. *Circ Res* **90:** 317–324.

Bazzi H, Getz A, Mahoney MG, Ishida-Yamamoto A, Langbein L, Wahl KK III, Christiano AM. 2006. Desmoglein 4 is expressed in highly differentiated keratinocytes and trichocytes in human epidermis and hair follicle. *Differentiation* **74:** 129–140.

Behrens J, Birchmeier W, Goodman SL, Imhof BA. 1985. Dissociation of Madin-Darby canine kidney epithelial cells by the monoclonal antibody anti-arc-1: Mechanistic aspects and identification of the antigen as a component related to uvomorulin. *J Cell Biol* **101:** 1307–1315.

Behrens J, Mareel MM, Van Roy FM, Birchmeier W. 1989. Dissecting tumour cell invasion: Epithelial cells acquire invasive properties after the loss of uvomorulin-mediated cell–cell adhesion. *J Cell Biol* **108:** 2435–2447.

Behrens J, Vakaet L, Friis R, Winterhager E, Van Roy F, Mareel MM, Birchmeier W. 1993. Loss of epithelial morphotype and gain of invasiveness correlates with tyrosine phosphorylation of the E-cadherin/β-catenin complex in cells transformed with a temperature-sensitive v-*src* gene. *J Cell Biol* **120:** 757–766.

Berndorff D, Gessner R, Kreft B, Schnoy N, Lajous-Petter A-M, Loch N, Reutter W, Hortsch M, Tauber R. 1994. Liver-intestine cadherin: Molecular cloning and characterization of a novel Ca^{2+}-dependent cell adhesion molecule expressed in liver and intestine. *J Cell Biol* **125:** 1353–1369.

Berx G, van Roy F. 2009. Involvement of members of the cadherin superfamily in cancer. *Cold Spring Harb Perspect Biol* **1:** a003129.

Beyer EC, Goodenough DA, Paul DL. 1990. Connexin family of gap-junction proteins. *J Membr Biol* **116:** 187–194.

Beyer EC, Paul DL, Goodenough DA. 1987. Connexin43: A protein from rat heart homologous to a gap junction protein from liver. *J Cell Biol* **105:** 2621–2629.

Beyer EC, Kistler J, Paul DL, Goodenough DA. 1989. Antisera directed against connexin43 peptides react with a 43-kD protein localized to gap junctions in myocardium and other tissues. *J Cell Biol* **108:** 595–605.

Beyer EC, Kanter HL, Rup DM, Westphale EM, Reed KE, Larson DM, Saffitz JE. 1993. Expression of multiple connexins by cells of the cardiovascular system and lens. In *Gap junctions* (eds. J.E. Hall, G.A. Zampighi, R.M. Davis), *Progr Cell Res.* Vol. **3**, pp. 171–175. Elsevier, Amsterdam.

Bierkamp C, McLaughlin KJ, Schwarz H, Huber O, Kemler R. 1996. Embryonic heart and skin defects in mice lacking plakoglobin. *Dev Biol* **180:** 780–785.

Birchmeier W, Hülsken J, Behrens J. 1995. E-cadherin as an invasion suppressor. In *Junctional complexes of epithelial cells*, Ciba Found. Symp. **125**, pp. 124–141. J. Wiley and Sons, Chichester.

Birchmeier W, Weidner KM, Hülsken J, Behrens J. 1993. Molecular mechanisms leading to cell junction (cadherin) deficiency in invasive carcinomas. *Sem Cancer Biol* **4:** 231–239.

Boller K, Vestweber D, Kemler R. 1985. Cell-adhesion molecule uvomorulin is localized in the intermediate junctions of adult intestinal epithelial cells. *J Cell Biol* **100:** 327–332.

Bonné S, van Hengel J, Nollet F, Kools P, van Roy F. 1999. Plakophilin-3, a novel armadillo-like protein present in nuclei and desmosomes of epithelial cells. *J Cell Sci* **112:** 2265–2276.

Bonné S, Gilbert B, Hatzfeld M, Chen X, Green KJ, van Roy F. 2003. Defining desmosomal plakophilin-3 interactions. *J Cell Biol* **161:** 403–416.

Bornslaeger EA, Corcoran CM, Stappenbeck TS, Green KJ. 1996. Breaking the connection: Displacement of the desmosomal plaque protein desmoplakin from cell–cell interfaces disrupts anchorage of intermediate filament bundles and alters intercellular junction assembly. *J Cell Biol* **134:** 985–1001.

Borrmann CM, Mertens C, Schmidt A, Langbein L, Kuhn C, Franke WW. 2000. Molecular diversity of plaques of epithelial adhering junctions. *Ann NY Acad Sci* **915:** 144–150.

Brabletz T. 2004. In vivo functions of catenins. In *Cell adhesion* (eds. J. Behrens, W.J. Nelson), Handbook Exp. Pharmacol., Vol. **165**, pp. 105–135. Springer, Berlin.

Brandner JM, Kief S, Grund C, Rendl M, Houdek P, Kuhn C, Tschachler E, Franke WW, Moll I. 2002. Organization and formation of the tight junction-system in human epidermis and cultured keratinocytes. *Eur J Cell Biol* **81:** 253–263.

Bretscher A. 1986. Purification of the intestinal microvillus cytoskeletal proteins villin, fimbrin, and ezrin. *Meth Enzymol* **134:** 24–37.

Brill S, Li S, Lyman CW, Church DM, Wasmuth JJ, Weissbach L, Bernards A, Snijders AJ. 1996. The Ras GTPase-activating-protein-related human protein IQGAP2 harbors a potential actin binding domain and interacts with calmodulin and Rho family GTPases. *Mol Cell Biol* **16:** 4869–4878.

Bruzzone R, White TW, Paul DL. 1996. Connections with connexins: The molecular basis of direct intercellular signalling. *Eur J Biochem* **238:** 1–27.

Bussemakers MJG, van Bokhoven A, Mees SGM, Kemler R, Schalken JA. 1993. Molecular cloning and characterization of the human E-cadherin cDNA. *Mol Biol Rep* **17:** 123–128.

Buxton RS, Magee AI. 1992. Structure and interactions of desmosomal and other cadherins. *Semin Cell Biol* **3:** 157–167.

Buxton RS, Cowin P, Franke WW, Garrod DR, Green KJ, King IA, Koch PJ, Magee AI, Rees DA, Stanley JR, et al. 1993. Nomenclature of the desmosomal cadherins. *J Cell Biol* **121:** 481–483.

Cadigan KM, Peifer M. 2009. Wnt signaling from development to disease: insights from model systems. *Cold Spring Harb Perspect Biol* **1:** a002881.

Campbell RD, Campbell JH. 1971. Origin and continuity of desmosomes. In *Origin and continuity of cell organelles* (eds. J. Reinert, H. Ursprung), pp. 261–298. Springer Verlag, Berlin.

Caspar D, Goodenough D, Makowski L, Phillips W. 1977. Gap junction structures. I. Correlated electron microscopy and x-ray diffraction. *J Cell Biol* **74:** 605–628.

Cereijido M, Anderson J. (eds.) 2001. *Tight Junctions* 2. edn. CRC Press, Boca Raton, FL.

Chen X, Bonné S, Hatzfeld M, van Roy F, Green KJ. 2002. Protein binding and functional characterization of plakophilin 2. *J Biol Chem* **277:** 10512–10522.

Chitaev NA, Troyanovsky SM. 1997. Direct Ca^{2+}-dependent heterophilic interaction between desmosomal cadherins, desmoglein and desmocollin, contributes to cell–cell adhesion. *J Cell Biol* **138:** 193–201.

Chitaev NA, Leube RE, Troyanovsky RB, Eshkind LG, Franke WW, Troyanovsky SM. 1996. The binding of plakoglobin to desmosomal cadherins: Patterns of binding sites and topogenic potential. *J Cell Biol* **133:** 359–369.

Choate KA, Lu Y, Lifton RP. 2001. Claudins mediate specific paracellular fluxes *in vivo*: Paracellin-1 is required for paracellular Mg^{2+} flux. In *Tight junctions* (eds. M. Cereijido, J. Anderson), pp. 483–492. CRC Press, Boca Raton, FL.

Choi H-J, Weis WI. 2004. Structural aspects of adherens junctions and desmosomes. In *Cell adhesion* (eds. J. Behrens, W.J. Nelson), Handbook Exp. Pharmacol., Vol. **165**, pp. 23–52. Springer, Berlin.

Citi S. 2001. The cytoplasmic plaque proteins of the tight junction. In *Tight junctions* (eds. M. Cereijido, J. Anderson), pp. 231–264. CRC Press, Boca Raton, FL.

Citi S, Sabanay H, Jakes R, Geiger B, Kendrick-Jones J. 1988. Cingulin, a new peripheral component of tight junctions. *Nature* **333:** 272–276.

Citi S, Sabanay H, Kendrick-Jones J, Geiger B. 1989. Cingulin: Characterization and localization. *J Cell Sci* **93:** 107–122.

Citi S, Amorosi A, Franconi F, Giotti A, Zampi G. 1991. Cingulin, a specific protein component of tight junctions, is expressed in normal and neoplastic human epithelial tissues. *Am J Pathol* **138:** 781–789.

Claude P. 1978. Morphological factors influencing transepithelial permeability: A model for the resistance of the zonula occludens. *J Membr Biol* **10:** 219–232.

Claude P, Goodenough DA. 1973. Fracture faces of zonulae occludentes from "tight" and "leaky" epithelia. *J Cell Biol* **58:** 390–400.

Cohen SM, Gorbsky GJ, Steinberg MS. 1983. Immunochemical characterization of related families of glycoproteins in desmosomes. *J Biol Chem* **258:** 2621–2627.

Colaco CALS, Evans WH. 1981. A biochemical dissection of the cardiac intercalated disk: Isolation of subcellular fractions containing fascia adherentes and gap junctions. *J Cell Sci* **52:** 313–325.

Collins JE, Legan PK, Kenny TP, MacGarvie J, Holton JL, Garrod DR. 1991. Cloning and sequence analysis of desmosomal glycoproteins 2 and 3 (desmocollins): Cadherin-like desmosomal adhesion molecules with heterogeneous cytoplasmic domains. *J Cell Biol* **113:** 381–391.

Cordenonsi M, D'Atri F, Hammar E, Parry DA, Kendrick-Jones J, Shore D, Citi S. 1999. Cingulin contains globular and coiled-coil doamins and interacts with ZO-1, ZO-2, ZO-3, and myosin. *J Cell Biol* **147:** 1569–1582.

Cowin P, Garrod DR. 1983. Antibodies to epithelial desmosomes show wide tissue and species cross-reactivity. *Nature* **302:** 148–150.

Cowin P, Kapprell H-P, Franke WW. 1985b. The complement of desmosomal plaque proteins in different cell types. *J Cell Biol* **101:** 1442–1454.

Cowin P, Mattey DM, Garrod DR. 1984a. Distribution of desmosomal components in the tissues of vertebrates, studied by fluorescent antibody staining. *J Cell Sci* **66:** 119–132.

Cowin P, Mattey DM, Garrod DR. 1984b. Identification of desmosomal surface components (desmocollins) and inhibition of desmosome formation by specific Fab. *J Cell Sci* **70:** 41–60.

Cowin P, Franke WW, Grund C, Kapprell HP, Kartenbeck J. 1985a. The desmosome-intermediate filament complex. In *The cell in contact* (eds. G.M. Edelman, J.-P. Thiery), pp. 427–460. John Wiley and Sons, New York.

Cowin P, Kapprell H-P, Franke WW, Tamkun J, Hynes RO. 1986. Plakoglobin: A protein common to different kinds of intercellular adhering junctions. *Cell* **46:** 1063–1073.

Cunningham BA. 1985. Structures of cell adhesion molecules. In *The cell in contact* (eds. G.M. Edelman, J.-P. Thiery), pp. 197–217. John Wiley and Sons, New York.

Cunningham BA, Leutzinger Y, Gallin WJ, Forkin BC, Edelman GM. 1984. Linear organization of the liver cell adhesion molecule L-CAM. *Proc Natl Acad Sci* **81:** 5787–5791.

Damsky CH, Richa J, Solter D, Knudsen K, Buck CA. 1983. Identification and purification of a cell surface glycoprotein mediating intercellular adhesion in embryonic and adult tissue. *Cell* **34:** 455–466.

Damsky CH, Richa J, Wheelock M, Damjanov I, Buck CA. 1985. Characterization of cell-CAM 120/80 and the role of surface membrane adhesion glycoproteins in early events in mouse embryo morphogenesis. In *The cell in contact* (eds. G.M. Edelman, J.-P. Thiery), pp. 233–255. John Wiley and Sons, New York.

Dbouk HA, Mroue RM, El-Sabban MW, Talhouk RS. 2009. Connexins: A myriad of functions extending beyond assembly of gap junction channels. *Cell Commun Signal* **7:** 4.

Deguchi M, Iizuka T, Hata Y, Nishimury W, Hirao K, Yao I, Kawabe H, Takai Y. 2000. PAPIN. A novel multiple PSD95/Dlg-A/ZO-1 protein interacting with neural plakophilin-related armadillo repeat protein/δ-catenin and p0071. *J Biol Chem* **275:** 29875–29880.

Dejana E. 1996. Endothelial adherens junctions: Implications in the control of vascular permeability and angiogenesis. *J Clin Invest* **98:** 1949–1953.

Dejana E, Lampugnani MG, Martinez-Estrada O, Bazzoni G. 2000. The molecular organization of endothelial junctions and their functional role in vascular morphogenesis and permeability. *Int J Dev Biol* **44:** 743–748.

Delva E, Tucker DK, Kowalczyk AP. 2009. The desmosome. Cold *Spring Harb Perspect Biol* **1:** a002543.

Demlehner MP, Schäfer S, Grund C, Franke WW. 1995. Continual assembly of half-desmosomal structures in the absence of cell contacts and their frustrated endocytosis: A coordinated Sisyphus cycle. *J Cell Biol* **131:** 745–760.

Denk H, Lackinger E, Cowin P, Franke WW. 1985a. Maintenance of desmosomes in mouse hepatocytes after drug-induced rearrangement of cytokeratin filament material. *Exp Cell Res* **161:** 161–171.

Denk H, Weybora W, Ratschek M, Sohar R, Franke WW. 1985b. Distribution of vimentin, cytokeratins, and desmosomal-plaque proteins in human nephroblastomas revealed by specific antibodies: Co-existence of cell groups of different degrees of epithelial differentiation. *Differentiation* **29:** 88–97.

Dermietzel R, Leibstein A, Frixen U, Janssen-Timmen U, Traub O, Willecke K. 1984. Gap junctions in several tissues share antigenic determinants with liver gap junctions. *EMBO J* **3:** 2261–2270.

Dewey MM, Barr L. 1964. A study of the structure and distribution of the nexus. *J Cell Biol* **23:** 553–585.

Diamond JM. 1977. The epithelial junction: Bridge, gate, and fence. *Physiologist* **20:** 10–18.

Drees F, Pokutta S, Yamada S, Nelson WJ, Weis WI. 2005. α-Catenin is a molecular switch that binds E-cadherin-β-catenin and regulates actin-filament assembly. *Cell* **123:** 903–915.

Drochmans P, Freudenstein C, Wanson J-C, Laurent L, Keenan TW, Stadler J, Leloup R, Franke WW. 1978. Structure and biochemical composition of desmosomes and tonofilaments isolated from calf muzzle epidermis. *J Cell Biol* **79:** 427–443.

Duden R, Franke WW. 1988. Organization of desmosomal plaque proteins in cells growing at low calcium concentrations. *J Cell Biol* **107:** 1049–1063.

Edelman GM. 1985. Specific cell adhesion in histogenesis and morphogenesis. In *The cell in contact* (eds. G.M. Edelman, J.-P. Thiery), pp. 139–168. John Wiley and Sons, New York.

Edelman GM, Gallin WJ, Delouvée A, Cunningham BA, Thiery J-P. 1983. Early epochal maps of two different cell adhesion molecules. *Proc Natl Acad Sci* **80:** 4384–4388.

Fanning AS. 2001. Organization and regulation of the tight junction by the actin-myosin cytoskeleton. In *Tight junctions* (eds. M. Cereijido, J. Anderson), pp. 265–284. CRC Press, Boca Raton, FL.

Fanning AS, Ma TY, Anderson JM. 2002. Isolation and functional characterization of the actin binding region in the tight junction protein ZO-1. *FASEB J* **16:** 1835–1837.

Fanning AS, Jameson BJ, Jesaitis LA, Anderson JM. 1998. The tight junction protein ZO-1 establishes a link between the transmembrane protein occludin and the actin cytoskeleton. *J Biol Chem* **273:** 29745–29753.

Farquhar MG, Palade GE. 1963. Junctional complexes in various epithelia. *J Cell Biol* **17:** 375–412.

Fawcett DW, McNutt S. 1969. The ultrastructure of the cat myocardium. I. Ventricular papillary muscle. *J Cell Biol* **42:** 1–45.

Ferreri DM, Vincent PA. 2008. Signaling to and through the endothelial adherens junction. In *Cell Junctions* (eds. S.E. La Flamme, A.P. Kowalczyk), pp. 169–195. Wiley-VCH, Weinheim.

Finbow ME. 1997. Vertebrate and invertebrate gap junctions: A common molecular basis? *Cell Biol Int* **21:** 329–331.

Finbow ME, Pitts JD. 1993. Is the gap junction channel—*the connexon*—made of connexin or ductin? *J Cell Sci* **106:** 463–472.

Finbow ME, Buultjens TEJ, John S, Kam E, Meagher L, Pitts JD. 1985. Molecular structure of the gap junctional channel. In *Junctional complexes of epithelial cells*, Ciba Found. Symp. **125**, pp. 92–107. J. Wiley and Sons, Chichester.

Förster C. 2008. Tight junctions and the modulation of barrier function in disease. *Histochem Cell Biol* **130:** 55–70.

Fouquet B, Zimbelmann R, Franke WW. 1992. Identification of plakoglobin in oocytes and early embryos of *Xenopus laevis*: Maternal expression of a gene encoding a junctional plaque protein. *Differentiation* **51:** 187–194.

Franke WW, Kapprell H-P, Cowin P. 1987b. Immunolocalization of plakoglobin in endothelial junctions: Identification as a special type of *Zonulae adhaerentes*. *Biol Cell* **59:** 205–218.

Franke WW, Borrmann CM, Grund C, Pieperhoff S. 2006. The *area composita* of adhering junctions connecting heart muscle cells of vertebrates. I. Molecular definition in intercalated disks of cardiomyocytes by immunoelectron microscopy of desmosomal proteins. *Eur J Cell Biol* **85:** 69–82.

Franke WW, Cowin P, Schmelz M, Kapprell H-P. 1987a. The desmosomal plaque and the cytoskeleton. In *Junctional complexes of epithelial cells*, Ciba Found. Symp. **125**, pp. 26–44. J. Wiley and Sons, Chichester.

Franke WW, Goldschmidt MD, Zimbelmann R, Mueller HM, Schiller DL, Cowin P. 1989. Molecular cloning and amino acid sequence of human plakoglobin, the common junctional plaque protein. *Proc Natl Acad Sci* **86:** 4027–4031.

Franke WW, Grund C, Kuhn C, Jackson BW, Illmensee K. 1982a. Formation of cytoskeletal elements during mouse embryogenesis. III. Primary mesenchymal cells and the first appearance of vimentin filaments. *Differentiation* **23:** 43–59.

Franke WW, Moll R, Müller H, Schmid E, Kuhn C, Krepler R, Artlieb U, Denk H. 1983a. Immunocytochemical identification of epithelium-derived human tumors with antibodies to desmosomal plaque proteins. *Proc Natl Acad Sci* **80:** 543–547.

Franke WW, Mueller H, Mittnacht S, Kapprell H-P, Jorcano JL. 1983b. Significance of two desmosome plaque-associated polypeptides of molecular weights 75,000 and 83,000. *EMBO J* **2:** 2211–2215.

Franke WW, Rickelt S, Barth M, Pieperhoff S. 2009. The junctions that don't fit the scheme: Special symmetrical junctions of their own kind. *Cell Tissue Res*, in press.

Franke WW, Schmid E, Grund C, Müller H, Engelbrecht I, Moll R, Stadler J, Jarasch E-D. 1981. Antibodies to high molecular weight polypeptides of desmosomes: Specific localization of a class of junctional proteins in cells and tissues. *Differentiation* **20:** 217–241.

Franke WW, Schmid E, Schiller DL, Winter S, Jarasch E-D, Moll R, Denk H, Jackson BW, Illmensee K. 1982b. Differentiation-related patterns of expression of proteins of intermediate-size filaments in tissues and cultured cells. *Cold Spring Harb Symp Quant Biol* **46:** 431–453.

Franke WW, Troyanovsky SM, Koch PJ, Troyanovsky R, Fouquet B, Leube RE. 1992. Desmosomal proteins—mediators of intercellular coupling and intermediate filament anchorage. *Cold Spring Harbor Symp Quant Biol* **57:** 643–652.

Friend DS, Gilula NB. 1972. Variations in tight and gap-junctions in mammalian tissues. *J Cell Biol* **53:** 758–776.

Frixen U, Behrens J, Sachs M. 1991. E-cadherin-mediated cell–cell adhesion prevents invasiveness of human carcinoma cells. *J Cell Biol* **113:** 173–185.

Fromaget C, Al Aoumari A, Dupont E, Briand JP, Gros D. 1990. Changes in the expression of Cx43, a cardiac gap-junctional protein, during mouse heart development. *J Mol Cell Cardiol* **22:** 1245–1258.

Fromaget C, El Aoumari A, Gros D. 1992. Distribution pattern of connexin 43, a gap junctional protein, during the differentiation of mouse heart myocytes. *Differentiation* **51:** 9–20.

Fromaget C, El Aoumari A, Jarry T, Briand J-P, Maurice M, Feldmann G, Maro B, Gros D. 1993. Expression of Cx43 in rat and mouse liver. In *Gap junctions* (eds. J.E. Hall, G.A. Zampighi, R.M. Davis), *Progr Cell Res* Vol. **3**, pp. 25–31. Elsevier, Amsterdam.

Funayama N, Nagafuchi A, Sato N, Tsukita S, Tsukita S. 1991. Radixin is a novel member of the band 4.1 family. *J Cell Biol* **115:** 1039–1048.

Furuse M, Sasaki H, Tsukita Sh. 1999. Manner of interaction of heterogenous claudin species within and between tight junction strands. *J Cell Biol* **147:** 891–903.

Furuse M, Fujimoto K, Sato N, Hirase T, Tsukita Sa, Tsukita Sh. 1996. Overexpression of occludin, a tight junction-associated integral membrane protein, induces the formation of intracellular multilamellar bodies bearing tight junction-like structures. *J Cell Sci* **109:** 429–435.

Furuse M, Fujita K, Hiiragi T, Fujimoto K, Tsukita Sh. 1998a. Claudin-1 and -2: Novel integral membrane proteins localizing at tight junctions with no sequence similarity to occludin. *J Cell Biol* **141:** 1539–1550.

Furuse M, Hata M, Furuse K, Yoshida Y, Haratake A, Sugitani Y, Noda T, Kubo A, Tsukita Sh. 2002. Claudin-based tight junctions are crucial for the mammalian epidermal barrier: A lesson from claudin-1-deficient mice. *J Cell Biol* **156:** 1099–1111.

Furuse M, Hirase T, Itoh M, Nagafuchi A, Yonemura S, Tsukita Sa, Tsukita Sh. 1993. Occludin: A novel integral membrane protein localizing at tight junctions. *J Cell Biol* **123:** 1777–1788.

Furuse M, Itoh M, Hirase T, Nagafuchi A, Yonemura S, Tsukita Sa, Tsukita Sh. 1994. Direct association of occludin with ZO-1 and its possible involvement in the localization of occludin at tight junctions. *J Cell Biol* **127:** 1617–1626.

Furuse M, Sasaki H, Fujimoto K, Tsukita Sh. 1998b. A single gene product, claudin-1 or -2, reconstitutes tight junction strands and recruits occludin in fibroblasts. *J Cell Biol* **143:** 391–401.

Gallin WJ, Edelman GM, Cunningham BA. 1983. Characterization of L-CAM, a major cell adhesion molecule from embryonic liver cells. *Proc Natl Acad Sci* **80:** 1038–1042.

Gallin WJ, Sorkin BC, Edelman GM, Cunningham BA. 1987. Sequence analysis of a cDNA clone encoding the liver cell adhesion molecule, L-CAM. *Proc Natl Acad Sci* **84:** 2808–2812.

Gallicano GI, Bauer C, Fuchs E. 2001. Rescuing desmoplakin function in extra-embryonic extoderm reveals the importance of this protein in embryonic heart, neuroepithelium, skin and vasuclature. *Development* **128:** 929–941.

Gallicano GI, Kouklis P, Bauer C, Yin M, Vasioukhin V, Degenstein L, Fuchs E. 1998. Desmoplakin is required early in development for assembly of desmosomes and cytoskeletal linkage. *J Cell Biol* **143:** 2009–2022.

Gary R, Bretscher B. 1993. Heterotypic and homotypic associations between ezrin and moesin, two putative membrane-cytoskeletal linking proteins. *Proc Natl Acad Sci* **90:** 10846–10850.

Garrod D, Chidgey M, North A. 1996. Desmosomes: Differentiation, development, dynamics and disease. *Curr Opin Cell Biol* **8:** 670–678.

Garrod DR, Merritt AJ, Nie Z. 2002. Desmosomal cadherins. *Curr Opin Cell Biol* **14:** 537–545.

Geiger B, Schmid E, Franke WW. 1983. Spatial distribution of proteins specific for desmosomes and adhaerens junctions in epithelial cells demonstrated by double immunofluorescence microscopy. *Differentiation* **23:** 189–205.

Geiger B, Volk T, Volberg T. 1985b. Molecular heterogeneity of adherens junctions. *J Cell Biol* **101:** 1523–1531.

Geiger B, Avnur Z, Volberg T, Volk T. 1985a. Molecular domains of adherens junctions. In *The cell in contact* (eds. G.M. Edelman, J.-P. Thiery), pp. 461–489. John Wiley and Sons, New York.

Gerull B, Heuser A, Wichter T, Paul M, Basson CT, McDermott DA, Lerman BB, Markowitz SM, Ellinor PT, MacRae CA et al. 2004. Mutations in the desmosomal protein plakophilin-2 are common in arrhythmogenic right ventricular cardiomyopathy. *Nat Genet* **36:** 1162–1164.

Giagtzoglou N, Ly CV, Bellen HJ. 2009. Cell adhesion, the backbone of the synapse: "Vertebrate" and "invertebrate" perspectives. *Cold Spring Harb Perspect Biol* **1:** a003079.

Giepmans BN, Molenaar WH. 1998. The gap junction protein connexin 43 interacts with the second PDZ domain of the zona occludes-1 protein. *Curr Biol* **8:** 931–934.

Giepmans BN, Verlaan I, Moolenaar WH. 2001. Connexin-43 interactions with ZO-1 and α- and β-tubulin. *Cell Commun Adhes* **8:** 213–217.

Gilula NB. 1985. Gap junctional contact between cells. In *The cell in contact* (eds. G.M. Edelman, J.-P. Thiery), pp. 395–409. John Wiley and Sons, New York.

Gilula NB. 1987. Topology of gap junction protein and channel function. In *Junctional complexes of epithelial cells*, Ciba Found. Symp. 125, pp. 128–139. J. Wiley and Sons, Chichester.

Gilula NB. 1990. Gap junction protein and its organization into the cell–cell channel. In *Morphoregulatory molecules* (eds. G.M. Edelman, B.A. Cunningham, Thiery J.P.), pp. 371–382. John Wiley and Sons, New York.

Giudice GJ, Cohen SM, Patel N, Steinberg MS. 1984. Immunological comparison of desmosomal components from several bovine tissues. *J Cell Biochem* **26:** 35–45.

Godsel LM, Getsios S, Huen AC, Green KJ. 2004. The molecular composition and function of desmosomes. In *Cell adhesion* (eds. J. Behrens, W.J. Nelson), Handbook Exp. Pharmacol., Vol. **165**, pp. 137–193. Springer, Berlin.

Golenhofen N, Drenckhahn D. 2000. The catenin, p120ctn, is a common membrane-associated protein in various epithelial cells and tissues. *Histochem Cell Biol* **114:** 147–155.

González-Mariscal L, Betanzos S, Nava P, Jaramillo BE. 2003. Tight junction proteins. *Progr Biophys Mol Biol* **81:** 1–44.

Goodenough DA. 1974. Bulk isolation of mouse hepatocyte gap junctions. Characterizations of the principal protein, connexin. *J Cell Biol* **61:** 557–563.

Goodenough DA. 1990. Gap junctions and intercellular communication. In *Morphoregulatory molecules* (eds. G.M. Edelman, B.A. Cunningham, J.P. Thiery), pp. 357–369. John Wiley and Sons, New York.

Goodenough DA, Revel JP. 1970. A fine structural analysis of intercellular junctions in the mouse liver. *J Cell Biol* **45:** 272–290.

Goodenough DA, Paul DL. 2009. Gap Junctions. *Cold Spring Harb Perspect Biol* **1:** a002576.

Goodenough DA, Revel JP. 1971. The permeability of isolated and in situ mouse hepatic gap junctions studied with enzymatic tracers. *J Cell Biol* **50:** 81–91.

Goodenough DA, Goliger JA, Paul DL. 1996. Connexins, Connexons, and intercellular communication. *Annu Rev Biochem* **65:** 475–502.

Goodwin L, Hill JE, Raynor K, Raszi L, Manabe M, Cowin P. 1990. Desmoglein shows extensive homology to the cadherin family of cell adhesion molecules. *Biochem Biophys Res Commun* **173:** 1224–1230.

Goossens S, Janssens B, Bonne S, De Rycke R, Braet F, van Hengel J, van Roy F. 2007. A unique and specific interaction between αT-catenin and plakophilin-2 in the area composita, the mixed-type junctional structure of cardiac intercalated discs. *J Cell Sci* **120:** 2126–2136.

Gorbsky G, Steinberg MS. 1981. Isolation of the intercellular glycoproteins of desmosomes. *J Cell Biol* **90:** 243–248.

Gorbsky G, Cohen SM, Shida H, Giudice GJ, Steinberg MS. 1985. Isolation of the non-glycosylated proteins of desmosomes and immunolocalization of a third plaque protein: Desmoplakin III. *Proc Natl Acad Sci* **82:** 810–814.

Gottardi CJ, Niessen CM. 2008. Tight junctions in simple and stratified epithelium. In *Cell Junctions* (eds. S.E. La Flamme, A.P. Kowalczyk), pp. 217–233. Wiley-VCH, Weinheim.

Gottardi CJ, Arpin M, Fanning AS, Louvard D. 1996. The junction-associated protein, zonula occludens-1, localizes to the nucleus before the maturation and during the remodeling of cell–cell contacts. *Proc Natl Acad Sci* **93:** 10779–10784.

Gould KL, Bretscher A, Esch FS, Hunter T. 1989. cDNA cloning and sequencing of the protein-tyrosine kinase substrate, ezrin, reveals homology to band 4.1. *EMBO J* **8:** 4133–4142.

Graw J, Löster J, Doewarto D, Fuchs H, Meyer B, Reis A, Wolf E, Balling R, Hrabé de Angelis M. 2001. Characterization of a mutation in the lens-specific MP70 encoding gene of the mouse leading to a dominant cataract. *Exp Eye Res* **73:** 867–876.

Green KJ, Gaudry CA. 2000. Are desmosomes more than tethers for intermediate filaments? *Nat Rev Mol Cell Biol* **1:** 208–216.

Green KJ, Goldman RD, Chisholm RL. 1988. Isolation of cDNAs encoding desmosomal plaque proteins: Evidence that bovine desmoplakins I and II are derived from two mRNAs and a single gene. *Proc Natl Acad Sci* **85:** 2613–1627.

Green KJ, Parry DAD, Steinert PM, Virata MLA, Wagner RM, Angst BD, Nilles LA. 1990. Structure of the human desmoplakins. Implications for function in the desmosomal plaque. *J Biol Chem* **265:** 2603–2612.

Green KJ, Getsios S, Troyanovsky S, Godsel LM. 2009. Intercellular junction assembly, dynamics and homeostasis. *Cold Spring Harb Perspect Biol* **2:** a000125.

Grossmann KS, Grund C, Huelsken J, Behrend M, Erdmann B, Franke WW, Birchmeier W. 2004. Requirement of plakophilin 2 for heart morphogenesis and cardiac junction formation. *J Cell Biol* **167:** 149–160.

Gumbiner BM. 1993. Breaking through the tight junction barrier. *J Cell Biol* **123:** 1631–1633.

Gumbiner B. 1996. Cell adhesion: The molecular basis of tissue architecture and morphogenesis. *Cell* **84:** 345–357.

Gumbiner B, Lowenkopf T, Apatira D. 1991. Identification of a 160-kDa polypeptide that binds to the tight junction protein ZO-1. *Proc Natl Acad Sci* **88:** 3460–3464.

Günzel D, Stuiver M, Kausalya PJ, Haisch L, Krug SM, Rosenthal R, Meij UC, Hunziker W, Fromm M, Müller D. 2009. Claudin-10 exists in six alternatively spliced isoforms that exhibit distinct localization and function. *J Cell Sci* **122:** 1507–1517.

Halbleib JM, Nelson WJ. 2006. Cadherins in development: Cell adhesion, sorting, and tissue morphogenesis. *Genes Dev* **20:** 3199–3214.

Hart MJ, Callow MG, Souza B, Polakis P. 1996. IQGAP1, a calmodulin-binding protein with a rasGAP-related domain, is a potential effector for cdc42Hs. *EMBO J* **15:** 2997–3005.

Haskins J, Lijie G, Wittchen ES, Hibbard J, Stevenson BR. 1998. ZO-3, a novel member of the MAGUK protein family found at the tight junction, interacts with ZO-1 and occludin. *J Cell Biol* **141:** 199–208.

Hatta K, Takeichi M. 1986. Expression of N-cadherin adhesion molecules associated with early morphogenetic events in chick development. *Nature* **320:** 447–449.

Hatta K, Nose A, Nagafuchi A, Takeichi M. 1988. Cloning and expression of cDNA encoding a neural calcium-dependent cell adhesion molecule: Its identity in the cadherin gene family. *J Cell Biol* **106:** 873–881.

Hatta K, Takagi S, Fujisawa H, Takeichi M. 1987. Spatial and temporal expression pattern of N-cadherin cell adhesion molecules correlated with morphogenetic processes of chicken embryo. *Dev Biol* **120:** 215–227.

Hatzfeld M. 1999. The armadillo family of structural proteins. *Int Rev Cytol* **186:** 179–224.

Hatzfeld M. 2005. The p120 family of cell adhesion molecules. *Eur J Cell Biol* **84:** 205–214.

Hatzfeld M, Nachtsheim C. 1996. Cloning and characterization of a new armadillo family member, p0071, associated with the junctional plaque: Evidence for a subfamily of closely related proteins. *J Cell Sci* **109:** 2767–2778.

Hatzfeld M, Kristjansson GI, Plessmann U, Weber K. 1994. Band 6 protein, a major constituent of desmosomes from stratified epitehlia, is a novel member of the armadillo multigene family. *J Cell Sci* **107:** 2259–2270.

Hecht G. 2001. Microbial pathogens that affect tight junctions. In *Tight junctions* (eds. M. Cereijido, J. Anderson), pp. 493–515. CRC Press, Boca Raton, FL.

Henderson D, Eibl H, Weber K. 1979. Structure and biochemistry of mouse hepatic gap junctions. *J Mol Biol* **132:** 193–218.

Herrenknecht K, Ozawa M, Eckerskorn C, Lottspeich F, Lenter M, Kemler R. 1991. The uvomorulin-anchorage protein α catenin is a vinculin homologue. *Proc Natl Acad Sci* **88:** 9156–9160.

Hertzberg EL. 1984. A detergent-independent procedure for the isolation of gap junctions from rat liver. *J Biol Chem* **259:** 9936–9943.

Hertzberg EL, Gilula NB. 1979. Isolation and characterization of gap junctions from rat liver. *J Biol Chem* **254:** 2138–2147.

Hertzberg EL, Skibbens RV. 1984. A protein homologous to the 27,000 dalton liver gap junction protein is present in a wide variety of species and tissues. *Cell* **39:** 61–69.

Hertzberg AL, Spray DC, Bennett MVL. 1985. Reduction of gap junctional conductance by microinjection of antibodies against the 27-kDa liver gap junction polypeptide. *Proc Natl Acad Sci* **82:** 2412–2416.

Hertzberg EL, Anderson DJ, Friedlander M, Gilula NB. 1982. Comparative analysis of the major polypeptides from liver gap junctions and lens fiber junctions. *J Cell Biol* **92:** 53–59.

Heuberger J, Birchmeier W. 2009. Interplay of cadherin-mediated cell adhesion and canonical Wnt signaling. *Cold Spring Harb Perspect Biol* **2:** 002915.

Heynkes R, Kozjek G, Traub O, Willecke K. 1986. Identification of a rat liver cDNA and mRNA coding for the 28 kDa gap junction. *FEBS Lett* **205:** 56–60.

Hinz B, Pittet P, Smith-Clerc J, Chaponnier C, Meister J-J. 2004. Myofibroblast development is characterized by specific cell–cell adherens junctions. *Mol Biol Cell* **15:** 4310–4320.

Hirao M, Sato N, Kondo T, Yonemura S, Monden M, Sasaki T, Takai Y, Tsukita Sh, Tsukita Sa. 1996. Regulation mechanism of ERM (Ezrin/Radixin/Moesin) protein/plasma membrane association: Possible involvement of phosphatidylinositol turnover and Rho-dependent signaling pathway. *J Cell Biol* **135:** 37–51.

Ho C, Zhou J, Medina M, Goto T, Jacobson M, Bhide PG, Kosik KS. 2000. δ-Catenin is a nervous system-specific adherens junction protein which undergoes dynamic relocalization during development. *J Comp Neurol* **420:** 261–276.

Hobmayer E, Hatta M, Fischer R, Fujisawa T, Holstein TW, Sugiyama T. 1996. Identification of a Hydra homologue of the β-catenin/plakoglobin/armadillo gene family. *Gene* **172:** 155–159.

Hobmayer B, Rentzsch F, Kuhn K, Happel CM, Cramer von Laue C, Snyder P, Rothbächer U, Holstein TW. 2000. WNT signalling molecules act in axis formation in the diploblastic metazoan *Hydra*. *Nature* **407:** 186–189.

Hofmann I, Kuhn C, Franke WW. 2008. Protein p0071, a major plaque protein of non-desmosomal adhering junctions, is a selective cell-type marker. *Cell Tissue Res* **334:** 381–399.

Hofmann I, Mertens C, Brettel M, Nimmrich V, Schnölzer M, Herrmann H. 2000. Interaction of plakophilins with desmoplakin and intermediate filament proteins: An in vitro analysis. *J Cell Sci* **113:** 2471–2483.

Hofmann I, Schlechter T, Kuhn C, Hergt M, Franke WW. 2009. Protein p0071 – an armadillo plaque protein that characterizes a specific subtype of adherens junctions. *J Cell Sci* **122:** 21–24.

Holtfreter J. 1939. Gewebeaffinität, ein Mittel der embryonalen Formbildung. *Arch Exp Zellforsch* **23:** 169–209.

Holthöfer B, Windoffer R, Troyanovsky S, Leube RE. 2007. Structure and function of desmosomes. *Int Rev Cytol* **264:** 65–163.

Holton JL, Kenny TP, Legan PK, Collins JE, Keen JN, Sharma R, Garrod DR. 1990. Desmosomal glycoproteins 2 and 3 (desmocollins) show N-terminal similarity ot calcium-dependent cell–cell adhesion molecules. *J Cell Sci* **97:** 239–246.

Horwarth AG, Hughes MR, Stevenson BR. 1992. Detection of the tight junction-associated protein ZO-1 in astrocytes and other nonepithelial cell types. *Am J Physiol* **262:** C461–C469.

Hsu S-N, Yonekura S, Ting C-Y, Robertson HM, Iwai Y, Uemura T, Lee C-H, Chiba A. 2009. Conserved alternative splicing and expression patterns of arthropod N-cadherin. *PLoS Genet* **5:** e1000441.

Huber AH, Nelson WJ, Weis WI. 1997. Three-dimensional structure of the armadillo repeat region of β-catenin. *Cell* **90:** 871–882.

Hull BE, Staehelin LA. 1976. Functional significance of the variations in the geometrical organization of tight junction networks. *J Cell Biol* **68:** 688–704.

Hülsken J, Birchmeier W, Behrens J. 1994. E-cadherin and APC compete for the interaction with β-catenin and the cytoskeleton. *J Cell Biol* **127:** 2061–2069.

Hyafil F, Morello D, Babinet C, Jacob F. 1980. A cell surface glycoprotein involved in the compaction of embryonal carcinoma cells and cleavage stage embryos. *Cell* **21:** 927–934.

Hyafil F, Babinet C, Jacob F. 1981. Cell–cell interactions in early embryogenesis: A molecular approach to the role of calcium. *Cell* **26:** 447–454.

Ikenouchi J, Furuse M, Furuse K, Sasaki H, Tsukita Sa, Tsukita Sh. 2005. Tricellulin constitutes a novel barrier at tricellular contacts of epithelial cells. *J Cell Biol* **171:** 939–945.

Imhof BA, Vollmers HP, Goodman SL, Birchmeier W. 1983. Cell–cell interaction and polarity of epithelial cells: Specific perturbation using a monoclonal antibody. *Cell* **36:** 667–675.

Itoh M, Morita K, Tsukita Sh. 1999. Characterization of ZO-2 as a MAGUK family member associated with tight as well as adherens junctions with a binding affinity to occludin and α catenin. *J Biol Chem* **274:** 5981–5986.

Itoh M, Nagafuchi A, Yonemura S, Kitani-Yasuda T, Tsukita S, Tsukita S. 1993. The 200-kD protein colocalizing with cadherins in non-epithelial cells is identical to ZO-1, a tight junction-associated protein in epithelial cells: cDNA cloning and immunoelectron microscopy. *J Cell Biol* **131:** 391–502.

Itoh M, Yonemura S, Nagafuchi A, Tsukita S, Tsukita S. 1991. A 220-kD undercoat-constitutive protein: Its specific localization at cadherin-based cell–cell adhesion sites. *J Cell Biol* **115:** 1449–1462.

Irie K, Shimizu K, Sakisaka T, Ikeda W, Takai Y. 2004. Roles of nectins in cell adhesion, signaling and polarization. In *Cell adhesion* (eds. J. Behrens, W.J. Nelson), Handbook Exp. Pharmacol., Vol. **165**, pp. 343–372. Springer, Berlin.

Iyer S, Ferreri DM, DeCocco NC, Minnear FL, Vincent PA. 2004. VE-cadherin-p120 interaction is required for maintenance of endothelial barrier function. *Am J Physiol Lung Cell Mol Physiol* **286:** L1143–L1153.

Izumi G, Sakisaka T, Baba T, Tanaka S, Morimoto K, Takai Y. 2004. Endocytosis of E-cadherin regulated by Rac and Cdc42 small G proteins through IQGAP1 and actin filaments. *J Cell Biol* **166:** 237–248.

Janssen-Timmen U, Dermietzel R, Frixen U, Leibstein A, Traub O, Willecke K. 1983. Immunocytochemical localization of the gap junction 26 K protein in mouse liver plasma membranes. *EMBO J* **2:** 295–302.

Janssens B, Goossens S, Staes K, Gilbert B, van Hengel J, Colpaert C, Bruyneel E, Mareel M, van Roy F. 2001. αT-catenin: A novel tissue-specific β-catenin-binding protein mediating strong cell–cell adhesion. *J Cell Sci* **114:** 3177–3188.

Jin C, Martyn KD, Kurata WE, Warn-Cramer BJ, Lau AF. 2004. Connexin43 PDZ2 binding domain mutants create functional gap junctions and exhibit altered phosphorylation. *Cell Commun Adhes* **11:** 67–87.

Kanai Y, Oda T, Shimoyama Y, Ochiai A, Oyama T, Yoshiura K, Akimoto S, Yamada T, Hirohashi S. 1994. Alterations of the cadherin-catenin cell adhesion system in cancers. In *Molecular and cellular basis for cell to cell interaction: Its significance in cancer* (eds. S. Hirohashi, H.L. Moses, E. Ruoslahti, T. Sugimura, M. Takeichi, M. Terada; *Proc. 24th Int. Symp. Princess Takamatsu Cancer Res. Fund*), pp. 51–62. Princeton Scientific Publishing, Princeton, NJ.

Kanter HL, Saffitz JE, Beyer EC. 1991. Canine cardiac myocytes express multiple gap-junction proteins. *Clin Res* **39:** p193A.

Kapprell H-P, Cowin P, Franke WW. 1987. Biochemical characterization of the soluble form of the junctional plaque protein, plakoglobin, from different cell types. *Eur J Biochem* **166:** 505–517.

Kapprell H-P, Owaribe K, Franke WW. 1988. Identification of a basic protein of Mr 75,000 as an accessory desmosomal plaque protein in stratified and complex epithelia. *J Cell Biol* **106:** 1679–1691.

Kartenbeck J, Franke WW, Moser JG, Stoffels U. 1983. Specific attachment of desmin filaments to desmosomal plaques in cardiac myocytes. *EMBO J* **2:** 735–742.

Kartenbeck J, Schmid E, Franke WW, Geiger B. 1982. Different modes of internalization of proteins associated with adhaerens junctions and desmosomes: Experimental separation of lateral contacts induces endocytosis of desmosomal plaque material. *EMBO J* **1:** 725–732.

Kartenbeck J, Schwechheimer K, Moll R, Franke WW. 1984. Attachment of vimentin filaments to desmosomal plaques in human meningiomal cells and arachnoidal tissue. *J Cell Biol* **98:** 1072–1081.

Kartenbeck J, Schmelz M, Franke WW, Geiger B. 1991. Endocytosis of junctional cadherins in bovine kidney epithelial (MDBK) cells cultured in low Ca^{2+} ion medium. *J Cell Biol* **113:** 881–892.

Kaufmann U, Zuppinger C, Waibler Z, Rudiger M, Urbich C, Martin B, Jockusch BM, Eppenberger H, Starzinski-Powitz A. 2000. The armadillo repeat region targets ARVCF to cadherin-based cellular junctions. *J Cell Sci* **113:** 4121–4135.

Kausalya PJ, Phua DCY, Hunziker W. 2004. Association of ARVCF with zonula occludens (ZO)-1 and ZO-2: Binding to PDZ-domain proteins and cell–cell adhesion regulate plasma membrane and nuclear localization of ARVCF. *Mol Biol Cell* **15:** 5503–5514.

Kausalya PJ, Reichert M, Hunziker W. 2001. Connexin45 directly binds to ZO-1 and localizes to the tight junction region in epithelial MDCK cells. *FEBS Lett* **505:** 92–96.

Kemler R, Babinet C, Eisen H, Jacob F. 1977. Surface antigen in early differentiation. *Proc Natl Acad Sci* **74:** 4449–4452.

King N, Hittinger CT, Carroll SB. 2003. Evolution of key cell signaling and adhesion protein families predates animal origins. *Science* **301:** 361–363.

King IA, O'Brien TJ, Buxton RS. 1996. Expression of the "skin-type" desmosomal cadherin DSC1 is closely linked to the keratinization of epithelial tissues during mouse development. *J Invest Dermatol* **107:** 531–538.

King IA, Angst BD, Hunt DM, Kruger M, Arnemann J, Buxton RS. 1997. Hierarchical expression of desmosomal cadherins during stratified epithelial morphogenesis in the mouse. *Differentiation* **62:** 83–96.

King IA, Arnemann J, Spurr NK, Buxton RS. 1993a. Cloning of the cDNA (DSC1) coding for human type 1 desmocollin and its assignment to chromosome 18. *Genomics* **18:** 185–194.

King IA, Sullivan KH, Bennett Jr, Buxton RS. 1995. The desmocollins of human foreskin epidermis: Identification and chromosomal assignment of a third gene and expression patterns of the three isoforms. *J Invest Dermatol* **105:** 314–321.

King IA, Tabiowo A, Purkis P, Leigh I, Magee AI. 1993b. Expression of distinct desmocollin isoforms in human epidermis. *J Invest Dermatol* **100:** 373–379.

Kljuic A, Bazzi H, Sundberg JP, Martinze-Mir A, S'shaughnessy R, Mahoney MG, Levy M, Montagutelli X, Ahmad W, Aita VM, et al. 2003. Desmoglein 4 in hair follicle differentiation and epidermal adhesion: Evidence from inherited hypotrichosis and acquired pemphigus vulgaris. *Cell* **113:** 249–260.

Knudsen KA, Wheelock MJ. 1992. Plakoglobin, or an 83-kD homologue distinct from β-catenin, interacts with E-cadherin and N-cadherin. *J Cell Biol* **118:** 671–679.

Koch PJ, Franke WW. 1994. Desmosomal cadherins: Another growing multigene family of adhesion molecules. *Curr Opin Cell Biol* **6:** 682–687.

Koch PJ, Goldschmidt MD, Walsh MJ, Zimbelmann R, Franke WW. 1991a. Complete amino acid sequence of the epidermal desmoglein precursor polypeptide and identification of a second type of desmoglein gene. *Eur J Cell Biol* **55:** 200–208.

Koch PJ, Goldschmidt MD, Walsh MJ, Zimbelmann R, Schmelz M, Franke WW. 1991b. Amino acid sequence of bovine muzzle epithelial desmocollin derived from cloned cDNA: A novel subtype of desmosomal cadherins. *Differentiation* **47:** 29–36.

Koch PJ, Goldschmidt MD, Zimbelmann R, Troyanovsky R, Franke WW. 1992. Complexity and expression patterns of the desmosomal cadherins. *Proc Natl Acad Sci* **89:** 353–357.

Koch PJ, Walsh MJ, Schmelz M, Goldschmidt MD, Zimbelmann R, Franke WW. 1990. Identification of desmoglein, a constitutive desmosomal glycoprotein, as a member of the cadherin family of cell adhesion molecules. *Eur J Cell Biol* **53:** 1–12.

Koeser J, Troyanovsky SM, Grund C, Franke WW. 2003. De novo formation of desmosomes in cultured cells upon transfection of genes encoding specific desmosomal components. *Exp Cell Res* **285:** 114–130.

Korman NJ, Eyre RW, Klaus-Kovtun V, Stanley JR. 1989. Demonstration of an adhering-junction molecule (plakoglobin) in the autoantigens of *Pemphigus foliaceus* and *Pemphigus vulgaris*. *N Engl J Med* **321:** 631–635.

Kosik KS, Donahue CP, Israely I, Liu X, Ochiishi T. 2005. δ-Catenin at the synaptic-adherens junction. *Trends Cell Biol* **15:** 172–178.

Koslov ER, Maupin P, Pradhan D, Morrow JS, Rimm DL. 1997. α-Catenin can form asymmetric homodimeric complexes and/or heterodimeric complexes with β-catenin. *J Biol Chem* **272:** 27301–27306.

Kottke MD, Delva E, Kowalczyk AP. 2006. The desmosome: Cell science lessons from human diseases. *J Cell Sci* **119:** 797–806.

Koval M. 2008. Gap junctions: Connexin functions and roles in human disease. In *Cell Junctions* (eds. S.E. La Flamme, A.P. Kowalczyk), pp. 197–216. Wiley-VCH, Weinheim.

Kowalczyk AP, Bornslaeger EA, Borgwardt JE, Palka HL, Dhaliwal AS, Corcoran CM, Denning MF, Green KJ. 1997. The amino-terminal domain of desmoplakin binds to plakoglobin and clusters desmosomal cadherin-plakoglobin complexes. *J Cell Biol* **139:** 773–784.

Kowalczyk AP, Navarro P, Dejana E, Bornslaeger EA, Green KJ, Kopp DS, Borgwardt JE. 1998. VE-cadherin and desmoplakin are assembled into dermal microvascular endothelial intercellular junctions: A pivotal role for plakoglobin in the recruitment of desmoplakin to the intercellular junctions. *J Cell Sci* **111:** 3045–3057.

Kowalczyk AP, Palka HL, Luu HH, Nilles LA, Anderson JE, Wheelock MJ, Green KJ. 1994. Posttranslational regulation of plakoglobin expression. Influence of the desmosomal cadherins on plakoglobin metabolic stability. *J Biol Chem* **269:** 31214–31223.

Kreft B, Berndorff D, Böttinger A, Finnemann S, Wedlich D, Hortsch M, Tauber R, Geßner R. 1997. LI-Cadherin-mediated cell–cell adhesion does not require cytoplasmic interactions. *J Cell Biol* **136:** 1109–1121.

Kumar NM, Gilula NB. 1986. Cloning and characterization of human and rat liver cDNAs coding for a gap junction protein. *J Cell Biol* **103:** 767–776.

Kumar N, Gilula NB. 1996. The gap junction communication channel. *Cell* **84:** 381–388.

Kuroda S, Fukata M, Kobayashi K, Nakafuku M, Nomura N, Iwamatsu A, Kaibuchi K. 1996. Identification of IQGAP as a putative target for the small GTPases, Cdc42 and Rac1. *J Biol Chem* **271:** 23363–23367.

Kuroda S, Fukata M, Nakagawa M, Fujii K, Nakamura T, Ookubo T, Izawa I, Nagase T, Nomura N, Tani H, et al. 1998. Role of IQGAP1, a target of the small GTPases Cdc42 and Rac1, in regulation of E-cadherin-mediated cell–cell adhesion. *Science* **281:** 832–835.

Laing JG, Manley-Markowski RN, Koval M, Civitelli R, Steinberg TH. 2001a. Connexin45 interacts with zonula occludens-1 and connexin43 in osteoblastic cells. *J Biol Chem* **276:** 23051–23055.

Laing JG, Manley-Markowski RN, Koval M, Civitelli R, Steinberg TH. 2001b. Connexin45 interacts with zonula occludens-I in osteoblastic cells. *Cell Commun Adhes* **8:** 209–212.

Lampugnani MG, Corada M, Caveda L, Breviario F, Ayalon O, Geiger B, Dejana E. 1995. The molecular organization of endothelial cell to cell junctions: Differential association of plakoglobin, β-catenin, and α-catenin with vascular endothelial cadherin (VE-cadherin). *J Cell Biol* **129:** 203–217.

Langbein L, Grund C, Kuhn C, Praetzel S, Kartenbeck J, Brandner JM, Moll I, Franke WW. 2002. Tight junctions and compositionally related junctional structures in mammalian stratified epithelia and cell cultures derived therefrom. *Eur J Cell Biol* **81:** 419–435.

Langbein L, Pape U-F, Grund C, Kuhn C, Praetzel S, Moll I, Moll R, Franke WW. 2003. Tight junction-related structures in the absence of a lumen: Occludin, claudins and tight junction plaque proteins in densely packed cell formations of stratified epithelia and squamous cell carcinomas. *Eur J Cell Biol* **82:** 385–400.

Legan PK, Yue KKM, Chidgey MAJ, Holton JL, Wilkinson RW, Garrod DR. 1994. The bovine desmocollin family: A new gene and expression patterns reflecting epithelial cell proliferation and differentiation. *J Cell Biol* **126:** 507–518.

Lehtonen S, Ryan JJ, Kudlicka K, Iino N, Zhou H, Farquhar MG. 2005. Cell junction-associated proteins IQGAP1, MAGI-2, CASK, spectrins, and α-actinin are components of the nephrin multiprotein complex. *Proc Natl Acad Sci* **102:** 9814–9819.

Leloup R, Laurent L, Ronveaux M-F, Drochmans P, Wanson J-C. 1979. Desmosomes and desmogenesis in the epidermis of calf muzzle. *Biol Cell* **34**, 137–152.

Lewis-Tuffin JJ, Anastasiadis PZ. 2008. Armadillo repeat proteins at epithelial adherens junctions. In *Cell Junctions* (eds. S.E. La Flamme, A.P. Kowalczyk), pp. 153–167. Wiley-VCH, Weinheim.

Loewenstein WR. 1981. Junctional intercellular communication: The cell-to-cell membrane channel. *Physiol Rev* **61:** 829–913.

Loh YH, Christoffels A, Brenner S, Hunziker W, Venkatesh B. 2004. Extensive expansion of the claudin gene family in the teleost fish, *Fugu rubripes*. *Genome Res* **14:** 1248–1257.

Lu Q, Paredes M, Medina M, Zhou J, Cavallo R, Peifer M, Orecchio L, Kosik KS. 1999. δ-Catenin, and adhesive junction-associated protein which promotes cell scattering. *J Cell Biol* **144:** 519–532.

Lui WY, Mruk DD, Cheng CY. 2005. Interactions among IQGAP1, Cdc42, and the cadherin/catenin protein complex regulate Sertoli-germ cell adherens junction dynamics in the testis. *J Cell Phsiol* **202:** 49–66.

Madara JL, Dharmsathaphorn K. 1985. Occluding junction structure-function relationships in a cultured epithelial monolayer. *J Cell Biol* **101:** 2124–2133.

Mandai K, Nakanishi H, Satoh A, Obaishi H, Wada M, Nishioka H, Itoh M, Mizoguchi A, Aoki T, Fujimoto T, Matsuda Y, Tsukita S, Takai Y. 1997. Afadin: A novel actin filament-binding protein with one PDZ domain localized at cadherin-based cell-to-cell adherens junction. *J Cell Biol* **139:** 517–528.

Marcozzi C, Burdett IDJ, Buxton RS, Magee AI. 1998. Coexpression of both types of desmosomal cadherin and plakoglobin confers strong intercellular adhesion. *J Cell Sci* **111:** 495–509.

Mareel MM, Behrens J, Birchmeier W, De Bruyne GK, Vleminckx K, Hoogewijs A, Fiers WC, Van Roy FM. 1991. Downregulation of E-cadherin expression in Madin Darby canine kidney (MDCK) cells inside nude mice tumors. *Int J Cancer* **47:** 922–928.

Mareel M, Vleminckx K, Vermeulen S, Yan G, Bracke M, van Roy F. 1994. Downregulation in vivo of the invasion-suppressor molecule E-cadherin in experimental and clinical cancer. In *Molecular and cellular basis for cell to cell interaction: Its significance in cancer* (eds. S. Hirohashi, HL Moses, E. Ruoslahti, T. Sugimura, M. Takeichi, M. Terada; Proc. 24th Int. Symp. Princess Takamatsu Cancer Res. Fund), pp. 63–80. Princeton Scientific Publishing, Princeton, NJ.

Mariner DJ, Sirotkin H, Daniel JM, Lindman BR, Mernaugh RL, Patten AK, Thoreson MA, Reynolds AB. 1999. Production and characterization of monoclonal antibodies to ARVCF. *Hybridoma* **18:** 343–349.

Mariner DJ, Wang J, Reynolds AB. 2000. ARVCF localizes to the nucleus and adherens junction and is mutually exclusive with p120$^{(ctn)}$ in E-cadherin complexes. *J Cell Sci* **113:** 1481–1490.

Matsunaga M, Hatta K, Takeichi M. 1988. Role of N-cadherin cell adhesion molecules in the histogenesis of neural retina. *Neuron* **1:** 289–295.

Matter K, Balda M. 1999. Occludin and the function of tight junctions. *Int Rev Cytol* **186:** 117–146.

McCallum SJ, Su WJ, Cerione RA. 1996. Identification of a putative effector for Cdc42Hs with high sequence similarity to the RasGAP-related protein IQGAP1 aqnd Cdc42Hs binding partner with similarity to IQGAP2. *J Biol Chem* **271:** 21732–21737.

McCrea PD, Gumbiner BM. 1991. Purification of a 92-kDa cytoplasmic protein tightly associated with the cell–cell adhesion molecule E-cadherin (uvomorulin). Characterization and extractability of the protein complex from the cell cytostructure. *J Biol Chem* **266:** 4514–4520.

McCrea PD, Turck CW, Gumbiner B. 1991. A homolog of the armadillo protein in *Drosophila* (plakoglobin) associated with E-cadherin. *Science* **254:** 1359–1361.

McNutt NS, Weinstein RS. 1970. The ultrastructure of the nexus: A correlated thin-section and freeze-cleave study. *J Cell Biol* **47:** 666–687.

Mechanic S, Raynor K, Hill JE, Cowin P. 1991. Desmocollins form a distinct subset of the cadherin family of cell adhesion molecules. *Proc Natl Acad Sci* **88:** 4476–4480.

Meng W, Takeichi M. 2009. Adherens junction: Molecular architecture and regulation. *Cold Spring Harb Perspect Biol* **1:** 002899.

Mertens C, Kuhn C, Franke WW. 1996. Plakophilins 2a and 2b: Constitutive proteins of dual location in the karyoplasm and the desmosomal plaque. *J Cell Biol* **135:** 1009–1025.

Mertens C, Kuhn C, Moll R, Schwetlick I, Franke WW. 1999. Desmosomal plakophilin 2 as a differentiation marker in normal and malignant tissues. *Differentiation* **64:** 277–290.

Mertens C, Hofmann I, Wang Z, Teichmann M, Sepehri Chong S, Schnölzer M, Franke WW. 2001. Nuclear particles containing RNA polymerase III complexes associated with the junctional plaque protein plakophilin 2. *Proc Natl Acad Sci* **98:** 7795–7800.

Meşe G, Richard G, White TW. 2007. Gap junctions: Basic structure and function. *J Invest Dermatol* **127:** 2516–2524.

Mitic LL, Anderson JM. 1998. Molecular architecture of tight junctions. *Annu Rev Physiol* **60:** 121–142.

Moll R, Cowin P, Kapprell H-P, Franke WW. 1986. Biology of disease. Desmosomal proteins: New markers for identification and classification of tumors. *Lab Invest* **54:** 4–25.

Montesano R, Friend DS, Perrelet A, Orci L. 1975. In vivo assembly of tight junctions in fetal rat liver. *J Cell Biol* **67:** 310–319.

Morita K, Itoh M, Saitou M, Ando-Akatsuka Y, Furuse M, Yoneda K, Imamura S, Tsukita Sh. 1998. Subcellular distribution of tight junction-associated proteins (occludin, ZO-1 and ZO-2) in rodent skin changes. *J Invest Dermatol* **110:** 862–866.

Morita K, Furuse M, Fujimoto K, Tsukita Sh. 1999a. Claudin multigene family encoding four-transmembrane domain protein components of tight junction strands. *Proc Natl Acad Sci* **96:** 511–516.

Morita K, Sasaki H, Fujimoto K, Furuse M, Tsukita Sh. 1999b. Claudin-11/OSP-based tight junctions of myelin sheaths in brain and Sertoli cells in testis. *J Cell Biol* **145:** 579–588.

Morita K, Sasaki H, Furuse M, Tsukita Sh. 1999c. Endothelial claudin: Claudin-5/TMVCF constitutes tight junction strands in endothelial cells. *J Cell Biol* **147:** 185–194.

Morita K, Furuse M, Yoshida Y, Itoh M, Sasaki H, Tsukita Sh, Miyachi Y. 2002. Molecular architecture of tight junctions of periderm differs from that of the maculae occludentes of epidermis. *J Invest Dermatol* **118:** 1073–1079.

Moscona AA. 1962. Analysis of cell recombinations in experimental synthesis of tissues in vitro. *J Cell Comp Physiol* **60** (Suppl. 1): 65–80.

Mueller H, Franke WW. 1983. Biochemical and immunological characterization of desmoplakins I and II, the major polypeptides of the desmosomal plaque. *J Mol Biol* **163:** 647–671.

Nagafuchi A, Tsukita S. 1994. The loss of the expression of α catenin, the 102 kD cadherin associated protein, in central nervous tissues during development. *Dev Growth Differ* **36:** 59–71.

Nagafuchi A, Takeichi M, Tsukita S. 1991. The 102 kD cadherin-associated protein: Similarity to vinculin and posttranscriptional regulation of expression. *Cell* **65:** 849–857.

Nagafuchi A, Ishihara S, Tsukita S. 1994. The roles of catenins in the cadherin-mediated cell adhesion: Functional analysis of E-cadherin-α catenin fusion molecules. *J Cell Biol* **127:** 235–245.

Näthke IS, Hinck L, Swedlow JR, Papkoff J, Nelson WJ. 1994. Defining interactions and distributions of cadherin and catenin complexes in polarized epithelial cells. *J Cell Biol* **125:** 1341–1352.

Nichols SA, Dirks W, Pearse JS, King N. 2006. Early evolution of animal cell signaling and adhesion genes. *Proc Natl Acad Sci* **103:** 12451–12456.

Nicholson BJ, Gros DB, Kent SBH, Hood LE, Revel J-P. 1985. The M_r 28,000 gap junction proteins from rat heart and liver are different but related. *J Biol Chem* **260:** 6514–6517.

Nielsen PA, Baruch A, Giepmans BN, Kumar NM. 2001. Characterization of the assocation of connexins and ZO-1 in the lens. *Cell Commun Aches* **8:** 219–223.

Nielsen PA, Baruch A, Shestopalov VI, Giepmans BN, Dunia I, Benedetti EL, Kumar NM. 2003. Lens connexins α3Cx46 and α8Cx50 interact with zonula occludens protein-1 (ZO-1). *Mol Biol Cell* **14:** 2470–2481.

Nielsen PA, Geahm DL, Giepmans BN, Baruch A, Hall JE, Kuman NM. 2002. Molecular cloning, functional expression, and tissue distribution of a novel human gap junction-forming protein, connexin-31.9. Interaction with zona occludens protein-1. *J Biol Chem* **277:** 38272–38283.

Nieset JE, Redfield AR, Jin F, Knudsen KA, Johnson KR, Wheelock MJ. 1997. Characterization of the interactions of α-catenin with α-actinin and β-catenin/plakoglobin. *J Cell Sci* **110:** 1013–1022.

Nilles LA, Parry DA, Powers EE, Angst BD, Wagner RM, Green KJ. 1991. Structural analysis and expression of human desmoglein: A cadherin-like component of the desmosome. *J Cell Sci* **99:** 809–821.

Nollet F, Kools P, van Roy F. 2000. Phylogenetic analysis of the cadherin superfamily allows identification of six major subfamilies besides several solitary members. *J Mol Biol* **299:** 551–572.

Noritake J, Fukata M, Sato K, Nakagawa M, Watanabe T, Izumi N, Wang S, Fukata Y, Kaibuchi K. 2004. Positive role of IQGAP1, an effector of Rac1, in actin-meshwork formation at sites of cell–cell contact. *Mol Biol Cell* **15:** 1065–1076.

North AJ, Bardsley WG, Hyam J, Bornslaeger EA, Cordingley HC, Trinnaman B, Hatzfeld M, Green KJ, Magee AI, Garrod DR. 1999. Moleclar map of the desmosomal plaque. *J Cell Sci* **112:** 4325–4336.

Nose A, Takeichi M. 1986. A novel cadherin cell adhesion molecule: Its expression patterns associated with implantation and organogenesis of mouse embryos. *J Cell Biol* **103:** 2649–2658.

Nose A, Nagafuchi A, Takeichi M. 1987. Isolation of placental cadherin cDNA: Identification of a novel gene family of cell–cell adhesion molecules. *EMBO J* **6:** 3655–3661.

Nuber UA, Schäfer S, Schmidt A, Koch PJ, Franke WW. 1995. The widespread human desmocollin Dsc2 and tissue-specific patterns of synthesis of various desmocollin subtypes. *Eur J Cell Biol* **66:** 69–74.

Nuber UA, Schäfer S, Stehr S, Rackwitz H-R, Franke WW. 1996. Patterns of desmocollin synthesis in human epithelia: Immunolocalization of desmocollins 1 and 3 in special epithelia and in cultured cells. *Eur J Cell Biol* **71:** 1–13.

Ohkubo T, Ozawa M. 1999. $p120^{ctn}$ binds to the membrane-proximal region of the E-cadherin cytoplasmic domain and is involved in modulation of adhesion activity. *J Biol Chem* **274:** 21409–21415.

Okada TS. 1996. The path leading to the discovery of cadherin: In retrospect. *Develop Growth Differ* **38:** 583–596.

Ozawa M, Kemler R. 1992. Molecular organization of the uvomorulin-catenin complex. *J Cell Biol* **116:** 989–996.

Ozawa M, Baribault H, Kemler R. 1989. The cytoplasmic domain of the cell adhesion molecules uvomorulin associates with three independent proteins structurally related in different species. *EMBO J* **8:** 1711–1717.

Ozawa M, Ringwald M, Kemler R. 1990b. Uvomorulin-catenin complex formation is regulated by a specific domain in the cytoplasmic region of the cell adhesion molecule. *Proc Natl Acad Sci* **87:** 4246–4250.

Ozawa M, Hoschützky M, Herrenknecht K, Kemler R. 1990a. A possible new adhesive site in the cell-adhesion molecule uvomorulin. *Mech Dev* **33:** 49–56.

Paffenholz R, Franke WW. 1997. Identification and localization of a neurally expressed member of the plakoglobin/*armadillo* multigene family. *Differentiation* **61:** 293–304.

Paffenholz R, Kuhn C, Grund C, Stehr S, Franke WW. 1999. The *arm*-repeat protein NPRAP (neurojungin) is a constituent of the plaques of the outer limiting zone in the retina, defining a novel type of adhering junction. *Exp Cell Res* **250:** 452–464.

Palka HL, Green KJ. 1997. Roles of plakoglobin end domains in desmosome assembly. *J Cell Sci* **110:** 2359–2371.

Park DJ, Wallick CJ, Martyn KD, Lau AF, Jin C, Warn-Cramer BJ. 2007. Akt phosphorylates connexin43 on Ser373, a "mode-1" binding site for 14-3-3. *Cell Commun Adhes* **14:** 211–226.

Parker AE, Wheeler GN, Arnemann J, Pidsley SC, Ataliotis P, Thomas CL, Rees DA, Magee AI, Buxton RS. 1991. Desmosomal glycoproteins II and III. Cadherin-like junctional molecules generated by alternative splicing. *J Biol Chem* **266:** 10438–10445.

Parrish EP, Garrod DR, Mattey DL, Hand L, Steart P, Weller RO. 1986. Mouse antisera specific for desmosomal adhesion molecules of suprabasal skin cells, meninges, and meningioma. *Proc Natl Acad Sci* **83:** 2657–2661.

Parrish EP, Marston JE, Mattey DL, Measures HR, Venning R, Garrod DR. 1990. Size heterogeneity, phosphorylation and transmembrane organisation of desmosomal glycoproteins 2 and 3 (desmocollins) in MDCK cells. *J Cell Sci* **96:** 239–248.

Parrish EP, Steart PV, Garrod DR, Weller RO. 1987. Antidesmosomal monoclonal antibody in the diagnosis of intracranial tumours. *J Pathol* **153:** 265–273.

Pasdar M, Nelson WJ. 1988a. Kinetics of desmosome assembly in Madin-Darby canine kidney epithelial cells: Temporal and spatial regulation of desmoplakin organization and stabilization upon cell–cell contact. I. Biochemical analysis. *J Cell Biol* **106:** 677–685.

Pasdar M, Nelson WJ. 1988b. Kinetics of desmosome assembly in Madin-Darby canine kidney epithelial cells: Temporal and spatial regulation of desmoplakin organization and stabilization upon cell–cell contact. II. Morphological analysis. *J Cell Biol* **109:** 163–177.

Paul DL. 1985. Antibody against liver gap junction 27-kD protein is tissue-specific and cross-reacts with a 54-kD protein. In *Gap junctions* (eds. M.V.L. Bennett, D.C. Spray), pp. 107–122. Cold Spring Habor Laboratory, Cold Spring Harbor, New York.

Paul D. 1986. Molecular cloning of cDNA for rat liver gap junction protein. *J Cell Biol* **103:** 123–134.

Paulson AF, Mooney E, Fang X, Ji H, McCrea PD. 2000. Xarvcf, *Xenopus* member of the p120 catenin subfamily associating with cadherin juxtamembrane region. *J Biol Chem* **275:** 30124–30131.

Peifer M, Wieschaus E. 1990. The segment polarity gene armadillo encodes a functionally modular protein that is the *Drosophila* homolog of human plakoglobin. *Cell* **63:** 1167–1176.

Peifer M, Berg S, Reynolds AB. 1994. A repeating amino acid motif shared by proteins with diverse cellular roles. *Cell* **76:** 789–791.

Peifer M, McCrea PD, Green KJ, Wieschaus E, Gumbiner BM. 1992. The vertebrate adhesive junction proteins β-catenin and plakoglobin and the *Drosophila* segment polarity gene armadillo form a multigene family with similar properties. *J Cell Biol* **118:** 681–691.

Peracchia C. (ed.) 2000. *Gap Junctions. Molecular Basis of Cell Communication and Health and Disease*. Curr. Top. in Membranes, Vol. **49**. Academic Press, San Diego.

Perez TD, Nelson WJ. 2004. Cadherin adhesion: mechanisms and molecular interactions. In *Cell adhesion* (eds. J. Behrens, W.J. Nelson), Handbook Exp. Pharmacol., Vol. **165**, pp. 3–21. Springer, Berlin.

Peyrieras N, Hyafil F, Louvard D, Ploegh H, Jacob F. 1983. Uvomorulin: A nonintegral membrane protein of early mouse embryo. *Proc Natl Acad Sci* **80:** 6274–6277.

Prochnow N, Dermietzel R. 2008. Connexons and cell adhesion: A romantic phase. *Histochem Cell Biol* **130:** 71–77.

Pummi K, Malminen M, Aho H, Karvonen S-L, Peltonen J, Peltonen S. 2001. Epidermal tight junctions: ZO-1 and occludin are expressed in mature, developing, and affected skin and in vitro differentiating keratinocytes. *J Invest Dermatol* **117:** 1050–1058.

Ranscht B, Dours-Zimmermann MT. 1991. T-cadherin, a novel cadherin cell adhesion molecule in the nervous system lacks the conserved cytoplasmic region. *Neuron* **7:** 391–402.

Revel JP, Yancey SB, Nicholson B. 1985. Biology of gap junction molecules and development. In *The cell in contact* (eds. G.M. Edelman, J.-P. Thiery), pp. 411–425. John Wiley and Sons, New York.

Revel JP, Yancey SB, Nicholson B, Hoh J. 1987. Sequence diversity of gap junction proteins. In *Junctional complexes of epithelial cells*, Ciba Found. Symp. **165**, pp. 108–127. J. Wiley and Sons, Chichester.

Reynolds AB, Daniel J, McCrea PD, Wheelock MJ, Wu J, Zang Z. 1994. Identification of a new catenin: the tyrosine kinase substrate p120cas associates with E-cadherin complexes. *Mol Cell Biol* **14:** 8333–8342.

Reynolds AB, Daniel JM, Mo Y-Y, Wu J, Zhang Z. 1996. The novel catenin p120cas binds classical cadherins and induces an unusual morphological phenotype in NIH3T3 fibroblasts. *Exp Cell Res* **225:** 328–337.

Reynolds AB, Herbert L, Cleveland JL, Berg ST, Gaut JR. 1992. p120, a novel substrate of protein tyrosine kinase receptors and p60v-src, is related to cadherin-binding factors β-catenin, plakoglobin and armadillo. *Oncogene* **7:** 2439–2445.

Rickelt S, Franke WW, Doerflinger Y, Goerdt S, Brandner JM, Peitsch WK. 2008. Subtypes of melanocytes and melanoma cells distinguished by their intercellular contacts: Heterotypic adherens junctions, adhesive associations, and dispersed desmoglein 2 glycoproteins. *Cell Tissue Res* **334:** 401–422.

Rickelt S, Winter-Simanowski S, Noffz E, Kuhn C, Franke WW. 2009. Upregulation of plakophilin-2 and its acquisition to adherens junctions identifies a novel molecular ensemble of cell–cell-attachment characteristic for transformed mesenchymal cells. *Int J Cancer*, in press; published online May 4. 2009; DOI: 10.1002/ijc.24552.

Rimm DL, Koslov ER, Kebriaei P, Cianci CD, Morrow JS. 1995. α_1(E)-catenin is an actin-binding and -bundling protein mediating the attachement of F-actin to the membrane adhesion complex. *Proc Natl Acad Sci* **92:** 8813–8817.

Ringwald M, Schuh R, Vestweber D, Eistetter H, Lottspeich F, Engel J, Dölz R, Jähnig F, Epplen J, Mayer S et al., 1987. The structure of cell adhesion molecule uvomorulin. Insights into the molecular mechanism of Ca^{2+}-dependent cell adhesion. *EMBO J* **6:** 3647–3653.

Ruhrberg C, Watt FM. 1997. The plakin family: Versatile organizers of cytoskeletal architecture. *Curr Opin Gen Dev* **7:** 392–397.

Ruiz P, Brinkmann V, Ledermann B, Behrend M, Grund C, Thalhammer C, Vogel F, Birchmeier C, Günthert U, Franke WW, Birchmeier W. 1996. Targeted mutation of plakoglobin in mice reveals essential functions of desmosomes in the embryonic heart. *J Cell Biol* **135:** 215–225.

Sahakibara A, Furuse M, Saitou M, Ando-Akatsuka Y, Tsukita Sh. 1997. Possible involvement of phosphorylation of occludin in tight junction formation. *J Cell Biol* **137:** 1393–1401.

Saitou M, Akatsuka-Ando Y, Itoh M, Furuse M, Inazawa J, Fujimoto K, Tsukita Sh. 1997. Mammalian occludin in epithelial cells: Its expression and subcellular distribution. *Eur J Cell Biol* **73:** 222–231.

Saitou M, Fujimoto K, Doi Y, Itoh M, Fujimoto T, Furuse M, Takano H, Noda T, Tsukita Sh. 1998. Occludin-deficient embryonic stem cells can differentiate into polarized epithelial cells bearing tight junctions. *J Cell Biol* **141:** 397–408.

Saitou M, Furuse M, Sasaki H, Schulzke J, Fromm M, Takano H, Noda T, Tsukita Sh. 2000. Complex phenotype of mice lacking occludin, a component of tight junction strands. *Mol Biol Cell* **11:** 4131–4142.

Sato N, Funayama N, Nagafuchi A, Yonemura S, Tsukita S, Tsukita S. 1992. A gene family consisting of ezrin, radixin and moesin. Its specific localization at actin filament/plasma membrane association sites. *J Cell Sci* **103:** 131–143.

Sato N, Yonemura S, Obinata T, Tsukita S, Tsukita S. 1991. Radixin, a barbed end-capping actin-modulating protein, is concentrated at the cleavage furrow during cytokinesis. *J Cell Biol* **113:** 321–330.

Satoh-Horikawa K, Nakanishi H, Takahashi K, Miyahara M, Nishimura M, Tachibana K, Mizoguchi A, Takai Y. 2000. Nectin-3, a new member of immunoglobulin like cell adhesion molecules that shows homophilic and heterophilic cell–cell adhesion activities. *J Biol Chem* **275:** 10291–10299.

Schäfer S, Koch PJ, Franke WW. 1994. Identification of the ubiquitous human desmoglein, Dsg2, and the expression catalogue of the desmoglein subfamily of desmosomal cadherins. *Exp Cell Res* **211:** 391–399.

Schäfer S, Stumpp S, Franke WW. 1996. Immunological identification and characterization of the desmosomal cadherin Dsg2 in coupled and uncoupled epithelial cells and in human tissues. *Differentiation* **60:** 99–108.

Schäfer S, Troyanovsky SM, Heid HW, Eshkind LG, Koch PJ, Franke WW. 1993. Cytoskeletal architecture and epithelial differentiation: Molecular determinants of cell interaction and cytoskeletal filament anchorage. *C R Acad Sci III Sci Vie* **316:** 1316–1323.

Schlüter H, Moll I, Wolburg H, Franke WW. 2007. The different structures containing tight junction proteins in epidermal and other stratified epithelial cells, including squamous cell metaplasia. *Eur J Cell Biol* **86:** 645–655.

Schlüter H, Wepf R, Moll I, Franke WW. 2004. Sealing the live part of the skin: The integrated meshwork of desmosomes, tight junctions and curvilinear ridge structures in the cells of the uppermost granular layer of the human epidermis. *Eur J Cell Biol* **83:** 655–665.

Schmelz M, Duden R, Cowin P, Franke WW. 1986a. A constitutive transmembrane glycoprotein of M_r 165,000 (desmoglein) in epidermal and non-epidermal desmosomes: I. Biochemical identification of the polypeptide. *Eur J Cell Biol* **42:** 177–183.

Schmelz M, Duden R, Cowin P, Franke WW. 1986b. A constitutive transmembrane glycoprotein of M_r 165,000 (desmoglein) in epidermal and non-epidermal desmosomes: II. Immunolocalization and microinjection studies. *Eur J Cell Biol* **42:** 184–199.

Schmidt A, Koch PJ. 2008. Desmosomes in development and disease. In *Cell Junctions* (eds. S.E. La Flamme, A.P. Kowalczyk), pp. 235–249. Wiley-VCH, Weinheim.

Schmidt A, Heid HW, Schäfer S, Nuber UA, Zimbelmann R, Franke WW. 1994. Desmosomes and cytoskeletal architecture in epithelial differentiation: Cell type-specific plaque components and intermediate filament anchorage. *Eur J Cell Biol* **65:** 229–245.

Schmidt A, Langbein L, Prätzel S, Rode M, Rackwitz H-R, Franke WW. 1999. Plakophilin 3 – a novel cell-type-specific desmosomal plaque protein. *Differentiation* **64:** 291–306.

Schmidt A, Langbein L, Rode M, Prätzel S, Zimbelmann R, Franke WW. 1997. Plakophilins 1a and 1b: Widespread nuclear proteins recruited in specific epithelial cells as desmosomal plaque components. *Cell Tissue Res* **290:** 481–499.

Schmitt CJ, Franke WW, Goerdt S, Falkowska-Hansen B, Rickelt S, Peitsch WK. 2007. Homo- and heterotypic cell contacts in malignant melanoma cells and desmoglein 2 as a novel solitary surface glyocprotein. *J Invest Dermatol* **127**, 2191–2206.

Schneeberger EE, Lynch RD. 1992. Structure, function and regulation of cellular tight junctions. *Am J Physiol* **262:** L647–L661.

Schuh R, Vestweber D, Riede I, Ringwald M, Rosenberg UB, Jäckle H, Kemler R. 1986. Molecular cloning of the mouse cell adhesion molecule uvomorulin: cDNA contains a B1-related sequence. *Proc Natl Acad Sci* **83:** 1364–1368.

Schwechheimer K, Kartenbeck J, Moll R, Franke WW. 1984. Vimentin filament-desmosome cytoskeleton of diverse types of human meningiomas. A distinctive diagnostic feature. *Lab Invest* **51:** 584–591.

Setzer SV, Calkins CC, Garner J, Summers S, Green KJ, Kowalczyk AP. 2004. Comparative analysis of armadillo family proteins in the regulation of A431 epithelial cell junction assembly, adhesion and migration. *J Invest Dermatol* **123:** 426–433.

Shibamoto S, Hayakawa M, Takeuchi K, Hori T, Miyazawa K, Kitamura N, Johnson KR, Wheelock MJ, Matsuyoshi N, Takeichi M, Ito F. 1995. Association of p120, a tyrosine kinase substrate, with E-cadherin/catenin complexes. *J Cell Biol* **128:** 949–957.

Simon DB, Lu Y, Choate KA, Velazquez H, Al-Sabban E, Praga M, Casari G, Bettinelli A, Colussi G, Rodriguez-Soriano J, McCredie D, Milford D, Sanjad S, Lifton RP. 1990. Paracellin-1, a renal tight junction protein required for paracellular Mg^{2+} resorption. *Science* **285:** 103–106.

Simons K. 1990. The epithelial tight junction: Occluding barrier and fence. In *Morphoregulatory molecules* (eds. G.M. Edelman, B.A. Cunningham, J.P. Thiery), pp. 341–356. John Wiley and Sons, New York.

Singh D, Solan JL, Taffet SM, Javier R, Lampe PD. 2005. Connexin43 interacts with zona occludens-I and -2 proteins in a cell cycle stage-specific manner. *J Biol Chem* **280:** 30416–30421.

Sirotkin H, O'Donnell H, DasGupta R, Halford S, St. Jore B, Puech A, Parimoo S, Morrow S, Skoultchi A, Weissman SM, et al. 1997. Identification of a new human catenin gene family member (ARVCF) from the region deleted in velo-cardiofacial syndrome. *Genomics* **41:** 75–83.

Skerrow CJ. 1986. Desmosomal proteins. In *Biology of the integument 2 Vertebrates* (eds. J. Bereiter-Hahn, A.G. Matoltsy, K.S. Richard), pp. 762–787. Springer-Verlag, Berlin.

Skerrow CJ, Matoltsy AG. 1974a. Isolation of epidermal desmosomes. *J Cell Biol* **63:** 515–523.

Skerrow CJ, Matoltsy AG. 1974b. Chemical characterization of epidermal desmosomes. *J Cell Biol* **63:** 524–530.

Smith EA, Fuchs E. 1998. Defining the interactions between intermediate filaments and desmosomes. *J Cell Biol* **141:** 1229–1241.

Sonoda N, Furuse M, Sasaki H, Yonemura S, Katahira J, Horiguchi A, Tsukita Sh. 1999. Clostridium perfringens enterotoxin fragment removes specific claudins from tight junction strands: Evidence for direct involvement of claudins in tight junction barrier. *J Cell Biol* **147:** 195–204.

Sosinsky G. 2000. Gap junction structure: New structures and new insights. In *Gap Junctions. Molecular Basis of Cell Communication and Health and Disease* (ed. C. Peracchia), Curr. Top. Membr., Vol. **49**, pp. 1–22. Academic Press, San Diego.

Staehelin LA. 1974. Structure and function of intercellular junctions. *Int Rev Cytol* **39:** 191–283.

Stappenbeck TS, Bornslaeger EA, Corcoran CM, Luu HH, Virata ML, Green KJ. 1993. Functional analysis of desmoplakin domains: Specification of the interaction with keratin versus vimentin intermediate filament networks. *J Cell Biol* **123:** 691–705.

Stappert J, Kemler R. 1994. A short core region of E-cadherin is essential for catenin binding and is highly phosphorylated. *Cell Adhes Commun* **2:** 319–327.

Stauffer KA, Kumar NM, Gilula NB, Unwin N. 1993. Biochemistry of gap-junction channels. In *Gap junctions* (eds. J.E. Hall, G.A. Zampighi, R.M. Davis), *Progr. Cell Res.* Vol. **3**, pp. 57–59. Elsevier, Amsterdam.

Steinberg M. 1958. On the chemical bonds between animal cells. A mechanism for type-specific association. *Am Nat* **92:** 65–81.

Stevenson BR, Siliciano JD, Mooseker JD, Goodenough DA. 1986. Identification of ZO-1: A high molecular weight polypeptide associated with the tight junction (zonula occludens) in a variety of epithelia. *J Cell Biol* **103:** 755–766.

Stokes DL. 2007. Desmosomes from a structural perspective. *Curr Opin Cell Biol* **19:** 565–571.

Strumane K, Berx G, Van Roy F. 2004. Cadherins in cancer. In *Cell adhesion* (eds. J. Behrens, W.J. Nelson), Handbook Exp. Pharmacol., Vol. **165**, pp. 69–103. Springer, Berlin.

Tachibana K, Nakanishi H, Mandai K, Ozaki K, Ideda W, Yamamoto Y, Nagafuchi A, Tsukita S, Takai Y. 2000. Two cell adhesion molecules, nectin and cadherin, interact through their cytoplasmic domain-associated proteins. *J Cell Biol* **150:** 1161–1176.

Takahashi K, Nakanishi H, Miyahara M, Mandai K, Satoh K, Satoh Y, Nishioka H, Aoki J, Nomoto A, Mizoguchi A, et al. 1999. Nectin/PRR: An immunoglobulin-like cell adhesion molecule recruited to cadherin-based adherens junctions through interaction with afadin, a PDZ domain-containing protein. *J Cell Biol* **145:** 539–549.

Takai Y, Nakanishi H. 2003. Nectin and afadin: Novel organizers of intercellular junctions. *J Cell Sci* **116:** 17–27.

Takai Y, Irie K, Shimizu K, Sakisaka T, Ikeda W. 2003. Nectins and nectin-like molecules: Roles in cell adhesion, migration, and polarization. *Cancer Sci* **94:** 655–667.

Takai Y, Ikeda W, Ogita H, Rikitake Y. 2008a. The immunoglobulin-like cell adhesion molecule nectin and its associated protein afadin. *Annu Rev Cell Dev Biol* **24:** 309–342.

Takai Y, Miyoshi J, Ideda W, Ogita H. 2008b. Nectins and nectin-like molecules: Roles in contact inhibition of cell movement and proliferation. *Nat Rev Mol Cell Biol* **9:** 603–615.

Takeda H, Shimoyama Y, Nagafuchi A, Hirohashi S. 1999. E-cadherin functions as a *cis*-dimer at the cell–cell adhesive interface *in vivo*. *Nat Struct Biol* **6:** 310–312.

Takeichi M. 1977. Functional correlation between cell adhesive properties and some cell surface proteins. *J Cell Biol* **75:** 464–474.

Takeichi M. 1988. The cadherins: Cell–cell adhesion molecules controlling animal morphogenesis. *Development* **102:** 639–655.

Takeichi M. 1990. Cadherins: A molecular family important in selective cell–cell adhesion. *Annu Rev Biochem* **59:** 237–252.

Takeichi M, Yoshida-Noro C, Shirayoshi Y, Hatta K. 1985. Calcium-dependent cell–cell adhesion system: Its molecular nature, cell type specificity, and morphogenetic role. In *The cell in contact* (eds. G.M. Edelman, J.-P. Thiery), pp. 219–232. John Wiley and Sons, New York.

Talhouk RS, Mroue RM, Mokalled M, Abi-Mosleh L, Nehme R, Imsail A, Khalil A, Zaatari M, El-Sabban ME. 2008. Heterocellular interaction enhances recruitment of α and β-catenins and ZO-2 into functional gap junction complexes and gap junction-dependent differentiation of mammary epithelial cells. *Exp Cell Res* **314:** 3275–3291.

Tanihara H, Sano K, Heimark RL, John T St, Suzuki S. 1994. Cloning of five human cadherins clarifies characteristic features of cadherin extracellular domain and provides further evidence for two structurally different types of cadherin. *Cell Adhes Commun* **2:** 15–26.

Theis DG, Koch PJ, Franke WW. 1993. Differential synthesis of type 1 and type 2 desmocollin mRNAs in human stratified epithelia. *Int J Dev Biol* **37:** 101–110.

Thiery J-P, Duband J-L, Delouvée A. 1985. The role of cell adhesion in morphogenetic movements during early embryogenesis. In *The cell in contact* (eds. G.M. Edelman, J.-P. Thiery), pp. 169–196. John Wiley and Sons, New York.

Thiery JP, Delouvée A, Gallin W, Cunningham BA, Edelman GM. 1984. Ontogenetic expression of cell adhesion molecules: L-CAM is found in epithelia derived from the three primary germ layers. *Dev Biol* **102:** 61–78.

Thomson RB, Igarashi P, Biemesderfer D, Kim R, Abu-Alfa A, Soleimani M, Aronson PS. 1995. Isolation and cDNA cloning of Ksp-cadherin, a novel kidney-specific member of the cadherin multigene family. *J Biol Chem* **270:** 17594–17601.

Toyofuku T, Akamatsu Y, Zhang H, Kuzuya T, Tada M, Hori M. 2001. c-SRC regulates the interaction between connexin-43 and ZO-1 in cardiac myocytes. *J Biol Chem* **276:** 1780–1788.

Toyofuku T, Yabuki M, Otsu K, Kuzuya T, Hori M, Tada M. 1998. Direct association of the gap junction protein connexin-43 with ZO-1 in cardiac myocytes. *J Biol Chem* **273:** 12725–12731.

Traub O, Look J, Dermietzel R, Brummer F, Hülser D, Willecke K. 1989. Comparative characterization of the 21-kDa and 26-kDa gap-junction proteins in murine liver and cultured hepatocytes. *J Cell Biol* **108:** 1039–1051.

Troyanovsky SM. 2005. Cadherin dimers in cell–cell adhesion. *Eur J Cell Biol* **84:** 225–233.

Troyanovsky RB, Chitaev NA, Troyanovsky SM. 1996. Cadherin binding sites of plakoglobin: Localization, specificity and role in targeting to adhering junctions. *J Cell Sci* **109:** 3069–3078.

Troyanovsky SM, Eshkind LG, Troyanovsky RB, Leube RE, Franke WW. 1993. Contributions of cytoplasmic domains of desmosomal cadherins to desmosome assembly and intermediate filament anchorage. *Cell* **72:** 561 574.

Troyanovsky SM, Troyanovsky RB, Eshkind LG, Krutovskikh VA, Leube RE, Franke WW. 1994a. Identification of the plakoglobin-binding domain in desmoglein and its role in plaque assembly and intermediate filament anchorage. *J Cell Biol* **127:** 151–160.

Troyanovsky SM, Troyanovsky RB, Eshkind LG, Leube RE, Franke WW. 1994b. Identification of amino acid sequence motifs in desmocollin, a desmosomal glycoprotein, that are required for plakoglobin binding and plaque formation. *Proc Natl Acad Sci* **91:** 10790–10794.

Tsukita Sh, Furuse M. 1998. Overcoming barries in the study of tight junction functions: From occludin to claudin. *Genes Cells* **3:** 569–573.

Tsukita Sh, Furuse M. 2000. Pores in the wall: Claudins constitute tight junction strands containing aqueous pores. *J Cell Biol* **149:** 13–16.

Tsukita Sh, Furuse M. 2002. Claudin-based barrier in simple and stratified cellular sheets. *Curr Opin Cell Biol* **14:** 531–536.

Tsukita Sh, Tsukita Sa. 1989. Isolation of cell-to-cell adherens junctions from rat liver. *J Cell Biol* **108:** 31–41.

Tsukita Sh, Furuse M, Itoh M. 2001. Multi-functional strands in tight junctions. *Nat Rev Mol Cell Biol* **2:** 285–293.

Tsukita Sa, Hieda Y, Tsukita Sh. 1989. A new 82-kD barbed end-capping protein (radixin) localized in the cell-to-cell adherens junction: Purification and characterization. *J Cell Biol* **108:** 2369–2382.

Tsukita Sa, Yonemura S, Tsukita Sh. 1997a. ERM (ezrin/radixin/moesin) family: From cytoskeleton to signal transduction. *Curr Opin Cell Biol* **9:** 70–75.

Tsukita Sa, Yonemura S, Tsukita Sh. 1997b. ERM proteins: Head-to-tail regulation of actin-plasma membrane interaction. *Trends Biochem Sci* **22:** 53–58.

Tsukita Sa, Oishi K, Sato N, Sagara J, Kawai A, Tsukita Sh. 1994. ERM family members as molecular linkers between the cell surface glycoprotein CD44 and actin-based cytoskeleton. *J Cell Biol* **126:** 391–401.

Uchida N, Honjo Y, Kohnson KR, Wheelock MJ, Takeichi M. 1996. The catenin/cadherin adhesion system is localized in synaptic junctions bordering trnsmitter release zones. *J Cell Biol* **135:** 767–779.

Unwin PNT. 1987. Gap junction structure and the control of cell to cell communication. In *Junctional complexes of epithelial cells*, Ciba Found. Symp. **125**, pp. 78–91. J. Wiley and Sons, Chichester.

Unwin PNT, Zampighi G. 1980. Structure of the junctions between communicating cells. *Nature* **283:** 545–549.

Van Itallie CM, Anderson JM. 2006. Claudins and epithelial paracellular transport. *Annu Rev Physiol* **68:** 403–429.

Vasioukhin V, Bauer C, Yin M, Fuchs E. 2000. Directed actin polymerization is the driving force for epithelial cell–cell adhesion. *Cell* **100:** 209–219.

Vestweber D, Kemler R. 1984. Rabbit antiserum against a purified surface glycoprotein decompacts mouse preimplantation embryos and reacts with specific adult tissues. *Exp Cell Res* **152:** 169–178.

Vestweber D, Kemler R. 1985. Identification of a putative cell adhesion domain of uvomorulin. *EMBO J* **4:** 3393–3398.

Vilela MJ, Parrish EP, Wright DH, Garrod DR. 1987. Monoclonal antibody to desmosomal glycoprotein 1 – A new epithelial marker for diagnostic pathology. *J Pathol* **153:** 365–375.

Virata MLA, Wagner RM, Parry DAD, Green KJ. 1992. Molecular structgure of the human desmoplakin I and II amino terminus. *Prod Natl Acad Sci* **89:** 544–548.

Vleminckx K, Vakaet LJr, Mareel M, Fiers W, Van Roy F. 1991. Genetic manipulation of E-cadherin expression by epithelial tumor cells reveals an invasion suppressor role. *Cell* **66:** 107–119.

Volk T, Geiger B. 1984. A 135-kD membrane protein of intercellular adherens junctions. *EMBO J* **3:** 2249–2260.

Wahl JK, Sacco PA, McGranahan-Sadler TM, Sauppe LM, Wheelock MJ, Johnson KR. 1996. Plakoglobin domains that define its association with the desmosomal cadherins and the classical cadherins: Identification of unique shared domains. *J Cell Sci* **109:** 1143–1154.

Waibler Z, Schäfer A, Starzinski-Powitz A. 2001. mARVCF cellular localisation and binding to cadherins is influenced by the cellular context but not by alternative splicing. *J Cell Sci* **114:** 3873–3884.

Walter B, Schlechter T, Hergt M, Berger I, Hofmann I. 2008. Differential expression pattern of protein ARVCF in nephron segments of human and mouse kidney. *Histochem Cell Biol* **130:** 943–956.

Warner AE, Guthrie SC, Gilula NB. 1984. Antibodies to gap junctional protein selectively disrupt junctional communication in the early amphibian embryo. *Nature* **311:** 127–131.

Waschke J. 2008. The desmosome and pemphigus. *Histochem Cell Biol* **130:** 130–121.

Watabe-Uchida M, Uchida N, Imamura Y, Nagafuchi A, Fujimoto K, Uemura T, Vermeulen S, van Roy F, Adamson ED, Takeichi M. 1998. α-catenin-vinculin interaction functions to organize the apical junctional complex in epithelial cells. *J Cell Biol* **142:** 847–857.

Watt FM, Mattey DL, Garrod DR. 1984. Calcium-induced reorganization of desmosomal components in cultured human keratinocytes. *J Cell Biol* **99:** 2211–2215.

Weiss EE, Kroemker M, Rüdiger A-H, Jockusch BM, Rüdiger M. 1998. Vinculin is part of the cadherin-catenin junctional complex: Complex formation between α-catenin and vinculin. *J Cell Biol* **141:** 755–764.

Wheeler GN, Buxton RS, Parker AE, Arnemann J, Rees DA, King IA, Magee AI. 1991a. Desmosomal glycoproteins I, II and III: Novel members of the cadherin superfamily. *Biochem Soc Trans* **19:** 1060–1064.

Wheeler GN, Parker AE, Thomas CL, Ataliotis P, Poynter D, Arnemann J, Rutman AJ, Pidsley SC, Watt FM, Rees DA, et al. 1991b. Desmosomal glycoprotein DGI, component of intercellular desmosome junctions, is related to the cadherin family of cell adhesion molecules. *Proc Natl Acad Sci* **88:** 4796–4800.

Wheelock MJ, Johnson KR. 2003. Cadherins as modulators of cellular phenotype. *Annu Rev Cell Dev Biol* **19:** 207–235.

Wheelock MJ, Shintani Y, Maeda M, Fukumoto Y, Johnson KR. 2008. Cadherin switching. *J Cell Sci* **121:** 727–735.

Willecke K, Dermietzel R, Drüge PM, Frixen U, Janssen-Timmen U, Schäfer R, Traub O. 1982. Biochemical and genetic investigations on gap junctions from mammalian cells. *Biophys Struct Mech* **9:** 103–107.

Willecke K, Hennemann H, Dahl E, Jungbluth S. 1993. The mouse connexin family. In *Gap junctions* (eds. J.E. Hall, G.A. Zampighi, R.M. Davis), *Progr. Cell Res.* Vol. 3, pp. 33–37. Elsevier, Amsterdam.

Willott E, Balda MS, Fanning AS, Jameson B, Van Itallie C, Anderson JM. 1993. The tight junction protein ZO-1 is homologous to the *Drosophila* discs-large tumor suppressor protein of septate junctions. *Proc Natl Acad Sci* **90:** 7834–7838.

Wilson HV. 1907. On some phenomena of coalescence and regeneration in sponges. *J Exp Zool* **5:** 245–258.

Witcher LL, Collins R, Puttagunta S, Mechanic SE, Munson M, Gumbiner B, Cowin P. 1996. Desmosomal cadherin binding domains of plakoglobin. *J Biol Chem* **271:** 10904–10909.

Wittchen ES, Haskins J, Stevenson BR. 1999. Protein interaction at the tight junction. Actin has multiple binding partners, and ZO-1 forms independent complexes with ZO-2 and ZO-3. *J Biol Chem* **274:** 35179–35185.

Yeager M. 1993. Structure and design of cardiac gap-junction membranes channels. In *Gap junctions* (eds. J.E. Hall, G.A. Zampighi, R.M. Davis), *Progr. Cell Res.* Vol. 3, pp. 47–55. Elsevier, Amsterdam.

Yonemura S, Itoh M, Nagafuchi A, Tsukita S. 1995. Cell-to-cell adherens junction formation and actin filament organization: similarities and differences between non-polarized fibroblasts and polarized epithelial cells. *J Cell Sci* **108:** 127–142.

Yoshida-Noro C, Takeichi M. 1982. Teratocarcinoma cell adhesion: Identification of cell surface protein involved in calcium-dependent cell aggregation. *Cell* **28:** 217–224.

Yoshida-Noro C, Suzuki N, Takeichi M. 1984. Molecular nature of the calcium-dependent cell–cell adhesion sstem in mouse teratocarcinoma and embryonic cells studied with a monoclonal antibody. *Dev Biol* **101:** 19–27.

Zhang M-T, Nicholson B. 1989. Sequence and tissue distribution of a second protein of hepatic gap junctions, Cx26, as deduced from its cDNA. *J Cell Biol* **109:** 3391–3401.

Zhou J, Liyanage U, Medina M, Ho C, Simmons AD, Lovett M, Kosiks KS. 1997. Presenilin 1 interaction in the brain with a novel member of the *armadillo* family. *NeuroReport* **8:** 1489–1494.

Zhou X, Stuart A, Dettin LE, Rodriguez G, Hoel B, Gallicano GI. 2004. Desmoplakin is required for microvascular tube formation in culture. *J Cell Sci* **117:** 3129–3140.

Zhurinsky J, Shtutman M, Ben-Ze'ev A. 2000. Plakoglobin and β-catenin: Protein interactions, regulation and biological roles. *J Cell Sci* **113:** 3127–3139.

Molecular Basis of the Core Structure of Tight Junctions

Mikio Furuse

Division of Cell Biology, Department of Physiology and Cell Biology, Kobe University Graduate School of Medicine, 7-5-1 Kusunoki-cho Chuo-ku, Kobe 650-0017, Japan

Correspondence: furuse@med.kobe-u.ac.jp

The morphological feature of tight junctions (TJs) fits well with their functions. The core of TJs is a fibril-like proteinaceous structure within the lipid bilayer, the so-called TJ strands. TJ strands in apposing plasma membranes associate with each other to eliminate the intercellular space. A network of paired TJ strands generates a continuous belt that circumscribes each cell to establish the diffusion barrier to the solutes in the paracellular pathway throughout the cellular sheet. Identification and characterization of TJ-associated proteins during the last two decades has unveiled the nature of TJ strands and how they are spatially organized. The interplay between integral membrane proteins, claudins, and cytoplasmic plaque proteins, ZO-1/ZO-2, is critical for TJ formation and function.

Tight junctions (TJs) are fascinating structures in terms of their function and morphology. In 1963, using ultrathin-section electron microscopy, Farquhar and Palade described the fine structure of TJs together with adherens junctions (AJs) and desmosomes at the most luminal side of the lateral membrane (Farquhar and Palade 1963). In addition, they demonstrated insightfully that TJs function as permeability seals for mass tracers. Indeed, the structure of TJs observed in electron microscopy indicates that TJs could physically restrict the leak of solutes through the intercellular space. However, physiological studies at the same time revealed that solute transport occurred via the intercellular space in a variety of epithelial cells. A resolution of these different views of TJ function comprises the current concept that the TJ regulates the diffusion of solutes with size and charge selectivity and that it is functionally different in physiologically diverse epithelial cell types (Powell 1981; Anderson and Cereijido 2001). To understand the molecular mechanism controlling TJ structure and function, it is important to determine their molecular composition and organization.

Although purification of TJs is difficult, Stevenson and Goodenough developed an isolation method for a TJ-enriched plasma membrane fraction from rodent liver. They discovered the first TJ-associated protein, ZO-1, in 1986 by generating monoclonal antibodies against this fraction (Stevenson et al. 1986). Since then, many molecular components of TJs have been identified using immunological approaches or searches for binding proteins

Cite this article as *Cold Spring Harb Perspect Biol* 2009;1:a002907

with known molecules, which have enabled detailed molecular cell biological analyses of TJs. Among the TJ-associated proteins, the claudin family of membrane proteins identified in 1998 by the Tsukita group are key molecules in the architecture and barrier function of TJs (Furuse et al. 1998a). Functional analyses of claudins have allowed remarkable progress in the development of a comprehensive understanding of the molecular basis of the ultrastructure and physiological characteristics of TJs (Van Itallie and Anderson 2006; Furuse and Tsukita 2006; Angelow et al. 2008). In addition, the cytoplasmic plaque proteins associated with TJs are important in regulating TJ architecture (Guillemot et al. 2008).

In this article, we present the molecular basis for the core structure of TJs based on recent progress in functional analyses of TJ-associated proteins. The current molecular basis of TJ physiology is covered in detail in Anderson and Van Itallie (2009).

MORPHOLOGY OF TJS

Electron Microscopic Observations of TJs

TJs occur in epithelial cell types, including epithelial and endothelial cells, in vertebrate species and tunicates (Lane 2001). TJs are located at the most apical part of junctional complexes, next to AJs in epithelial cells. In contrast, TJs and AJs are spatially intermingled in endothelial cells (Schulze and Firth 1993). TJs are also located in myelin sheaths formed by oligodendrocytes (Dermietzel 1974) and Schwann cells (Tetzlaff 1978).

In ultrathin-section electron microscopy, TJs are visualized as a zone in which adjacent plasma membranes are closely apposed, that circumscribes the cell as a belt together with AJs. At high magnification, TJs are characterized as focal attachments of adjacent cell membranes that exclude the intercellular gap (Fig. 1). This appearance of TJs suggests their function as rate-limiting barriers to the diffusion of solutes, and this idea has been supported by electron microscopic observations using electron-opaque tracers (Farquhar and Palade 1963; Brightman and Reese 1969; Goodenough and Revel 1970).

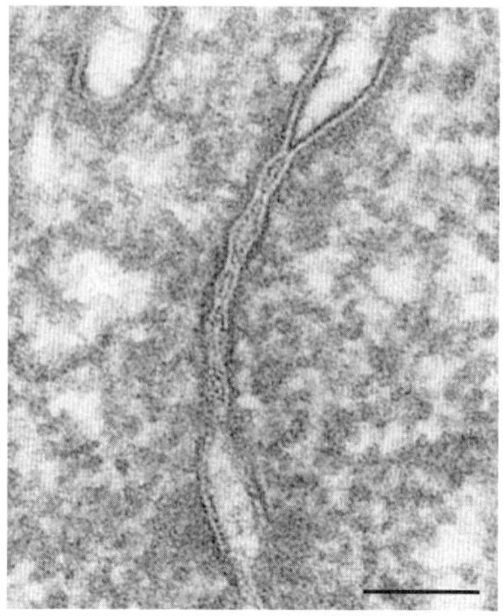

Figure 1. Thin section of TJs of mouse epididymis epithelial cells. TJs are defined as close contacts between plasma membranes of adjacent cells. Bar, 100 nm.

TJs have been intensely analyzed by freeze-fracture electron microscopy, in which the morphology of the central hydrophobic plane of the lipid bilayer, including the particles and pits, can be observed. In general, TJs in glutaraldehyde-fixed epithelial cells appear as anastomosing linear fibrils (TJ strands) on the cytoplasmic leaflet (P-face) and grooves on the exoplasmic leaflet (E-face) surrounding the cell (Fig. 2) (Staehelin 1973, 1974). In some cell types, however, TJs exhibit discontinuous particle strands on the P-face and grooves

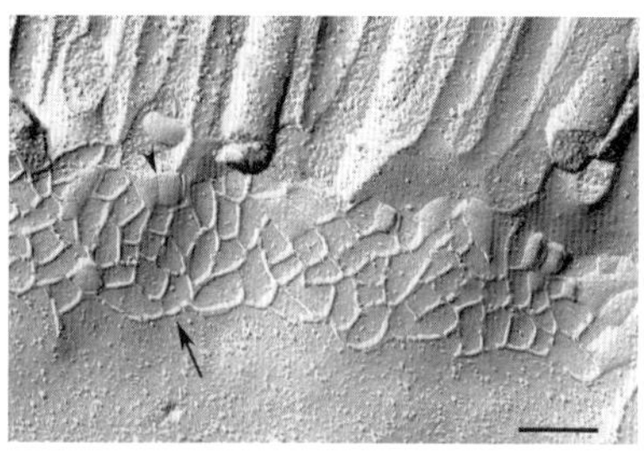

Figure 2. Freeze-fracture replica of TJs of mouse intestinal epithelial cells. TJs appear as anastomosing fibrils, namely, TJ strands, on P-face (arrow) or complementary grooves on E-face (arrowhead) in glutaraldehyde-fixed specimens. Bar, 200 nm.

containing many particles on the E-face (Simionescu et al. 1975; van Deurs and Koehler 1979). Furthermore, specimens that are not fixed with glutaraldehyde before freezing also reveal particle strands on the E-face (Staehelin 1973). Detailed analyses of double replicas have revealed that these strands and grooves are complementary between the P- and E-face (Chalcroft and Bullivant 1970; van Deurs and Koehler 1979). It is widely accepted that TJ strands correspond to the focal attachments of adjacent plasma membranes observed in ultrathin sections. Freeze-fracture electron microscopy has also revealed that the number and network pattern of TJ strands varies between cell types (Friend and Gilula 1972; Claude and Goodenough 1973). By combining morphological observations and physiological measurements in various epithelial cell types, the correlation between the structure and barrier function of TJs has been discussed (Claude 1978). In several cases, however, the correlation was not perfect. At that time, there was no information about the molecular nature of TJs, and they hypothesized that each TJ strand in various cell types was equal. Currently, we know that the combination of caudin types is an important factor to determine the barrier property of TJs.

Two models have been proposed for the structural nature of the TJ strand based on morphological analyses using electron microscopy. The most plausible model is the protein model, in which proteinaceous attachment elements, i.e., adhesion molecules, hold the apposing plasma membranes together (Staehelin 1973). Detailed characterization of TJ-associated integral membrane proteins, including claudins as described in the following section, strongly support this model. The other model, proposed in the 1980s, is the lipid model in which TJ strands are cylindrical inverted lipid micelles formed by the fusion of apposed plasma membranes, resulting in the continuity of outer leaflets of plasma membranes between adjacent cells (Pinto da Silva and Kachar 1982). The observations that a fluorescent-labeled lipid that was introduced in the outer leaflet of the apical membrane of an epithelial cell never moved to the next cell (van Meer et al. 1986) and TJ strands are detergent resistant (Stevenson and Goodenough 1984) do not conform to this model, although specific lipids may contribute to the formation of TJs.

Tricellular TJs: Specialized TJs at Tricellular Corners

Considering that TJs function as continuous barriers throughout the cellular sheet, it is interesting to understand how tricellular contacts formed by the joining of three cells are sealed. Because TJs are thought to be zipper-like structures formed between two adjacent cells, it is not simple to see how they could exclude the extracellular space between three plasma membranes. Specialized structures, namely tricellular TJs (tTJs), are observed in these regions by freeze-fracture electron microscopy (Staehelin et al. 1969; Friend and Gilula 1972; Staehelin 1973; Wade and Karnovsky 1974). At tTJs, the most apical elements of the horizontal TJ strands from both sides attach and turn together to extend vertically in the basal direction (Fig. 3). These vertical TJ strands link to several short horizontal TJ strands that connect with the main portion of the TJ strands. As a result, three sets of parallel and closely spaced TJ strands extend vertically to form a narrow tube at the extracellular space of the tricellular contact region. This tube is thought to reduce to a negligible amount the free diffusion of solutes through the intercellular space (Staehelin 1973).

MOLECULAR ORGANIZATION OF THE CORE STRUCTURE OF TJS

Overview of TJ-associated Proteins

Many molecular components of TJs have been identified and characterized. Like other intercellular junctions, TJs are protein complexes composed of integral membrane proteins, cytoplasmic plaque proteins, and cytoskeletal proteins. Among these, the claudin family

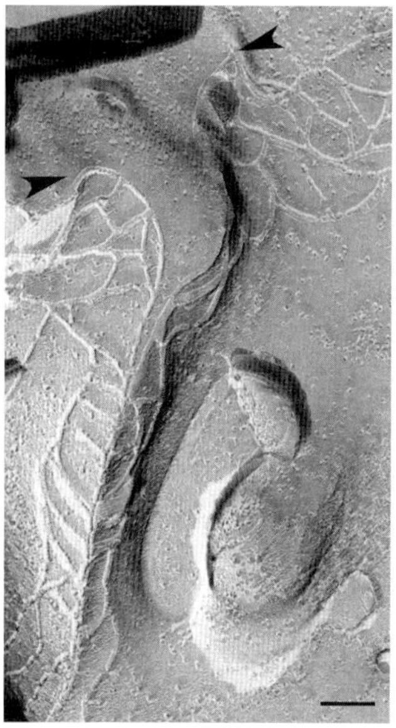

Figure 3. Freeze-fracture replica of tTJs in MDCK cells. At tricellular corners, the most apical elements of the horizontal TJ strands from both sides (arrowheads) attach to form the central seal and extend vertically in the basal direction. Short, horizontal strands connect with the central seal. Bar, 100 nm.

membrane proteins are key components for the structure and function of TJs (Van Itallie and Anderson 2006; Furuse and Tsukita 2006; Angelow et al. 2008). Claudin molecules interact with each other between cells and probably polymerize within the plasma membrane as fibrils to generate TJ strands. Occludin and tricellulin, comprising other types of four transmembrane domain-containing proteins with related amino acid sequences, are incorporated into or localized very close to claudin-based TJ strands (Furuse et al. 1993; Fujimoto 1995; Ikenouchi et al. 2005). In addition, immunoglobulin superfamily membrane proteins with two extracellular Ig-like domains, including JAM-A, JAM4, coxsackie adenovirus receptor (CAR), and endothelial cell-selective adhesion molecule (ESAM), are also localized at TJs (Martin-Padura et al. 1998; Hirabayashi et al. 2003; Cohen et al. 2001; Nasdala et al. 2002).

On the cytoplasmic side of TJ strands, the membrane-associated guanylate kinase (MAGUK) family proteins ZO-1, ZO-2, and ZO-3 (Stevenson et al. 1986; Gumbiner et al. 1991; Haskins et al. 1998) bind to the C-terminal cytoplasmic domain of claudins, occludin, tricellulin, and JAM-A. In addition, MAGI-1, MAGI-3, MUPP1, and PATJ are known to be PDZ (PSD-95, discs-large, ZO-1) domain-containing proteins that directly bind to claudins or other TJ-associated membrane proteins (Ide et al. 1999; Laura et al. 2002; Hamazaki et al. 2002; Lemmers et al. 2002; Roh et al. 2002; Jeansonne et al. 2003). Other cytoplasmic proteins with coiled-coil domains, such as cingulin and JACOP/paracingulin, are concentrated relatively far from the plasma membrane of TJs (Citi et al. 1988; Ohnishi et al. 2004). Of these proteins, at least the ZO family of proteins and cingulin interact with F-actin (Itoh et al. 1997, 1999b; Fanning et al. 1998, 2002; Wittchen et al. 1999; D'Atri and Citi. 2001), which is present on the cytoplasmic side of TJs (Madara 1987). Signaling molecules, including the aPKC-Par3-Par6 complex, which is involved in epithelial polarity formation, are also localized on the cytoplasmic side of TJs and regulate mature TJ formation (Izumi et al. 1998; Suzuki and Ohno 2006; Ebnet et al. 2008). Furthermore, ZONAB, a transcription factor involved in cell growth, has been reported to be localized at TJs (Balda et al. 2003).

Among these TJ-associated proteins, claudins and ZO proteins are directly involved in the formation of TJ strands in epithelial cells. Two other membrane proteins with four transmembrane domains, occludin and tricellulin, appear to have intimate relationships with claudins and modulate TJ strand formation. These proteins are highlighted in this section as constituents of the core structure of TJs.

Claudins

Claudin Family

Claudins were originally identified as membrane proteins that cofractionated with occludin, the first identified integral membrane protein of TJs, by sucrose density gradient centrifugation of a sonicated TJ/AJ-enriched membrane fraction isolated from chick liver (Furuse et al. 1998a). Claudins comprise a multigene family with approximately 24 members in humans and mice (Van Itallie and Anderson 2006; Furuse and Tsukita 2006; Angelow et al. 2008). Claudins are 21–28-kD proteins and consist of four transmembrane domains, two extracellular loops, amino- and carboxy-terminal cytoplasmic domains, and a short cytoplasmic turn. Most claudins have the conserved motif GLWxxC(8-10 aa)C in the first extracellular loop, and a PDZ domain-binding motif at the carboxy-terminal, which binds to the TJ-associated PDZ domain-containing plaque proteins ZO-1, ZO-2, and ZO-3 (Fig. 4) (Itoh et al. 1999a). Claudins are key components for the structure and function of TJs. To investigate the roles of claudins in the barrier/channel property of TJs, overexpression and RNAi-mediated suppression of particular claudins in epithelial cell lines followed by physiological measurements have been carried out (see Anderson and Van Itallie 2009). In addition, many claudin knockout mice have been generated and analyzed to clarify the roles of the individual claudins in vivo (Furuse 2009).

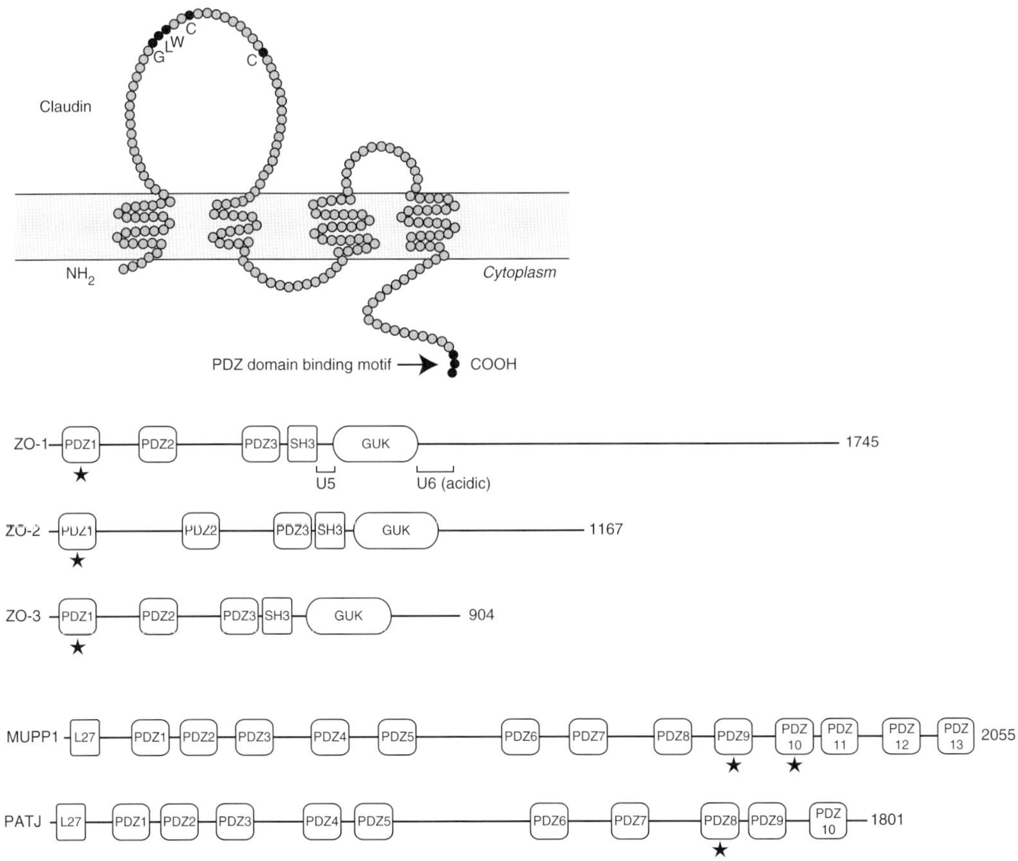

Figure 4. The membrane-spanning model of claudins and the domain organization of claudin-binding plaque proteins. Claudins contain four transmembrane domains. The conserved amino acid residues in the first extracellular loop are indicated. The PDZ domain binding motif is located at the carboxyl terminus, which is known to bind to PDZ-domains in ZO-1, ZO-2, ZO-3, MUPP1, and PATJ (stars). (GUK) guanylate kinase domain; (L27) LIN2, 7 homology domain; (U5) unique 5 region; (U6) unique 6 region. U6 contains the acidic region.

TJ Strand Formation by Claudins

The most obvious function of claudins is to generate TJ strands, the structural and functional core of TJs within the plasma membrane. When claudins are overexpressed in mouse L fibroblasts lacking endogenous TJs, the exogenous claudins concentrate into cell–cell contacts between adjacent cells and generate TJ strands with the induction of cell adhesive activity (Furuse et al. 1998b; Kubota et al. 1999). Immunoelectron and immunofluorescence microscopy analyses revealed that claudins themselves generate TJ strands as structural components (Furuse et al. 1998b; Sasaki et al. 2003). Importantly, addition to the culture medium of the carboxy-terminal of *Clostridium perfringens* enterotoxin, which binds to the second extracellular loop of some claudin types and functions as an inhibitor of these claudins, results in the removal of some claudin types from TJs accompanied by a reduction in the number of TJ strands and loss-of-barrier function (Sonoda et al. 1999). Furthermore, in claudin-11-deficient mice, TJ strands disappear in Sertoli cells, oligodendrocytes (Gow et al. 1999), and marginal cells of the inner ear (Gow et al. 2004; Kitajiri et al. 2004), where claudin-11 is the only known claudin type to be expressed. These observations from gain-of-function and loss-of-function studies confirm that claudins are involved in the structure of TJ strands. Deletion of the carboxy-terminal cytoplasmic region of claudin-1 did not abolish TJ strand reconstitution in L cells (Furuse et al. 1998b), also indicating that the ability to form TJ strands is mediated by claudins. In epithelial cells, however, the TJ-associated plaque proteins ZO-1 or ZO-2 are required for TJ strand formation by claudins (Umeda et al. 2006) (see Section ZO-1, ZO-2, and ZO-3).

TJ Strand as a Mosaic of Multiple Claudin Types

Most cell types coexpress multiple claudin subtypes in vivo in combinations and proportions that vary among different cell types. This complex expression pattern of claudins creates diversity in the barrier/channel property of TJs, depending on each epithelium (see Anderson and Van Itallie 2009 for details). The manner in which multiple claudins generate TJs between cells has been analyzed using L cells overexpressing various claudin types (Furuse et al. 1999). Coculture of L cells expressing claudin-1, -2, or -3 revealed that the combinations of claudin-1/-3 and claudin-2/-3, but not claudin-1/-2, could form heterotypic TJ strands. Furthermore, coexpression of two of these three claudins in L cells revealed that these claudins could copolymerize into single TJ strands in a heteromeric manner in individual cells. In other analyses using HeLa cells, claudin-3 and claudin-4 were unable to assemble heterotypically between cells, but their heteromeric assembly is possible (Daugherty et al. 2007). These experiments suggest that TJ strands are generally comprised of mosaics of different claudins that form homotypic and heterotypic interactions between cells and homomeric and heteromeric complexes in individual cells, although their compatibility is dependent on the right claudin combination. It is important to remember this complex architecture of claudin-based TJ strands when the barrier property of TJs is considered in terms of claudin content. Note, however, that claudin types may also be separated within TJs in the same cell surface in vivo. For example, TJ strands can be divided into two subdomains between outer hair cells and supporting Deiter cells in the organ of Corti in the inner ear; claudin-14 is concentrated in the apical TJ strands, whereas claudin-6 and claudin-9 are localized at lateral TJ strands (Nunes et al. 2006).

Structural Units of TJ Strands

Freeze-fracture replica electron microscopy studies with glutaraldehyde-fixed samples revealed that the E-face/P-face distributions of the intramembrane particles of TJ strands vary among different cell types. This variation appears to depend on the claudin types forming the TJ strands. For instance, TJ strands generated in L cells with claudin-1 and fixed with glutaraldehyde exhibit continuous rod-shaped strands on the P-face, whereas those in L cells with claudin-5 show thin and shallow grooves on the P-face and many particles on the E-face (Furuse et al. 1998b; Morita et al. 1999). An attractive hypothesis is that the P-face distribution of TJ strands is because of the interaction of the intramembrane components of TJs with cytoplasmic components (Balda and Anderson 1993). This does not appear to be the case because carboxy-terminal-deleted claudin-1, which cannot associate with cytoplasmic plaque proteins, still exhibits a P-face distribution (Furuse et al. 1999).

Because quick-freeze samples show chains of E-face-associated particles, it is reasonable to think that this is the unit particle structure of TJ strands. However, it remains unclear how this unit particle is constituted from claudin molecules. Interestingly, an intramembrane particle of TJs is similar to the diameter of a connexon, the basic unit of gap junctions consisting of the hexamer of connexins (Tsukita and Furuse 1999; Kumar and Gilula 1996). Both claudins and connexins have four transmembrane domains. Therefore, each E-face-associated particle of TJs may be composed of a claudin oligomer, similar to the hexamer of connexins, rather than claudin monomer. So far, there is no direct evidence to demonstrate this idea, although a crosslinking experiment indirectly implied the possibility of claudin oligomer formation in claudin-4-overexpressing insect cells (Mitic et al. 2003).

Modification of Claudins

Claudins are posttranslationally modified by palmitoylation and phosphorylation. All claudins have membrane proximal CxxC motifs near the ends of the second and fourth transmembrane domains. Both of these regions in claudin-14 are palmitoylated (Van Itallie et al. 2005). Site-directed mutagenesis revealed that this modification of claudin-14 is required for its efficient localization at TJs, but not the stability of claudin-14 protein or the assembly of TJ strands.

Ser or Thr residues in the carboxy-terminal cytoplasmic regions of several claudins, including claudin-3, -4, -5, and -16, are phosphorylated (D'souza et al. 2005; Aono and Hirai. 2008; Ishizaki et al. 2003; Ikari et al. 2006). This modification seems to regulate the localization of claudins and the barrier property of TJs, but the molecular mechanisms remain unknown.

Occludin

Occludin, the first-reported integral membrane protein of TJs, was identified by screening with monoclonal antibodies raised against an AJ/TJ-enriched plasma membrane fraction isolated from chick liver (Furuse et al. 1993). Occludin is an approximately 65-kD protein with four transmembrane domains, two extracellular loops, an intracellular turn, and carboxy- and amino-terminal cytoplasmic domains, but does not exhibit any sequence similarities with claudins. The first extracellular loop has a characteristic amino acid content ($\sim$60% of the amino acids are glycine and tyrosine residues), whose physiological significance remains elusive. The carboxy-terminal cytoplasmic region of occludin binds to ZO-1, ZO-2, and ZO-3, comprising the MAGUK family plaque proteins of TJs (Furuse et al. 1994; Fanning et al. 1998; Itoh et al. 1999a; Haskins et al. 1998). Multiple Ser and Thr residues in this region are phosphorylated, and highly-phoshorylated occludin is selectively concentrated at TJs (Sakakibara et al. 1997).

Overexpression studies in cultured cells revealed some activities of occludin in TJ formation and cell adhesion, although these activities are much weaker than those of claudins. When overexpressed in mouse L fibroblasts, which lack TJs as well as cadherin-mediated cell–cell adhesion, exogenous mouse occludin assembles into cell–cell contacts formed by adjacent cells as dots observed by immuno fluorescence microscopy (Furuse et al. 1998b). Short TJ strandlike structures containing occludin are observed in these regions by freeze-fracture electron microscopy. When human occludin is expressed in fibroblasts, the cells acquire weak cell–cell adhesive activity (Van Itallie and Anderson 1997). In these cells, addition of synthetic peptides corresponding to the first extracellular loop inhibits cell–cell adhesion. Overexpression of chicken occludin in MDCK cells increased the number of TJ strands (McCarthy et al. 1996), whereas overexpression of human occludin induced side-to-side aggregation of TJ strands (Medina et al. 2000). Furthermore, overexpression of occludin affected the transepithelial electric resistance as well as the paracellular flux in various epithelial cell lines (McCarthy et al. 1996; Balda et al. 1996; Bamforth et al. 1999). These observations suggest that occludin has some accessory functions in TJ formation. However, occludin-deficient epithelial cells exhibit no significant abnormalities in the morphology and physiology of their TJs, and occludin-deficient mice are viable (Saitou et al. 1998; Saitou et al. 2000). Occludin-related proteins, such as tricellulin, may have redundant functions, thereby obscuring the function of occludin.

Tricellulin

Tricellulin is the only known protein that is highly concentrated at tTJs, the vertically oriented TJ strands at tricellular contacts. Tricellulin was originally identified as an epithelium-specific transcript in a screening for putative targets of the Snail transcription repressor, which determines epithelial–mesenchymal transition (Ikenouchi et al. 2005). Tricellulin gene was also identified as a causative gene for recessive, nonsyndromic deafness DFNB49 (Riazuddin et al. 2006). Tricellulin contains four transmembrane domains and has sequence similarity with occludin in the carboxy-terminal cytoplasmic domain of approximately 130 amino acids. This region of occludin is responsible for its association with ZO-1, and tricellulin also binds ZO-1 (Riazuddin et al. 2006). RNAi-mediated suppression of tricellulin in epithelial cells results in abnormal organization of tTJs and impairment of epithelial barrier function (Ikenouchi et al. 2005). Furthermore, tricellulin knockdown also appears to affect bicellular TJ formation, as evaluated by immunostaining of occludin. Interestingly, occludin knockdown in cultured epithelial cells causes mislocalization of tricellulin to bicellular TJs, indicating a role of occludin in the concentration of tricellulin at tTJs (Ikenouchi et al. 2008).

When tricellulin is expressed in claudin-1-overexpressing L cells, it is colocalized with claudin-1-based reconstituted TJ strands and induces their crosslinking (Ikenouchi et al. 2008), suggesting that tricellulin influences the morphology of TJ strands. Although tricellulin is highly concentrated at tTJs, it is also detected at bicellular TJs and the extent of the bicellular distribution of tricellulin depends on the epithelial cell type (Ikenouchi et al. 2005). Therefore, it will be of interest to analyze the relationship between the amount of tricellulin within bicellular TJs and the morphology of TJ strand branching in different cell-types. Despite the critical role of tricellulin in tTJ formation, the function of tricellulin in the organization of tTJs together with claudins remains largely unknown.

ZO-1, ZO-2, and ZO-3

TJ MAGUKs

Just beneath the plasma membrane of TJs, PDZ domain-containing cytoplasmic proteins directly interact with TJ-associated integral membrane proteins and form cytoplasmic plaques, which function as scaffolds to recruit other structural proteins, signaling proteins, and the actin cytoskeleton to the underlying surface of TJs. Three structurally related proteins, ZO-1, ZO-2, and ZO-3, are included in these cytoplasmic plaques. ZO-1 was the first reported TJ-associated molecule. It was identified by generating monoclonal antibodies against a TJ-enriched membrane fraction isolated from mouse liver (Stevenson et al. 1986). ZO-2 and ZO-3 were later identified as coprecipitated proteins with ZO-1 from extracts of cultured epithelial cells (Gumbiner et al. 1991;

Haskins et al. 1998). All of these proteins belong to the MAGUK family and consist of three PDZ domains, a SH3 domain, and a guanylate kinase (GUK) domain from their amino-termini (Guillemot et al. 2008). ZO-1 and ZO-2 have acidic domains next to their GUK domains (Fig. 4). Multiple domains of ZO-1, ZO-2, and ZO-3 are involved in complex interactions with other TJ-associated molecules. The first PDZ domains of ZO-1, ZO-2, and ZO-3 directly bind to the carboxy-termini of claudins in vitro (Itoh et al. 1999b). ZO-1 also binds to the TJ-associated integral membrane proteins occludin and JAM-A (Furuse et al. 1994; Fanning et al. 1998; Bazzoni et al. 2000; Itoh et al. 2001; Li et al. 2005), and the cytoplasmic proteins cingulin and ZONAB (Corndenonsi et al. 1999; Balda et al. 2003). ZO-1, ZO-2, and ZO3 can interact with actin filaments (Itoh et al. 1997, 1999b; Fanning et al. 1998, 2002; Wittchen et al. 1999). ZO-1 and ZO-2 form heterodimers through their second PDZ domains (Utepbergenov et al. 2006). ZO-1 and ZO-3 also form heterodimers, but trimer formation between ZO-1, ZO-2, and ZO-3 has not been detected (Wittchen et al. 1999).

Roles of ZO-1 and ZO-2 in TJ Formation

Among the three TJ MAGUK proteins, ZO-1 and ZO-2 have been shown to be indispensable for TJ formation in epithelial cells. Targeted disruption of the ZO-1 gene and RNAi-mediated depletion of ZO-2 protein in mouse epithelial cells without ZO-3 expression results in deficient TJ formation (Umeda et al. 2006). Using these cells, the role of TJ MAGUKs in TJ strand formation was further characterized. Re-expression of either ZO-1 or ZO-2 recovered normal TJ formation in these cells, whereas the introduction of ZO-3 did not. A deletion construct of ZO-1 containing only three PDZ domains did not rescue the phenotype, but a longer construct containing the three PDZ domains, SH3 domain, GUK domain, and the following acidic domain induced normal TJ formation. However, a slightly shorter construct without the acidic domain induced aberrant TJ strand formation on lateral membranes.

These observations indicate that the SH3+GUK+acidic domains of ZO-1 play important roles in the correct localization of ZO-1 as well as the induction of claudin polymerization into TJ strands. Indeed, this region containing the SH3+unique-5 (U5)+GUK+unique-6 (U6) domains is highlighted as the entire SH3-GUK module in terms of the function in TJ formation (Fanning et al. 2007). U5 is the domain that links the SH3 domain and the GUK domain, whereas the U6 domain is immediately adjacent to the GUK domain. The GUK+U6 domains can bind to the SH3 domain, suggesting that the SH3-GUK module forms intramolecular interactions. A ZO-1 mutant lacking the U5 domain expressed in cultured epithelial cells was unable to localize to TJs and was distributed in the cytoplasm. On the other hand, a ZO-1 mutant lacking the U6 domain induced ectopically extended TJ strands to the lateral plasma membrane (Fanning et al. 2007).

Despite critical roles of the SH3-GUK module containing U5 and U6 domains, it remains unclear how ZO-1 is recruited to the correct region of the lateral membrane to induce TJ strand formation. The interaction of the SH3-GUK module of ZO-1 with AJ-associated molecules such as α-catenin and afadin might be implicated in this recruitement (Itoh et al. 1997; Yokoyama et al. 2001). Interestingly, in ZO-1/ZO-2/ZO-3-depleted epithelial cells, artificial anchoring of the three PDZ domains of ZO-1 to the plasma membrane by myristylation at the amino terminus did not induce TJ formation, whereas dimerization of this construct induced TJ formation throughout the lateral membrane (Umeda et al. 2006). Therefore, dimerization of ZO-1 (or ZO-2) just beneath the plasma membrane may trigger TJ formation, for instance by inducing claudin polymerization. The SH3-GUK module of ZO-1 may be involved in this dimerization, as proposed for the other MAGUK proteins hDlg and hCASK (Nix et al. 2000).

The roles of ZO-1 and ZO-2 are not limited to TJ formation. In calcium switch assays, ZO-1/ZO-2/ZO-3-depleted epithelial cells exhibit a delay in the transition from spotlike

AJs to beltlike AJs with linear actin cables. Depletion of ZO-1 in epithelial cells causes a delay in TJ formation (Umeda et al. 2004; McNeil et al. 2006), but this phenotype may also be because of a delay in AJ formation. ZO-1 and ZO-2 are required for Rac1 activation and myosin II recruitment during linear AJ formation (Ikenouchi et al. 2007; Yamazaki et al. 2008). Therefore, ZO-1 and ZO-2 appear to be involved in sequential steps in the assembly of AJ and TJ as well as their physical segregation in the membrane.

CONCLUDING REMARKS

The major molecular components of the core structure of TJs have been identified and characterized. Claudins are the main constituent of TJ strands, and cytoplasmic plaque proteins ZO-1 and ZO-2 directly regulate TJ strand formation and location in epithelial cells. However, many issues concerning the detailed structure of TJ strands remain unsolved.

A more complete understanding of the fine structure of TJ strands will require analysis by structural biological approaches, which may also provide important information about the size and charge selectivity of TJs (see review by Anderson and Van Itallie 2009). Furthermore, little is known about the mechanism of the ZO-1/ZO-2-mediated control of claudin polymerization within membranes. Recent studies indicate that dimerization of the claudin-interacting domains of ZO-1 and ZO-2 triggers claudin polymerization. On the other hand, it is unlikely that all claudins in the TJ bind to ZO-1 and ZO-2. Further research into the SH3-GUK module (the regulatory domains for TJ MAGUKs) of ZO family members will be key to understanding this problem. Interestingly, claudins lacking the carboxy-terminal cytoplasmic region, which interacts with ZO-1 and ZO-2, reconstitute TJ strands in mouse L fibroblasts, whereas TJs are not formed in epithelial cells lacking ZO-1 and ZO-2. Clarification of the molecular mechanisms behind this discrepancy in claudin-based TJ formation between epithelial cells and fibroblasts will provide important information about the polymerization of claudins within the plasma membrane.

A new research area of TJs is how tTJs are formed. Although tricellulin is required for the formation of normal tTJs and the epithelial barrier function, it is unknown how tricellulin functions in tTJs and how it is recruited there. Further functional analyses of tricellulin as well as the identification of novel components associated with tTJs will lead to better understanding of the molecular mechanism of tTJ formation and biological significance for tTJ sealing.

Finally, the function of occludin, the first identified TJ-associated integral membrane protein, is still not well understood although it was identified more than 15 years ago. It must have an important function because occludin-deficient mice cannot produce offspring (Saitou et al. 2000). Because occludin exhibits structural similarity with tricellulin, these two proteins may share redundant functions, although it is still unknown whether they are incorporated into or localized very closed to claudin-based TJ strands. Further investigation of the relationship between occludin and tricellulin may clarify the common role of these molecules in TJ formation and regulation.

ACKNOWLEDGMENTS

I would like to thank Drs. Hiroyuki Sasaki (The Jikei University School of Medicine) and Toshio Imai (KAN Research Institute, Inc.) for providing electron micrographs. M.F. is supported by a Grant-in-Aid for Cancer Research and the National Project on Targeted Protein Research Program (TPRP) from the Ministry of Education, Culture, Sports, Science and Technology of Japan.

REFERENCES

Anderson JM, Cereijido M. 2001. Evolution of ideas on the tight junction. In *Tight junctions*, 2nd ed. (ed. Cereijido M, Anderson J), pp. 1–18. CRC Press, Boca Raton.

Anderson JM, Van Itallie CM. 2009. Physiology and function of the tight junction. *Cold Spring Harb Perspect Biol* **1:** 002584.

Angelow S, Ahlstrom R, Yu AS. 2008. Biology of claudins. *Am J Physiol Renal Physiol* **295:** 867–876.

Aono S, Hirai Y. 2008. Phosphorylation of claudin-4 is required for tight junction formation in a human keratinocyte cell line. *Exp Cell Res* **314:** 3326–3339.

Balda MS, Anderson JM. 1993. Two classes of tight junctions are revealed by ZO-1 isoforms. *Am J Physiol* **264:** 918–924.

Balda MS, Garrett MD, Matter K. 2003. The ZO-1-associated Y-box factor ZONAB regulates epithelial cell proliferation and cell density. *J Cell Biol* **160:** 423–432.

Balda MS, Whitney JA, Flores C, Gonzalez S, Cereijido M, Matter K. 1996. Functional dissociation of paracellular permeability and transepithelial electrical resistance and disruption of the apical-basolateral intramembrane diffusion barrier by expression of a mutant tight junction membrane protein. *J Cell Biol* **134:** 1031–1049.

Bamforth SD, Kniesel U, Wolburg H, Engelhardt B, Risau W. 1999. A dominant mutant of occludin disrupts tight junction structure and function. *J Cell Sci* **112:** 1879–1888.

Bazzoni G, Martinez-Estrada OM, Orsenigo F, Cordenonsi M, Citi S, Dejana E. 2000. Interaction of junctional adhesion molecule with the tight junction components ZO-1, cingulin, and occludin. *J Biol Chem* **275:** 20520–20526.

Brightman MW, Reese TS. 1969. Junctions between intimately apposed cell membranes in the vertebrate brain. *J Cell Biol* **40:** 648–677.

Citi S, Sabanay H, Jakes R, Geiger B, Kendrick-Jones J. 1988. Cingulin, a new peripheral component of tight junctions. *Nature* **333:** 272–276.

Chalcroft JP, Bullivant S. 1970. An interpretation of liver cell membrane and junction structure based on observation of freeze-fracture replicas of both sides of the fracture. *J Cell Biol* **47:** 49–60.

Claude P. 1978. Morphological factors influencing transepithelial permeability: A model for the resistance of the zonula occludens. *J Membr Biol* **39:** 219–232.

Claude P, Goodenough DA. 1973. Fracture faces of zonulae occludentes from "tight" and "leaky" epithelia. *J Cell Biol* **58:** 390–400.

Cohen CJ, Shieh JT, Pickles RJ, Okegawa T, Hsieh JT, Bergelson JM. 2001. The coxsackievirus and adenovirus receptor is a transmembrane component of the tight junction. *Proc Natl Acad Sci* **98:** 15191–15196.

Cordenonsi M, D'Atri F, Hammar E, Parry DA, Kendrick-Jones J, Shore D, Citi S. 1999. Cingulin contains globular and coiled-coil domains and interacts with ZO-1, ZO-2, ZO-3, and myosin. *J Cell Biol* **147:** 1569–1582.

D'Atri F, Citi S. 2001. Cingulin interacts with F-actin in vitro. *FEBS Lett* **507:** 21–24.

D'souza T, Agarwal R, Morin PJ. 2005. Phosphorylation of claudin-3 at threonine 192 by cAMP-dependent protein kinase regulates tight junction barrier function in ovarian cancer cells. *J Biol Chem* **280:** 26233–26240.

da Silva PP, Kacher B. 1982. On tight-junction structure. *Cell* **28:** 441–450.

Daugherty BL, Ward C, Smith T, Ritzenthaler JD, Koval M. 2007. Regulation of heterotypic claudin compatibility. *J Biol Chem* **282:** 30005–30013.

Dermietzel R. 1974. Junctions in the central nervous system of the cat. I. Membrane fusion in central myelin. *Cell Tissue Res* **148:** 565–576.

Ebnet K, Iden S, Gerke V, Suzuki A. 2008. Regulation of epithelial and endothelial junctions by PAR proteins. *Front Biosci* **13:** 6520–6536.

Fanning AS, Ma TY, Anderson JM. 2002. Isolation and functional characterization of the actin binding region in the tight junction protein ZO-1. *Faseb J* **16:** 1835–1837.

Fanning AS, Jameson BJ, Jesaitis LA, Anderson JM. 1998. The tight junction protein ZO-1 establishes a link between the transmembrane protein occludin and the actin cytoskeleton. *J Biol Chem* **273:** 29745–29753.

Fanning AS, Little BP, Rahner C, Utepbergenov D, Walther Z, Anderson JM. 2007. The unique-5 and -6 motifs of ZO-1 regulate tight junction strand localization and scaffolding properties. *Mol Biol Cell* **18:** 721–731.

Farquhar MG, Palade GE. 1963. Junctional complexes in various epithelia. *J Cell Biol* **17:** 375–412.

Friend DS, Gilula NB. 1972. Variations in tight and gap junctions in mammalian tissues. *J Cell Biol* **53:** 758–776.

Fujimoto K. 1995. Freeze-fracture replica electron microscopy combined with SDS digestion for cytochemical labeling of integral membrane proteins. Application to the immunogold labeling of intercellular junctional complexes. *J Cell Sci* **108:** 3443–3449.

Furuse M. 2009. Knockout animals and natural mutations as experimental and diagnostic tool for studying tight junction functions in vivo. *Biochim Biophys Acta* **1788:** 813–819.

Furuse M, Tsukita S. 2006. Claudins in occluding junctions of humans and flies. *Trends Cell Biol* **16:** 181–188.

Furuse M, Sasaki H, Tsukita S. 1999. Manner of interaction of heterogeneous claudin species within and between tight junction strands. *J Cell Biol* **147:** 891–903.

Furuse M, Sasaki H, Fujimoto K, Tsukita S. 1998b. A single gene product, claudin-1 or -2, reconstitutes tight junction strands and recruits occludin in fibroblasts. *J Cell Biol* **143:** 391–401.

Furuse M, Fujita K, Hiiragi T, Fujimoto K, Tsukita S. 1998a. Claudin-1 and -2: Novel integral membrane proteins localizing at tight junctions with no sequence similarity to occludin. *J Cell Biol* **141:** 1539–1550.

Furuse M, Hirase T, Itoh M, Nagafuchi A, Yonemura S, Tsukita S, Tsukita S. 1993. Occludin: A novel integral membrane protein localizing at tight junctions. *J Cell Biol* **123:** 1777–1788.

Furuse M, Itoh M, Hirase T, Nagafuchi A, Yonemura S, Tsukita S, Tsukita S. 1994. Direct association of occludin with ZO-1 and its possible involvement in the localization of occludin at tight junctions. *J Cell Biol* **127:** 1617–1626.

Goodenough DA, Revel JP. 1970. A fine structural analysis of intercellular junctions in the mouse liver. *J Cell Biol* **45:** 272–290.

Gow A, Davies C, Southwood CM, Frolenkov G, Chrustowski M, Ng L, Yamauchi D, Marcus DC, Kachar B. 2004. Deafness in Claudin 11-null mice reveals the critical contribution of basal cell tight junctions to stria vascularis function. *J Neurosci* **24:** 7051–7062.

Gow A, Southwood CM, Li JS, Pariali M, Riordan GP, Brodie SE, Danias J, Bronstein JM, Kachar B, Lazzarini RA. 1999. CNS myelin and sertoli cell tight junction strands are absent in Osp/claudin-11 null mice. *Cell* **99:** 649–659.

Guillemot L, Paschoud S, Pulimeno P, Foglia A, Citi S. 2008. The cytoplasmic plaque of tight junctions: A scaffolding and signalling center. *Biochim Biophys Acta* **1778:** 601–613.

Gumbiner B, Lowenkopf T, Apatira D. 1991. Identification of a 160-kDa polypeptide that binds to the tight junction protein ZO-1. *Proc Natl Acad Sci* **88:** 3460–3464.

Hamazaki Y, Itoh M, Sasaki H, Furuse M, Tsukita S. 2002. Multi-PDZ domain protein 1 (MUPP1) is concentrated at tight junctions through its possible interaction with claudin-1 and junctional adhesion molecule. *J Biol Chem* **277:** 455–461.

Haskins J, Gu L, Wittchen ES, Hibbard J, Stevenson BR. 1998. ZO-3, a novel member of the MAGUK protein family found at the tight junction, interacts with ZO-1 and occludin. *J Cell Biol* **141:** 199–208.

Hirabayashi S, Tajima M, Yao I, Nishimura W, Mori H, Hata Y. 2003. JAM4, a junctional cell adhesion molecule interacting with a tight junction protein, MAGI-1. *Mol Cell Biol* **23:** 4267–4282.

Ide N, Hata Y, Nishioka H, Hirao K, Yao I, Deguchi M, Mizoguchi A, Nishimori H, Tokino T, Nakamura Y, Takai Y. 1999. Localization of membrane-associated guanylate kinase (MAGI)-1/BAI-associated protein (BAP) 1 at tight junctions of epithelial cells. *Oncogene* **18:** 7810–7815.

Ikari A, Matsumoto S, Harada H, Takagi K, Hayashi H, Suzuki Y, Degawa M, Miwa M. 2006. Phosphorylation of paracellin-1 at Ser217 by protein kinase A is essential for localization in tight junctions. *J Cell Sci* **119:** 1781–1789.

Ikenouchi J, Furuse M, Furuse K, Sasaki H, Tsukita S, Tsukita S. 2005. Tricellulin constitutes a novel barrier at tricellular contacts of epithelial cells. *J Cell Biol* **171:** 939–945.

Ikenouchi J, Sasaki H, Tsukita S, Furuse M, Tsukita S. 2008. Loss of occludin affects tricellular localization of tricellulin. *Mol Biol Cell* **19:** 4687–4693.

Ikenouchi J, Umeda K, Tsukita S, Furuse M, Tsukita S. 2007. Requirement of ZO-1 for the formation of belt-like adherens junctions during epithelial cell polarization. *J Cell Biol* **176:** 779–786.

Ishizaki T, Chiba H, Kojima T, Fujibe M, Soma T, Miyajima H, Nagasawa K, Wada I, Sawada N. 2003. Cyclic AMP induces phosphorylation of claudin-5 immunoprecipitates and expression of claudin-5 gene in blood-brain-barrier endothelial cells via protein kinase A-dependent and -independent pathways. *Exp Cell Res* **290:** 275–288.

Itoh M, Morita K, Tsukita S. 1999b. Characterization of ZO-2 as a MAGUK family member associated with tight as well as adherens junctions with a binding affinity to occludin and α catenin. *J Biol Chem* **274:** 5981–5986.

Itoh M, Furuse M, Morita K, Kubota K, Saitou M, Tsukita S. 1999a. Direct binding of three tight junction-associated MAGUKs, ZO-1, ZO-2, and ZO-3, with the COOH termini of claudins. *J Cell Biol* **147:** 1351–1363.

Itoh M, Nagafuchi A, Moroi S, Tsukita S. 1997. Involvement of ZO-1 in cadherin-based cell adhesion through its direct binding to α catenin and actin filaments. *J Cell Biol* **138:** 181–192.

Itoh M, Sasaki H, Furuse M, Ozaki H, Kita T, Tsukita S. 2001. Junctional adhesion molecule (JAM) binds to PAR-3: A possible mechanism for the recruitment of PAR-3 to tight junctions. *J Cell Biol* **154:** 491–497.

Izumi Y, Hirose T, Tamai Y, Hirai S, Nagashima Y, Fujimoto T, Tabuse Y, Kemphues KJ, Ohno S. 1998. An atypical PKC directly associates and colocalizes at the epithelial tight junction with ASIP, a mammalian homologue of *Caenorhabditis elegans* polarity protein PAR-3. *J Cell Biol* **143:** 95–106.

Jeansonne B, Lu Q, Goodenough DA, Chen YH. 2003. Claudin-8 interacts with multi-PDZ domain protein 1 (MUPP1) and reduces paracellular conductance in epithelial cells. *Cell Mol Biol* **49:** 13–21.

Kitajiri S, Miyamoto T, Mineharu A, Sonoda N, Furuse K, Hata M, Sasaki H, Mori Y, Kubota T, Ito J, et al. 2004. Compartmentalization established by claudin-11-based tight junctions in stria vascularis is required for hearing through generation of endocochlear potential. *J Cell Sci* **117:** 5087–5096.

Kubota K, Furuse M, Sasaki H, Sonoda N, Fujita K, Nagafuchi A, Tsukita S. 1999. Ca(2+)-independent cell-adhesion activity of claudins, a family of integral membrane proteins localized at tight junctions. *Curr Biol* **9:** 1035–1038.

Kumar NM, Gilula NB. 1996. The gap junction communication channel. *Cell* **84:** 381–388.

Lane JN. 2001. Tight junctions in invertebrates. In *Tight junctions*, 2nd ed. (ed. Cereijido M and Anderson J), pp. 39–59, CRC Press, Boca Raton.

Laura RP, Ross S, Koeppen H, Lasky LA. 2002. MAGI-1: A widely expressed, alternatively spliced tight junction protein. *Exp Cell Res* **275:** 155–170.

Lemmers C, Medina E, Delgrossi MH, Michel D, Arsanto JP, Le Bivic A. 2002. hINADl/PATJ, a homolog of discs lost, interacts with crumbs and localizes to tight junctions in human epithelial cells. *J Biol Chem* **277:** 25408–25415.

Li Y, Fanning AS, Anderson JM, Lavie A. 2005. Structure of the conserved cytoplasmic C-terminal domain of occludin: Identification of the ZO-1 binding surface. *J Mol Biol* **352:** 151–164.

Madara JL. 1987. Intestinal absorptive cell tight junctions are linked to cytoskeleton. *Am J Physiol* **253:** 171–175.

Martin-Padura I, Lostaglio S, Schneemann M, Williams L, Romano M, Fruscella P, Panzeri C, Stoppacciaro A, Ruco L, Villa A, et al. 1998. Junctional adhesion molecule, a novel member of the immunoglobulin superfamily that distributes at intercellular junctions and modulates monocyte transmigration. *J Cell Biol* **142:** 117–127.

McCarthy KM, Skare IB, Stankewich MC, Furuse M, Tsukita S, Rogers RA, Lynch RD, Schneeberger EE. 1996. Occludin is a functional component of the tight junction. *J Cell Sci* **109:** 2287–2298.

McNeil E, Capaldo CT, Macara IG. 2006. Zonula occludens-1 function in the assembly of tight junctions in Madin-Darby canine kidney epithelial cells. *Mol Biol Cell* **17:** 1922–1932.

Medina R, Rahner C, Mitic LL, Anderson JM, Van Itallie CM. 2000. Occludin localization at the tight junction requires the second extracellular loop. *J Membr Biol* **178:** 235–247.

Mitic LL, Unger VM, Anderson JM. 2003. Expression, solubilization, and biochemical characterization of the tight junction transmembrane protein claudin-4. *Protein Sci* **12:** 218–227.

Morita K, Sasaki H, Furuse M, Tsukita S. 1999. Endothelial claudin: Claudin-5/TMVCF constitutes tight junction strands in endothelial cells. *J Cell Biol* **147:** 185–194

Nasdala I, Wolburg-Buchholz K, Wolburg H, Kuhn A, Ebnet K, Brachtendorf G, Samulowitz U, Kuster B, Engelhardt B, Vestweber D, et al. 2002. A transmembrane tight junction protein selectively expressed on endothelial cells and platelets. *J Biol Chem* **277:** 16294–16303.

Nix SL, Chishti AH, Anderson JM, Walther Z. 2000. hCASK and hDlg associate in epithelia, and their src homology 3 and guanylate kinase domains participate in both intramolecular and intermolecular interactions. *J Biol Chem* **275:** 41192–41200.

Nunes FD, Lopez LN, Lin HW, Davies C, Azevedo RB, Gow A, Kachar B. 2006. Distinct subdomain organization and molecular composition of a tight junction with adherens junction features. *J Cell Sci* **119:** 4819–4827.

Ohnishi H, Nakahara T, Furuse K, Sasaki H, Tsukita S, Furuse M. 2004. JACOP, a novel plaque protein localizing at the apical junctional complex with sequence similarity to cingulin. *J Biol Chem* **279:** 46014–46022.

Powell DW. 1981. Barrier function of epithelia. *Am J Physiol* **241:** 275–288.

Riazuddin S, Ahmed ZM, Fanning AS, Lagziel A, Kitajiri S, Ramzan K, Khan SN, Chattaraj P, Friedman PL, Anderson JM, et al. 2006. Tricellulin is a tight-junction protein necessary for hearing. *Am J Hum Genet* **79:** 1040–1051.

Roh MH, Liu CJ, Laurinec S, Margolis B. 2002. The carboxyl terminus of zona occludens-3 binds and recruits a mammalian homologue of discs lost to tight junctions. *J Biol Chem* **277:** 27501–27509.

Saitou M, Fujimoto K, Doi Y, Itoh M, Fujimoto T, Furuse M, Takano H, Noda T, Tsukita S. 1998. Occludin-deficient embryonic stem cells can differentiate into polarized epithelial cells bearing tight junctions. *J Cell Biol* **141:** 397–408.

Saitou M, Furuse M, Sasaki H, Schulzke JD, Fromm M, Takano H, Noda T, Tsukita S. 2000. Complex phenotype of mice lacking occludin, a component of tight junction strands. *Mol Biol Cell* **11:** 4131–4142.

Sakakibara A, Furuse M, Saitou M, Ando-Akatsuka Y, Tsukita S. 1997. Possible involvement of phosphorylation of occludin in tight junction formation. *J Cell Biol* **137:** 1393–1401.

Sasaki H, Matsui C, Furuse K, Mimori-Kiyosue Y, Furuse M, Tsukita S. 2003. Dynamic behavior of paired claudin strands within apposing plasma membranes. *Proc Natl Acad Sci* **100:** 3971–3976.

Schulze C, Firth JA. 1993. Immunohistochemical localization of adherens junction components in blood-brain barrier microvessels of the rat. *J Cell Sci* **104:** 773–782.

Simionescu M, Simionescu N, Palade GE. 1975. Segmental differentiations of cell junctions in the vascular endothelium. The microvasculature. *J Cell Biol* **67:** 863–885.

Sonoda N, Furuse M, Sasaki H, Yonemura S, Katahira J, Horiguchi Y, Tsukita S. 1999. Clostridium perfringens enterotoxin fragment removes specific claudins from tight junction strands: Evidence for direct involvement of claudins in tight junction barrier. *J Cell Biol* **147:** 195–204.

Staehelin LA. 1973. Further observations on the fine structure of freeze-cleaved tight junctions. *J Cell Sci* **13:** 763–786.

Staehelin LA. 1974. Structure and function of intercellular junctions. *Int Rev Cytol* **39:** 191–283.

Staehelin LA, Mukherjee TM, Williams AW. 1969. Freeze-etch appearance of the tight junctions in the epithelium of small and large intestine of mice. *Protoplasma* **67:** 165–184.

Stevenson BR, Goodenough DA. 1984. Zonulae occludentes in junctional complex-enriched fractions from mouse liver: Preliminary morphological and biochemical characterization. *J Cell Biol* **98:** 1209–1221.

Stevenson BR, Siliciano JD, Mooseker MS, Goodenough DA. 1986. Identification of ZO-1: A high molecular weight polypeptide associated with the tight junction (zonula occludens) in a variety of epithelia. *J Cell Biol* **103:** 755–766.

Suzuki A, Ohno S. 2006. The PAR-aPKC system: Lessons in polarity. *J Cell Sci* **119:** 979–987.

Tetzlaff W. 1978. The development of a zonula occludens in peripheral myelin of the chick embryo. A freeze-fracture study. *Cell Tissue Res* **189:** 187–201.

Tsukita S, Furuse M. 1999. Occludin and claudins in tight-junction strands: Leading or supporting players? *Trend Cell Biol* **9:** 268–273.

Umeda K, Ikenouchi J, Katahira-Tayama S, Furuse K, Sasaki H, Nakayama M, Matsui T, Tsukita S, Furuse M, Tsukita S. 2006. ZO-1 and ZO-2 independently determine where claudins are polymerized in tight-junction strand formation. *Cell* **126:** 741–754.

Umeda K, Matsui T, Nakayama M, Furuse K, Sasaki H, Furuse M, Tsukita S. 2004. Establishment and characterization of cultured epithelial cells lacking expression of ZO-1. *J Biol Chem* **279:** 44785–44794.

Utepbergenov DI, Fanning AS, Anderson JM. 2006. Dimerization of the scaffolding protein ZO-1 through the second PDZ domain. *J Biol Chem* **281:** 24671–24677.

van Deurs B, Koehler JK. 1979. Tight junctions in the choroid plexus epithelium. A freeze-fracture study including complementary replicas. *J Cell Biol* **80:** 662–673.

Van Itallie CM, Anderson JM. 1997. Occludin confers adhesiveness when expressed in fibroblasts. *J Cell Sci* **110:** 1113–1121.

Van Itallie CM, Anderson JM. 2006. Claudins and epithelial paracellular transport. *Annu Rev Physiol* **68:** 403–429.

Van Itallie CM, Gambling TM, Carson JL, Anderson JM. 2005. Palmitoylation of claudins is required for efficient tight-junction localization. *J Cell Sci* **118:** 1427–1436.

van Meer G, Gunbiner B, Simons K. 1986. The tight junction does not allow lipid molecules to diffuse from one epithelial cell to the next. *Nature* **322:** 639–641.

Wade JB, Karnovsky MJ. 1974. The structure of the zonula occludens. A single fibril model based on freeze-fracture. *J Cell Biol* **60:** 168–180.

Wittchen ES, Haskins J, Stevenson BR. 1999. Protein interactions at the tight junction. Actin has multiple binding partners, and ZO-1 forms independent complexes with ZO-2 and ZO-3. *J Biol Chem* **274:** 35179–35185.

Yamazaki Y, Umeda K, Wada M, Nada S, Okada M, Tsukita S, Tsukita S. 2008. ZO-1- and ZO-2-dependent integration of myosin-2 to epithelial zonula adherens. *Mol Biol Cell* **19:** 3801–3811.

Yokoyama S, Tachibana K, Nakanishi H, Yamamoto Y, Irie K, Mandai K, Nagafuchi A, Monden M, Takai Y. 2001. α-Catenin-independent recruitment of ZO-1 to nectin-based cell-cell adhesion sites through afadin. *Mol Biol Cell* **12:** 1595–1609.

Adherens Junction: Molecular Architecture and Regulation

Wenxiang Meng and Masatoshi Takeichi

RIKEN Center for Developmental Biology, Chuo-ku, Kobe 650-0047, Japan

Correspondence: takeichi@cdb.riken.jp

The adherens junction (AJ) is an element of the cell–cell junction in which cadherin receptors bridge the neighboring plasma membranes via their homophilic interactions. Cadherins associate with cytoplasmic proteins, called catenins, which in turn bind to cytoskeletal components, such as actin filaments and microtubules. These molecular complexes further interact with other proteins, including signaling molecules, rendering the AJs into highly dynamic and regulatable structures. The AJs of such nature contribute to the physical linking of cells, as well as to the regulation of cell–cell contacts, which is essential for morphogenesis and remodeling of tissues and organs. Thus, elucidating the molecular architecture of the AJs and their regulatory mechanisms are crucial for understanding how the multicellular system is organized.

The adherens junction (AJ) is a form of cell–cell adhesion structure observed in a variety of cell types, as well as in different animal species. It is characterized by a pair of plasma membranes apposed with a distance of 10–20 nm between them, whose intercellular space is occupied by rod-shaped molecules bridging the membranes (Hirokawa and Heuser 1981; Miyaguchi 2000), and the cytoplasmic side of the AJ is associated with condensed actin filaments. In polarized epithelia of vertebrates, the AJ is part of the tripartite junctional complex localized at the juxtaluminal region, which comprises the tight junction (zonula occludens), AJ, and desmosome (macula adherens) aligned in this order from the apical end of the junction (Farquhar and Palade 1963). In this type of epithelia, the AJ is specifically termed the "zonula adherens" or "adhesion belt," as it completely encloses the cells along with the F-actin lining, called the circumferential actin belt (Fig. 1). The AJs in other cell types assume different morphologies: For example, the AJs in fibroblastic cells are spotty and discontinuous (Yonemura et al. 1995), and those in neurons are organized into tiny puncta as a constituent of the synaptic junctions (Uchida et al. 1996).

A major function of AJs is to maintain the physical association between cells, as disruption of them causes loosening of cell–cell contacts, leading to disorganization of tissue architecture. Calcium chelators such as EDTA and EGTA are widely used as a reagent to promote the dissociation of cells in tissues or monolayer cultures. A major target of these chelators is the

 doi: 10.1101/cshperspect.a002899
Cite this article as *Cold Spring Harb Perspect Biol* 2009;1:a002899

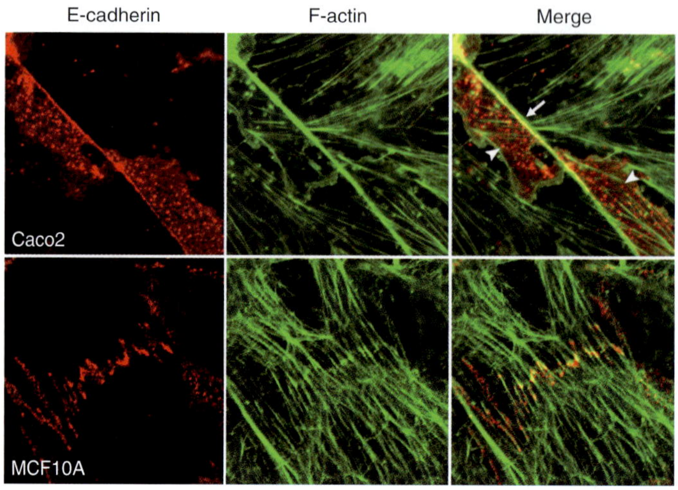

Figure 1. Morphological variations of the adherens junction. In Caco2 cells (colonic carcinoma line), E-cadherin is localized along the actin circumferential belt to organize the zonula adherens (arrow). At the lateral portions of cell junction (arrowheads), E-cadherin signals are punctate, only occasionally overlapping with actin signals in this specific sample. The lateral patterns of cadherin and actin distribution, however, vary with cellular conditions. In MCF10A cells (mammary epithelial line), spotty adherens junctions are seen, where actin filaments perpendicularly terminate at E-cadherin puncta.

AJ, as this is a calcium-sensitive structure; although, calcium removal is generally insufficient for the complete dispersion of cells because of the presence of calcium-independent cell–cell adhesion mechanisms (Takeichi et al. 1977). Early studies to search for the molecules responsible for the calcium-dependent junctions resulted in the identification of a group of type-I transmembrane proteins, and its founding member was termed cadherin (Yoshida and Takeichi 1982; Yoshida-Noro et al. 1984). Related molecules identified were also called by various names, such as uvomorulin (Peyrieras et al. 1983), LCAM (Gallin et al. 1983), and ACAM (Volk and Geiger 1984). Later studies revealed that the cadherins form a superfamily, and therefore, the original cadherins are now called "classic" cadherins.

Another series of studies have identified nectins, a family of immunoglobulin-like transmembrane proteins, as an AJ component. Nectins function in a calcium-independent way to promote cell–cell adhesion (Nakanishi and Takai 2004). In this article, we overview the molecular organization of the AJs constructed with these membrane proteins, as well as the regulatory mechanisms that operate to sustain or remodel these junctions, paying much attention to the linkages between the AJ and cytoskeletal or signaling proteins.

CADHERINS

The classic cadherin family comprises approximately 20 members that share a common domain organization. The members are called E-cadherin (cdh1), N-cadherin (cdh2), and so on, each of which shows a distinct tissue distribution pattern (Takeichi 1988). Their extracellular domain is divided into five repetitive subdomains, called cadherin repeats or EC domains, and each subdomain contains calcium-binding sequences (Overduin et al. 1995). The interaction of calcium ions with these sequences controls the conformation of the extracellular domain (Pokutta et al. 1994), switching its adhesive function "off" and "on." On its association with calcium, the extracellular domain of cadherins on a cell undergoes homophilic interaction with that of cadherins present on the apposed cells (Fig. 2). The precise mechanism for this homophilic interaction is still controversial (Troyanovsky 2005), although current studies support the

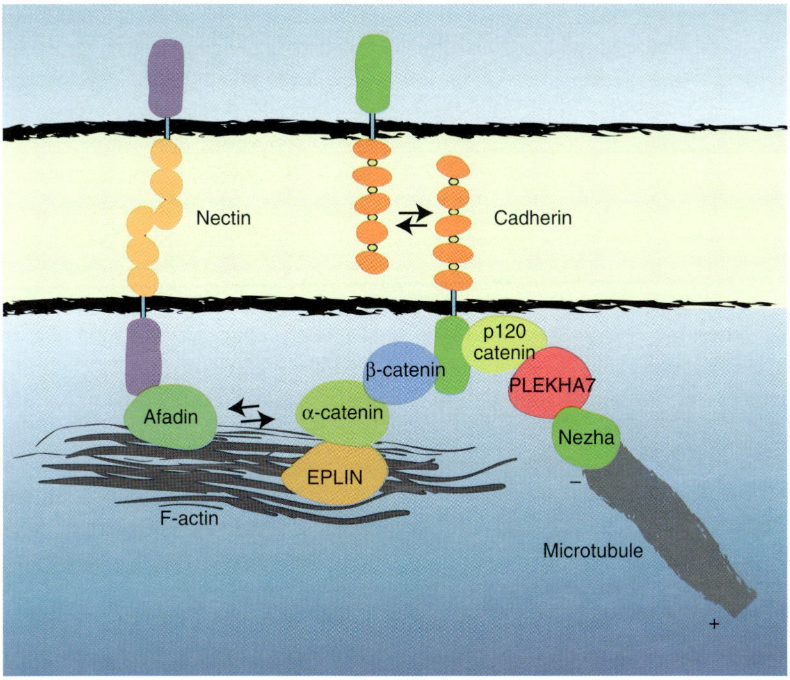

Figure 2. Representative molecular constituents of the zonula adherens. These constituents vary with the types of adherens junction.

idea that a *trans* binding between cadherin monomers via their EC1 domains initiates their interactions (Zhang et al. 2009).

The calcium-sensitive sequences are highly conserved among the family members, whose sequences are the hallmarks for this family, and all classic cadherins show a similar calcium dependence. Other sequences in the extracellular domain, however, vary, and this sequence variation confers the adhesive specificity on each member (Nose et al. 1988). For example, E-cadherin preferentially binds E-cadherin, and N-cadherin does so to N-cadherin; although, the degree of the selectivity changes depending on the partners (Shimoyama et al. 2000; Patel et al. 2006). This nature of classic cadherins has been implicated in the sorting of different cell types (Takeichi 1988).

AJs occur in a wide variety of animal species, and classic cadherin-type molecules have been identified in many species (Oda et al. 2005). However, their molecular organization is not perfectly conserved. For example, although *Drosophila* E-cadherin (DE-cadherin) is a component of the AJ, it has seven EC domains, instead of the five in the vertebrate classic cadherins (Oda et al. 1994). The other *Drosophila* classic cadherin, DN-cadherin, has even 17 EC domains (Iwai et al. 1997). These *Drosophila* cadherins also contain other domains, such as EGF-like and laminin globular-like domains, inserted between the set of EC domains and the transmembrane domain. As a result, these *Drosophila* cadherins are much larger in size than the vertebrate ones, despite the similar appearance of their AJs. The functional importance of AJs also differs among species. For example, *Caenorhabditis elegans* has a classic-cadherin-type molecule (HMR-1), but its role in cell junction formation is limited: HMR-1-deficient animals fail to enclose the epidermal sheets, but their general cell junctions look normal even in the absence of this cadherin (Cox et al. 2004). The organization of apical junctional complexes is also different between vertebrates and invertebrates. In *Drosophila*, the relative positions of the AJ and tight junction are inverse; that is, the AJ is located above the septate junction, which is considered as a structure equivalent to the vertebrate tight junction. Irrespective of these species-dependent diversifications, the key structure and function of the cadherin cytoplasmic domain is highly conserved among different species; that is, the classic cadherins from any species interact with their conserved cytoplasmic partners (see the following).

After the discovery of the classic cadherins, a number of other molecules that share the conserved EC domains but have divergent cytoplasmic sequences were identified, and these are collectively called nonclassic cadherins. These include desmosomal cadherins (Wheeler et al. 1991; Buxton and Magee 1992), protocadherins (Redies et al. 2005; Morishita and Yagi 2007), Fat and Dachsous cadherins (Saburi and McNeill 2005; Tanoue and Takeichi 2005), and Flamingo/Celsr (Takeichi 2007) (see McNeill et al. 2009). Among them, desmosomal cadherins are the closest to the classic cadherins, and desmosomes, in fact, are similar in appearance to AJs, though not identical from a number of aspects (Holthofer et al. 2007) (see Green et al. 2009; Delva et al. 2009). Except for the desmosomal cadherins, other nonclassic cadherins appear not to organize specialized junctions, nor to be the components essential for AJ formation; although, they generally can undergo homophilic interactions at the cell–cell interfaces. These nonclassic cadherins seem to have acquired unique molecular roles, rather than that for the physical linking of cells. For example, some of the protocadherins negatively regulate the classic cadherin-dependent adhesion (Chen and Gumbiner 2006), and Fat (Nollet et al. 2000; Strutt and Strutt 2005; Tanoue and Takeichi 2005) and Flamingo (Usui et al. 1999) regulate planar cellular polarity as well as other forms of cellular interactions.

CADHERIN–CATENIN COMPLEX

The cytoplasmic domains are highly conserved among the classic cadherin members, and they bind common cytoplasmic molecules, collectively called catenins (Fig. 2). The juxtamembrane portion of the cytoplasmic domain associates with p120-catenin, which belongs to a subfamily of the armadillo proteins (Reynolds et al. 1992; Shibamoto et al. 1995; Hatzfeld 2005). The carboxy-terminal half of the cytoplasmic domain, on the other hand, binds β-catenin or plakoglobin (γ-catenin), which are close relatives of each other (Ozawa et al. 1989; Knudsen and Wheelock 1992; Ozawa and Kemler 1992). As mentioned above, these molecular partners for the cytoplasmic domain are well conserved among different animal species; e.g., its invertebrate versions can bind both p120-catenin (JAC-1 in *C. elegans*) and β-catenin (Armadillo in *Drosophila*; HMP-2 in *C. elegans*) (Peifer and Wieschaus 1990; Cox et al. 2004). These catenins in turn associate with a variety of other molecules, including cytoskeletal proteins and their regulators. These cytoplasmic components of AJ affect the adhesive action of the extracellular domain of cadherins in various ways, leading to alterations in the strength and stability of cell–cell contacts.

INTERACTIONS WITH THE ACTIN CYTOSKELETON

The AJ is morphologically associated with actin filaments, posing the questions of how this association is established and what role the actin plays in AJ organization and function. A key player is thought to be α-catenin, a molecule similar to vinculin (Herrenknecht et al. 1991; Nagafuchi et al. 1991). The α-catenin binds β-catenin, resulting in the formation of the cadherin–β-catenin–α-catenin complex. Early biochemical studies showed that α-catenin can interact with actin filaments (Rimm et al. 1995), giving rise to the general belief that α-catenin acts as a linker between the cadherin–β-catenin complex and F-actin. However, this concept was challenged by the finding that the α-catenin complexed with cadherin and β-catenin cannot bind F-actin in vitro and that only free α-catenins can do so (Drees et al. 2005; Yamada et al. 2005). This finding suggested the possibility that the cadherin–β-catenin–α-catenin complex might interact with F-actin via some other mediator(s). In fact, α-catenin has been shown to associate with actin-binding proteins such as formin (Kobielak et al. 2004) and vinculin (Watabe-Uchida et al. 1998), and a very recent study identified another actin-binding protein, EPLIN (epithelial protein lost in neoplasm; also known as Lima-1), as an α-catenin partner (Abe and Takeichi 2008). EPLIN is known to enhance the bundling of actin filaments and to stabilize them by suppressing F-actin depolymerization (Maul et al. 2003). The EPLIN can bind α-catenin when the latter is associated with the cadherin–β-catenin complex, and this entire complex binds F-actin. This finding illustrates a novel pathway for the interaction between cadherin and F-actin (Fig. 2).

It should be noted that the morphology of actin-AJ association differs with the cell types or cellular conditions (Yonemura et al. 1995). Although actin filaments run parallel along the ZA in simple cuboidal or columnar epithelia, these filaments often perpendicularly terminate at cell–cell borders in many other junctions, e.g., in those of stratified epithelium-derived cells and fibroblastic cells, and also in immature junctions of most cell types (Fig. 1). It is highly possible that different molecular mechanisms operate for the linking of F-actin to AJs in such different types of junction. In fact, even in the absence of EPLIN, F-actin still morphologically associates with AJs (see the following). Meanwhile, the actin-associated cadherins do not necessarily form a static domain. In some cell lines, actin filaments are aligned from the basal to apical end of cell junctions, and these actins display a type of retrograde flow. Cadherins are tethered to these actin filaments in an α-catenin-dependent manner and move together with the actins, displaying "cadherin flow" (Kametani and Takeichi 2007); although, the biological role of this flow is not understood yet. In mature epithelia, α-catenin is also important for regulating the mobility of cadherins (Cavey et al. 2008) (see Stepniak et al. 2009).

What happens if the linkage between the cadherin–catenin complex and F-actin is disrupted? When α-catenin is removed, AJ organization is disrupted, and the apical actin belt becomes segregated from the cadherin–catenin complex (Watabe-Uchida et al. 1998). If this occurs in neural epithelia in vivo, their architecture is seriously damaged (Vasioukhin et al. 2001). Removal of neural α-catenin (αN-catenin) from synaptic AJs destabilizes synaptic contacts (Abe et al. 2004). On the other hand, EPLIN loss in epithelia results in different types of defects at the junctions. The circumferential actin belt disappears, being converted to radially oriented actin filaments, indicating that EPLIN is important not only for the linkage between cadherin and F-actin but also for stabilizing this unique configuration of actin fibers (Abe and Takeichi 2008). Importantly, the actin filaments, rearranged as a result of EPLIN loss, still target cadherins, which now assume a spotty localization as seen in fibroblasts or immature epithelial junctions. This finding suggests that EPLIN is not the sole linker between cadherin and F-actin and that other linkers must be present and also suggests the possibility that the shape of AJs could be altered by the type of cadherin–actin linkers.

Because the actin filaments are essential for AJ assembly, regulators of these filaments would be expected to affect it. In fact, Rho-family small GTPases, such as RhoA, Rac1, and Cdc42, as well as their GEFs and GAPs, have been shown to regulate AJ formation and integrity (Braga 2000). Because these molecules are used for a wide variety of cell behavior, precise dissection of whether a given GTPase targets the actin filaments directly involved in junction formation or those involved in other processes, such as cell motility, is important for our correct understanding of the roles of these enzymes. aPKC has also been shown to be essential for the circumferential actin belt formation in epithelial cells. When aPKC is inactivated, the circumferential actin belt is lost, and AJs become spot-like (Suzuki et al. 2002). In vivo, aPKC knockout results in the loss of adherens junctions in neuroepithelial cells (Imai et al. 2006). It has been proposed that the action of aPKC is to antagonize the myosin-II-driven centripetal contraction of the circumferential actin cables (Kishikawa et al. 2008).

The contractility of AJ-associated circumferential actin belts is used for morphogenesis. A well-known example is the Shroom3-dependent constriction of the zonula adherens (ZA) in epithelial layers, where constriction plays an important role in epithelial folding or bending (Haigo et al. 2003; Hildebrand 2005). In this case, the actin-binding protein Shroom3 recruits Rho kinases (ROCKs) near the ZA, and activates myosin-II, inducing the contraction of the circumferential actin belts (Nishimura and Takeichi 2008). The actomyosin-dependent regulation of the AJ is also implicated in cell intercalation during early morphogenesis of *Drosophila* embryos (Bertet et al. 2004; Zallen and Wieschaus 2004; Blankenship et al. 2006).

INTERACTIONS WITH MICROTUBULES

As compared with the actin cytoskeleton, less attention has been paid to microtubules (MTs) with reference to the structure and function of the AJ. Actually, the MTs do not show specialized condensation at the AJ, contrasted with the unique bundling of actin fibers along the AJ. However, MTs have occasionally been observed to be located in a close proximity to the AJ, running parallel to this junction. Moreover, the radially extending MTs are targeted to the AJs with their plus ends in a CLIP-170-dependent manner, and blocking of the MT extension toward the AJs causes a reduction in the accumulation of junctional E-cadherin (Stehbens et al. 2006). In addition, dynein was found to bind β-catenin, and this β-catenin-associated dynein was proposed to tether MTs to cell junctions (Ligon et al. 2001; Shaw et al. 2007). These observations suggest a potential interaction of AJ with MT plus ends, and their interaction seems to have some roles in AJ assembly. In support of this idea, reagents that depolymerize MTs are known to disrupt the integrity of the AJs (Waterman-Storer et al. 2000), and even inhibit the disassembly of cell junctions (Ivanov et al. 2006)

Recent studies have also revealed that MT minus ends interact with AJs via p120-catenin (p120). This catenin recognizes and binds a specific sequence located in the juxtamembrane region of the cadherin cytoplasmic domain. The p120-cadherin binding is known to play a central role in the stability of cadherin-mediated junctions; i.e., when p120 is removed, the plasma membrane-associated cadherins become endocytosed, leading to reduced cell–cell associations (see the following discussion). This action of p120 requires the armadillo domain occupying its central region (Liu et al. 2007). Other series of studies have found that p120 can associate with MTs (Chen et al. 2003; Roczniak-Ferguson and Reynolds 2003; Franz and Ridley 2004; Yanagisawa et al. 2004), and this ability of p120 requires its carboxy-terminal domain (Ichii and Takeichi 2007), suggesting that the amino-terminal and armadillo domains of this catenin have separate functions. Further studies on the amino-terminal domain identified a new partner for p120 (Meng et al. 2008). This is PLEKHA7, which binds the amino-terminal domain of p120. Intriguingly, this protein localizes specifically along the ZA, but not in other portions of the junctions in epithelial sheets, despite the ubiquitous distribution of p120 and cadherins at the cell–cell contacts. Depletion of PLEKHA7 exclusively disrupts the ZA, but not the entire cell–cell junctions, suggesting that this protein is specifically required to maintain the ZA. Subsequently, PLEKHA7 was found to bind another protein, termed Nezha, which is again distributed along the ZA; although Nezha was also detected as punctate signals in the cytoplasm. Nezha displays an important property: It binds the minus ends of MTs, and tethers them to the ZA, allowing their extension and retraction from the cell junctions.

These studies have uncovered the presence of a novel population of MTs, whose minus ends are anchored at the ZA via the PLEKHA7–Nezha complex (Fig. 2). The functions of these MTs are not fully understood yet, but preliminary observations suggest that minus-end directed kinesin motors, such as KIFC3, use these MTs to transport themselves to the ZA. Depletion of PLEKHA7, Nezha, KIFC3, and MTs results in similar defects in ZA organization. Thus, these molecules appear to work together by forming a complex to sustain the ZA architecture. It should be re-emphasized that this novel molecular complex appears to be important only for the ZA, but not for the entire cadherin-mediated cell–cell contacts, confirming that the ZA is a specialized domain of the cadherin-mediated junctions/AJs. The mechanisms of how PLEKHA7 interacts with p120 only at the ZA remains unknown.

In sum, evidence is accumulating that MTs play roles in AJ assembly, where both the plus and minus ends of MTs have been suggested to be involved. Although the molecular mechanisms for the plus-end interactions with AJs have not been determined yet, the two populations of MTs with the opposite polarity might cooperate together for AJ regulation.

COOPERATION BETWEEN CADHERIN AND NECTIN

Nectins are a family of immunoglobulin-like molecules, consisting of four members (Nakanishi and Takai 2004). They are accumulated at the AJ, colocalizing with cadherins (Fig. 2). The cytoplasmic domain of nectins associates with AF6/afadin via the carboxy-terminal PDZ-binding motif of the former. Afadin, on the other hand, was shown to interact with α-catenin, suggesting that a physical association might occur between the cadherin–catenin and nectin–afadin complexes (Tachibana et al. 2000). Because afadin is an actin-binding protein (Mandai et al. 1997; Takahashi et al. 1999), this system may also play a role in the linking of AJ to actin filaments.

Nectins interact with other nectins in either a homophilic or heterophilic way. Different from the classic cadherins, nectins prefer heterotypic partners to homotypic ones, and their heterophilic binding produces stronger cell–cell adhesion than the homophilic interactions (Fabre et al. 2002; Yasumi et al. 2003; Martinez-Rico et al. 2005). Importantly, during the process of early cell–cell contacts, nectins first accumulate at the contacts, and then cadherins follow them (Takai et al. 2003), suggesting that the former may guide the latter in their junctional localization. This can be seen in the following example: When cells expressing nectin-1 and nectin-3 are mixed, these nectins preferentially accumulate at the heterotypic interfaces of the cells. In these cells, cadherins also become predominantly concentrated at the heterotypic nectin-positive cell boundaries (Togashi et al. 2006). Thus, this form of nectin interaction serves for recruiting cadherins to heterotypic cell–cell borders, which are otherwise distributed throughout cell–cell borders. This ability of nectins is used for recruiting cadherins to the synaptic contacts formed between two distinct domains of hippocampal neurons, i.e., axons and dendrites, which express nectin-1 and nectin-3, respectively (Togashi et al. 2006). Thus, nectins show important cooperativity with classic cadherins in generating heterotypic cell–cell contacts.

INTERACTIONS WITH CELL POLARITY REGULATORS

How is the apical junctional complex located apically? The apico-basal polarity of cells is regulated by a couple of molecular complexes, including the aPKC-Par6-Par3 complex localized at a subapical region of the cell junction (Margolis and Borg 2005). Genetic analysis using *Drosophila* embryos showed that the *Drosophila* homolog of Par3 (Bazooka) could establish apical complexes in the absence of AJs, indicating that Bazooka acts upstream of AJ formation (Harris and Peifer 2005). Some reports, on the other hand, suggest the AJs have a physical interaction with these polarity factors: Par3 and Par6, but not aPKC, coprecipitate with VE-cadherin, the endothelial-specific cadherin (Iden et al. 2006). Par3/Bazooka colocalizes with cadherins in epithelial junctions (Harris and Peifer 2005; Afonso and Henrique 2006). These observations illustrate a possible pathway by which the apical membrane domain is primarily determined by those apical determinants, and, in turn, the AJ/ZA is recruited to this domain, possibly via interactions with the pre-existing Par complex. Cadherins may also interact with another polarity regulator, the Scrib-Dlg-Lgl complex localized at the lateral membrane (Reuver and Garner 1998; Navarro et al. 2005). However, the mechanisms responsible for the clustering of the three junctions, i.e., tight junction, AJ, and desmosomes, to form the apical junctional complex and their alignment in a specific order remain unresolved.

CADHERIN EXPRESSION AND RECYCLING

The AJ in mature tissues appears to be a static structure, but actually, cadherin molecules are turning over, and their surface levels are controlled by various mechanisms. Elucidating these mechanisms is important for a molecular understanding of the homeostatic nature of cell junctions. In epithelial cells, the newly synthesized cadherins are transferred from the Golgi to AJs via an exocyst-dependent mechanism (Yeaman et al. 2004). For recycling, E-cadherin is transported to recycling endosomes, and then trafficked to late endosomes for return to the cell surface. DE-cadherin trafficking depends on the interaction of Rab11 and β-catenin with exocyst components Sec15 and Sec10, respectively (Langevin et al. 2005).

The cell surface-located cadherins are stabilized by their homophilic interactions. When cell–cell junctions are artificially disrupted by depletion of extracellular calcium or by other means, the cadherins are actively internalized (Kartenbeck et al. 1991). Under the physiological situation, p120 plays a critical role in cadherin stability (Reynolds 2007). It has been proposed that the attachment of p120 to cadherin masks a dileucine motif on the juxtamembrane region of the cytoplasmic domain, which is sensitive to endocytotic signals (Miyashita and Ozawa 2007), and thereby stabilizes the cadherins. The p120-cadherin binding is strengthened by the interaction of p120 with nectin-associated afadin in a way depending on Rap1, a small GTPase known to be important for AJ formation (Kooistra et al. 2007), and this results in the suppression of E-cadherin endocytosis (Hoshino et al. 2005). A component of the tight junction, PALS1, can also regulate the cadherin trafficking: In PALS1-knocked-down epithelial cells, the exocyst complex is mislocalized, and E-cadherin puncta accumulate in the cell periphery (Wang et al. 2007). Recently, the Cdc42-Par6-aPKC pathway was reported to stabilize the AJ via the control of Arp2/3-dependent endocytosis (Georgiou et al. 2008; Leibfried et al. 2008). Thus, the cadherin stability is regulated in a variety of ways.

The level of cadherins on the cell surface is also controlled by transcriptional and posttranscriptional regulators. Many zinc finger family transcription factors have been implicated in the control of cadherin expression. For example, the zinc finger transcription factor "Snail" is considered as a repressor of E-cadherin transcription, and the expression of Snail inversely correlates with that of E-cadherin (Cano et al. 2000). Other zinc-finger-family transcription factors, such as SIP1, δEF1, Slug, Twist, and E12, similarly act as cadherin transcription repressors through their interaction with the E-box (Remacle et al. 1999; Comijn et al. 2001; Perez-Moreno et al. 2001; Hajra et al. 2002; Yang et al. 2004). Recently, a family of micro RNAs (miRNAs), such as miR-200, was reported to control the expression level of E-cadherin during the epithelial–mesenchymal transition (EMT). Ectopic expression of miR-200 in cell lines upregulates the expression of E-cadherin. These microRNAs act on E-cadherin transcriptional repressors ZEB1/δEF1 and ZEB2/SIP1, and thereby regulate EMT (Gregory et al. 2008; Park et al. 2008). Another miRNA, miR-373, was found to induce E-cadherin expression by recognizing a target site in the promoter of the gene for E-cadherin (Li et al. 2006; Place et al. 2008). These findings provide novel insights into the regulation of cadherin gene expression, which shows highly complex patterns during development (Takeichi 1988).

CONCLUDING REMARKS

The AJs do not display highly specialized ultrastructures, except for having the actin undercoats, as compared with the tight junction, desmosome, and gap junction, each of which shows uniquely decorated plasma membranes or intercellular architecture. The relatively simple structure of the AJ may reflect its dynamic nature and flexibility. In fact, the overall morphology of AJs varies with the cell type and changes during morphogenetic cell rearrangements such as convergent-extension and EMT. The observations of AJ remodeling during *Drosophila* germ band extension suggest that the AJs function not only as a physical ligand between cells but also as an active regulator for cell rearrangement (Lecuit and Lenne 2007). Under these circumstances, studies on AJs will continue for answering at least two lines of questions: How are the AJs maintained or disrupted? And, how does the regulation of AJs contribute to normal morphogenetic cell behavior as well as to the pathogenic one, such as cancer invasion and metastasis?

The integrity of AJs is sustained by cytoplasmic components, including catenins and associated molecules, actin filaments, and microtubules, as well as by the recycling machinery. Although the entire system is apparently complicated, one of the crucial mechanisms to maintain the AJs obviously underlies the interplay of cadherin and the cytoskeleton. We have not obtained a complete answer to the long-standing question of how actin regulates the adhesive function of cadherins. Now, a new question has been posed: What role do MTs have in AJ assembly? Future studies are needed to resolve these problems, through the analysis of the roles of actin regulators and MT-associated proteins.

To understand the morphogenetic roles of AJs, physical and mathematical modeling is becoming a powerful tool (Honda et al. 2008). For example, anisotropy of cortical tension at the AJs has been shown to be sufficient to drive tissue elongation (Rauzi et al. 2008). On the experimental biology side, it will be important to delineate the cooperative mechanisms between AJs and other morphogenetic regulators that work at cell–cell borders, for example, planar cell polarity (PCP) signaling (Fanto and McNeill 2004), as AJs alone would not be sufficient for responding to so many morphogenetic signals.

Other important issues yet unanswered include how the morphology of AJs differs between cell types, and the roles of different classic cadherin members in AJ formation. Each cadherin subtype, which shows a unique tissue distribution, might confer some kind of tissue specificity on the structure and functions of AJs, but this concept has not been tested. The roles of nonclassic cadherins in AJ regulation is another interesting issue, as some protocadherins can down-regulate the functions of classic cadherins (Chen and Gumbiner 2006; Nakao et al. 2008). There must be a number of unknown interactions between the classic and nonclassic cadherin systems, as both of their components accumulate at cell–cell contacts. All these studies targeted on the AJ give us deeper insights into the problem of how individual cells, which can freely move when isolated, can regulate themselves via cell–cell contacts to generate highly ordered multicellular systems.

ACKNOWLEDGMENTS

We thank Katsutoshi Taguchi for photographs. Work in our laboratory was supported by the program Grants-in-Aid for Specially Promoted Research of the Ministry of Education, Science, Sports, and Culture of Japan.

REFERENCES

Abe K, Chisaka O, Van Roy F, Takeichi M. 2004. Stability of dendritic spines and synaptic contacts is controlled by α N-catenin. *Nat Neurosci* **7:** 357–363.

Abe K, Takeichi M. 2008. EPLIN mediates linkage of the cadherin catenin complex to F-actin and stabilizes the circumferential actin belt. *Proc Natl Acad Sci* **105:** 13–19.

Afonso C, Henrique D. 2006. PAR3 acts as a molecular organizer to define the apical domain of chick neuroepithelial cells. *J Cell Sci* **119:** 4293–4304.

Bertet C, Sulak L, Lecuit T. 2004. Myosin-dependent junction remodelling controls planar cell intercalation and axis elongation. *Nature* **429:** 667–671.

Blankenship JT, Backovic ST, Sanny JS, Weitz O, Zallen JA. 2006. Multicellular rosette formation links planar cell polarity to tissue morphogenesis. *Dev Cell* **11:** 459–470.

Braga V. 2000. Epithelial cell shape: Cadherins and small GTPases. *Exp Cell Res* **261:** 83–90.

Buxton RS, Magee AI. 1992. Structure and interactions of desmosomal and other cadherins. *Semin Cell Biol* **3:** 157–167.

Cano A, Perez-Moreno MA, Rodrigo I, Locascio A, Blanco MJ, del Barrio MG, Portillo F, Nieto MA. 2000. The transcription factor snail controls epithelial-mesenchymal transitions by repressing E-cadherin expression. *Nat Cell Biol* **2:** 76–83.

Cavey M, Rauzi M, Lenne PF, Lecuit T. 2008. A two-tiered mechanism for stabilization and immobilization of E-cadherin. *Nature* **453:** 751–756.

Chen X, Gumbiner BM. 2006. Paraxial protocadherin mediates cell sorting and tissue morphogenesis by regulating C-cadherin adhesion activity. *J Cell Biol* **174:** 301–313.

Chen X, Kojima S, Borisy GG, Green KJ. 2003. p120 catenin associates with kinesin and facilitates the transport of cadherin-catenin complexes to intercellular junctions. *J Cell Biol* **163:** 547–557.

Comijn J, Berx G, Vermassen P, Verschueren K, van Grunsven L, Bruyneel E, Mareel M, Huylebroeck D, van Roy F. 2001. The two-handed E box binding zinc finger protein SIP1 downregulates E-cadherin and induces invasion. *Mol Cell* **7:** 1267–1278.

Cox EA, Tuskey C, Hardin J. 2004. Cell adhesion receptors in C. elegans. *J Cell Sci* **117:** 1867–1870.

Delva E, Tucker DK, Kowalczyk AP. 2009. The Desmosome. *Cold Spring Harb Perspect Biol* **1:** a002543.

Drees F, Pokutta S, Yamada S, Nelson WJ, Weis WI. 2005. α-Catenin is a molecular switch that binds E-cadherin-β-catenin and regulates actin-filament assembly. *Cell* **123:** 903–915.

Fabre S, Reymond N, Cocchi F, Menotti L, Dubreuil P, Campadelli-Fiume G, Lopez M. 2002. Prominent role of the Ig-like V domain in trans-interactions of nectins. Nectin3 and nectin 4 bind to the predicted C-C′-C″-D β-strands of the nectin1 V domain. *J Biol Chem* **277:** 27006–27013.

Fanto M, McNeill H. 2004. Planar polarity from flies to vertebrates. *J Cell Sci* **117:** 527–533.

Farquhar MG, Palade GE. 1963. Junctional complexes in various epithelia. *J Cell Biol* **17:** 375–412.

Franz CM, Ridley AJ. 2004. p120 catenin associates with microtubules: Inverse relationship between microtubule binding and Rho GTPase regulation. *J Biol Chem* **279:** 6588–6594.

Gallin WJ, Edelman GM, Cunningham BA. 1983. Characterization of L-CAM, a major cell adhesion molecule from embryonic liver cells. *Proc Natl Acad Sci* **80:** 1038–1042.

Georgiou M, Marinari E, Burden J, Baum B. 2008. Cdc42, Par6, and aPKC regulate Arp2/3-mediated endocytosis to control local adherens junction stability. *Curr Biol* **18:** 1631–1638.

Green KJ, Getsios S, Troyanovsky S, Godsel LM. 2009. Intercellular junction assembly, dynamics and homeostasis. *Cold Spring Harb Perspect Biol* **2:** a000125.

Gregory PA, Bracken CP, Bert AG, Goodall GJ. 2008. MicroRNAs as regulators of epithelial-mesenchymal transition. *Cell Cycle* **7:** 3112–3118.

Haigo SL, Hildebrand JD, Harland RM, Wallingford JB. 2003. Shroom induces apical constriction and is required for hingepoint formation during neural tube closure. *Curr Biol* **13:** 2125–2137.

Hajra KM, Chen DY, Fearon ER. 2002. The SLUG zinc-finger protein represses E-cadherin in breast cancer. *Cancer Res* **62:** 1613–1618.

Harris TJ, Peifer M. 2005. The positioning and segregation of apical cues during epithelial polarity establishment in *Drosophila*. *J Cell Biol* **170:** 813–823.

Hatzfeld M. 2005. The p120 family of cell adhesion molecules. *Eur J Cell Biol* **84:** 205–214.

Herrenknecht K, Ozawa M, Eckerskorn C, Lottspeich F, Lenter M, Kemler R. 1991. The uvomorulin-anchorage protein α catenin is a vinculin homologue. *Proc Natl Acad Sci* **88:** 9156–9160.

Hildebrand JD. 2005. Shroom regulates epithelial cell shape via the apical positioning of an actomyosin network. *J Cell Sci* **118:** 5191–5203.

Hirokawa N, Heuser JE. 1981. Quick-freeze, deep-etch visualization of the cytoskeleton beneath surface differentiations of intestinal epithelial cells. *J Cell Biol* **91:** 399–409.

Holthofer B, Windoffer R, Troyanovsky S, Leube RE. 2007. Structure and function of desmosomes. *Int Rev Cytol* **264:** 65–163.

Honda H, Nagai T, Tanemura M. 2008. Two different mechanisms of planar cell intercalation leading to tissue elongation. *Dev Dyn* **237:** 1826–1836.

Hoshino T, Sakisaka T, Baba T, Yamada T, Kimura T, Takai Y. 2005. Regulation of E-cadherin endocytosis by nectin through afadin, Rap1, and p120ctn. *J Biol Chem* **280:** 24095–24103.

Ichii T, Takeichi M. 2007. p120-catenin regulates microtubule dynamics and cell migration in a cadherin-independent manner. *Genes Cells* **12:** 827–839.

Iden S, Rehder D, August B, Suzuki A, Wolburg-Buchholz K, Wolburg H, Ohno S, Behrens J, Vestweber D, Ebnet K. 2006. A distinct PAR complex associates physically with VE-cadherin in vertebrate endothelial cells. *EMBO Rep* **7:** 1239–1246.

Imai F, Hirai S, Akimoto K, Koyama H, Miyata T, Ogawa M, Noguchi S, Sasaoka T, Noda T, Ohno S. 2006. Inactivation of aPKCλ results in the loss of adherens junctions in neuroepithelial cells without affecting neurogenesis in mouse neocortex. *Development* **133:** 1735–1744.

Ivanov AI, McCall IC, Babbin B, Samarin SN, Nusrat A, Parkos CA. 2006. Microtubules regulate disassembly of epithelial apical junctions. *BMC Cell Biol* **7:** 12.

Iwai Y, Usui T, Hirano S, Steward R, Takeichi M, Uemura T. 1997. Axon patterning requires DN-cadherin, a novel neuronal adhesion receptor, in the *Drosophila* embryonic CNS. *Neuron* **19:** 77–89.

Kametani Y, Takeichi M. 2007. Basal-to-apical cadherin flow at cell junctions. *Nat Cell Biol* **9:** 92–98.

Kartenbeck J, Schmelz M, Franke WW, Geiger B. 1991. Endocytosis of junctional cadherins in bovine kidney epithelial (MDBK) cells cultured in low Ca2+ ion medium. *J Cell Biol* **113:** 881–892.

Kishikawa M, Suzuki A, Ohno S. 2008. aPKC enables development of zonula adherens by antagonizing centripetal contraction of the circumferential actomyosin cables. *J Cell Sci* **121:** 2481–2492.

Knudsen KA, Wheelock MJ. 1992. Plakoglobin, or an 83-kD homologue distinct from β-catenin, interacts with E-cadherin and N-cadherin. *J Cell Biol* **118:** 671–679.

Kobielak A, Pasolli HA, Fuchs E. 2004. Mammalian formin-1 participates in adherens junctions and polymerization of linear actin cables. *Nat Cell Biol* **6:** 21–30.

Kooistra MR, Dube N, Bos JL. 2007. Rap1: A key regulator in cell-cell junction formation. *J Cell Sci* **120:** 17–22.

Langevin J, Morgan MJ, Sibarita JB, Aresta S, Murthy M, Schwarz T, Camonis J, Bellaiche Y. 2005. *Drosophila* exocyst components Sec5, Sec6, and Sec15 regulate DE-Cadherin trafficking from recycling endosomes to the plasma membrane. *Dev Cell* **9:** 365–376.

Lecuit T, Lenne PF. 2007. Cell surface mechanics and the control of cell shape, tissue patterns and morphogenesis. *Nat Rev Mol Cell Biol* **8:** 633–644.

Leibfried A, Fricke R, Morgan MJ, Bogdan S, Bellaiche Y. 2008. *Drosophila* Cip4 and WASp define a branch of the Cdc42-Par6-aPKC pathway regulating E-cadherin endocytosis. *Curr Biol* **18:** 1639–1648.

Li LC, Okino ST, Zhao H, Pookot D, Place RF, Urakami S, Enokida H, Dahiya R. 2006. Small dsRNAs induce transcriptional activation in human cells. *Proc Natl Acad Sci* **103:** 17337–17342.

Ligon LA, Karki S, Tokito M, Holzbaur EL. 2001. Dynein binds to beta-catenin and may tether microtubules at adherens junctions. *Nat Cell Biol* **3:** 913–917.

Liu H, Komiya S, Shimizu M, Fukunaga Y, Nagafuchi A. 2007. Involvement of p120 carboxy-terminal domain in cadherin trafficking. *Cell Struct Funct* **32:** 127–137.

Mandai K, Nakanishi H, Satoh A, Obaishi H, Wada M, Nishioka H, Itoh M, Mizoguchi A, Aoki T, Fujimoto T, Matsuda Y, Tsukita S, Takai Y. 1997. Afadin: A novel actin filament-binding protein with one PDZ domain localized at cadherin-based cell-to-cell adherens junction. *J Cell Biol* **139:** 517–528.

Margolis B, Borg JP. 2005. Apicobasal polarity complexes. *J Cell Sci* **118:** 5157–5159.

Martinez-Rico C, Pincet F, Perez E, Thiery JP, Shimizu K, Takai Y, Dufour S. 2005. Separation force measurements reveal different types of modulation of E-cadherin-based adhesion by nectin-1 and -3. *J Biol Chem* **280:** 4753–4760.

Maul RS, Song Y, Amann KJ, Gerbin SC, Pollard TD, Chang DD. 2003. EPLIN regulates actin dynamics by cross-linking and stabilizing filaments. *J Cell Biol* **160:** 399–407.

McNeill H. 2009. Planar cell polarity: Keeping hairs straight is not so simple. *Cold Spring Harb Perspect Biol* **2:** a003376.

Meng W, Mushika Y, Ichii T, Takeichi M. 2008. Anchorage of microtubule minus ends to adherens junctions regulates epithelial cell-cell contacts. *Cell* **135:** 948–959.

Miyaguchi K. 2000. Ultrastructure of the zonula adherens revealed by rapid-freeze deep-etching. *J Struct Biol* **132:** 169–178.

Miyashita Y, Ozawa M. 2007. A dileucine motif in its cytoplasmic domain directs β-catenin-uncoupled E-cadherin to the lysosome. *J Cell Sci* **120:** 4395–4406.

Morishita H, Yagi T. 2007. Protocadherin family: Diversity, structure, and function. *Curr Opin Cell Biol* **19:** 584–592.

Nagafuchi A, Takeichi M, Tsukita S. 1991. The 102 kd cadherin-associated protein: Similarity to vinculin and posttranscriptional regulation of expression. *Cell* **65:** 849–857.

Nakanishi H, Takai Y. 2004. Roles of nectins in cell adhesion, migration and polarization. *Biol Chem* **385:** 885–892.

Nakao S, Platek A, Hirano S, Takeichi M. 2008. Contact-dependent promotion of cell migration by the OL-protocadherin-Nap1 interaction. *J Cell Biol* **182:** 395–410.

Navarro C, Nola S, Audebert S, Santoni MJ, Arsanto JP, Ginestier C, Marchetto S, Jacquemier J, Isnardon D, Le Bivic A, Birnbaum D, Borg JP. 2005. Junctional recruitment of mammalian Scribble relies on E-cadherin engagement. *Oncogene* **24:** 4330–4339.

Nishimura T, Takeichi M. 2008. Shroom3-mediated recruitment of Rho kinases to the apical cell junctions regulates epithelial and neuroepithelial planar remodeling. *Development* **135:** 1493–1502.

Nollet F, Kools P, van Roy F. 2000. Phylogenetic analysis of the cadherin superfamily allows identification of six major subfamilies besides several solitary members. *J Mol Biol* **299:** 551–572.

Nose A, Nagafuchi A, Takeichi M. 1988. Expressed recombinant cadherins mediate cell sorting in model systems. *Cell* **54:** 993–1001.

Oda H, Uemura T, Harada Y, Iwai Y, Takeichi M. 1994. A *Drosophila* homolog of cadherin associated with armadillo and essential for embryonic cell-cell adhesion. *Dev Biol* **165:** 716–726.

Oda H, Tagawa K, Akiyama-Oda Y. 2005. Diversification of epithelial adherens junctions with independent reductive changes in cadherin form: Identification of potential molecular synapomorphies among bilaterians. *Evol Dev* **7:** 376–389.

Overduin M, Harvey TS, Bagby S, Tong KI, Yau P, Takeichi M, Ikura M. 1995. Solution structure of the epithelial cadherin domain responsible for selective cell adhesion. *Science* **267:** 386–389.

Ozawa M, Kemler R. 1992. Molecular organization of the uvomorulin-catenin complex. *J Cell Biol* **116:** 989–996.

Ozawa M, Baribault H, Kemler R. 1989. The cytoplasmic domain of the cell adhesion molecule uvomorulin associates with three independent proteins structurally related in different species. *EMBO J* **8:** 1711–1717.

Park SM, Gaur AB, Lengyel E, Peter ME. 2008. The miR-200 family determines the epithelial phenotype of cancer cells by targeting the E-cadherin repressors ZEB1 and ZEB2. *Genes Dev* **22:** 894–907.

Patel SD, Ciatto C, Chen CP, Bahna F, Rajebhosale M, Arkus N, Schieren I, Jessell TM, Honig B, Price SR, Shapiro L. 2006. Type II cadherin ectodomain structures: Implications for classical cadherin specificity. *Cell* **124:** 1255–1268.

Peifer M, Wieschaus E. 1990. The segment polarity gene armadillo encodes a functionally modular protein that is the *Drosophila* homolog of human plakoglobin. *Cell* **63:** 1167–1176.

Perez-Moreno MA, Locascio A, Rodrigo I, Dhondt G, Portillo F, Nieto MA, Cano A. 2001. A new role for E12/E47 in the repression of E-cadherin expression and epithelial-mesenchymal transitions. *J Biol Chem* **276:** 27424–27431.

Peyrieras N, Hyafil F, Louvard D, Ploegh HL, Jacob F. 1983. Uvomorulin: A nonintegral membrane protein of early mouse embryo. *Proc Natl Acad Sci* **80:** 6274–6277.

Place RF, Li LC, Pookot D, Noonan EJ, Dahiya R. 2008. MicroRNA-373 induces expression of genes with complementary promoter sequences. *Proc Natl Acad Sci* **105:** 1608–1613.

Pokutta S, Herrenknecht K, Kemler R, Engel J. 1994. Conformational changes of the recombinant extracellular domain of E-cadherin upon calcium binding. *Eur J Biochem* **223:** 1019–1026.

Rauzi M, Verant P, Lecuit T, Lenne PF. 2008. Nature and anisotropy of cortical forces orienting Drosophila tissue morphogenesis. *Nat Cell Biol* **10:** 1401–1410.

Redies C, Vanhalst K, Roy F. 2005. δ-Protocadherins: Unique structures and functions. *Cell Mol Life Sci* **62:** 2840–2852.

Remacle JE, Kraft H, Lerchner W, Wuytens G, Collart C, Verschueren K, Smith JC, Huylebroeck D. 1999. New mode of DNA binding of multi-zinc finger transcription factors: δEF1 family members bind with two hands to two target sites. *EMBO J* **18:** 5073–5084.

Reuver SM, Garner CC. 1998. E-cadherin mediated cell adhesion recruits SAP97 into the cortical cytoskeleton. *J Cell Sci* **111:** 1071–1080.

Reynolds AB. 2007. p120-catenin: Past and present. *Biochim Biophys Acta* **1773:** 2–7.

Reynolds AB, Herbert L, Cleveland JL, Berg ST, Gaut JR. 1992. p120, a novel substrate of protein tyrosine kinase receptors and of p60v-src, is related to cadherin-binding factors beta-catenin, plakoglobin and armadillo. *Oncogene* **7:** 2439–2445.

Rimm DL, Koslov ER, Kebriaei P, Cianci CD, Morrow JS. 1995. α1(E)-catenin is an actin-binding and -bundling protein mediating the attachment of F-actin to the membrane adhesion complex. *Proc Natl Acad Sci* **92:** 8813–8817.

Roczniak-Ferguson A, Reynolds AB. 2003. Regulation of p120-catenin nucleocytoplasmic shuttling activity. *J Cell Sci* **116:** 4201–4212.

Saburi S, McNeill H. 2005. Organising cells into tissues: New roles for cell adhesion molecules in planar cell polarity. *Curr Opin Cell Biol* **17:** 482–488.

Shaw RM, Fay AJ, Puthenveedu MA, von Zastrow M, Jan YN, Jan LY. 2007. Microtubule plus-end-tracking proteins target gap junctions directly from the cell interior to adherens junctions. *Cell* **128:** 547–560.

Shibamoto S, Hayakawa M, Takeuchi K, Hori T, Miyazawa K, Kitamura N, Johnson KR, Wheelock MJ, Matsuyoshi N, Takeichi M, et al. 1995. Association of p120, a tyrosine kinase substrate, with E-cadherin/catenin complexes. *J Cell Biol* **128:** 949–957.

Shimoyama Y, Tsujimoto G, Kitajima M, Natori M. 2000. Identification of three human type-II classic cadherins and frequent heterophilic interactions between different subclasses of type-II classic cadherins. *Biochem J* **349:** 159–167.

Stehbens SJ, Paterson AD, Crampton MS, Shewan AM, Ferguson C, Akhmanova A, Parton RG, Yap AS. 2006. Dynamic microtubules regulate the local concentration of E-cadherin at cell-cell contacts. *J Cell Sci* **119:** 1801–1811.

Stepniak E, Radice GL, Vasioukhin V. 2009. Adhesive and signaling functions of cadherins and catenins in vertebrate development. *Cold Spring Harb Perspect Biol* **1:** a002949.

Strutt H, Strutt D. 2005. Long-range coordination of planar polarity in *Drosophila*. *Bioessays* **27:** 1218–1227.

Suzuki A, Ishiyama C, Hashiba K, Shimizu M, Ebnet K, Ohno S. 2002. aPKC kinase activity is required for the asymmetric differentiation of the premature junctional complex during epithelial cell polarization. *J Cell Sci* **115:** 3565–3573.

Tachibana K, Nakanishi H, Mandai K, Ozaki K, Ikeda W, Yamamoto Y, Nagafuchi A, Tsukita S, Takai Y. 2000. Two cell adhesion molecules, nectin and cadherin, interact through their cytoplasmic domain-associated proteins. *J Cell Biol* **150:** 1161–1176.

Takahashi K, Nakanishi H, Miyahara M, Mandai K, Satoh K, Satoh A, Nishioka H, Aoki J, Nomoto A, Mizoguchi A, Takai Y. 1999. Nectin/PRR: An immunoglobulin-like cell adhesion molecule recruited to cadherin-based adherens junctions through interaction with Afadin, a PDZ domain-containing protein. *J Cell Biol* **145:** 539–549.

Takai Y, Irie K, Shimizu K, Sakisaka T, Ikeda W. 2003. Nectins and nectin-like molecules: Roles in cell adhesion, migration, and polarization. *Cancer Sci* **94:** 655–667.

Takeichi M. 1988. The cadherins: Cell-cell adhesion molecules controlling animal morphogenesis. *Development* **102:** 639–655.

Takeichi M. 2007. The cadherin superfamily in neuronal connections and interactions. *Nat Rev Neurosci* **8:** 11–20.

Takeichi S, Otsuka H, Kimura S. 1977. Studies on tumors produced by cells transformed with herpes simplex virus type 2. *Gann* **68:** 653–661.

Tanoue T, Takeichi M. 2005. New insights into Fat cadherins. *J Cell Sci* **118:** 2347–2353.

Togashi H, Miyoshi J, Honda T, Sakisaka T, Takai Y, Takeichi M. 2006. Interneurite affinity is regulated by heterophilic nectin interactions in concert with the cadherin machinery. *J Cell Biol* **174:** 141–151.

Troyanovsky S. 2005. Cadherin dimers in cell-cell adhesion. *Eur J Cell Biol* **84:** 225–233.

Uchida N, Honjo Y, Johnson KR, Wheelock MJ, Takeichi M. 1996. The catenin/cadherin adhesion system is localized in synaptic junctions bordering transmitter release zones. *J Cell Biol* **135:** 767–779.

Usui T, Shima Y, Shimada Y, Hirano S, Burgess RW, Schwarz TL, Takeichi M, Uemura T. 1999. Flamingo, a seven-pass transmembrane cadherin, regulates planar cell polarity under the control of Frizzled. *Cell* **98:** 585–595.

Vasioukhin V, Bauer C, Degenstein L, Wise B, Fuchs E. 2001. Hyperproliferation and defects in epithelial polarity upon conditional ablation of α-catenin in skin. *Cell* **104:** 605–617.

Volk T, Geiger B. 1984. A 135-kd membrane protein of intercellular adherens junctions. *EMBO J* **3:** 2249–2260.

Wang Q, Chen XW, Margolis B. 2007. PALS1 regulates E-cadherin trafficking in mammalian epithelial cells. *Mol Biol Cell* **18:** 874–885.

Watabe-Uchida M, Uchida N, Imamura Y, Nagafuchi A, Fujimoto K, Uemura T, Vermeulen S, van Roy F, Adamson ED, Takeichi M. 1998. α-Catenin-vinculin interaction functions to organize the apical junctional complex in epithelial cells. *J Cell Biol* **142:** 847–857.

Waterman-Storer CM, Salmon WC, Salmon ED. 2000. Feedback interactions between cell-cell adherens junctions and cytoskeletal dynamics in newt lung epithelial cells. *Mol Biol Cell* **11:** 2471–2483.

Wheeler GN, Buxton RS, Parker AE, Arnemann J, Rees DA, King IA, Magee AI. 1991. Desmosomal glycoproteins I, II and III: Novel members of the cadherin superfamily. *Biochem Soc Trans* **19:** 1060–1064.

Yamada S, Pokutta S, Drees F, Weis WI, Nelson WJ. 2005. Deconstructing the cadherin-catenin-actin complex. *Cell* **123:** 889–901.

Yanagisawa M, Kaverina IN, Wang A, Fujita Y, Reynolds AB, Anastasiadis PZ. 2004. A novel interaction between kinesin and p120 modulates p120 localization and function. *J Biol Chem* **279:** 9512–9521.

Yang J, Mani SA, Donaher JL, Ramaswamy S, Itzykson RA, Come C, Savagner P, Gitelman I, Richardson A, Weinberg RA. 2004. Twist, a master regulator of morphogenesis, plays an essential role in tumor metastasis. *Cell* **117:** 927–939.

Yasumi M, Shimizu K, Honda T, Takeuchi M, Takai Y. 2003. Role of each immunoglobulin-like loop of nectin for its cell-cell adhesion activity. *Biochem Biophys Res Commun* **302:** 61–66.

Yeaman C, Grindstaff KK, Nelson WJ. 2004. Mechanism of recruiting Sec6/8 (exocyst) complex to the apical junctional complex during polarization of epithelial cells. *J Cell Sci* **117:** 559–570.

Yonemura S, Itoh M, Nagafuchi A, Tsukita S. 1995. Cell-to-cell adherens junction formation and actin filament organization: Similarities and differences between non-polarized fibroblasts and polarized epithelial cells. *J Cell Sci* **108:** 127–142.

Yoshida C, Takeichi M. 1982. Teratocarcinoma cell adhesion: Identification of a cell-surface protein involved in calcium-dependent cell aggregation. *Cell* **28:** 217–224.

Yoshida-Noro C, Suzuki N, Takeichi M. 1984. Molecular nature of the calcium-dependent cell-cell adhesion system in mouse teratocarcinoma and embryonic cells studied with a monoclonal antibody. *Dev Biol* **101:** 19–27.

Zallen JA, Wieschaus E. 2004. Patterned gene expression directs bipolar planar polarity in Drosophila. *Dev Cell* **6:** 343–355.

Zhang Y, Sivasankar S, Nelson WJ, Chu S. 2009. Resolving cadherin interactions and binding cooperativity at the single-molecule level. *Proc Natl Acad Sci* **106:** 109–114.

Gap Junctions

Daniel A. Goodenough[1] and David L. Paul[2]

[1]Department of Cell Biology, Harvard Medical School, Boston, Massachusetts 02115
[2]Department of Neurobiology, Harvard Medical School, Boston, Massachusetts 02115
Correspondence: dgoodenough@hms.harvard.edu

Gap junctions are aggregates of intercellular channels that permit direct cell–cell transfer of ions and small molecules. Initially described as low-resistance ion pathways joining excitable cells (nerve and muscle), gap junctions are found joining virtually all cells in solid tissues. Their long evolutionary history has permitted adaptation of gap-junctional intercellular communication to a variety of functions, with multiple regulatory mechanisms. Gap-junctional channels are composed of hexamers of medium-sized families of integral proteins: connexins in chordates and innexins in precordates. The functions of gap junctions have been explored by studying mutations in flies, worms, and humans, and targeted gene disruption in mice. These studies have revealed a wide diversity of function in tissue and organ biology.

Gap junctions are clusters of intercellular channels that allow direct diffusion of ions and small molecules between adjacent cells. The intercellular channels are formed by head-to-head docking of hexameric assemblies (connexons) of tetraspan integral membrane proteins, the connexins (Cx) (Goodenough et al. 1996). These channels cluster into polymorphic maculae or plaques containing a few to thousands of units (Fig. 1). The close membrane apposition required to allow the docking between connexons sterically excludes most other membrane proteins, leaving a narrow ~2 nm extracellular "gap" for which the junction is named (Fig. 2). Gap junctions in prechordates are composed of innexins (Phelan et al. 1998; Phelan 2005). In chordates, connexins arose by convergent evolution (Alexopoulos et al. 2004), to expand by gene duplication (Cruciani and Mikalsen 2007) into a 21-member gene family. Three innexin-related proteins, called pannexins, have persisted in vertebrates, although it is not clear if they form intercellular channels (Panchin et al. 2000; Bruzzone et al. 2003). 7Å-resolution electron crystallographic structures of intercellular channels composed of either a carboxy-terminal truncation of Cx43 (Unger et al. 1999; Yeager and Harris 2007) or an M34A mutant of Cx26 (Oshima et al. 2007) are available. The overall pore morphologies are similar with the exception of a "plug" in the Cx26 channel pore. The density of this plug is substantively decreased by deletion of amino acids 2–7, suggesting that the amino-terminus contributes to this structure (Oshima et al. 2008). A 3.5-Å X-ray crystallographic structure has visualized the amino-terminus of Cx26 folded into the mouth of the channel without forming a plug, thought to be an image of the open channel

 doi: 10.1101/cshperspect.a002576
Cite this article as *Cold Spring Harb Perspect Biol* 2009;1:a002576

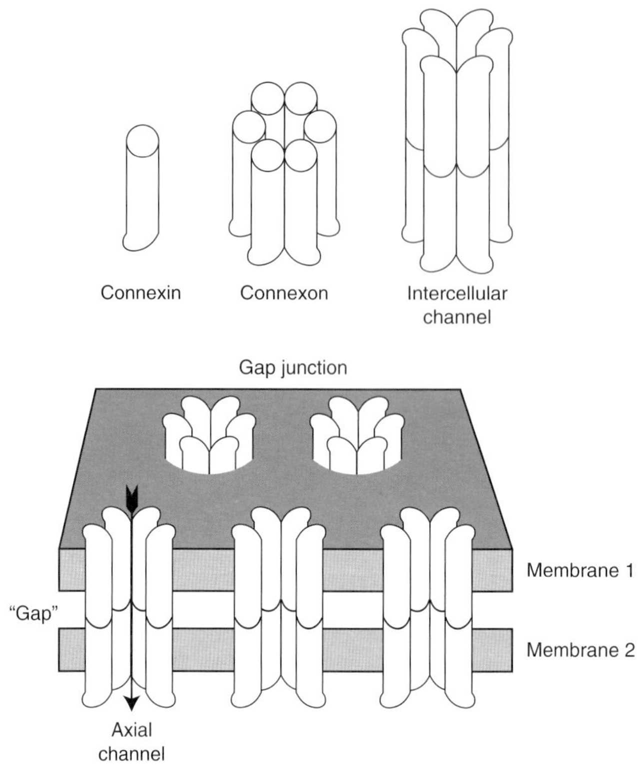

Figure 1. A diagram showing the multiple levels of gap junction structure. Individual connexins assemble intracellularly into hexamers, called connexons, which then traffic to the cell surface. There, they dock with connexons in an adjacent cell, assembling an axial channel spanning two plasma membranes and a narrow extracellular "gap."

conformation (Maeda et al. 2009). The amino-terminus has been physiologically implicated in voltage-gating of the Cx26 and Cx32 channels (Purnick et al. 2000; Oh et al. 2004), lending support to a role for the amino-terminus as a gating structure. However, Cx43 also shows voltage-gating, and its lack of any structure resembling a plug remains unresolved. A comparison of a 1985 intercellular channel structure (Makowski 1985) with the 2009 3.5Å structure (Maeda et al. 2009) summarizes a quarter-century of X-ray progress (Fig. 3).

Most cells express multiple connexins. These may co-oligomerize into the same (homomeric) or mixed (heteromeric) connexons, although only certain combinations are permitted (Falk et al. 1997; Segretain and Falk 2004). A connexon may dock with an identical connexon to form a homotypic intercellular channel or with a connexon containing different connexins to form a heterotypic channel (Dedek et al. 2006). Although only some assembly combinations are permitted (White et al. 1994), the number of possible different intercellular channels formed by this 21-member family is astonishingly large. This diversity has significance because intercellular channels composed of different connexins have different physiological properties, including single-channel conductances and multiple conductance states (Takens-Kwak and Jongsma 1992), as well as permeabilities to experimental tracers (Elfgang et al. 1995) and to biologically relevant permeants (Gaunt and Subak-Sharpe 1979; Veenstra et al. 1995; Bevans et al. 1998; Gong and Nicholson 2001; Goldberg et al. 2002; Ayad et al. 2006; Harris 2007).

Opening of extrajunctional connexons in the plasma membrane, described as "hemichannel" activity, can be experimentally induced in a

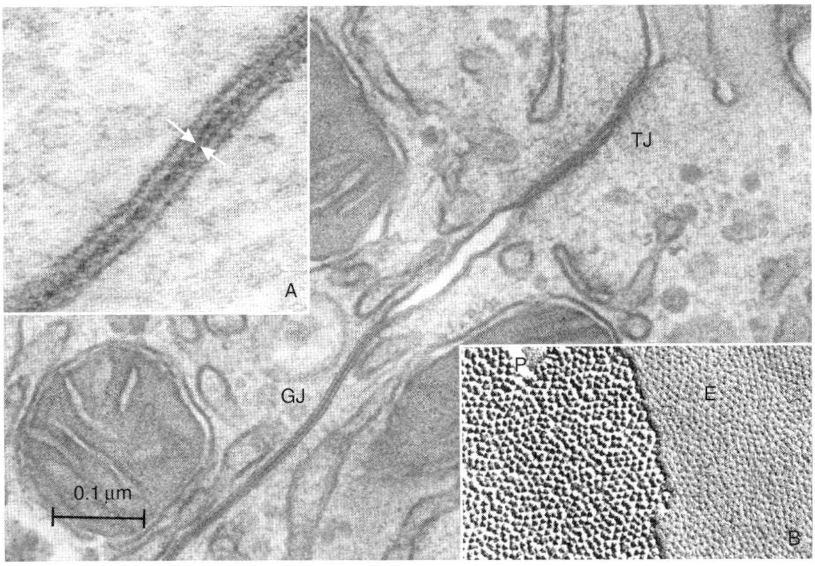

Figure 2. Electron microscopy of gap junctions joining adjacent hepatocytes in the mouse. The gap junction (GJ) is seen as an area of close plasma membrane apposition, clearly distinct from the tight junction (TJ) joining these cells. (*Inset A*) A high magnification view of the gap junction revealing the 2–3 nm "gap" (white arrows) separating the plasma membranes. (*Inset B*) A freeze-fracture replica of a gap junction showing the characteristic particles on the protoplasmic (P) fracture face and pits on the ectoplasmic (E) fracture face. The particles and pits show considerable disorder in their packing with an average 9-nm center-to-center spacing.

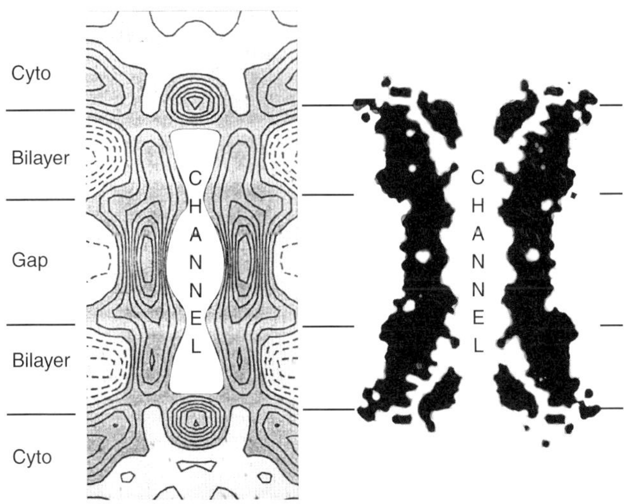

Figure 3. A comparison of axial sections through gap-junction structures deduced from X-ray diffraction. The 1985 data (Makowski 1985) were acquired from gap junctions isolated biochemically from mouse liver containing mixtures of Cx32 and Cx26. The intercellular channel (CHANNEL) is blocked at the two cytoplasmic surfaces by electron density at the channel mouths along the sixfold symmetry axis. The 2009 data (Maeda et al. 2009), acquired from three-dimensional crystals of recombinant Cx26, resolve this density at the channel opening as the amino-termini of the connexin proteins, the 2009 model possibly showing an open channel structure.

variety of cell types. Because first observations of hemichannel activity were in an oocyte expression system (Paul et al. 1991) and dissociated retinal horizontal cells (DeVries and Schwartz 1992), the possible functions of hemichannels composed of connexins and pannexins has enjoyed vigorous investigation (Goodenough and Paul 2003; Bennett et al. 2003; Locovei et al. 2006; Evans et al. 2006; Srinivas et al. 2007; Schenk et al. 2008; Thompson and MacVicar 2008; Anselmi et al. 2008; Goodenough and Paul 2003). Hemichannels have been implicated in various forms of paracrine signaling, for example in providing a pathway for extracellular release of ATP (Cotrina et al. 1998; Kang et al. 2008), glutamate (Ye et al. 2003), NAD^+ (Bruzzone et al. 2000), and prostaglandins (Jiang and Cherian 2003).

GAP JUNCTIONAL INTERCELLULAR CHANNELS ARE DYNAMICALLY REGULATED

Communication via intercellular channels is regulated at multiple levels. The most rapid timescales involve changing the unitary conductance of single channels or altering their probability of opening. Slower regulation is achieved by altering the number of channels present in the membrane by changing rates of synthesis and assembly, posttranslational modification and/or protein degradation. The mechanisms of regulation can overlap between these different time frames, for example, phosphorylation is involved both in changing single channel conductance and in protein trafficking to the cell surface and degradation. The different timescales will be considered in turn.

Rapid Regulation

On the shortest time scale, it is known that gap-junction channels are gated by voltage and can display multiple voltage-dependent conductance states (Turin and Warner 1977; Spray et al. 1979; Neyton and Trautmann 1985; Chen and DeHaan 1992; Bukauskas and Weingart 1993). Voltage-gating is a common property of connexins, although they show substantive differences in their sensitivities. Voltage-gating could explain the rectifying neuronal synapses observed in crayfish (Furshpan and Potter 1959), *Drosophila* (Allen et al. 2006), and hatchetfish (Auerbach and Bennett 1969; Hall et al. 1985), in which action potentials are permitted to pass orthodromically but not antidromically. This behavior requires a structural asymmetry that could be most simply modeled by a heterotypic intercellular channel in which one connexon showed fast voltage-dependent closure whereas the other did not. Indeed, rectification was observed in heterotypic junctions formed between connexins expressed in paired *Xenopus* oocytes (Dahl et al. 1987), but the time scale was too slow to completely explain rectifying synapses (Swenson et al. 1989). In addition to rapid closure of a channel in response to postsynaptic depolarization, rectification at an electrical synapse could also be achieved by opening channels in response to presynaptic depolarization. However, it requires at least 9.5 ms to reopen a closed Cx40 channel in this manner, which is also too slow to account for synaptic rectification (Bukauskas et al. 1995). Although rectifying synapses require near-instantaneous rectification of current, somewhat slower voltage inactivation may be functional in other contexts. For example, Cx45/Cx43 heterotypic junctions may rectify fast enough to influence dendro–dendritic interactions in the central nervous system or may modulate re-entry circuits in myocardium (Bukauskas et al. 2002a).

Fast rectification has been shown using Cx32/Cx26 heterotypic channels (Oh et al. 1999). However, neither connexin displays particularly fast homotypic voltage-dependent gating and thus the rectification observed cannot be predicted from the properties of the individual channels. A model is that the asymmetry of the heterotypic channel results in a separation of fixed positive and negative charges across the two junctional membranes and that rectification of ionic currents occurs within the channel rather than resulting from voltage-induced connexin conformational changes. Regardless, Cx26 and Cx32 are not typically found in excitable cells and are unlikely to participate in rectifying synapses.

Recently, Phelan et al. 2008 have explored innexin composition and physiology of rectifying synapses in the *Drosophila* giant fiber system. These rectifying synapses were shown to be composed of heterotypic channels formed from two different products of the *shaking-B* innexin gene: Shaking-B (neural+16) and Shaking-B (lethal). The former innexin is expressed in the presynaptic neuron and the latter in postsynaptic cell. Although technical limitations did not permit direct electrophysiological measurements in vivo, the two innexins were expressed in the *Xenopus* paired-oocyte system that allowed the characterization of both homotypic and heterotypic innexin interactions. Homotypic intercellular channels composed of Shaking-B (lethal) were highly voltage-dependent compared with those composed of Shaking-B (neural+16). However, in neither case did homotypic channels display rectification. In contrast, Shaking-B (neural+16) and Shaking-B (lethal) assembled heterotypic junctions that rectified. Importantly, channel closure was complete within 5 ms of the application of a transjunctional voltage, and displayed the appropriate gating polarity seen in vivo. However, crayfish junctions in vivo show channel gating within 0.8–1 ms (Furshpan and Potter 1959; Giaume et al. 1987) fivefold faster than the values measured using innexin channels in paired oocytes. Because it is not known how fast channels rectify in the fly, it is not yet possible to conclude that innexin composition explains the entire phenomenon. Regardless, this study provides the first molecular in vivo model to explain part of this 40-year-old conundrum.

Other than rectification, voltage gating of gap-junction channels may not be an important mode of channel regulation in vivo (Harris 2002). However, experimental manipulation of transjunctional voltage reveals a range of conductance states that are likely stabilized by other forms of channel regulation. For example, phosphorylation may function to favor one conductance state more than another, and hence be of great importance in terms of channel selectivity. Although phosphorylation is observed in most members of the connexin family (Lampe and Lau 2000; Lampe and Lau 2004; Laird 2005), most studies have focused on Cx43, which contains 21 serine and two tyrosine residues that are targets of phosphorylation by protein kinase A (PKA), protein kinase C (PKC), p34(cdc2)/cyclin B kinase, casein kinase 1, mitogen-activated protein kinase (MAPK), and pp60 (src) kinase (review Solan and Lampe 2005). Phosphorylation of Cx43 changes the shape of the current voltage relationship (Moreno et al. 1994). In particular, phosphorylation of serine368 (Lampe et al. 2000) by PKC results in a ~50% reduction in unitary conductance. The change in conductance state likely reflects significant changes in channel permeation. For example, driving intercellular channels into subconductance states with transjunctional voltage has been shown to produce a change in charge selectivity (Bukauskas et al. 2002b) or a block of intercellular cAMP and dye-transfer (Qu and Dahl 2002) with little effect on macroscopic electrical coupling. Phosphorylation effects on permeation have also been noted with Cx43 hemichannels where dephosphorylation was correlated with increased channel permeability in liposome reconstitution studies (Kim et al. 1999).

Cx45 has been shown to change its open probability in response to activation of cAMP-dependent protein kinases (van Veen et al. 2000). Activation of pp60$^{v\text{-}src}$ is correlated with tyrosine phosphorylation of Cx43 and concomitant channel inactivation (Swenson et al. 1990; Lampe et al. 2000; Lampe and Lau 2000; Lin et al. 2001; for a review see Pahujaa et al. 2007), although recent studies suggest this regulation may be complex, as src activation also led to phosphorylation of MAPK and PKC sites in Cx43 (Solan and Lampe 2008).

Although many studies have shown changes in channel conductance with phosphorylation in cell culture, there are also in vivo studies documenting this role. For example, during reinitiation of meiosis by luteinizing in developing mouse ovarian follicles, Cx43 is multiply serine phosphorylated via MAPK (Norris et al.

2008), resulting in closure of gap junctional channels between mural granulosa cells, and internalization of gap junctions (Gilula et al. 1978). Another example in which connexin phosphorylation has a clear physiological relevance is in light–dark adaptation, which is globally regulated in the retina by the extrasynaptic release of dopamine (Puopolo et al. 2001). Dopamine acts on most if not all retinal neurons to adjust the gain of neural networks so that sensitivity to contrast can be maintained as the intensity of background illumination changes. In the outer retina, dopamine release rapidly and reversibly leads to a decrease in junctional coupling between horizontal cells (Lasater and Dowling 1985; DeVries and Schwartz 1989; Xin and Bloomfield 1999), which among other actions decreases the size of their receptive field (i.e., restricts the response of a given horizontal cell to a smaller number of photoreceptors), with the overall effect being an improvement in contrast sensitivity. In the inner retina, dopamine has similar effects on junctional coupling between amacrine cells, particularly the AII amacrine, which expresses Cx36 and is a critical part of the rod photoreceptor signaling pathway. D1 dopamine receptor activation in mouse AII amacrine cells leads to a PKA-mediated phosphorylation of Cx36 correlating with a decrease of dye coupling in vivo (Urschel et al. 2006). In the teleost retina, it was shown using phospho-specific antibodies that the natural stimulus of dark-adaptation dramatically increased the levels of Cx35 (the teleost ortholog of Cx36) phosphorylation (Kothmann et al. 2007). Furthermore, these phosphorylation events occurred at sites shown to regulate Cx35 channel gating using in vitro expression studies (O'Brien et al. 2004).

Slow Regulation

A slower temporal level of regulation involves connexin biosynthesis and junctional plaque assembly and turnover (Segretain and Falk 2004). Connexins can show a remarkably rapid turnover rate for a membrane protein. For example, the in vivo half-life of Cx32 in gap junctional plaques from rodent hepatocytes is less than 5 hours (Fallon and Goodenough 1981) and turnover of Cx43 in tissue culture cells is even faster (Musil and Goodenough 1991; Laird et al. 1991). Gap junctions have been shown to turn over by addition of subunits at the edges and removal of subunits from the center of plaques (Gaietta et al. 2002; Lauf et al. 2002). Accretion of connexons at the edges of pre-existing plaques could require nothing more than lateral diffusion in the plasma membrane, but it is not at all clear how the selective removal of connexins/connexons/intercellular channels from the center of a plaque might be orchestrated. Gap junctions are also removed from the cell surface by gross internalization of the entire plaque, leaving large double-membrane vesicles in the cytoplasm (Albertini and Anderson 1975; Larsen et al. 1979; Jordan et al. 2001). Studies with cultured cells suggest that internalization is a clathrin-mediated process (Piehl et al. 2007; Nickel et al. 2008). The relationship between the removal of connexins from the center of pre-existing junctional plaques and the clathrin-dependent endocytosis of whole junctional plaques remains unclear.

Gap junction assembly is associated with multiple phosphorylation steps (Musil and Goodenough 1991). Cx43 is phosphorylated soon after synthesis, and trafficking of the protein through the Golgi to the plasma membrane is accompanied by phosphorylation of specific residues, suggesting a requirement for these modifications in protein transport (Solan and Lampe 2007). Consistent with this notion, chemical or temperature blockade of trafficking in the ER or Golgi results in incomplete Cx43 phosphorylation (Musil and Goodenough 1993). Phosphorylation is also used by different connexins to both block and enhance degradation (Laird et al. 1995). For example, it has been shown that phosphorylation protects Cx32 from calpain digestion (Elvira et al. 1993), while serine phosphorylation of Cx45.6, the chick lens counterpart of Cx50, stimulates protein turnover (Yin et al. 2008). Cx43 can be degraded by both the proteosomal and lysosomal pathways, although no ubiquitin ligase has been shown to specifically associate with a connexin (Laing and Beyer 1995; Berthoud et al. 2004). Proteosome inhibitors block connexin degradation and up-regulate both gap junction assembly and intercellular dye transfer, demonstrating control of gap-junctional intercellular communication (GJIC) via the degradation pathway (Musil et al. 2000). Cx43 dephosphorylation has been associated with disassembly of gap junctions in cells treated with the gap junction blocking agent 18 β-glycyrrhetinic acid (Guan et al. 1996).

Assembly is also affected by interaction with connexin binding partners. A Cx43-interacting protein, CIP85 can induce the turnover of Cx43 through the lysosomal pathway (Lan et al. 2005). Another important interactor is ZO-1, which colocalizes with Cx43 in myocardium and links this connexin to α-spectrin in HEK293 cells in culture (Toyofuku et al. 1998). Cx43 binds to the second PDZ domain of ZO-1 (Giepmans and Moolenaar 1998). Mutations in Cx43 that alter the consensus PDZ binding domain do not inhibit the formation of gap junctions or the activity of intercellular channels. However, there is a dramatic deregulation of plaque size and abnormally large gap junctions are observed (Falk 2000; Hunter et al. 2005). The size expansion results from increased accretion of cytoplasmic pools of Cx43 connexons to the edges of existing junctional plaques and not from de novo synthesis or inhibited degradation. ZO-1 is preferentially associated with the periphery of gap junctional plaques in cells expressing Cx43, suggesting that ZO-1 is a negative regulator of accretion. It has been proposed that accretion is suppressed by a ZO-1 mediated association with filamentous actin (Hunter and Gourdie 2008). In addition, Cx43 may directly associate with tubulin (Giepmans et al. 2001), possibly explaining the observed transport of Cx43 along microtubule tracks (Lauf et al. 2002) that in turn may influence the rate or location of plaque assembly.

The myriad forms of regulation of gap junction function seem surprisingly diverse in comparison to other membrane channels. As reviewed in the following section, the multiple cellular, tissue, and organ functions that have adapted gap-junctional communication as part of their mechanisms have developed a diverse set of regulatory strategies to provide the spatial and temporal controls required in different contexts. Indeed, in some cases, the evolution of multiple connexin genes may have occurred in part because of requirements for unique mechanisms of regulation. In other cases, for example with Cx43, which is used by many different cell types in specialized contexts, multiple regulatory mechanisms are needed to provide specialized control. It is clear from this diversity that the regulation of gap junctional intercellular communication must be experimentally determined on a case-by-case basis as different mechanisms have evolved to subserve this function in different cellular contexts.

UNIVERSAL FUNCTIONS OF GAP JUNCTIONS

The ability of adjacent cells to share ions through low-resistance pathways is fundamental to the function of electrically excitable cells, such as neurons, heart, and smooth muscle. Indeed, gap junctions (electrical synapses) were first discovered in myocardium and nerve because of their properties of electrical transmission between adjacent cells (Weidmann 1952; Furshpan and Potter 1957). In these contexts, connecting cells with gap junctions provides both increased speed in synaptic transmission and the ability to synchronize groups of cells for coordinated electrical and mechanical output.

In addition to electrically excitable cells, virtually all cells in solid tissues are joined by gap junctions. A core function of GJIC is to share metabolic demands across groups of cells and thereby buffer spatial gradients of nutrients or signaling molecules. For example, targeted deletion of Cx32 in mice has been shown to result in a loss of responsiveness to sympathetic stimulation, resulting in an impaired mobilization of glucose from glycogen stores. Postganglionic sympathetic axons terminate at the edges of the liver lobules and thus can only directly stimulate a fraction of the hepatocytes. Presumably, the remainder of the lobule is stimulated indirectly by diffusion of second messengers through gap junctions (Stümpel et al. 1998). Gap junctions may also function as suppressors of somatic cell mutations so that loss of a critical metabolic enzyme or ion channel in one cell might be compensated by its neighbors. For example, Lesch-Nyhan syndrome results from impaired activity of hypoxanthine phosphoribosyltransferase (HGPRTase), a key enzyme in the nucleotide salvage pathway. Impaired HGPRTase results in an elevated concentration of phosphoribosyl pyrophosphate, a marked increase in the rate of purine biosynthesis, and an overproduction of urate. Mutant fibroblasts from patients with Lesch-Nyhan syndrome can be metabolically rescued in cell culture by gap junction formation with normal cells (Cox et al. 1970), a process termed metabolic cooperation (Subak-Sharpe et al. 1969). Furthermore, metabolic cooperation likely accounts for the lack of symptoms in heterozygous female Lesch-Nyhan carriers. As HGPRTase is located on the X chromosome, random X-inactivation results in a mosaic of mutant and normal cells. Thus, individuals are asymptomatic because of metabolic rescue of mutant cells by adjacent nonmutant cells.

SPECIALIZED FUNCTIONS REVEALED BY CONNEXIN MUTATIONS

Human Mutations

Given the long phylogenetic history of gap junctions in metazoans (Fraser and Bode 1981; Potenza et al. 2002; Starich et al. 2003; Nogi and Levin 2005), it is not surprising that this method of cell–cell communication has been adapted to subserve a wide variety of physiological functions in different cell types. Many cell- and tissue-specific functions of GJIC have been brought to light by human mutations and targeted connexin gene deletion in mice (for reviews see Simon and Goodenough 1998; White and Paul 1999; Gerido and White 2004; Dobrowolski and Willecke 2008). In humans, mutations in Cx32 underlie X-linked Charcot-Marie-Tooth syndrome, a common peripheral demyelination neuropathy (Bergoffen et al. 1993), and mutations in Cx47 result in a central demyelinating condition called Pelizaeus-Merzbacher-Like-Disease (Uhlenberg et al. 2004). More than half of all profound hereditary deafness results from mutations in Cx26, which are often syndromic and involve skin disorders (Kelsell et al. 1997; Denoyelle et al. 1997). Similarly, although usually less severe, disorders of the skin and the auditory system accompany mutations in Cx31 and Cx30 (Common et al. 2002; Abrams et al. 2006; Yang et al. 2007; Apps et al. 2007; Yum et al. 2007). Familial cataracts are commonly associated with mutations in either Cx46 or Cx50, whose expression is largely restricted to the ocular lens (Gong et al. 2007; Richard 2005; van Steensel 2004; Vreeburg et al. 2007; Mese et al. 2007). Finally, mutations in Cx43 give rise to oculodentodigital dysplasia, a pleomorphic, syndromic condition affecting a large number of cell types (Paznekas et al. 2003).

Targeted Mutations in Mice

In mice, targeted mutations of connexins have uncovered a wide variety of gap-junction functions in various organs. In many of these cases, a given connexin occupies a particular niche, supplying an essential function that is not compensated by another connexin. For example, Cx26 deletion is embryonic lethal because of a disruption of glucose transport between syncytiotrophoblast I and II in the labyrinth layer of the placenta, which are coupled by gap junctions (Gabriel et al. 1998). In contrast, the human placenta contains only one giant syncytiotrophoblast and so is not vulnerable to Cx26 mutations. Cx45 deletions are also embryonic lethal (Kruger et al. 2000; Willecke et al. 2002), in this case likely the result of myocardial arrhythmia shortly after the heart begins to beat (Nishii et al. 2003). Cx37 knockouts are female sterile from a failure of ovarian follicle development at the antral stage. Presumably, loss of communication between oocyte and cumulus cells leads to premature resumption of meiosis and luteinization (Simon et al. 1997). The loss of Cx40, prevalent in the His-Purkinje system, results in cardiac arrhythmias resembling right-bundle-branch block in humans (Simon et al. 1998; Kirchhoff et al. 1998).

Unique roles played by some connexins have been shown by knockin experiments. The Cx43 coding sequence was replaced in three separate mouse lines with Cx32, Cx40, or Cx26 coding regions. All three animal lines showed new functional defects unique to each connexin, revealing that the three connexins were not able to substitute for Cx43 in all contexts (Plum et al. 2000; Winterhager et al. 2007). Although none of the lines displayed the pulmonary outflow defects seen in the Cx43KO mouse (Reaume et al. 1995), a knockin of Cx31 into the Cx43 locus did show the defect (Zheng-Fischhofer et al. 2006). Thus, connexins may have both unique and redundant functions.

SURPRISING AND PUZZLING RESULTS FROM CONNEXIN MUTATIONS

Other functions that emerge from connexin deletions may result from the loss of a complex interplay of multiple connexin-family members in an incompletely defined network, producing unexpected and unexplained outcomes. Some of these examples are explored here in more detail.

Gap Junctions in the Vascular System

Arterioles are composed of a longitudinal layer of endothelial cells facing the blood, which is separated by a basal lamina from a layer of circular smooth muscle cells that control lumen diameter. There is a surprising complexity of connexin expression in the arteriolar layers. Smooth muscle cells express mainly Cx43 (Gabriels and Paul 1998) and endothelial cells mainly Cx40 (Little et al. 1995; van Kempen and Jongsma 1999), although both cell types express both connexins. Cx32 expression has been reported in endothelial cells (Okamoto et al. 2009). Smooth muscle cells uniquely express Cx45 (Kruger et al. 2000), whereas only the endothelium contains Cx37 (Gabriels and Paul 1998; van Kempen and Jongsma 1999). In addition, there can be significant regional variations in the relative abundance of these connexins in the vessel wall. As an example, endothelial Cx43 is dramatically up-regulated at the expense of the other connexins in areas that experience shear stresses such as vessel branch points (Gabriels and Paul 1998). Not only are gap junctions formed within arteriolar layers, but junctions are also formed between smooth muscle and endothelial cells. The connexin content of the myoendothelial junctions is not yet clear, although in vitro studies suggest that the endothelial side contains largely if not exclusively Cx40 (Isakson and Duling 2005).

Gap junctions have been strongly implicated in the conducted spread of vasodilation. Local endothelial stimulation initiates a rapidly propagated, bidirectional wave of relaxation along the vessel axis (Welsh and Segal 1998; Figueroa et al. 2003; de Wit et al. 2006). An intact endothelium is required for conducted vasodilation, which does not decay with distance and so must contain a self-regenerative component. The propagation of vasomotor activity is significantly depressed in Cx40 KO but not Cx37 KO animals (Figueroa et al. 2003; de Wit et al. 2000). While it was initially surprising that the loss of Cx37, which is co-expressed in endothelial cells, had no effect on propagation, this could be explained by the fact that loss of Cx40 causes a dramatic (>20-fold) reduction in the levels of endothelial Cx37, while loss of Cx37 results in only a mild (~fourfold) reduction in the levels of Cx40 (Simon and McWhorter 2003).

A simple model for the role of gap junctions in propagation is that endothelial stimulation results in a change in membrane potential that is passively conducted along the endothelial layer through gap junctions, critically those containing Cx40. However, this model does not explain self-propagation. Even more problematic, knockin of Cx45 into the Cx40 locus does not rescue the Cx40 KO phenotype, suggesting that ionic spread of membrane potential changes through endothelial–endothelial gap junctions is not a critical factor (Wolfle et al. 2007). On the other hand, studies using connexin-mimetic peptides to selectively inhibit junctional communication in rabbit iliac arteries suggest that although Cx40 is required for endothelium-dependent smooth muscle hyperpolarization, Cx43 is required for spread of that hyperpolarization within the smooth muscle layer (Chaytor et al. 2005). Taken together, these observations suggest another model in which propagation requires both myoendothelial gap junctions as well as gap junctions joining smooth muscle cells. In the first phase, endothelial stimulation leads to release of an endothelium-derived hyperpolarizing factor (EDHF), causing hyperpolarization of immediately adjacent smooth muscle. It has been suggested that EDHF signaling requires myoendothelial junctions (Griffith 2007), which are permeable to inositol trisphosphate and Ca^{2+} (Isakson et al. 2007). A second phase might involve electrotonic spread of hyperpolarization within the smooth muscle layer through gap junctions composed of Cx43. The extent of this spread would be modest as electrical coupling in this layer is relatively weak. In the third phase, smooth muscle must restimulate endothelial cells distal to the site of initial stimulus, regenerating additional rounds of EDHF release. Relaxation of smooth muscle accompanies release of a second factor, endothelium-derived relaxation factor (likely nitric oxide), which can move from endothelium to smooth muscle in the absence of gap junctions. This model is consistent with the loss of conducted vasodilation in the Cx40 KO, but not Cx37 KO, and predicts a Cx40 KO phenocopy in a smooth muscle-specific Cx43 KO, which has not yet been evaluated.

In addition to vasomotor responses, connexin knockouts can dramatically impact systemic blood pressure. Conditional disruption of Cx43 in vascular endothelial cells results in hypotension and bradycardia (Liao et al. 2001), accompanied by elevated plasma levels of nitric oxide because of increased activity of endothelial nitric oxide synthase. These phenotypes are currently without explanation and are not seen in another model of vascular deletion of Cx43 (Theis et al. 2001). In contrast to the hypotension accompanying vascular loss of Cx43, constitutive deletion of Cx40 results in hypertension (de Wit et al. 2006). In this case, disregulation of angiotensin levels may be responsible. In these animals, renin-producing cells are anatomically displaced during development (Kurtz et al. 2007) and are also less responsive to feedback inhibition by plasma angiotensin, leading to increased plasma levels of renin (Wagner et al. 2007). Why the loss of Cx40 results in this cellular localization defect is not known. Interestingly, although knockin of Cx45 into the Cx40 locus is unable to rescue propagation of the vasomotor activity (Wolfle et al. 2007), it abrogates the hyperreninemia, partially attenuating the systemic hypertension and restoring angiotensin-suppression of renin release (Schweda et al. 2008). Parenthetically, Cx45 deletion from smooth muscle in the juxtaglomerular apparatus later in development also results in increased renin secretion and significant blood pressure elevation (Hanner et al. 2008; Yao et al. 2008).

The double knockout (dKO) of Cx37 and Cx40 displays an additional phenotype not seen in either individual knockout. dKO animals die perinatally with dramatic vascular abnormalities. By E18.5, numerous hemorrhages are visible through the skin and internally in the testes, lungs, and intestines. Vasculogenesis is aberrant in the testis and in the connective tissues of the small bowel, but seemingly unaffected in other organs (Simon and McWhorter 2002; Simon and McWhorter 2003). It is not known if these new pathologies result from a combination of the individual regulation and selectivities of the individual connexins, or if this is because of unique properties exhibited by heteromeric or heterotypic intercellular channels.

Gap Junctions in the Ocular Lens

During development, the optic vesicle induces the overlying ectoderm to invaginate and pinch off a hollow sphere of cells, the lens vesicle. The posterior cells of the vesicle then elongate anteriorly as lens fibers, which contact the anterior cells occluding the vesicle lumen. The lens thus becomes a solid cyst of cells, with an anterior epithelium and posterior fibers. The organ eventually loses an enveloping basket of blood vessels, becoming totally avascular and therefore dependent on the aqueous humor for all metabolic needs. The lens continues to grow in volume throughout the life of the organism by appositional growth, differentiating new lens fibers from a stem cell population at the equatorial surface. The older fibers do not turn over, remaining in the lens interior. To achieve a high refractive index and transparency, the differentiating fibers synthesize high concentrations of soluble proteins, the crystallins, and then undergo a limited apoptosis, destroying their nuclei and all light-scattering organelles. Thus, the lens fibers are metabolically dependent on the anterior epithelial cells that retain their organelles. The lens fibers are joined to each other and to the epithelial cells by large numbers of gap junctions (Goodenough 1992). The asymmetric location of the Na^+K^+ATPase in the epithelium results in a translenticular potential and a DC current flow (Candia et al. 1970), modeled as the circulatory system of the lens (Rae 1979; Mathias 1985; Mathias and Rae 1989). As the high concentration of the crystallins requires a tight control of ionic balance to remain in solution, the ionic syncytium created by the gap junctions is essential for lens transparency.

Cx43, 46, and 50 are expressed in the lens. Cx43 and 50 are found abundantly in the lens epithelium (Beyer et al. 1987; Jiang et al. 1995; Martinez-Wittinghan et al. 2003). Cx46 and 50 are found joining the lens fibers where they colocalize to the same junctional plaques (Paul et al. 1991) and have been shown to co-oligomerize into the same connexons and intercellular channels (Konig and Zampighi 1995; Jiang and Goodenough 1996). Indeed, immunofluorescence studies have shown colocalization of Cx46 and 50 in all junctional plaques joining the fibers. Given this anatomical overlap, it is surprising that targeted deletion of Cx46 and 50 result in distinctly different phenotypes (Gong et al. 1997; White et al. 1998). First, both cause cataracts but with differences in timing of onset and in morphology. Second, deletion of Cx50, but not Cx46, results in a slower postnatal growth rate with concomitant decrease in lens size and microphthalmia (White et al. 1998). Interestingly, the normal growth rate is uniquely dependent on Cx50 because replacing the coding region of Cx50 with that of Cx46 ($Cx50^{46/46}$) does not fully rescue the lens mitotic rate (White 2002; Sellitto et al. 2004). The identity of the Cx50-dependent signal controlling mitosis is not known (White et al. 2007). The Cx46/Cx50 double knockout shows a phenotype more severe but predictable as the sum of the two individual connexin deletions (Xia et al. 2006).

$Cx50^{46/46}$ animals are completely free of cataracts (White 2002), suggesting that this pathology could be prevented by simply restoring adequate numbers of junctional channels. Thus, it is surprising that mice heterozygous for Cx46 and Cx50 at the Cx50 locus ($Cx50^{+/46}$) develop a cataract (Martinez-Wittinghan et al. 2003). Furthermore, this cataract is morphologically different from those in either Cx46KO or Cx50KO lenses. Although the latter two are primarily nuclear, the $Cx50^{+/46}$ cataract is largely subepithelial. Additional crosses show that the $Cx50^{+/46}$ cataract is insensitive to dosage of Cx46 at the Cx46 locus, proving that this unexpected phenotype is the result of changes in connexin stoichiometry in the epithelium, where Cx46 is not normally detected. Importantly, the phenotype only occurs when Cx50 and Cx46 are coexpressed in the epithelium, because no cataract is observed in the homozygous ($Cx50^{46/46}$) knockin (White 2002). In addition to the cataract, $Cx50^{+/46}$ lenses display impaired dye transfer both within the epithelial plane and between epithelium and underlying fibers (Martinez-Wittinghan et al. 2003). Why mixing of Cx46 and Cx50 in the epithelium should depress dye transfer and cause a novel cataract is completely without explanation because those connexins functionally interact in heterotypic and heteromeric configurations both in vivo and in expression systems (White et al. 1994; Jiang and Goodenough 1996; Hopperstad et al. 2000).

Demonstration of mechanisms underlying the specificity of connexin intercellular channels in these contexts is still missing. It was shown that fiber–fiber conductance was lower in the $Cx50^{46/46}$ knockin than WT (Martinez-Wittinghan et al. 2004), thus the knockin approach may provide equal numbers of channels but does not provide equal levels of coupling. Regardless, the relationship between coupling level and differential mitotic rates remains obscure. We favor the notion that differential permeability of intercellular channels may play a more important role, as connexin-dependent differences in small molecule permeability have been observed in several studies (Harris 2007). For example, Cx43 channel permeability to cAMP is approximately three times higher than Cx26 and approximately five times higher than Cx40 (Kanaporis et al. 2008), providing a conceptual framework for the observed differences in knockin phenotypes (Harris 2008).

Gap Junctions in Myelin and the Central Nervous System

Mutations in Cx32 associated with the X-linked form of Charcot-Marie-Tooth syndrome result in a peripheral neuropathy associated with myelin failure in Schwann cells. Cx32 forms "reflexive" gap junctions that the Schwann cell makes with itself at the paranodal membranes and incisures of Schmidt-Lantermann. This anatomy suggests that the reflexive junctions in myelin are essential for communication between perinuclear and adaxonal Schwann cell cytoplasm. Measurements of the rate of diffusion between these two cytoplasmic compartments in individual Schwann cells support this notion (Balice-Gordon et al. 1998). However, there is no significant difference between diffusion rates in WT and Cx32 KO animals. To explain this discrepancy, it was hypothesized that Cx29, which is equally abundant although with a somewhat different intracellular distribution, might substitute for the loss of Cx32. However, Cx29 does not accumulate in gap junctional plaques in vivo in oligodendrocytes or Schwann cells (Altevogt et al. 2002; Nagy et al. 2003; Altevogt and Paul 2004) or form function gap junctions when expressed in tissue culture cells (Altevogt et al. 2002). On the other hand, the Cx29 KO does show a myelin defect but one that is restricted to cell bodies of the spiral ganglion neurons in the organ of Corti (Tang et al. 2006).

An additional surprising role for connexins has been shown in the developing neocortex (Elias et al. 2007). Cx26 and Cx43 protein expression was substantively knocked down by electroporation of shRNAs into E16 embryonic cortex. Connexin knockdown resulted in the stalling of migration of neurons along radial glia in the intermediate zone and a loss of cells arriving in the lower and upper cortical plates. Further experiments showed that normal migration was dependent on neuronal rather than glial expression of connexins (Elias et al. 2007). Connexin knockdown neurons showed normal timing of exit from mitosis and no detectable changes in apoptosis, which is unexpected because changes in cell–cell communication and hemichannel involvement in Ca^{2+} waves have been correlated with stages of the mitotic cycle (Bittman et al. 2007). Surprisingly, a channel-dead mutant (Beahm et al. 2006) rescued the migration defect, whereas mutations that resulted in both the loss of connexon pairing (but not hemichannel activity) and the loss of interaction with cytoplasmic partners (C-terminal truncations) were unable to rescue (Elias et al. 2007). These data led to the conclusion that the adhesive properties of connexins, rather than channel activity, were required for correct neuronal migration. In this context, it is of interest that Cx43 hemichannels can confer adhesivity between HeLa and C6 glioma cells in culture (Cotrina et al. 2008).

In summary, connexins and innexins are universally used to promote intercellular interactions between cells in solid tissues and circulating elements of the blood (Wong et al. 2006). They show multiple levels of regulation from instantaneous to hours. Genetic studies have shown that gap junctions are involved in a wide variety of functions in homeostasis, regulation, regeneration, and development. Given that a complex spectrum of small molecules within a cell can potentially diffuse through gap-junctional channels into neighbors, the identification of the relevant small molecules subserving each function has been difficult. Connexons, the hexameric precursor to the gap-junction channel, can function as a hemichannel in nonjunctional membranes promoting paracrine signaling. Even without channel function, the adhesivity of connexons can provide critical migratory cues. Unraveling the multiple functions of connexins and innexins and the contributions to these functions controlled by channel selectivity and regulation, is fundamental to understanding many aspects of collective cellular behavior.

ACKNOWLEDGMENTS

The authors gratefully acknowledge support from grants EY02430 (DAG) and GM37751 (DLP).

REFERENCES

Abrams CK, Freidin MM, Verselis VK, Bargiello TA, Kelsell DP, Richard G, Bennett MV, Bukauskas FF. 2006. Properties of human connexin 31, which is implicated in hereditary dermatological disease and deafness. *Proc Natl Acad Sci* **103:** 5213–5218.

Albertini DF, Anderson E. 1975. Structural modifications of lutein cell gap junctions during pregnancy in the rat and the mouse. *Anat Rec* **181:** 171–194.

Alexopoulos H, Bottger A, Fischer S, Levin A, Wolf A, Fujisawa T, Hayakawa S, Gojobori T, Davies JA, David CN, et al. 2004. Evolution of gap junctions: The missing link? *Curr Biol* **14:** R879–R880.

Allen MJ, Godenschwege TA, Tanouye MA, Phelan P. 2006. Making an escape: Development and function of the Drosophila giant fibre system. *Semin Cell Dev Biol* **17:** 31–41.

Altevogt BM, Paul DL. 2004. Four classes of intercellular channels between glial cells in the CNS. *J Neurosci* **24:** 4313–4323.

Altevogt BM, Kleopa KA, Postma FR, Scherer SS, Paul DL. 2002. Connexin29 is uniquely distributed within myelinating glial cells of the central and peripheral nervous systems. *J Neurosci* **22:** 6458–6470.

Anselmi F, Hernandez VH, Crispino G, Seydel A, Ortolano S, Roper SD, Kessaris N, Richardson W, Rickheit G, Filippov MA, et al. 2008. ATP release through connexin hemichannels and gap junction transfer of second messengers propagate Ca2+ signals across the inner ear. *Proc Natl Acad Sci* **105:** 18770–18775.

Apps SA, Rankin WA, Kurmis AP. 2007. Connexin 26 mutations in autosomal recessive deafness disorders: A review. *Int J Audiol* **46:** 75–81.

Auerbach AA, Bennett MV. 1969. A rectifying electrotonic synapse in the central nervous system of a vertebrate. *J Gen Physiol* **53:** 211–237.

Ayad WA, Locke D, Koreen IV, Harris AL. 2006. Heteromeric, but not homomeric, connexin channels are selectively permeable to inositol phosphates. *J Biol Chem* **281:** 16727–16739.

Balice-Gordon RJ, Bone LJ, Scherer SS. 1998. Functional gap junctions in the Schwann cell myelin sheath. *J Cell Biol* **142:** 1095–1104.

Beahm DL, Oshima A, Gaietta GM, Hand GM, Smock AE, Zucker SN, Toloue M, Chandrasekhar A, Nicholson BJ, Sosinsky GE. 2006. Mutation of a conserved threonine in the third transmembrane helix of α- and β-connexins creates a dominant negative closed gap junction channel. *J Biol Chem* **281:** 7994–8009.

Bennett MV, Contreras JE, Bukauskas FF, Saez JC. 2003. New roles for astrocytes: Gap junction hemichannels have something to communicate. *Trends Neurosci* **26:** 610–617.

Bergoffen J, Scherer SS, Wang S, Scott MO, Bone LJ, Paul DL, Chen K, Lensch MW, Chance PF, Fischbeck KH. 1993. Connexin mutations in X-linked Charcot-Marie-Tooth disease. *Science* **262:** 2039–2042.

Berthoud VM, Minogue PJ, Laing JG, Beyer EC. 2004. Pathways for degradation of connexins and gap junctions. *Cardiovasc Res* **62:** 256–267.

Bevans CG, Kordel M, Rhee SK, Harris AL. 1998. Isoform composition of connexin channels determines selectivity among second messengers and uncharged molecules. *J Biol Chem* **273:** 2808–2816.

Beyer EC, Paul DL, Goodenough DA. 1987. Connexin43: A protein from rat heart homologous to a gap junction protein from liver. *J Cell Biol* **105:** 2621–2629.

Bittman K, Owens DF, Kriegstein AR, LoTurco JJ. 1997. Cell coupling and uncoupling in the ventricular zone of developing neocortex. *J Neurosci* **17:** 7037–7044.

Bruzzone S, Guida L, Zocchi E, Franco L, De Flora A. 2000. Connexin 43 hemichannels mediate Ca2+-regulated transmembrane NAD+ fluxes in intact cells. *FASEB J* **15:** 10–12.

Bruzzone R, Hormuzdi SG, Barbe MT, Herb A, Monyer H. 2003. Pannexins, a family of gap junction proteins expressed in brain. *Proc Natl Acad Sci USA* **100:** 13644–13649.

Bukauskas FF, Weingart R. 1993. Multiple conductance states of newly formed single gap junction channels between insect cells. *Pflügers Arch* **423:** 152–154.

Bukauskas FF, Elfgang C, Willecke K, Weingart R. 1995. Biophysical properties of gap junction channels formed by mouse connexin40 in induced pairs of transfected human HeLa cells. *Biophys J* **68:** 2289–2298.

Bukauskas FF, Angele AB, Verselis VK, Bennett MV. 2002a. Coupling asymmetry of heterotypic connexin 45/ connexin 43-EGFP gap junctions: Properties of fast and slow gating mechanisms. *Proc Natl Acad Sci* **99:** 7113–7118.

Bukauskas FF, Bukauskiene A, Verselis VK. 2002b. Conductance and permeability of the residual state of connexin43 gap junction channels. *J Gen Physiol* **119:** 171–186.

Candia OA, Bentley PJ, Mills CD, Toyofuku H. 1970. Asymmetrical distribution of the potential difference in the toad lens. *Nature* **227:** 852–853.

Chaytor AT, Bakker LM, Edwards DH, Griffith TM. 2005. Connexin-mimetic peptides dissociate electrotonic EDHF-type signalling via myoendothelial and smooth muscle gap junctions in the rabbit iliac artery. *Br J Pharmacol* **144:** 108–114.

Chen Y-H, DeHaan RL. 1992. Multiple-channel conductance states and voltage regulation of embryonic chick cardiac gap junctions. *J Membr Biol* **127:** 95–111.

Common JE, Becker D, Di WL, Leigh IM, O'Toole EA, Kelsell DP. 2002. Functional studies of human skin disease- and deafness-associated connexin 30 mutations. *Biochem Biophys Res Commun* **298:** 651–656.

Cotrina ML, Lin JH, Nedergaard M. 2008. Adhesive properties of connexin hemichannels. *Glia* **56:** 1791–1798.

Cotrina ML, Lin JHC, Alves-Rodrigues A, Liu S, Li J, Azmi-Ghadimi H, Kang J, Naus CCG, Nedergaard M. 1998. Connexins regulate calcium signaling by controlling ATP release. *Proc Natl Acad Sci* **95:** 15735–15740.

Cox RP, Krauss ME, Balis ME, Dancis J. 1970. Evidence for transfer of enzyme product as the basis of metabolic cooperation between tissue culture fibroblasts of Lesch-Nyhan disease and normal cells. *Proc Natl Acad Sci* **67:** 1573–1579.

Cruciani V, Mikalsen SO. 2007. Evolutionary selection pressure and family relationships among connexin genes. *Biol Chem* **388:** 253–264.

Dahl G, Miller T, Paul D, Voellmy R, Werner R. 1987. Expression of functional cell-cell channels from cloned rat liver gap junction complementary DNA. *Science* **236:** 1290–1293.

Dedek K, Schultz K, Pieper M, Dirks P, Maxeiner S, Willecke K, Weiler R, Janssen-Bienhold U. 2006. Localization of heterotypic gap junctions composed of connexin45 and connexin36 in the rod pathway of the mouse retina. *Eur J Neurosci* **24:** 1675–1686.

Denoyelle F, Weil D, Maw MA, Wilcox SA, Lench NJ, Allen-Powell DR, Osborn AH, Dahl HH, Middleton A, Houseman MJ, et al. 1997. Prelingual deafness: High prevalence of a 30delG mutation in the connexin 26 gene. *Hum Mol Genet* **6:** 2173–2177.

DeVries SH, Schwartz EA. 1989. Modulation of an electrical synapse between solitary pairs of catfish horizontal cells by dopamine and second messengers. *J Physiol* **414:** 351–375.

DeVries SH, Schwartz EA. 1992. Hemi-gap-junction channels in solitary horizontal cells of the catfish retina. *J Physiol* **445:** 201–230.

de Wit C, Wolfle SE, Hopfl B. 2006. Connexin-dependent communication within the vascular wall: Contribution to the control of arteriolar diameter. *Adv Cardiol* **42:** 268–283.

de Wit C, Roos F, Bolz SS, Kirchhoff S, Kruger O, Willecke K, Pohl U. 2000. Impaired conduction of vasodilation along arterioles in connexin40- deficient mice. *Circ Res* **86:** 649–655.

Dobrowolski R, Willecke K. 2008. Connexin-caused genetic diseases and corresponding mouse models. *Antioxid Redox Signal* **11:** 283–295.

Elfgang C, Eckert R, Lichtenberg-Fraté H, Butterweck A, Traub O, Klein RA, Hülser DF, Willecke K. 1995. Specific permeability and selective formation of gap junction channels in connexin-transfected HeLa cells. *J Cell Biol* **129:** 805–817.

Elias LA, Wang DD, Kriegstein AR. 2007. Gap junction adhesion is necessary for radial migration in the neocortex. *Nature* **448:** 901–907.

Elvira M, Diez JA, Wang KKW, Villalobo A. 1993. Phosphorylation of connexin-32 by protein kinase C prevents its proteolysis by mu-calpain and m-calpain. *J Biol Chem* **268:** 14294–14300.

Evans WH, De Vuyst E, Leybaert L. 2006. The gap junction cellular internet: Connexin hemichannels enter the signalling limelight. *Biochem J* **397:** 1–14.

Falk MM. 2000. Connexin-specific distribution within gap junctions revealed in living cells. *J Cell Sci* **113:** 4109–4120.

Falk MM, Buehler LK, Kumar NM, Gilula NB. 1997. Cell-free synthesis and assembly of connexins into functional gap junction membrane channels. *EMBO J* **16:** 2703–2716.

Fallon RF, Goodenough DA. 1981. Five hour half-life of mouse liver gap-junction protein. *J Cell Biol* **90:** 521–526.

Figueroa XF, Paul DL, Simon AM, Goodenough DA, Day KH, Damon DN, Duling BR. 2003. Central role of connexin40 in the propagation of electrically activated vasodilation in mouse cremasteric arterioles in vivo. *Circ Res* **92:** 793–800.

Fraser SE, Bode HR. 1981. Epithelial cells of Hydra are dye-coupled. *Nature* **294:** 356–358.

Furshpan EJ, Potter DD. 1957. Mechanism of nerve-impulse transmission at a crayfish synapse. *Nature* **180:** 342–343.

Furshpan EJ, Potter DD. 1959. Transmission at the giant motor synapses of the crayfish. *J Physiol* **145:** 289–325.

Gabriel H-D, Jung D, Bützler C, Temme A, Traub O, Winterhager E, Willecke K. 1998. Transplacental uptake of glucose is decreased in embryonic lethal connexin26-deficient mice. *J Cell Biol* **140:** 1453–1461.

Gabriels JE, Paul DL. 1998. Connexin43 is highly localized to sites of disturbed flow in rat aortic endothelium but connexin37 and connexin40 are more uniformly distributed. *Circ Res* **83:** 636–643.

Gaietta G, Deerinck TJ, Adams SR, Bouwer J, Tour O, Laird DW, Sosinsky G, Tsien RY, Ellisman MH. 2002. Multicolor and electron microscopic imaging of connexin trafficking. *Science* **296:** 503–507.

Gaunt SJ, Subak-Sharpe JH. 1979. Selectivity in metabolic cooperation between cultured mammalian cells. *Exp Cell Res* **120:** 307–320.

Gerido DA, White TW. 2004. Connexin disorders of the ear, skin, and lens. *Biochim Biophys Acta* **1662:** 159–170.

Giaume C, Kado RT, Korn H. 1987. Voltage-clamp analysis of a crayfish rectifying synapse. *J Physiol* **386:** 91–112.

Giepmans BN, Moolenaar WH. 1998. The gap junction protein connexin43 interacts with the second PDZ domain of the zona occludens-1 protein. *Curr Biol* **8:** 931–934.

Giepmans BN, Verlaan I, Hengeveld T, Janssen H, Calafat J, Falk MM, Moolenaar WH. 2001. Gap junction protein connexin-43 interacts directly with microtubules. *Curr Biol* **11:** 1364–1368.

Gilula NB, Epstein ML, Beers WH. 1978. Cell-to-cell communication and ovulation. A study of the cumulus-oocyte complex. *J Cell Biol* **78:** 58–75.

Goldberg GS, Moreno AP, Lampe PD. 2002. Gap junctions between cells expressing connexin 43 or 32 show inverse permselectivity to adenosine and ATP. *J Biol Chem* **277:** 36725–36730.

Gong XQ, Nicholson BJ. 2001. Size selectivity between gap junction channels composed of different connexins. *Cell Adhes Commun* **8:** 187–192.

Gong X, Li E, Klier G, Huang Q, Wu Y, Lei H, Kumar N, Horwitz J, Gilula NB. 1997. Disruption of α_3 connexin gene leads to proteolysis and cataractogenesis in mice. *Cell* **91:** 833–843.

Gong X, Cheng C, Xia CH. 2007. Connexins in lens development and cataractogenesis. *J Membr Biol* **218:** 9–12.

Goodenough DA. 1992. The crystalline lens: A system networked by gap junctional intercellular communication. *Semin Cell Biol* **3:** 49–58.

Goodenough DA, Paul DL. 2003. Beyond the gap: Functions of unpaired connexon channels. *Nat Rev Mol Cell Biol* **4:** 285–295.

Goodenough DA, Goliger JA, Paul DL. 1996. Connexins, connexons, and intercellular communication. *Annu Rev Biochem* **65:** 475–502.

Griffith TM. 2007. Which connexins connect? *Circ Res* **101:** 1219–1221.

Guan XJ, Wilson S, Schlender KK, Ruch RJ. 1996. Gap-junction disassembly and connexin 43 dephosphorylation induced by 18-β-glycyrrhetinic acid. *Mol Carcinogenesis* **16:** 157–164.

Hall DH, Gilat E, Bennett MV. 1985. Ultrastructure of the rectifying electrotonic synapses between giant fibres and pectoral fin adductor motor neurons in the hatchetfish. *J Neurocytol* **14:** 825–834.

Hanner F, von Maltzahn J, Maxeiner S, Toma I, Sipos A, Kruger O, Willecke K, Peti-Peterdi J. 2008. Connexin45 is expressed in the juxtaglomerular apparatus and is involved in the regulation of renin secretion and blood pressure. *Am J Physiol Regul Integr Comp Physiol* **295:** R371–R380.

Harris AL. 2002. Voltage-sensing and substate rectification: Moving parts of connexin channels. *J Gen Physiol* **119:** 165–170.

Harris AL. 2007. Connexin channel permeability to cytoplasmic molecules. *Prog Biophys Mol Biol* **94:** 120–143.

Harris AL. 2008. Connexin specificity of second messenger permeation: Real numbers at last. *J Gen Physiol* **131:** 287–292.

Hopperstad MG, Srinivas M, Spray DC. 2000. Properties of gap junction channels formed by Cx46 alone and in combination with Cx50. *Biophys J* **79:** 1954–1966.

Hunter AW, Gourdie RG. 2008. The second PDZ domain of zonula occludens-1 is dispensable for targeting to connexin 43 gap junctions. *Cell Commun Adhes* **15:** 55–63.

Hunter AW, Barker RJ, Zhu C, Gourdie RG. 2005. ZO-1 alters connexin43 gap junction size and organization by influencing channel accretion. *Mol Biol Cell* **16:** 5686–5698.

Isakson BE, Duling BR. 2005. Heterocellular contact at the myoendothelial junction influences gap junction organization. *Circ Res* **97:** 44–51.

Isakson BE, Ramos SI, Duling BR. 2007. Ca2+ and 1,4,5-trisphosphate-mediated signaling across the myoendothelial junction. *Circ Res* **100:** 246–254.

Jiang JX, Cherian PP. 2003. Hemichannels formed by connexin 43 play an important role in the release of prostaglandin e(2) by osteocytes in response to mechanical strain. *Cell Commun Adhes* **10:** 259–264.

Jiang JX, Goodenough DA. 1996. Heteromeric connexons in lens gap junction channels. *Proc Natl Acad Sci* **93:** 1287–1291.

Jiang JX, White TW, Goodenough DA. 1995. Changes in connexin expression and distribution during chick lens development. *Dev Biol* **168:** 649–661.

Jordan K, Chodock R, Hand AR, Laird DW. 2001. The origin of annular junctions: A mechanism of gap junction internalization. *J Cell Sci* **114:** 763–773.

Kanaporis G, Mese G, Valiuniene L, White TW, Brink PR, Valiunas V. 2008. Gap junction channels exhibit connexin-specific permeability to cyclic nucleotides. *J Gen Physiol* **131:** 293–305.

Kang J, Kang N, Lovatt D, Torres A, Zhao Z, Lin J, Nedergaard M. 2008. Connexin 43 hemichannels are permeable to ATP. *J Neurosci* **28:** 4702–4711.

Kelsell DP, Dunlop J, Stevens HP, Lench NJ, Liang JN, Parry G, Mueller RF, Leigh IM. 1997. Connexin 26 mutations in hereditary non-syndromic sensorineural deafness. *Nature* **387:** 80–83.

Kim DY, Kam Y, Koo SK, Joe CO. 1999. Gating connexin 43 channels reconstituted in lipid vesicles by mitogen-activated protein kinase phosphorylation. *J Biol Chem* **274:** 5581–5587.

Kirchhoff S, Nelles E, Hagendorff A, Krüger O, Traub O, Willecke K. 1998. Reduced cardiac conduction velocity and predisposition to arrhythmias in connexin40-deficient mice. *Current Biol* **8:** 299–302.

Konig N, Zampighi G. 1995. Purification of bovine lens cell-to-cell channels composed of connexin44 and connexin50. *J Cell Sci* **108:** 3091–3098.

Kothmann WW, Li X, Burr GS, O'Brien J. 2007. Connexin 35/36 is phosphorylated at regulatory sites in the retina. *Vis Neurosci* **24:** 363–375.

Kruger O, Plum A, Kim J, Winterhager E, Maxeiner S, Hallas G, Kirchhoff S, Traub O, Lamers WH, Willecke K. 2000. Defective vascular development in connexin 45-deficient mice. *Development* **127:** 4179–4193.

Kurtz L, Schweda F, de Wit C, Kriz W, Witzgall R, Warth R, Sauter A, Kurtz A, Wagner C. 2007. Lack of connexin 40 causes displacement of renin-producing cells from afferent arterioles to the extraglomerular mesangium. *J Am Soc Nephrol* **18:** 1103–1111.

Laing JG, Beyer EC. 1995. The gap junction protein connexin43 is degraded via the ubiquitin proteasome pathway. *J Biol Chem* **270:** 26399–26403.

Laird DW. 2005. Connexin phosphorylation as a regulatory event linked to gap junction internalization and degradation. *Biochim Biophys Acta* **1711:** 172–182.

Laird DW, Puranam KL, Revel JP. 1991. Turnover and phosphorylation dynamics of connexin43 gap junction protein in cultured cardiac myocytes. *Biochem J* **273:** 67–72.

Laird DW, Castillo M, Kasprzak L. 1995. Gap junction turnover, intracellular trafficking, and phosphorylation of connexin43 in brefeldin A-treated rat mammary tumor cells. *J Cell Biol* **131:** 1193–1203.

Lampe PD, Lau AF. 2000. Regulation of gap junctions by phosphorylation of connexins. *Arch Biochem Biophys* **384:** 205–215.

Lampe PD, Lau AF. 2004. The effects of connexin phosphorylation on gap junctional communication. *Int J Biochem Cell Biol* **36:** 1171–1186.

Lampe PD, Tenbroek EM, Burt JM, Kurata WE, Johnson RG, Lau AF. 2000. Phosphorylation of connexin43 on serine368 by protein kinase C regulates gap junctional communication. *J Cell Biol* **149:** 1503–1512.

Lan Z, Kurata WE, Martyn KD, Jin C, Lau AF. 2005. Novel Rab GAP-like protein, CIP85, interacts with connexin43 and induces its degradation. *Biochemistry* **44:** 2385–2396.

Larsen WJ, Tung H-N, Murray S, Swenson CA. 1979. Evidence for the participation of actin microfilaments and bristle coats in the internalization of gap junction membrane. *J Cell Biol* **83:** 576–587.

Lasater EM, Dowling JE. 1985. Dopamine decreases conductance of the electrical junctions between cultured retinal horizontal cells. *Proc Natl Acad Sci* **82:** 3025–3029.

Lauf U, Giepmans BN, Lopez P, Braconnot S, Chen SC, Falk MM. 2002. Dynamic trafficking and delivery of connexons to the plasma membrane and accretion to gap junctions in living cells. *Proc Natl Acad Sci* **99:** 10446–10451.

Liao Y, Day KH, Damon DN, Duling BR. 2001. Endothelial cell-specific knockout of connexin 43 causes hypotension and bradycardia in mice. *Proc Natl Acad Sci* **98:** 9989–9994.

Lin R, Warn-Cramer BJ, Kurata WE, Lau AF. 2001. v-Src phosphorylation of connexin 43 on Tyr247 and Tyr265 disrupts gap junctional communication. *J Cell Biol* **154:** 815–828.

Little TL, Beyer EC, Duling BR. 1995. Connexin 43 and connexin 40 gap junctional proteins are present in arteriolar smooth muscle and endothelium in vivo. *Am J Physiol* **268:** H729–H739.

Locovei S, Bao L, Dahl G. 2006. Pannexin 1 in erythrocytes: Function without a gap. *Proc Natl Acad Sci* **103:** 7655–7659.

Maeda S, Nakagawa S, Suga M, Yamashita E, Oshima A, Fujiyoshi Y, Tsukihara T. 2009. Structure of the connexin 26 gap junction channel at 3.5 A resolution. *Nature* **458:** 597–602.

Makowski L. 1985. Structural domains in gap junctions: Implications for the control of intercellular communication. In *Gap junction* (ed. M.V.L. Bennett, et al.), pp. 5–12. Cold Spring Harbor Laboratory, Cold Spring Harbor, NY.

Martinez-Wittinghan FJ, Sellitto C, Li L, Gong X, Brink PR, Mathias RT, White TW. 2003. Dominant cataracts result from incongruous mixing of wild-type lens connexins. *J Cell Biol* **161:** 969–978.

Martinez-Wittinghan FJ, Sellitto C, White TW, Mathias RT, Paul D, Goodenough DA. 2004. Lens gap junctional coupling is modulated by connexin identity and the locus of gene expression. *Invest Ophthalmol Vis Sci* **45:** 3629–3637.

Mathias RT. 1985. Steady-state voltages, ion fluxes and volume regulation in syncytial tissues. *Biophys J* **48:** 435–448.

Mathias RT, Rae JL. 1989. Cell to cell communication in the lens. In *Cell interactions and gap junctions* (ed. N. Sperelakis, et al.), pp. 29–50. CRC Press, Boca Raton, FL.

Mese G, Richard G, White TW. 2007. Gap junctions: Basic structure and function. *J Invest Dermatol* **127:** 2516–2524.

Moreno AP, Saez JC, Fishman GI, Spray DC. 1994. Human connexin43 gap junction channels - regulation of unitary conductances by phosphorylation. *Circ Res* **74:** 1050–1057.

Musil LS, Goodenough DA. 1991. Biochemical analysis of connexin43 intracellular transport, phosphorylation, and assembly into gap junctional plaques. *J Cell Biol* **115:** 1357–1374.

Musil LS, Goodenough DA. 1993. Multisubunit assembly of an integral plasma membrane channel protein, gap junction connexin43, occurs after exit from the ER. *Cell* **74:** 1065–1077.

Musil LS, Le AC, VanSlyke JK, Roberts LM. 2000. Regulation of connexin degradation as a mechanism to increase gap junction assembly and function. *J Biol Chem* **275:** 25207–25215.

Nagy JI, Ionescu AV, Lynn BD, Rash JE. 2003. Coupling of astrocyte connexins Cx26, Cx30, Cx43 to oligodendrocyte Cx29, Cx32, Cx47: Implications from normal and connexin32 knockout mice. *Glia* **44:** 205–218.

Neyton J, Trautmann A. 1985. Single-channel currents of an intercellular junction. *Nature* **317:** 331–335.

Nickel BM, Defranco BH, Gay VL, Murray SA. 2008. Clathrin and Cx43 gap junction plaque endoexocytosis. *Biochem Biophys Res Commun* **374:** 679–682.

Nishii K, Kumai M, Egashira K, Miwa T, Hashizume K, Miyano Y, Shibata Y. 2003. Mice lacking connexin45 conditionally in cardiac myocytes display embryonic lethality similar to that of germline knockout mice without endocardial cushion defect. *Cell Commun Adhes* **10:** 365–369.

Nogi T, Levin M. 2005. Characterization of innexin gene expression and functional roles of gap-junctional communication in planarian regeneration. *Dev Biol* **287:** 314–335.

Norris RP, Freudzon M, Mehlmann LM, Cowan AE, Simon AM, Paul DL, Lampe PD, Jaffe LA. 2008. Luteinizing hormone causes MAP kinase-dependent phosphorylation and closure of connexin 43 gap junctions in mouse ovarian follicles: One of two paths to meiotic resumption. *Development* **135:** 3229–3238.

O'Brien J, Nguyen HB, Mills SL. 2004. Cone photoreceptors in bass retina use two connexins to mediate electrical coupling. *J Neurosci* **24:** 5632–5642.

Oh S, Rubin JB, Bennett MV, Verselis VK, Bargiello TA. 1999. Molecular determinants of electrical rectification of single channel conductance in gap junctions formed by connexins 26 and 32. *J Gen Physiol* **114:** 339–364.

Oh S, Rivkin S, Tang Q, Verselis VK, Bargiello TA. 2004. Determinants of gating polarity of a connexin 32 hemichannel. *Biophys J* **87:** 912–928.

Okamoto T, Akiyama M, Takeda M, Gabazza EC, Hayashi T, Suzuki K. 2009. Connexin32 is expressed in vascular endothelial cells and participates in gap junction intercellular communication. *Biochem Biophys Res Commun* **382:** 264–268.

Oshima A, Tani K, Hiroaki Y, Fujiyoshi Y, Sosinsky GE. 2007. Three-dimensional structure of a human connexin26 gap junction channel reveals a plug in the vestibule. *Proc Natl Acad Sci* **104:** 10034–10039.

Oshima A, Tani K, Hiroaki Y, Fujiyoshi Y, Sosinsky GE. 2008. Projection structure of a N-terminal deletion mutant of connexin 26 channel with decreased central pore density. *Cell Commun Adhes* **15:** 85–93.

Pahujaa M, Anikin M, Goldberg GS. 2007. Phosphorylation of connexin43 induced by Src: Regulation of gap junctional communication between transformed cells. *Exp Cell Res* **313:** 4083–4090.

Panchin Y, Kelmanson I, Matz M, Lukyanov K, Usman N, Lukyanov S. 2000. A ubiquitous family of putative gap junction molecules. *Curr Biol* **10:** R473–R474.

Paul DL, Ebihara L, Takemoto LJ, Swenson KI, Goodenough DA. 1991. Connexin46, a novel lens gap junction protein, induces voltage-gated currents in nonjunctional plasma membrane of *Xenopus* oocytes. *J Cell Biol* **115:** 1077–1089.

Paznekas WA, Boyadjiev SA, Shapiro RE, Daniels O, Wollnik B, Keegan CE, Innis JW, Dinulos MB, Christian C, Hannibal MC, et al. 2003. Connexin 43 (GJA1) mutations cause the pleiotropic phenotype of oculodentodigital dysplasia. *Am J Hum Genet* **72:** 408–418.

Phelan P. 2005. Innexins: Members of an evolutionarily conserved family of gap-junction proteins. *Biochim Biophys Acta* **1711:** 225–245.

Phelan P, Bacon JP, Davies JA, Stebbings LA, Todman MG, Avery L, Baines RA, Barnes TM, Ford C, Hekimi S, et al. 1998. Innexins: A family of invertebrate gap-junction proteins. *Trends Genet* **14:** 348–349.

Phelan P, Goulding LA, Tam JL, Allen MJ, Dawber RJ, Davies JA, Bacon JP. 2008. Molecular mechanism of rectification at identified electrical synapses in the Drosophila giant fiber system. *Curr Biol* **18:** 1955–1960.

Piehl M, Lehmann C, Gumpert A, Denizot JP, Segretain D, Falk MM. 2007. Internalization of large double-membrane intercellular vesicles by a clathrin-dependent endocytic process. *Mol Biol Cell* **18:** 337–347.

Plum A, Hallas G, Magin TM, Dombrowski F, Hagendorff A, Schumacher B, Wolpert C, Kim J-S, Lamers WH, Evert M, Meda P, et al. 2000. Unique and shared functions of different connexins in mice. *Curr Biol* **10:** 1083–1091.

Potenza N, del Gaudio R, Rivieccio L, Russo GM, Geraci G. 2002. Cloning and molecular characterization of the first innexin of the phylum annelida-expression of the gene during development. *J Mol Evol* **54:** 312–321.

Puopolo M, Hochstetler SE, Gustincich S, Wightman RM, Raviola E. 2001. Extrasynaptic release of dopamine in a retinal neuron: Activity dependence and transmitter modulation. *Neuron* **30:** 211–225.

Purnick PE, Oh S, Abrams CK, Verselis VK, Bargiello TA. 2000. Reversal of the gating polarity of gap junctions by negative charge substitutions in the N-terminus of connexin 32. *Biophys J* **79:** 2403–2415.

Qu Y, Dahl G. 2002. Function of the voltage gate of gap junction channels: Selective exclusion of molecules. *Proc Natl Acad Sci* **99:** 697–702.

Rae JL. 1979. The electrophysiology of the crystalline lens. *Curr Topics in Eye Research* **1:** 37–90.

Reaume AG, De Sousa PA, Kulkarni S, Langille BL, Zhu D, Davies TC, Juneja SC, Kidder GM, Rossant J. 1995. Cardiac malformation in neonatal mice lacking connexin43. *Science* **267:** 1831–1834.

Richard G. 2005. Connexin disorders of the skin. *Clin Dermatol* **23:** 23–32.

Schenk U, Westendorf AM, Radaelli E, Casati A, Ferro M, Fumagalli M, Verderio C, Buer J, Scanziani E, Grassi F. 2008. Purinergic control of T cell activation by ATP released through pannexin-1 hemichannels. *Sci Signal* **1:** ra6.

Schweda F, Kurtz L, de Wit C, Janssen-Bienhold U, Kurtz A, Wagner C. 2008. Substitution of connexin40 with connexin45 prevents hyperreninemia and attenuates hypertension. *Kidney Int* **75:** 482–490.

Segretain D, Falk MM. 2004. Regulation of connexin biosynthesis, assembly, gap junction formation, and removal. *Biochim Biophys Acta* **1662:** 3–21.

Sellitto C, Li L, White TW. 2004. Connexin50 is essential for normal postnatal lens cell proliferation. *Invest Ophthalmol Vis Sci* **45:** 3196–3202.

Simon AM, Goodenough DA. 1998. Diverse functions of vertebrate gap junctions. *Trends Cell Biol* **8:** 477–482.

Simon AM, McWhorter AR. 2002. Vascular abnormalities in mice lacking the endothelial gap junction proteins connexin37 and connexin40. *Dev Biol* **251:** 206–220.

Simon AM, McWhorter AR. 2003. Role of connexin37 and connexin40 in vascular development. *Cell Commun Adhes* **10:** 379–385.

Simon AM, Goodenough DA, Li E, Paul DL. 1997. Female infertility in mice lacking connexin 37. *Nature* **385:** 525–529.

Simon AM, Goodenough DA, Paul DL. 1998. Mice lacking connexin40 have cardiac conduction abnormalities characteristic of atrioventricular block and bundle branch block. *Current Biol* **8:** 295–298.

Solan JL, Lampe PD. 2005. Connexin phosphorylation as a regulatory event linked to gap junction channel assembly. *Biochim Biophys Acta* **1711:** 154–163.

Solan JL, Lampe PD. 2007. Key connexin 43 phosphorylation events regulate the gap junction life cycle. *J Membr Biol* **217:** 35–41.

Solan JL, Lampe PD. 2008. Connexin 43 in LA-25 cells with active v-src is phosphorylated on Y247, Y265, S262,

S279/282, and S368 via multiple signaling pathways. *Cell Commun Adhes* **15:** 75–84.

Spray DC, Harris AL, Bennett MVL. 1979. Voltage dependence of junctional conductance in early amphibian embryos. *Science* **204:** 432–434.

Srinivas M, Kronengold J, Bukauskas FF, Bargiello TA, Verselis VK. 2007. Correlative studies of gating in Cx46 and Cx50 hemichannels and gap junction channels. *Biophys J* **88:** 1725–1739.

Starich TA, Miller A, Nguyen RL, Hall DH, Shaw JE. 2003. The caenorhabditis elegans innexin INX-3 is localized to gap junctions and is essential for embryonic development. *Dev Biol* **256:** 403–417.

Stümpel F, Ott T, Willecke K, Jungermann K. 1998. Connexin 32 gap junctions enhance stimulation of glucose output by glucagon and noradrenaline in mouse liver. *Hepatology* **28:** 1616–1620.

Subak-Sharpe H, Burk RR, Pitts JD. 1969. Metabolic cooperation between biochemically marked mammalian cells in tissue culture. *J Cell Sci* **4:** 353–367.

Swenson KI, Jordan JR, Beyer EC, Paul DL. 1989. Formation of gap junctions by expression of connexins in *Xenopus* oocyte pairs. *Cell* **57:** 145–155.

Swenson KI, Piwnica-Worms H, McNamee H, Paul DL. 1990. Tyrosine phosphorylation of the gap junction protein connexin43 is required for the pp60v-src-induced inhibition of communication. *Cell Regul* **1:** 989–1002.

Takens-Kwak BR, Jongsma HJ. 1992. Cardiac gap junctions: Three distinct single channel conductances and their modulation by phosphorylating treatments. *Pflügers Arch* **422:** 198–200.

Tang W, Zhang Y, Chang Q, Ahmad S, Dahlke I, Yi H, Chen P, Paul DL, Lin X. 2006. Connexin29 is highly expressed in cochlear Schwann cells, and it is required for the normal development and function of the auditory nerve of mice. *J Neurosci* **26:** 1991–1999.

Theis M, de Wit C, Schlaeger TM, Eckardt D, Kruger O, Doring B, Risau W, Deutsch U, Pohl U, Willecke K. 2001. Endothelium-specific replacement of the connexin43 coding region by a lacZ reporter gene. *Genesis* **29:** 1–13.

Thompson RJ, MacVicar BA. 2008. Connexin and pannexin hemichannels of neurons and astrocytes. Channels (Austin) **2.** (epub ahead of print).

Toyofuku T, Yabuki M, Otsu K, Kuzuya T, Hori M, Tada M. 1998. Direct association of the gap junction protein connexin-43 with ZO-1 in cardiac myocytes. *J Biol Chem* **273:** 12725–12731.

Turin L, Warner A. 1977. Carbon dioxide reversibly abolishes ionic communication between cells of early amphibian embryos. *Nature* **270:** 56–57.

Uhlenberg B, Schuelke M, Ruschendorf F, Ruf N, Kaindl AM, Henneke M, Thiele H, Stoltenburg-Didinger G, Aksu F, Topaloglu H, et al. 2004. Mutations in the gene encoding gap junction protein alpha 12 (connexin 46.6) cause Pelizaeus-Merzbacher-like disease. *Am J Hum Genet* **75.**

Unger VM, Kumar NM, Gilula NB, Yeager M. 1999. Three-dimensional structure of a recombinant gap junction membrane channel. *Science* **283:** 1176–1180.

Urschel S, Hoher T, Schubert T, Alev C, Sohl G, Worsdorfer P, Asahara T, Dermietzel R, Weiler R, Willecke K. 2006. Protein kinase A mediated phosphorylation of connexin36 in mouse retina results in decreased gap junctional communication between AII amacrine cells. *J Biol Chem* **281:** 33163–33171.

van Kempen MJ, Jongsma HJ. 1999. Distribution of connexin37, connexin40 and connexin43 in the aorta and coronary artery of several mammals. *Histochem Cell Biol* **112:** 479–486.

van Steensel MAM. 2004. Gap junction diseases of the skin. *Am J Med Genet C Semin Med Genet Genet* **131C:** 12–19.

van Veen TA, Van Rijen HV, Jongsma HJ. 2000. Electrical conductance of mouse connexin45 gap junction channels is modulated by phosphorylation. *Cardiovasc Res* **46:** 496–510.

Veenstra RD, Wang HZ, Beblo DA, Chilton MG, Harris AL, Beyer EC, Brink PR. 1995. Selectivity of connexin-specific gap junctions does not correlate with channel conductance. *Circ Res* **77:** 1156–1165.

Vreeburg M, Zwart-Storm EA, Schouten MI, Nellen RG, Marcus-Soekarman D, Devies M, van Geel M, van Steensel MA. 2007. Skin changes in oculo-dento-digital dysplasia are correlated with C-terminal truncations of connexin 43. *Am J Med Genet A* **143A:** 360–363.

Wagner C, de Wit C, Kurtz L, Grunberger C, Kurtz A, Schweda F. 2007. Connexin40 is essential for the pressure control of renin synthesis and secretion. *Circ Res* **100:** 556–563.

Weidmann S. 1952. The electrical constants of Purkinje fibres. *J Physiol* **118:** 348–362

Weissman TA, Riquelme PA, Ivic L, Flint AC, Kriegstein AR. 2004. Calcium waves propagate through radial glial cells and modulate proliferation in the developing neocortex. *Neuron* **43:** 647–661.

Welsh DG, Segal SS. 1998. Endothelial and smooth muscle cell conduction in arterioles controlling blood flow. *Am J Physiol* **274:** H178–H186.

White TW. 2002. Unique and redundant connexin contributions to lens development. *Science* **295:** 319–320.

White TW, Paul DL. 1999. Genetic diseases and gene knockouts reveal diverse connexin functions. *Annu Rev Physiol* **61:** 283–310.

White TW, Bruzzone R, Wolfram S, Paul DL, Goodenough DA. 1994. Selective interactions among the multiple connexin proteins expressed in the vertebrate lens: The second extracellular domain is a determinant of compatibility between connexins. *J Cell Biol* **125:** 879–892.

White TW, Goodenough DA, Paul DL. 1998. Targeted ablation of connexin50 in mice results in microphthalmia and zonular pulverulent cataracts. *J Cell Biol* **143:** 815–825.

White TW, Gao Y, Li L, Sellitto C, Srinivas M. 2007. Optimal lens epithelial cell proliferation is dependent on the connexin isoform providing gap junctional coupling. *Invest Ophthalmol Vis Sci* **48:** 5630–5637.

Willecke K, Eiberger J, Degen J, Eckardt D, Romualdi A, Güldenagel M, Deutsch U, Söhl G. 2002. Structural and functional diversity of connexin genes in the mouse and human genome. *Biol Chem* **383:** 725–737.

Winterhager E, Pielensticker N, Freyer J, Ghanem A, Schrickel JW, Kim JS, Behr R, Grummer R, Maass K, Urschel S, et al. 2007. Replacement of connexin43 by connexin26 in transgenic mice leads to dysfunctional reproductive organs and slowed ventricular conduction in the heart. *BMC Dev Biol* **7:** 26.

Wolfle SE, Schmidt VJ, Hoepfl B, Gebert A, Alcolea S, Gros D, de Wit C. 2007. Connexin45 cannot replace the function of connexin40 in conducting endothelium-dependent dilations along arterioles. *Circ Res* **101:** 1292–1299.

Wong CW, Christen T, Roth I, Chadjichristos CE, Derouette JP, Foglia BF, Chanson M, Goodenough DA, Kwak BR. 2006. Connexin37 protects against atherosclerosis by regulating monocyte adhesion. *Nat Med* **12:** 950–954.

Xia CH, Cheng C, Huang Q, Cheung D, Li L, Dunia I, Benedetti LE, Horwitz J, Gong X. 2006. Absence of alpha3 (Cx46) and alpha8 (Cx50) connexins leads to cataracts by affecting lens inner fiber cells. *Exp Eye Res* **83:** 688–696.

Xin D, Bloomfield SA. 1999. Dark- and light-induced changes in coupling between horizontal cells in mammalian retina. *J Comp Neurol* **405:** 75–87.

Yang JJ, Huang SH, Chou KH, Liao PJ, Su CC, Li SY. 2007. Identification of mutations in members of the connexin gene family as a cause of nonsyndromic deafness in Taiwan. *Audiol Neurootol* **12:** 198–208.

Yao J, Oite T, Kitamura M. 2008. Gap junctional intercellular communication in the juxtaglomerular apparatus. *Am J Physiol Renal Physiol* **296:** F939–F946

Ye ZC, Wyeth MS, Baltan-Tekkok S, Ransom BR. 2003. Functional hemichannels in astrocytes: A novel mechanism of glutamate release. *J Neurosci* **23:** 3588–3596.

Yeager M, Harris AL. 2007. Gap junction channel structure in the early 21st century: Facts and fantasies. *Curr Opin Cell Biol* **19:** 521–528.

Yin X, Liu J, Jiang JX. 2008. Lens fiber connexin turnover and caspase-3-mediated cleavage are regulated alternately by phosphorylation. *Cell Commun Adhes* **15:** 1–11.

Yum SW, Zhang J, Valiunas V, Kanaporis G, Brink PR, White TW, Scherer S. 2007. Human connexin26 and connexin30 form functional heteromeric and heterotypic channels. *Am J Physiol Cell Physiol* **293:** C1032–C1048.

Zheng-Fischhofer Q, Ghanem A, Kim JS, Kibschull M, Schwarz G, Schwab JO, Nagy J, Winterhager E, Tiemann K, Willecke K. 2006. Connexin31 cannot functionally replace connexin43 during cardiac morphogenesis in mice. *J Cell Sci* **119:** 693–701.

The Desmosome

Emmanuella Delva, Dana K. Tucker, and Andrew P. Kowalczyk

Departments of Cell Biology and Dermatology, Emory University School of Medicine, Atlanta, Georgia 30322

Correspondence: akowalc@emory.edu

Desmosomes are intercellular junctions that tether intermediate filaments to the plasma membrane. Desmogleins and desmocollins, members of the cadherin superfamily, mediate adhesion at desmosomes. Cytoplasmic components of the desmosome associate with the desmosomal cadherin tails through a series of protein interactions, which serve to recruit intermediate filaments to sites of desmosome assembly. These desmosomal plaque components include plakoglobin and the plakophilins, members of the armadillo gene family. Linkage to the cytoskeleton is mediated by the intermediate filament binding protein, desmoplakin, which associates with both plakoglobin and plakophilins. Although desmosomes are critical for maintaining stable cell–cell adhesion, emerging evidence indicates that they are also dynamic structures that contribute to cellular processes beyond that of cell adhesion. This article outlines the structure and function of the major desmosomal proteins, and explores the contributions of this protein complex to tissue architecture and morphogenesis.

The desmosome is an adhesive intercellular junction that is crucial to tissues that experience mechanical stress, such as the myocardium, bladder, gastrointestinal mucosa, and skin (Getsios et al. 2004b; Holthofer et al. 2007). The desmosome was first observed in the spinous layer of epidermis by the Italian pathologist Giulio Bizzozero (1846–1901). Bizzozero's observations of these small dense nodules, subsequently named "nodes of Bizzozero," led him to the insightful interpretation of these structures as adhesive cell–cell contact points. The term desmosome was later coined by Josef Schaffer in 1920 and is derived from the Greek words "desmo," meaning bond or fastening, and "soma," meaning body (Wells 2005; Calkins and Setzer 2007). The introduction of electron microscopy yielded a series of advances by Porter, Odland, and Kelly in the 1950s and 1960s, which revealed desmosome organization at the ultrastructural level. These studies and others indicated that the desmosome can be divided into three morphologically identifiable zones: the extracellular core region (desmoglea), the outer dense plaque (ODP), and the inner dense plaque (IDP) (Fig. 1A) (Kowalczyk et al. 1994; Schmidt et al. 1994; Green and Jones 1996; North et al. 1999; Garrod and Chidgey 2008).

In the mid 1970s, Skerrow and Matoltsy (Skerrow and Matoltsy 1974a; Skerrow and Matoltsy 1974b) advanced the field by isolating desmosomes using biochemical approaches (Bass-Zubek and Green 2007).

 doi: 10.1101/cshperspect.a002543
Cite this article as *Cold Spring Harb Perspect Biol* 2009;1:a002543

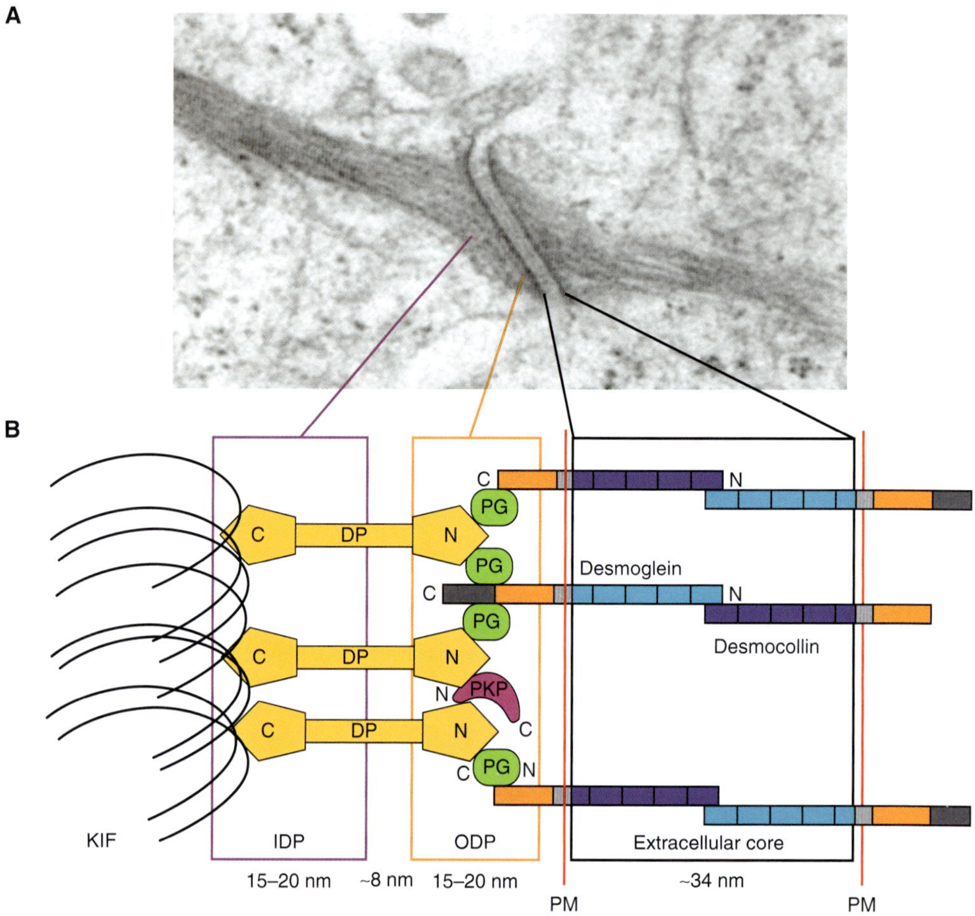

Figure 1. A model for the structure of desmosomes. (*A*) Electron micrograph of a desmosome. (*B*) Schematic of desmosomal proteins and relative distance from the plasma membrane (PM). The desmosomal cadherins, the desmogleins and desmocollins, extend into extracellular core and outer dense plaque (ODP) to establish contact and adhere to neighboring cells in a Ca^{2+}-dependent manner. The cadherin cytoplasmic tails associate linker proteins, plakoglobin (PG), the plakophilins (PKP), and desmoplakin (DP). DP binds to keratin intermediate filaments (KIF) within the inner dense plaque (IDP), serving to tether the intermediate filaments to the plasma membrane. (Adapted with permission from Kottke et al. 2006.)

These landmark studies provided a foundation for the Franke and Steinberg laboratories to characterize the transmembrane glycoproteins and cytoplasmic plaque proteins that linked the structure to the intermediate filament cytoskeleton, and to develop immunological tools for localizing specific components (Franke et al. 1981; Kapprell et al. 1985; Steinberg et al. 1987). Collectively, these and other studies shaped our current view of how desmosomal components are organized.

The transmembrane glycoproteins, termed desmogleins and desmocollins (Garrod and Chidgey 2008), represent separate subfamilies of the cadherin superfamily of calcium dependent adhesion molecules. The extracellular domains of the desmogleins and desmocollins mediate adhesion, whereas the cytoplasmic tails of these cadherins associate with the desmosomal plaque proteins. The outer dense plaque consists of the cytoplasmic tails of the desmosomal cadherins, which bind to members of the armadillo and plakin family of linker proteins (Kowalczyk et al. 1994; Getsios et al. 2004b; Garrod and Chidgey 2008). Plakoglobin, a member of the armadillo family, binds directly to the cytoplasmic tails of both the desmogleins and the desmocollins (Wahl et al. 1996; Witcher et al. 1996). Desmoplakin, a member of the plakin family, interacts with both plakoglobin and another subgroup of armadillo family proteins, the plakophilins (Cowin and Burke 1996). Finally, the interaction between desmoplakin and the keratin filaments forms the inner dense plaque, tethering the cytoskeletal network to the adhesion complex (Fig. 1B) (Kowalczyk et al. 1994; Getsios et al. 2004b; Garrod and Chidgey 2008).

The following sections of this article describe the structural and functional characteristics of the major desmosomal proteins. In addition, we discuss differences in tissue expression patterns of desmosomal proteins and the role of desmosomes in human disease. A comprehensive review of additional proteins found to regulate or associate with desmosomes is provided elsewhere (Holthofer et al. 2007) and discussion of desmosome dynamics is provided in Green et al. 2009.

THE DESMOSOMAL CADHERINS

Desmogleins and Desmocollins: Structure and Function

Desmogleins (Dsgs) and desmocollins (Dscs) are the two types of desmosomal cadherins (Nollet et al. 2000). In humans, there are three isoforms of desmocollins (Dsc 1–3) and four isoforms of desmogleins (Dsg 1–4) (Green and Simpson 2007). All three desmocollin gene products undergo alternative splicing, resulting in the generation of the Dsc "a" form and a shorter Dsc "b" form of the proteins, which differ in the length of their respective carboxy-terminal domains (Fig. 2) (Collins et al. 1991; Parker et al. 1991). Evolutionarily and structurally, the desmocollin genes are more closely related to the classical cadherins (i.e., E- and N-cadherin) than they are to the desmogleins (Kljuic et al. 2004).

The desmogleins and desmocollins both contain four extracellular cadherin homology repeats and a fifth domain termed the extracellular anchor (Fig. 2). Cadherin repeats are approximately 110 amino acids each and are separated by calcium binding motifs. Extracellular calcium supports cadherin mediated

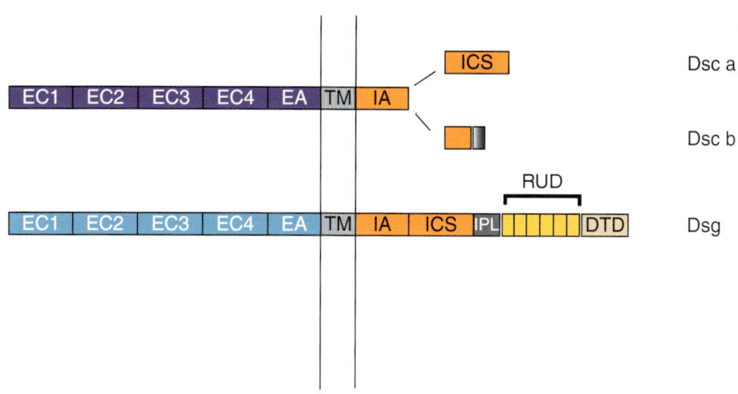

Figure 2. Domains of the desmosomal cadherins. Desmogleins (Dsg) and desmocollins (Dsc) contain four highly conserved extracellular (EC) domains, a more variable extracellular anchor (EA), a single transmembrane domain (TM), and an intracellular anchor (IA), which is followed by additional cytoplasmic domains. Desmocollin "a" isoforms and desmogleins contain an intracellular cadherin-like sequence (ICS) that binds plakoglobin. Desmogleins contain the additional intracellular proline-rich linker (IPL), a variable number of repeat unit domains (RUD), and a desmoglein terminal domain (DTD). The numbers of RUDs contained in each desmoglein isoform are as follows: Dsg 1, 5 repeats; Dsg 2, 6 repeats; Dsg 3, 2 repeats; and Dsg 4, 3 repeats.

adhesion by allowing the cadherin extracellular domain to assume a rigidified and functional conformation (Pokutta and Weis 2007).

A single pass transmembrane domain is followed by an intracellular anchor (IA) located at the cytoplasmic face of the plasma membrane (Kowalczyk et al. 1999a; Green and Simpson 2007; Holthofer et al. 2007). Desmogleins and the longer Dsc "a" form contain an intracellular cadherin-like sequence (ICS), which binds plakoglobin (Mathur et al. 1994; Troyanovsky et al. 1994; Roh and Stanley 1995). The desmogleins also harbor other unique motifs that are specific to this subfamily of the cadherins, including an intracellular proline-rich linker (IPL) domain, a variable number of repeat unit domain (RUD), and a glycine-rich desmoglein terminal domain (DTD) (Fig. 2) (Holthofer et al. 2007; Garrod and Chidgey 2008).

Desmosomes show Ca^{2+}-dependent assembly and adhesion, although the precise mechanisms of desmosomal cadherin adhesion and specificity are not fully understood. Classical cadherins typically show homophilic interactions, which support cell–cell adhesion and tissue patterning. Most studies to date suggest that both desmocollins and desmogleins are required for strong cell–cell interactions (Marcozzi et al. 1998; Tselepis et al. 1998; Getsios et al. 2004a). However, it remains unclear whether homophilic or heterophilic interactions are primarily responsible for desmosome adhesion. Heterophilic interactions between Dsg2 and Dsc1a were observed in human fibrosarcoma cells (HT-1080) (Chitaev and Troyanovsky 1997) and between Dsg2 and Dsc2 using recombinant polypeptides encompassing the first two extracellular domains of the proteins (Syed et al. 2002).

Other studies using atomic force microscopy suggest homophilic interactions mediated by the desmogleins (Waschke et al. 2005; Heupel et al. 2008). Additional studies are required to understand precisely how these cadherins associate and assemble into the highly organized adhesive interface observed in mature desmosomes (Al-Amoudi and Frangakis 2008).

Role of Desmosomal Cadherins in Epithelial Differentiation

The desmosomal cadherins show complex developmental and differentiation patterns of expression (Holthofer et al. 2007). Dsg2 and Dsc2 are expressed widely and in all desmosome bearing tissues, whereas the other desmosomal cadherins are predominantly expressed in stratifying epithelia. In the epidermis, all seven desmosomal cadherins are present. Furthermore, these genes are differentially expressed as keratinocytes and undergo terminal differentiation (Fig. 3) (Kottke et al. 2006; Holthofer et al. 2007). Dsg2 and Dsg3 are mainly distributed throughout the lower layers of the epidermis, whereas Dsg1 is expressed at higher levels in the upper layers. Dsg4 is primarily expressed in the hair follicle and in the granular layer. With respect to the desmocollins, Dsc2 and Dsc3 are present in the basal and spinous layers, whereas Dsc1 is expressed in the granular layer. The differential expression pattern of the desmosomal cadherins implies that desmosomes within different tissues are biochemically, and presumably functionally, distinct.

The precise role for the tissue-specific expression patterns of desmosomal cadherins is not fully understood, but manipulation of desmosomal cadherin expression suggests that tight regulation of their expression pattern is critical to tissue homeostasis. Misexpression of Dsg3 using the keratin 1 promoter to drive expression in the suprabasal epidermal layers in transgenic mice resulted in epidermal hyperproliferation and abnormal differentiation (Merritt et al. 2002). Furthermore, expressing Dsg3 in the upper layers of the epidermis using the involucrin promoter resulted in an even more severe phenotype, including a resemblance to oral mucosa, a reduction in epidermal barrier function, and early postnatal lethality because of extensive water loss (Elias et al. 2001). More recently, Dsg2 misexpression in the differentiated layers of the epidermis was found to cause hyperproliferation and susceptibility to chemically induced carcinogenesis (Brennan et al. 2007). Gene ablation studies have also revealed key roles for desmosomal

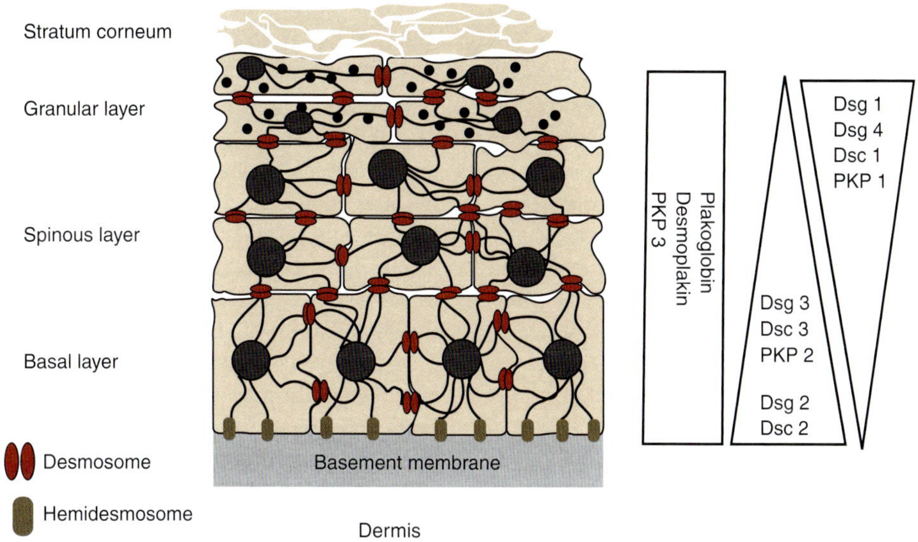

Figure 3. Expression patterns of the desmosomal cadherins in the epidermis. The epidermis is a stratified epithelium comprised of four distinct layers—the basal layer, spinous layer, granular layer, and the stratum corneum. Keratin filaments are shown connecting to desmosomes at sites of cell–cell contact and to hemidesmosomes at the basement membrane. The relative expression profiles of the various desmosomal cadherins and plaque proteins in the epidermal layers are depicted on the *right*. (Adapted with permission from Kottke et al. 2006.)

cadherins in development and in adult tissue function. For example, loss of mouse Dsg3 resulted in animals with compromised cell–cell adhesion in the oral mucosa, with separation of keratinocytes and weakened desmosomal adhesion (Koch et al. 1997). Ablation of Dsg2, on the other hand, results in embryonic lethality shortly after implantation (Eshkind et al. 2002). Interestingly, this lethality on loss of Dsg2 was attributed in part to decreased embryonic stem cell proliferation, suggesting roles of Dsg2 that transcend desmosomal adhesive functions.

Desmocollins also play an important role in the structural integrity of the epidermis and in the control of keratinocyte proliferation and differentiation. Mice lacking Dsc1, which is normally expressed in the upper layers of the epidermis, display loss of cell–cell adhesion in the granular layer as well as hyperproliferation, hair follicle degeneration, and increased expression of the wound keratins, 6 and 16 (Chidgey et al. 2001). Similarly, the conditional ablation of DSC3 in mouse epidermis also resulted in epidermal fragility and hair loss (Chen et al. 2008).

Even more remarkable is that complete loss of Dsc3 expression resulted in preimplantation embryonic lethality (E2.5), strongly suggesting nondesmosomal roles for Dsc3 during development (Den et al. 2006). Lastly, misexpression of Dsc3 in the suprabasal epidermis results in keratinocyte hyperproliferation and abnormal differentiation. These defects were attributed to increased stability of β-catenin, an adherens junction protein known to regulate gene expression and epithelial proliferation (Hardman et al. 2005).

The studies outlined in this section using mouse genetic models implicate desmosomal cadherins in epithelial proliferation and differentiation, and in the robust cell–cell adhesion that is required for tissue integrity. As discussed next, a number of inherited and acquired human diseases confirm these conclusions.

Desmosomal Cadherins and Disease

During the past 10–15 years, a wide range of studies revealed roles for desmosomal cadherins in human diseases, particularly in disorders affecting the heart and the skin. Mutations in three of the desmoglein isoforms cause inherited disorders. Loss of DSG4 is associated with defective hair-follicle differentiation (Kljuic et al. 2003), whereas DSG1 haploinsufficiency leads to striate palmoplantar keratoderma, an epidermal-thickening disease (Rickman et al. 1999). The localized impact of these mutations in DSG1 and DSG4 on the skin is consistent with the tissue expression patterns of these genes (Fig. 3). In contrast, mutations in DSG2 result in arrhythmogenic right ventricular cardiomyopathy (ARVC) (Awad et al. 2006; Pilichou et al. 2006).

In this case, heart manifestation of disease most likely reflects a strict requirement for Dsg2 function in resisting mechanical stresses associated with cardiac contraction. To date, no examples of DSG3 mutations have been reported. However, Simpson and colleagues recently reported the first case of a human mutation in a desmocollin (DSC2) and its association with autosomal recessive ARVC (Simpson et al. 2008). Similar to the mouse genetic models described previously, these human disorders further reveal functions for desmosomal cadherins in both tissue integrity and differentiation.

In addition to inherited disorders, desmosomal cadherins are also targets of autoimmune disease. Pemphigus is a class of diseases in which autoantibodies target the desmosomal cadherins, predominantly Dsg1 and Dsg3. The term "pemphigus" is derived from the Greek word "pemphix," which means blister. These pathologies include pemphigus foliaceus (PF) and pemphigus vulgaris (PV) (Prajapati and Mydlarski 2008). PV is a potentially fatal autoimmune skin disease characterized by circulating autoantibodies (IgG) against Dsg3 and sometimes Dsg1 (Amagai et al. 1991; Amagai 1996; Amagai 1999; Amagai 2002; Payne et al. 2004). PV is characterized histologically by suprabasal acantholysis (loss of cell–cell adhesion) and clinically by mucous membrane erosion and epidermal blisters (Stanley and Amagai 2006). In contrast, PF patients show antibodies against Dsg1 and present clinically with superficial blisters of the epidermis and without mucous membrane involvement. The difference in the clinical presentation in PV and PF is thought to be because of the tissue distribution of Dsg1 and Dsg3 (Mahoney et al. 1999). Nonetheless, in both cases, autoantibodies directed against the desmosomal cadherins are known to cause loss of keratinocyte adhesion (Amagai et al. 2006).

Pemphigus is relatively rare, but these diseases have proven important in revealing functions of desmogleins as well as the mechanisms by which desmosome function is regulated. For example, the fact that IgG against Dsg3 and Dsg1 causes loss of epidermal integrity substantiates a key role for these proteins in epithelial cell–cell adhesion strength. Furthermore, advances in understanding the molecular mechanisms of pemphigus led Stanley and Amagai to correctly predict that bullous impetigo, a common infection caused by a staphylococcal bacterium, causes epidermal blistering by targeting Dsg1. The bacterium produces exfoliative toxin, a serine protease that cleaves the Dsg1 extracellular domain. The cleavage results in blistering in the granular layer that is histologically identical to the acantholysis caused by Dsg1 autoantibodies in PF patients (Amagai et al. 2000; Stanley and Amagai 2006). Thus, epidermal integrity is compromised by either autoantibodies or bacterial toxins that target desmogleins.

Although great strides have been made in identifying the pemphigus antigens, it remains unclear exactly how pemphigus antibodies cause loss of adhesion. One likely explanation is that antibody binding to the Dsg extracellular domain simply blocks adhesion by steric hindrance. However, this hypothesis is called into question by several observations. For example, PV IgG are unable to induce acantholysis in plakoglobin-null cells (Caldelari et al. 2001). In addition, human keratinocytes treated with PV IgG at 4°C do not show loss of cell–cell adhesion until the cells are shifted to 37°C, suggesting that keratinocyte responses are required in order for the antibodies to cause loss of adhesion (Calkins et al. 2006). Other studies have implicated a role for signal transduction in mediating PV-induced acantholysis.

Pemphigus IgG binding has been shown to cause activation of numerous cell signaling pathways.

Work by the Rubenstein laboratory has illustrated that PV IgG binding induced phosphorylation of heat shock protein 27 (HSP27) via p38 mitogen-activating protein kinase (p38MAPK) (Berkowitz et al. 2006; Berkowitz et al. 2007; Berkowitz et al. 2008). p38MAPK has also been shown to be involved in phosphorylation of Dsg3 in response to PV IgG (Kawasaki et al. 2006). Furthermore, inhibition of p38MAPK activity prevents keratin retraction, actin reorganization, and formation of epidermal blisters in a mouse model (Berkowitz et al. 2005; Berkowitz et al. 2006). Similarly, signaling through the c-myc pathway has also been implicated in PV pathogenesis (Williamson et al. 2006). These studies suggest that either the activation of signaling pathways on autoantibody binding to the desmosomal cadherins causes loss of adhesion, or alternatively, that manipulation of intracellular signaling pathways can bolster baseline adhesion strength and prevent blistering caused by pemphigus IgG. Either scenario implies that pemphigus patients could be treated by targeting these regulatory networks.

A number of observations have also illustrated that pemphigus IgG binding may interfere with the normal turnover of the desmogleins. Early studies found that PV IgG were internalized after binding to the surface of keratinocytes (Patel et al. 1984; Iwatsuki et al. 1989; Sato et al. 2000). Work from our lab has shown that PV IgG binding results in clathrin independent endocytosis of Dsg3 and subsequent routing of the cadherin to a lysosomal compartment for degradation (Calkins et al. 2006; Delva et al. 2008). Furthermore, blocking Dsg3 endocytosis (Delva et al. 2008) or up-regulating Dsg3 biosynthesis by exogenously expressing Dsg3 prevents keratinocyte loss of adhesion in response to PV IgG (JM Jennings and A. Kowalczyk, unpubl.).

Work performed by Payne and Kitajima also lends support to the impact of PV IgG on desmosomal assembly, as PV IgG causes internalization of newly synthesized pools of Dsg3 (Sato et al. 2000; Mao et al. 2008). Altogether, these findings suggest that pemphigus IgG disrupt the normal turnover and assembly of desmosomes. In fact, a growing body of evidence suggests that regulation of post Golgi trafficking of cadherins is a key mechanism by which cell adhesion is regulated during development and disease (Mosesson et al. 2008; Delva and Kowalczyk 2009).

ARMADILLO FAMILY OF PROTEINS

Cadherins, like other transmembrane adhesion proteins, require cytoplasmic interactions for strong adhesion and cytoskeletal attachment. In the desmosome, the cytoplasmic associations with the cadherins are partly mediated by the armadillo family proteins, plakoglobin, and the plakophilins. Armadillo was originally discovered as a regulator of segment polarity in *Drosophila*, and later found to be the homolog of the vertebrate junctional and signaling proteins, β-catenin and plakoglobin (Peifer et al. 1992). The armadillo proteins are characterized by the presence of a central domain, containing repeating units of a 42 amino acid sequence homology domain (arm repeats) (Peifer et al. 1992). These proteins include β-catenin, plakoglobin (γ-catenin), p120-catenin, p0071, ARVC, δ-catenin, and the plakophilins 1–3 (PKP 1–3) (Hatzfeld 2005; Hatzfeld 2007). The roles of these proteins in the desmosome appear to be to facilitate the tethering of desmoplakin and keratin intermediate filaments to the desmosome, and to regulate clustering of the desmosomal components. In addition to these structural roles, armadillo proteins also mediate important signal transduction pathways.

Plakoglobin

Plakoglobin is the best characterized armadillo protein in the desmosome. Structurally, the protein contains 12-arm repeats flanked by distinct amino- and carboxy-terminal domains (Fig. 4) (Peifer et al. 1992; Huber et al. 1997; Garrod and Chidgey 2008). High resolution structural studies of the desmosomal

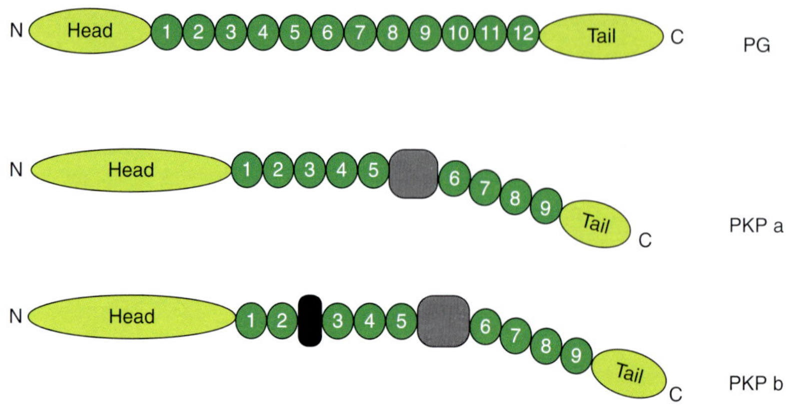

Figure 4. Armadillo protein family members. Plakoglobin (PG) contains 12-arm repeats that are flanked by distinct amino- and carboxy-termini. The plakophilins (PKP) contain 9-arm repeats with an insert between repeats 5 and 6 that introduces a bend into the overall structure. PKP1 and PKP2 exist as two isoforms, each generating a short "a" form and a longer "b" form, differing by the addition of 21 amino acids in arm repeat three (PKP1) and the addition of 44 amino acids in arm repeat four (PKP2).

cadherin–plakoglobin complex have not yet been reported, but deletion mutagenesis studies suggest that the plakoglobin amino-terminus and several arm repeats near the carboxy-terminus are key for desmosomal cadherin binding (Chitaev et al. 1996; Wahl et al. 1996; Witcher et al. 1996). Although plakoglobin localizes to desmosomes and adherens junctions, its affinity for desmosomal cadherins is several times greater than for E-cadherin (Chitaev et al. 1996).

The central armadillo domain of plakoglobin also interacts with desmoplakin, which tethers intermediate filaments to the desmosomal plaque (Kowalczyk et al. 1997; Smith and Fuchs 1998; Bornslaeger et al. 2001). These data have been used to generate models in which the desmosomal cadherins are linked to desmoplakin and the intermediate filament cytoskeleton in a simple linear chain. This type of linkage was also thought to be responsible for classical cadherin associations with actin, whereby β-catenin functioned in an analogous manner to link E-cadherin to actin binding proteins such as α-catenin. However, this model was called into question by studies from the Nelson and Weis laboratories (Drees et al. 2005; Yamada et al. 2005).

Thus, it remains speculative as to whether plakoglobin associates with the desmosomal cadherins and the desmoplakin–intermediate filament complex simultaneously. Nonetheless, plakoglobin clearly plays a key role in desmosome assembly as revealed by both human diseases and mouse genetic model systems (see the following).

A critical role for plakoglobin in desmosome assembly in vivo first emerged from plakoglobin knockout studies in mice. Plakoglobin null animals die because of fragility of the myocardium (Bierkamp et al. 1996; Ruiz et al. 1996). These mice also show acantholysis, indicative of compromised desmosome function (Acehan et al. 2008). In some genetic backgrounds, mouse pups are born but show serious epidermal fragility, heart defects, and early postnatal lethality. Consistent with these findings, Naxos disease, an autosomal recessive disease characterized by ARVC, woolly hair, and palmoplantar keratoderma, develops as a result of a plakoglobin gene mutation leading to the truncation of the carboxy-terminus of the protein (McKoy et al. 2000). The heart defect occurs as a result of fragility of the myocyte syncytium, which leads to its degradation and fibrofatty replacement (Kottke et al. 2006). Although the palmoplantar keratoderma phenotype may result from weakened cell adhesion, the epidermal manifestations seen in Naxos patients are not as severe as the plakoglobin-null mice (Kottke et al. 2006). Interestingly, Naxos disease patients also show woolly hair, in which the hair shaft is lighter and finer than unaffected individuals. The precise mechanism by which this hair defect occurs is unknown, but plakoglobin and β-catenin have both been implicated in hair follicle formation.

Both β-catenin and plakoglobin function as regulators of the T-cell factor (TCF) family of transcription factors and modulators of Wnt growth factor signaling, a pathway known to be important in hair follicle development (Huelsken and Behrens 2002; Chien et al. 2009). Interestingly, plakoglobin overexpression in mouse epidermis decreases keratinocyte proliferation and shortens the anagen phase of the hair cycle (Charpentier et al. 2000), whereas β-catenin overexpression causes hyperproliferation and hair follicle differentiation (Gat et al. 1998). These observations suggest that either the balance between plakoglobin and β-catenin signaling is misregulated in Naxos patients, or that the hair defect is caused by loss of proliferation and/or adhesion in the hair follicle and shaft.

Plakophilins

The other armadillo members represented in the desmosome are the plakophilins (PKPs). Based on the number and organization of the central armadillo repeat domain, the plakophilins are considered to be members of the p120-catenin subfamily of armadillo proteins. In contrast to p120-catenin, the plakophilins are predominantly found at desmosomes instead of adherens junctions, with PKP1 and 2 also localizing to the nucleus. Plakophilins 1–3 share approximately 55% sequence similarity within the armadillo repeats and approximately 50% sequence similarity with the p120-catenin arm domain (Hatzfeld 2007). Both PKP1 and 2 exist as two isoforms, each generating a short "a" form and a longer "b" form, differing in the addition of 21 amino acids in arm repeat three (PKP1) and the addition of 44 amino acids in arm repeat four (PKP2) (Mertens et al. 1996; Schmidt et al. 1997). Based on structural analysis, the plakophilins contain 9 arm repeat domains (Fig. 4) (Choi and Weis 2005).

Although the armadillo domain of plakoglobin, β-catenin, and other p120-catenin family members all function as a scaffold for numerous binding partners, plakophilin binding partners that have been identified to date associate with the plakophilin amino-terminal head domain. Thus, the precise role of the central armadillo domain of the plakophilins remains mysterious.

Similar to the desmosomal cadherins, the plakophilins show tissue and differentiation-specific patterns of expression. PKP1 is more highly expressed in the suprabasal layers of stratified epithelia, whereas PKP2 is expressed in simple epithelia, the lower layers of stratified epithelia, and in nonepithelial tissues such as cardiac muscle and lymph nodes (Heid et al. 1994; Mertens et al. 1996; Schmidt et al. 1997; Mertens et al. 1999; Franke et al. 2007). PKP3 resides in simple and stratified epithelia, showing uniform expression throughout the epidermis (Fig. 3) (Bonne et al. 1999; Schmidt et al. 1999). It is unclear how these expression patterns are integrated with potential differences in functions of the various plakophilin isoforms.

An obvious morphological difference between desmosomes and adherens junctions is the remarkably electron-dense and highly organized disk-shaped plaque that characterizes the desmosome. Plakophilins are likely to play a key role in the clustering that drives the formation of this highly ordered structure. This function is particularly evident in PKP1. The amino-terminal head domain of PKP1 associates with Dsg1, desmoplakin, and keratin intermediate filaments (Hatzfeld et al. 2000). The binding of the PKP1 head domain to desmoplakin appears particularly robust, and this binding event drives desmoplakin recruitment to cell–cell junctions (Kowalczyk et al. 1999b; Hatzfeld et al. 2000; Wahl 2005).

Based on these interactions, a model has been proposed in which plakophilins play a role in the lateral interactions between desmosomal plaque complexes (Kowalczyk et al. 1999b).

In support of this model, PKP1 is highly expressed in the suprabasal layers of the epidermis where desmosomes are larger and more numerous than in the lower layers of the tissue. In contrast, PKP3 overexpression does not result in larger desmosomes or an increase in desmosome number (Bonne et al. 1999; Schmidt et al. 1999; Bonne et al. 2003). However, PKP3 interacts with the largest number of desmosomal proteins, including desmoplakin, plakoglobin, all three desmoglein proteins, Dsc 3a and 3b, and Dsc 1a and 2a (Hatzfeld 2007). PKP2 is also less efficient than PKP1 at recruiting desmoplakin and other desmosomal proteins to the plasma membrane (Chen et al. 2002), suggesting differences in plakophilin isoform function. Recent studies from the Green lab indicate that PKP2 plays an important role in transport of desmoplakin to the plasma membrane during desmosome assembly. In this model system, PKP2 functions as a scaffold for PKC-α and thereby regulates desmoplakin association with intermediate filaments (Godsel et al. 2005; Bass-Zubek et al. 2008; see also Green et al. 2009). Additional studies will be needed to determine whether PKP1 and PKP3 also regulate assembly through PKC-α.

The importance of plakophilins in maintaining desmosomal integrity is underscored by various diseases and/or defective cellular processes that are associated with mutations in each of the plakophilins. For example, patients who suffer from a severe autosomal recessive ectodermal dysplasia and skin fragility syndrome were found to have mutations in PKP1 (McGrath et al. 1997; McGrath et al. 1999).

The dysplasia suggests that in addition to its role in regulating desmosome stability, PKP1 may also regulate epidermal morphogenesis. Because PKP2 is the only isoform expressed in the heart, mutations associated with this particular plakophilin cause ARVC (Gerull et al. 2004). In most cases, the mutations are found within the carboxy-terminus of the PKP2 protein, although other mutations have been observed. Based on experiments in which the gene encoding PKP2 is ablated in mice, the effects seen in ARVC patients are likely the result of mechanical fragility. These mice display mid-gestational embryonic lethality caused by cardiac patterning defects and fragility of the myocardium (Grossmann et al. 2004). PKP2-null mice also display retraction of intermediate filaments from the plasma membrane, further illustrating the importance of plakophilins in desmoplakin recruitment and intermediate filament tethering to the desmosome plaque. Although no human pathologies are associated with mutations in PKP3, conditional ablation of PKP3 in mouse epidermis resulted in defective hair follicle morphogenesis, increased keratinocyte proliferation, and desmoplakin mislocalization. These mice are also more susceptible to dermatitis (skin inflammation) and secondary alopecia (hair loss) (Sklyarova et al. 2008).

Emerging evidence indicates that plakophilins also function outside of the desmosome. Plakophilins, especially PKP1, show striking nuclear localization, although it is currently unclear what functions these proteins may play in gene regulation and/or nuclear structure (Mertens et al. 1996; Hatzfeld et al. 2000).

In addition, PKP2 was found in RNA polymerase III complexes and PKP3 in cytoplasmic particles containing RNA-binding proteins (Mertens et al. 2001; Hofmann et al. 2006). Additional studies are clearly warranted to understand how these associations and nuclear localization are integrated with function in the desmosome, and more broadly, with epithelial cell contributions to tissue architecture.

DESMOPLAKIN

Desmoplakin is the most abundant component of the desmosome (Mueller and Franke 1983) and serves as the key linker between intermediate filaments and the plasma membrane (Bornslaeger et al. 1996; Gallicano et al. 1998; Vasioukhin et al. 2001). Desmoplakin is among the founding members of a large family of proteins termed the plakin family of cytolinkers (Green et al. 1992b; Sonnenberg and Liem 2007). These large molecules link cytoskeletal networks to the plasma membrane and integrate actin, microtubules, and intermediate filaments. Desmoplakin contains globular amino- and carboxy-termini, connected by a central, α-helical coiled-coil rod domain. The amino-terminal head domain provides binding sites for plakoglobin and plakophilins, and thereby targets the protein to the cadherin-armadillo complex at the desmosome (Bornslaeger et al. 1996; Kowalczyk et al. 1999a; Bornslaeger et al. 2001; Holthofer et al. 2007). The carboxy-terminal tail contains three plakin repeat domains (PRDs; A, B, C), as well as a glycine-serine-arginine rich domain thought to regulate desmoplakin binding to intermediate filaments (Fig. 5) (Choi et al. 2002; Getsios et al. 2004b; Yin and Green 2004). There are two desmoplakin isoforms (I and II), with desmoplakin II missing approximately two-thirds of the rod domain (Green et al. 1992a).

Similar to the "a" and "b" forms of desmocollins, desmoplakin I and II are produced as a result of alternative RNA splicing. Although both isoforms of desmoplakin are widely expressed in numerous tissues, desmoplakin II is absent from the heart and its expression is lower in simple epithelia (Angst et al. 1990).

Several studies highlight the importance of desmoplakin in desmosome structure and function. Mice lacking desmoplakin die shortly after implantation at day E6.5 and show fewer desmosomes, compared with wild-type mice (Gallicano et al. 1998). Furthermore, desmosomes that are observed in desmoplakin-null embryos do not attach to intermediate filaments. These observations support in vitro studies and illustrate the importance of desmoplakin in linking intermediate filaments to the plasma membrane (Bornslaeger et al. 1996). There is also striking evidence that desmoplakin plays roles in tissue morphogenesis. Desmoplakin-null embryos fail to undergo the massive increase in cell proliferation normally observed at E5-6, and other studies have illustrated that desmoplakin is pivotal in the development of the epidermis, neuroepithelium, heart and blood vessels (Gallicano et al. 2001; Vasioukhin et al. 2001). Interestingly, loss of desmoplakin dramatically alters actin organization at the cell periphery, demonstrating that key aspects of cross-talk between desmosomes and adherens junctions are somehow mediated by desmoplakin during epithelial sheet formation (Vasioukhin et al. 2001).

Perhaps not surprisingly, there are a number of human genetic disorders associated with desmoplakin mutations, with varying degrees of severity, depending on the particular mutation generated (Lai Cheong et al. 2005). For example, desmoplakin haploinsufficiency causes striate palmoplantar keratoderma (Armstrong et al. 1999; Whittock et al. 1999).

Compound heterozygosity with amino-terminal missense mutations and carboxy-terminal nonsense mutations leads to a more severe keratoderma, skin fragility, and woolly hair/alopecia (Whittock et al. 2002). One patient harboring compound heterozygous mutations that truncated the desmoplakin

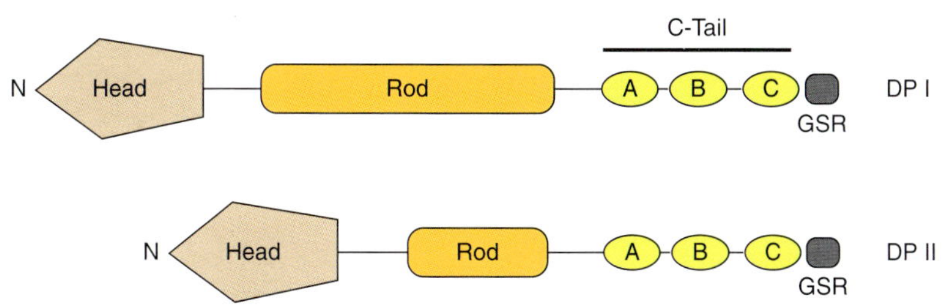

Figure 5. Desmoplakin. Shown above are the different domains that form the two desmoplakin isoforms, DP I and II. The two isoforms differ in the length of the central rod domain (Rod), with DP II missing approximately two-thirds of the Rod domain. The carboxy-terminal tail (C-Tail) contains three plakin repeat domains (A, B, C), as well as a glycine-serine-arginine rich domain (GSR) thought to regulate desmoplakin binding to intermediate filaments. The amino-terminal globular head domain (Head) mediates protein–protein interactions. See text for additional details.

carboxy-terminus presented with severe acantholytic epidermolysis bullosa. This lethal disorder included complete alopecia (lack of hair), neonatal teeth, nail loss, and death 10 days postpartum because of transcutaneous fluid loss as a result of extensive skin erosion (Jonkman et al. 2005). This finding illustrates that attachment of intermediate filaments to the desmosome (via the desmoplakin carboxy-terminus) is necessary for supporting strong cell–cell attachment, particularly in the epidermis.

CONCLUDING REMARKS

The desmosome is a remarkable protein complex that was appreciated by early microscopists as a key structural element in a wide range of epithelial tissues. Work in the 1970s and 1980s led the identification and localization of the major desmosomal components. This foundation led to the employment of molecular and genetic approaches in the 1990s to identify desmosomal protein binding partners and roles during development.

A number of exciting and important challenges have emerged with our growing understanding of desmosome biology and with the development of new tools and model systems. For example, desmosomes are clearly not static structures, even though they mediate robust adhesion.

Significant effort is being focused on the mechanisms by which individual desmosomal components traffic to and from the desmosome, and how these dynamics are coordinated with the assembly of other intercellular junctions, including adherens junctions, tight junctions, and gap junctions. One of the most exciting and perhaps important issues to address is how desmosomal proteins contribute to tissue patterning and epithelial homeostasis. As outlined above, mutations in many of the desmosomal genes result in keratodermas or ectodermal dysplasia. Whether these changes in epithelial structure are compensatory, or if they reflect direct loss- or gain-of-function phenotypes because of misregulation of signaling pathways in the absence of the wild-type gene is unclear. A related issue arises when one considers the number of desmosomal cadherins and plakophilin isoforms that show tissue-specific expression patterns. It is clear that the correct spatial and temporal patterns of expression of these isoforms is driven by tissue- and differentiation-specific functions of each molecule. The continued use of human and mouse genetics, in combination with emerging studies of desmosome dynamics, promises to advance our understanding of how specific desmosomal proteins function at the molecular, cellular, and tissue levels.

ACKNOWLEDGMENTS

The authors are grateful for advice and support provided by Drs. Kathleen Green, Spiro Getsios, and members of the Kowalczyk laboratory. Work in the Kowalczyk lab is supported by National Institute of Health (NIH) R01 AR048266 and R01 AR050501. ED was supported by NIH F31 CA110278 and DKT was supported by NIH T32 GM008367.

REFERENCES

Acehan D, Petzold C, Gumper I, Sabatini DD, Muller EJ, Cowin P, Stokes DL. 2008. Plakoglobin is required for effective intermediate filament anchorage to desmosomes. *J Invest Dermatol* **128:** 2665–2675.

Al-Amoudi A, Frangakis AS. 2008. Structural studies on desmosomes. *Biochem Soc Trans* **36:** 181–187.

Amagai M. 1996. Pemphigus: Autoimmunity to epidermal cell adhesion molecules. *Adv Dermatol* **11:** 319–352; discussion 353.

Amagai M. 1999. Autoimmunity against desmosomal cadherins in pemphigus. *J Dermatol Sci* **20:** 92–102.

Amagai M. 2002. Pemphigus as a paradigm of autoimmunity and cell adhesion. *Keio J Med* **51:** 133–139.

Amagai M, Ahmed AR, Kitajima Y, Bystryn JC, Milner Y, Gniadecki R, Hertl M, Pincelli C, Fridkis-Hareli M, Aoyama Y, et al. 2006. Are desmoglein autoantibodies essential for the immunopathogenesis of pemphigus vulgaris, or just 'witnesses of disease'? *Exp Dermatol* **15:** 815.

Amagai M, Klaus-Kovtun V, Stanley JR. 1991. Autoantibodies against a novel epithelial cadherin in pemphigus vulgaris, a disease of cell adhesion. *Cell* **67:** 869–877.

Amagai M, Matsuyoshi N, Wang ZH, Andl C, Stanley JR. 2000. Toxin in bullous impetigo and staphylococcal scalded-skin syndrome targets desmoglein 1. *Nat Med* **6:** 1275–1277.

Angst BD, Nilles LA, Green KJ. 1990. Desmoplakin II expression is not restricted to stratified epithelia. *J Cell Sci* **97:** 247–257.

Armstrong DK, McKenna KE, Purkis PE, Green KJ, Eady RA, Leigh IM, Hughes AE. 1999. Haploinsufficiency of desmoplakin causes a striate subtype of palmoplantar keratoderma. [erratum appears in Hum Mol Genet 1999 May;8(5):943]. *Human Mol Gen* **8:** 143–148.

Awad MM, Dalal D, Cho E, Amat-Alarcon N, James C, Tichnell C, Tucker A, Russell SD, Bluemke DA, Dietz HC, et al. 2006. DSG2 mutations contribute to arrhythmogenic right ventricular dysplasia/cardiomyopathy. *Am J Human Gen* **79:** 136–142.

Bass-Zubek AE, Green KJ. 2007. Biochemical characterization of the desmosome. *J Invest Dermatol* **127:** E4–E5.

Bass-Zubek AE, Hobbs RP, Amargo EV, Garcia NJ, Hsieh SN, Chen X, Wahl JK, Denning MF 3rd, Green KJ. 2008. Plakophilin 2: A critical scaffold for PKC alpha that regulates intercellular junction assembly. *JJ Cell Biol* **181:** 605–613.

Berkowitz P, Chua M, Liu Z, Diaz LA, Rubenstein DS. 2008. Autoantibodies in the autoimmune disease pemphigus foliaceus induce blistering via p38 mitogen-activated protein kinase-dependent signaling in the skin. *Am J Pathol* **173:** 1628–1636.

Berkowitz P, Diaz LA, Hall RP, Rubenstein DS. 2007. Induction of p38MAPK and HSP27 phosphorylation in pemphigus patient skin. *J Investig Dermatol* **128:** 738–740.

Berkowitz P, Hu P, Liu Z, Diaz LA, Enghild JJ, Chua MP, Rubenstein DS. 2005. Desmosome signaling. Inhibition of p38MAPK prevents pemphigus vulgaris IgG-induced cytoskeleton reorganization. *J Biol Chem* **280:** 23778–23784.

Berkowitz P, Hu P, Warren S, Liu Z, Diaz LA, Rubenstein DS. 2006. p38MAPK inhibition prevents disease in pemphigus vulgaris mice. *Proc Natl Acad Sci* **103:** 12855–12860.

Bierkamp C, McLaughlin KJ, Schwarz H, Huber O, Kemler R. 1996. Embryonic heart and skin defects in mice lacking plakoglobin. *Developmental biology* **180:** 780–785.

Bonne S, Gilbert B, Hatzfeld M, Chen X, Green KJ, van Roy F. 2003. Defining desmosomal plakophilin-3 interactions. *J Cell Biol* **161:** 403–416.

Bonne S, van Hengel J, Nollet F, Kools P, van Roy F. 1999. Plakophilin-3, a novel armadillo-like protein present in nuclei and desmosomes of epithelial cells. *J Cell Sci* **112:** 2265–2276.

Bornslaeger EA, Corcoran CM, Stappenbeck TS, Green KJ. 1996. Breaking the connection: Displacement of the desmosomal plaque protein desmoplakin from cell-cell interfaces disrupts anchorage of intermediate filament bundles and alters intercellular junction assembly. *J Cell Biol* **134:** 985–1001.

Bornslaeger EA, Godsel LM, Corcoran CM, Park JK, Hatzfeld M, Kowalczyk AP, Green KJ. 2001. Plakophilin 1 interferes with plakoglobin binding to desmoplakin, yet together with plakoglobin promotes clustering of desmosomal plaque complexes at cell-cell borders. *J Cell Sci* **114:** 727–738.

Brennan D, Hu Y, Joubeh S, Choi YW, Whitaker-Menezes D, O'Brien T, Uitto J, Rodeck U, Mahoney MG. 2007. Suprabasal Dsg2 expression in transgenic mouse skin confers a hyperproliferative and apoptosis-resistant phenotype to keratinocytes. *J Cell Sci* **120:** 758–771.

Caldelari R, de Bruin A, Baumann D, Suter MM, Bierkamp C, Balmer V, Muller E. 2001. A central role for the armadillo protein plakoglobin in the autoimmune disease pemphigus vulgaris. *J Cell Biol* **153:** 823–834.

Calkins CC, Setzer SV. 2007. Spotting desmosomes: The first 100 years. *J Invest Dermatol* **127:** E2–E3.

Calkins CC, Setzer SV, Jennings JM, Summers S, Tsunoda K, Amagai M, Kowalczyk AP. 2006. Desmoglein endocytosis and desmosome disassembly are coordinated responses to pemphigus autoantibodies. *J Biol Chem* **281:** 7623–7634.

Charpentier E, Lavker RM, Acquista E, Cowin P. 2000. Plakoglobin suppresses epithelial proliferation and hair growth in vivo. *J Cell Biol* **149:** 503–520.

Chen J, Den Z, Koch PJ. 2008. Loss of desmocollin 3 in mice leads to epidermal blistering. *J Cell Sci* **121:** 2844–2849.

Chen X, Bonne S, Hatzfeld M, van Roy F, Green KJ. 2002. Protein binding and functional characterization of plakophilin 2. Evidence for its diverse roles in desmosomes and beta-catenin signaling. *J Biol Chem* **277:** 10512–10522.

Chidgey M, Brakebusch C, Gustafsson E, Cruchley A, Hail C, Kirk S, Merritt A, North A, Tselepis C, Hewitt J, et al. 2001. Mice lacking desmocollin 1 show epidermal fragility accompanied by barrier defects and abnormal differentiation. *J Cell Biol* **155:** 821–832.

Chien AJ, Conrad WH, Moon RT. 2009. A Wnt survival guide: From flies to human disease. *J Invest Dermatol* doi: 10.1038/jid.2008.445.

Chitaev NA, Leube RE, Troyanovsky RB, Eshkind LG, Franke WW, Troyanovsky SM. 1996. The binding of plakoglobin to desmosomal cadherins: Patterns of binding sites and topogenic potential. *J Cell Biol* **133:** 359–369.

Chitaev NA, Troyanovsky SM. 1997. Direct Ca2+-dependent heterophilic interaction between desmosomal cadherins, desmoglein and desmocollin, contributes to cell-cell adhesion. *J Cell Biol* **138:** 193–201.

Choi HJ, Park-Snyder S, Pascoe LT, Green KJ, Weis WI. 2002. Structures of two intermediate filament-binding fragments of desmoplakin reveal a unique repeat motif structure. *Nat Structural Biol* **9:** 612–620.

Choi HJ, Weis WI. 2005. Structure of the armadillo repeat domain of plakophilin 1. *J Mol Biol* **346:** 367–376.

Collins JE, Legan PK, Kenny TP, MacGarvie J, Holton JL, Garrod DR. 1991. Cloning and sequence analysis of desmosomal glycoproteins 2 and 3 (desmocollins): Cadherin-like desmosomal adhesion molecules with heterogeneous cytoplasmic domains. *J Cell Biol* **113:** 381–391.

Cowin P, Burke B. 1996. Cytoskeleton-membrane interactions. *Current Opinion Cell Biol* **8:** 56–65.

Delva E, Jennings JM, Calkins CC, Kottke MD, Faundez V, Kowalczyk AP. 2008. Pemphigus vulgaris IgG-induced desmoglein-3 endocytosis and desmosomal disassembly are mediated by a clathrin- and dynamin-independent mechanism. *Biol Chem* **283:** 18303–18313.

Delva E, Kowalczyk AP. 2009. Regulation of cadherin trafficking. *Traffic* **10:** 259–267.

Den Z, Cheng X, Merched-Sauvage M, Koch PJ. 2006. Desmocollin 3 is required for pre-implantation development of the mouse embryo. *J Cell Sci* **119:** 482–489.

Drees F, Pokutta S, Yamada S, Nelson WJ, Weis WI. 2005. Alpha-catenin is a molecular switch that binds E-cadherin-beta-catenin and regulates actin-filament assembly. *Cell* **123:** 903–915.

Elias PM, Matsuyoshi N, Wu H, Lin C, Wang ZH, Brown BE, Stanley JR. 2001. Desmoglein isoform distribution affects stratum corneum structure and function. *J Cell Biol* **153:** 243–249.

Eshkind L, Tian Q, Schmidt A, Franke WW, Windoffer R, Leube RE. 2002. Loss of desmoglein 2 suggests essential functions for early embryonic development and proliferation of embryonal stem cells. *Eur J Cell Biol* **81:** 592–598.

Franke WW, Schmid E, Grund C, Muller H, Engelbrecht I, Moll R, Stadler J, Jarasch ED. 1981. Antibodies to high molecular weight polypeptides of desmosomes: Specific localization of a class of junctional proteins in cells and tissue. *Differentiation* **20:** 217–241.

Franke WW, Schumacher H, Borrmann CM, Grund C, Winter-Simanowski S, Schlechter T, Pieperhoff S, Hofmann I. 2007. The area composita of adhering junctions connecting heart muscle cells of vertebrates - III: Assembly and disintegration of intercalated disks in rat cardiomyocytes growing in culture. *Eur J Cell Biol* **86:** 127–142.

Gallicano GI, Bauer C, Fuchs E. 2001. Rescuing desmoplakin function in extra-embryonic ectoderm reveals the importance of this protein in embryonic heart, neuroepithelium, skin and vasculature. *Development* **128:** 929–941.

Gallicano GI, Kouklis P, Bauer C, Yin M, Vasioukhin V, Degenstein L, Fuchs E. 1998. Desmoplakin is required early in development for assembly of desmosomes and cytoskeletal linkage. *J Cell Biol* **143:** 2009–2022.

Garrod D, Chidgey M. 2008. Desmosome structure, composition and function. *Bioch biophys Acta* **1778:** 572–587.

Gat U, DasGupta R, Degenstein L, Fuchs E. 1998. De Novo hair follicle morphogenesis and hair tumors in mice expressing a truncated beta-catenin in skin. *Cell* **95:** 605–614.

Gerull B, Heuser A, Wichter T, Paul M, Basson CT, McDermott DA, Lerman BB, Markowitz SM, Ellinor PT, MacRae CA, et al. 2004. Mutations in the desmosomal protein plakophilin-2 are common in arrhythmogenic right ventricular cardiomyopathy. *Nat Gen* **36:** 1162–1164.

Getsios S, Amargo EV, Dusek RL, Ishii K, Sheu L, Godsel LM, Green KJ. 2004a. Coordinated expression of desmoglein 1 and desmocollin 1 regulates intercellular adhesion. *Differentiation* **72:** 419–433.

Getsios S, Huen AC, Green KJ. 2004b. Working out the strength and flexibility of desmosomes. *Nat Rev Mol Cell Biol* **5:** 271–281.

Godsel LM, Hsieh SN, Amargo EV, Bass AE, Pascoe-McGillicuddy LT, Huen AC, Thorne ME, Gaudry CA, Park JK, Myung K, et al. 2005. Desmoplakin assembly dynamics in four dimensions: Multiple phases differentially regulated by intermediate filaments and actin. *J Cell Biol* **171:** 1045–1059.

Green KJ, Jones JC. 1996. Desmosomes and hemidesmosomes: Structure and function of molecular components. *FASEB J* **10:** 871–881.

Green KJ, Simpson CL. 2007. Desmosomes: New perspectives on a classic. *J Investig Dermatol* **127:** 2499–2515.

Green KJ, Stappenbeck TS, Parry DA, Virata ML. 1992a. Structure of desmoplakin and its association with intermediate filaments. *J Dermatol* **19:** 765–769.

Green KJ, Virata ML, Elgart GW, Stanley JR, Parry DA. 1992b. Comparative structural analysis of desmoplakin, bullous pemphigoid antigen and plectin: Members of a new gene family involved in organization of intermediate filaments. *Int J Biol Macromolecules* **14:** 145–153.

Green KJ, Getsios S, Troyanovsky S, Godsel LM. 2009. Intercellular junction assembly, dynamics and homeostasis. *Cold Spring Harb Perspect Biol* **2:** a000125.

Grossmann KS, Grund C, Huelsken J, Behrend M, Erdmann B, Franke WW, Birchmeier W. 2004. Requirement of plakophilin 2 for heart morphogenesis and cardiac junction formation. *J Cell Biol* **167:** 149–160.

Hardman MJ, Liu K, Avilion AA, Merritt A, Brennan K, Garrod DR, Byrne C. 2005. Desmosomal cadherin misexpression alters beta-catenin stability and epidermal differentiation. *Mol Cell Biol* **25:** 969–978.

Hatzfeld M. 2005. The p120 family of cell adhesion molecules. *Eur J Cell Biol* **84:** 205–214.

Hatzfeld M. 2007. Plakophilins: Multifunctional proteins or just regulators of desmosomal adhesion? *Bioch Biophys Acta* **1773:** 69–77.

Hatzfeld M, Haffner C, Schulze K, Vinzens U. 2000. The function of plakophilin 1 in desmosome assembly and actin filament organization. *J Cell Biol* **149:** 209–222.

Heid HW, Schmidt A, Zimbelmann R, Schafer S, Winter-Simanowski S, Stumpp S, Keith M, Figge U, Schnolzer M, Franke WW. 1994. Cell type-specific desmosomal plaque proteins of the plakoglobin family: Plakophilin 1 (band 6 protein). *Differentiation* **58:** 113–131.

Heupel WM, Zillikens D, Drenckhahn D, Waschke J. 2008. Pemphigus vulgaris IgG directly inhibit desmoglein 3-mediated transinteraction. *J Immunol* **181:** 1825–1834.

Hofmann I, Casella M, Schnolzer M, Schlechter T, Spring H, Franke WW. 2006. Identification of the junctional plaque protein plakophilin 3 in cytoplasmic particles containing RNA-binding proteins and the recruitment of plakophilins 1 and 3 to stress granules. *Mol Biol Cell* **17:** 1388–1398.

Holthofer B, Windoffer R, Troyanovsky S, Leube RE. 2007. Structure and function of desmosomes. *Int Rev Cytol* **264:** 65–163.

Huber AH, Nelson WJ, Weis WI. 1997. Three-dimensional structure of the armadillo repeat region of beta-catenin. *Cell* **90:** 871–882.

Huelsken J, Behrens J. 2002. The Wnt signalling pathway. *J Cell Sci* **115:** 3977–3978.

Iwatsuki K, Takigawa M, Imaizumi S, Yamada M. 1989. In vivo binding site of pemphigus vulgaris antibodies and their fate during acantholysis. *J Am Acad Dermatol* **20:** 578–582.

Jonkman MF, Pasmooij AM, Pasmans SG, van den Berg MP, Ter Horst HJ, Timmer A, Pas HH. 2005. Loss of desmoplakin tail causes lethal acantholytic epidermolysis bullosa. *Am J Human Gen* **77:** 653–660.

Kapprell HP, Cowin P, Franke WW, Ponstingl H, Opferkuch HJ. 1985. Biochemical characterization of desmosomal proteins isolated from bovine muzzle epidermis: Amino acid and carbohydrate composition. *Eur J Cell Biol* **36:** 217–229.

Kawasaki Y, Aoyama Y, Tsunoda K, Amagai M, Kitajima Y. 2006. Pathogenic monoclonal antibody against desmoglein 3 augments desmoglein 3 and p38 MAPK phosphorylation in human squamous carcinoma cell line. *Autoimmunity* **39:** 587–590.

Kljuic A, Bauer RC, Christiano AM. 2004. Genomic organization of mouse desmocollin genes reveals evolutionary conservation. *DNA Seq* **15:** 148–152.

Kljuic A, Bazzi H, Sundberg JP, Martinez-Mir A, O'shaughnessy R, Mahoney MG, Levy M, Montagutelli X, Ahmad W, Aita VM, et al. 2003. Desmoglein 4 in hair follicle differentiation and epidermal adhesion: Evidence from inherited hypotrichosis and acquired pemphigus vulgaris. *Cell* **113:** 249–260.

Koch PJ, Mahoney MG, Ishikawa H, Pulkkinen L, Uitto J, Shultz L, Murphy GF, Whitaker-Menezes D, Stanley JR. 1997. Targeted disruption of the pemphigus vulgaris antigen (desmoglein 3) gene in mice causes loss of keratinocyte cell adhesion with a phenotype similar to pemphigus vulgaris. *J Cell Biol* **137:** 1091–1102.

Kottke MD, Delva E, Kowalczyk AP. 2006. The desmosome: Cell science lessons from human diseases. *J Cell Sci* **119:** 797–806.

Kowalczyk AP, Bornslaeger EA, Borgwardt JE, Palka HL, Dhaliwal AS, Corcoran CM, Denning MF, Green KJ. 1997. The amino-terminal domain of desmoplakin binds to plakoglobin and clusters desmosomal cadherin-plakoglobin complexes. *J Cell Biol* **139:** 773–784.

Kowalczyk AP, Bornslaeger EA, Norvell SM, Palka HL, Green KJ. 1999a. Desmosomes: Intercellular adhesive junctions specialized for attachment of intermediate filaments. *Int Rev Cytol* **185:** 237–302.

Kowalczyk AP, Hatzfeld M, Bornslaeger EA, Kopp DS, Borgwardt JE, Corcoran CM, Settler A, Green KJ. 1999b. The head domain of plakophilin-1 binds to desmoplakin and enhances its recruitment to desmosomes. Implications for cutaneous disease. *J Biol Chem* **274:** 18145–18148.

Kowalczyk AP, Stappenbeck TS, Parry DA, Palka HL, Virata ML, Bornslaeger EA, Nilles LA, Green KJ. 1994. Structure and function of desmosomal transmembrane core and plaque molecules. *Biophys Chem* **50:** 97–112.

Lai Cheong JE, Wessagowit V, McGrath JA. 2005. Molecular abnormalities of the desmosomal protein desmoplakin in human disease. *Clin Exp Dermatol* **30:** 261–266.

Mahoney MG, Wang Z, Rothenberger K, Koch PJ, Amagai M, Stanley JR. 1999. Explanations for the clinical and microscopic localization of lesions in pemphigus foliaccus and vulgaris. *J Clin Invest* **103:** 461–468.

Mao X, Choi EJ, Payne AS. 2008. Disruption of desmosome assembly by monovalent human pemphigus vulgaris monoclonal antibodies. *J Invest Dermatol* **129:** 908–918.

Marcozzi C, Burdett ID, Buxton RS, Magee AI. 1998. Coexpression of both types of desmosomal cadherin and plakoglobin confers strong intercellular adhesion. *J Cell Sci* **111:** 495–509.

Mathur M, Goodwin L, Cowin P. 1994. Interactions of the cytoplasmic domain of the desmosomal cadherin Dsg1 with plakoglobin. *J Biol Chem* **269:** 14075–14080.

McGrath JA, Hoeger PH, Christiano AM, McMillan JR, Mellerio JE, Ashton GH, Dopping-Hepenstal PJ, Lake BD, Leigh IM, Harper JI, et al. 1999. Skin fragility and hypohidrotic ectodermal dysplasia resulting from ablation of plakophilin 1. *Br J Dermatol* **140:** 297–307.

McGrath JA, McMillan JR, Shemanko CS, Runswick SK, Leigh IM, Lane EB, Garrod DR, Eady RA. 1997. Mutations in the plakophilin 1 gene result in ectodermal dysplasia/skin fragility syndrome. *Nat Gen* **17:** 240–244.

McKoy G, Protonotarios N, Crosby A, Tsatsopoulou A, Anastasakis A, Coonar A, Norman M, Baboonian C, Jeffery S, McKenna WJ. 2000. Identification of a deletion in plakoglobin in arrhythmogenic right ventricular cardiomyopathy with palmoplantar keratoderma and woolly hair (Naxos disease). *Lancet* **355:** 2119–2124.

Merritt AJ, Berika MY, Zhai W, Kirk SE, Ji B, Hardman MJ, Garrod DR. 2002. Suprabasal desmoglein 3 expression in the epidermis of transgenic mice results in hyperproliferation and abnormal differentiation. *Mol Cell Biol* **22:** 5846–5858.

Mertens C, Hofmann I, Wang Z, Teichmann M, Sepehri Chong S, Schnolzer M, Franke WW. 2001. Nuclear particles containing RNA polymerase III complexes associated with the junctional plaque protein plakophilin 2. *Proc Natl Acad Sci* **98:** 7795–7800.

Mertens C, Kuhn C, Franke WW. 1996. Plakophilins 2a and 2b: Constitutive proteins of dual location in the karyoplasm and the desmosomal plaque. *J Cell Biol* **135:** 1009–1025.

Mertens C, Kuhn C, Moll R, Schwetlick I, Franke WW. 1999. Desmosomal plakophilin 2 as a differentiation marker in normal and malignant tissues. *Differentiation* **64:** 277–290.

Mosesson Y, Mills GB, Yarden Y. 2008. Derailed endocytosis: An emerging feature of cancer. *Nat Rev Cancer* **8:** 835–850.

Mueller H, Franke WW. 1983. Biochemical and immunological characterization of desmoplakins I and II, the major polypeptides of the desmosomal plaque. *J Mol Biol* **163:** 647–671.

Nollet F, Kools P, van Roy F. 2000. Phylogenetic analysis of the cadherin superfamily allows identification of six major subfamilies besides several solitary members. *J Mol Biol* **299:** 551–572.

North AJ, Bardsley WG, Hyam J, Bornslaeger EA, Cordingley HC, Trinnaman B, Hatzfeld M, Green KJ, Magee AI, Garrod DR. 1999. Molecular map of the desmosomal plaque. *J Cell Sci* **112:** 4325–4336.

Parker AE, Wheeler GN, Arnemann J, Pidsley SC, Ataliotis P, Thomas CL, Rees DA, Magee AI, Buxton RS. 1991. Desmosomal glycoproteins II and III. Cadherin-like junctional molecules generated by alternative splicing. *J Biol Chem* **266:** 10438–10445.

Patel HP, Diaz LA, Anhalt GJ, Labib RS, Takahashi Y. 1984. Demonstration of pemphigus antibodies on the cell surface of murine epidermal cell monolayers and their internalization. *J Invest Dermatol* **83:** 409–415.

Payne AS, Hanakawa Y, Amagai M, Stanley JR. 2004. Desmosomes and disease: Pemphigus and bullous impetigo. *Current Opinion Cell Biol* **16:** 536–543.

Peifer M, McCrea PD, Green KJ, Wieschaus E, Gumbiner BM. 1992. The vertebrate adhesive junction proteins beta-catenin and plakoglobin and the Drosophila segment polarity gene armadillo form a multigene family with similar properties. *J Cell Biol* **118:** 681–691.

Pilichou K, Nava A, Basso C, Beffagna G, Bauce B, Lorenzon A, Frigo G, Vettori A, Valente M, Towbin J, et al. 2006. Mutations in desmoglein-2 gene are associated with arrhythmogenic right ventricular cardiomyopathy. *Circulation* **113:** 1171–1179.

Pokutta S, Weis WI. 2007. Structure and mechanism of cadherins and catenins in cell-cell contacts. *Ann Rev Cell Develop Biol* **23:** 237–261.

Prajapati V, Mydlarski PR. 2008. Advances in pemphigus therapy. *Skin Ther Lett* **13:** 4–7.

Rickman L, Simrak D, Stevens HP, Hunt DM, King IA, Bryant SP, Eady RA, Leigh IM, Arnemann J, Magee AI, et al. 1999. N-terminal deletion in a desmosomal cadherin causes the autosomal dominant skin disease striate palmoplantar keratoderma. *Human Mol Gen* **8:** 971–976.

Roh JY, Stanley JR. 1995. Plakoglobin binding by human Dsg3 (pemphigus vulgaris antigen) in keratinocytes requires the cadherin-like intracytoplasmic segment. *J Invest Dermatol* **104:** 720–724.

Ruiz P, Brinkmann V, Ledermann B, Behrend M, Grund C, Thalhammer C, Vogel F, Birchmeier C, Gunthert U, Franke WW, et al. 1996. Targeted mutation of plakoglobin in mice reveals essential functions of desmosomes in the embryonic heart. *J Cell Biol* **135:** 215–225.

Sato M, Aoyama Y, Kitajima Y. 2000. Assembly pathway of desmoglein 3 to desmosomes and its perturbation by pemphigus vulgaris-IgG in cultured keratinocytes, as revealed by time-lapsed labeling immunoelectron microscopy. *Lab Invest* **80:** 1583–1592.

Schmidt A, Heid HW, Schafer S, Nuber UA, Zimbelmann R, Franke WW. 1994. Desmosomes and cytoskeletal architecture in epithelial differentiation: Cell type-specific plaque components and intermediate filament anchorage. *Eur J Cell Biol* **65:** 229–245.

Schmidt A, Langbein L, Pratzel S, Rode M, Rackwitz HR, Franke WW. 1999. Plakophilin 3–a novel cell-type-specific desmosomal plaque protein. *Differentiation* **64:** 291–306.

Schmidt A, Langbein L, Rode M, Pratzel S, Zimbelmann R, Franke WW. 1997. Plakophilins 1a and 1b: Widespread nuclear proteins recruited in specific epithelial cells as desmosomal plaque components. *Cell Tissue Res* **290:** 481–499.

Simpson MA, Mansour S, Ahnood D, Kalidas K, Patton MA, McKenna WJ, Behr ER, Crosby AH. 2008. Homozygous mutation of desmocollin-2 in arrhythmogenic right ventricular cardiomyopathy with mild palmoplantar keratoderma and woolly hair. *Cardiology* **113:** 28–34.

Skerrow CJ, Matoltsy AG. 1974a. Chemical characterization of isolated epidermal desmosomes. *J Cell Biol* **63:** 524–530.

Skerrow CJ, Matoltsy AG. 1974b. Isolation of epidermal desmosomes. *J Cell Biol* **63**(2 Pt 1): 515–523.

Sklyarova T, Bonne S, D'Hooge P, Denecker G, Goossens S, De Rycke R, Borgonie G, Bosl M, van Roy F, van Hengel J. 2008. Plakophilin-3-deficient mice develop hair coat abnormalities and are prone to cutaneous inflammation. *J Invest Dermatol* **128:** 1375–1385.

Smith EA, Fuchs E. 1998. Defining the interactions between intermediate filaments and desmosomes. *J Cell Biol* **141:** 1229–1241.

Sonnenberg A, Liem RK. 2007. Plakins in development and disease. *Exp Cell Res* **313:** 2189–2203.

Stanley JR, Amagai M. 2006. Pemphigus, bullous impetigo, and the staphylococcal scalded-skin syndrome. *N Engl J Med* **355:** 1800–1810.

Steinberg MS, Shida H, Giudice GJ, Shida M, Patel NH, Blaschuk OW. 1987. On the molecular organization, diversity and functions of desmosomal proteins. *Ciba Found Symp* **125:** 3–25.

Syed SE, Trinnaman B, Martin S, Major S, Hutchinson J, Magee AI. 2002. Molecular interactions between desmosomal cadherins. *Biochem J* **362:** 317–327.

Troyanovsky SM, Troyanovsky RB, Eshkind LG, Leube RE, Franke WW. 1994. Identification of amino acid sequence motifs in desmocollin, a desmosomal glycoprotein, that are required for plakoglobin binding and plaque formation. *Proc Natl Acad Sci* **91:** 10790–10794.

Tselepis C, Chidgey M, North A, Garrod D. 1998. Desmosomal adhesion inhibits invasive behavior. *Proc Natl Acad Sci* **95:** 8064–8069.

Vasioukhin V, Bowers E, Bauer C, Degenstein L, Fuchs E. 2001. Desmoplakin is essential in epidermal sheet formation. *Nat Cell Biol* **3:** 1076–1085.

Wahl JK 3rd. 2005. A role for plakophilin-1 in the initiation of desmosome assembly. *J Cell Biochem* **96:** 390–403.

Wahl JK, Sacco PA, McGranahan-Sadler TM, Sauppe LM, Wheelock MJ, Johnson KR. 1996. Plakoglobin domains that define its association with the desmosomal cadherins and the classical cadherins: Identification of unique and shared domains. *J Cell Sci* **109:** 1143–1154.

Waschke J, Bruggeman P, Baumgartner W, Zillikens D, Drenckhahn D. 2005. Pemphigus foliaceus IgG causes dissociation of desmoglein 1-containing junctions without blocking desmoglein 1 transinteraction. *J Clin Invest* **115:** 3157–3165.

Wells WA. 2005. Defining junctional complexes. *J Cell Biol* **168:** 989.

Whittock NV, Ashton GH, Dopping-Hepenstal PJ, Gratian MJ, Keane FM, Eady RA, McGrath JA. 1999. Striate palmoplantar keratoderma resulting from desmoplakin haploinsufficiency. *J Invest Dermatol* **113:** 940–946.

Whittock NV, Wan H, Morley SM, Garzon MC, Kristal L, Hyde P, McLean WH, Pulkkinen L, Uitto J, Christiano AM, et al. 2002. Compound heterozygosity for non-sense and mis-sense mutations in desmoplakin underlies skin fragility/woolly hair syndrome. *J Invest Dermatol* **118:** 232–238.

Williamson L, Raess NA, Caldelari R, Zakher A, de Bruin A, Posthaus H, Bolli R, Hunziker T, Suter MM, Muller EJ. 2006. Pemphigus vulgaris identifies plakoglobin as key suppressor of c-Myc in the skin. *EMBO J* **25:** 3298–3309.

Witcher LL, Collins R, Puttagunta S, Mechanic SE, Munson M, Gumbiner B, Cowin P. 1996. Desmosomal cadherin binding domains of plakoglobin. *J Biol Chem* **271:** 10904–10909.

Yamada S, Pokutta S, Drees F, Weis WI, Nelson WJ. 2005. Deconstructing the cadherin-catenin-actin complex. *Cell* **123:** 889–901.

Yin T, Green KJ. 2004. Regulation of desmosome assembly and adhesion. *Sem Cell Develop Biol* **15:** 665–677.

Structure and Biochemistry of Cadherins and Catenins

Lawrence Shapiro[1] and William I. Weis[2]

[1]Department of Biochemistry and Molecular Biophysics and Edward S. Harkness Eye Institute, Columbia University, New York, New York 10032

[2]Departments of Structural Biology, Molecular and Cellular Physiology, and Photon Science, Stanford University School of Medicine, Stanford, California 94305

Correspondence: lss8@columbia.edu

Classical cadherins mediate specific adhesion at intercellular adherens junctions. Interactions between cadherin ectodomains from apposed cells mediate cell–cell contact, whereas the intracellular region functionally links cadherins to the underlying cytoskeleton. Structural, biophysical, and biochemical studies have provided important insights into the mechanism and specificity of cell–cell adhesion by classical cadherins and their interplay with the cytoskeleton. Adhesive binding arises through exchange of β strands between the first extracellular cadherin domains (EC1) of partner cadherins from adjacent cells. This "strand-swap" binding mode is common to classical and desmosomal cadherins, but sequence alignments suggest that other cadherins will bind differently. The intracellular region of classical cadherins binds to p120 and β-catenin, and β-catenin binds to the F-actin binding protein α-catenin. Rather than stably bridging β-catenin to actin, it appears that α-catenin actively regulates the actin cytoskeleton at cadherin-based cell–cell contacts.

Cadherins constitute a large family of cell surface proteins, many of which participate in Ca^{2+}-dependent cell adhesion that plays a fundamental role in the formation of solid tissues (Takeichi 1995; Tepass 1999; Gumbiner 2005). Many events in the development of multicellular assemblies are associated with changes in cadherin expression (Takeichi 1995; Honjo et al. 2000; Price et al. 2002). Expression of particular cadherins often correlates with formation of discrete tissue structures, and in mature tissues discrete cell layers or other cell assemblies are often demarcated by particular cadherins (Gumbiner 1996). Conversely, down-regulation or loss of cadherins correlates with an increased metastatic potential of the affected cells that arises from the loss of their adhesive properties (Hajra and Fearon 2002; Gumbiner 2005).

The cadherins of vertebrates, and some of their invertebrate homologs, are the most highly characterized. "Classical" cadherins, associated with the adherens junction, and the closely related desmosomal cadherins feature an amino-terminal extracellular region or ectodomain that is followed by a transmembrane anchor and a carboxy-terminal intracellular region. Interactions between ectodomains on

 doi: 10.1101/cshperspect.a003053
Cite this article as *Cold Spring Harb Perspect Biol* 2009;1:a003053

apposed cells mediate specific cell–cell contacts, whereas the intracellular region functionally links cadherins to the underlying cytoskeleton. This article focuses on structural, biophysical, and biochemical studies that have provided important mechanistic insights into the specificity of cell–cell adhesion and its interplay with the cytoskeleton (see also Meng and Takeichi 2009; Cavey and Lecuit 2009).

Cadherin Function in Specific Cell Adhesion

Early in vitro experiments with transfected cells demonstrated that cadherin expression can induce cells to aggregate into tissue-like clumps in the presence of Ca^{2+}, suggesting a Ca^{2+}-dependent adhesive function (Takeichi 1977; Takeichi et al. 1981; Takeichi 1988). When cells expressing different cadherin subtypes are mixed, the cells often segregate by cadherin subtype, suggesting that their adhesive function is selective (Takeichi 1990). This "homotypic" cell sorting behavior led to the idea that cadherin molecules of one subtype bind to other molecules of the same subtype but not to cadherins of a different subtype. A classic example occurs during the process of neurulation in vertebrates (Radice et al. 1997). The separation of the neural tube from the embryonic ectoderm layer is mediated in part by expression changes in two cadherins, E- and N-cadherin. Cells of the early ectoderm layer express E-cadherin, and thus adhere together homotypically to form an epithelial cell layer. Some of these cells then "switch" from expression of E-cadherin to expression of N-cadherin. These N-cadherin expressing cells, with their different adhesive properties, are thus displaced from the E-cadherin containing ectoderm to form a new cell layer, which adheres together homotypically via N-cadherin to form the presumptive neural tube (see also Stepniak et al. 2009).

The early notion that cadherin binding preferences are exclusively homophilic is not entirely correct. In vitro experiments that examined sorting of cadherin-transfected cells (Volk et al. 1987; Shimoyama et al. 2000; Duguay et al. 2003; Patel et al. 2006) and binding of such cells to immobilized cadherin ectodomains (Niessen and Gumbiner 2002) revealed substantial cross-binding among different cadherin subtypes, including between N- and E-cadherins. In addition, heterophilic interactions between the two subgroups of desmosomal cadherins, the desmogleins and desmocollins, have been reported (Chitaev and Tryoanovsky 1997).

Structural studies of cadherin ectodomains have revealed much about the molecular basis of Ca^{2+}-dependent adhesion by cadherins. However, as discussed later, understanding the physical basis for the specificity of tissue patterning by cadherins is still at an early stage.

THE CADHERIN ECTODOMAIN

Overall Structure of Cadherin Ectodomains

Cadherin ectodomains are characterized by a repeating amino acid sequence of about 110 residues, each of which corresponds to a protein module of immunoglobulin-like fold called an "extracellular cadherin" or "EC" domain (Overduin et al. 1995; Shapiro et al. 1995a; Nollet et al. 2000; Posy et al. 2008). A loose convention in the literature refers to any protein containing an EC domain as a "cadherin." By this definition, the cadherin superfamily includes structurally diverse proteins including single-pass transmembrane cell adhesion proteins, G protein coupled receptors, and others, and ranges phylogenetically from single celled organisms to man. Many of these EC domain-containing proteins have poorly characterized function, and the function of the EC domains is unclear. Here, we restrict our focus to cadherins that have been shown to mediate adhesive interactions at the intercellular junction structures they help form between apposed cells. These are the "classical" cadherins and the closely related desmosomal cadherins (Patel et al. 2003; Pokutta and Weis 2007), which mediate adhesion at adherens junctions and desmosomes, respectively.

Classical cadherins, such as E- and N-cadherin, were first identified in vertebrates (Takeichi 1988; Takeichi et al. 1988; Takeichi 1990) and later found to interact with cytoplasmic proteins called catenins (Ozawa et al. 1989). Subsequent studies identified EC domain-containing proteins in invertebrates, notably the *Drosophila* homologs DE- and DN-cadherin that appear to serve tissue patterning functions analogous to their vertebrate counterparts (Oda et al. 1994; Iwai et al. 1997). Although the cytoplasmic regions of DE- and DN-cadherins are very similar to their vertebrate counterparts, their ectodomains differ markedly, containing up to 15 EC repeats rather than five (Fig. 1). As described later, the mechanistic understanding of ectodomain-mediated adhesion that has been obtained for vertebrate classical cadherins is unlikely to apply to the invertebrate proteins or to other nonclassical members of the cadherin family. The ectodomains of the vertebrate desmosmal cadherins are closely related to classical cadherins, and as discussed below, sequence analysis suggests a similar adhesive mechanism. Desmosomal cadherins differ from classical cadherins in portions of their cytoplasmic regions, which enable their interaction with the intermediate filament system (see Delva et al. 2009).

The vertebrate classical cadherins are subdivided into two closely related families, the type I

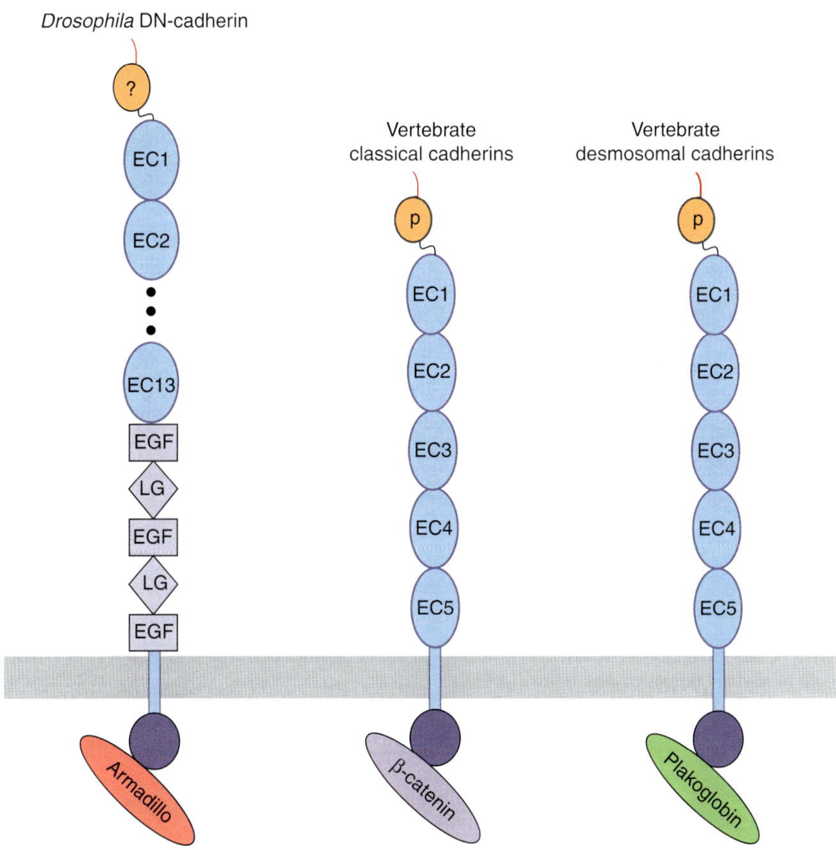

Figure 1. Domain structure of cadherins. The best understood cadherins are the "classical" cadherins of vertebrates, and the closely related desmosomal cadherins. These proteins contain a prodomain (P) immediately following the signal sequence, which is removed by proteolysis. Mature classical and desmosomal cadherins have ectodomains composed of five extracellular cadherin (EC) repeats, a single transmembrane region, and a cytoplasmic domain that interacts with either β-catenin (classical cadherins) or plakoglobin, also known as γ-catenin (desmosomal cadherins). Other cadherin family members have widely differing domain structures, as illustrated by the diagram of *Drosophila* DN-cadherin.

and type II cadherins. At least 19 classical cadherins are conserved in vertebrate genomes: six type I subfamily members and 13 type II cadherins (Nollet et al. 2000). Type I cadherins typically have broad distributions that are segregated by embryonic germ layer or tissue type (Nishimura et al. 1999). In contrast, type II cadherins show more fine-grained, and often overlapping patterns of expression, notably within the developing nervous system (Bekirov et al. 2002; Price et al. 2002).

Evidence Identifying the Adhesive Binding Sites of Classical Cadherins

The large majority of data on cadherins suggests that the adhesive binding site is localized to the membrane-distal EC1 domain. The evidence includes domain shuffling experiments in which chimeric cadherins show the specificity of the transplanted EC1 domain (Nose et al. 1990; Shan et al. 1999; Shan et al. 2000), cross-linking with engineered disulfides (Klingelhofer et al. 2002; Troyanovsky et al. 2003; Troyanovsky 2005), and crystallographic (Boggon et al. 2002; Haussinger et al. 2004; Patel et al. 2006; Parisini et al. 2007) and electron tomography (He et al. 2003; Al-Amoudi et al. 2005; Al-Amoudi et al. 2007; Al-Amoudi and Frangakis 2008), structural data described later. However, atomic force microscopy and force spectroscopy experiments have shown that under some conditions, when cadherin-coated surfaces are "pulled apart," force maxima at multiple distances can be observed (Sivasankar et al. 2001; Zhu et al. 2003; Bayas et al. 2006). The multiple steps were interpreted as representing discrete modes of adhesive interaction, with the strongest corresponding to full interdigitation of ectodomains in which EC1 domains would pair with EC5, EC2 with EC4, etc.; the EC1-EC1 interactions were suggested to represent an intermediate, low affinity interaction initially formed before a contact matures (Sivasankar et al. 2001). This discrepancy has caused some controversy in the field, despite the weight of evidence for EC1-EC1 interactions. Recently, a single-molecule fluorescence study by the current Secretary of Energy of the United States, Steven Chu, appears to have put this controversy to rest. Chu and colleagues labeled the EC1 and EC5 domains of cadherins expressed on cell surfaces with fluorescent dyes, and measured fluorescence resonance energy transfer (FRET) between individual cadherin pairs (Zhang et al. 2009). The results of these studies show that cadherins bind in *trans* (between cells) by association of their EC1 domains.

High-resolution Cadherin Ectodomain Structures

Vertebrate classical and desmosomal cadherins are single-pass transmembrane proteins that are encoded as precursor proteins containing a signal sequence and a prodomain that immediately precedes EC1 (Ozawa and Kemler 1990; Patel et al. 2003; Pokutta and Weis 2007). Proteolytic removal of the prodomain by furin-like proteolysis yields the mature ectodomain, comprised of five EC domains repeated in tandem, EC1 to EC5 (Figs. 1 and 2A). The adhesive capacity of classical cadherins depends on removal of the prodomain (Ozawa and Kemler 1990; Haussinger et al. 2004). High-resolution structural studies have been performed on a number of classical cadherin ectodomains and fragments. Each EC domain is composed of seven β-strands arranged as two opposed β-sheets with carboxy- and amino-termini at opposite ends, so they can be efficiently repeated in tandem (Overduin et al. 1995a,b; Shapiro et al. 1995a,b). The Greek-key topology of cadherins is identical to that of immunoglobulin constant domains. In keeping with the convention for immunoglobulin superfamily (IgSF) domains, successive strands are labeled A–G, with strands A, G, F, and C forming one sheet, and strands D, E, and B forming the other. The EC1 domains of type I (but not type II) cadherins contain a small loop between the C and D strands, designated a "pseudo-β helix" in which the backbone torsion angles are β-like (Patel et al. 2006). This feature is positioned similarly to the C' and C' strands of Ig variable domains. Thus, type I cadherins are remarkably similar to some IgSF

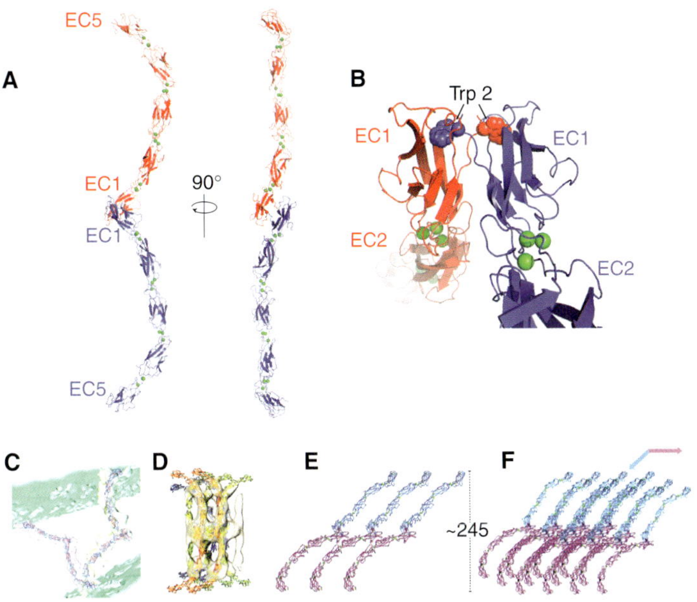

Figure 2. Structures of cadherins. (*A* and *B*) Crystal structure of the ectodomain from C-cadherin. (*A*) Adhesive dimer, joined through the strand-swap interface formed between EC1 domains. (*B*) Close-up rotated view of the strand-swap interface, highlighting the conserved "anchor" residue Trp 2. Bound calcium ions are shown as green spheres. (*C–F*) Possible cadherin junction structures. Cryoelectron tomography has been used to acquire three-dimensional images of desmosome junctions in two cases. One of these (*C*) by He and colleagues (He et al. 2003) examined plastic-embedded samples and revealed a poorly ordered structure. Another (*D*) by Al-Amoudi and colleagues (Al-Amoudi et al. 2007) examined frozen hydrated samples and used averaging procedures, and revealed a uniformly repeating structure. Crystals of C-cadherin (*D* and *E*) are arranged in lattice layers suggestive of a junction structure. These structures are made up of intersecting lines of cadherins (*D*), which together form a two-dimensional array (*E*).

adhesion proteins that have membrane distal variable-like domains atop a stalk of constant-like domains. Despite this similarity, the hydrophobic cores of cadherin and IgSF proteins are distinct (Shapiro et al. 1995b; Posy et al. 2008), and it is unclear whether cadherins and Ig superfamily adhesion proteins evolved from a common precursor.

Cell adhesion by cadherins depends on the presence of Ca^{2+}. Crystal structures of ectodomain regions containing multiple EC domains show that the connections between successive domains are rigidified by Ca^{2+} coordination (Nagar et al. 1996; Boggon et al. 2002). Three Ca^{2+} are coordinated by conserved amino acids contributed from the base of one domain, the top of the next, and the linker region between them (Fig. 2). These Ca^{2+} binding sites are among the most highly conserved sequence features of cadherins across all species (Nollet et al. 2000; Posy et al. 2008). Structures of the vertebrate proteins show that, in the Ca^{2+}-bound state, successive domains are held at well-defined orientations that impart a strong curvature to the full-length ectodomain. The curvature of the Ca^{2+}-bound ectodomain, which places the long axes of the EC1 and EC5 domains at a nearly right angle, is evident in the C-cadherin crystal structure (Boggon et al. 2002), electron micrographs of purified E-cadherin ectodomains (Pokutta et al. 1994), and EM tomograms of desmosome junctions between cells (He et al. 2003; Al-Amoudi et al. 2007). Removal of Ca^{2+} leads to a disordering of interdomain orientations, as can be seen by electron microscopy (Pokutta et al. 1994), increased sensitivity to proteolysis, and increased motion between successive domains (Haussinger et al. 2002).

The Strand-swapped Adhesive Interface

Crystallographic studies of type I cadherins have provided detailed information on the nature of the adhesive dimer interface (Shapiro et al. 1995a; Boggon et al. 2002; Haussinger et al. 2004). Its primary feature is the "swapping" of the amino-terminal β-strands of paired EC1 domains (the A strands) so that the A strand of one monomer replaces the A strand of the other (Fig. 2B). A key element of this twofold symmetric interaction is the insertion of the side chain from the conserved residue Trp-2 of the A strand of one protomer into a pocket extending into the hydrophobic core of the adhesive partner. Mutations of Trp-2 or its acceptor pocket residues result in the loss of adhesive function for all cadherins tested thus far (Tamura et al. 1998; May et al. 2005). A number of studies have also revealed structures for monomeric cadherins. In some of these structures, the A strand is disordered near Trp-2 (Overduin et al. 1995; Tamura et al. 1998), but in one structure, the side chain of Trp-2 inserts into the core of its own rather than a partner protomer (Pertz et al. 1999).

Exchange of A strands between two partner cadherins is characteristic of the general phenomenon of "3D domain swapping" (Bennett et al. 1995; Liu and Eisenberg 2002; Chen et al. 2005). Consistent with the inherently symmetric nature of this binding mode, the EC1 domains of cadherin pairs presented from apposing cells bind in a parallel fashion even though the overall orientation of the two cadherins is antiparallel (Figs. 2 and 3). The parallel orientation is crucial to achieve a true domain-swapped mode of interaction in which near-equivalent interfaces can be achieved in both the monomer and dimer states, and is enabled by the highly curved shape of the cadherin ectodomain.

Crystal structures of type II cadherin ectodomain fragments reveals similar strand-swap binding (Fig. 3). In type II cadherins, Trp-2 plays the same role as it does in type I cadherins but in addition Trp-4, which is conserved in all type II cadherins, also inserts into a hydrophobic pocket of the dimer mate EC1 domain (Patel et al. 2006). Moreover, the EC1–EC1 interface of type II cadherins features a much larger interaction surface arising in part from an extended region of nonpolar contacts that runs along the entire length of the interface. The substantial differences between the binding interfaces of type I and type II

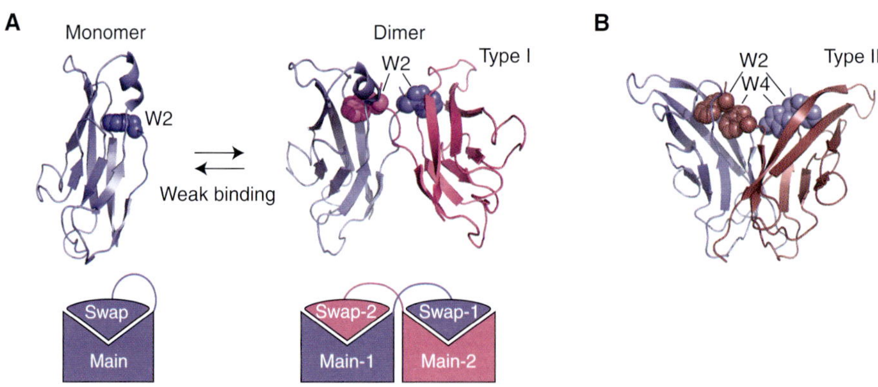

Figure 3. Strand-swap binding by classical cadherins. (*A*) 3D domain swapping by cadherin EC1 domains. 3D domain swapping provides a simple mechanism for constructing homophilic interfaces. A protein made up of a "main" domain and a "swapping" domain connected by a flexible region can form a "closed" monomer, or a multimer (a dimer in the case of cadherins). These two molecular configurations compete with one another, leading to weak binding affinities. (*B*) Comparison of strand-swap interfaces of type I and type II classical cadherins. Although folding topology is identical, these two subfamilies have incompatible binding interfaces. Type II cadherins include two conserved Trp anchor residues, rather than one, and form a hydrophobic interface that runs the length of the EC1 domain.

cadherins provide a simple explanation for the lack of adhesion observed between members of these different subfamilies (Patel et al. 2006). In contrast, the conservation of the central elements of the adhesive interface (for example, the buried Trp side chains) provides a structural explanation for the cross-binding found among members of the same subfamily (Patel et al. 2003). This high level of similarity suggests that, whereas members of different subfamilies will not interact, members of the same subfamily are likely to have graded levels of affinity in cross-adhesion.

Although high-resolution structures have been determined only for classical cadherin ectodomains, sequence analysis strongly suggests that desmosomal cadherins will engage in a similar form of strand-swap binding (Posy et al. 2008). Among the features suggestive of strand-swap binding are a shortened length for the A-strand that engages in strand swapping, and conservation of the anchor residue Trp2 and three residues from the Trp2 binding pocket, Ile24, Tyr36, and Ala80. In contrast, other cadherin families, such as the clustered protocadherins, do not share these features. These observations suggest that desmosomal cadherins, but not others, will engage in strand-swap binding similar to that of classical cadherins.

Unique Properties of Strand-swap Binding

Although affinity differences lie at the heart of interaction specificity, there have been only a few reports of binding affinities between cadherins, mostly focused on E-cadherin. Surprisingly, a wide range of dissociation constants for dimerization of E-cadherin, ranging from 80–720 μM (Koch et al. 1997; Haussinger et al. 2004), have been reported. Structural data provide some insight into the basis of these disparate results. The dimerization interface for classical cadherins, as described above, depends on swapping of the amino-terminal β-strand. Mutation analysis has shown that the native amino terminus (i.e., that generated by cleavage of the prodomain) is required for cell adhesion, and crystal structures reveal a salt bridge between the amino-terminal amino group and a conserved glutamic acid at position 89. The salt bridge is also conserved in type II cadherins (Patel et al. 2006). This salt bridge has been shown to play a key role in forming the stand swapped interface (Harrison et al. 2005). However, many of the reported affinity measurements used bacterially produced proteins that contain extensions at their amino-termini. For example, one of these mutant E-cadherins included an extra Met-Arg dipeptide at the amino terminus and gave $K_D = 170$ μM (Alattia et al. 1997). On the other hand, a study of a "native" E-cadherin gave extremely weak binding measurements, determined by NMR spectroscopy, with $K_D = 720$ μM (Haussinger et al. 2004). However, this protein was produced under denaturing conditions with 8M urea and subsequently refolded, providing a possible explanation for this outlying affinity measurement.

Domain swapping provides a general mechanism of forming homophilic interfaces (Liu and Eisenberg 2002). 3D domain swapping, in general, requires a protomer consisting of a "main" domain and a "swapped" domain, connected by a flexible hinge loop. In this way, symmetric dimers can be formed simply by changing the conformation of the hinge loop. All molecular contacts between "main" and "swapped" domains are locally identical in the dimer and monomer forms, except that they are *intra*molecular in the monomer, and *inter*molecular in the dimer. The A strand, containing Trp-2, as shown in Figure 3, constitutes the "swapped" domain for classical cadherins.

Because the A strand can bind to the body of its own protomer, there is competition for dimeric A strand binding (Fig. 3), which lowers the binding affinity (Chen et al. 2005). This reduction of binding affinity via competition with self-binding appears to be common to many domain swapped multimers. In particular, a study of swapped and nonswapped multimers in the PDB showed that binding affinities for assemblies with swapped interfaces were weaker than nonswapped assemblies with similar interfacial buried surface area

(Chen et al. 2005). It has been proposed that the lowered affinities provided by domain swapping are critical to cadherin specificity in cell adhesion (Chen et al. 2005).

Strand-swap Binding Between Cadherins from Apposed Cells

Why does the cadherin strand-swap interface preferentially form between cadherins from apposed cells (*trans* binding), rather than between cadherin molecules from the same cell surface (*cis* binding)? A substantial body of data has accumulated to suggest that type I cadherins exist as either *cis* dimers or *trans* dimers, formed via the same strand-swapped EC1 interface, and that these are mutually exclusive configurations.

Immunoprecipitation studies with differently tagged cadherins show that in the presence of calcium, *trans* dimers form between apposed cells, whereas *cis* dimers from cadherins on the same cell predominate in the absence of calcium (Pokutta et al. 1994; Chitaev and Troyanovsky 1998; Shan et al. 2000). Importantly, mutagenesis studies reveal that both *cis* and *trans* dimers depend on the strand-swapping interaction (Klingelhofer et al. 2002). Furthermore, site-directed cross-linking studies (Troyanovsky et al. 2003) show identical cross-link formation for both *cis* and *trans* dimers, indicating that they form via the same interface: the strand exchanged dimer of EC1. This behavior can be explained by the rigidification of cadherin ectodomains in the presence of calcium (Pokutta et al. 1994). The geometry of the strand-swap binding interface places orientational constraints on dimer formation. When calcium is present and cadherin ectodomains are rigid, cadherins from apposed cell membranes are oriented so as to easily interact through the strand-swap interface, whereas cadherins from the same cell surface are unable to orient appropriately, hindering *cis* dimer formation. When calcium is depleted, cadherin ectodomains lose their rigidity and orientation constraints are removed, thus enabling strand-swap dimers to form in *cis*. Because this *cis* binding of EC1 domains occurs only in the absence of calcium ions, a condition not encountered in vivo, it seems likely that *cis* strand-swap binding represents a configuration that has no biological relevance. This is supported by recent single molecule FRET studies, which show that *cis* dimerization (between cadherins on the same cell surface) is not required for these *trans* interactions (Zhang et al. 2009).

Assembly of Junctions: Other Potential Lateral (*cis*) Interactions

Cadherins assemble between apposed adherent cells into junctional structures (Fig. 2C–F), adherens junctions and desmosomes (Garrod 1986; Geiger et al. 1987). These structures, which have been studied by classical EM and tomography methods (He et al. 2003; Al-Amoudi et al. 2007), are densely packed with cadherins, raising the question of whether specific lateral interactions of cadherins on the same cell contribute to the higher-order structure of the junctions. Furthermore, EM studies of liposomes coated with the ectodomain of the type II VE-cadherin reveal junction-like structures (Taveau et al. 2008), suggesting that the ectodomain alone may be sufficient to specify junction structure. These observations have prompted the search for "*cis*" or lateral interfaces between cadherin ectodomains, but so far, few solid answers have arisen.

The two extant cryo-EM tomography studies of desmosomes (He et al. 2003; Al-Amoudi et al. 2007) yield conflicting results on lateral interactions. Although some *cis* contacts were identified between cadherin ectodomains in both studies, the proposed interaction sites do not agree. The study by Al-Amoudi et al. presents a lattice-like array of cadherins with *cis* interactions involving a part of the EC1 domain distal from the strand-swapping region (Al-Amoudi et al. 2007). In contrast, the desmosome structure determined by Stokes and collaborators is far less ordered, yet shows some *cis* interactions in the EC5 region (He et al. 2003). These discrepancies likely arise from the different sample preparation techniques used in these studies. Because the resolution in these tomograms is low, it is not possible to ascribe specific molecular contacts to these putative *cis* interactions.

Another potential *cis* interaction has been identified based on crystal packing interactions in the crystal structure of the EC1-EC5 C-cadherin ectodomain (Boggon et al. 2002). For this interaction, the EC1 domain of one molecule contacts another near the EC2-EC3 linker region. This interaction has also been found in several two-domain structures of E-cadherin (Nagar et al. 1996; Pertz et al. 1999; Boggon et al. 2002; Haussinger et al. 2004), but is absent in structures of N-cadherin (Tamura et al. 1998) and type II cadherins (Patel et al. 2006). In this potential *cis* interface, the side chain of Asp 44 approaches the Ca^{2+} binding sites. Mutation of Asp 44 in type I R-cadherin diminishes adhesiveness (Kitagawa et al. 2000), suggesting that these interactions may be relevant. Nevertheless, it is clear that more work needs to be performed to identify and verify potential *cis* interfaces among classical and desmosomal cadherin ectodomains, and to determine whether such interfaces play a role in junction formation.

Molecular Basis of Cell Adhesive Specificity

Classical cadherin ectodomains within a subfamily (e.g., Type I cadherins) are remarkably similar. Sequences of the EC1 domains of E- and N-cadherin are 57% identical, and only a few residues within the strand-swap interface differ. Yet N- and E-cadherins mediate the homotypic sorting of cells in aggregation experiments (Tamura et al. 1998; Shan et al. 1999). Furthermore, cell sorting specificity resides within the EC1 domain (Nose et al. 1990; Shan et al. 1999). How then can small differences between cadherin EC1 domains determine their cellular specificity? Unfortunately, to date there have been no systematic studies of cadherin binding affinities, so the most important parameters required to address this question remain unknown. Nonetheless, it is clear that multiple interactions can amplify small affinity differences at the molecular level into large differences in intercellular adhesive strength (Chen et al. 2005).

Theoretical frameworks have been proposed to explain how differing cellular affinities can lead to the formation of defined tissue patterns. The "differential adhesion hypothesis" proposed by Malcolm Steinberg in the 1960s posits that mixtures of cells behave like molecules in immiscible liquids, with their sorting properties determined by the adhesion molecules expressed on their surfaces (Steinberg 1962; Steinberg 1975; Steinberg 2007). Thus, cell separations, like that observed for E- and N-cadherin expressing cells, can be thought of as analogous to the separation of oil and water. With the wide range of specificities potentially available to cell adhesion molecules, it is possible that complex tissue patterns could be created by such a mechanism.

THE INTRACELLULAR SIDE OF ADHERENS JUNCTIONS

Immunoprecipitation of detergent-solubilized cell extracts with anticadherin antibodies identified three major cytoplasmic partners of E-cadherin: α-catenin, β-catenin, and plakoglobin (also known as γ-catenin) (Ozawa et al. 1989). Subsequent characterization revealed that the cytoplasmic region of E-, N-, and P-cadherins binds to β-catenin or plakoglobin, which in turn binds to α-catenin (Ozawa et al. 1990; Ozawa and Kemler 1992; Nagafuchi et al. 1991). The high level of sequence identity in the cadherin cytoplasmic region suggests that all classical cadherins interact with the catenins in this fashion. α-Catenin has a number of binding partners, including actin (Kobielak and Fuchs 2004). In addition, the protein p120 catenin binds to the juxtamembrane portion of the cadherin cytoplasmic domain (Ozawa and Kemler 1998; Yap et al. 1998; Ohkubo and Ozawa 1999; Thoreson et al. 2000).

Tissues form through cell–cell contacts that are initiated by the cadherin ectodomains, and which strengthen over a time scale of minutes because of rearrangements of the cytoskeleton near the developing junction (Chu et al. 2004). Fluorescence microscopy and time-lapse imaging of developing Madin-Darby canine kidney (MDCK) and keratinocyte epithelium has shown that highly dynamic lamellopodial protrusions from opposing cells form initial contacts that contain cell-surface cadherins, β-catenin, and α-catenin (Adams et al. 1996; Adams et al. 1998; Vaezi et al. 2002). Cadherin–catenin complexes then cluster into punctate structures spatially coincident with actin filaments that emanate from underlying cortical actin belt (Adams et al. 1996; Adams et al. 1998; Vasioukhin et al. 2000). The subsequent change from lamellopodial contact to a continuously sealed adhesion is driven by reorganization of the actin cytoskeleton. Studies of α-catenin-null keratinocytes demonstrate that this process depends on α-catenin (Vasioukhin et al. 2000).

In addition to the initial formation of cell–cell contacts, tissue remodeling during development requires cells to remain attached to one another while undergoing shape changes that enable morphogenetic movements. Conversely, disassembly of cell–cell contacts occurs during normal development and in metastasis of transformed cells. All of these processes require close coordination of cadherin-mediated adhesive contacts with changes in the underlying cytoskeleton. In the adherens junction, the cytoplasmic complex of cadherins with catenins and elements of the actin cytoskeleton serve this essential role. In this section, we discuss the molecular properties of the cadherin–catenin complex and their implications for the regulated assembly of cell–cell contacts.

Cadherin Cytoplasmic Domain

The cytoplasmic region of classical cadherins, roughly 150 amino acids long, is the most highly conserved portion of these proteins (Nollet et al. 2000). The juxtamembrane region binds to p120, and the carboxy-terminal ca. 100 amino acids bind to β-catenin and to plakoglobin. Sequences homologous to the β-catenin/plakoglobin-binding region are also present in the desmosomal cadherins.

The cadherin cytoplasmic domain is unstructured in the absence of a binding partner (Huber et al. 2001), and is an example of an intrinsically unstructured protein (Dyson and Wright 2005; Tompa 2005). Crystal structures of the full 151-residue E-cadherin domain bound to β-catenin show that the last 100 residues become ordered in the complex (Huber et al. 2001) (Fig. 4). A portion of the

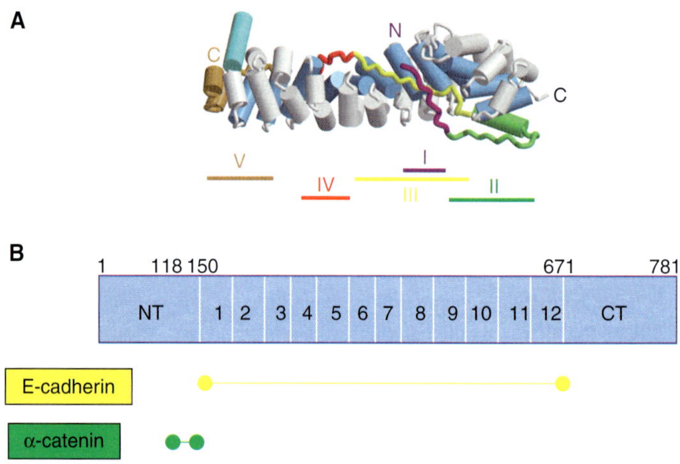

Figure 4. β-Catenin structure and its complex with E-cadherin cytoplasmic domain. (*A*) Three-dimensional structure of the β-catenin arm repeat region in complex with the E-cadherin cytoplasmic domain (Huber and Weis 2001). The arm repeats are formed by three helices, colored gray (H1 and H2) and blue (H3). Residues 134–161, which include the α-catenin binding site and a portion of the first arm repeat, form a single helix in this particular crystal structure (cyan). E-cadherin is divided into five regions of primary structure that are indicated in distinct colors. (*B*) The β-catenin primary structure. Binding sites for cadherin and α-catenin are indicated.

juxtamembrane cadherin region presumably becomes ordered on forming a stable complex with p120, but this has yet to be verified experimentally.

The lack of structure of E-cadherin in the absence of binding partners may be important to its regulated turnover. Newly synthesized E-cadherin associates with β-catenin while still in the endoplasmic reticulum, and the two proteins move together to the cell surface (Hinck et al. 1994). Impairing β-catenin binding leads to proteosomal destruction of cadherin (Chen et al. 1999). The catenin-binding region of cadherins features a "PEST" sequence motif that is recognized by ubiquitin ligases but would be inaccessible in the complex with β-catenin (Huber et al. 2001; Huber and Weis 2001). Thus, it appears that β-catenin prevents proteosomal destruction of cadherin, insuring delivery of cadherin–β-catenin complexes to the cell surface.

The interaction of E-cadherin with β-catenin is very strong and is modulated by phosphorylation (Lickert et al. 2000). In the absence of phosphorylation, the two proteins bind with a K_D of 36 nM (Choi et al. 2006). Phosphorylation of cadherin by the Ser/Thr kinase CKII increases the affinity to 52 pM (Choi et al. 2006). Mutation of these phosphorylation sites reduces cell adhesiveness (Lickert et al. 2000), but the mechanistic connection between the affinity modulation by phosphorylation and changes in cell adhesiveness is not clear. The phosphorylation increases the on rate and decreases the off rate of the interaction (Choi et al. 2006), and causes more of the PEST region to be structured (Huber and Weis 2001). These observations suggest that one role of phosphorylation could be to further reduce the likelihood that cadherin will be recognized by the cellular destruction machinery.

β-Catenin

As part of adherens junctions, β-catenin binds to the cytoplasmic portion of cadherins and to α-catenin. It is also a transcriptional coactivator central to the Wnt signaling pathway that determines cell fate during embryogenesis and tissue renewal in the adult (see Heuberger and Birchmeier 2009; Cadigan and Peifer 2009). Unlike the multitude of cadherins and several tissue-specific α-catenin variants, a single β-catenin protein is present in vertebrates and insects. The protein is highly conserved, with, for example, only six amino acids different between the human and *Xenopus* proteins, and it is 67% identical to the *Drosophila* homolog armadillo. The primary structure of the 781 amino acid β-catenin consists of an amino-terminal region of about 150 amino acids, a central ~520 residue domain composed of 12 armadillo (arm) repeats, and a carboxy-terminal 100 residue region (Fig. 4B). E-cadherin binds to the arm domain, whereas α-catenin binds to residues 118-149, just before the start of the arm domain (Aberle et al. 1996) (Fig. 4B). The arm domain is an elongated superhelical structure formed by the successive packing of helical arm repeats (Huber et al. 1997) (Fig. 4A). The superhelix features a groove that forms part of the binding site for β-catenin ligands (see Choi et al. 2006 and references therein).

The entire β-catenin arm domain interacts with cadherin (Fig. 4A). For convenience, we divide the cadherin sequence into five β-catenin interaction regions. Region I includes a β strand that pairs with region III and forms several direct polar contacts with β-catenin. Cadherin region II includes a helix that interacts with the carboxy-terminal arm repeats. In contrast to the effect of Ser/Thr phosphorylation of cadherin, phosphorylation of β-catenin Tyr654 (repeat 12) by Src kinase produces a sixfold reduction in affinity (Roura et al. 1999). The phenolic hydroxyl group of the tyrosine side chain forms a hydrogen bond with an aspartate from cadherin (Huber and Weis 2001); phosphorylation would clearly disrupt this interaction. This modification is correlated with a change in cell contacts and the development of a more invasive cell, consistent with dysregulation of the junctional complex, but direct mechanistic data are lacking.

Moving toward the amino terminus of the arm domain, cadherin region III forms an extended peptide that binds to the basic groove formed by repeats five and nine (Fig. 4A). This is the site common to Wnt-pathway ligands Tcfs, APC, and the transcription inhibitor ICAT. Region IV is a 10-residue stretch that contains the target serine residues of CKII and GSK-3β. This region binds to repeats three to five only when it is phosphorylated; it is disordered in the structure of the complex of the nonphosphorylated cadherin and the β-catenin arm domain (Huber and Weis 2001). The extreme carboxyl terminus of cadherin region V features two small helices that "cap" the amino-terminal arm repeats.

The amino- and carboxy-terminal regions or "tails" of β-catenin that flank the arm repeats are less well characterized than the arm domain, but their proteolytic sensitivity indicates that these regions are flexibly linked to the arm domain and probably unstructured in the absence of a binding partner. Indeed, these regions are disordered in the crystal structure of full-length β-catenin (Xing et al. 2008). Although there have been reports that the tails, both of which are acidic, influence ligand binding by competing for binding to the positively charged groove of the arm domain (Piedra et al. 2001; Castaño et al. 2002; Gottardi and Gumbiner 2004), accurate calorimetric affinity measurements showed no significant differences of purified full-length, arm only, N + arm, or arm + C β-catenin constructs binding to E-cadherin, the transcription factor Lef-1, or the transcriptional inhibitor ICAT (Choi et al. 2006). Binding to the lower affinity ligands adenomatous polyposis coli and Axin is influenced by the presence of the amino- and carboxy-terminal tails, but the molecular basis for this effect is not clear (Choi et al. 2006).

The carboxy-terminal tail is highly acidic and functions as a transcriptional activator in Wnt signaling. Analysis of the *Drosophila* protein indicates that this region is dispensible for adhesion (Cox et al. 1999). In contrast, the amino-terminal tail contains the binding site for α-catenin, specifically residues 118–149, which lie just amino-terminal to the arm domain (Fig. 4B), and is discussed later.

p120 Catenin

p120 Catenin (p120) is an armadillo-family protein whose arm repeat domain binds to the juxtamembrane region of the cadherin cytoplasmic tail (Daniel and Renynolds 1995; Ohkubo and Ozawa 1999; Thoreson et al. 2000). The arm domain is preceded by an amino-terminal sequence that varies in length in the four splice variants of the protein (Anastasiadis and Reynonds 2000). There appear to be two principal roles for p120. First, it may function to regulate cadherin levels by controlling the rate of cadherin turnover (Ireton et al. 2002; Davis et al. 2003; Xiao et al. 2003; Hoshino et al. 2005; Xiao et al. 2007). Second, p120 controls small GTPases and their effectors that regulate the actin cytoskeleton (Anastasiadis 2007). The interaction with cadherin is thought to localize the small GTPase regulatory activity of p120 at adherens junctions to affect actin organization.

The crystal structure of the arm domain of the desmosomal p120 homolog plakophilin-1 shows that this armadillo protein subfamily contains nine arm repeats; a sequence in the middle of the domain that had predicted to form a 10th repeat is instead a large insertion whose ends form a wedge that introduces a pronounced kink into the domain, making it quite distinct from β-catenin (Choi and Weis 2005). Nonetheless, the superhelical structure imparted by packing of successive arm repeats gives rise to a prominent groove. By analogy to β-catenin, the groove may represent the cadherin-binding site, but this has not been verified experimentally.

α-Catenin

α-Catenin is a roughly 900-amino-acid protein that is homologous to the focal adhesion protein vinculin. Three isoforms of α-catenin have been described, E(epithelial), N(neuronal), and T (found in heart and testis). Both α-E- and α-N-catenin are coexpressed in a number of tissues, often at different levels and at different developmental stages. Alternative splicing generates two variants of α-N-catenin, the significance of which is unknown. Most of the biochemical and structural data reported to date are for α-E-catenin. A number of α-catenin interaction partners have been reported (Kobielak and Fuchs 2004), but not all have been verified biochemically with pure proteins. Distinct regions of primary structure are associated with binding to different partners (Fig. 5A).

1. Dimerization and β-catenin binding—α-E-catenin can homodimerize, but the homodimer dissociates on binding to β-catenin (Koslov et al. 1997; Pokutta and Weis 2000). The dimerization/β-catenin binding region lies near the amino terminus of the protein, in residues 55–264. The crystal structure of the dimerization domain defined by proteolysis, residues 82–264, starts as a 30-residue α helix, followed by an antiparallel 50-residue helix, then three more helices (Fig. 5B) (Pokutta and Weis 2000). The first half of the long helix pairs with the first helix, and the second half forms part of an antiparallel four-helix bundle at the carboxyl terminus of the domain. The two antiparallel amino-terminal helices pair with the corresponding helices in the other protomer to form an antiparallel four-helix bundle (Fig. 5B).

Alanine scanning mutagenesis demonstrated that β-catenin residues 118–149 bind to α-catenin largely as an α helix (Aberle et al. 1996). α-Catenin residues 57–81, which appear to be flexibly linked to the following dimerization domain, are also predicted to form an α helix (Pokutta and Weis 2000).

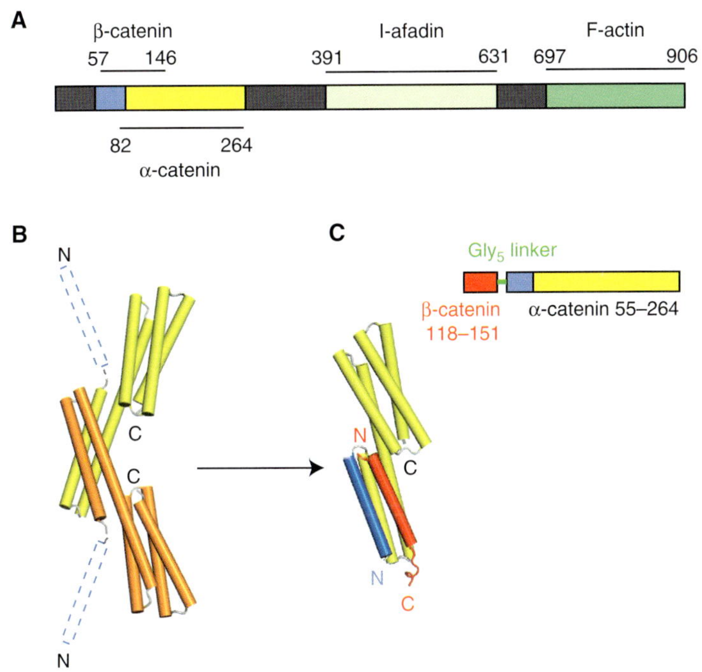

Figure 5. α-Catenin structure and β-catenin binding. (*A*) The primary structure of α-catenin, showing interaction sites mapped by direct binding and/or crystallography. The dimerization domain is shown in yellow; the preceding helix required for β-catenin binding is shown in blue. The M-domain is shown in light green, and the actin-binding domain in dark green. (*B*) The three-dimensional structure of the α-catenin dimerization region 82–264 (Pokutta and Weis 2000). The protomers of the α-catenin dimer are shown in yellow and orange. Residues 57–81, which were not in the crystallized construct and are flexibly linked to the dimerization domain, are schematically illustrated by the dashed blue box. (*C*) The βα-catenin chimera primary structure is shown to the right of the crystal structure. The helices of the dimerization region 82–264 are colored in yellow, and the helix formed by 57–81 in blue. The α-catenin binding region of β-catenin is shown in red.

These findings suggested a model for the α-catenin–β-catenin interaction in which the β-catenin residues 118–149 and α-catenin residues 57–79 form two helices that replace one protomer of the α-catenin homodimer (Fig. 5B,C) (Pokutta and Weis 2000). To test this model, a chimera in which the α-catenin binding site of β-catenin is fused to residues 57–264 of α-catenin by a flexible linker was constructed to effect a high local concentration of β-catenin that outcompetes the homodimerization interaction (Pokutta and Weis 2000). The crystal structure of this chimera (Fig. 5C) confirmed the expected overall arrangement of helices. However, the β-catenin helix and α-catenin 57–79 helix pack against the other two α-catenin helices at a different angle than those observed in the homodimer. This creates a slightly larger interface, which may account for the observation that when the two proteins are mixed, the β–α heterodimer is favored (Koslov et al. 1997). Also, the β-catenin helix is interrupted at residue 141, with residues 142–144 bound in an extended conformation. β-catenin Tyr142 is buried in the interface. This tyrosine has been found to be a substrate for Met kinase, and phosphorylation of this residue diminishes the β–α interaction (Brembeck et al. 2004).

The bulk of endogenous α-E-catenin found in MDCK cell extracts is monomeric, but small amounts of both α-E-catenin homodimer and β–α-E-catenin heterodimer are present (Drees et al. 2005). The cytosolic concentration of α-E-catenin was estimated to be approximately 0.6 μM (Drees et al. 2005). Mixtures of α-E-catenin monomer and dimer can be found in purified recombinant material soon after purification when applied to a gel filtration column at a concentration of 2 μM. These observations suggest that the K_D for α-catenin homodimer formation lies in the single micromolar range. Gel filtration experiments performed with reasonably pure monomer suggested that the K_D for β-catenin binding is roughly 1 μM (Drees et al. 2005), similar to the estimate given from surface plasmon resonance measurements (Koslov et al. 1997).

2. *M domain*—A second, proteolytically defined region in the middle of α-catenin, designated the M domain, spans residues 377–633 (Yang et al. 2001; Pokutta et al. 2002). This region comprises two four-helix bundles connected by a short linker. Structures of two distinct crystal forms, each containing several crystallographically independent copies, revealed that the relative positions of the two four-helix bundles can vary by as much as 56° (Yang et al. 2001; Pokutta et al. 2002). This interdomain flexibility may be important for transmitting conformational changes between the amino-terminal β-catenin-binding and carboxy-terminal actin-binding regions, as discussed later.

The M domain is the binding site for l-afadin (Pokutta et al. 2002), an actin-binding protein associated with the Ig-superfamily adhesion molecule nectin (Takai et al. 2008). This adhesion system is thought to be involved in early establishment of cell contacts (Takai et al. 2008). Both M subdomains are required for the interaction (Pokutta et al. 2002), which may imply that afadin binds to a particular arrangement of the two domains.

3. *Actin-binding domain*—α-Catenin and vinculin bind to filamentous (F-)actin. They differ in that α-catenin binds actin constitutively, whereas vinculin is autoinhibited by an interaction between the head and tail regions and therefore requires activation to bind actin. The highest level of sequence homology between these two proteins is found in their carboxy-terminal actin-binding domains (Rimm et al. 1995; Bakolitsa et al. 1999; Pokutta et al. 2002). The vinculin domain is a five-helix bundle (Bakolitsa et al. 1999), and electron microscopy of the vinculin domain bound to F-actin has identified a putative interaction surface (Jannsen et al. 2006). The five-helix architecture and the actin-binding surface residues appear to be conserved in α-catenin (Bakolitsa et al. 1999). However, a fragment of α-catenin corresponding to the vinculin actin-binding domain does not bind to actin; a 42-amino acid carboxy-terminal extension unique to α-catenin is also required, indicating that there are likely to be differences

in how these two proteins interact with actin (Pokutta et al. 2002).

Comparison of α-catenin monomer and dimer binding to actin revealed that the latter binds more strongly to actin (Drees et al. 2005). This is likely an avidity effect because the dimer has two binding sites.

Role of α-Catenin in the Cadherin–Catenin Complex

The ability of α-catenin to bind to both β-catenin and F-actin led to the conclusion that α-catenin links the cadherin/β-catenin complex to actin, providing a stable mechanical linkage between the extracellular contact and the underlying cytoskeleton. However, experiments with purified, recombinant proteins demonstrated that β-catenin inhibits binding of α-catenin to actin (Yamada et al. 2005). The inhibition of α-catenin actin-binding activity by β-catenin was independently confirmed using a different assay (Abe and Takeichi 2008).

Fluorescence recovery after photobleaching experiments performed on fluorescently tagged E-cadherin, β-catenin, α-catenin, and actin showed that the cadherins and catenins have very similar dynamics on the membrane, as expected if they associate as a complex, whereas actin associated with cell–cell contacts diffused more rapidly and was more mobile (Yamada et al. 2005). Deletion of the cadherin cytoplasmic domain or the actin-binding domain of α-catenin, either of which would break the putative link between cadherin and actin, did not significantly alter the dynamics of cadherin (Yamada et al. 2005). These results indicate the absence of a stable linkage between the cadherin–catenin complex and actin at cell–cell contacts, consistent with the inhibition of α-catenin actin binding activity by β-catenin. Higher-resolution imaging experiments revealed that in fact there are relatively immobile clusters of E-cadherin and actin in addition to the more dynamic cadherin and actin present at cell–cell contacts, but α-catenin is not required for the stability of these clusters (Cavey et al. 2008). In contrast, epithelial architecture appears to be determined by a dynamic actin network that controls the mobility of these adhesive clusters, and this feature requires α-catenin. It was suggested that α-catenin interacts transiently with actin and β-catenin to achieve this dynamic architecture, or it does so through interactions with other actin-binding proteins (Cavey et al. 2008).

Several lines of evidence indicate that α-catenin regulates actin assembly and dynamics in cell–cell contacts. Reduction of α-N-catenin in dendritic spines produces enhanced membrane activity, whereas overexpression results in decreased activity (Abe et al. 2004). Deletion of α-E-catenin in keratinocytes results in increased cell migration, a shortened cell cycle, and increased proliferation rather than simple loss of cell–cell adhesion (Vasioukhin et al. 2001; Lien et al. 2006). It is known that the Arp2/3 complex, which nucleates branched actin networks that drive lamellipodial movement needed for cell migration (Pollard and Borisy 2003), interacts with the cadherin complex in nascent cell contacts (Kovacs et al. 2002), and that lamellipodial activity decays as the contacts form (Ehrlich et al. 2002; Vaezi et al. 2002). In vitro actin polymerization experiments demonstrated that α-catenin can inhibit the polymerization of actin into branched networks by the Arp2/3 complex, starting at concentrations of about 1 μM, above the α-catenin concentration of 0.6 μM in bulk cytosol (Drees et al. 2005).

These live cell and in vitro observations can be rationalized in a model that explains the change in membrane dynamics and the organization of the actin cytoskeleton during cell contact formation (Fig. 6). Transient contacts between cadherins present on lamellipodia lead to cadherin clustering, which produces a locally high concentration of α-catenin at the developing junction. The concentration of α-catenin at these sites becomes high enough so that when it dissociates from β-catenin, it can suppress Arp2/3-stimulated branched actin polymerization, leading to the cessation of lamellipodial movement and formation of a stable cell–cell contact with concomitant reorganization of actin into linear cables

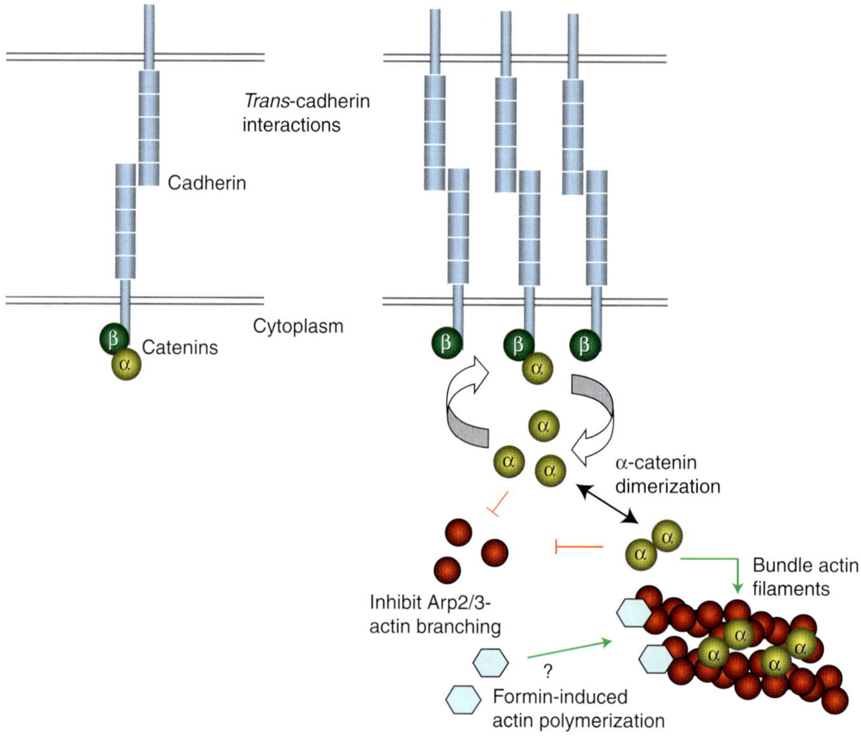

Figure 6. Model of α-catenin function in actin polymerization and reorganization in developing cell–cell contacts. See text for details. From Pokutta et al. 2008.

(Drees et al. 2005). The ability of α-catenin to bundle actin (Rimm et al. 1995) may indicate that it also has a role in reorganizing actin into the bundles observed in mature contacts. Moreover, it has been reported that formin-1, which promotes formation of linear actin cables, can be positively regulated by α-catenin (Kobielak et al. 2004). Thus, α-catenin might actively mediate the switch between Arp2/3-stimulated branched actin polymerization and formin-simulated linear actin cable assembly.

As noted previously, Lecuit and colleagues found a portion of actin associated with immobile cadherin clusters, but this association is independent of α-catenin. In contrast, α-catenin is important for the dynamic architecture of the junctions (Cavey et al. 2008). These authors point out that this "two-tiered" mechanism provides mechanical linkage between adhesion molecules and actin, while allowing dynamic remodeling of the junctions. Because α-catenin does not seem to mediate stable linkage between cadherin and actin, it is likely that other proteins fulfill this role. Although there are candidates (see Weis and Nelson 2006), the components of these immobile clusters are not known. Identifying these molecules will be important to furthering our understanding of adherens junctions and their role in providing mechanical strength to tissues as well as tissue dynamics.

CONCLUSIONS

Vertebrate classical cadherin ectodomains mediate intercellular adhesion by stereotyped strand-swap binding in their membrane-distal EC1 domains. Differences in the details of the strand-swap interface likely accounts for the lack of binding between type I and type II cadherins. Sequence analysis suggests that desmosomal cadherins will share this binding mode, but other cadherins—including all cadherins from invertebrates—likely bind in a different manner.

Clustering of classical cadherin ectodomains play an important role in junction assembly, in part by providing a high local concentration of catenins on the intracellular side of the contact. The chain of binary interactions between cadherin and β-catenin, between β-catenin and α-catenin, and α-catenin and F-actin were thought to indicate that the catenins stably link clustered cadherins to actin, but recent biochemical data indicate that α-catenin interactions with actin are negatively regulated by β-catenin. Instead of serving as a passive linker between the cadherin–β-catenin complex and actin, it appears that α-catenin has an active role in organizing actin near the membrane. The precise role of α-catenin and also p120 in organizing and regulating the cytoskeleton at developing and mature contacts remains an important challenge.

ACKNOWLEDGMENTS

Work in the authors' laboratories is supported by grants GM56169 (W.I.W.) and GM062270 (L.S.) from the National Institutes of Health.

REFERENCES

Abe K, Takeichi M. 2008. EPLIN mediates linkage of the cadherin–catenin complex to F-actin and stabilizes the circumferential actin belt. *Proc Natl Acad Sci* **105:** 13–19.

Abe K, Chisaka O, Van Roy F, Takeichi M. 2004. Stability of dendritic spines and synaptic contacts is controlled by α-N-catenin. *Nat Neurosci* **7:** 357–363.

Aberle H, Schwartz H, Hoschuetzky H, Kemler R. 1996. Single amino acid substitutions in proteins of the *armadillo* gene family abolish their binding to α-catenin. *J Biol Chem* **271:** 1520–1526.

Adams CL, Nelson WJ, Smith SJ. 1996. Quantitative analysis of cadherin-catenin-actin reorganization during development of cell-cell adhesion. *J Cell Biol* **135:** 1899–1911.

Adams CL, Chen Y-T, Smith SJ, Nelson WJ. 1998. Mechanisms of epithelial cell-cell adhesion and cell compaction revealed by high-resolution tracking of E-cadherin-green fluorescent protein. *J Cell Biol* **142:** 1105–1119.

Al-Amoudi A, Frangakis AS. 2008. Structural studies on desmosomes. *Biochem Soc Trans* **36:** 181–187.

Al-Amoudi A, Dubochet J, Norlen L. 2005. Nanostructure of the epidermal extracellular space as observed by cryo-electron microscopy of vitreous sections of human skin. *J Invest Dermatol* **124:** 764–777.

Al-Amoudi A, Diez DC, Betts MJ, Frangakis AS. 2007. The molecular architecture of cadherins in native epidermal desmosomes. *Nature* **450:** 832–837.

Alattia JR, Ames JB, Porumb T, Tong KI, Heng YM, Ottensmeyer P, Kay CM, Ikura M. 1997. Lateral self-assembly of E-cadherin directed by cooperative calcium binding. *FEBS Lett* **417:** 405–408.

Anastasiadis PZ. 2007. p120-ctn: A nexus for contextual signaling via Rho GTPases. *Biochim Biophys Acta* **1773:** 34–46.

Anastasiadis PZ, Reynonds AB. 2000. The p120 catenin family: Complex roles in adhesion, signaling and cancer. *J Cell Sci* **113:** 1319–1334.

Bakolitsa C, de Pereda JM, Bagshaw CR, Critchley DR, Liddington RC. 1999. Crystal structure of the vinculin tail suggests a pathway for activation. *Cell* **99:** 603–613.

Bayas MV, Leung A, Evans E, Leckband D. 2006. Lifetime measurements reveal kinetic differences between homophilic cadherin bonds. *Biophys J* **90:** 1385–1395.

Bekirov IH, Needleman LA, Zhang W, Benson DL. 2002. Identification and localization of multiple classic cadherins in developing rat limbic system. *Neuroscience* **115:** 213–227.

Bennett MJ, Schlunegger MP, Eisenberg D. 1995. 3D domain swapping: A mechanism for oligomer assembly. *Protein Sci* **4:** 2455–2468.

Boggon T, Murray J, Chappuis-Flament S, Wong E, Gumbiner B, Shapiro L. 2002. C-cadherin ectodomain structure and implications for cell adhesion mechanisms. *Science* **296:** 1308–1313.

Brembeck FH, Schwarz-Romond Tl., Bakkers J, Wilhelm S, Hammerschmidt M, Birchmeier W. 2004. Essential role of BCL9–2 in the switch between β-catenin's adhesive and transcriptional functions. *Genes Dev* **18:** 2225–2230.

Castaño J, Raurell I, Piedra JA, Miravet S, Duñach M, García de Herreros A. 2002. β-Catenin N- and C-terminal tails modulate the coordinated binding of adherens junction proteins to β-catenin. *J Biol Chem* **277:** 31541–31550.

Cavey M, Lecuit T. 2009. Molecular bases of cell–cell junctions stability and dynamics. *Cold Spring Harb Perspect Biol* **1:** a002998.

Cavey M, Rauzi M, Lenne PF, Lecuit T. 2008. A two-tiered mechanism for stabilization and immobilization of E-cadherin. *Nature* **453:** 751–756.

Chen Y-T, Stewart DB, Nelson WJ. 1999. Coupling assembly of the E-cadherin/b-catenin complex to efficient endoplasmic reticulum exit and basal-lateral membrane targeting of E-cadherin in polarized MDCK cells. *J Cell Biol* **144:** 687–699.

Chen CP, Posy S, Ben-Shaul A, Shapiro L, Honig BH. 2005. Specificity of cell-cell adhesion by classical cadherins: Critical role for low-affinity dimerization through β-strand swapping. *Proc Natl Acad Sci* **102:** 8531–8536.

Chitaev NA, Troyanovsky SM. 1997. Direct Ca^{2+}-dependent heterophilic interaction between desmosomal cadherins, desmoglein and desmocollin, contributes to cell-cell adhesion. *J Cell Biol* **138:** 193–201.

Chitaev NA, Averbakh AZ, Troyanovsky RB, Troyanovsky SM. 1998. Molecular organization of the desmoglein-plakoglobin complex. *J Cell Sci* **111:** 1941–1949.

Choi H-J, Weis WI. 2005. Structure of the armadillo repeat domain of plakophilin 1. *J Mol Biol* **346:** 367–376.

Choi H-J, Huber AH, Weis WI. 2006. Thermodynamics of β-catenin–ligand interactions. The roles of the N- and C-terminal tails in modulating binding affinity. *J Biol Chem* **281:** 1027–1038.

Chu YS, Thomas WA, Eder O, Pincet F, Perez E, Thiery JP, Dufour S. 2004. Force measurements in E-cadherin-mediated cell doublets reveal rapid adhesion strengthened by actin cytoskeleton remodeling through Rac and Cdc42. *J Cell Biol* **167:** 1183–1194.

Cox RT, Pai LM, Kirkpatrick C, Stein J, Peifer M. 1999. Roles of the C-terminus of armadillo in Wingless signaling in *Drosophila*. *Genetics* **153:** 319–332.

Daniel JM, Renynolds AB. 1995. The tyrosine kinase substrate p120^{cas} binds directly to E-cadherin but not to the Adenomatous Polyposis Coli protein or α-catenin. *Mol Cell Biol* **15:** 4819–4824.

Davis MA, Ireton RC, Reynolds AB. 2003. A core function for p120-catenin in cadherin turnover. *J Cell Biol* **163:** 525–534.

Delva E, Tucker DK, Kowalczyk AP. 2009. The desmosome. *Cold Spring Harb Perspect Biol* **1:** a002543.

Drees F, Pokutta S, Yamada S, Nelson WJ, Weis WI. 2005. α-Catenin is a molecular switch that binds E-cadherin/β-catenin and regulates actin filament assembly. *Cell* **123:** 903–915.

Duguay D, Foty RA, Steinberg MS. 2003. Cadherin-mediated cell adhesion and tissue segregation: Qualitative and quantitative determinants. *Dev Biol* **253:** 309–323.

Dyson HJ, Wright PE. 2005. Intrinsically unstructured proteins and their functions. *Nat Rev Mol Cell Biol* **6:** 197–208.

Ehrlich JS, Hansen MD, Nelson WJ. 2002. Spatio-temporal regulation of Rac1 localization and lamellipodia dynamics during epithelial cell-cell adhesion. *Dev Cell* **3:** 259–270.

Garrod DR. 1986. Desmosomes, cell adhesion molecules and the adhesive properties of cells in tissues. *J Cell Sci Suppl* **4:** 221–237.

Geiger B, Volk T, Volberg T, Bendori R. 1987. Molecular interactions in adherens-type contacts. *J Cell Sci Suppl* **8:** 251–272.

Gottardi CJ, Gumbiner BM. 2004. Distinct molecular forms of β-catenin are targeted to adhesive or transcriptional complexes. *J Cell Biol* **167:** 339–349.

Gumbiner BM. 1996. Cell adhesion: The molecular basis of tissue architecture and morphogenesis. *Cell* **84:** 345–357.

Gumbiner BM. 2005. Regulation of cadherin-mediated adhesion in morphogenesis. *Nat Rev Mol Cell Biol* **6:** 622–634.

Hajra KM, Fearon ER. 2002. Cadherin and catenin alterations in human cancer. *Genes Chromosomes Cancer* **34:** 255–268.

Harrison O, Corps E, Kilshaw P. 2005. Cadherin adhesion depends on a salt bridge at the N-terminus. *J Cell Sci* **118:** 4123–4130.

Haussinger D, Ahrens T, Aberle T, Engel J, Stetefeld J, Grzesiek S. 2004. Proteolytic E-cadherin activation followed by solution NMR and X-ray crystallography. *EMBO J* **23:** 1699–1708.

Haussinger D, Ahrens T, Sass H, Pertz O, Engel J, Grzesiek S. 2002. Calcium-dependent homoassociation of E-cadherin by NMR spectroscopy: Changes in mobility, conformation and mapping of contact regions. *J Mol Biol* **324:** 823–839.

He W, Cowin P, Stokes DL. 2003. Untangling desmosomal knots with electron tomography. *Science* **302:** 109–113.

Hinck L, Nathke IS, Papkoff J, Nelson WJ. 1994. Dynamics of cadherin/catenin complex formation: Novel protein interactions and pathways of complex assembly. *J Cell Biol* **125:** 1327–1340.

Honjo M, Tanihara H, Suzuki S, Tanaka T, Honda Y, Takeichi M. 2000. Differential expression of cadherin adhesion receptors in neural retina of the postnatal mouse. *Invest Ophthalmol Vis Sci* **41:** 546–551.

Hoshino T, Sakisaka T, Baba T, Yamada T, Kimura T, Takai Y. 2005. Regulation of E-cadherin endocytosis by nectin through afadin, Rap1, and p120ctn. *J Biol Chem* **280:** 24095–24103.

Huber AH, Weis WI. 2001. The structure of the β-catenin/E-cadherin complex and the molecular basis of diverse ligand recognition by β-catenin. *Cell* **105:** 391–402.

Huber AH, Nelson WJ, Weis WI. 1997. Three-dimensional structure of the armadillo repeat region of β-catenin. *Cell* **90:** 871–882.

Huber AH, Stewart DB, Laurents DV, Nelson WJ, Weis WI. 2001. The cadherin cytoplasmic domain is unstructured in the absence of β-catenin: A possible mechanism for regulating cadherin turnover. *J Biol Chem* **276:** 12301–12309.

Ireton RC, Davis MA, van Hengel J, Mariner DJ, Barnes K, Thoreson MA, Anastasiadis PZ, Matrisian L, Bundy LM, Sealy L et al. 2002. A novel role for p120 catenin in E-cadherin function. *J Cell Biol* **159:** 465–476.

Iwai Y, Usui T, Hirano S, Steward R, Takeichi M, Uemura T. 1997. Axon patterning requires DN-cadherin, a novel neuronal adhesion receptor, in the *Drosophila* embryonic CNS. *Neuron* **19:** 77–89.

Jannsen ME, Kim E, Liu H, Fujimoto LM, Bobkov AA, Volkmann N, Hanein D. 2006. Three-dimensional structure of vinculin bound to actin filaments. *Mol Cell* **21:** 271–281.

Kitagawa M, Natori M, Murase S, Hirano S, Taketani S, Suzuki S. 2000. Mutation analysis of cadherin-4 reveals amino acid residues of EC1 important for the structure and function. *Biochem Biophys Res Commun* **271:** 358–363.

Klingelhofer J, Laur OY, Troyanovsky RB, Troyanovsky SM. 2002. Dynamic interplay between adhesive and lateral E-cadherin dimers. *Mol Cell Biol* **22:** 7449–7458.

Kobielak A, Fuchs E. 2004. α-catenin: At the junction of intercellular adhesion and actin dynamics. *Nat Rev Mol Cell Biol* **5:** 614–625.

Kobielak A, Pasolli HA, Fuchs E. 2004. Mammalian formin-1 participates in adherens junctions and polymerization of linear actin cables. *Nat Cell Biol* **6:** 21–30.

Koch AW, Pokutta S, Lustig A, Engel J. 1997. Calcium binding and homoassociation of E-cadherin domains. *Biochemistry* **36:** 7697–7705.

Koslov ER, Maupin P, Pradhan D, Morrow JS, Rimm DL. 1997. α-catenin can form asymmetric homodimeric complexes and/or heterodimeric complexes with β-catenin. *J Biol Chem* **272:** 27301–27306.

Kovacs EM, Goodwin M, Ali RG, Paterson AD, Yap AS. 2002. Cadherin-directed actin assembly: E-cadherin physically associates with the Arp2/3 complex to direct actin assembly in nascent adhesive contacts. *Curr Biol* **12:** 379–382.

Lickert H, Bauer A, Kemler R, Stappert J. 2000. Casein Kinase II phosphorylation of E-cadherin Increases E-cadherin/β-catenin interaction and strengthens cell–cell adhesion. *J Biol Chem* **275:** 5090–5095.

Lien WH, Klezovitch O, Fernandez TE, Delrow J, Vasioukhin V. 2006. a-E-catenin controls cerebral cortical size by regulating the hedgehog signaling pathway. *Science* **311:** 1609–1612.

Liu Y, Eisenberg D. 2002. 3D domain swapping: As domains continue to swap. *Protein Sci* **11:** 1285–1299.

May C, Doody JF, Abdullah R, Balderes P, Xu X, Chen CP, Zhu Z, Shapiro L, Kussie P, Hicklin DJ et al. 2005. Identification of a transiently exposed VE-cadherin epitope that allows for specific targeting of an antibody to the tumor neovasculature. *Blood* **105:** 4337–4344.

Meng W, Takeichi M. 2009. Adherens junction: Molecular architecture and regulation. *Cold Spring Harb Perspect Biol* **1:** a002899.

Nagafuchi A, Takeichi M, Tsukita S. 1991. The 102 kD cadherin-associated protein: Similarity to vinculin and posttranscriptional regulation of expression. *Cell* **65:** 849–857.

Nagar B, Overduin M, Ikura M, Rini JM. 1996. Structural basis of calcium-induced E-cadherin rigidification and dimerization. *Nature* **380:** 360–364.

Niessen CM, Gumbiner BM. 2002. Cadherin-mediated cell sorting not determined by binding or adhesion specificity. *J Cell Biol* **156:** 389–399.

Nishimura EK, Yoshida H, Kunisada T, Nishikawa SI. 1999. Regulation of E- and P-cadherin expression correlated with melanocyte migration and diversification. *Dev Biol* **215:** 155–166.

Nollet F, Kools P, van Roy F. 2000. Phylogenetic analysis of the cadherin superfamily allows identification of six major subfamilies besides several solitary members. *J Mol Biol* **299:** 551–572.

Nose A, Tsuji K, Takeichi M. 1990. Localization of specificity determining sites in cadherin cell adhesion molecules. *Cell* **61:** 147–155.

Oda T, Kanai Y, Oyama K, Yoshiura K, Shimoyama Y, Birchmeier W, Sugimura T, Hirohashi S. 1994. E-cadherin gene mutations in human gastric carcinoma cell lines. *Proc Natl Acad Sci* **91:** 1858–1862.

Ohkubo T, Ozawa M. 1999. p120^{ctn} binds to the membrane-proximal region of the E-cadherin cytoplasmic domain and is involved in modulation of adhesion activity. *J Biol Chem* **274:** 21409–21415.

Overduin M, Harvey TS, Bagby S, Tong KI, Yau P, Takeichi M, Ikura M. 1995. Solution structure of the epithelial cadherin domain responsible for selective cell adhesion. *Science* **267:** 386–389.

Ozawa M, Kemler R. 1992. Molecular organization of the uvomorulin-catenin complex. *J Cell Biol* **116:** 989–996.

Ozawa M, Kemler R. 1990. Correct proteolytic cleavage is required for the cell adhesion function of uvomorulin. *J Cell Biol* **111:** 1645–1650.

Ozawa M, Kemler R. 1998. Altered cell adhesion activity by pervanadate due to the dissociation of a-catenin from the E-cadherin catenin complex. *J Biol Chem* **273:** 6166–6170.

Ozawa M, Baribault H, Kemler R. 1989. The cytoplasmic domain of the cell adhesion molecule uvomorulin associates with three independent proteins structurally related in different species. *EMBO J* **8:** 1711–1717.

Ozawa M, Ringwald M, Kemler R. 1990. Uvomorulin-catenin complex formation is regulated by a specific domain in the cytoplasmic region of the cell adhesion molecule. *Proc Natl Acad Sci* **87:** 4246–4250.

Parisini E, Higgins JM, Liu JH, Brenner MB, Wang JH. 2007. The Crystal Structure of Human E-cadherin Domains 1 and 2, and Comparison with other Cadherins in the Context of Adhesion Mechanism. *J Mol Biol* **373:** 401–411.

Patel SD, Chen CP, Bahna F, Honig B, Shapiro L. 2003. Cadherin-mediated cell-cell adhesion: Sticking together as a family. *Curr Opin Struct Biol* **13:** 690–698.

Patel SD, Ciatto C, Chen CP, Bahna F, Rajebhosale M, Arkus N, Schieren I, Jessell TM, Honig B, Price SR, et al. 2006. Type II cadherin ectodomain structures: Implications for classical cadherin specificity. *Cell* **124:** 1255–1268.

Pertz O, Bozic D, Koch AW, Fauser C, Brancaccio A, Engel J. 1999. A new crystal structure, Ca^{2+} dependence and mutational analysis reveal molecular details of E-cadherin homoassociation. *EMBO J* **18:** 1738–1747.

Piedra J, Martínez D, Castaño J, Miravet S, Duñach M, García de Herreros A. 2001. Regulation of β-catenin structure and activity by tyrosine phosphorylation. *J Biol Chem* **276:** 20436–20443.

Pokutta S, Weis WI. 2000. Structure of the dimerization and β-catenin binding region of α-catenin. *Mol Cell* **5:** 533–543.

Pokutta S, Weis WI. 2007. Structure and mechanism of cadherins and catenins in cell-cell contacts. *Annu Rev Cell Dev Biol* **23:** 237–261.

Pokutta S, Drees F, Yamada S, Nelson WJ, Weis WI. 2008. Biochemical and structural analysis of alpha-catenin in cell–cell contacts. *Biochem Soc Trans* **36:** 141–147.

Pokutta S, Drees F, Takai Y, Nelson WJ, Weis WI. 2002. Biochemical and structural definition of the l-afadin- and actin-binding sites of α-catenin. *J Biol Chem* **277:** 18868–18874.

Pokutta S, Herrenknecht K, Kemler R, Engel J. 1994. Conformational changes of the recombinant extracellular domain of E-cadherin upon calcium binding. *Eur J Biochem* **223:** 1019–1026.

Pollard TD, Borisy GG. 2003. Cellular motility driven by assembly and disassembly of actin filaments. *Cell* **112:** 453–465.

Posy S, Shapiro L, Honig B. 2008. Sequence and structural determinants of strand swapping in cadherin domains:

Do all cadherins bind through the same adhesive interface? *J Mol Biol* **378:** 954–968.

Price SR, De Marco Garcia NV, Ranscht B, Jessell TM. 2002. Regulation of motor neuron pool sorting by differential expression of type II cadherins. *Cell* **109:** 205–216.

Radice GL, Rayburn H, Matsunami H, Knudsen KA, Takeichi M, Hynes RO. 1997. Developmental defects in mouse embryos lacking N-cadherin. *Dev Biol* **181:** 64–78.

Rimm DL, Koslov ER, Kebriaei P, Cianci CD, Morrow JS. 1995. α_1(E)-catenin is an actin-binding and -bundling protein mediating the attachment of F-actin to the membrane adhesion complex. *Proc Natl Acad Sci* **92:** 8813–8817.

Roura S, Miravet S, Piedra J, García de Herreros A, Dunach M. 1999. Regulation of E-cadherin/catenin association by tyrosine phosphorylation. *J Biol Chem* **274:** 36734–36740.

Shan WS, Koch A, Murray J, Colman DR, Shapiro L. 1999. The adhesive binding site of cadherins revisited. *Biophys Chem* **82:** 157–163.

Shan W, Tanaka H, Phillips G, Arndt K, Yoshida M, Colman D, Shapiro L. 2000. Functional cis-heterodimers of N- and R-cadherins. *J Cell Biol* **148:** 579–590.

Shapiro L, Fannon AM, Kwong PD, Thompson A, Lehmann MS, Grubel G, Legrand J-F, Als-Nielsen J, Colman DR, Hendrickson WA. 1995a. Structural basis of cell-cell adhesion by cadherins. *Nature* **374:** 327–337.

Shapiro L, Kwong PD, Fannon AM, Colman DR, Hendrickson WA. 1995b. Considerations on the folding topology and evolutionary origin of cadherin domains. *Proc Natl Acad Sci USA* **92:** 6793–6797.

Shimoyama Y, Tsujimoto G, Kitajima M, Natori M. 2000. Identification of three human type-II classic cadherins and frequent heterophilic interactions between different subclasses of type-II classic cadherins. *Biochem J* **349:** 159–167.

Sivasankar S, Gumbiner B, Leckband D. 2001. Direct measurements of multiple adhesive alignments and unbinding trajectories between cadherin extracellular domains. *Biophys J* **80:** 1758–1768.

Steinberg MS. 1962. On the Mechanism of Tissue Reconstruction by Dissociated Cells, Iii. Free Energy Relations and the Reorganization of Fused, Heteronomic Tissue Fragments. *Proc Natl Acad Sci* **48:** 1769–1776.

Steinberg MS. 1975. Adhesion-guided multicellular assembly: A commentary upon the postulates, real and imagined, of the differential adhesion hypothesis, with special attention to computer simulations of cell sorting. *J Theor Biol* **55:** 431–443.

Steinberg MS. 2007. Differential adhesion in morphogenesis: A modern view. *Curr Opin Genet Dev* **17:** 281–286.

Stepniak E, Radice GL, Vasioukhin V. 2009. Adhesive and signaling functions of cadherins and catenins in vertebrate development. *Cold Spring Harb Perspect Biol* **1:** a002949.

Takai Y, Ikeda W, Ogita H, Rikitake Y. 2008. The immunoglobulin-like cell adhesion molecule nectin and its associated protein afadin. *Annu Rev Cell Dev Biol* **24:** 309–342.

Takeichi M. 1977. Functional correlation between cell adhesive properties and some cell surface proteins. *J Cell Biol* **75:** 464–474.

Takeichi M. 1988. The cadherins: Cell-cell adhesion molecules controlling animal morphogenesis. *Development* **102:** 639–655.

Takeichi M. 1990. Cadherins: A molecular family important in selective cell-cell adhesion. *Annu Rev Biochem* **59:** 237–252.

Takeichi M, Hatta K, Nose A, Nagafuchi A. 1998. Identification of a gene family of cadherin cell adhesion molecules. *Cell Differ Dev* **25:** 91–94.

Takeichi M. 1995. Morphogenetic roles of classic cadherins. *Curr Opin Cell Biol* **7:** 619–627.

Takeichi M, Atsumi T, Yoshida C, Uno K, Okada TS. 1981. Selective adhesion of embyonal carcinoma cells and differentiated cells by Ca^{2+}-dependent sites. *Dev Biol* **87:** 340–350.

Tamura K, Shan W, Hendrickson W, Colman D, Shapiro L. 1998. Structure-function analysis of cell adhesion by neural (N-) cadherin. *Neuron* **20:** 1153–1163.

Taveau JC, Dubois M, Le Bihan O, Trepout S, Almagro S, Hewat E, Durmort C, Heyraud S, Gulino-Debrac D, Lambert O. 2008. Structure of artificial and natural VE-cadherin-based adherens junctions. *Biochemical Society transactions* **36:** 189–193.

Tepass U. 1999. Genetic analysis of cadherin function in animal morphogenesis. *Curr Opin Cell Biol* **11:** 540–548.

Thoreson MA, Anastasiadis PZ, Daniel JM, Ireton RC, Wheelock MJ, Johnson KR, Hummingbird DK, Reynolds AB. 2000. Selective uncoupling of $p120^{ctn}$ from E-cadherin disrupts strong adhesion. *J Cell Biol* **148:** 189–201.

Tompa P. 2005. The interplay between structure and function in intrinsically unstructured proteins. *FEBS Lett* **579:** 3346–3354.

Troyanovsky S. 2005. Cadherin dimers in cell-cell adhesion. *Eur J Cell Biol* **84:** 225–233.

Troyanovsky R, Sokolov E, Troyanovsky S. 2003. Adhesive and lateral E-cadherin dimers are mediated by the same interface. *Mol Cell Biol* **23:** 7965–7972.

Vaezi A, Bauer C, Vasioukhin V, Fuchs E. 2002. Actin cable dynamics and Rho/Rock orchestrate a polarized cytoskeletal architecture in the early steps of assembling a stratified epithelium. *Dev Cell* **3:** 367–381.

Vasioukhin V, Bauer C, Begenstein L, Wise B, Fuchs E. 2001. Hyperproliferation and defects in epithelial polarity upon conditional ablation of α-catenin in skin. *Cell* **104:** 605–617.

Vasioukhin V, Bauer C, Yin M, Fuchs E. 2000. Directed actin polymerization is the driving force for epithelial cell-cell adhesion. *Cell* **100:** 209–219.

Volk T, Cohen O, Geiger B. 1987. Formation of heterotypic adherens-type junctions between L-CAM-containing liver cells and A-CAM-containing lens cells. *Cell* **50:** 987–994.

Weis WI, Nelson WJ. 2006. Re-solving the cadherin-catenin-actin conundrum. *J Biol Chem* **281:** 35593–35597.

Xiao K, Allison DF, Buckley KM, Kottke MD, Vincent PA, Faundez V, Kowalczyk AP. 2003. Cellular levels of p120 catenin function as a set point for cadherin expression levels in microvascular endothelial cells. *J Cell Biol* **163:** 535–545.

Xiao K, Oas RG, Chiasson CM, Kowalczyk AP. 2007. Role of p120-catenin in cadherin trafficking. *Biochim Biophys Acta* **1773:** 8–16.

Xing Y, Takemaru K, Liu J, Berndt JD, Zheng JJ, Moon RT, Xu W. 2008. Crystal structure of a full-length β-catenin. *Structure* **16:** 478–487.

Yamada S, Pokutta S, Drees F, Weis WI, Nelson WJ. 2005. Deconstructing the cadherin-catenin-actin complex. *Cell* **123:** 889–901.

Yang J, Dokurno P, Tonks NK, Barford D. 2001. Crystal structure of the M-fragment of α-catenin: Implications for modulation of cell adhesion. *EMBO J* **20:** 3645–3656.

Yap AS, Niessen CM, Gumbiner BM. 1998. The juxtamembrane region of the cadherin cytoplasmic tail supports lateral clustering, adhesive strengthening, and interaction with p120^{ctn}. *J Cell Biol* **141:** 779–789.

Zhang Y, Sivasankar S, Nelson WJ, Chu S. 2009. Resolving cadherin interactions and binding cooperativity at the single-molecule level. *Proc Natl Acad Sci* **106:** 109–114.

Zhu B, Chappuis-Flament S, Wong E, Jensen IE, Gumbiner BM, Leckband D. 2003. Functional analysis of the structural basis of homophilic cadherin adhesion. *Biophys J* **84:** 4033–4042.

Physiology and Function of the Tight Junction

James M. Anderson[1] and Christina M. Van Itallie[2]

[1]Department of Cell and Molecular Physiology, University of North Carolina at Chapel Hill, Chapel Hill, North Carolina 27599-7545

[2]Department of Medicine, Division of Gastroenterology and Hepatology, University of North Carolina at Chapel Hill, Chapel Hill, North Carolina 27599-7545

Correspondence: jandersn@med.unc.edu

Understanding of tight junctions has evolved from their historical perception as inert solute barriers to recognition of their physiological and biochemical complexity. Many proteins are specifically localized to tight junctions, including cytoplasmic actin-binding proteins and adhesive transmembrane proteins. Among the latter are claudins, which are critical barrier proteins. Current information suggests that the paracellular barrier is most usefully modeled as having two physiologic components: a system of charge-selective small pores, 4 Å in radius, and a second pathway created by larger discontinuities in the barrier, lacking charge or size discrimination. The first pathway is influenced by claudin expression patterns and the second is likely controlled by different proteins and signals. Recent information on claudin function and disease-causing mutations have led to a more complete understanding of their role in barrier formation, but progress is impeded by lack of high resolution structural information.

Tight junctions form the continuous intercellular barrier between epithelial cells, which is required to separate tissue spaces and regulate selective movement of solutes across the epithelium. Although there are now >40 proteins (Schneeberger and Lynch 2004; Yamazaki et al. 2008) identified within the tight junction, the claudin family of transmembrane proteins, named from the Latin *claudere* to close, has emerged as the most critical for defining tight junction selectivity. Here, we review evidence that claudins regulate permselectivity (including size, electrical resistance, and ionic charge preference) derived from studies in cultured epithelial cell models and the phenotypes of knockout mice and human mutants. We highlight the physiologic relevance of selectivity but only briefly discuss how it might be physiologically regulated and altered in pathologic situations. We develop the perspective that the barrier is usefully described as having two pathways: first a system of charge-selective claudin-based pores that are 4 Å in radius and a second pathway created by larger discontinuities in the barrier and that lacks charge and size discrimination. The two pathways may be controlled by different proteins and signals. This article focuses on claudins and physiology and is meant to be read as a companion to the article in this collection contributed by

 doi: 10.1101/cshperspect.a002584
Cite this article as *Cold Spring Harb Perspect Biol* 2009;1:a002584

M. Furuse, which focuses on the molecular structure, proteins, and cell biology of the tight junction (Furuse 2009). The reader is also referred to comprehensive reviews on physiology (Diamond 1978; Powell 1981; Van Itallie and Anderson 2006), pathophysiology (Turner 2006; Schmitz et al. 1999; Nusrat et al. 2001), regulation (Tsukita et al. 2008; Gonzalez-Mariscal et al. 2008), and molecular components of the junction (Schneeberger and Lynch 2004; Krause et al. 2008; Gonzalez-Mariscal et al. 2003).

EVOLUTION OF IDEAS ABOUT PARACELLULAR TRANSPORT AND THE RELEVANCE OF TIGHT JUNCTION PERMSELECTIVITY

Explicit description of a sealing contact between epithelial cells can first be found in the biologic literature in the latter part of the 19th century (Cereijido and Anderson 2001). Staining of epithelial tissues, such as the intestine, with vital dyes revealed a distinct region at the apical end of the lateral cell interspaces referred to as the "terminal bar." This was thought to be an absolute barrier preventing anything from passing between cells. By the early 20th century, studies began to acknowledge that some material, for example macrophages and water, could cross epithelia through the paracellular space.

Our current understanding of how material selectively crosses epithelia originates with the pioneering work of Hans Ussing and associates, beginning in the 1940s. Ussing addressed the question of how sodium ions are moved in a directional fashion across the epithelium of frog skin, resulting in a steady-state electrical potential across the epithelium (Koefeld-Johnsen and Ussing 1958; Ussing and Zerahn 1951). Their first conceptual breakthrough, the so-called "two membrane model," stated that the apical and basal membrane surfaces had different conductance properties, namely, that sodium enters the cell across the apical membrane down its concentration gradient (later shown to be through Na channels) but was transported in an energy-dependent fashion out the basal lateral surface (later shown to be by the NaK-ATPase). Ironically, they chose to study an epithelium with one of the electrically tightest junctions in nature, leading them initially to discount the possibility of ion movements between cells. Subsequent inconsistencies between their model and the data led to the realization that Cl^- ions must be following Na^+ in a passive fashion through the tight junction (based on electrical circuit modeling, they called this the shunt pathway) coupled to the electrical gradient generated by active Na^+ transport (Ussing and Windhager 1964). This was the beginning of understanding about how transcellular and paracellular transport are physiologically coupled (Fig. 1).

In the 1960s, investigators began to study transport across other epithelia, such as the gallbladder, where the paracellular resistance was much lower than in frog skin (reviewed in Diamond 1977). Because the transepithelial electrical resistance (TER) was so much lower than frog skin, there was initial concern that these tissues were damaged during experimental preparation and this explained their low resistance. Gradually, it was accepted that intact normal epithelia display a wide range of electrical resistances (Table 1). Because the cell membranes are generally of very high resistance

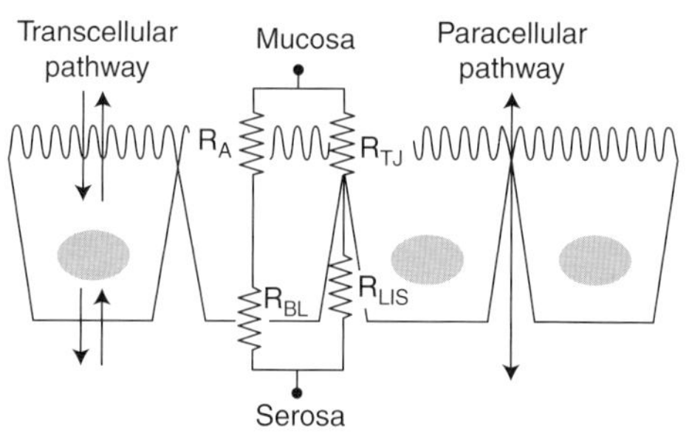

Figure 1. Electrical circuit model of the series and paracellular resistances across trans- and paracellular pathways of an epithelial cell monolayer. Transcellular transport is controlled by transporters in the apical and basolateral surfaces. The resistance of these series elements is typically much higher than that of the parallel elements of the paracellular pathway. Thus, the overall resistance of an epithelium is defined by R_{TJ}, which is defined by the composition of claudins in the tight junction (TJ).

Table 1. Electrical characteristics of some epithelia and endothelia

Tissues[a]	Species	R_{cell} $\Omega \times cm^2$	$R_{paracellular}$ $\Omega \times cm^2$	P_{Na}/P_{Cl}[b]
Proximal tubule	dog	–	6–7	1.4
Gallbladder	rabbit	229	21	3.3
Duodenum	rat	–	98	–
Jejunum	rat	67	51	10.0
Ileum	rabbit	115	100	2.5
Distal colon	rabbit	730	385	0.6
	mouse surface	132	3,200	–
	crypt	429	–	–
Brain endothelium		1,300		
Gastric fundus	Necturus	2,826	10,573	–
Urinary bladder	rabbit	160,000	300,000	–
Cell lines[c]				
Caco-2	human colon	125–250	–	3.0
LLC-PK_1	pig prox. tubule	100	–	0.6
MDCK	dog	60–4000	–	10.0

[a]All values can be found in Powell (Powell 1981) or Crone (Crone and Christensen 1981).

[b]Permeability ratio of Na^+ versus Cl^-. P_{Na}/P_{Cl} in free solute is 0.66. Paracellular pathways with ratios above this value are more permeable for Na^+ than Cl^-, i.e., cation-selective.

[c]Values for cell lines are the personal observations of Dr. C. Van Itallie. Modified, with permission, from Van Itallie and Anderson (2004) (© American Thoracic Society).

(R_{apical} and $R_{basolateral}$), it is the TJ (R_{TJ}) that determines whether the transepithelial resistance is high or low (Fig. 1). Theoretically, the lateral intercellular space (R_{lis}) could contribute a resistance in series with the tight junction, but there is little evidence that this is physiologically significant.

TER differs by several orders of magnitude between so-called "tight" and "leaky" epithelia (Table 1). For example, the epithelium of the mammalian proximal tubule is only about 6 $\Omega \cdot cm^2$ compared with the toad urinary bladder at 300,000 $\Omega \cdot cm^2$ (Powell 1981; Van Itallie and Anderson 2006). The relevance of TER is that epithelia with "tight" tight junctions can maintain the high electrochemical gradients produced by active transcellular transport. This configuration is used to produce either highly concentrated or diluted secretions, as in the distal nephron, which can produce urine with osmolarity several fold higher or lower than plasma. In contrast, epithelia with "leaky" tight junctions move large amounts of isosmotic fluids. A good example is provided by the human gastrointestinal tract, which secretes and then reabsorbs about 10 L of fluid each day (Boron and Boulpaep 2005). Most of the gastrointestinal track is of low TER, except in the distal colon where steeper electrochemical gradients are required to reabsorb NaCl and water and form solid stool.

Study of leaky epithelia revealed another interesting and variable property, namely that tight junctions (TJs) have ionic charge selectivity. Charge selectivity is most relevant in leaky tight junctions where higher amounts of ions flow. Almost all leaky TJs show a preference for Na^+ over Cl^-, and the permeability ratio (expressed as P_{Na}/P_{Cl}) ranges from about 10 to 0.1 among different epithelia and experimental cultured cell models (Table 1). This represents only a modest ability to discriminate compared with membrane ion channels. For comparison, some amiloride-sensitive Na^+ channels (ENaC) show a 1000-fold preference for Na^+ over K^+, an identically charged ion of similar size (Hille 2001). However, even a 10-fold cation to anion discrimination has significant physiologic implications. For example, epithelia that secrete salt and water initiated by an apical Cl^- channel (such as CFTR in the airway, lacrimal glands, etc.) are found to have a threefold to 10-fold preference for Na^+ over Cl^-. This allows paracellular passage of Na^+

to follow secreted Cl^- while relatively discriminating against back diffusion of the Cl^-, which would defeat the ability to secrete NaCl and water.

The TJ was first visualized at the ultrastructural level in 1963 by Farquhar and Palade (Farquhar and Palade 1963). They were able for the first time to resolve the apical junction complex (terminal bar) into three morphologically discrete contacts from the apical-most occluding or TJ (zonula occludens) followed by the cadherin-based adherens junction and the more basally positioned desmosomes. Gap junction plaques are also commonly found within the TJ contacts. The TJ zone is formed by a variable number of close cell–cell contacts or "kisses," which range from a single continuous contact (e.g., some endothelial cells) to a half dozen (jejunum) or dozens of contacts in the most extreme mammalian example (Sertoli cells). Subsequent visualization of the tight junction in freeze fracture EM images revealed that membrane kisses correspond to continuous rows of 10-nm transmembrane particles, which are presumed to contain an oligomerized array of claudins. The physical barrier to paracellular diffusion occurs where rows of particles adhere and seal across the intercellular space. The existence of multiple strands is speculated to provide redundancy to the barrier. Although there is no experimental proof for this idea, it seems very reasonable to assume that multiple barriers in series would provide a fail-safe barrier during cell dynamics, such as when cells move relative to each other (Matsuda et al. 2004) or single cells leave the epithelia sheet during apoptosis (Madara et al. 1980). In the early 1970s, Philippe Claude proposed that TER has a logarithmic dependence on the number of strands in series, the so-called Claude hypothesis (Claude 1978). By a quirk of etymology, the Claude hypothesis is no longer accepted but the TER is now thought to depend on the profile of different claudins expressed.

To summarize, the field of TJs before the discovery of claudins was well described at a physiologic level. Comparing epithelia: Cell-specific TJ barriers allow the passage of varying levels of electrical current and noncharged solutes. Based on how paracellular ion selectivity varied with extracellular pH, several insightful investigators in the 1970s had already proposed that the junction was created by pore-forming proteins with variable side-chain chemistries (Wright and Diamond 1968). A major transformation (in fact a resurrection) of the field occurred with discovery of the barrier and pore-forming proteins.

TRANSMEMBRANE TJ PROTEINS AND THE CLAUDIN FAMILY

The TJ has a surprisingly complex protein composition compared with other cell–cell junctions and is composed of at least 40 different proteins (Schneeberger and Lynch 2004; Gonzalez-Mariscal et al. 2003; Yamazaki et al. 2008). This complexity is a consequence of its many interrelated roles in cell polarity (Cereijido et al. 1998), signaling (Gonzalez-Mariscal et al. 2008; Van Itallie and Anderson 2006), transcriptional regulation, and cell cycle (Balda and Matter 2003; Tsukita et al. 2008) and vesicle trafficking (Yeaman et al. 2004), in addition to creating the paracellular barrier. Excellent reviews of the cytoplasmic plaque proteins and nonclaudin transmembrane proteins can be found in Gonzalez-Mariscal et al. (2003) and Schneeberger and Lynch (2004). A recent proteomics study suggests that there are many more transmembrane proteins yet to be characterized (Yamazaki et al. 2008). The role of these other transmembrane proteins remains an active area of investigation; however, all current evidence supports a central role for claudins in defining electrical resistance and permselectivity.

Nonclaudin Transmembrane Proteins

In addition to claudin family members, there are currently three other transmembrane proteins localized within the adhesive barrier strands and which in theory might directly influence the barrier. Occludin (Furuse et al. 1993) and tricellulin (Ikenouchi et al. 2005) are related tetraspanning proteins of currently unknown function. Occludin knockdown (KD) cell lines (Yu et al. 2005) and even knockout (KO) mice have no definable barrier defects (Saitou et al. 2000). KD cell lines reform TJs more slowly after a Ca-switch and have elevated levels of active GTP-bound RhoA (Yu et al. 2005), leading to speculation of a role for occludin in cytoskeletal dynamics. KO mice have a collection of defects, but there are no obvious barrier changes when assessed by TER, steady-state potential difference, electrical impedance, or solute tracer flux (Schulzke et al. 2005). An alternative role for occludin in coordinating transmembrane signaling is suggested by the findings that it binds the TGF-β type III receptor (Barrios-Rodiles et al. 2005) and manipulation of occludin in cultured cells affects Raf-1 signaling (Wang et al. 2005). Human tricellulin mutations result in a non-syndromic form of deafness by an unknown mechanism (Riazuddin et al. 2006). There is a third protein in the human database homologous to occludin and tricellulin (MARVELD3, Genebank ID: 91862, personal observation). Perhaps functional redundancy among these proteins has obscured the role of the individual proteins. Finally, there are several IgG superfamily members within the stand contacts (CAR [Coyne and Bergelson 2005], CLP24 [Kearsey et al. 2004]); the best studied being JAM-A, which appears to stabilize the barrier (Bazzoni 2003). In support of this idea, KD of JAM-A in cultured epithelia monolayers is reported to induce leakiness for large paracellular tracer molecules (Liu et al. 2000). The intestinal epithelium of JAM-A KO mice is extremely sensitive to disruption in inflammatory models (Laukoetter et al. 2007; Vetrano et al. 2008).

The Claudin Family

The first claudins were discovered in 1998 by Furuse and Tsukita through traditional biochemical fractionation of membrane fractions from liver (Furuse et al. 1998). When expressed in claudin-null fibroblasts, claudins formed the characteristic linear strands of 10-nm particles observed in TJs and conferred cell-to-cell adhesion (Kubota et al. 1999). Mammalian claudins range from 20 to 27 kDa and have four transmembrane helices: a short internal amino-terminal sequence (2–6 residues), two extracellular domains (loop 1 is 49–52 residues and loop 2 is 16–33 residues), and a longer and more variable cytoplasmic tail (21–63 residues) (Fig. 2B). Claudins are recognized by the signature residues W-GLW-C-C in the first extracellular loop, although the function of these highly conserved residues remains unknown (Fig. 2A). The carboxy-termini of claudins contain a PDZ-binding motif and several have been shown to bind PDZ domains within the cytoplasmic scaffolding proteins ZO-1,-2, and -3 (Itoh et al. 1999), MUPP1 (Hamazaki et al. 2002; Jeansonne et al. 2003) and PATJ (Roh et al. 2002). ZO-1 has three PDZ domains, MUPP1 13, and PATJ 10, suggesting that there exists a dense Velcro-like trap of PDZ interactions under the claudin strands. However, there is actually no clear evidence that PDZ binding is required for targeting of claudins to the TJ in epithelial cells or for claudins to confer physiologic effects when expressed in cultured cell models. In a single case, when claudin-1 was expressed in cultured MDCK cells without a functional PDZ motif, this resulted in ectopic strands on the lateral cell surface (McCarthy et al. 2000).

Claudins are members of the much larger pfam00822 or PMP-22/EMP/MP20/Claudin family. These claudin relatives share the tetra-spanning topology and W-GLW-C-C signature motif in the first extracellular loop. Beyond these structural similarities, their functions appear highly divergent and only some are believed to create intercellular barriers. The most homologous to claudins are MP20 (eye lens specific membrane protein) (Steele et al. 1998); epithelial membrane proteins (EMP-1, -2, -3) (Jetten and Suter 2000), and peripheral myelin protein 22 (PMP22) (Notterpek et al. 2001). PMP-22 is highly expressed in Schwann cells and required for myelin formation (Bronstein 2000). Several forms of human peripheral polyneuropathies arise from PMP22 mutations, deletions, or gene duplications (Brancolini et al. 2000) (Table 2). Surprisingly,

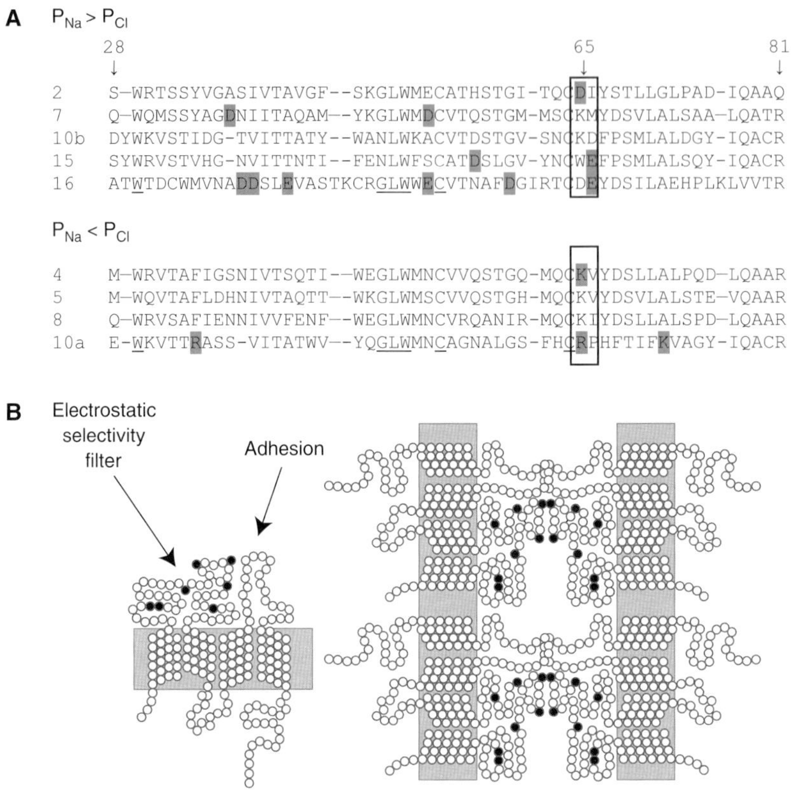

Figure 2. (*A*) The first extracellular loops of several claudins, which have been tested for charge selectivity. Residue numbering is based on the sequence of claudin-2. Numerous shaded positions alter selectivity when mutated. Alternatively, the boxed residues around residue 65 have been suggested to be the most critical in determining selectivity. Signature W-GLW-C-C residues are underlined. (*B*) Conceptual model of the claudin-based TJ barrier. The first extracellular loop contains the electrostatic selectivity filter of the pore and the second loop the cell–cell adhesion sites. In an unknown way, claudin monomers assemble into continuous strands within each cell membrane and adhere across the intercellular space to create a barrier with size and charge-selective pores.

although PMP22 is only 19% identical to human claudin-1, it has been found in tight junctions in liver, intestine (Notterpek et al. 2001), and the blood-brain barrier (Roux et al. 2004), and when expressed in MDCK cells, it increases TER (Roux et al. 2005). More distant members of the pfam00822 family include the subfamily of γ subunits of voltage-dependent calcium channels. These are required for proper membrane delivery of channel complexes (Tomita et al. 2004). One of these, stargazin, which is an AMPA receptor regulator, was recently shown to mediate cell–cell adhesion when expressed in fibroblasts, suggesting some distantly related members of this family may have retained claudin's cell–cell adhesive property (Price et al. 2005).

Study of invertebrate claudins supports a role in barrier formation, although unlike vertebrates, they lack tight junctions. Their epithelial barriers are formed by septate junctions with wide intercellular gaps, very different from the near fusions at tight junctions. *Drosophila* has six claudin sequences, two of which, Megatrachea (Mega) (Behr et al. 2003) and Sinuous (Sinu) (Wu et al. 2004) are located at septate junctions; mutations in Mega disrupt the barrier. Mutations of either claudin also result in developmental defects in the size and shape of the tracheal epithelium.

Table 2. Genetic diseases of tight junction proteins

Gene	Disease	Pathology/Mechanism	Ref.
Cldn-1	Ichthyosis, sclerosing cholangitis	Affects skin and bile ducts	(Hadj-Rabia et al. 2004)
Cldn-14	Nonsyndromic deafness, *DFNB29*	Cochlear hair cell degeneration	(Ben Yosef et al. 2003)
Cldn-16			
Human	HHNC[a]	Defective renal Mg^{++} reabsorption	(Simon et al. 1999)
Bovine	interstitial nephritis		(Hirano et al. 2000)
Cldn-19	Renal Mg^{++} loss and vision loss	Similar to Cldn-16	(Konrad et al. 2006)
PMP22	Peripheral polyneuropathies	Demyelinization	(Gabreels-Festen and Wetering 1999)
	HNPP[b]	Gene deletion	
	Charcot-Marie-Tooth Type 1A	Gene duplication	
	Dejerine-Sottas syndrome	Point mutations	
ZO-2	Familial hypercholanemia	Defective PDZ-claudin binding	(Carlton et al. 2003)
Tricellulin	Nonsyndromic deafness	Loss of ZO-1 binding	(Riazuddin et al. 2006)

[a]Hypomagnesemia hypercalciuria with nephrocalcinosis

[b]Hereditary neuropathy with liability to pressure palsies. Modified, with permission, from Van Itallie and Anderson 2006 (© Annual Reviews www.annualreviews.org).

Five claudin-like sequences have been identified in *Caenorhabditis elegans* (Asano et al. 2003) and RNAi-mediated ablation of claudin-like protein 1 disrupted the barrier between epithelial cells of the hypodermis.

Claudin Cell–Cell Adhesion and Assembly

There remains almost no information about how individual claudins assemble into higher order structures to create 10-nm membrane particles, polymerize into the strands observed in freeze-fracture EM, or adhere across the cell–cell interspace. Currently, two papers suggest claudins, like connexins in the gap junction, form hexamers. This is based on a study of human claudin-4 expressed in insect cells (Mitic et al. 2003) and on purified native MP20, the distant claudin relative found in the lens (Jarvis and Louis 1995). There is consistent evidence that claudins can form homotypic adhesive plaques, and two studies provide evidence that this homotypic cell–cell adhesion can occur through residues in the center of the second extracellular loop (Daugherty et al. 2007; Piontek et al. 2007). Molecular modeling of claudin-5 complemented by mutagenesis suggests that adhesion requires an interaction of complementing hydrophobic residues along a helix in the center of the second loop (Piontek et al. 2007). A model of the barrier in Figure 2B is based on evidence that claudins are cell-to-cell adhesion molecules and form small pores through the barrier, but there is currently insufficient evidence to speculate on how they oligomerize to form the particles seen in freeze fracture EM images.

Most cell types express several different claudins, which assemble into the same adhesive strands, yet the generality of heterotypic binding across the intercellular space remains an unresolved question. There is limited evidence that a subset of claudins can interact heterotypically across cell contacts; for example, claudin 1 can bind claudin-3 but not claudin-2 (Furuse et al. 1999), but there is currently no information that allows generalizations about which claudins can interact or if adhesion involves additional proteins (Fig. 2A,B). Some claudins are observed to coimmunoprecipitate with a list of proteins (e.g., claudin-7 with EpCAM [Le Naour and Zoller 2008]; claudin-11 with β1 integrin and tetraspanins [Tiwari-Woodruff et al. 2001]), although the biologic implications remain unknown. Further understanding of how various functions (selectivity, adhesion, and assembly) are organized within the protein is severely limited by a current lack of structural information.

PERMSELECTIVITY AND THE TWO PATHWAY MODEL: SMALL SELECTIVE PORES AND NONSELECTIVE BREAKS

The term "permselectivity" is used to describe variations among TJs in electrical resistance, ionic charge and size discrimination, and the magnitude of solute permeability (Table 1) (Powell 1981). TER and solute flux are in part a function of cell geometry. For example, a monolayer with smaller cells has more cell–cell contact length per unit area through which electrical current and solutes can pass. Assuming the same claudins are expressed, a monolayer with smaller cells will show lower TER and higher solute flux. Geometry of the individual cell–cell contacts can also affect TJ length per unit area. Some cells touch with straight junction contacts, whereas others make elaborate interdigitations. For example, those between cells in mammalian thin ascending limb of the loop of Henle are extremely redundant, which increases TJ length per unit area of the epithelium at least 10-fold over strength contacts. Presumably, this arrangement enhances the space for paracellular transport without requiring larger epithelium. However, at the protein level, it is now recognized that a critical determinant of permselectivity is the profile and levels of different claudins expressed in a given tight junction.

Size

In general, description of paracellular size selectivity in the cell biology and physiology literature has been rather imprecise, in contrast to studies in the pharmaceutical literature, which have specifically focused on how transepithelial drug absorption is influenced by solute size and chemistry. Research in the former fields is often concerned only with TJ assembly or disassembly, rather than with subtleties of the permselectivity. In most studies, a single tracer size is used to report paracellular flux, which limits the ability to determine experimental changes in the size-dependence of permeability. Tracer charge is usually ignored as is a possible transcellular contribution from transcytosis. These subtleties are important because information about alterations in size and charge dependence may have important pathologic implications, for example in determining if specific cytokines or bacterial toxins might have increased access across the paracellular space.

Now that the actual barrier-forming proteins have been discovered, it is instructive to characterize permeability as a continuous function of solute size (Watson et al. 2001; Watson et al. 2005; Van Itallie et al. 2008). This is performed by characterizing the permeability for solutes of progressively larger molecular radii. Apparent Permeability [$P_{app} = (dQ/dt)/$ (Concentration Gradient $\times$ Area)] takes into account the chemical driving force for each solute and surface area, and, unlike simple flux (dQ/dt), can be compared among tissues and laboratories. When size profiling is performed using a noncharged solute like graded polyethylene polymers (PEGs), it becomes obvious that permeability has two components (Fig. 3). There is a higher capacity pathway with a steep size dependence for solutes less than about 4 Å in radius (Knipp et al. 1997; Watson et al. 2001; Van Itallie et al. 2008). This pore-pathway shows ionic charge discrimination and its magnitude varies among epithelia. This is the pathway that carries most of the

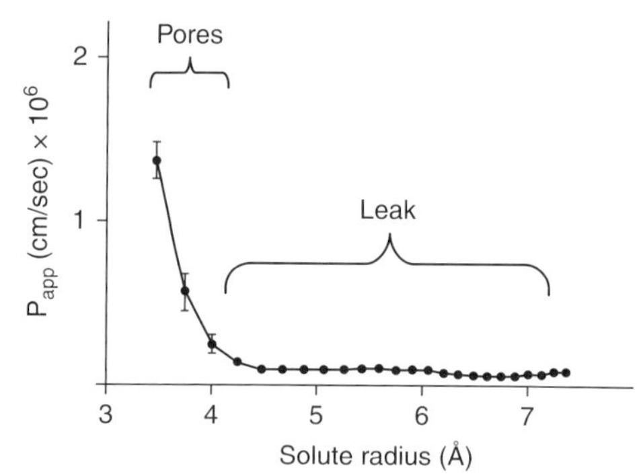

Figure 3. The two pathway model. Idealized data show the permeability of noncharged polyethylene glycol molecules of different sizes across an epithelium. The pathway for molecules below 4 Å is formed by claudin pores that are size and charge selective. The pathway for molecule above 4 Å in diameter shows no charge selectivity and appears to be defined by the status of cell-to-cell adhesion and the cytoskeleton.

electrical current during physiologic transport and during experimentally imposed electrical field (reflected in the TER), thus its magnitude and charge selectivity are key determinants of physiologic transport. As described below, claudins confer the charge selectivity, but there is less information on what determines the magnitude of flux through the pores. However, different claudins may control the level of porosity because it has been observed that expression of claudin-2 in cultured epithelial cells selectively increases the permeability for solutes smaller than 4 Å, whereas claudins 4 and 18 do not (Van Itallie et al. 2008). In vivo, deletion of claudin-5 from mice results in a size-selective increase in permeability of TJs in blood vessels (Nitta et al. 2003). It is important to recognize that the permeability for noncharged and charged solutes are not necessarily directly correlated. Because pores in different epithelia can have different charge discrimination, one can observe an apparent paradox in which the solute permeability is very high but the ion permeability is very low (high TER). Taken together, the present data suggest that claudins create the system of small discriminating pores, and by an unknown mechanism regulate the magnitude of flux (Fig. 3).

Solutes that are larger than 4 Å can also get through intact TJs. The magnitude of this component varies among epithelia and shows no charge discrimination (Knipp. et al. 1997; Artursson et al. 1993). The distinction between the pore and nonpore pathways has been overlooked in most of the literature because permeability is typically measured only by tracers, which are larger than the pores, namely mannitol (4.2 Å in radius), inulin (15 Å), and a graded series of fluorescently labeled dextrans (4 kDa, 10 kDa, 40 kDa, etc). This larger pathway is speculated to represent small temporary breaks in the otherwise continuous TJ contacts. Interestingly, flux through this pathway can be enhanced by proinflammatory cytokines without altering the pore pathway (Watson et al. 2005). The nonpore pathway is sensitive to cytoskeletal disruption (Bruewer et al. 2004; Ivanov et al. 2005) and any form of cellular injury (Nusrat et al. 2000). There is significant evidence that the "intactness" of the TJ is controlled by cytoskeletal dynamics (Hartsock and Nelson 2008), myosin light chain activity (Ma et al. 2005), and any factor affecting cell homeostasis. As investigators dissect the role of individual proteins and signals in regulating the barrier, it will be useful to characterize barriers in terms of the pore and nonpore pathways.

Claudin Charge Selectivity and TER

All of the available data supports the idea that the first extracellular loop of claudins creates an "electrostatic selectivity filter," controlling overall resistance and charge selectivity of the small pores (Fig. 2). In contrast, admittedly limited data suggest that the second loop is involved in cell–cell adhesion (Blasig et al. 2006; Piontek et al. 2007). Charged amino acid side chains on the claudins limit similarly charged ions in solution without opposing permeability of oppositely charged ions. Most of these studies have been performed by expressing a single foreign claudin in the background of all the other claudins expressed in a cultured epithelial cell monolayer, most often MDCK and LLC-PK_1 cells or by siRNA knockdown of selected claudins (Hou et al. 2006). A change from the baseline TER and dilution potential (a measure of cation/anion selectivity) is interpreted to reveal selectivity of the transfected or silenced claudin compared with the background. These studies are somewhat qualitative. Lacking the electrophysiologic tools used to characterize membrane channels, such as patch clamping, it has been impossible to assign a specific conductance to individual claudins.

A total of 15 claudins have now been tested with good consensus. Claudins 2 (Furuse et al. 2001; Colegio et al. 2002; Amasheh et al. 2002) and 10 (Van Itallie et al. 2006) tend to make tight monolayers leakier. Claudins 1 (Inai et al. 1999; McCarthy et al. 2000), 4 (Van Itallie et al. 2001), 5 (Wen et al. 2004; Amasheh et al. 2005), 7 (Alexandre et al. 2005), 8 (Angelow et al. 2006; Yu et al. 2003; Jeansonne et al. 2003), 11 (Van Itallie et al. 2003), 14 (Ben Yosef et al. 2003), 15 (Colegio et al. 2003; Van Itallie et al. 2003), 16 (Hou et al. 2005; Ikari et al. 2008), 18 (Jovov et al. 2007), and 19 (Hou et al. 2008) tend to make leaky monolayers tighter. Claudins 2, 15 (Amasheh et al. 2002; Colegio et al. 2003), 16, and 19 (Hou et al. 2008) have a preference for cations over anions. Claudin-10a (Van Itallie et al. 2006) prefers anions. Where tested, the effect on TER is dose-dependent and each claudin reaches a saturating influence on TER (Van Itallie et al. 2001), although the reason for this is not clear. Most reports simply express the change in barrier permselectivity that results from the maximal induction level. Some studies suggest that the transfected claudin adds to those already present, whereas a single study convincingly shows that claudin-8 can replace endogenous claudin-2 (Yu et al. 2003). Potencies vary, with claudins 14 (Ben Yosef et al. 2003) and 18 (Jovov et al. 2007) producing the electrically tightest barriers. A single study has simultaneously expressed two different claudins, showing that claudins 16 and 19 exert properties in a cooperative and not just additive fashion, consistent with the idea that claudins might form heterotypic pores (Hou et al. 2008).

There is a general relationship between the chemistry of the first loop and charge selectivity (Fig. 2A). However, the relative influence of specific positions is still under debate. For example, claudin-16 has the highest proportion of negatively charged residues (Fig. 2A, shaped positions) and when expressed in cultured epithelial cell monolayers, results in a strong enhancement in permeability for Na^+ and Mg^{++} but has no affect on P_{Cl} (Hou et al. 2005). In the opposite direction, claudin-10a has more basic residues and reduces P_{Na}, while simultaneously increasing P_{Cl} (Van Itallie et al. 2006). Each claudin has shown a signature pattern of whether it will increase or decrease TER and increase, decrease, or not alter the individual P_{Na} and P_{Cl}. The ability to increase TER can be based on limiting permeability for just cations (claudin-4 [Van Itallie et al. 2001]) or both cations and anions (claudin-7 [Alexandre et al. 2005]). There is a reassuring correlation between the physiology of particular tissues, the claudins they express, and the physiology of those claudin as determined experimentally (Holmes et al. 2006). This is best documented along the nephron, where TER and ion selectivity are very well documented. For example, "leaky" claudin-2 is the dominant form in the leaky proximal tubule and is excluded from tighter segments; "tight" claudin-4 and -8 are restricted to the tight collecting ducts (Angelow et al. 2008).

The role of individual positions in influencing charge selectivity has most often been tested by expressing charge-reversing mutations. This approach has led to the conclusion that positions along much of the first loop can influence selectivity and thus this loop is folded to line the pore space through which soluble ions pass. This has been most extensively studied for claudins 15 (Van Itallie et al. 2003) and 16 (Hou et al. 2005). In the case of claudin-15, three widely spaced and negatively charged residues were sequentially reversed and shown to have additive effects (Van Itallie et al. 2003). The "distributed filter" model has been criticized because it is based on introducing abnormal charges rather than simply neutralizing them. Extensive characterization of charge-neutralizing mutations of claudin-2 convincingly showed that the only charged residue that provides the high P_{Na} is the aspartic acid at position 65 (Fig. 2A) (Yu et al. 2009). Resolution of the mechanism of selectivity must await high-resolution structural information.

The TJ pore is slightly wider than that of transmembrane pores and consequently shows less ionic discrimination. For comparison, some Na^+ channels (ENaC) show a permeability ratio for Na^+ to K^+ (P_{Na}/P_K) as high as 1000 to 1 (Hille 2001). In contrast, claudin pores show almost no discrimination between K^+ and Na^+ and the P_{Cl}/P_{Na} ratio ranges only from about 10 to 0.1 when compared among various in vivo epithelia and cultured cell monolayers. Membrane channels are of a size where the associated water molecules must be stripped off so the ion can fit through the pore (Hille 2001; Yu et al. 2009). The modest ability of TJs in MDCK cells to discriminate among the alkali metal cations (K > Rb > Na > Li > Cs, so-called Eisenman selectivity sequence V–VIII) does actually suggest that they are being partially dehydrated as they pass through the pore. The observed Eisenman sequence is slightly different from the rank order of their free mobilities in solution, leading to the conclusion that the pore radius is about 3.6 Å (Yu et al. 2009). This is concordant with the 4 Å radius estimated using noncharged solutes (Watson et al. 2001). It is remarkable that this degree of size and charge selectivity can be maintained while cells in an epithelial sheet continuously move relative to one another.

Several claudins have now been deleted from mice and the resulting phenotypes are consistent with a barrier role for claudins 1 (Ladwein et al. 2005), 5 (Nitta et al. 2003), 11 (Gow et al. 2000), 14 (Ben Yosef et al. 2003), and 15 (Tamura et al. 2008), although the phenotypes are not easily explained as changes in permselectivity. In contrast, deletions of 16 (Himmerkus et al. 2008) and 19 (Hou et al. 2008) phenocopy their human disease mutants and are more easily explained as ion selectivity defects, as described in the following section.

In zebrafish, reduction in claudin-15 through morphlino treatment produces a developmental intestinal defect, which very likely results from loss of ion transport selectivity (Bagnat et al. 2007). In the absence of claudin-15, the gut remains as multiple lumens, which fail to fuse. It is proposed that active ion and water secretion is required for the lumens to enlarge, come into contact, and fuse. In the absence of claudin-15, there may be a back leak of Na^+, preventing fluid accumulation and expansion. Interestingly, claudin-15 KO mice develop a megaintestine, but this phenotype appears to be a proliferation and not a permeability defect (Tamura et al. 2008).

Claudin-16 Mutants and Other Inherited Diseases of Tight Junction Proteins

Currently, there are seven human diseases known to be caused by mutations in genes encoding tight junction proteins (Table 2). The basis for ZO-2 (Carlton et al. 2003) and tricellulin-based defects (Riazuddin et al. 2006) remains unclear, but the phenotypes of several of the recessive claudin mutants provide additional evidence that claudins provide selectivity for the barrier. We describe the claudin-16 defect in some detail because it provides the best evidence that claudins in vivo create ion selective pores.

The first disease-causing claudin mutation was identified through positional cloning of the locus responsible for a rare form of renal magnesium loss, hypomagnesemia hypercalcemia with nephrocalcinosis (HHNC) (Simon et al. 1999; Kaushansky et al. 2007). Patients with this defect show increased urinary loss of Mg^{++} and Ca^{++}, and reduced plasma Mg^{++} levels, leading to weakness and seizures. The gene responsible encodes claudin-16 (originally referred to a Paracellin-1). Claudin-16 is highly restricted to the thick ascending limb of the loop of Henle, the segment of the nephron most involved in reabsorbing filtered Mg^{++}. Other claudins are also expressed in the same TJs. Paracellular reabsorption is driven by an intralumenal positive electrical potential with respect to the peritubular interstitium. These observations led to an initial hypothesis that claudin-16 is a Mg^{++} pore and when absent, Mg^{++} remains in the tubule and is lost in the urine. When expressed in cultured monolayers, wild-type claudin-16 does form a highly cation-selective pore (Hou et al. 2005). Consistent with the electrostatic claudin pore model, the first loop contains many negatively charged residues (Fig. 2), and neutralizing mutations reduce cation permeability (Hou et al. 2005). Many of the human mutations are missense mutations, which fail to traffic properly to the plasma membrane when expressed in cultured cell models (Kausalya et al. 2006) suggesting the defect results from the selective absence of claudin-16, leaving other claudins to define the barrier's ion selectivity. Recently, mutations in the human claudin-19 gene were shown to cause renal Mg^{++} wasting by a very similar mechanism. In addition, claudin-19 is expressed in the retina and affected individuals have defects in retinal development and vision loss (Konrad et al. 2006) (Table 2).

The Mg-pore model is probably too simplistic. The intralumenal positive potential driving Mg^{++} results from a paracellular back diffusion of Na^+ down its concentration gradient into the lumen. If Na^+ entry is limited in the absence of claudin-16 pores, there would be a reduced electrical gradient to drive Mg^{++} out. Although this can not be tested in humans, study of isolated perfused nephron segments from claudin-16 KO mice is consistent with a defect in Na^+ permeability (Himmerkus et al. 2008). Whatever the full explanation, claudin-16 is clearly a cation pore and HHNC remains the best example of a disease caused by an ion-selective TJ defect.

It is tempting to also explain hearing loss in the human claudin-14 mutants as an ion selectivity defect, but the evidence is only circumstantial. Claudin-14 is expressed in TJs lining the intrachoclear space, which contains fluid uniquely high in K^+. This high gradient promotes rapid entry of K^+ into the outer hair cells; this is required for their depolarization during acoustic mechanotransduction. Other forms of deafness are caused by mutations in the transporters that normally secrete K^+ into this space, suggesting that the loss of claudin-14 might cause deafness by allowing the K^+ gradient to dissipate. In fact, when claudin-14 is expressed in cultured MDCK cells, the junctions become very tight by specifically restricting cation permeability (Riazuddin et al. 2006). Finally, claudin-1 mutation results in neonatal ichthyosis-sclerosing cholangitis syndrome (NISCH), but whether this pathophysiology results from a permeability defect is unclear (Hadj-Rabia et al. 2004). Although claudin-1 is very widely expressed among all epithelia, it is difficult to rationalize why patients with these mutations develop predominately a scaling skin disease and obliteration of their bile ducts.

CONCLUSION

Significant progress has been made over the past decade in understanding the role of claudins in regulation of TER and paracellular ionic selectivity. In addition, the recognition of the existence of two pathways for paracellular solute flux should allow a more sophisticated analysis of the roles individual TJ proteins play in physiologic and pathologic regulation of permeability of nonionic solutes. However, further insights are hampered by the lack of a three-dimensional model of TJ structure. Studies on the synaptic junction may provide clues and methods for studying TJ organization, but ultimately we will need structural information about the integral membrane proteins of the TJ to ask informative questions about how this complex is organized and physiologically regulated.

ACKNOWLEDGMENTS

This work was supported by grants from the National Institutes of Health, DK45134 (JMA, CVI), P30 DK034987, and DK61397 (JMA).

REFERENCES

Alexandre MD, Lu Q, Chen YH. 2005. Overexpression of claudin-7 decreases the paracellular Cl^- conductance and increases the paracellular Na^+ conductance in LLC-PK1 cells. *J Cell Sci* **118:** 2683–2693.

Amasheh S, Meiri N, Gitter AH, Schoneberg T, Mankertz J, Schulzke JD, Fromm M. 2002. Claudin-2 expression induces cation-selective channels in tight junctions of epithelial cells. *J Cell Sci* **115:** 4969–4976.

Amasheh S, Schmidt T, Mahn M, Florian P, Mankertz J, Tavalali S, Gitter AH, Schulzke JD, Fromm M. 2005. Contribution of claudin-5 to barrier properties in tight junctions of epithelial cells. *Cell Tissue Res* **321:** 89–96.

Angelow S, Ahlstrom R, Yu AS. 2008. Biology of claudins. *Am J Physiol Renal Physiol* **295:** F867–F876.

Angelow S, Kim KJ, Yu AS. 2006. Claudin-8 modulates paracellular permeability to acidic and basic ions in MDCK II cells. *J Physiol* **571:** 15–26.

Artursson P, Ungell AL, Lofroth JE. 1993. Selective paracellular permeability in two models of intestinal absorption: Cultured monolayers of human intestinal epithelial cells and rat intestinal segments. *Pharm Res* **10:** 1123–1129.

Asano A, Asano K, Sasaki H, Furuse M, Tsukita S. 2003. Claudins in Caenorhabditis elegans: Their distribution and barrier function in the epithelium. *Curr Biol* **13:** 1042–1046.

Bagnat M, Cheung ID, Mostov KE, Stainier DY. 2007. Genetic control of single lumen formation in the zebrafish gut. *Nat Cell Biol* **9:** 954–960.

Balda MS, Matter K. 2003. Epithelial cell adhesion and the regulation of gene expression. *Trends Cell Biol* **13:** 310–318.

Barrios-Rodiles M, Brown KR, Ozdamar B, Bose R, Liu Z, Donovan RS, Shinjo F, Liu Y, Dembowy J, Taylor IW, et al. 2005. High-throughput mapping of a dynamic signaling network in mammalian cells. *Science* **307:** 1621–1625.

Bazzoni G. 2003. The JAM family of junctional adhesion molecules. *Curr Opin Cell Biol* **15:** 525–530.

Behr M, Riedel D, Schuh R. 2003. The claudin-like megatrachea is essential in septate junctions for the epithelial barrier function in Drosophila. *Dev Cell* **5:** 611–620.

Ben Yosef T, Belyantseva IA, Saunders TL, Hughes ED, Kawamoto K, Van Itallie CM, Beyer LA, Halsey K, Gardner DJ, Wilcox ER, et al. 2003. Claudin 14 knockout mice, a model for autosomal recessive deafness DFNB29, are deaf due to cochlear hair cell degeneration. *Hum Mol Genet* **12:** 2049–2061.

Blasig IE, Winkler L, Lassowski B, Mueller SL, Zuleger N, Krause E, Krause G, Gast K, Kolbe M, Piontek J. 2006. On the self-association potential of transmembrane tight junction proteins. *Cell Mol Life Sci* **63:** 505–514.

Boron W, Boulpaep EL. 2005. *Medical physiology. A cellular and molecular approach*, pp. 931–946. Elsevier Saunders, Philidelphia.

Brancolini C, Edomi P, Marzinotto S, Schneider C. 2000. Exposure at the cell surface is required for Gas3/PMP22 to regulate both cell death and cell spreading: Implication for the Charcot-Marie-Tooth type 1A and Dejerine-Sottas diseases. *Mol Biol Cell* **11:** 2901–2914.

Bronstein JM. 2000. Function of tetraspan proteins in the myelin sheath. *Curr Opin Neurobiol* **10:** 552–557.

Bruewer M, Hopkins AH, Hobert ME, Nusrat A, Madara JL. 2004. RhoA, Rac1, and Cdc42 exert distinct effects on epithelial barrier via selective structural and biochemical modulation of junctional proteins and F-actin. *Am J Physiol Cell Physiol* **287:** C327–C335.

Carlton VE, Harris BZ, Puffenberger EG, Batta AK, Knisely AS, Robinson DL, Strauss KA, Shneider BL, Lim WA, Salen G, et al. 2003. Complex inheritance of familial hypercholanemia with associated mutations in TJP2 and BAAT. *Nat Genet* **34:** 91–96.

Cereijido M, Anderson JM. 2001. Introduction: Evolution of ideas on the tight junction. In *Tight junctions* (ed. Cereijido M, Anderson JM), pp. 1–18. CRC Press, Boca Raton.

Cereijido M, Valdes J, Shoshani L, Contreras RG. 1998. Role of tight junctions in establishing and maintaining cell polarity. *Annu Rev Physiol* **60:** 161–177.

Claude P. 1978. Morphological factors influencing transepithelial permeability: A model for the resistance of the zonula occludens. *J Membr Biol* **39:** 219–232.

Colegio OR, Van Itallie CM, Mccrea HJ, Rahner C, Anderson JM. 2002. Claudins create charge-selective channels in the paracellular pathway between epithelial cells. *Am J Physiol Cell Physiol* **283:** C142–C147.

Colegio OR, Van Itallie C, Rahner C, Anderson JM. 2003. Claudin extracellular domains determine paracellular charge selectivity and resistance but not tight junction fibril architecture. *Am J Physiol Cell Physiol* **284:** C1346–C1354.

Coyne CB, Bergelson JM. 2005. CAR: A virus receptor within the tight junction. *Adv Drug Deliv Rev* **57:** 869–882.

Crone C, Christensen O. 1981. Electrical resistance of a capillary endothelium. *J Gen Physiol* **77:** 349–371.

Daugherty BL, Ward C, Smith T, Ritzenthaler JD, Koval M. 2007. Regulation of heterotypic claudin compatibility. *J Biol Chem* **282:** 30005–30013.

Diamond JM. 1977. Twenty-first Bowditch lecture. The epithelial junction: Bridge, gate, and fence. *Physiologist* **20:** 10–18.

Diamond JM. 1978. Channels in epithelial cell membranes and junctions. *Fed Proc* **37:** 2639–2644.

Farquhar MG, Palade GE. 1963. Junctional complexes in various epithelia. *J Cell Biol* **17:** 375–412.

Furuse M. 2009. Molecular basis of the core structure of tight junctions. *Cold Spring Harb Perspect Biol* **2:** a002907.

Furuse M, Hirase T, Itoh M, Nagafuchi A, Yonemura S, Tsukita S, Tsukita S. 1993. Occludin - A Novel Integral Membrane-Protein Localizing at Tight Junctions. *J Cell Biol* **123:** 1777–1788.

Furuse M, Fujita K, Hiiragi T, Fujimoto K, Tsukita S. 1998. Claudin-1 and -2: Novel integral membrane proteins localizing at tight junctions with no sequence similarity to occludin. *J Cell Biol* **141:** 1539–1550.

Furuse M, Sasaki H, Tsukita S. 1999. Manner of interaction of heterogeneous claudin species within and between tight junction strands. *J Cell Biol* **147:** 891–903.

Furuse M, Furuse K, Sasaki H, Tsukita S. 2001. Conversion of Zonulae occludentes from tight to leaky strand type by introducing claudin-2 into Madin-Darby Canine Kidney I cells. *J Cell Biol* **153:** 263–272.

Gabreels-Festen A, Wetering RV. 1999. Human nerve pathology caused by different mutational mechanisms of the PMP22 gene. *Ann NY Acad Sci* **883:** 336–343.

Gonzalez-Mariscal L, Betanzos A, Nava P, Jaramillo BE. 2003. Tight junction proteins. *Prog Biophys Mol Biol* **81:** 1–44.

Gonzalez-Mariscal L, Tapia R, Chamorro D. 2008. Crosstalk of tight junction components with signaling pathways. *Biochim Biophys Acta* **1778:** 729–756.

Gow A, Southwood CM, Li JS, Pariali M, Riordan GP, Danias J, Bronstein JM, Brodie SE, Kachar B, Lazzarini RA. 2000. CNS myelin and sertoli cell tight junction stands are absent in osp/claudin 11-null mice. *J Neurochem* **74:** S35.

Hadj-Rabia S, Baala L, Vabres P, Hamel-Teillac D, Jacquemin E, Fabre M, Lyonnet S, De Prost Y, Munnich A, Hadchouel M, et al. 2004. Claudin-1 gene mutations in neonatal sclerosing cholangitis associated with ichthyosis: A tight junction disease. *Gastroenterology* **127:** 1386–1390.

Hamazaki Y, Itoh M, Sasaki H, Furuse M, Tsukita S. 2002. Multi-PDZ domain protein 1 (MUPP1) is concentrated at tight junctions through its possible interaction with claudin-1 and junctional adhesion molecule. *J Biol Chem* **277:** 455–461.

Hartsock A, Nelson WJ. 2008. Adherens and tight junctions: Structure, function and connections to the actin cytoskeleton. *Biochim Biophys Acta* **1778:** 660–669.

Hille B. 2001. *Ion channels of excitable membranes*, pp. 573–574. Sinauer Associates, Sunderland, Massachusetts.

Himmerkus N, Shan Q, Goerke B, Hou J, Goodenough DA, Bleich M. 2008. Salt and acid-base metabolism in claudin-16 knockdown mice: Impact for the pathophysiology of FHHNC patients. *Am J Physiol Renal Physiol* **295:** F1641–F1647.

Hirano T, Kobayashi N, Itoh T, Takasuga A, Nakamaru T, Hirotsune S, Sugimoto Y. 2000. Null mutation of PCLN-1/Claudin-16 results in bovine chronic interstitial nephritis. *Genome Res* **10:** 659–663.

Holmes JL, Van Itallie CM, Rasmussen JE, Anderson JM. 2006. Claudin profiling in the mouse during postnatal intestinal development and along the gastrointestinal tract reveals complex expression patterns. *Gene Expr Patterns* **6:** 581–588.

Hou J, Paul DL, Goodenough DA. 2005. Paracellin-1 and the modulation of ion selectivity of tight junctions. *J Cell Sci* **118:** 5109–5118.

Hou J, Gomes AS, Paul DL, Goodenough A. 2006. Study of claudin function by RNA interference. *J Biol Chem* **281:** 36117–36123.

Hou J, Renigunta A, Konrad M, Gomes AS, Schneeberger EE, Paul DL, Waldegger S, Goodenough DA. 2008. Claudin-16 and claudin-19 interact and form a cation-selective tight junction complex. *J Clin Invest* **118:** 619–628.

Ikari A, Ito M, Okude C, Sawada H, Harada H, Degawa M, Sakai H, Takahashi T, Sugatani J, Miwa M. 2008. Claudin-16 is directly phosphorylated by protein kinase A independently of a vasodilator-stimulated phosphoprotein-mediated pathway. *J Cell Physiol* **214:** 221–229.

Ikenouchi J, Furuse M, Furuse K, Sasaki H, Tsukita S, Tsukita S. 2005. Tricellulin constitutes a novel barrier at tricellular contacts of epithelial cells. *J Cell Biol* **171:** 939–945.

Inai T, Kobayashi J, Shibata Y. 1999. Claudin-1 contributes to the epithelial barrier function in MDCK cells. *Eur J Cell Biol* **78:** 849–855.

Itoh M, Furuse M, Morita K, Kubota K, Saitou M, Tsukita S. 1999. Direct binding of three tight junction-associated MAGUKs, ZO-1, ZO-2 and ZO-3, with the COOH termini of claudins. *J Cell Biol* **147:** 1351–1363.

Ivanov AI, Hunt D, Utech M, Nusrat A, Parkos CA. 2005. Differential roles for actin polymerization and a myosin II motor in assembly of the epithelial apical junctional complex. *Mol Biol Cell* **16:** 2636–2650.

Jarvis LJ, Louis CF. 1995. Purification and oligomeric state of the major lens fiber cell membrane proteins. *Curr Eye Res* **14:** 799–808.

Jeansonne B, Lu Q, Goodenough DA, Chen YH. 2003. Claudin-8 interacts with multi-PDZ domain protein 1 (MUPP1) and reduces paracellular conductance in epithelial cells. *Cell Mol Biol(Noisy-le-grand)* **49:** 13–21.

Jetten AM, Suter U. 2000. The peripheral myelin protein 22 and epithelial membrane protein family. *Prog Nucl Acid Res Mol Biol* **64:** 97–129.

Jovov B, Van Itallie CM, Shaheen NJ, Carson JL, Gambling TM, Anderson JM, Orlando RC. 2007. Claudin-18: A Dominant Tight Junctional Protein in Barrett's Esophagus and Potential Contributor to its Acid Resistance. *Am J Physiol GI* **293:** G1106–G1113.

Kausalya PJ, Amasheh S, Gunzel D, Wurps H, Muller D, Fromm M, Hunziker W. 2006. Disease-associated mutations affect intracellular traffic and paracellular Mg transport function of Claudin-16. *J Clin Invest* **116:** 878–891.

Kaushansky N, Hemo R, Eisenstein M, Ben Nun A. 2007. OSP/claudin-11-induced EAE in mice is mediated by pathogenic T cells primarily governed by OSP192Y residue of major encephalitogenic region OSP179-207. *Eur J Immunol* **37:** 2018–2031.

Kearsey J, Petit S, De Oliveira C, Schweighoffer F. 2004. A novel four transmembrane spanning protein, CLP24. *Eur J Biochem* **271:** 2584–2592.

Knipp GT, Ho NF, Barsuhn CL, Borchardt RT. 1997. Paracellular diffusion in Caco-2 cell monolayers: Effect of perturbation on the transport of hydrophilic compounds that vary in charge and size. *J Pharm Sci* **86:** 1105–1110.

Koefeld-Johnsen V, Ussing HH. 1958. The nature of the frog skin potential. *Acta Physiol Scand* **42:** 298–308.

Konrad M, Schaller A, Seelow E, Pandey AV, Waldegger S, Lesslauer A, Vitzthum H, Suzuki Y, Luk JM, Becker C, et al. 2006. Mutations in the tight-junction gene claudin 19 (CLDN19) are associated with renal magnesium wasting, renal failure, and severe ocular involvement. *Am J Hum Genet* **79:** 949–957.

Krause G, Winkler L, Mueller SL, Haseloff F, Piontek J, Blasig IE. 2008. Structure and function of claudins. *Biochim Biophys Acta* **1778:** 631–645.

Kubota K, Furuse M, Sasaki H, Sonoda N, Fujita K, Nagafuchi A, Tsukita S. 1999. Ca^{2+}-independent cell-adhesion activity of claudins, a family of integral membrane proteins localized at tight junctions. *Curr Biol* **9:** 1035–1038.

Ladwein M, Pape UF, Schmidt DS, Schnolzer M, Fiedler S, Langbein L, Franke WW, Moldenhauer G, Zoller M. 2005. The cell-cell adhesion molecule EpCAM interacts directly with the tight junction protein claudin-7. *Exp Cell Res* **309:** 345–357.

Laukoetter MG, Nava P, Lee WY, Severson EA, Capaldo CT, Babbin BA, Williams IR, Koval M, Peatman E, Campbell JA, et al. 2007. JAM-A regulates permeability and inflammation in the intestine in vivo. *J Exp Med* **204:** 3067–3076.

Le Naour F, Zoller M. 2008. The tumor antigen EpCAM: Tetraspanins and the tight junction protein claudin-7, new partners, new functions. *Front Biosci* **13:** 5847–5865.

Liu Y, Nusrat A, Schnell FJ, Reaves TA, Walsh S, Pochet M, Parkos CA. 2000. Human junction adhesion molecule regulates tight junction resealing in epithelia 18. *J Cell Sci* **113:** 2363–2374.

Ma TY, Boivin MA, Ye D, Pedram A, Said HM. 2005. Mechanism of TNF-{alpha} modulation of Caco-2 intestinal epithelial tight junction barrier: Role of myosin light-chain kinase protein expression. *Am J Physiol Gastrointest Liver Physiol* **288:** G422–G430.

Madara JL, Trier JS, Neutra MR. 1980. Structural-Changes in the Plasma-Membrane Accompanying Differentiation of Epithelial-Cells in Human and Monkey Small-Intestine. *Gastroenterology* **78:** 963–975.

Matsuda M, Kubo A, Furuse M, Tsukita S. 2004. A peculiar internalization of claudins, tight junction-specific adhesion molecules, during the intercellular movement of epithelial cells. *J Cell Sci* **117:** 1247–1257.

McCarthy KM, Francis SA, McCormack JM, Lai J, Rogers RA, Skare IB, Lynch RD, Schneeberger EE. 2000. Inducible expression of claudin-1-myc but not occludin-VSV-G results in aberrant tight junction strand formation in MDCK cells. *J Cell Sci* **113:** 3387–3398.

Mitic LL, Unger VM, Anderson JM. 2003. Expression, solubilization, and biochemical characterization of the tight junction transmembrane protein claudin-4. *Protein Sci* **12:** 218–227.

Nitta T, Hata M, Gotoh S, Seo Y, Sasaki H, Hashimoto N, Furuse M, Tsukita S. 2003. Size-selective loosening of the blood-brain barrier in claudin-5-deficient mice. *J Cell Biol* **161:** 653–660.

Notterpek L, Roux KJ, Amici SA, Yazdanpour A, Rahner C, Fletcher BS. 2001. Peripheral myelin protein 22 is a constituent of intercellular junctions in epithelia. *Proc Natl Acad Sci* **98:** 14404–14409.

Nusrat A, Turner JR, Madara JL. 2000. Molecular physiology and pathophysiology of tight junctions. IV. Regulation of tight junctions by extracellular stimuli: Nutrients, cytokines, and immune cells. *Am J Physiol Gastrointest Liver Physiol* **279:** G851–G857.

Nusrat A, Eichel-Streiber C, Turner JR, Verkade P, Madara JL, Parkos CA. 2001. Clostridium difficile toxins disrupt epithelial barrier function by altering membrane microdomain localization of tight junction proteins. *Infect Immun* **69:** 1329–1336.

Piontek J, Winkler L, Wolburg H, Muller SL, Zuleger N, Piehl C, Wiesner B, Krause G, Blasig IE. 2007. Formation of tight junction: Determinants of homophilic interaction between classic claudins. *FASEB J* **22:** 146–158.

Powell DW. 1981. Barrier function of epithelia. *Am J Physiol* **241:** G275–G288.

Price MG, Davis CF, Deng F, Burgess DL. 2005. The alpha-amino-3-hydroxyl-5-methyl-4-isoxazolepropionate receptor trafficking regulator "stargazin" is related to the claudin family of proteins by Its ability to mediate cell-cell adhesion. *J Biol Chem* **280:** 19711–19720.

Riazuddin S, Ahmed ZA, Fanning AS, Lagziel A, Kitajiri S, Ramzan K, Khan SN, Chattaraj P, Friedman PL, Anderson JM, et al. 2006. Tricellulin is a tight-junction protein necessary for hearing. *Am J Hum Genet* **79:** 1040–1051.

Roh MH, Liu CJ, Laurinec S, Margolis B. 2002. The carboxyl terminus of zona occludens-3 binds and recruits a mammalian homologue of discs lost to tight junctions. *J Biol Chem* **277:** 27501–27509.

Roux KJ, Amici SA, Notterpek L. 2004. The temporospatial expression of peripheral myelin protein 22 at the developing blood-nerve and blood-brain barriers. *J Comp Neurol* **474:** 578–588.

Roux KJ, Amici SA, Fletcher BS, Notterpek L. 2005. Modulation of epithelial morphology, monolayer permeability, and cell migration by growth arrest specific 3/peripheral myelin protein 22. *Mol Biol Cell* **16:** 1142–1151.

Saitou M, Furuse M, Sasaki H, Schulzke JD, Fromm M, Takano H, Noda T, Tsukita S. 2000. Complex phenotype of mice lacking occludin, a component of tight junction strands. *Mol Biol Cell* **11:** 4131–4142.

Schmitz H, Barmeyer C, Fromm M, Runkel N, Foss HD, Bentzel CJ, Riecken EO, Schulzke JD. 1999. Altered tight junction structure contributes to the impaired epithelial barrier function in ulcerative colitis. *Gastroenterology* **116:** 301–309.

Schneeberger EE, Lynch RD. 2004. The tight junction: A multifunctional complex. *Am J Physiol-Cell Physiol* **286:** C1213–C1228.

Schulzke JD, Gitter AH, Mankertz J, Spiegel S, Seidler U, Amasheh S, Saitou M, Tsukita S, Fromm M. 2005. Epithelial transport and barrier function in occludin-deficient mice. *Biochim Biophys Acta* **1669:** 34–42.

Simon DB, Lu Y, Choate KA, Velazquez H, Al Sabban E, Praga M, Casari C, Bettinelli C, Colussi C, Rodriguez-Soriano J, et al. 1999. Paracellin-1, a renal tight junction protein required for paracellular Mg^{2+} resorption. *Science* **285:** 103–106.

Steele EC, Lyon MF, Favor J, Guillot PV, Boyd Y, Church RL. 1998. A mutation in the connexin 50 (Cx50) gene is a candidate for the No2 mouse cataract. *Curr Eye Res* **17:** 883–889.

Tamura A, Kitano Y, Hata M, Katsuno T, Moriwaki K, Sasaki H, Hayashi H, Suzuki Y, Noda T, Furuse M, et al. 2008. Megaintestine in claudin-15-deficient mice. *Gastroenterology* **134:** 523–534.

Tiwari-Woodruff SK, Buznikov AG, Vu TQ, Micevych PE, Chen K, Kornblum HI, Bronstein JM. 2001. OSP/claudin-11 forms a complex with a novel member of the tetraspanin super family and beta 1 integrin and regulates proliferation and migration of oligodendrocytes. *J Cell Biol* **153:** 295–305.

Tomita S, Fukata M, Nicoll RA, Bredt DS. 2004. Dynamic interaction of stargazin-like TARPs with cycling AMPA receptors at synapses. *Science* **303:** 1508–1511.

Tsukita S, Yamazaki Y, Katsuno T, Tamura A, Tsukita S. 2008. Tight junction-based epithelial microenvironment and cell proliferation. *Oncogene* **27:** 6930–6938.

Turner JR. 2006. Molecular basis of epithelial barrier regulation: From basic mechanisms to clinical application. *Am J Pathol* **169:** 1901–1909.

Ussing HH, Windhager EE. 1964. Nature of shunt path and active sodium transport path through frog skin epithelium. *Acta Physiol Scand* **61:** 484–504.

Ussing HH, Zerahn K. 1951. Active transport of sodium as the source of electric current in the short-circuited isolated frog skin. *Acta Physiol Scand* **23:** 110–127.

Van Itallie CM, Fanning AS, Anderson JM. 2003. Reversal of charge selectivity in cation or anion-selective epithelial lines by expression of different claudins. *Am J Physiol Renal Physiol* **285:** F1078–F1084.

Van Itallie C, Rahner C, Anderson JM. 2001. Regulated expression of claudin-4 decreases paracellular conductance through a selective decrease in sodium permeability. *J Clin Invest* **107:** 1319–1327.

Van Itallie CM, Anderson JM. 2004. The Role of Claudins in Determining Paracellular Charge Selectivity. *Proc Am Thorac Soc* **1:** 38–41.

Van Itallie CM, Anderson JM. 2006. Claudins and Epithelial Paracellular Transport. *Annu Rev Physiol* **68:** 403–429.

Van Itallie CM, Rogan S, Yu A, Vidal LS, Holmes JL, Anderson JM. 2006. Two splice variants of claudin-10 in the kidney create paracellular pores with different ion selectivities 1. *Am J Physiol Renal Physiol* **291:** F1288–F1299.

Van Itallie CM, Holmes J, Bridges A, Gookin JL, Coccaro MR, Proctor W, Colegio OR, Anderson JM. 2008. The density of small tight junction pores varies among cell types and is increased by expression of claudin-2. *J Cell Sci* **121:** 298–305.

Vetrano S, Rescigno M, Cera MR, Correale C, Rumio C, Doni A, Fantini M, Sturm A, Borroni E, Repici A, et al. 2008. Unique role of junctional adhesion molecule-a in maintaining mucosal homeostasis in inflammatory bowel disease. *Gastroenterology* **135:** 173–184.

Wang Z, Mandell KJ, Parkos CA, Mrsny RJ, Nusrat A. 2005. The second loop of occludin is required for suppression of Raf1-induced tumor growth. *Oncogene* **24:** 4412–4420.

Watson CJ, Rowland M, Warhurst G. 2001. Functional modeling of tight junctions in intestinal cell monolayers using polyethylene glycol oligomers. *Am J Physiol-Cell Physiol* **281:** C388–C397.

Watson CJ, Hoare CJ, Garrod DR, Carlson GL, Warhurst G. 2005. Interferon-gamma selectively increases epithelial permeability to large molecules by activating different populations of paracellular pores. *J Cell Sci* **118:** 5221–5230.

Wen HJ, Watry DD, Marcondes MCG, Fox HS. 2004. Selective decrease in paracellular conductance of tight junctions: Role of the first extracellular domain of claudin-5. *Mol Cell Biol* **24:** 8408–8417.

Wright EM, Diamond JM. 1968. Effects of pH and polyvalent cations on the selective permeability of gallbladder epithelium to monovalent ions. *Biochim Biophys Acta* **163:** 57–74.

Wu VM, Schulte J, Hirschi A, Tepass U, Beitel GJ. 2004. Sinuous is a Drosophila claudin required for septate junction organization and epithelial tube size control. *J Cell Biol* **164:** 313–323.

Yamazaki Y, Okawa K, Yano T, Tsukita S, Tsukita S. 2008. Optimized proteomic analysis on gels of cell-cell adhering junctional membrane proteins. *Biochem* **47:** 5378–5386.

Yeaman C, Grindstaff KK, Nelson WJ. 2004. Mechanism of recruiting Sec6/8 (exocyst) complex to the apical junctional complex during polarization of epithelial cells. *J Cell Sci* **117:** 559–570.

Yu AS, Enck AH, Lencer WI, Schneeberger EE. 2003. Claudin-8 expression in Madin-Darby canine kidney cells augments the paracellular barrier to cation permeation. *J Biol Chem* **278:** 17350–17359.

Yu AS, McCarthy KM, Francis SA, McCormack JM, Lai J, Rogers RA, Lynch RD, Schneeberger EE. 2005. Knock down of occludin expression leads to diverse phenotypic alterations in epithelial cells. *Am J Physiol Cell Physiol* **288:** C1231–C1241.

Yu AS, Cheng MH, Angelow S, Gunzel D, Kanzawa SA, Schneeberger EE, Fromm M, Coalson RD. 2009. Molecular Basis for Cation Selectivity in Claudin-2-based Paracellular Pores: Identification of an Electrostatic Interaction Site. *J Gen Physiol* **133:** 111–127.

Molecular Bases of Cell–Cell Junctions Stability and Dynamics

Matthieu Cavey and Thomas Lecuit

IBDML, UMR6216 CNRS-Université de la Méditerranée, Campus de Luminy case 907, 13288 Marseille Cedex 09, France

Correspondence: lecuit@ibdml.univ-mrs.fr

Epithelial cell–cell junctions are formed by apical adherens junctions (AJs), which are composed of cadherin adhesion molecules interacting in a dynamic way with the cortical actin cytoskeleton. Regulation of cell–cell junction stability and dynamics is crucial to maintain tissue integrity and allow tissue remodeling throughout development. Actin filament turnover and organization are tightly controlled together with myosin-II activity to produce mechanical forces that drive the assembly, maintenance, and remodeling of AJs. In this review, we will discuss these three distinct stages in the lifespan of cell–cell junctions, using several developmental contexts, which illustrate how mechanical forces are generated and transmitted at junctions, and how they impact on the integrity and the remodeling of cell–cell junctions.

Cell–cell junction formation and remodeling occur repeatedly throughout development. Epithelial cells are linked by apical adherens junctions (AJs) that rely on the cadherin-catenin-actin module. Cadherins, of which epithelial E-cadherin (E-cad) is the most studied, are Ca^{2+}-dependent transmembrane adhesion proteins forming homophilic and heterophilic bonds in *trans* between adjacent cells. Cadherins and the actin cytoskeleton are mutually interdependent (Jaffe et al. 1990; Matsuzaki et al. 1990; Hirano et al. 1992; Oyama et al. 1994; Angres et al. 1996; Orsulic and Peifer 1996; Adams et al. 1998; Zhang et al. 2005; Pilot et al. 2006). This has long been attributed to direct physical interaction of E-cad with β-catenin (β-cat) and of α-catenin (α-cat) with actin filaments (for reviews, see Gumbiner 2005; Leckband and Prakasam 2006; Pokutta and Weis 2007). Recently, biochemical and protein dynamics analyses have shown that such a link may not exist and that instead, a constant shuttling of α-cat between cadherin/β-cat complexes and actin may be key to explain the dynamic aspect of cell–cell adhesion (Drees et al. 2005; Yamada et al. 2005). Regardless of the exact nature of this link, several studies show that AJs are indeed physically attached to actin and that cadherins transmit cortical forces exerted by junctional acto-myosin networks (Costa et al. 1998; Sako et al. 1998; Pettitt et al. 2003; Dawes-Hoang et al. 2005; Cavey et al. 2008; Martin et al. 2008; Rauzi et al. 2008). In addition, physical

 doi: 10.1101/cshperspect.a002998
Cite this article as *Cold Spring Harb Perspect Biol* 2009;1:a002998

association depends in part on α-cat (Cavey et al. 2008) and additional intermediates have been proposed to represent alternative missing links (Abe and Takeichi 2008) (reviewed in Gates and Peifer 2005; Weis and Nelson 2006). Although further work is needed to address the molecular nature of cadherin/actin dynamic interactions, association with actin is crucial all throughout the lifespan of AJs. In this article, we will review our current understanding of the molecular mechanisms at work during three different developmental stages of AJs biology: assembly, stabilization, and remodeling, with special emphasis on the mechanical forces controlling AJs integrity and development.

CELL JUNCTIONS FORMATION

Cell junctions form in two contexts during development. (1) Migrating cells undergo mesenchymal-epithelial transitions (MET) during which they establish membrane contacts with neighbors and initiate assembly of AJs at these sites. Subsequently, cell–cell contacts expand and newly formed AJs serve as landmarks for establishing tissue polarity. (2) In primary embryonic epithelia, cell junctions are formed in a subregion of pre-existing cell contacts, which is defined by upstream polarity cues.

Cell–cell junction formation during MET has been extensively studied in cell cultures: (1) formation of junctions after membrane contacts or after Ca^{2+} switch (activating cadherin adhesive function by raising the extracellular Ca^{2+} concentration); (2) adhesion of cells on cadherin-coated substrata. These studies have provided considerable insights into in vivo epitheliogenesis and epithelial sheet sealing processes, which occur during embryogenesis and wound healing. In this section, we first review how cell junctions assemble in cell cultures, focusing on the role of forces generated by actin polymerization and acto-myosin tension. We then examine how these forces are used during epithelial sheet sealing in whole organisms. Finally, we summarize how cell junctions form in a primary embryonic epithelium, i.e., without MET.

Mesenchymal-Epithelial Transitions (MET): Cell Culture Studies

Cell–cell junction formation during MET suggests the following steps:

1. Membrane protrusions explore the environment to generate initial cell contacts. Cadherin molecules diffusing in the plasma membrane engage in homophilic interactions and form clusters.
2. Homophilic ligation of cadherins triggers actin cytoskeleton rearrangement by directly controlling the recruitment and activity of several actin regulators.
3. Actin reorganization drives contact expansion (an increase of the surface of contact between two cells), and is also linked to the stabilization of adhesive interfaces.

Formation of Initial Cell–Cell Contacts

The first step is an opportunistic event resulting from the exploratory behavior of cells extending actin-based protrusions (lamellipodia and membrane ruffles) (Fig. 1A) (reviewed in Vasioukhin and Fuchs 2001). Immunofluoresence and the use of GFP-fusions show that bright cadherin puncta rapidly form where protrusions touch (Fig. 1B). These puncta are thought to represent clusters of homophilic cadherin dimers, a view supported by the fact that they contain cadherin molecules from both contacting cells (Kametani and Takeichi 2007). However, cadherin clustering mechanisms are still not fully understood and remain debated. They may include lateral interactions in the cadherin extracellular domain and intracellular interactions including interactions with actin (reviewed in Leckband and Prakasam 2006; Pokutta and Weis 2007). Upon contact formation, cadherin molecules previously diffusing "freely" in the membrane become immobilized (Adams et al. 1998), presumably by anchoring to actin (Sako et al. 1998; Iino et al. 2001; Lambert et al. 2002). Within a few tens of minutes, cadherin puncta grow, suggesting a continuous addition of new molecules

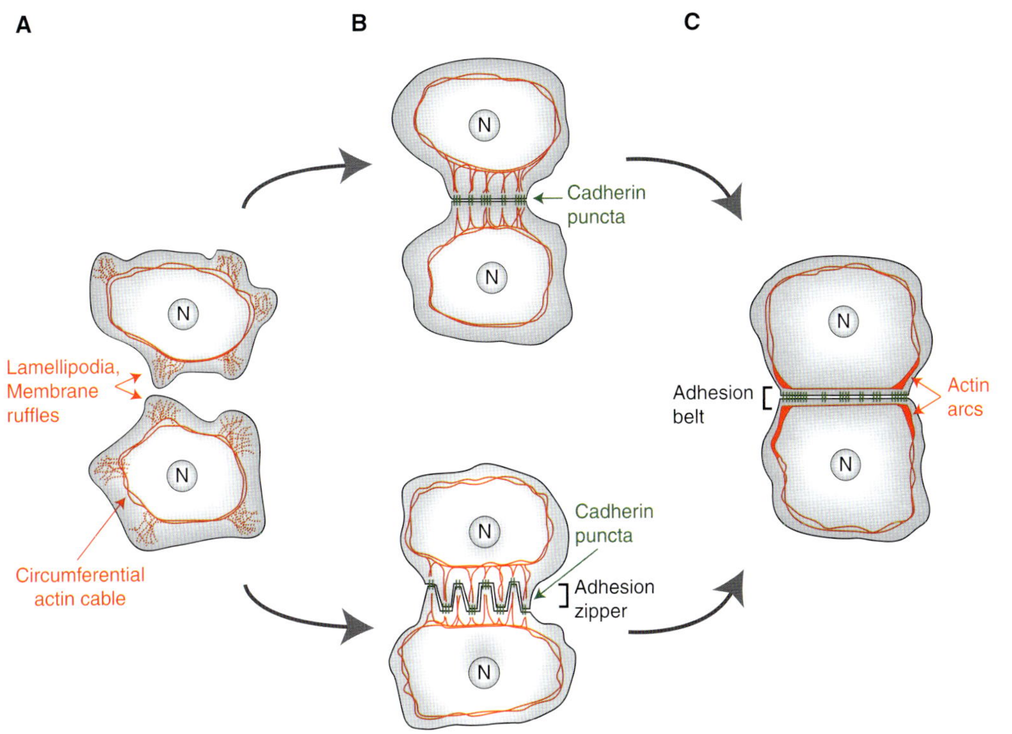

Figure 1. Actin reorganization during cell–cell junction formation in cell culture. Actin is shown in red, cadherin in green. (*A*) Before cell–cell contact, epithelial cells extend protrusions (lamellipodia and membrane ruffles). (*B*) Cadherin puncta form at the tips of these projections and are connected to the circumferential actin cable via radial actin bundles. (*B, bottom*) Contacting keratinocytes adopt an intermediate configuration known as the "adhesion zipper" as a consequence of myosin-mediated tension pulling inward on cadherin puncta. (*C*) As cell–cell contacts expand and mature, actin arcs focus on the edges of the contact. Actin remodeling along the contact results in formation of the adhesion belt.

(Yonemura et al. 1995; Adams et al. 1996; Adams et al. 1998).

In the *Drosophila* embryo, a similar punctate organization of cadherins has been reported (Muller and Wieschaus 1996; Harris and Peifer 2004). Moreover, puncta were recently shown to represent *bona fide* sites of immobilized E-cad clusters with slower dynamics than outside puncta (Cavey et al. 2008), suggesting that they are equivalent to the spot adherens junctions (SAJs) structures observed at the electron microscopy level (Tepass and Hartenstein 1994; Oda et al. 1998). Therefore the puncta observed in cell cultures and in several epithelia in vivo likely represent the same structures, namely sites of homophilic cadherin dimers enrichment.

Actin Reorganization

Before cell contacts, actin forms concentric ring(s) (the "circumferential actin cable/ring") and a dense meshwork between the ring and the plasma membrane (Fig. 1A) (Yonemura et al. 1995; Adams et al. 1996; Gloushankova et al. 1997; Adams et al. 1998; Krendel and Bonder 1999; Ehrlich et al. 2002; Vaezi et al. 2002; Ivanov et al. 2005a). As contacts form, cadherin puncta are connected to the actin ring via radial actin bundles (Fig. 1B). Subsequently, actin bundles are replaced by finer ones under the region of contact (the "perijunctional actin belt/adhesion belt") resembling actin organization in epithelia in vivo (Hirokawa et al. 1983), and thick bundles ("actin arcs") focus on the contact edges (Fig. 1C). This cytoskeletal reorganization is triggered and controlled by cadherins whose homophilic ligation can directly recruit and activate actin regulators including Rac1, Cdc42 (Nakagawa et al. 2001; Noren et al. 2001; Kovacs et al. 2002a), Abl kinase (Zandy et al. 2007), Arp2/3 (Kovacs et al. 2002b; Verma et al. 2004), Cortactin (Helwani et al. 2004), N-WASP (Ivanov et al. 2005a), Formin1 (Kobielak et al. 2004), and Ena/VASP (Vasioukhin et al. 2000) (for reviews, see Bershadsky 2004; Braga and Yap 2005). Cell–cell contact formation is dependent on actin polymerization (Braga et al. 1997; Adams et al. 1998; Vasioukhin et al. 2000; Ivanov et al. 2005a; Zhang et al. 2005; Yamada and Nelson 2007) and adhesion strength (measured as the force required to detach cell doublets) increases in an F-actin- (Angres et al. 1996), Cdc42-, and Rac1-dependent manner (Chu et al. 2004). Cytoskeletal reorganization likely serves two related purposes examined below: expansion and stabilization of the adhesive interface.

Cell–Cell Contact Expansion

Two forces are coordinated to expand cell contacts: (1) Actin polymerization produces membrane protrusions to generate new sites of contact, and (2) acto-myosin tension focusing on contact edges generates a pulling force to facilitate contact expansion.

(1) Lamellipodia and membrane ruffles initiating cell–cell contacts (Gloushankova et al. 1997; Krendel and Bonder 1999) are generated by branched actin networks (Vaezi et al. 2002; Bershadsky 2004). New sites of contact are then generated in adjacent regions. In keratinocytes, the adhesive interface develops in an "adhesion zipper" structure, because of myosin-mediated tension pulling inward on cadherin puncta (Fig. 1B, *bottom*). Actin polymerization from the tip of radial actin bundles could provide the force necessary to resolve the two rows of cadherin puncta into a single belt of mature junctions, but the underlying mechanism is not understood (Vasioukhin et al. 2000; Kobielak et al. 2004). In other cell types, lamellipodial activity is initially distributed evenly around the cell periphery but becomes restricted to the region of contact and subsequently propagates to adjacent regions in waves generating new sites of membrane apposition (Ehrlich et al. 2002) (Fig. 2A). Supporting this idea, factors promoting branched actin polymerization are enriched at the leading edge of cells spreading on cadherin-coated substrata (Kovacs et al. 2002a; Helwani et al. 2004) and Rac1 is specifically activated in regions of contact expansion (Yamada and Nelson 2007). Moreover, interfering with actin branching impairs cell spreading or contact expansion (Ehrlich et al. 2002; Kovacs et al. 2002a; Helwani et al. 2004; Verma et al. 2004; Ivanov et al. 2005a; Zandy et al. 2007). Conversely, Rac1 constitutive activation increases cell contact expansion rate and adhesion strength (Ehrlich et al. 2002). Together, these studies show that actin polymerization in lamellipodia is specifically concentrated in regions adjacent to the site of initial contact and provides a pushing force required for membrane apposition. These fluctuating contacts are then ligated and rectified by cadherin homophilic dimers.

(2) Myo-II-mediated tension is also important for contact expansion. Myo-II localizes to peripheral actin bundles in epithelial cells (Krendel et al. 1999; Krendel and Bonder 1999; Bertet et al. 2004; Zallen and Wieschaus 2004; Shewan et al. 2005; Zhang et al. 2005; Yamada and Nelson 2007) and is required to bundle actin filaments (Vaezi et al. 2002; Ivanov et al. 2005a; Shewan et al. 2005; Zhang et al. 2005). Cell adhesion induces Myo-II activation (Ivanov et al. 2005a), which can be directly triggered by cadherin homophilic ligation (Shewan et al. 2005). In keratinocytes, Myo-II appears to act negatively on contact expansion by generating the adhesion zipper structure (Vaezi et al. 2002). However, in other cell types Myo-II acts positively on contact expansion (Yamada and Nelson 2007). How can Myo-II activity participate in contact expansion?

As contacts expand, cadherin puncta (Adams et al. 1998) and ConA-coated beads

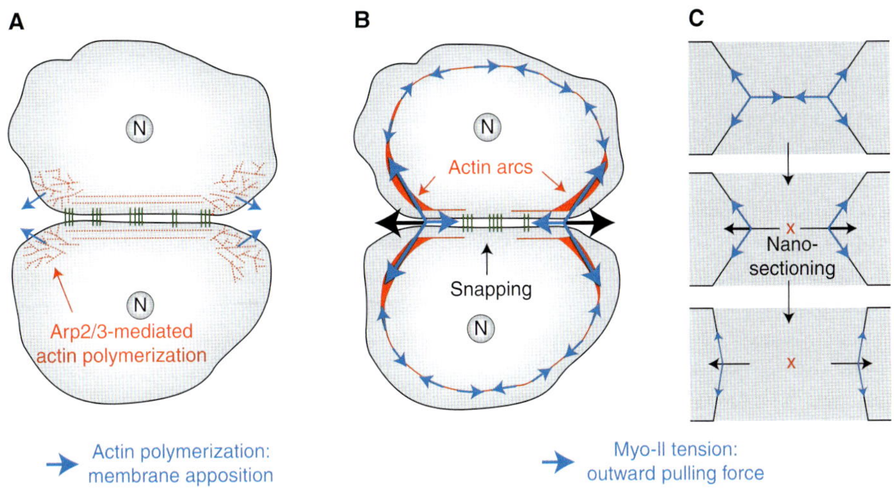

Figure 2. Forces driving cell–cell contact expansion in cell culture. (*A*) Polymerization of branched actin networks at the edges of a contact generates a pushing force (blue arrows) necessary to create new sites of membrane apposition, which are then ligated by homophilic cadherin dimers. (*B,C*) Analogy for the role of myosin-II-mediated tension (blue arrows) during contact expansion in cell culture (*B*) and after laser nano-sectioning in an epithelium (*C*). In both cases, the sum of tension forces applied at the contact edges is initially null, but once actin bundles are sectioned by snapping along the cell–cell contact (*B*) or laser nano-sectioning (*C*), these forces are not balanced anymore. This produces a net outward pulling force (black arrows), which drives contact expansion.

that mark plasma membrane proteins (Gloushankova et al. 1997) move toward the contact edges (i.e., tangentially) at similar speeds, revealing a flow of cortical material toward the edges. The cortical forces responsible for this flow likely result from acto-myosin contractility. Activated Myo-II and activated Rho are enriched on the actin arcs, which contract at the edges of cell contacts (Krendel and Bonder 1999; Yamada and Nelson 2007) (Fig. 2B). In addition, during contact expansion, actin bundles along the contact zone break and retract toward the edges of the contact (i.e., tangentially), in a process termed "actin bundle snapping" (Krendel and Bonder 1999) (Fig. 2B). This suggests a model whereby the local unbalance of cortical acto-myosin forces at contact edges drives expansion. Actin bundles along the contact have to resist the outward pulling forces generated on the actin arcs. However, actin bundle snapping along the contact alleviates such a resistance, resulting in a net outward force pulling on the contact edges that drives contact expansion (Fig. 2B). Snapping along cell contacts could be a consequence of increased tension generated on actin arcs, combined with actin remodeling triggered by cell–cell adhesion. Moreover, retraction of these bundles after snapping could drag cadherin puncta toward the edges (Adams et al. 1998). Such a mechanism would be analogous to laser nano-sectioning experiments performed in live epithelial cells of *Drosophila* embryos (Cavey et al. 2008; Rauzi et al. 2008). Indeed, nano-sectioning of a tensile acto-myosin cortical network along a cell–cell contact causes the actin network to retract, resulting in a net expansion of the contact (Fig. 2C). Actin retraction induces E-cad puncta redistribution away from the region of sectioning by tethering E-cad puncta to the contractile acto-myosin network (Cavey et al. 2008; Rauzi et al. 2008).

Junction Formation by MET in Embryos

Interfering with AJs components in embryonic tissues results in tissue collapse at varying developmental stages (Larue et al. 1994; Riethmacher et al. 1995; Tepass et al.

1996; Uemura et al. 1996; Torres et al. 1997; Carmeliet et al. 1999; Vasioukhin et al. 2001; De Vries et al. 2004). One interesting phenomenon is when cell–cell contacts are increased during compaction of late 8-cell mouse embryos. Interfering with cell adhesion using anti-E-cad antibodies and Ca^{2+} depletion both block compaction (Kemler et al. 1977; Hyafil et al. 1980; Pratt et al. 1982; Shirayoshi et al. 1983; Johnson et al. 1986). Compaction may be driven by similar forces as contact expansion in cell culture. Compaction requires actin polymerization (Pratt et al. 1982; Fleming et al. 1986; Clayton et al. 1999) and involves a redistribution of E-cad to the baso-lateral domain where cell contacts expand (Vestweber et al. 1987). Interestingly, numerous membrane protrusions (microvilli) form bridges between cells along regions of contacts (Calarco and Epstein 1973; Fleming et al. 1986), suggesting a role analogous to that of lamellipodia in cell cultures. In later stages, microvilli are excluded from regions of cell contacts (Fleming et al. 1986), as membrane protrusive activity may not be compatible with stabilization of the interface (see below). This may be controlled in part by the ERM (Ezrin-Radixin-Moesin) protein Ezrin, which organizes actin networks to form microvilli and has to be excluded from regions of cell–cell contacts for compaction to be completed (Dard et al. 2001; Dard et al. 2004). A role for acto-myosin tension in driving compaction has not been directly investigated yet, but Rho inhibition (using C3-transferase) disrupts actin and E-cad organization, blocking compaction (Clayton et al. 1999).

Cell–cell contact establishment also occurs during the sealing of epithelial sheets during embryogenesis and wound healing. At the end of embryogenesis, morphogenetic rearrangements leave holes in the dorsal epidermis of *Drosophila* embryos and in the ventral hypodermis of *C. elegans* embryos. Similar holes are created when embryonic or adult tissues are wounded. In all cases, migrating epithelial sheets cover and eventually seal these holes (Fig. 3) (reviewed in Jacinto et al. 2001; Martin and Parkhurst 2004). As for the establishment of cell–cell contacts in cell culture, actin protrusions and acto-myosin cables are both involved in epithelial sheet sealing. An acto-myosin cable

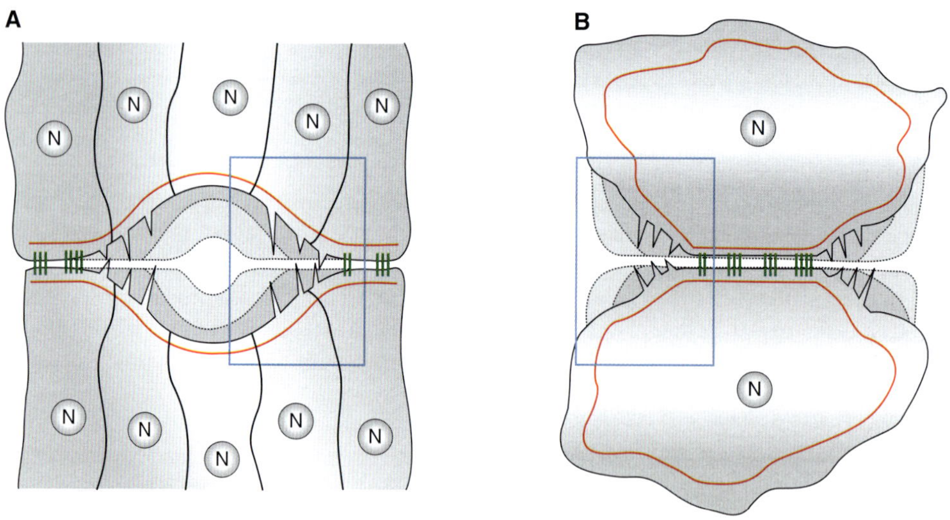

Figure 3. Parallel between epithelial sheet sealing and contact expansion in cell culture. (*A*) Epithelial sheet sealing at the end of embryogenesis and during wound healing and (*B*) cell–cell contact expansion in cell culture. More advanced stages are shown by dashed lines. (*A*) A contracting acto-myosin cable (red) closes the hole and the final sealing step is facilitated by actin-based protrusions. Note the similarity of structures involved in the two systems (blue rectangles).

assembled at the periphery of the hole/wound provides contractile force to progressively close the hole in flies (Harden et al. 1999; Magie et al. 1999; Kiehart et al. 2000; Bloor and Kiehart 2002; Jacinto et al. 2002) during wound healing (Wood et al. 2002)(reviewed in Jacinto et al. 2001; Redd et al. 2004) and possibly in worms as well (Williams-Masson et al. 1997; Raich et al. 1999). Filopodia establish connections over the holes and are required during the final phase to seal the two sheets together. They may exert pulling forces to help sealing (Williams-Masson et al. 1997; Raich et al. 1999; Jacinto et al. 2000; Bloor and Kiehart 2002; Jacinto et al. 2002; Soto et al. 2002; Wood et al. 2002; Gates et al. 2007; Sheffield et al. 2007; Millard and Martin 2008). Electron microscopy studies have revealed interdigitated filopodia during sealing (Redd et al. 2004) harboring AJs puncta at their tips (Raich et al. 1999; Vaezi et al. 2002). This structure appears analogous to the adhesion zipper observed in keratinocytes (Vasioukhin et al. 2000).

Cell–Cell Junctions Formation by Non-MET Processes

In the examples mentioned so far, AJs formation defines a spatial cue for organizing the cytoskeleton and recruiting apical–basal polarity complexes. In turn, these complexes stabilize and maintain AJs, possibly via regulation of the junctional actin cytoskeleton (reviewed in Knust and Bossinger 2002; Nelson 2003; Ebnet 2008). However, in the *Drosophila* embryonic primary epithelium, AJs formation occurs in a small region of the surface of contact between cells, which is defined by already-present apical–basal polarity cues. The polarity protein Par3/Bazooka (Baz) is the first component recruited to the apical region, in a microtubule- and Dynein-dependent process (Harris and Peifer 2004; Harris and Peifer 2005). Par3/Baz then initiates AJs assembly and recruits further polarity components, which maintain AJs in later stages (Muller and Wieschaus 1996; Bilder et al. 2003; Tanentzapf and Tepass 2003; Harris and Peifer 2004; Hutterer et al. 2004; Harris and Peifer 2005). Par3/Baz oligomers could serve as structural adaptors for AJs integrity (Benton and St Johnston 2003) or could link AJs to actin via the nectin/afadin system (Wei et al. 2005). Polarity complexes are also probably implicated in regulating actin polymerization via the small GTPases. In mammals, Par3 serves as a platform connecting Rho signaling to Rac1 regulation to control front/rear polarity in migrating cells. Rac1 is activated by its GEF Tiam1/2, which is in a complex with Par3/aPKC/Cdc42 (Nishimura et al. 2005). Phosphorylation of Par3 by the RhoA-ROCK pathway separates Rac1 from Tiam1/2, resulting in Rac1 inactivation (Nakayama et al. 2008). In flies, Par3 has also been implicated in organizing the cortical actin cytoskeleton by acting on the recruitment of Moesin, an ERM protein (Pilot et al. 2006) and in the control of phosphoinositides levels (von Stein et al. 2005), which can impact on actin dynamics and organization in many ways (for reviews, see Zheng 2001; Yin and Janmey 2003).

MAINTENANCE OF CELL–CELL JUNCTIONS

Once cell contacts have been established, actin polymerization and acto-myosin tension are required for the stabilization and maturation of adhesive interfaces. Actin network organization and Myo-II-mediated tension control the organization of adhesion molecules at the cell surface and the dynamics of the plasma membrane, which are both crucial for maintaining stable cell interfaces and tissue architecture.

Immobilization of E-cadherin Clusters

The organization of cadherins at cell interfaces depends on the integrity of the actin cytoskeleton (Pilot et al. 2006). In *Drosophila* embryos, apical–basal polarity cues (Par3/Baz and the phosphoinoside derivative PIP2) recruit the synaptotagmin-like protein Bitesize, which in turn recruits Moesin to the apical region.

These factors organize the cortical actin network and ensure its integrity, which is crucial for a homogeneous distribution of adhesion foci at cell interfaces (Pilot et al. 2006). In fact, cortical actin is composed of two intermixed populations of filaments, which control E-cad distribution at two levels (Cavey et al. 2008). (1) Locally, actin filaments with a low turnover control the stability of E-cad molecules within puncta. (2) A peripheral contractile network of acto-myosin controls the mobility (displacement) of these puncta along cell–cell contacts. Myo-II-mediated tension is crucial for effectively tethering E-cad puncta in the plasma membrane and thus for controlling their spatial distribution along cell–cell contacts. This suggests that stabilization of adhesion requires the regulated immobilization of homophilic E-cad clusters by a tensile cortical actin network, independently of cluster stabilization per se (Cavey et al. 2008). These results shed light on previous reports, which implicated Myo-II in the spatial organization of cadherin clusters in various cell types (Gloushankova et al. 1998; Krendel et al. 1999; Vaezi et al. 2002; Conti et al. 2004; Shewan et al. 2005; Zhang et al. 2005).

E-cadherin Clusters Stability

Actin depolymerization studies suggest that actin turnover is reduced in mature cell junctions compared to younger ones (Adams et al. 1998; Braga et al. 1999; Ivanov et al. 2005a). Actin turnover is also specifically lower at cadherin puncta compared to neighboring regions in *Drosophila* epithelial cells (Wood et al. 2002; Cavey et al. 2008). In mammalian cells, cadherin stability and actin stability may be directly coupled by Eplin, a protein recruited by E-cad/β-cat/α-cat complexes and which is required for stabilizing actin filaments associated with adhesion complexes (Abe and Takeichi 2008). However, recycling of actin filaments associated with adhesion complexes is likely to occur because small GTPases are required to maintain adhesion. Alternatively, GTPase activity may reflect the need to remodel junctions in dynamic epithelia (reviewed in Braga and Yap 2005; Kooistra et al. 2007; Yamazaki et al. 2007).

Role of Junctional Actin Architecture

Adhesion strengthening seems to involve regulation of actin cytoskeleton organization: Branched networks associated with lamellipodial protrusions are replaced by parallel contractile bundles. This transition probably takes place very shortly after the initial clustering of cadherins and appears to be controlled by AJs components. Factors promoting branched actin polymerization (Rac1, Arp2/3, and cortactin) are relatively depleted from "older" regions of cell contacts (Helwani et al. 2004; Verma et al. 2004; Yamada and Nelson 2007). α-cat was proposed to directly control this transition by repressing Arp2/3 activity (Drees et al. 2005). As AJs mature, progressive enrichment of α-cat would result in local inhibition of Arp2/3 (Perez-Moreno and Fuchs 2006; Pokutta and Weis 2007). In addition, factors promoting unbranched F-actin elongation can be recruited to AJs, such as Formin1 and Ena/VASP in keratinocytes (Vasioukhin et al. 2000; Kobielak et al. 2004). Another formin, Diaphanous (Dia), is recruited to cell contacts and required for junction maintenance (Sahai and Marshall 2002; Carramusa et al. 2007; Homem and Peifer 2008). Dia may be directly recruited at junctions by α-cat and p120, which recruit its activator Rho1 (Magie et al. 2002). Bundling factors at nascent AJs may also contribute to the transition of actin organization (for reviews, see Adams 2004; Broderick and Winder 2005; Mege et al. 2006). In this context, direct observation of actin network architecture at AJs is an important, albeit challenging, avenue.

Tension and Rigidity of the Interface

Linear arrays of unbranched actin filaments favor Myo-II tension, which can stabilize cell–cell junctions in two ways.

(1) The protrusive activity of cell membranes is thought to destabilize adhesive

interfaces (Gloushankova et al. 1997; Sahai and Marshall 2002; Zhang et al. 2005). Myo-II-mediated tension inhibits the formation of protrusions by its ability to align actin filaments parallel to the cell membrane (Gloushankova et al. 1997). In cell cultures, protrusive activity decreases after cell–cell contact formation (Gloushankova et al. 1997; Ehrlich et al. 2002) and increases upon Myo-II inhibition in keratinocytes (Vaezi et al. 2002; Zhang et al. 2005). Similar results were observed in *Drosophila* (Bloor and Kiehart 2002; Jacinto et al. 2002). Interfering with Dia function (polymerization of unbranched actin filaments) increases the membrane protrusive activity and creates gaps in cell contacts (Sahai and Marshall 2002). The formation of a continuous belt of adhesion is compromised upon Myo-II inhibition in keratinocytes and MCF-7 cells but whether this is a direct consequence of increased protrusions is unclear (Shewan et al. 2005; Zhang et al. 2005).

(2) Myo-II-mediated tension may also affect the distribution of cadherin molecules at junctions. For instance, applying tension on the cell membrane is sufficient to drive the clustering of cadherins independently of actin. The current model proposes that membrane tension along a cell contact brings the two cell membranes in close proximity, thereby favoring new homophilic cadherin interactions to form and thus trapping these molecules in the region under tension (Delanoe-Ayari et al. 2004).

DYNAMICS AND REMODELING OF CELL–CELL JUNCTIONS

Throughout embryonic development and in adults, morphogenetic processes, which shape tissues and organisms, require constant remodeling of cell junctions. What underlies the dynamics of cell–cell junctions? They first require a constant turnover of AJs components at the cell surface. Second, they result from the regulated balance of two forces at the cell surface: adhesion and cortical tension. Adhesion tends to increase the surface of contacts with neighbors, whereas cortical tension tends to decrease it (for a more in-depth discussion, see Lecuit and Lenne 2007).

Regulation of Adhesion

Cadherin Endocytosis and Recycling

The turnover of AJs components is achieved by endocytosis and recycling of cadherins to the cell surface (reviewed in D'souza-Schorey 2005; Ivanov et al. 2005b; Yap et al. 2007). Indeed, several morphogenetic processes involving junction remodeling are blocked upon inhibition of endocytosis (Jarrett et al. 2002; Classen et al. 2005; Ulrich et al. 2005; Shaye et al. 2008). Several studies point to a role for actin in mediating cadherin endocytosis (Le et al. 2002; Ivanov et al. 2004) (reviewed in Kaksonen et al. 2006). Myosin may facilitate endocytosis in Ca^{2+}-depleted cells by assembling a contractile acto-myosin ring, which could provide mechanical force for vesicle formation at AJs (Ivanov et al. 2004). On the other hand, association with actin has been proposed to protect E-cad from endocytosis based on experiments suggesting that homophilically engaged E-cad is not endocytosed, whereas free E-cad is (Izumi et al. 2004). However, this issue remains controversial because cadherin homophilic dimers—which are presumably associated to actin—can be dissociated by endocytosis (Troyanovsky et al. 2006). Recently, the Par polarity proteins Cdc42, aPKC, and Par6 (but not Par3/Baz) have been implicated in regulating AJs stability in *Drosophila* by controlling E-cad endocytosis. Cdc42 is thought to couple actin polymerization (via WASp and Arp2/3) with vesicle scission (via Cip4 and Dynamin) to promote E-cad endocytosis and hence turnover (Georgiou et al. 2008; Leibfried et al. 2008; Harris and Tepass 2008).

E-cadherin recycling is equally important for the regulation of AJs dynamics. In *Drosophila* thoracic epithelial cells, members of the exocyst complex (Sec5, Sec6, and Sec15) directly control the recycling of E-cad to AJs via interactions with β-cat (Langevin et al. 2005). In the *Drosophila* pupal wing epithelium, irregularly arranged cells remodel their contacts to pack into a highly regular hexagonal array (Classen et al. 2005) (Fig. 4A). This process requires E-cad endocytosis (Rab5), recycling (Rab11), and exocytosis (Sec5). E-cad exocytosis may be under the control of the planar cell polarity pathway, which may bias E-cad recycling to specific cortical locations to promote regular hexagonal packing (Fig. 4A) (Classen et al. 2005).

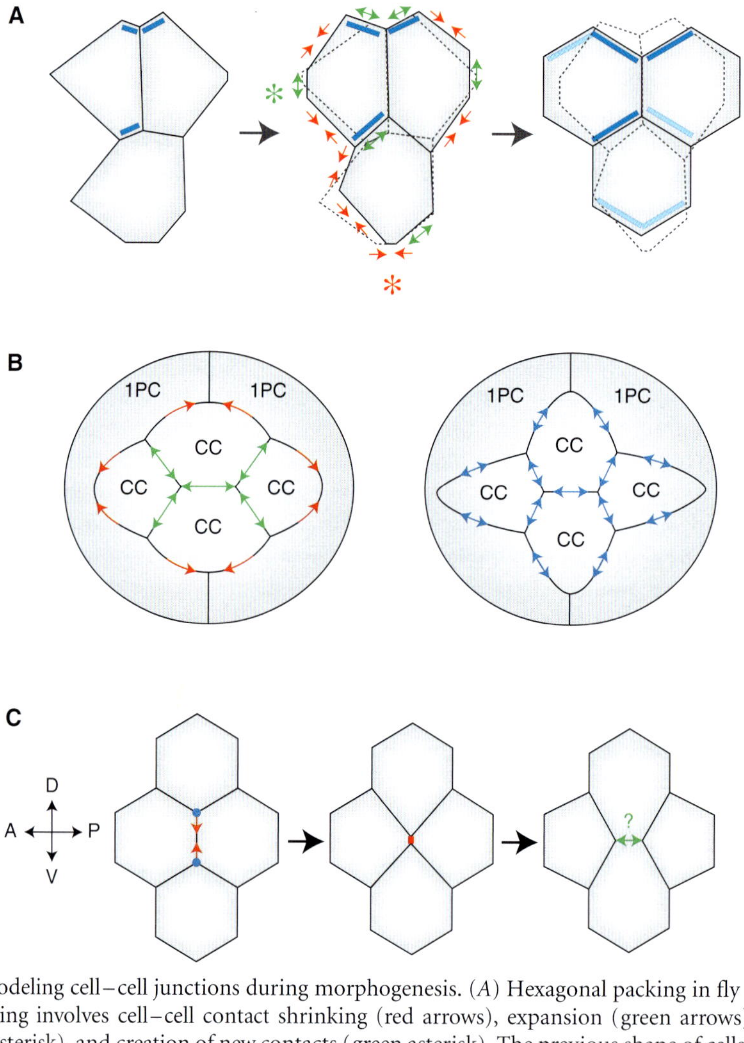

Figure 4. Remodeling cell–cell junctions during morphogenesis. (*A*) Hexagonal packing in fly wing epithelial cells. Remodeling involves cell–cell contact shrinking (red arrows), expansion (green arrows), loss of some contacts (red asterisk), and creation of new contacts (green asterisk). The previous shape of cells is indicated by dashed lines in the *middle* and *right* panels. Flamingo enrichment (dark blue rectangles) and emergent polarity (light blue rectangles) may spatially bias exocytic delivery of E-cad to promote hexagonal packing. (*B*) Pattern formation in fly retina. (*Left* panel) Wild type ommatidial cluster with four cone cells (CC) surrounded by two primary pigment cells (1PC). Strong adhesion (N-cad + E-cad) increases inter-CC contacts (green arrows) at the expense of weaker adhesion (E-cad only) with 1PCs (red arrows). (*Right* panel) Uniform adhesion in a mutant ommatidium (blue arrows) distorts the cell pattern. (*C*) Cell intercalation during germ band elongation in fly embryos. The T1 transition involves shrinking of junctions between A/P neighbors (vertical junctions) and creation of new junctions between D/V neighbors (horizontal junctions). Shrinking is triggered by increased acto-myosin tension along A/P junctions (red arrows; vertices are indicated by blue dots). Expansion of the new D/V interface could rely on adhesion (green arrows), as in cell culture systems.

Generation of Tissue Patterns by Differential Adhesion

The concept of differential adhesion states that differences in adhesion strength between different cell types induce cell sorting, that is, the segregation of groups of cells based on their relative affinity for each other (Steinberg 1963). The strength of interaction is hypothesized to be different between homophilic and heterophilic cadherin dimers. However, cell sorting can happen with similar adhesion strengths (Niessen and Gumbiner 2002; Prakasam et al. 2006) and likely depends on additional parameters including the kinetics of the interactions and the actin cytoskeleton (discussed in Leckband and Prakasam 2006; Lecuit and Lenne 2007). Regardless of the exact mechanisms underlying differential adhesion, the idea that it can drive cell sorting has been largely confirmed in vivo (Godt and Tepass 1998; Gonzalez-Reyes and St Johnston 1998). This concept can be extended to the generation of cell patterns, which do not involve the complete separation of different cell types but subtle cell shape changes. One striking example comes from the morphogenesis of the *Drosophila* ommatidia in the developing retina (Fig. 4B). Ommatidia are composed of four cone cells (CCs) located in the center of a cluster and expressing both N-cad and E-cad, surrounded by two primary pigment cells (1PCs) expressing only E-cad. Genetic analyses have shown that increased adhesion strength in CCs because of N-cad expression maximizes inter-CCs contacts at the expense of CC-1PCs contacts and is responsible for establishing the specific geometry of CCs (Hayashi and Carthew 2004). Another aspect of ommatidial geometry is dictated by differential adhesion, relying on the immunoglobulin (Ig)-domain cell adhesion molecules Hibris and Roughest (Bao and Cagan 2005).

An increase in contact surface is hypothesized to result from increased adhesion strength along the interface. Adhesion strength depends on the surface levels of cadherins (Angres et al. 1996; Duguay et al. 2003; Chu et al. 2004), on the inherent properties (strength, kinetics) of homophilic versus heterophilic interactions (reviewed in Prakasam et al. 2006), as well as on the interaction of cadherins with actin and on actin dynamics (Angres et al. 1996; Imamura et al. 1999; Vasioukhin et al. 2000; Drees et al. 2005; Yamada et al. 2005; Zhang et al. 2005; Cavey et al. 2008). How does an increase in adhesion strength lead to an increase in contact surface? In an ideal system in which no forces resist the deformations induced by changes in contact surface, a zipping-like mechanism could induce contact surface expansion. However, living cells represent a much more complex system in which the deformation of cell shape produces a restoring force—cortical elasticity—which resists to contact expansion. Stronger adhesion alone is thus unlikely to be sufficient to expand the contact surface. As we have seen in the first section, the expansion of cell–cell contacts relies on actin-based forces (protrusions mediated by actin polymerization, and acto-myosin tension), which cooperate to bring cell membranes in close apposition and thereby counteract deformation-induced resistance to contact expansion. Such forces are very likely to be required to increase contact surfaces in the examples mentioned above. One could even imagine that increased adhesion strength only acts via actin remodeling to promote contact surface expansion. The magnitude of adhesion strength (i.e., amount and/or adhesive strength of cadherin dimers) may thus not be directly relevant to contact expansion but may simply reflect the degree of mobilization of actin-based forces. Alternatively, adhesion may regulate the kinetics of contact expansion but not the equilibrium geometry of cell contacts.

Cortical Tension and Junction Dynamics

Recent studies have characterized some important features of junctional mechanics underlying morphogenetic processes (Farhadifar et al. 2007; Kafer et al. 2007; Rauzi et al. 2008). The models were inspired from work done with soap films in which bubbles assemble in geometries almost identical to that of cells in an epithelium and which can be explained as resulting from progressive exploration of local surface energy minima (for details, see Carthew 2005; Lecuit and Lenne 2007). These studies suggest that the extent of cell–cell contacts is dictated by (1) local forces contributed by adhesion and acto-myosin tension, and (2) global forces generated by acto-myosin tension and dependent on the state of cell deformation.

Explaining Steady-State Cell Patterns

This paradigm has been used to explain pattern formation in the fly retina. A simple model that only incorporates the action of adhesive forces cannot recapitulate the patterns observed in wild type ommatidia and mutants in which the expression levels of cadherins are altered. In contrast, these different patterns are faithfully reproduced when including cortical tension as a force resisting cell adhesion (Kafer et al. 2007). Similarly, the steady state geometry of the wing imaginal disc epithelium before hexagonal packing can be modeled in terms of adhesive and elastic tension forces (Farhadifar et al. 2007). Comparing computer simulations to in vivo data yielded estimates of the relative contributions of elastic forces and line tensions at work at cell contacts and suggested that in this epithelium, local acto-myosin contractility predominates over adhesion (Farhadifar et al. 2007).

Tensile Networks Regulating Junction Dynamics

Following similar physical working hypotheses, a recent study probed the spatial distribution of junctional forces underlying the dynamics of tissue elongation (Rauzi et al. 2008). In gastrulating *Drosophila* embryos, epithelial cells of the ventro-lateral tissue (the germ band) intercalate to promote germ band elongation (GBE). This process involves the regulated disassembly of cell contacts between antero-posterior (A/P) neighbors and the creation of new contacts between dorso-ventral (D/V) neighbors (Bertet et al. 2004) (Fig. 4C). An anisotropy in Myo-II localization, more specifically, Myo-II enrichment along shrinking A/P junctions, is required for intercalation (Bertet et al. 2004), suggesting that polarized Myo-II distribution could generate anisotropic tension that would drive cell intercalation. Computer simulations and comparisons with in vivo data indicated that anisotropic tension could be sufficient to drive GBE. Laser-nanodissection of cortical actin measured the anisotropy at AJs and its dependence on Myo-II (Rauzi et al. 2008).

It is worth noting that tensile forces, which shape cells, need not be generated directly at the cell cortex. This was recently illustrated by the observation that apical cell constriction, which underlies tissue bending and invagination in several developmental contexts (reviewed in Lecuit and Lenne 2007), is driven by contraction of a centrally located acto-myosin network in the plane of AJs (Martin et al. 2008). This contrasts with other models stating that an acto-myosin purse string physically attached to AJs is responsible for constriction. During *Drosophila* mesoderm invagination, acto-myosin aggregates located in the medial part of cells undergo pulses and pull on the cell cortex to drive constriction as a ratchet (Martin et al. 2008). These recent results add another level of complexity to the description of junctional mechanics controlling cell contacts and cell shape.

Interplay of Adhesion and Cortical Tension

As illustrated throughout this review, actin plays a lead role in the regulation of cell contacts, by sustaining adhesion and cortical tension, whose balance dictates the extent of cell contacts. The cadherin/catenin system provides one among several other mechanical links between the actin skeletons of adjacent cells and thereby integrates intra- and intercellular forces to the scale of the whole tissue during morphogenesis. How such coupling impacts on the amplitude and the spatial range of force transmission at the cell cortex is one of the most important and open fields of investigation now, in which the contributions of different candidate linkers will have to be evaluated in detail. The extent of frictions between cadherins and contacting tensile actin networks is also likely to define the kinetics of cell shape changes. There is evidence that actin filaments are subdivided in functionally distinct populations, which are dedicated to specific purposes (Zhang et al. 2005; Cavey et al. 2008). Future work will have to focus on the mechanisms controlling the functional subdivision of the actin cytoskeleton in subcellular domains. It will be equally important to understand how the balance of forces is fine-tuned to reach equilibrium and obtain stable cell patterns in a tissue. Many experimental and pathological conditions show that misregulating the central players of adhesion and cortical tension leads to disequilibrium and has dramatic consequences on tissue integrity. For instance, excess tension can rupture cell–cell contacts (Sahai and Marshall 2002; Diogon et al. 2007), whereas the loss of cell–cell adhesion triggers epithelia–mesenchymal transitions (EMT) and is associated with cancer (reviewed in Thiery and Sleeman 2006; Baum et al. 2008).

Vertices as Central Regulatory Units of Epithelial Remodeling

Vertices are geometrical points in a tissue where three or more cells meet and can define both cell shape and cell dynamics. The number of neighbors a cell has simply reflects the number of vertices it is part of and the length of cell–cell contacts reflects the inter-vertices distances. Thus, epithelial remodeling can be explained as the displacement of vertices and most importantly the subsequent exchange of neighbors at vertices. Understanding how these two processes are regulated is thus key to understanding epithelial remodeling.

A drastic change in inter-vertices distance occurs during GBE in *Drosophila* embryos, when A/P junctions shrink during T1 transitions (Fig. 4C) (Bertet et al. 2004; Rauzi et al. 2008). Theoretically, anchoring of acto-myosin contractile bundles to the cell membrane specifically at vertices could be sufficient to remodel cell contacts. However, acto-myosin bundles are anchored to AJs puncta all along cell–cell junctions and not simply at vertices (Cavey et al. 2008). Moreover, vertices impose structural limits on the movement of stable E-cad clusters that cannot “cross” a vertex. Consistent with this, nano-ablation experiments indicate that vertices represent physical barriers to the lateral displacement of retracting actin bundles and E-cad clusters (Cavey et al. 2008; Rauzi et al. 2008), supporting the idea that vertices represent special sites of attachment of acto-myosin bundles to the cortex.

The regulation of AJs stability specifically at vertices also requires further understanding. Anchoring of acto-myosin bundles to the cortex depends on AJs (Dawes-Hoang et al. 2005; Martin et al. 2008) suggesting that the stability of AJs complexes, namely homophilic E-cad clusters, at vertices is essential to drive junction shrinkage. However, T1 transitions, the process whereby four cells meeting at a vertex after junction shrinkage produce two new three-way vertices (Fig. 4C), requires local remodeling of homophilic E-cad clusters. How E-cad stability is regulated during this step is currently unknown. There may be specific mechanisms targeted to vertices to regulate AJs integrity, which remain to be characterized.

ACKNOWLEDGMENTS

M.C. was supported by a fellowship from Association pour la Recherche sur le Cancer (ARC) and the Fondation Schlumberger pour l’Education et la Recherche (FSER). T.L. is supported by the CNRS, the ANR, the Fondation Recherche Médicale (équipe FRM), a subvention fixe from ARC, and a program grant from the HFSP.

REFERENCES

Abe K, Takeichi M. 2008. EPLIN mediates linkage of the cadherin catenin complex to F-actin and stabilizes the circumferential actin belt. *Proc Natl Acad Sci* **105:** 13–19.

Adams JC. 2004. Roles of fascin in cell adhesion and motility. *Current Opinion in Cell Biol* **16:** 590–596.

Adams CL, Nelson WJ, Smith SJ. 1996. Quantitative analysis of cadherin-catenin-actin reorganization during development of cell-cell adhesion. *J Cell Biol* **135:** 1899–1911.

Adams CL, Chen YT, Smith SJ, Nelson WJ. 1998. Mechanisms of epithelial cell-cell adhesion and cell compaction revealed by high-resolution tracking of E-cadherin-green fluorescent protein. *J Cell Biol* **142:** 1105–1119.

Angres B, Barth A, Nelson WJ. 1996. Mechanism for transition from initial to stable cell-cell adhesion: Kinetic analysis of E-cadherin-mediated adhesion using a quantitative adhesion assay. *J Cell Biol* **134:** 549–557.

Bao S, Cagan R. 2005. Preferential adhesion mediated by Hibris and Roughest regulates morphogenesis and patterning in the *Drosophila* eye. *Develop Cell* **8:** 925–935.

Baum B, Settleman J, Quinlan MP. 2008. Transitions between epithelial and mesenchymal states in development and disease. *Sem Cell Develop Biol* **19:** 294–308.

Benton R, St Johnston D. 2003. *Drosophila* PAR-1 and 14-3-3 inhibit Bazooka/PAR-3 to establish complementary cortical domains in polarized cells. *Cell* **115:** 691–704.

Bershadsky A. 2004. Magic touch: How does cell-cell adhesion trigger actin assembly? *Trends Cell Biol* **14:** 589–593.

Bertet C, Sulak L, Lecuit T. 2004. Myosin-dependent junction remodelling controls planar cell intercalation and axis elongation. *Nature* **429:** 667–671.

Bilder D, Schober M, Perrimon N. 2003. Integrated activity of PDZ protein complexes regulates epithelial polarity. *Nat Cell Biol* **5:** 53–58.

Bloor JW, Kiehart DP. 2002. *Drosophila* RhoA regulates the cytoskeleton and cell-cell adhesion in the developing epidermis. *Development* **129:** 3173–3183.

Braga VM, Yap AS. 2005. The challenges of abundance: Epithelial junctions and small GTPase signalling. *Current Opinion Cell Biol* **17:** 466–474.

Braga VM, Machesky LM, Hall A, Hotchin NA. 1997. The small GTPases Rho and Rac are required for the establishment of cadherin-dependent cell-cell contacts. *J Cell Biol* **137:** 1421–1431.

Braga VM, Del Maschio A, Machesky L, Dejana E. 1999. Regulation of cadherin function by Rho and Rac: Modulation by junction maturation and cellular context. *Mol Biol Cell* **10:** 9–22.

Broderick MJ, Winder SJ. 2005. Spectrin, α-actinin, and dystrophin. *Adv Protein Chem* **70:** 203–246.

Calarco PG, Epstein CJ. 1973. Cell surface changes during preimplantation development in the mouse. *Develop Biol* **32:** 208–213.

Carmeliet P, Lampugnani MG, Moons L, Breviario F, Compernolle V, Bono F, Balconi G, Spagnuolo R, Oostuyse B, Dewerchin M. 1999. Targeted deficiency or cytosolic truncation of the VE-cadherin gene in mice impairs VEGF-mediated endothelial survival and angiogenesis. *Cell* **98:** 147–157.

Carramusa L, Ballestrem C, Zilberman Y, Bershadsky AD. 2007. Mammalian diaphanous-related formin Dia1 controls the organization of E-cadherin-mediated cell-cell junctions. *J Cell Sci* **120:** 3870–3882.

Carthew RW. 2005. Adhesion proteins and the control of cell shape. *Current Opinion Gen Develop* **15:** 358–363.

Cavey M, Rauzi M, Lenne PF, Lecuit T. 2008. A two-tiered mechanism for stabilization and immobilization of E-cadherin. *Nature* **453:** 751–756.

Chu YS, Thomas WA, Eder O, Pincet F, Perez E, Thiery JP, Dufour S. 2004. Force measurements in E-cadherin-mediated cell doublets reveal rapid adhesion strengthened by actin cytoskeleton remodeling through Rac and Cdc42. *J Cell Biol* **167:** 1183–1194.

Classen AK, Anderson KI, Marois E, Eaton S. 2005. Hexagonal packing of *Drosophila* wing epithelial cells by the planar cell polarity pathway. *Develop Cell* **9:** 805–817.

Clayton L, Hall A, Johnson MH. 1999. A role for Rho-like GTPases in the polarisation of mouse eight-cell blastomeres. *Develop Cell* **205:** 322–331.

Conti MA, Even-Ram S, Liu C, Yamada KM, Adelstein RS. 2004. Defects in cell adhesion and the visceral endoderm following ablation of nonmuscle myosin heavy chain II-A in mice. *J Biol Chem* **279:** 41263–41266.

Costa M, Raich W, Agbunag C, Leung B, Hardin J, Priess JR. 1998. A putative catenin-cadherin system mediates morphogenesis of the *Caenorhabditis elegans* embryo. *J Cell Biol* **141:** 297–308.

D'souza-Schorey C. 2005. Disassembling adherens junctions: Breaking up is hard to do. *Trends Cell Biol* **15:** 19–26.

Dard N, Louvet S, Santa-Maria A, Aghion J, Martin M, Mangeat P, Maro B. 2001. In vivo functional analysis of ezrin during mouse blastocyst formation. *Develop Biol* **233:** 161–173.

Dard N, Louvet-Vallee S, Santa-Maria A, Maro B. 2004. Phosphorylation of ezrin on threonine T567 plays a crucial role during compaction in the mouse early embryo. *Develop Biol* **271:** 87–97.

Dawes-Hoang RE, Parmar KM, Christiansen AE, Phelps CB, Brand AH, Wieschaus EF. 2005. folded gastrulation, cell shape change and the control of myosin localization. *Development* **132:** 4165–4178.

De Vries WN, Evsikov AV, Haac BE, Fancher KS, Holbrook AE, Kemler R, Solter D, Knowles BB. 2004. Maternal β-catenin and E-cadherin in mouse development. *Development* **131:** 4435–4445.

Delanoe-Ayari H, Al Kurdi R, Vallade M, Gulino-Debrac D, Riveline D. 2004. Membrane and acto-myosin tension promote clustering of adhesion proteins. *Proc Natl Acad Sci* **101:** 2229–2234.

Diogon M, Wissler F, Quintin S, Nagamatsu Y, Sookhareea S, Landmann F, Hutter H, Vitale N, Labouesse M. 2007. The RhoGAP RGA-2 and LET-502/ROCK achieve a balance of actomyosin-dependent forces in *C. elegans* epidermis to control morphogenesis. *Development* **134:** 2469–2479.

Drees F, Pokutta S, Yamada S, Nelson WJ, Weis WI. 2005. α-catenin is a molecular switch that binds E-cadherin-β-catenin and regulates actin-filament assembly. *Cell* **123:** 903–915.

Duguay D, Foty RA, Steinberg MS. 2003. Cadherin-mediated cell adhesion and tissue segregation:

Qualitative and quantitative determinants. *Develop Biol* **253:** 309–323.

Ebnet K. 2008. Organization of multiprotein complexes at cell-cell junctions. *Histochem Cell Biol* **130:** 1–20.

Ehrlich JS, Hansen MD, Nelson WJ. 2002. Spatio-temporal regulation of Rac1 localization and lamellipodia dynamics during epithelial cell-cell adhesion. *Develop Cell* **3:** 259–270.

Farhadifar R, Roper JC, Aigouy B, Eaton S, Julicher F. 2007. The influence of cell mechanics, cell-cell interactions, and proliferation on epithelial packing. *Curr Biol* **17:** 2095–2104.

Fleming TP, Pickering SJ, Qasim F, Maro B. 1986. The generation of cell surface polarity in mouse 8-cell blastomeres: The role of cortical microfilaments analysed using cytochalasin D. *J Embryol Exp Morph* **95:** 169–191.

Gates J, Peifer M. 2005. Can 1000 reviews be wrong? Actin, α-Catenin, and adherens junctions. *Cell* **123:** 769–772.

Gates J, Mahaffey JP, Rogers SL, Emerson M, Rogers EM, Sottile SL, Van Vactor D, Gertler FB, Peifer M. 2007. Enabled plays key roles in embryonic epithelial morphogenesis in *Drosophila. Development* **134:** 2027–2039.

Georgiou M, Marinari E, Burden J, Baum B. 2008. Cdc42, Par6, and aPKC Regulate Arp2/3-Mediated Endocytosis to Control Local Adherens Junction Stability. *Curr Biol* **18:** 1631–1638.

Gloushankova NA, Alieva NA, Krendel MF, Bonder EM, Feder HH, Vasiliev JM, Gelfand IM. 1997. Cell-cell contact changes the dynamics of lamellar activity in nontransformed epitheliocytes but not in their ras-transformed descendants. *Proc Natl Acad Sci* **94:** 879–883.

Gloushankova NA, Krendel MF, Alieva NO, Bonder EM, Feder HH, Vasiliev JM, Gelfand IM. 1998. Dynamics of contacts between lamellae of fibroblasts: Essential role of the actin cytoskeleton. *Proc Natl Acad Sci* **95:** 4362–4367.

Godt D, Tepass U. 1998. *Drosophila* oocyte localization is mediated by differential cadherin-based adhesion. *Nature* **395:** 387–391.

Gonzalez-Reyes A, St Johnston D. 1998. The *Drosophila* AP axis is polarised by the cadherin-mediated positioning of the oocyte. *Development* **125:** 3635–3644.

Gumbiner BM. 2005. Regulation of cadherin-mediated adhesion in morphogenesis. *Nature Rev* **6:** 622–634.

Harden N, Ricos M, Ong YM, Chia W, Lim L. 1999. Participation of small GTPases in dorsal closure of the *Drosophila* embryo: Distinct roles for Rho subfamily proteins in epithelial morphogenesis. *J Cell Sci* **112:** 273–284.

Harris TJ, Peifer M. 2004. Adherens junction-dependent and -independent steps in the establishment of epithelial cell polarity in *Drosophila. J Cell Biol* **167:** 135–147.

Harris TJ, Peifer M. 2005. The positioning and segregation of apical cues during epithelial polarity establishment in *Drosophila. J Cell Biol* **170:** 813–823.

Harris KP, Tepass U. 2008. Cdc42 and Par proteins stabilize dynamic adherens junctions in the *Drosophila* neuroectoderm through regulation of apical endocytosis. *J Cell Biol* **183:** 1129–1143.

Hayashi T, Carthew R. 2004. Surface mechanics mediate pattern formation in the developing retina. *Nature* **431:** 647–652.

Helwani FM, Kovacs EM, Paterson AD, Verma S, Ali RG, Fanning AS, Weed SA, Yap AS. 2004. Cortactin is necessary for E-cadherin-mediated contact formation and actin reorganization. *J Cell Biol* **164:** 899–910.

Hirano S, Kimoto N, Shimoyama Y, Hirohashi S, Takeichi M. 1992. Identification of a neural α-catenin as a key regulator of cadherin function and multicellular organization. *Cell* **70:** 293–301.

Hirokawa N, Keller TC 3rd, Chasan R, Mooseker MS. 1983. Mechanism of brush border contractility studied by the quick-freeze, deep-etch method. *J Cell Biol* **96:** 1325–1336.

Homem CC, Peifer M. 2008. Diaphanous regulates myosin and adherens junctions to control cell contractility and protrusive behavior during morphogenesis. *Development* **135:** 1005–1018.

Hutterer A, Betschinger J, Petronczki M, Knoblich JA. 2004. Sequential roles of Cdc42, Par-6, aPKC, and Lgl in the establishment of epithelial polarity during *Drosophila* embryogenesis. *Develop Cell* **6:** 845–854.

Hyafil F, Morello D, Babinet C, Jacob F. 1980. A cell surface glycoprotein involved in the compaction of embryonal carcinoma cells and cleavage stage embryos. *Cell* **21:** 927–934.

Iino R, Koyama I, Kusumi A. 2001. Single molecule imaging of green fluorescent proteins in living cells: E-cadherin forms oligomers on the free cell surface. *Biophys J* **80:** 2667–2677.

Imamura Y, Itoh M, Maeno Y, Tsukita S, Nagafuchi A. 1999. Functional domains of α-catenin required for the strong state of cadherin-based cell adhesion. *J Cell Biol* **144:** 1311–1322.

Ivanov AI, McCall IC, Parkos CA, Nusrat A. 2004. Role for actin filament turnover and a myosin II motor in cytoskeleton-driven disassembly of the epithelial apical junctional complex. *Mol Biol Cell* **15:** 2639–2651.

Ivanov AI, Hunt D, Utech M, Nusrat A, Parkos CA. 2005a. Differential roles for actin polymerization and a myosin II motor in assembly of the epithelial apical junctional complex. *Mol Biol Cell* **16:** 2636–2650.

Ivanov AI, Nusrat A, Parkos CA. 2005b. Endocytosis of the apical junctional complex: Mechanisms and possible roles in regulation of epithelial barriers. *Bioessays* **27:** 356–365.

Izumi G, Sakisaka T, Baba T, Tanaka S, Morimoto K, Takai Y. 2004. Endocytosis of E-cadherin regulated by Rac and Cdc42 small G proteins through IQGAP1 and actin filaments. *J Cell Biol* **166:** 237–248.

Jacinto A, Wood W, Balayo T, Turmaine M, Martinez-Arias A, Martin P. 2000. Dynamic actin-based epithelial adhesion and cell matching during *Drosophila* dorsal closure. *Curr Biol* **10:** 1420–1426.

Jacinto A, Martinez-Arias A, Martin P. 2001. Mechanisms of epithelial fusion and repair. *Nat Cell Biol* **3:** E117–123.

Jacinto A, Wood W, Woolner S, Hiley C, Turner L, Wilson C, Martinez-Arias A, Martin P. 2002. Dynamic Analysis of Actin Cable Function during *Drosophila* Dorsal Closure. *Curr Biol* **12:** 1245.

Jaffe SH, Friedlander DR, Matsuzaki F, Crossin KL, Cunningham BA, Edelman GM. 1990. Differential effects of the cytoplasmic domains of cell adhesion molecules on cell aggregation and sorting-out. *Proc Natl Acad Sci* **87:** 3589–3593.

Jarrett O, Stow JL, Yap AS, Key B. 2002. Dynamin-dependent endocytosis is necessary for convergent-extension movements in *Xenopus* animal cap explants. *Int J Develop Biol* **46:** 467–473.

Johnson MH, Maro B, Takeichi M. 1986. The role of cell adhesion in the synchronization and orientation of polarization in 8-cell mouse blastomeres. *J Embryol Exp Morph* **93:** 239–255.

Kafer J, Hayashi T, Maree AF, Carthew RW, Graner F. 2007. Cell adhesion and cortex contractility determine cell patterning in the *Drosophila* retina. *Proc Natl Acad Sci* **104:** 18549–18554.

Kaksonen M, Toret CP, Drubin DG. 2006. Harnessing actin dynamics for clathrin-mediated endocytosis. *Nature Rev* **7:** 404–414.

Kametani Y, Takeichi M. 2007. Basal-to-apical cadherin flow at cell junctions. *Nat Cell Biol* **9:** 92–98.

Kemler R, Babinet C, Eisen H, Jacob F. 1977. Surface antigen in early differentiation. *Proc Natl Acad Sci* **74:** 4449–4452.

Kiehart DP, Galbraith CG, Edwards KA, Rickoll WL, Montague RA. 2000. Multiple forces contribute to cell sheet morphogenesis for dorsal closure in *Drosophila*. *J Cell Biol* **149:** 471–490.

Knust E, Bossinger O. 2002. Composition and formation of intercellular junctions in epithelial cells. *Science* **298:** 1955–1959.

Kobielak A, Pasolli HA, Fuchs E. 2004. Mammalian formin-1 participates in adherens junctions and polymerization of linear actin cables. *Nat Cell Biol* **6:** 21–30.

Kooistra MR, Dube N, Bos JL. 2007. Rap1: A key regulator in cell-cell junction formation. *J Cell Sci* **120:** 17–22.

Kovacs EM, Ali RG, McCormack AJ, Yap AS. 2002a. E-cadherin homophilic ligation directly signals through Rac and phosphatidylinositol 3-kinase to regulate adhesive contacts. *J Biol Chem* **277:** 6708–6718.

Kovacs EM, Goodwin M, Ali RG, Paterson AD, Yap AS. 2002b. Cadherin-directed actin assembly: E-cadherin physically associates with the Arp2/3 complex to direct actin assembly in nascent adhesive contacts. *Curr Biol* **12:** 379–382.

Krendel MF, Bonder EM. 1999. Analysis of actin filament bundle dynamics during contact formation in live epithelial cells. *Cell Mot Cytoskel* **43:** 296–309.

Krendel M, Gloushankova NA, Bonder EM, Feder HH, Vasiliev JM, Gelfand IM. 1999. Myosin-dependent contractile activity of the actin cytoskeleton modulates the spatial organization of cell-cell contacts in cultured epitheliocytes. *Proc Natl Acad Sci* **96:** 9666–9670.

Lambert M, Choquet D, Mege RM. 2002. Dynamics of ligand-induced, Rac1-dependent anchoring of cadherins to the actin cytoskeleton. *J Cell Biol* **157:** 469–479.

Langevin J, Morgan MJ, Sibarita JB, Aresta S, Murthy M, Schwarz T, Camonis J, Bellaiche Y. 2005. *Drosophila* exocyst components Sec5, Sec6, and Sec15 regulate DE-Cadherin trafficking from recycling endosomes to the plasma membrane. *Develop Cell* **9:** 355–376.

Larue L, Ohsugi M, Hirchenhain J, Kemler R. 1994. E-cadherin null mutant embryos fail to form a trophectoderm epithelium. *Proc Natl Acad Sci* **91:** 8263–8267.

Le TL, Joseph SR, Yap AS, Stow JL. 2002. Protein kinase C regulates endocytosis and recycling of E-cadherin. *Am J Physiol* **283:** C489–499.

Leckband D, Prakasam A. 2006. Mechanism and dynamics of cadherin adhesion. *Ann Rev Biomed Eng* **8:** 259–287.

Lecuit T, Lenne PF. 2007. Cell surface mechanics and the control of cell shape, tissue patterns and morphogenesis. *Nat Rev* **8:** 633–644.

Leibfried A, Fricke R, Morgan MJ, Bogdan S, Bellaiche Y. 2008. *Drosophila* Cip4 and WASp define a branch of the Cdc42-Par6-aPKC pathway regulating E-cadherin endocytosis. *Curr Biol* **18:** 1639–1648.

Magie CR, Meyer MR, Gorsuch MS, Parkhurst SM. 1999. Mutations in the Rho1 small GTPase disrupt morphogenesis and segmentation during early *Drosophila* development. *Development* **126:** 5353–5364.

Magie CR, Pinto-Santini D, Parkhurst SM. 2002. Rho1 interacts with p120ctn and α-catenin, and regulates cadherin-based adherens junction components in *Drosophila*. *Development* **129:** 3771–3782.

Martin P, Parkhurst SM. 2004. Parallels between tissue repair and embryo morphogenesis. *Development* **131:** 3021–3034.

Martin AC, Kaschube M, Wieschaus EF. 2008. Pulsed contractions of an actin-myosin network drive apical constriction. *Nature* **457:** 495–499.

Matsuzaki F, Mege RM, Jaffe SH, Friedlander DR, Gallin WJ, Goldberg JI, Cunningham BA, Edelman GM. 1990. cDNAs of cell adhesion molecules of different specificity induce changes in cell shape and border formation in cultured S180 cells. *J Cell Bio* **110:** 1239–1252.

Mege RM, Gavard J, Lambert M. 2006. Regulation of cell-cell junctions by the cytoskeleton. *Current Opinion Cell Biol* **18:** 541–548.

Millard TH, Martin P. 2008. Dynamic analysis of filopodial interactions during the zippering phase of *Drosophila* dorsal closure. *Development* **135:** 621–626.

Muller HA, Wieschaus E. 1996. armadillo, bazooka, and stardust are critical for early stages in formation of the zonula adherens and maintenance of the polarized blastoderm epithelium in *Drosophila*. *J Cell Biol* **134:** 149–163.

Nakagawa M, Fukata M, Yamaga M, Itoh N, Kaibuchi K. 2001. Recruitment and activation of Rac1 by the formation of E-cadherin-mediated cell-cell adhesion sites. *J Cell Sci* **114:** 1829–1838.

Nakayama M, Goto TM, Sugimoto M, Nishimura T, Shinagawa T, Ohno S, Amano M, Kaibuchi K. 2008. Rho-kinase phosphorylates PAR-3 and disrupts PAR complex formation. *Develop Cell* **14:** 205–215.

Nelson WJ. 2003. Adaptation of core mechanisms to generate cell polarity. *Nature* **422:** 766–774.

Niessen CM, Gumbiner BM. 2002. Cadherin-mediated cell sorting not determined by binding or adhesion specificity. *J Cell Biol* **156:** 389–399.

Nishimura T, Yamaguchi T, Kato K, Yoshizawa M, Nabeshima Y, Ohno S, Hoshino M, Kaibuchi K. 2005. PAR-6-PAR-3 mediates Cdc42-induced Rac activation through the Rac GEFs STEF/Tiam1. *Nat Cell Biol* **7:** 270–277.

Noren NK, Niessen CM, Gumbiner BM, Burridge K. 2001. Cadherin engagement regulates Rho family GTPases. *J Biol Chem* **276:** 33305–33308.

Oda H, Tsukita S, Takeichi M. 1998. Dynamic behavior of the cadherin-based cell-cell adhesion system during *Drosophila* gastrulation. *Develop Biol* **203:** 435–450.

Orsulic S, Peifer M. 1996. An in vivo structure-function study of armadillo, the β-catenin homologue, reveals both separate and overlapping regions of the protein required for cell adhesion and for wingless signaling. *J Cell Biol* **134:** 1283–1300.

Oyama T, Kanai Y, Ochiai A, Akimoto S, Oda T, Yanagihara K, Nagafuchi A, Tsukita S, Shibamoto S, Ito F. 1994. A truncated β-catenin disrupts the interaction between E-cadherin and α-catenin: A cause of loss of intercellular adhesiveness in human cancer cell lines. *Cancer Res* **54:** 6282–6287.

Perez-Moreno M, Fuchs E. 2006. Catenins: Keeping cells from getting their signals crossed. *Develop Cell* **11:** 601–612.

Pettitt J, Cox EA, Broadbent ID, Flett A, Hardin J. 2003. The *Caenorhabditis elegans* p120 catenin homologue, JAC-1, modulates cadherin-catenin function during epidermal morphogenesis. *J Cell Biol* **162:** 15–22.

Pilot F, Philippe J, Lemmers C, Lecuit T. 2006. Spatial control of actin organization at adherens junctions by a synaptotagmin like protein Btsz. *Nature* **442:** 580–584.

Pokutta S, Weis WI. 2007. Structure and mechanism of cadherins and catenins in cell-cell contacts. *Ann Rev Cell Develop Biol* **23:** 237–261.

Prakasam AK, Maruthamuthu V, Leckband DE. 2006. Similarities between heterophilic and homophilic cadherin adhesion. *Proc Natl Acad Sci* **103:** 15434–15439.

Pratt HP, Ziomek CA, Reeve WJ, Johnson MH. 1982. Compaction of the mouse embryo: An analysis of its components. *J Embryol Exp Morph* **70:** 113–132.

Raich WB, Agbunag C, Hardin J. 1999. Rapid epithelial-sheet sealing in the *Caenorhabditis elegans* embryo requires cadherin-dependent filopodial priming. *Curr Biol* **9:** 1139–1146.

Rauzi M, Verant P, Lecuit T, Lenne PF. 2008. Nature and anisotropy of cortical forces orienting *Drosophila* tissue morphogenesis. *Nat Cell Biol* **10:** 1401–1410.

Redd MJ, Cooper L, Wood W, Stramer B, Martin P. 2004. Wound healing and inflammation: Embryos reveal the way to perfect repair. *Phil Trans Roy Soc London* **359:** 777–784.

Riethmacher D, Brinkmann V, Birchmeier C. 1995. A targeted mutation in the mouse E-cadherin gene results in defective preimplantation development. *Proc Natl Acad Sci* **92:** 855–859.

Sahai E, Marshall CJ. 2002. ROCK and Dia have opposing effects on adherens junctions downstream of Rho. *Nat Cell Biol* **4:** 408–415.

Sako Y, Nagafuchi A, Tsukita S, Takeichi M, Kusumi A. 1998. Cytoplasmic regulation of the movement of E-cadherin on the free cell surface as studied by optical tweezers and single particle tracking: Corralling and tethering by the membrane skeleton. *J Cell Biol* **140:** 1227–1240.

Shaye DD, Casanova J, Llimargas M. 2008. Modulation of intracellular trafficking regulates cell intercalation in the *Drosophila* trachea. *Nat Cell Biol* **10:** 964–970.

Sheffield M, Loveless T, Hardin J, Pettitt J. 2007. *C. elegans* Enabled exhibits novel interactions with N-WASP, Abl, and cell-cell junctions. *Curr Biol* **17:** 1791–1796.

Shewan AM, Maddugoda M, Kraemer A, Stehbens SJ, Verma S, Kovacs EM, Yap AS. 2005. Myosin 2 is a key Rho kinase target necessary for the local concentration of E-cadherin at cell-cell contacts. *Mol Biol Cell* **16:** 4531–4542.

Shirayoshi Y, Okada TS, Takeichi M. 1983. The calcium-dependent cell-cell adhesion system regulates inner cell mass formation and cell surface polarization in early mouse development. *Cell* **35:** 631–638.

Soto MC, Qadota H, Kasuya K, Inoue M, Tsuboi D, Mello CC, Kaibuchi K. 2002. The GEX-2 and GEX-3 proteins are required for tissue morphogenesis and cell migrations in *C. elegans*. *Genes Develop* **16:** 620–632.

Steinberg MS. 1963. Reconstruction of tissues by dissociated cells. Some morphogenetic tissue movements and the sorting out of embryonic cells may have a common explanation. *Science* **141:** 401–408.

Tanentzapf G, Tepass U. 2003. Interactions between the crumbs, lethal giant larvae and bazooka pathways in epithelial polarization. *Nat Cell Biol* **5:** 46–52.

Tepass U, Hartenstein V. 1994. The development of cellular junctions in the *Drosophila* embryo. *Develop Biol* **161:** 563–596.

Tepass U, Gruszynski-DeFeo E, Haag TA, Omatyar L, Torok T, Hartenstein V. 1996. shotgun encodes *Drosophila* E-cadherin and is preferentially required during cell rearrangement in the neurectoderm and other morphogenetically active epithelia. *Genes Develop* **10:** 672–685.

Thiery JP, Sleeman JP. 2006. Complex networks orchestrate epithelial-mesenchymal transitions. *Nat Rev* **7:** 131–142.

Torres M, Stoykova A, Huber O, Chowdhury K, Bonaldo P, Mansouri A, Butz S, Kemler R, Gruss P. 1997. An α-E-catenin gene trap mutation defines its function in preimplantation development. *Proc Natl Acad Sci* **94:** 901–906.

Troyanovsky RB, Sokolov EP, Troyanovsky SM. 2006. Endocytosis of Cadherin from Intracellular Junctions Is the Driving Force for Cadherin Adhesive Dimer Disassembly. *Mol Biol Cell* **17:** 3484–3493.

Uemura T, Oda H, Kraut R, Hayashi S, Kotaoka Y, Takeichi M. 1996. Zygotic *Drosophila* E-cadherin expression is required for processes of dynamic epithelial cell rearrangement in the *Drosophila* embryo. *Genes Develop* **10:** 659–671.

Ulrich F, Krieg M, Schotz EM, Link V, Castanon I, Schnabel V, Taubenberger A, Mueller D, Puech PH, Heisenberg CP. 2005. Wnt11 functions in gastrulation by controlling cell cohesion through Rab5c and E-cadherin. *Develop Cell* **9:** 555–564.

Vaezi A, Bauer C, Vasioukhin V, Fuchs E. 2002. Actin cable dynamics and Rho/Rock orchestrate a polarized cytoskeletal architecture in the early steps of assembling a stratified epithelium. *Develop Cell* **3:** 367–381.

Vasioukhin V, Fuchs E. 2001. Actin dynamics and cell-cell adhesion in epithelia. *Current Opinion Cell Biol* **13:** 76–84.

Vasioukhin V, Bauer C, Yin M, Fuchs E. 2000. Directed actin polymerization is the driving force for epithelial cell-cell adhesion. *Cell* **100:** 209–219.

Vasioukhin V, Bauer C, Degenstein L, Wise B, Fuchs E. 2001. Hyperproliferation and defects in epithelial polarity upon conditional ablation of α-catenin in skin. *Cell* **104:** 605–617.

Verma S, Shewan AM, Scott JA, Helwani FM, den Elzen NR, Miki H, Takenawa T, Yap AS. 2004. Arp2/3 activity is necessary for efficient formation of E-cadherin adhesive contacts. *J Biol Chem* **279:** 34062–34070.

Vestweber D, Gossler A, Boller K, Kemler R. 1987. Expression and distribution of cell adhesion molecule uvomorulin in mouse preimplantation embryos. *Develop Biol* **124:** 451–456.

von Stein W, Ramrath A, Grimm A, Muller-Borg M, Wodarz A. 2005. Direct association of Bazooka/PAR-3 with the lipid phosphatase PTEN reveals a link between the PAR/aPKC complex and phosphoinositide signaling. *Development* **132:** 1675–1686.

Wei SY, Escudero LM, Yu F, Chang LH, Chen LY, Ho YH, Lin CM, Chou CS, Chia W, Modolell J. 2005. Echinoid is a component of adherens junctions that cooperates with DE-Cadherin to mediate cell adhesion. *Develop Cell* **8:** 493–504.

Weis WI, Nelson WJ. 2006. Re-solving the cadherin-catenin-actin conundrum. *J Biol Chem* **281:** 35593–35597.

Williams-Masson EM, Malik AN, Hardin J. 1997. An actin-mediated two-step mechanism is required for ventral enclosure of the *C. elegans* hypodermis. *Development* **124:** 2889–2901.

Wood W, Jacinto A, Grose R, Woolner S, Gale J, Wilson C, Martin P. 2002. Wound healing recapitulates morphogenesis in *Drosophila* embryos. *Nat Cell Biol* **4:** 907–912.

Yamada S, Nelson WJ. 2007. Localized zones of Rho and Rac activities drive initiation and expansion of epithelial cell-cell adhesion. *J Cell Biol* **178:** 517–527.

Yamada S, Pokutta S, Drees F, Weis WI, Nelson WJ. 2005. Deconstructing the cadherin-catenin-actin complex. *Cell* **123:** 889–901.

Yamazaki D, Oikawa T, Takenawa T. 2007. Rac-WAVE-mediated actin reorganization is required for organization and maintenance of cell-cell adhesion. *J Cell Sci* **120:** 86–100.

Yap AS, Crampton MS, Hardin J. 2007. Making and breaking contacts: The cellular biology of cadherin regulation. *Current Opinion Cell Biol* **19:** 508–514.

Yin HL, Janmey PA. 2003. Phosphoinositide regulation of the actin cytoskeleton. *Ann Rev Physiol* **65:** 761–789.

Yonemura S, Itoh M, Nagafuchi A, Tsukita S. 1995. Cell-to-cell adherens junction formation and actin filament organization: Similarities and differences between non-polarized fibroblasts and polarized epithelial cells. *J Cell Sci* **108:** 127–142.

Zallen JA, Wieschaus E. 2004. Patterned gene expression directs bipolar planar polarity in *Drosophila*. *Develop Cell* **6:** 343–355.

Zandy NL, Playford M, Pendergast AM. 2007. Abl tyrosine kinases regulate cell-cell adhesion through Rho GTPases. *Proc Natl Acad Sci* **104:** 17686–17691.

Zhang J, Betson M, Erasmus J, Zeikos K, Bailly M, Cramer LP, Braga VM. 2005. Actin at cell-cell junctions is composed of two dynamic and functional populations. *J Cell Sci* **118:** 5549–5562.

Zheng Y. 2001. Dbl family guanine nucleotide exchange factors. *Trends Biochem Sci* **26:** 724–732.

Intercellular Junction Assembly, Dynamics, and Homeostasis

Kathleen J. Green[1], Spiro Getsios[2], Sergey Troyanovsky[3], and L.M. Godsel[1]

[1]Northwestern University Feinberg School of Medicine, Departments of Pathology and Dermatology, R.H. Lurie Comprehensive Cancer Center, Chicago, Illinois 60611

[2]Departments of Dermatology and Cell and Molecular Biology, R.H. Lurie Comprehensive Cancer Center, Chicago, Illinois 60611

[3]Department of Dermatoogy, R.H. Lurie Comprehensive Cancer Center, Chicago, Illinois 60611

Correspondence: kgreen@northwestern.edu

Intercellular anchoring junctions are highly specialized regions of the plasma membrane where members of the cadherin family of transmembrane adhesion molecules on opposing cells interact through their extracellular domains, and through their cytoplasmic domains serve as a platform for organizing cytoskeletal anchors and remodelers. Here we focus on assembly of so-called "anchoring" or "adhering" junctions—adherens junctions (AJs) and desmosomes (DSMs), which associate with actin and intermediate filaments, respectively. We will examine how the assembly and function of AJs and DSMs are intimately connected during embryogenesis and in adult cells and tissues, and in some cases even form specialized "mixed" junctions. We will explore signaling and trafficking machineries that drive assembly and remodeling and how these mechanisms are co-opted in human disease.

In multicellular organisms intercellular junctions serve to maintain cell and tissue polarity and integrity. In vertebrates these organelles include: gap junctions, which chemically and electrically couple neighboring cells; tight junctions, which are essential for establishing the epithelial barrier; and "adhering junctions," which organize the cortical cytoskeleton beneath the plasma membrane to modulate cell and tissue behavior. Together, these organelles integrate intra- and intercellular signaling, including regulation of nuclear functions and transcriptional pathways.

Calcium-dependent, cadherin-based anchoring junctions known as adherens junctions (AJs) and desmosomes (DSMs) organize and tether microfilaments and intermediate filaments (IF) to the plasma membrane, respectively. In polarized cells, AJs appear as continuous zonula, subjacent to the tight junctions, and are uniformly distributed along the plasma membrane in a belt that is closely opposed to an actin-rich mat underneath the plasma membrane (Niessen and Gottardi 2008). DSMs appear as "spot welds" distributed subjacent to the zonula and scattered throughout the

Cite this article as *Cold Spring Harb Perspect Biol* 2009;1:a000125

lateral border, where they distribute forces of mechanical stress via interactions with the IF cytoskeleton (Holthofer et al. 2007; Garrod and Chidgey 2008). These junction types appeared at different times during the evolution of multicellular organisms, but share common blueprints, linked assembly mechanisms and synergistic functions.

Building Blocks for Assembly of Intercellular Adhering Junctions

The structure of AJs and DSMs is roughly parallel in blueprint and can be broken down into three major components: the (1) transmembrane cadherins, the tails of which provide a scaffold for (2) armadillo family members, which help build a cortical platform for additional (3) cytoskeletal adapter proteins (see also Meng and Takeichi 2009). Together these components associate with and organize actin and IF, respectively, at the plasma membrane.

Cadherins and Nectins: Classical cadherins (e.g., E- and N- and P-cadherin) are single-pass transmembrane proteins (Hulpiau and van Roy 2009). Their extracellular cadherin domains contain a series of five conserved cadherin repeats, which extend from the cell surfaces and bind to cadherins present on adjacent cells (see Shapiro and Weis 2009). The cytoplasmic domains contain binding sites for associated catenins (see the following discussion) and other regulatory proteins (Perez-Moreno and Fuchs 2006). The desmosomal cadherins, desmogleins and desmocollins, have a similar ectodomain structure but more divergent cytoplasmic domains (Getsios et al. 2004b). In particular, the desmogleins have extended cytoplasmic tails containing variable numbers of RUD (repeating unit domain) motifs, whose function is unknown.

Nectins are more recently identified calcium-independent adhesion molecules in the IgG superfamily (Irie et al. 2004). They are thought to prime adhesive interactions for cadherins, thus initiating AJ assembly. Notably, the nectin cytoplasmic domain interacts with an actin binding protein called afadin/AF6, which has been shown to bind to α-catenin and has thus been implicated in recruiting/clustering cadherin complexes in nascent junctions, and as a possible alternative way to link actin to the plasma membrane, as described below (Tachibana et al. 2000).

Armadillo Proteins: Both desmosomal and classic cadherins contain conserved intracellular domains that bind members of the armadillo family (Hatzfeld 2007). Junctional armadillo proteins fall into two subclasses that are each characterized by a central domain built from a series of ~42 amino acid armadillo repeats. Armadillo drives the transcriptional program responsible for segment polarity in the fly, in addition to being a key component of fly AJs (Brembeck et al. 2006). Intercellular adhering junctions contain members from two armadillo subfamilies: One family contains armadillo and its close relative plakoglobin; the other group contains p120catenin, δ-catenin/NPRAP (neural plakophilin related arm protein, NPRAP), ARVCF (armadillo repeat gene deleted in Velo-Cardio-Facial syndrome), and plakophilins (PKPs) 1–3 and p0071/PKP4 (Hatzfeld 2007).

Cytoskeletal Adaptor Proteins: The cytoplasmic cadherin tails and associated armadillo proteins are embedded within a cortical meshwork containing cytoskeletal adaptor proteins and associated cytoskeletal fibers that make up the electron dense junctional "plaque." Contained within this plaque region are the β-catenin binding protein α-catenin, and the plakoglobin- and PKP-binding protein desmoplakin (Green and Simpson 2007; Niessen and Gottardi 2008). In previous textbook models, these adaptors are final links in a chain that connect cadherins to their respective cytoskeletons, actin and IF. However, cadherin-β–α-catenin and α-catenin–actin interactions have recently been shown to be mutually exclusive in vitro (Yamada et al. 2005). Thus, the identity or identities of the real actin anchors are uncertain. α-Catenin may still contribute to the effort indirectly via associated proteins such as afadin, α-actinin, vinculin, ZO-1, or EPLIN (Abe and Takeichi 2008; Stemmler 2008). Data from AJ studies also prompt new consideration of the model proposed for

DSMs, whereby desmoplakin directly links IF to the desmosomal cadherin complex.

These core junctional components are assisted by additional cell-type specific and accessory molecules including actin remodeling proteins, IF binding and cornified envelope proteins, tyrosine kinases and phosphatases, growth factor receptors, and their associated proteins, which cooperate to tailor junctions structurally and functionally (Perez-Moreno and Fuchs 2006; Holthofer et al. 2007).

EMBRYOGENESIS AND JUNCTION ASSEMBLY

An Evolutionary Perspective

The cadherin superfamily comprises over one hundred members (Hulpiau and van Roy 2009). Cadherins first appeared in single-celled organisms, such as the choanoflagellate *Monosiga brevicollis*, where they are speculated to have served a role in binding bacterial prey (Abedin and King 2008; King et al. 2008). With the acquisition of multicellularity, cadherins recruited other structural and signaling proteins into primitive AJs. This ancient "assembly" step advanced AJ functions in regulating tissue morphogenesis through junctional and nonjunctional pathways, the latter through armadillo proteins that moved between the surface and the nucleus (Grimson et al. 2000; Schneider et al. 2003). With the appearance of vertebrates came the desmosomal cadherins. Their expanded diversity allowed tailoring of adhesive junctions for new structural and morphogenetic requirements, providing complex tissues with mechanical integrity whose importance is underscored by human diseases of the desmosome and corresponding animal models (reviewed by Kowalczyk and colleagues [Delva et al. 2009]) (Cheng and Koch 2004; Lai-Cheong et al. 2007). Consistent with the mantra "ontogeny recapitulates phylogeny," AJs appear first during embryonic development. A rich body of literature regards AJs as master regulators of TJ and DSM assembly, suggesting that their assembly routes may have been overlaid on pre-existing cytoskeletal and signaling pathways.

De Novo Junction Assembly in the Preimplantation Embryo

Intercellular junctions are initially formed during compaction of the 8-cell stage embryo in a process that culminates with the differentiation of the first epithelial layer, the trophectoderm (Fig. 1) (Vestweber et al. 1987; Fleming et al. 1991; Fleming et al. 1993; Barcroft et al. 1998). Up until the early 8-cell stage, the embryo consists of loosely adherent blastomeres. On compaction, these relatively spherical blastomeres are converted into highly polarized cells that flatten to maximize contact points and begin to elaborate AJs along with focal tight and gap junctions (Fleming et al. 1993; Eckert and Fleming 2008). In contrast, DSMs anchoring intermediate filaments to cell–cell contacts are only found in the mature trophectodermal epithelium at the 32-cell stage when a blastocoel cavity has formed and tight junctions have adopted a more zonular distribution. This sequence of developmental events places E-cadherin upstream in a pathway that allows for the formation and organization of other anchoring junctions.

E-cadherin-based adhesion complexes are required for embryo compaction and the subsequent formation of a trophectodermal epithelium (Larue et al. 1994; Riethmacher et al. 1995; Torres et al. 1997; De Vries et al. 2004). The mere presence of E-cadherin or its cytoplasmic partners (α-, β-, γ-catenin) is not sufficient to trigger compaction because all of these proteins are diffusely expressed along the cell surface in blastomeres of 4-cell embryos (Kidder and McLachlin 1985; Vestweber et al. 1987; Fleming et al. 1991; Fleming et al. 1993; Collins et al. 1995; Ohsugi et al. 1996; Barcroft et al. 1998). Although the process that activates the cadherin complex during compaction is not known, PKCα is redistributed to cell–cell contacts in 8-cell blastomeres and its activation can trigger compaction and β-catenin phosphorylation (Sefton et al. 1992; Pauken and Capco 1999). Because depletion of β-catenin

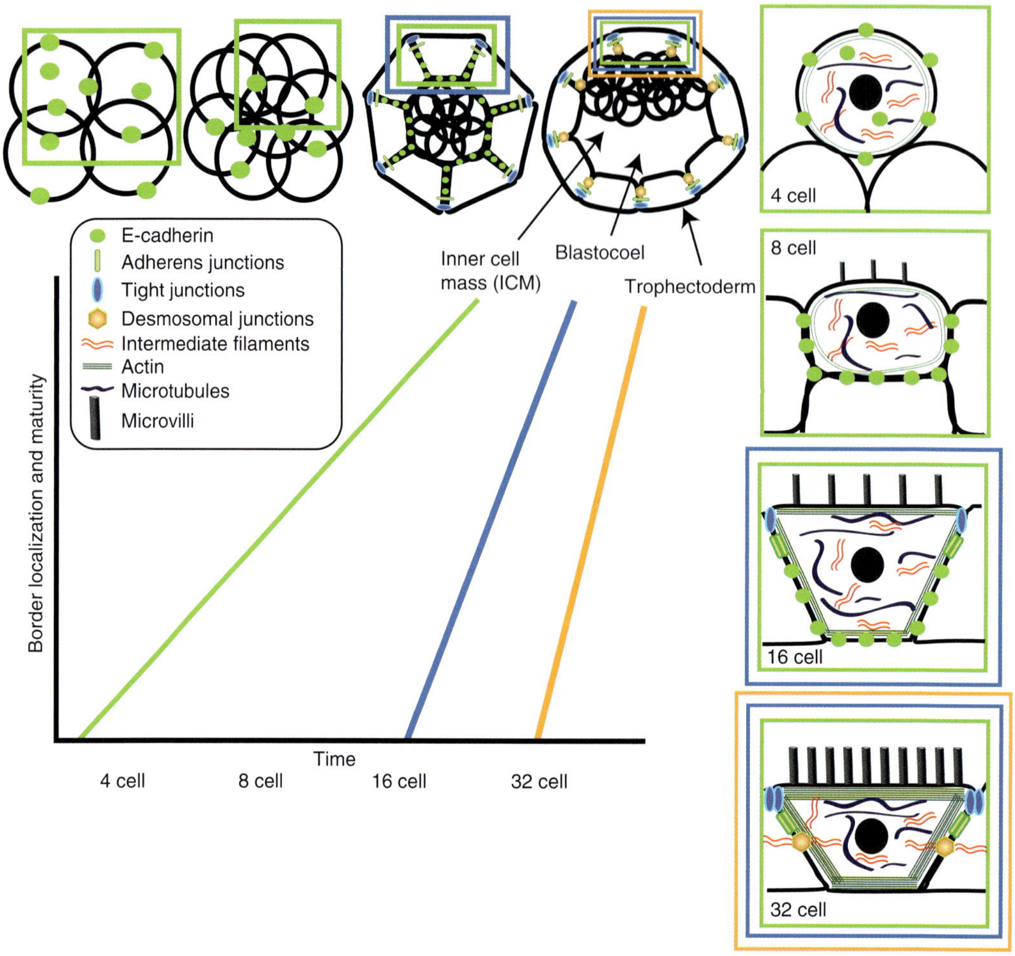

Figure 1. Junction biogenesis in preimplantation mammalian embryos. Following the initial cleavage event of the zygote and throughout the entire 4-cell stage, blastomeres are loosely connected to one another and exhibit a diffuse distribution of E-cadherin. Embryo compaction at the 8-cell stage stabilizes E-cadherin along increasing areas of cell–cell contact and initiates a process of junction assembly, cytoskeletal (actin microfilaments and microtubules) reorganization, and increased polarization of apico-basolateral membrane domains (apical microvilli). Asymmetric cell divisions that give rise to a 16-cell embryo allocate inner daughter cells lacking epithelial characteristics and an outer epithelial cell layer that gradually elaborates actin cytoskeleton-associated adherens and tight junctions (reviewed in Eckert and Fleming 2008). A functional tight junction barrier triggers cavitation and the formation of a fluid-filled blastocoel that contacts the inner cell mass; this process is coordinated with the abrupt assembly of intermediate filament-associated desmosomes in the outer trophectodermal epithelium.

results in postimplantation loss of embryos (Haegel et al. 1995), this armadillo family member is not the only target for PKC-mediated compaction events. Vezatin, which bridges the unconventional myosin VIIa motor to the cadherin–catenin complex, is a potential candidate because it maintains the levels of E-cadherin and β-catenin to allow for compaction (Hyenne et al. 2007). Accordingly, compaction also requires active actin remodeling and contraction (Kabir et al. 1996; Clayton et al. 1999).

Desmosomes form fairly abruptly in the trophectodermal epithelium at the 32-cell stage, in contrast to the more gradual maturation and redistribution of tight junctions into a zonular pattern at the apical interface. Although plakoglobin and Dsc3 are present at the time of compaction, most other desmosome proteins, including membrane (Dsg2, Dsc2) and plaque (desmoplakin) components, are first detected in the early blastocyst upon formation of an outer trophectoderm layer that surrounds the inner cell mass (Jackson et al. 1980; Fleming et al. 1991; Den et al. 2006). The induction of Dsc2 mRNA transcripts is tightly coordinated with the appearance of desmosomes and subsequent blastocoel formation (Fleming et al. 1991; Collins et al. 1995). The developmental signal that triggers Dsc2 expression appears to be E-cadherin-independent because Dsc2 and desmoplakin can be induced in the presence of function blocking antibodies for this classical cadherin. Moreover, desmosomal cadherins are normally distributed at the cell surface of E-cadherin-null blastocysts that have undergone compaction owing to the persistence of maternally derived E-cadherin or its ectopic replacement by transgenic N-cadherin expression (Ohsugi et al. 1997; Kan et al. 2007). Although Dsc2 induction appears to be a key initiator of desmosome assembly, E-cadherin is uniquely required for trophectoderm maturation leading to the death of these embryos during the preimplantation stages.

Embryos completely lacking zygotic and maternal E-cadherin cannot form DSMs and never reach the 32-cell stage because of a failure in compaction (De Vries et al. 2004; Kan et al. 2007). These results are mirrored by the observation that depletion of classic cadherins in keratinocytes results in a failure to form DSMs in vitro (Tinkle et al. 2008; Michels et al. 2009). These findings suggest that the temporal pattern of junction assembly is linked to their functional interdependence, a concept that will be further explored in the following section.

IN VITRO ANALYSIS OF CELL JUNCTION ASSEMBLY AND DYNAMICS

As during evolution and embryogenesis, AJ formation precedes DSM formation in vitro; however, rather than being drawn out over days of embryogenesis, events occur within minutes of de novo cell contact or elevation of extracellular calcium (Figs. 2 and 3). AJs and DSMs exhibit interdependence from the initiation of nascent complexes through segregation of components into specific domains. Functionally, mature junctions synergize to maintain epithelial polarity and integrity. Further, junction homeostasis is a balance of biosynthetic and endocytic pathways, each step subject to post-translational modifications and pathogenic input.

Assembly of the Adherens Junction and Desmosome Adhesive Interface

Trafficking behavior of cadherins and associated armadillo proteins: Desmosomal and classic cadherins are constantly synthesized and delivered to the plasma membrane (Fig. 4) (Penn et al. 1987; Pasdar and Nelson 1989; Shore and Nelson 1991). In response to de novo contact or elevation of extracellular calcium, cadherins cluster and become stabilized, and, in most cell types, segregate into distinct domains.

Cadherins follow a classic secretory route beginning with synthesis in the endoplasmic reticulum (ER), followed by transport to the Golgi for processing, sorting into post-Golgi carriers, and long range trafficking via MT-dependent motors (reviewed in Bryant and Stow 2004). Transport of tubulovesicular carriers containing E-cadherin out of the trans-Golgi network is dependent on the tethering protein Golgin-97 (Lock et al. 2005) and association with ankyrin-G and β-2-spectrin (Kizhatil et al. 2007). The ankyrin-G–spectrin complex later stabilizes E-cadherin by directly connecting it to the spectrin/actin cytoskeleton (Kizhatil et al. 2007), and these functions are both critical for coordinating cadherin trafficking with formation of cell–cell contacts in both cultured cells and developing embryos (Kizhatil et al. 2007). Studies linking desmosomal PKPs to vesicle trafficking have begun to emerge, with the discovery that PKP3 associates with the dynamin-like protein DNM1L, which is involved in vesicle trafficking within the

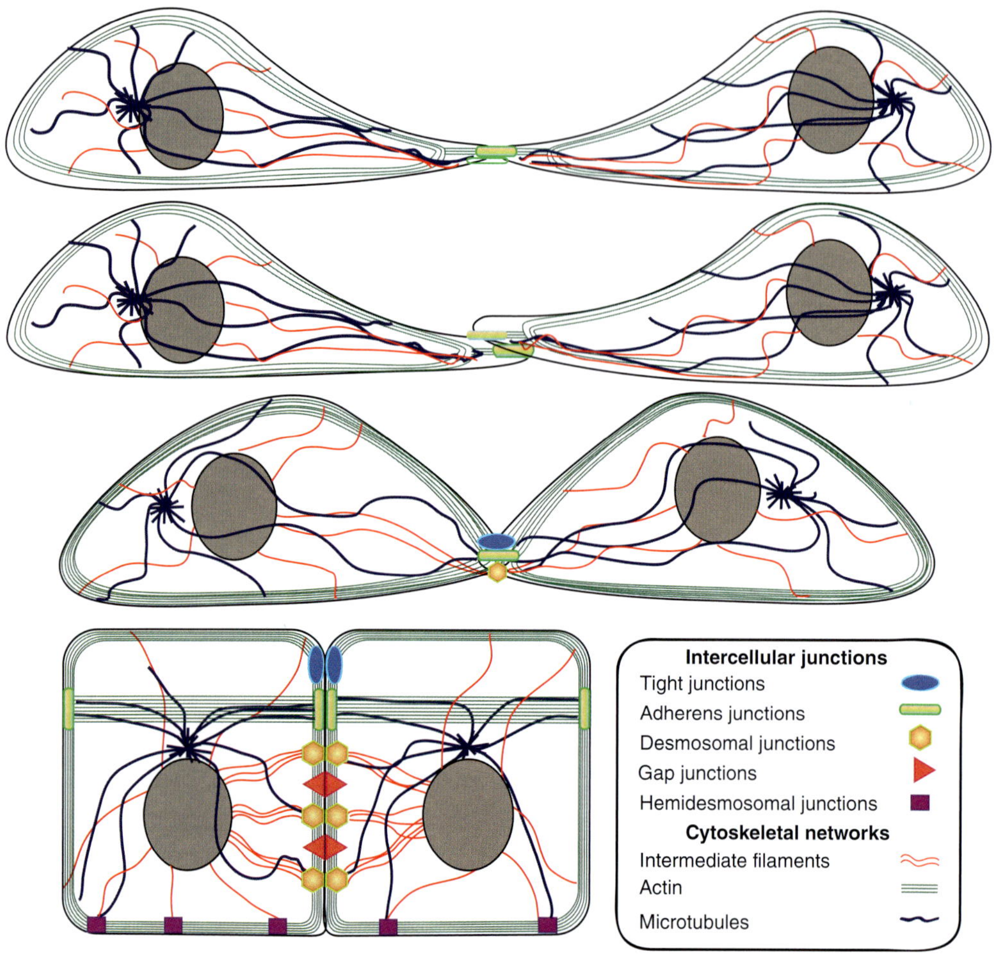

Figure 2. Junction biogenesis in vitro and associated cytoskeletal arrangements. Following cell contact in vitro, initial cadherin engagement triggers reorganization of the cytoskeleton, coordinated with downstream assembly of tight junctions and desmosomes. In polarized simple epithelial cells, the maturation of junctional contacts is accompanied by the formation of a circumferential zone of parallel microfilament bundles near the apical region of the cell, below the tight junction barrier. Desmosomes are subjacent to this complex and also appear as disclike structures that extend along the entire lateral border, where they anchor IF to the plasma membrane. MTs also anchor at AJs and DSMs.

secretory pathway in lung carcinoma cells (Furukawa et al. 2005).

Both classic and desmosomal counterparts are synthesized as precursor propeptides and processed via a furin-dependent mechanism before reaching the cell surface (Muller et al. 2004; Yokouchi et al. 2009). The presence of the propeptide is thought to prevent premature adhesion inside the cell (Ozawa and Kemler 1990). Glycosylation does not appear to be required for efficient processing of the precursor or transport of E-cadherin to the cell surface, but in the case of desmosomal cadherins, may contribute to their notable insolubility (Shore and Nelson 1991).

The E-cadherin tail is unstructured in the absence of its binding partners β-catenin and p120catenin, which protect it from degradation (Huber et al. 2001; Davis et al. 2003) and shepherd the cadherin from the ER to the plasma membrane (Chen et al. 1999). A soluble pool of plakoglobin, which is capable of associating

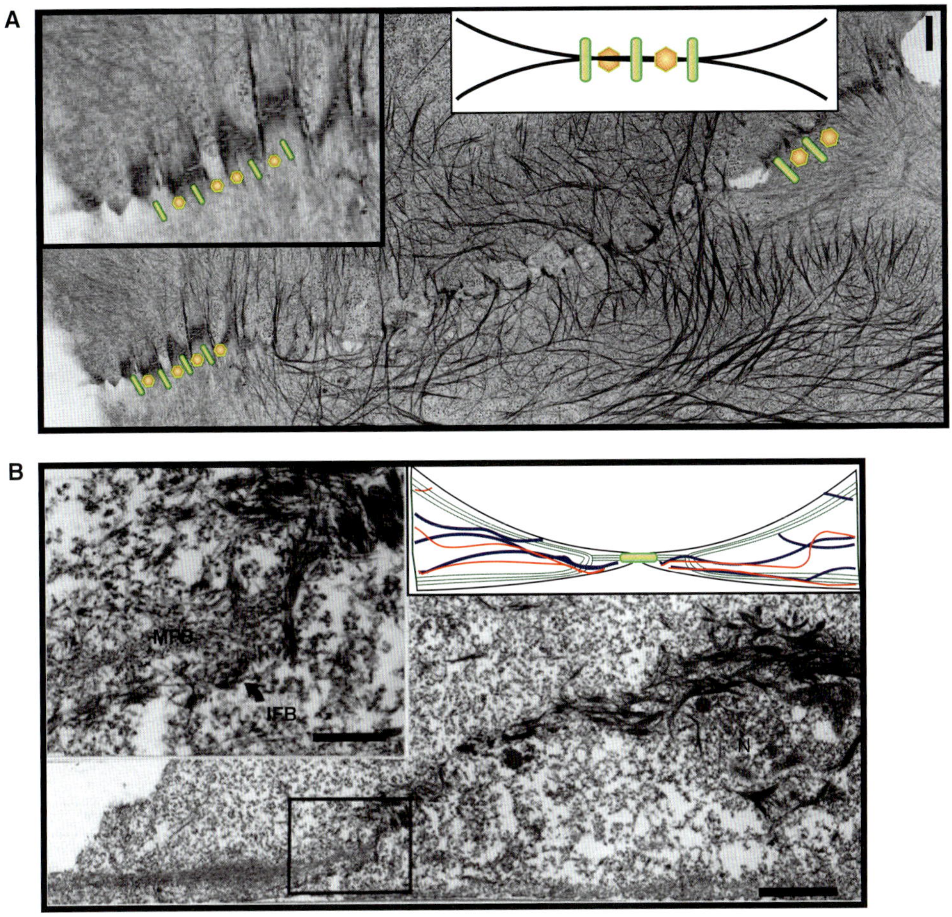

Figure 3. Ultrastructural relationships between microfilaments, IFs, and associated adhering junctions at keratinocyte cell–cell contacts and during contact formation. (*A*) Transverse section through cell contact zone between two mouse keratinocytes showing alternating arrangement of AJs and DSMs at the cell margins with their associated actin and IF networks. (*B*) Cross section of mouse keratinocyte during contact initiation shows the intimate connection between keratin IF bundles, which emanate from the nuclear surface and dive down towards and integrate within actin bundles near the substrate (Reprinted, with permission, from Green et al. 1987).

with both classic and desmosomal cadherins, initially associates primarily with E-cadherin and may serve a similar role (Hinck et al. 1994). An insoluble pool of newly synthesized plakoglobin associated with desmoglein is required for its stabilization and incorporation into DSMs (Pasdar et al. 1995). The functional implications of a potential temporal switch in partnership between classic and desmosomal cadherins was indicated by experiments suggesting that desmosome assembly in A431 cells depends on the presence of plakoglobin associated with E-cadherin (Lewis et al. 1997). Whether plakoglobin serves a chaperone role for desmosomal cadherins has not been directly tested, although desmogleins are also unstructured in the absence of plakoglobin binding (Kami et al. 2009), and truncated desmogleins deficient in plakoglobin binding fail to properly localize to plasma membranes (Andl and Stanley 2001).

In polarized epithelial cells, α-catenin and p120 associate with E-cadherin as it reaches the plasma membrane (Hinck et al. 1994;

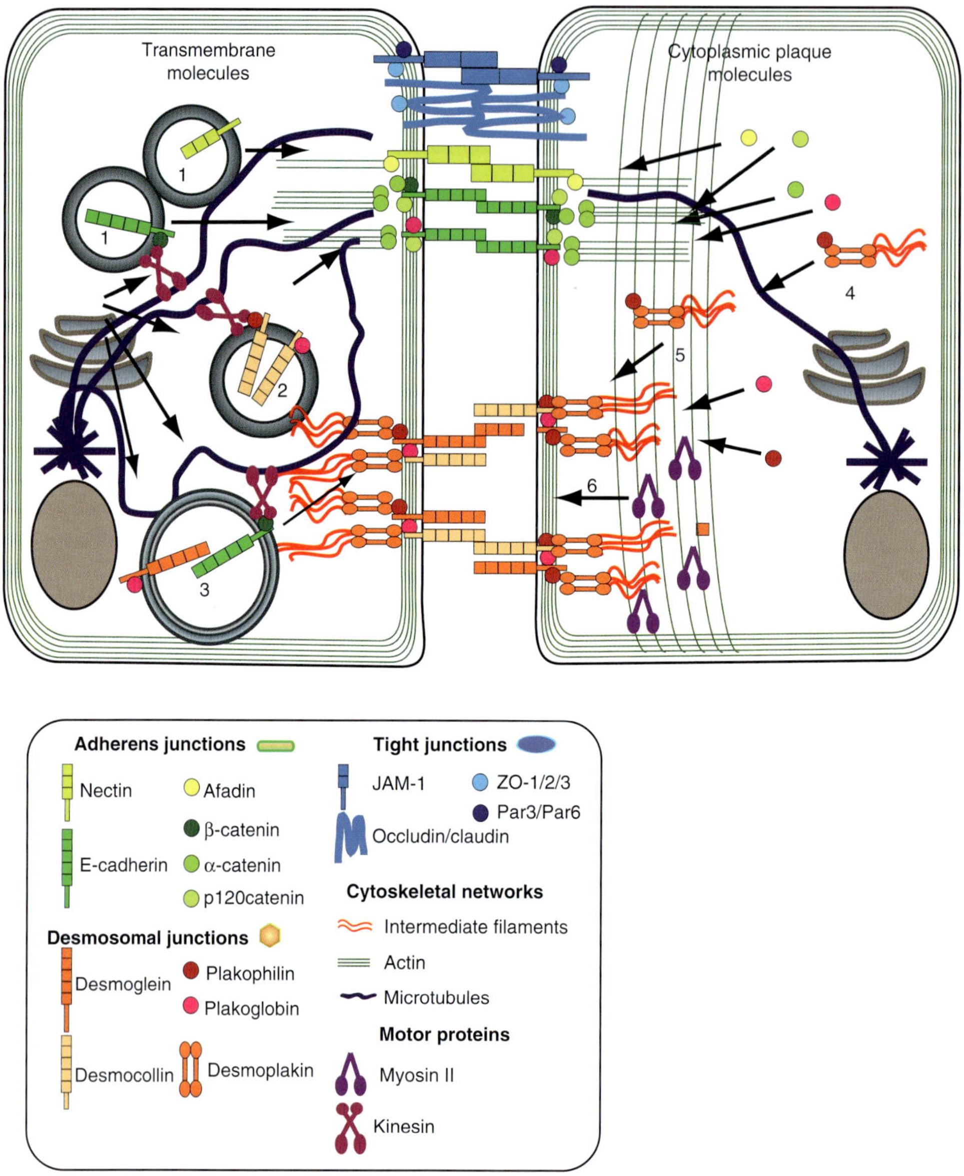

Figure 4. Model depicting possible molecular dynamics of intercellular junction assembly. Molecular players in assembly of the adhesive core (*left* cell) and junctional plaques (*right* cell) are depicted, with emphasis on AJs and DSMs assembly. Transmembrane molecules are transported in a MT-dependent fashion toward the plasma membrane, possibly via molecular motors in the kinesin family. Cadherins and nectins (1) initiate the process of assembly of the AJ adhesive interface, whereas initial recruitment of desmocollins (2) is followed by stabilization at sites of junction assembly by desmogleins (3). Cargo loading may be facilitated through armadillo family members in some cases. Junctional plaque assembly occurs through dynamic remodeling of the cortical cytoskeleton, initiated by the cadherin-dependent concentration of actin remodeling proteins at sites of junction assembly to drive actin polymerization and reorganization from branched to bundled actin. In coordination with these changes, desmosomal precursors containing desmoplakin and plakophilin are formed in the cytoplasm in close association with IF. Desmosome precursors (4) subsequently translocate to sites of desmosome assembly in an actomyosin-dependent manner to reinforce the plaque (5), in conjunction with actin rearrangement depending on myosin II (6).

Miranda et al. 2003). Here p120 promotes junction formation by stabilizing cadherins through interfering with endocytic internalization (Davis et al. 2003; Kowalczyk and Reynolds 2004) and association with the actin-associated protein afadin in a Rap1 GTPase-dependent manner (Hoshino et al. 2005).

Cadherin Dynamics and Role of Microtubules

Intercellular adhesions depend on the presence of an intact microtubule (MT) cytoskeleton (Grindstaff et al. 1998; Waterman-Storer et al. 2000; Stehbens et al. 2006; Kizhatil et al. 2007; Stehbens et al. 2009). MTs facilitate long-range trafficking of classic and desmosomal cadherins including their anterograde transport to the plasma membrane (Pasdar et al. 1991; Mary et al. 2002; Chen et al. 2003). Evidence is consistent with transport via kinesins, possibly through associated armadillo proteins (Mary et al. 2002; Chen et al. 2003; Teng et al. 2005) (Fig. 4).

Established junctions interact with both plus- and minus-end-directed MTs and associated motor proteins. These interactions regulate junction assembly state and provide transport routes to sites of cell adhesion (Ligon and Holzbaur 2007; Shaw et al. 2007). β-catenin-dynein (Ligon et al. 2001), α-E-catenin-dynactin (Lien et al. 2008) and Nezha/PLEKHA7-KIFC3 (Meng et al. 2008) associations have been revealed as critical for MT interaction and AJ stability. Further, dynamic MTs at adhesion sites promotes active myosin II recruitment, which in turn increases junctional E-cadherin (Stehbens et al. 2006). Intercellular junctions also control the spatial distribution of MTs. Notably, DSMs are required for reorganization of MTs to cell–cell contacts in differentiating keratinocytes through interactions with the centrosomal protein ninein, whose redistribution to cell–cell contacts along with MTs requires desmoplakin (Lechler and Fuchs 2007). Collectively, these data paint a picture of cooperative interactions between cadherin-based junctions and MTs in junction assembly and regulation of MT dynamics and organization.

Role of Rabs and the Exocyst

Cadherin trafficking and junction assembly is tailored through small GTPases called Rabs which recruit specific effectors in the secretion process (Caviston and Holzbaur 2006). E-cadherin has been reported to associate with an intermediate Rab11-positive recycling compartment in MDCK cells (Lock and Stow 2005). Further, Rab11 is associated with the exocyst, a multisubunit complex responsible for correct targeting of proteins to the plasma membrane. Loss of function of the exocyst components sec 5, 6, and 15 results in accumulation of *Drosophila* E-cadherin in a Rab11 compartment, inhibiting its delivery to the membrane (Langevin et al. 2005). Interfering with the tight junction scaffold protein associated with Lin seven 1 (PALS1) results in mislocalization of the exocyst and impairs E-cadherin delivery to the plasma membrane (Wang et al. 2007), and coordination between these junctions is facilitated by the balance of activities of Rab13 and Rab8 and their associated proteins (Yamamura et al. 2008).

How desmosomal and classic cadherins are coordinated during their egress out to the cell surface is not well understood. Based on the existence of "mixed" junctions that occur naturally or because of genetic interference, it seems plausible that cotrafficking of classic and desmosomal cadherins may be coupled with mechanisms to remodel or segregate components into distinct junctions in epithelial tissues (Bornslaeger et al. 1995; Ruiz et al. 1996; Borrmann et al. 2006). Temporal analysis of vesicular trafficking following a calcium switch suggests that 60-nm vesicles containing mostly Dsc2 are first observed to transport to the entire cell surface (Burdett 1998), consistent with the possibility that desmocollins serve as initiators of junction assembly in vitro (Hanakawa et al. 2000) and during embryogenesis. Larger vesicles containing desmoglein and E-cadherin were later observed to target basolaterally where they presumably associate with desmocollins and stabilize at sites of junction assembly (Wollner et al. 1992; Burdett 1998). Electron microscope pulse-chase experiments demonstrated that in keratinocytes, desmogleins first form small clusters on the surface lacking IF attachment, and later found associated with IF-attached desmosomal plaques (Sato et al. 2000). Collectively, these data suggest both a temporal and spatial staging of cadherin targeting and incorporation into junctions.

Establishing Functional Adhesion

Two models have been proposed to explain how intercellular adhesion is established once cadherins have arrived at the plasma membrane. The first depends on the initial formation of lateral or "cis" cadherin dimers, which then come together on adjacent cells to interact in trans, followed by a zippering of the adhesive interface (Shapiro et al. 1995; Pokutta and Weis 2007). The second suggests that cadherin monomers interact in *trans* to form a "strand dimer" (Boggon et al. 2002), an arrangement supported by site-specific cross-linking experiments (Troyanovsky et al. 2003) and visualized by tomographic analysis of DSMs (He et al. 2003). Additional support for this idea comes from recent single-molecule structural (fluorescence resonance energy transfer) and functional (atomic force microscopy) assays that identified *trans* but not *cis* interactions (Zhang et al. 2009).

In contrast to classic cadherins, robust desmosomal cadherin-mediated adhesion requires the presence of both desmogleins and desmocollins in the correct ratio (Amagai et al. 1994; Chidgey et al. 1996; Kowalczyk et al. 1996; Marcozzi et al. 1998; Tselepis et al. 1998; Getsios et al. 2004a). Biochemical and atomic force microscopy (AFM) suggest that desmosomal cadherins are capable of interacting in both homophilic and heterophilic configurations (Syed et al. 2002; Waschke et al. 2005; Heupel et al. 2008) and desmocollin and desmoglein from neighboring cells can be coprecipitated (Chitaev and Troyanovsky 1997). However, the organization of desmosomal cadherins in situ is still unknown.

It has been suggested that weak *trans* interactions (Chen et al. 2005) are subsequently strengthened by lateral cadherin clustering and cytoskeleton attachment (Gumbiner 1996; Kusumi et al. 1999). But the possible existence of higher affinity interactions is supported by the isolation of cadherin dimers from cells (Chitaev and Troyanovsky 1997) and the observation that cadherin mutants lacking catenin-binding sites can support cell–cell adhesion (Ozawa and Kemler 1998). The discrepancy between in vivo and in vitro data could arise from incomplete unfolding of the cadherin EC1 domain in vitro that makes strand dimer formation impossible (Troyanovsky et al. 2007). The observed range of mechanical strengths between *trans*-bonded E-cadherin pairs is also consistent with the possibility that longer lived associations can exist (Perret et al. 2004).

Assembly of the Junctional Plaque: Actin Remodeling and Desmosome Assembly are Linked

Although one can debate the existence of weak versus strong cadherin–cadherin dimers, the cortical actin cytoskeleton and myosin II activity are clearly required for adhesion strengthening, junctional plaque assembly and maintenance (Chu et al. 2004; Mege et al. 2006; Miyake et al. 2006). Classic cadherin ligation triggers rearrangements and polymerization of cortical actin that drive formation of the AJ plaque (Yonemura et al. 1995; Perez-Moreno and Fuchs 2006; Pokutta et al. 2008). During this time, the process of DSM assembly and establishment of IF attachments is set into motion, the latter stages occurring in an actin-dependent manner (Figs. 3 and 4) (Green et al. 1987; Godsel et al. 2005). A functional link between actin rearrangements and desmosome assembly may help explain why interfering with classic cadherins or α-catenin suppresses desmosome formation (Gumbiner et al. 1988; Watabe et al. 1994; Amagai et al. 1995; Taniguchi et al. 2005). Here, we will discuss actin remodeling pathways triggered by cadherin-dependent cell contact and then explore how these events may couple with downstream assembly of DSMs.

Classic Cadherins Trigger Rho GTPase-Dependent Actin Polymerization and Contractile Signaling

Cell contact occurs through the extension of lamellopodia driven by the polymerization of branched actin filaments at the leading edge of simple polarized epithelial cells (Drees et al. 2005), or filopodial projections driven by polymerization of linear actin bundles, as has been described in primary keratinocytes and developing *Drosophila*/*Caenorhabditis elegans* during tissue morphogenesis (Raich et al. 1999; Vasioukhin et al. 2000; Bloor and Kiehart 2002). The extent to which cells use one or both of these actin-dependent processes to drive the initial stages of adhesion in specific cell types is not known. It seems likely that a combination of mechanisms and actin modulators tailor remodeling of the cell cortex to contextual and cell type differences (Gates and Peifer 2005).

In the contacting lamellopodia of MDCK cells, a highly mobile fraction of E-cadherin (Adams et al. 1998) converts into immobile puncta and ultimately into larger plaques whose formation and stability requires an intact actin cytoskeleton (Adams et al. 1996). Actin rearrangements occur in concert with alterations in Rac1, Cdc42, and RhoA signaling triggered by nectin and cadherin-mediated contacts (Noren et al. 2001; Irie et al. 2004). Rac-dependent recruitment of arp2/3 (Kovacs et al. 2002), which promotes polymerization of branched actin filaments, collaborates with actin binding proteins such as cortactin, to facilitate expansion of the contact zone (Helwani et al. 2004). Actin remodeling progresses through the local accumulation and dimerization of α-catenin near contact sites. In its dimerized form, α-catenin inhibits Arp2/3, converting actin from branched to linear, bundled arrays typical of maturing AJs (Yamada et al. 2005).

Rho activity facilitates contact expansion by activating actomyosin contractility in the flanking regions of the junctional contact zone (Yamada and Nelson 2007). Rho also regulates formins, actin-associated proteins that mediate polymerization of linear bundles. The Rho effector Dia is required for junction stability and attenuation of protrusive activity in vertebrates and *Drosophila* (Sahai and Marshall 2002; Homem and Peifer 2008), depending on myosin II activity for reinforcement of cell–cell junctions (Carramusa et al. 2007). Formins are also recruited by α-catenin to E-cadherin-tipped filopodia that project into neighboring keratinocytes to draw them together and initiate AJ assembly (Kobielak et al. 2004). Rho activity can also destabilize junctions through alternate pathways (Sahai and Marshall 2002; Zandy et al. 2007; Fang et al. 2008). The ability of the cadherin-associated protein p120catenin to inhibit Rho through p190RhoGAP may ensure that Rho activity is properly titrated at cell–cell junctions (Wildenberg et al. 2006). The continued activities of Rho and its downstream effector ROCK help to transform keratinocyte sheets to stratifying tissues (Vaezi et al. 2002) and increase the height of polarized epithelial cells (Zhang et al. 2005).

AJ Maturation and Assembly of the Desmosomal Plaque

Shortly after calcium-induced cell contact in keratinocytes, electron-dense desmosome precursors associated with keratin IF align along perpendicular actin bundles that appear to emanate from newly formed AJ plaques (Green et al. 1987; Godsel et al. 2005). These precursors eventually disappear from the cytoplasm as DSMs mature (Jones and Goldman 1985; Pasdar and Nelson 1988), frequently alternating with AJs along contacting cell borders (Green et al. 1987) (Fig. 3). Live cell imaging of desmoplakin-GFP suggests that these precursors represent part of a three phase process that occurs in response to cell–cell contact in cultured keratinocytes at the leading edge of a wound. Desmosomal plaque assembly begins with the rapid accumulation of desmoplakin at contacting borders followed by a second phase (within ~20–30 min of cell contact) in which fluorescent bright nonmembrane bound desmoplakin-containing particles form in the cortical region of the cell. The subsequent translocation of particles to nascent junctions to reinforce the plaque, is coordinated with actin rearrangements and dependent on myosin II activity (Godsel et al. 2005 and unpublished results). Once formed, desmosomes continue to exhibit dynamic fusion events as well as internalization and retrograde transport (Windoffer et al. 2002).

How is the desmoplakin-IF system coupled to AJ assembly and the actin-based machinery? Although plakoglobin has been the focus of early studies as important for junctional cross talk (Lewis et al. 1997), inducible expression of the p120-related protein PKP1 has also been reported to initiate DSM assembly in a classic cadherin-dependent fashion (Wahl 2005). In addition to interacting with desmoplakin and IF, PKPs can associate with actin and cause p120-like rearrangements of the actin cytoskeleton (Hatzfeld et al. 2000) (Chen, Bass-Zubek, and Green, unpublished). PKP2 deficiency also results in major alterations in the actin cytoskeleton and rhoA signaling, likely interfering with AJ-dependent actin remodeling driving DSM assembly (Bass-Zubek, Godsel and Green, unpublished). Thus, PKPs may provide a link between the desmoplakin-IF network and an actin-dependent phase of desmosome assembly.

Loss of PKP2 also results in retention of desmoplakin on IF, preventing its incorporation into DSMs. This "beads on a string" appearance of desmoplakin is reminiscent of the distribution of desmoplakin in PKCα knockdown cells (Bass-Zubek et al. 2008) and a phosphorylation deficient desmoplakin mutant (S2849G) that exhibits delayed outward trafficking (Godsel et al. 2005) because of its higher affinity for, and aberrant retention on, IF (Stappenbeck et al. 1994; Meng et al. 1997; Fontao et al. 2003). Our data suggest a model whereby PKP2 scaffolding of PKCα promotes desmoplakin phosphorylation and assembly competence (Bass-Zubek et al. 2008). Consistent with this, PKC restores DSM assembly in α-catenin-null colon carcinoma cells and can promote DSM assembly even in reduced calcium (Sheu et al. 1989; Hengel et al. 1997). PKPs also collaborate with plakoglobin to mediate desmosomal cadherin clustering into ultrastructurally identifiable plaques (Bornslaeger et al. 2001; Koeser et al. 2003).

Like AJs, DSMs undergo a period of maturation, ultimately attaining a hyper-adhesive state which can be reversed by the calcium-dependent kinase PKCα (Watt et al. 1984; Wallis et al. 2000; Garrod et al. 2005; Kimura et al. 2006). The fact that PKCα also promotes desmosome assembly supports the idea that this kinase is important for maintaining a dynamic state, which would allow rapid desmosome remodeling in response to environmental cues. Desmosomes are also defective in patients with Darier's disease who have impaired calcium homeostasis because of mutations in the sarco(endo)plasmic reticulum Ca^{2+}-ATPase (SERCA) isoform 2 pump (Sakuntabhai et al. 1999; Dhitavat et al. 2003). Although these abnormalities were suggested to arise from improper folding of desmoplakin, it also seems possible that alterations in calcium-dependent kinase activity could contribute to the observed defects.

Achieving maximum mechanical integrity of epithelial cell sheets requires proper attachments of junctions to both IF and cortical actin, as intercellular adhesion is synergistically strengthened by the presence of IF and actin attachments (Huen et al. 2002). Further, AJs are unable to mature properly in cells lacking desmoplakin and normal IF attachments suggesting bi-directional regulation between AJs and DSMs (Vasioukhin et al. 2001b).

JUNCTION HOMEOSTASIS: ROLE OF ENDOCYTOSIS

AJs and DSMs can be modified posttranslationally by physiological stimuli resulting in changes in phosphorylation state and proteolytic processing (Gumbiner 2005; Green and Simpson 2007). These processes have traditionally been proposed to alter interactions with the cytoskeleton, thus weakening adhesive strength. However, modification of the cadherins and their associated proteins can also regulate endocytic internalization and recycling (Bryant and

Stow 2004; Ivanov et al. 2005; Yap et al. 2007; Delva and Kowalczyk 2009). Derailed endocytosis of cadherins may contribute to tumor progression (Mosesson et al. 2008) and blister formation in the autoimmune disease pemphigus (Kottke et al. 2006).

Like other receptors, cadherins can be constitutively endocytosed by a clathrin-dependent pathway (Le et al. 1999; Palacios et al. 2002; Xiao et al. 2005), possibly via binding of the adaptor complex AP2 to the dileucine motif in the juxtamembrane region (Miyashita and Ozawa 2007). This motif may be exposed on dissociation of p120, consistent with earlier work demonstrating a role for p120 in E-cadherin stability (Davis et al. 2003; Xiao et al. 2003). However, other determinants must be responsible for clathrin-dependent endocytosis of VE-cadherin, because even though it is similarly regulated by p120 (Xiao et al. 2005), it does not possess such a motif. Further, N-cadherin intracellular regions lacking the dileucine motif facilitate endocytosis (Tai et al. 2007). Basolateral targeting of E-cadherin may also be regulated by the dileucine motif (Lock and Stow 2005), and assembly and recycling facilitated by the adapter mu1B subunit through a variant of PIPKγ (Ling et al. 2007).

Growth factors such as TGFβ (Janda et al. 2006), EGF (Lu et al. 2003; Hirao et al. 2006; Bryant et al. 2007), HGF (Palacios et al. 2001), and, FGF (Bryant et al. 2005) can promote internalization and loss of junctions. Further, cadherin molecules can physically complex with growth factor receptors including EGF receptor, ErbB-1 (Dumstrei et al. 2002; Fedor-Chaiken et al. 2003) HGF receptor, c-Met (Davies et al. 2001) and FGFR (Suyama et al. 2002). These associations can either promote internalization or stabilize cell surface retention, impacting both junction integrity and growth factor-dependent signaling.

Small molecule inhibitors of EGFR/ErB2 have been shown to block internalization of E-cadherin and desmoglein 2 in tumor cells, resulting in increased adhesive strength even under conditions of reduced calcium, highlighting the therapeutic value of EGFR inhibitors for patients with cancers that overexpress EGFR (Lorch et al. 2004; Klessner et al. 2009). The ADAM family of sheddases cooperates with EGFR in increasing the internalized pool of desmoglein 2, thus reducing adhesive strength (Klessner et al. 2009). Different endocytic pathways have been suggested to mediate growth factor-regulated endocytosis, including rac-dependent macropinocytosis (Akhtar and Hotchin 2001; Bryant et al. 2007) and caveolin-dependent endocytosis (Lu et al. 2003).

Growth factor stimulation has also been linked to activation of nonreceptor tyrosine kinases such as Src. Src-dependent phosphorylation of two tyrosines of E-cadherin (Y755 and Y756) produces a binding site for E3 ubiquitin-ligase Hakai, and ubiquitination of the E-cadherin–catenin complex facilitates clathrin-dependent internalization (Fujita et al. 2002) and targeting for lysosomal degradation (Palacios et al. 2005). This pathway likely explains cadherin internalization and dissolution of AJs shown to accompany the activation of ts-v-Src in MDCK cells (Warren and Nelson 1987; Palovuori et al. 2003) or reduced extracellular calcium via activation of Cdc42 (Shen et al. 2008).

Desmosomal cadherin internalization is a contributing factor in the autoimmune disease pemphigus (Kottke et al. 2006). Dsg3 is degraded in response to PV autoantibody binding resulting in Dsg3-depleted DSMs (Aoyama and Kitajima 1999), shown to be because of rapid clathrin- and dynamin-independent internalization of a soluble pool of PV-IgG-bound Dsg3 (Calkins et al. 2006; Delva et al. 2008). PV antibodies may impair assembly of new DSMs by targeting internalization of newly synthesized Dsg3 (Sato et al. 2000; Mao et al. 2009). Although it has been reported that plakoglobin, but not desmoplakin, remains associated with the cadherin (Calkins et al. 2006), others suggest that plakoglobin dissociates from phosphorylated Dsg3 after which it is degraded, depleting a nuclear pool required for attenuating c-myc transcription (Muller et al. 2006). PV induces alterations in both IF attachment (Caldelari et al. 2001; Calkins et al. 2006) and cortical actin (Waschke et al. 2006). The latter has been linked with a reduction in Rho A activity, suggesting another link between pathways regulating AJs and DSMs (Waschke et al. 2006).

Endocytosis can be envisioned to target two pools of cadherin molecules—engaged or unengaged in adhesion. In isolated epithelial MCF-7 cells, a pool of free cadherin is subjected to macropinocytosis (Paterson et al. 2003; Bryant et al. 2007) and clathrin-dependent endocytosis was suggested to target extrajunctional cadherin (Izumi et al. 2004; Hoshino et al. 2005). However, it is also plausible that cadherins are internalized directly from the intrajunctional region, forcefully breaking apart cadherin adhesive bonds. This would be consistent with the observed association of endocytic vesicles with AJs and supported by the protective effect of endocytosis inhibitors, which lead to the rapid expansion of the junctional interface (Troyanovsky et al. 2006). Thus, cadherin endocytosis may be essential to counterbalance a continuous assembly of cadherin dimers in AJs. Supporting this idea, defects in endocytosis induced by the lack of dynamin or proteins involved in Arp2/3-dependent actin polymerization prevented completion of endocytosis in *Drosophilia* epithelial cells (Georgiou et al. 2008; Leibfried et al. 2008). Together the data suggest that different endocytic routes collaborate to regulate the assembly state of adhering junctions in normal tissues and pathogenic states.

JUNCTION ASSEMBLY IN COMPLEX TISSUES: CHALLENGES FOR THE FUTURE

Onto these basic modes of junction assembly and remodeling are overlaid additional complexities in the case of tissues such as the epidermis. This stratified squamous epithelium does not exhibit the typical apico-basolateral polarity found in simple epithelial cells. Yet, junctional components are integrated into specialized membrane complexes in a polarized fashion as basal layer keratinocytes stratify and differentiate (Green and Simpson 2007; Niessen 2007; Fuchs 2008). For example, adherens junction-enriched basal keratinocytes are accompanied by increasing numbers of DSMs, which change in molecular composition in the spinous layers and granular layers (Fig. 5 and Fig. 3 of Delva et al. 2009). Although many

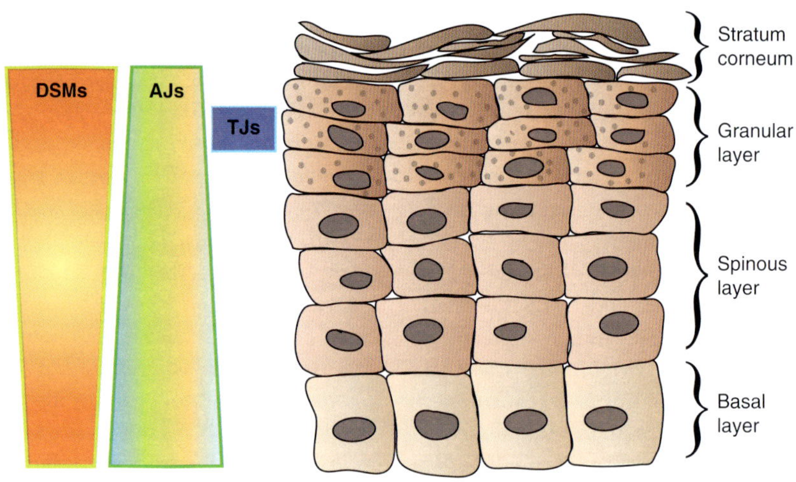

Figure 5. Organization of intercellular junctions in the epidermis. Desmosomes and AJs are found throughout the epidermis but AJs enriched in E- and P-cadherin are found in the basal layers whereas the number and surface area occupied by DSMs increases during differentiation, during which time P-cadherin is lost. See Kowalczyk and colleagues (Delva et al. 2009, Fig. 3) for the expression patterns of DSM components during differentiation in epidermis. The TJ barrier is in the granular layer of the epidermis, with structurally definable junctions restricted to the second cell layer.

tight junction components (e.g., claudin-1, occludin, and ZO-1) are expressed throughout the epidermis, a tight junction barrier is sealed only in granular cells before these eventually give rise to a cornified dead cell layer (Furuse et al. 2002). Thus, junction assembly in the epidermis is coordinated throughout multiple cell layers instead of within distinct basolateral membrane domains of a simple epithelial cell.

As in the case of compacting embryos, keratinocytes devoid of classical cadherins (E- and P-cadherin) fail to elaborate DSMs and tight junctions in vitro (Tinkle et al. 2008; Michels et al. 2009). Although depletion of classical cadherins interferes with epidermal cohesion and the tight junction barrier in vivo, desmosomes still form in epidermis lacking classical cadherins or alpha-catenin (Vasioukhin et al. 2001a; Tunggal et al. 2005; Tinkle et al. 2008). α-catenin also plays an important role in orienting the mitotic spindle in the basal layer for asymmetric cell divisions that promote stratification into the differentiated layers (Lechler and Fuchs 2005). The lack of desmoplakin has little influence on the place of cell division in basal keratinocytes, but this IF binding protein regulates MT organization in the suprabasal epidermis via interactions with ninein (Lechler and Fuchs 2007). These studies emphasize that elaboration of AJs and DSMs into a stratified epithelium not only holds the tissue together but also provides positional cues for targeting cytoskeletal, trafficking, and endocytic machineries.

How junction assembly and remodeling occurs on the molecular level in vertebrate tissues in vivo is a challenge for future study. During stratification, DSMs with different compositions reflecting differentiation-dependent expression patterns must either be made de novo or remodeled or both (North et al. 1996). In part this remodeling may be driven by transcriptional programs, such as that regulated by grainyhead transcription factors that drive expression of desmoglein 1 in differentiated layers of the epidermis (Wilanowski et al. 2008). Once expressed, whether newly synthesized cadherins are immediately incorporated into junctions or require additional regulatory cues or post-translational modifications, similar to those that occur during embryogenesis, is unknown. We also do not yet know whether, once assembly competent, molecules are inserted into pre-existing junctions or used to create new junctions. In the future, studies at the molecular level that address fundamental properties of junction dynamics, coupled with genetic analysis of in vivo models, should begin to reveal the complexity of mechanisms used in vivo.

ACKNOWLEDGMENTS

The authors would like to acknowledge Andrew Kowalczyk for helpful comments and discussion. Work in the authors' labs is supported by NIH R01's AR43380, AR41836, CA122151 (K.G.), a Dermatology Career Development Award and Zell Foundation Awards (S.G.), and R01 AR44016 (S.T.).

REFERENCES

Abe K, Takeichi M. 2008. EPLIN mediates linkage of the cadherin catenin complex to F-actin and stabilizes the circumferential actin belt. *Proc Natl Acad Sci* **105:** 13–19.

Abedin M, King N. 2008. The premetazoan ancestry of cadherins. *Science* **319:** 946–948.

Adams CL, Nelson WJ, Smith SJ. 1996. Quantitative analysis of cadherin-catenin-actin reorganization during development of cell-cell adhesion. *J Cell Biol* **135:** 1899–1911.

Adams CL, Chen Y-T, Smith SJ, Nelson WJ. 1998. Mechanisms of epithelial cell-cell adhesion and cell compaction revealed by high-resolution tracking of E-cadherin-green fluorescent protein. *J Cell Biol* **142:** 1105–1119.

Akhtar N, Hotchin NA. 2001. RAC1 regulates adherens junctions through endocytosis of E-cadherin. *Mol Biol Cell* **12:** 847–862.

Amagai M, Karpati S, Klaus-Kovtun V, Udey MC, Stanley JR. 1994. The extracellular domain of pemphigus vulgaris antigen (desmoglein 3) mediates weak homophilic adhesion. *J Invest Derm* **102:** 402–408.

Amagai M, Fujimori T, Masunaga T, Shimizu H, Nishikawa T, Shimizu N, Takeichi M, Hashimoto T. 1995. Delayed assembly of desmosomes in keratinocytes with disrupted classic-cadherin-mediated cell adhesion by a dominant negative mutant. *J Invest Derm* **104:** 27–32.

Andl CD, Stanley JR. 2001. Central role of the plakoglobin-binding domain for desmoglein 3 incorporation into desmosomes. *J Invest Dermatol* **117:** 1068–1074.

Aoyama Y, Kitajima Y. 1999. Pemphigus vulgaris-IgG causes a rapid depletion of desmoglein 3 (Dsg3) from the Triton

X-100 soluble pools, leading to the formation of Dsg3-depleted desmosomes in a human squamous carcinoma cell line, DJM-1 cells. *J Invest Dermatol* **112:** 67–71.

Barcroft LC, Hay-Schmidt A, Caveney A, Gilfoyle E, Overstrom EW, Hyttel P, Watson AJ. 1998. Trophectoderm differentiation in the bovine embryo: Characterization of a polarized epithelium. *J Reprod Fertil* **114:** 327–339.

Bass-Zubek AE, Hobbs RP, Amargo EV, Garcia NJ, Hsieh SN, Chen X, Wahl JK 3rd, Denning MF, Green KJ. 2008. Plakophilin 2: A critical scaffold for PKC α that regulates intercellular junction assembly. *J Cell Biol* **181:** 605–613.

Bloor JW, Kiehart DP. 2002. *Drosophila* RhoA regulates the cytoskeleton and cell-cell adhesion in the developing epidermis. *Development* **129:** 3173–3183.

Boggon TJ, Murray J, Chappuis-Flament S, Wong E, Gumbiner BM, Shapiro L. 2002. C-cadherin ectodomain structure and implications for cell adhesion mechanisms. *Science* **296:** 1308–1313.

Bornslaeger EA, Corcoran CM, Stappenbeck TS, Green KJ. 1995. Disruption of desmosome assembly and linkage to keratin intermediate filament bundles by an amino terminal domain of the desmosomal plaque protein desmoplakin. *Mol Biol Cell* **6:** p341a.

Bornslaeger E, Godsel L, Corcoran C, Park J, Hatzfeld M, Kowalczyk A, Green K. 2001. Plakophilin 1 interferes with plakoglobin binding to desmoplakin, yet together with plakoglobin promotes clustering of desmosomal plaque complexes at cell-cell borders. *J Cell Sci* **114:** 727–738.

Borrmann CM, Grund C, Kuhn C, Hofmann I, Pieperhoff S, Franke WW. 2006. The area composita of adhering junctions connecting heart muscle cells of vertebrates. II. Colocalizations of desmosomal and fascia adhaerens molecules in the intercalated disk. *Eur J Cell Biol* **85:** 469–485.

Brembeck FH, Rosario M, Birchmeier W. 2006. Balancing cell adhesion and Wnt signaling, the key role of β-catenin. *Curr Opin Genet Dev* **16:** 51–59.

Bryant DM, Stow JL. 2004. The ins and outs of E-cadherin trafficking. *Trends Cell Biol* **14:** 427–434.

Bryant DM, Wylie FG, Stow JL. 2005. Regulation of endocytosis, nuclear translocation, signaling of fibroblast growth factor receptor 1 by E-cadherin. *Mol Biol Cell* **16:** 14–23.

Bryant DM, Kerr MC, Hammond LA, Joseph SR, Mostov KE, Teasdale RD, Stow JL. 2007. EGF induces macropinocytosis and SNX1-modulated recycling of E-cadherin. *J Cell Sci* **120:** 1818–1828.

Burdett IDJ. 1998. Aspects of the structure and assembly of desmosomes. *Micron* **29:** 309–328.

Caldelari R, de Bruin A, Baumann D, Suter MM, Bierkamp C, Balmer V, Muller E. 2001. A central role for the armadillo protein plakoglobin in the autoimmune disease pemphigus vulgaris. *J Cell Biol* **153:** 823–834.

Calkins CC, Setzer SV, Jennings JM, Summers S, Tsunoda K, Amagai M, Kowalczyk AP. 2006. Desmoglein endocytosis and desmosome disassembly are coordinated responses to pemphigus autoantibodies. *J Biol Chem* **281:** 7623–7634.

Carramusa L, Ballestrem C, Zilberman Y, Bershadsky AD. 2007. Mammalian diaphanous-related formin Dia1 controls the organization of E-cadherin-mediated cell-cell junctions. *J Cell Sci* **120:** 3870–3882.

Caviston JP, Holzbaur EL. 2006. Microtubule motors at the intersection of trafficking and transport. *Trends Cell Biol* **16:** 530–537.

Chen Y-T, Stewart DB, Nelson WJ. 1999. Coupling assembly of the E-cadherin/b-catenin complex to efficient endoplasmic reticulum exit and basal-lateral membrane targeting of E-cadherin in polarized MDCK cells. *J Cell Biol* **144:** 687–699.

Chen X, Kojima S, Borisy GG, Green KJ. 2003. 120 catenin associates with kinesin and facilitates the transport of cadherin-catenin complexes to intercellular junctions. *J Cell Biol* **163:** 547–557.

Chen CP, Posy S, Ben-Shaul A, Shapiro L, Honig BH. 2005. Specificity of cell-cell adhesion by classical cadherins: Critical role for low-affinity dimerization through β-strand swapping. *Proc Natl Acad Sci* **102:** 8531–8536.

Cheng X, Koch PJ. 2004. In vivo function of desmosomes. *J Dermatol* **31:** 171–187.

Chidgey MAJ, Clarke JP, Garrod DR. 1996. Expression of full-length desmosomal glycoproteins (desmocollins) is not sufficient to confer strong adhesion on transfected L929 cells. *J Invest Derm* **106:** 689–695.

Chitaev NA, Troyanovsky SM. 1997. Direct $Ca2^+$-dependent heterophilic interaction between desmosomal cadherins, desmoglein and desmocollin, contributes to cell-cell adhesion. *J Cell Biol* **138:** 193–201.

Chu YS, Thomas WA, Eder O, Pincet F, Perez E, Thiery JP, Dufour S. 2004. Force measurements in E-cadherin-mediated cell doublets reveal rapid adhesion strengthened by actin cytoskeleton remodeling through Rac and Cdc42. *J Cell Biol* **167:** 1183–1194.

Clayton L, Hall A, Johnson MH. 1999. A role for Rho-like GTPases in the polarisation of mouse eight-cell blastomeres. *Dev Biol* **205:** 322–331.

Collins JE, Lorimer JE, Garrod DR, Pidsley SC, Buxton RS, Fleming TP. 1995. Regulation of desmocollin transcription in mouse preimplantation embryos. *Development* **121:** 743–753.

Davies G, Jiang WG, Mason MD. 2001. HGF/SF modifies the interaction between its receptor c-Met, the E-cadherin/catenin complex in prostate cancer cells. *Int J Mol Med* **7:** 385–388.

Davis MA, Ireton RC, Reynolds AB. 2003. A core function for 120-catenin in cadherin turnover. *J Cell Biol* **163:** 525–534.

De Vries WN, Evsikov AV, Haac BE, Fancher KS, Holbrook AE, Kemler R, Solter D, Knowles BB. 2004. Maternal β-catenin and E-cadherin in mouse development. *Development* **131:** 4435–4445.

Delva E, Kowalczyk AP. 2009. Regulation of cadherin trafficking. *Traffic* **10:** 259–267.

Delva E, Jennings JM, Calkins CC, Kottke MD, Faundez V, Kowalczyk AP. 2008. Pemphigus vulgaris IgG-induced desmoglein-3 endocytosis and desmosomal disassembly are mediated by a clathrin- and dynamin-independent mechanism. *J Biol Chem* **283:** 18303–18313.

Delva E, Tucker DK, Kowalczyk AP. 2009. The desmosome. *Cold Spring Harb Perspect Biol* **1:** 002543.

Den Z, Cheng X, Merched-Sauvage M, Koch PJ. 2006. Desmocollin 3 is required for pre-implantation development of the mouse embryo. *J Cell Sci* **119:** 482–489.

Dhitavat J, Cobbold C, Leslie N, Burge S, Hovnanian A. 2003. Impaired trafficking of the desmoplakins in cultured Darier's disease keratinocytes. *J Invest Dermatol* **121:** 1349–1355.

Drees F, Pokutta S, Yamada S, Nelson WJ, Weis WI. 2005. α-catenin is a molecular switch that binds E-cadherin-β-catenin and regulates actin-filament assembly. *Cell* **123:** 903–915.

Dumstrei K, Wang F, Shy D, Tepass U, Hartenstein V. 2002. Interaction between EGFR signaling and DE-cadherin during nervous system morphogenesis. *Development* **129:** 3983–3994.

Eckert JJ, Fleming TP. 2008. Tight junction biogenesis during early development. *Biochim Biophys Acta* **1778:** 717–728.

Fang WB, Ireton RC, Zhuang G, Takahashi T, Reynolds A, Chen J. 2008. Overexpression of EPHA2 receptor destabilizes adherens junctions via a RhoA-dependent mechanism. *J Cell Sci* **121:** 358–368.

Fedor-Chaiken M, Hein PW, Stewart JC, Brackenbury R, Kinch MS. 2003. E-cadherin binding modulates EGF receptor activation. *Cell Commun Adhes* **10:** 105–118.

Fleming TP, Garrod DR, Elsmore AJ. 1991. Desmosome biogenesis in the mouse preimplantation embryo. *Development* **112:** 527–539.

Fleming TP, Javed Q, Collins J, Hay M. 1993. Biogenesis of structural intercellular junctions during cleavage in the mouse embryo. *J Cell Sci Suppl* **17:** 119–125.

Fontao L, Favre B, Riou S, Geerts D, Jaunin F, Saurat JH, Green KJ, Sonnenberg A, Borradori L. 2003. Interaction of the bullous pemphigoid antigen 1 (BP230) and desmoplakin with intermediate filaments is mediated by distinct sequences within their COOH terminus. *Mol Biol Cell* **14:** 1978–1992.

Fuchs E. 2008. Skin stem cells: Rising to the surface. *J Cell Biol* **180:** 273–284.

Fujita Y, Krause G, Scheffner M, Zechner D, Leddy HE, Behrens J, Sommer T, Birchmeier W. 2002. Hakai, a c-Cbl-like protein, ubiquitinates and induces endocytosis of the E-cadherin complex. *Nat Cell Biol* **4:** 222–231.

Furukawa C, Daigo Y, Ishikawa N, Kato T, Ito T, Tsuchiya E, Sone S, Nakamura Y. 2005. Plakophilin 3 oncogene as prognostic marker and therapeutic target for lung cancer. *Cancer Res* **65:** 7102–7110.

Furuse M, Hata M, Furuse K, Yoshida Y, Haratake A, Sugitani Y, Noda T, Kubo A, Tsukita S. 2002. Claudin-based tight junctions are crucial for the mammalian epidermal barrier: A lesson from claudin-1-deficient mice. *J Cell Biol* **156:** 1099–1111.

Garrod DR, Berika MY, Bardsley WF, Holmes D, Tabernero L. 2005. Hyper-adhesion in desmosomes: Its regulation in wound healing and possible relationship to cadherin crystal structure. *J Cell Sci* **118:** 5743–5754.

Garrod D, Chidgey M. 2008. Desmosome structure, composition and function. *Biochim Biophys Acta* **1778:** 572–587.

Gates J, Peifer M. 2005. Can 1000 reviews be wrong? Actin, α-Catenin, adherens junctions. *Cell* **123:** 769–772.

Georgiou M, Marinari E, Burden J, Baum B. 2008. Cdc42, Par6, aPKC regulate Arp2/3-mediated endocytosis to control local adherens junction stability. *Curr Biol* **18:** 1631–1638.

Getsios S, Amargo EV, Dusek RL, Ishii K, Sheu L, Godsel LM, Green KJ. 2004a. Coordinated expression of desmoglein 1 and desmocollin 1 regulates intercellular adhesion. *Differentiation* **72:** 419–433.

Getsios S, Huen AC, Green KJ. 2004b. Working out the strength and flexibility of desmosomes. *Nat Rev Mol Cell Biol* **5:** 271–281.

Godsel LM, Hsieh SN, Amargo EV, Bass AE, Pascoe-McGillicuddy LT, Huen AC, Thorne ME, Gaudry CA, Park JK, Myung K, et al. 2005. Desmoplakin assembly dynamics in four dimensions: Multiple phases differentially regulated by intermediate filaments and actin. *J Cell Biol* **171:** 1045–1059.

Green KJ, Simpson CL. 2007. Desmosomes: New perspectives on a classic. *J Invest Dermatol* **127:** 2499–2515.

Green KJ, Geiger B, Jones JCR, Talian JC, Goldman RD. 1987. The relationship between intermediate filaments and microfilaments prior to and during the formation of desmosomes and adherens-type junctions in mouse epidermal keratinocytes. *J Cell Biol* **104:** 1389–1402.

Grimson MJ, Coates JC, Reynolds JP, Shipman M, Blanton RL, Harwood AJ. 2000. Adherens junctions and β-catenin-mediated cell signalling in a non-metazoan organism. *Nature* **408:** 727–731.

Grindstaff KK, Bacallao RL, Nelson WJ. 1998. Apiconuclear organization of microtubules does not specify protein delivery from the trans-Golgi network to different membrane domains in polarized epithelial cells. *Mol Biol Cell* **9:** 685–699.

Gumbiner BM. 1996. Cell adhesion: The molecular basis of tissue architecture and morphogenesis. *Cell* **84:** 345–357.

Gumbiner BM. 2005. Regulation of cadherin-mediated adhesion in morphogenesis. *Nat Rev Mol Cell Biol* **6:** 622–634.

Gumbiner B, Stevenson B, Grimaldi A. 1988. The role of the cell adhesion molecule uvomorulin in the formation and maintenance of the epithelial junctional complex. *J Cell Biol* **107:** 1575–1587.

Haegel H, Larue L, Ohsugi M, Fedorov L, Herrenknecht K, Kemler R. 1995. Lack of b-catenin affects mouse development at gastrulation. *Development* **121:** 3529–3537.

Hanakawa Y, Amagai M, Shirakata Y, Sayama K, Hashimoto K. 2000. Different effects of dominant negative mutants of desmocollin and desmoglein on the cell-cell adhesion of keratinocytes. *J Cell Sci* **113:** 1803–1811.

Hatzfeld M. 2007. Plakophilins: Multifunctional proteins or just regulators of desmosomal adhesion? *Biochim Biophys Acta* **1773:** 69–77.

Hatzfeld M, Haffner C, Schulze K, Vinzens U. 2000. The function of plakophilin 1 in desmosome assembly and actin filament organization. *J Cell Biol* **149:** 209–222.

He W, Cowin P, Stokes DL. 2003. Untangling desmosomal knots with electron tomography. *Science* **302:** 109–113.

Helwani FM, Kovacs EM, Paterson AD, Verma S, Ali RG, Fanning AS, Weed SA, Yap AS. 2004. Cortactin is necessary for E-cadherin-mediated contact formation and actin reorganization. *J Cell Biol* **164:** 899–910.

Hengel JV, Gohon L, Bruyneel E, Vermeulen S, Cornelissen M, Mareel M, Roy FV. 1997. Protein kinase C activation upregulates intercellular adhesion of a-catenin-negative human colon cancer cell variants via induction of desmosomes. *J Cell Biol* **137:** 1103–1116.

Heupel WM, Zillikens D, Drenckhahn D, Waschke J. 2008. Pemphigus vulgaris IgG directly inhibit desmoglein 3-mediated transinteraction. *J Immunol* **181:** 1825–1834.

Hinck L, Nathke IS, Papkoff J, Nelson WJ. 1994. Dynamics of cadherin/catenin complex formation: Ovel protein interactions and pathways of complex assembly. *J Cell Biol* **125:** 1327–1340.

Hirao T, Nanba D, Tanaka M, Ishiguro H, Kinugasa Y, Doki Y, Yano M, Matsuura N, Monden M, Higashiyama S. 2006. Overexpression of ADAM9 enhances growth factor-mediated recycling of E-cadherin in human colon cancer cell line HT29 cells. *Experimental cell research* **312:** 331–339.

Holthofer B, Windoffer R, Troyanovsky S, Leube RE. 2007. Structure and function of desmosomes. *Int Rev Cytol* **264:** 65–163.

Homem CC, Peifer M. 2008. Diaphanous regulates myosin and adherens junctions to control cell contractility and protrusive behavior during morphogenesis. *Development* **135:** 1005–1018.

Hoshino T, Sakisaka T, Baba T, Yamada T, Kimura T, Takai Y. 2005. Regulation of E-cadherin endocytosis by nectin through afadin, Rap1, 120ctn. *J Biol Chem* **280:** 24095–24103.

Huber AH, Stewart DB, Laurents DV, Nelson WJ, Weis WI. 2001. The cadherin cytoplasmic domain is unstructured in the absence of β- catenin. A possible mechanism for regulating cadherin turnover. *J Biol Chem* **276:** 12301–12309.

Huen AC, Park JK, Godsel LM, Chen X, Bannon LJ, Amargo EV, Hudson TY, Mongiu AK, Leigh IM, Kelsell DP, et al. 2002. Intermediate filament-membrane attachments function synergistically with actin-dependent contacts to regulate intercellular adhesive strength. *J Cell Biol* **159:** 1005–1017.

Hulpiau P, van Roy F. 2009. Molecular evolution of the cadherin superfamily. *Int J Biochem Cell Biol* **41:** 349–369.

Hyenne V, Souilhol C, Cohen-Tannoudji M, Cereghini S, Petit C, Langa F, Maro B, Simmler MC. 2007. Conditional knock-out reveals that zygotic vezatin-null mouse embryos die at implantation. *Mech Dev* **124:** 449–462.

Irie K, Shimizu K, Sakisaka T, Ikeda W, Takai Y. 2004. Roles and modes of action of nectins in cell-cell adhesion. *Semin Cell Dev Biol* **15:** 643–656.

Ivanov AI, Nusrat A, Parkos CA. 2005. Endocytosis of the apical junctional complex: Mechanisms and possible roles in regulation of epithelial barriers. *Bioessays* **27:** 356–365.

Izumi G, Sakisaka T, Baba T, Tanaka S, Morimoto K, Takai Y. 2004. Endocytosis of E-cadherin regulated by Rac and Cdc42 small G proteins through IQGAP1 and actin filaments. *J Cell Biol* **166:** 237–248.

Jackson BW, Grund C, Schmid E, Burki K, Franke WW, Illmensee K. 1980. Formation of cytoskeletal elements during mouse embryogenesis. *Differentiation* **17:** 161–179.

Janda E, Nevolo M, Lehmann K, Downward J, Beug H, Grieco M. 2006. Raf plus TGFβ-dependent EMT is initiated by endocytosis and lysosomal degradation of E-cadherin. *Oncogene* **25:** 7117–7130.

Jones JCR, Goldman RD. 1985. Intermediate filaments and the initiation of desmosome assembly. *J Cell Biol* **101:** 506–517.

Kabir N, Yamamura H, Takagishi Y, Inouye M, Oda S, Hidaka H. 1996. Regulation of preimplantation development of mouse embryos: Effects of inhibition of myosin light-chain kinase, a Ca^{2+}/calmodulin-dependent enzyme. *J Exp Zool* **274:** 101–110.

Kami K, Chidgey M, Dafforn T, Overduin M. 2009. The desmoglein-specific cytoplasmic region is intrinsically disordered in solution and interacts with multiple desmosomal protein partners. *J Mol Biol* **386:** 531–543.

Kan NG, Stemmler MP, Junghans D, Kanzler B, de Vries WN, Dominis M, Kemler R. 2007. Gene replacement reveals a specific role for E-cadherin in the formation of a functional trophectoderm. *Development* **134:** 31–41.

Kidder GM, McLachlin JR. 1985. Timing of transcription and protein synthesis underlying morphogenesis in preimplantation mouse embryos. *Dev Biol* **112:** 265–275.

Kimura TE, Merritt AJ, Garrod DR. 2006. Calcium-Independent Desmosomes of Keratinocytes are Hyper-Adhesive. *J Invest Dermatol.*

King N, Westbrook MJ, Young SL, Kuo A, Abedin M, Chapman J, Fairclough S, Hellsten U, Isogai Y, Letunic I, et al. 2008. The genome of the choanoflagellate *Monosiga brevicollis* and the origin of metazoans. *Nature* **451:** 783–788.

Kizhatil K, Davis JQ, Davis L, Hoffman J, Hogan BL, Bennett V. 2007. Ankyrin-G is a molecular partner of E-cadherin in epithelial cells and early embryos. *J Biol Chem* **282:** 26552–26561.

Klessner JL, Desai BV, Amargo EV, Getsios S, Green KJ. 2009. EGFR and ADAMs cooperate to regulate shedding and endocytic trafficking of the desmosomal cadherin desmoglein 2. *Mol Biol Cell* **20:** 328–337.

Kobielak A, Pasolli HA, Fuchs E. 2004. Mammalian formin-1 participates in adherens junctions and polymerization of linear actin cables. *Nat Cell Biol* **6:** 21–30.

Koeser J, Troyanovsky SM, Grund C, Franke WW. 2003. De novo formation of desmosomes in cultured cells upon transfection of genes encoding specific desmosomal components. *Experimental cell research* **285:** 114–130.

Kottke MD, Delva E, Kowalczyk AP. 2006. The desmosome: Cell science lessons from human diseases. *J Cell Sci* **119:** 797–806.

Kovacs EM, Goodwin M, Ali RG, Paterson AD, Yap AS. 2002. Cadherin-directed actin assembly. E-cadherin physically associates with the arp2/3 complex to direct actin assembly in nascent adhesive contacts. *Curr Biol* **12:** 379–382.

Kowalczyk AP, Reynolds AB. 2004. Protecting your tail: Regulation of cadherin degradation by p120-catenin. *Curr Opin Cell Biol* **16:** 522–527.

Kowalczyk AP, Borgwardt JE, Green KJ. 1996. Analysis of desmosomal cadherin-adhesive function and stoichiometry of desmosomal cadherin-plakoglobin complexes. *J Invest Derm* **107:** 293–300.

Kusumi A, Suzuki K, Koyasako K. 1999. Mobility and cytoskeletal interactions of cell adhesion receptors. *Curr Opin Cell Biol* **11:** 582–590.

Lai-Cheong JE, Arita K, McGrath JA. 2007. Genetic diseases of junctions. *J Invest Dermatol* **127:** 2713–2725.

Langevin J, Morgan MJ, Sibarita JB, Aresta S, Murthy M, Schwarz T, Camonis J, Bellaiche Y. 2005. *Drosophila* exocyst components Sec5, Sec6, Sec15 regulate DE-Cadherin trafficking from recycling endosomes to the plasma membrane. *Dev Cell* **9:** 355–376.

Larue L, Ohsugi M, Hirchenhain J, Kemler R. 1994. E-cadherin null mutant embryos fail to form a trophectoderm epithelium. *Proc Natl Acad Sci* **91:** 8263–8267.

Le TL, Yap AS, Stow JL. 1999. Recycling of E-cadherin: A potential mechanism for regulating cadherin dynamics. *J Cell Biol* **146:** 219–232.

Lechler T, Fuchs E. 2005. Asymmetric cell divisions promote stratification and differentiation of mammalian skin. *Nature* **437:** 275–280.

Lechler T, Fuchs E. 2007. Desmoplakin: An unexpected regulator of microtubule organization in the epidermis. *J Cell Biol* **176:** 147–154.

Leibfried A, Fricke R, Morgan MJ, Bogdan S, Bellaiche Y. 2008. *Drosophila* Cip4 and WASp define a branch of the Cdc42-Par6-aPKC pathway regulating E-cadherin endocytosis. *Curr Biol* **18:** 1639–1648.

Lewis JE, Wahl JK 3rd, Sass KM, Jensen PJ, Johnson KR, Wheelock MJ. 1997. Cross-talk between adherens junctions and desmosomes depends on plakoglobin. *J Cell Biol* **136:** 919–934.

Lien WH, Gelfand VI, Vasioukhin V. 2008. α-E-catenin binds to dynamitin and regulates dynactin-mediated intracellular traffic. *J Cell Biol* **183:** 989–997.

Ligon LA, Holzbaur EL. 2007. Microtubules tethered at epithelial cell junctions by dynein facilitate efficient junction assembly. *Traffic* **8:** 808–819.

Ligon LA, Karki S, Tokito M, Holzbaur EL. 2001. Dynein binds to β-catenin and may tether microtubules at adherens junctions. *Nat Cell Biol* **3:** 913–917.

Ling K, Bairstow SF, Carbonara C, Turbin DA, Huntsman DG, Anderson RA. 2007. Type I γ phosphatidylinositol phosphate kinase modulates adherens junction and E-cadherin trafficking via a direct interaction with μ1B adaptin. *J Cell Biol* **176:** 343–353.

Lock JG, Stow JL. 2005. Rab11 in recycling endosomes regulates the sorting and basolateral transport of E-cadherin. *Mol Biol Cell* **16:** 1744–1755.

Lock JG, Hammond LA, Houghton F, Gleeson PA, Stow JL. 2005. E-cadherin transport from the trans-Golgi network in tubulovesicular carriers is selectively regulated by golgin-97. *Traffic* **6:** 1142–1156.

Lorch JH, Klessner J, Park JK, Getsios S, Wu YL, Stack MS, Green KJ. 2004. Epidermal growth factor receptor inhibition promotes desmosome assembly and strengthens intercellular adhesion in squamous cell carcinoma cells. *J Biol Chem* **279:** 37191–37200.

Lu Z, Ghosh S, Wang Z, Hunter T. 2003. Downregulation of caveolin-1 function by EGF leads to the loss of E-cadherin, increased transcriptional activity of β-catenin, enhanced tumor cell invasion. *Cancer Cell* **4:** 499–515.

Mao X, Choi EJ, Payne AS. 2009. Disruption of desmosome assembly by monovalent human pemphigus vulgaris monoclonal antibodies. *J Invest Dermatol* **129:** 908–918.

Marcozzi C, Burdett IDJ, Buxton RS, Magee AI. 1998. Coexpression of both types of desmosomal cadherin and plakoglobin confers strong intercellular adhesion. *J Cell Sci* **111:** 495–509.

Mary S, Charrasse S, Meriane M, Comunale F, Travo P, Blangy A, Gauthier-Rouviere C. 2002. Biogenesis of N-cadherin-dependent cell-cell contacts in living fibroblasts is a microtubule-dependent kinesin-driven mechanism. *Mol Biol Cell* **13:** 285–301.

Mege RM, Gavard J, Lambert M. 2006. Regulation of cell-cell junctions by the cytoskeleton. *Curr Opin Cell Biol* **18:** 541–548.

Meng W, Takeichi M. 2009. Adherens junction: molecular architecture and regulation. *Cold Spring Harb Perspect Biol* **1:** 002899.

Meng J-J, Bornslaeger EA, Green KJ, Steinert PM, Ip W. 1997. Two hybrid analysis reveals fundamental differences in direct interactions between desmoplakin and cell type specific intermediate filaments. *J Biol Chem* **272:** 21495–21503.

Meng W, Mushika Y, Ichii T, Takeichi M. 2008. Anchorage of microtubule minus ends to adherens junctions regulates epithelial cell-cell contacts. *Cell* **135:** 948–959.

Michels C, Buchta T, Bloch W, Krieg T, Niessen CM. 2009. Classical cadherins regulate desmosome formation. *J Invest Dermatol.* Doi: 10.1038/jid.2009.17.

Miranda KC, Joseph SR, Yap AS, Teasdale RD, Stow JL. 2003. Contextual binding of 120ctn to E-cadherin at the basolateral plasma membrane in polarized epithelia. *J Biol Chem* **278:** 43480–43488.

Miyake Y, Inoue N, Nishimura K, Kinoshita N, Hosoya H, Yonemura S. 2006. Actomyosin tension is required for correct recruitment of adherens junction components and zonula occludens formation. *Experimental cell research* **312:** 1637–1650.

Miyashita Y, Ozawa M. 2007. A dileucine motif in its cytoplasmic domain directs β-catenin-uncoupled E-cadherin to the lysosome. *J Cell Sci* **120:** 4395–4406.

Mosesson Y, Mills GB, Yarden Y. 2008. Derailed endocytosis: An emerging feature of cancer. *Nat Rev Cancer* **8:** 835–850.

Muller EJ, Caldelari R, Posthaus H. 2004. Role of subtilisin-like convertases in cadherin processing or the conundrum to stall cadherin function by convertase inhibitors in cancer therapy. *J Mol Histol* **35:** 263–275.

Muller EJ, Caldelari R, Kolly C, Williamson L, Baumann D, Richard G, Jensen P, Girling P, Delprincipe F, Wyder M, et al. 2006. Consequences of depleted SERCA2-gated calcium stores in the skin. *J Invest Dermatol* **126:** 721–731.

Niessen CM. 2007. Tight junctions/adherens junctions: Basic structure and function. *J Invest Dermatol* **127:** 2525–2532.

Niessen CM, Gottardi CJ. 2008. Molecular components of the adherens junction. *Biochim Biophys Acta* **1778:** 562–571.

Noren NK, Niessen CM, Gumbiner BM, Burridge K. 2001. Cadherin engagement regulates Rho family GTPases. *J Biol Chem* **276:** 33305–33308.

North AJ, Chidgey MAJ, Clarke JP, Bardsley WG, Garrod DR. 1996. Distinct desmocollin isoforms occur in the same desmosomes and show reciprocally graded distributions in bovine nasal epidermis. *Proc Natl Acad Sci* **93:** 7701–7705.

Ohsugi M, Hwang SY, Butz S, Knowles BB, Solter D, Kemler R. 1996. Expression and cell membrane localization of catenins during mouse preimplantation development. *Dev Dyn* **206:** 391–402.

Ohsugi M, Larue L, Schwarz H, Kemler R. 1997. Cell-junctional and cytoskeletal organization in mouse blastocysts lacking E-cadherin. *Develop Biol* **185:** 261–271.

Ozawa M, Kemler R. 1990. Correct proteolytic cleavage is required for the cell adhesive function of uvomorulin. *J Cell Biol* **111:** 1645–1650.

Ozawa M, Kemler R. 1998. The membrane-proximal region of the E-cadherin cytoplasmic domain prevents dimerization and negatively regulates adhesion activity. *J Cell Biol* **142:** 1605–1613.

Palacios F, Price L, Schweitzer J, Collard JG, D'souza-Schorey C. 2001. An essential role for ARF6-regulated membrane traffic in adherens junction turnover and epithelial cell migration. *Embo J* **20:** 4973–4986.

Palacios F, Schweitzer JK, Boshans RL, D'souza-Schorey C. 2002. ARF6-GTP recruits Nm23-H1 to facilitate dynamin-mediated endocytosis during adherens junctions disassembly. *Nat Cell Biol* **4:** 929–936.

Palacios F, Tushir JS, Fujita Y, D'souza-Schorey C. 2005. Lysosomal targeting of E-cadherin: A unique mechanism for the down-regulation of cell-cell adhesion during epithelial to mesenchymal transitions. *Mol Cell Biol* **25:** 389–402.

Palovuori R, Sormunen R, Eskelinen S. 2003. SRC-induced disintegration of adherens junctions of madin-darby canine kidney cells is dependent on endocytosis of cadherin and antagonized by Tiam-1. *Lab Invest* **83:** 1901–1915.

Pasdar M, Nelson WJ. 1988. Kinetics of desmosome assembly in Madin-Darby canine kidney epithelial cells: Temporal and spatial regulation of desmoplakin organization and stabilization upon cell-cell contact I. Biochemical analysis. *J Cell Biol* **106:** 677–685.

Pasdar M, Nelson WJ. 1989. Regulation of desmosome assembly in epithelial cells: Kinetics of synthesis, transport, stabilization of desmoglein I, a major protein of the membrane core domain. *J Cell Biol* **109:** 163–177.

Pasdar M, Krzeminski KA, Nelson WJ. 1991. Regulation of desmosome assembly in MDCK epithelial cells: Coordination of membrane core and cytoplasmic plaque domain assembly at the plasma membrane. *J Cell Biol* **113:** 645–655.

Pasdar M, Li Z, Chlumecky V. 1995. Plakoglobin: Kinetics of synthesis, phosphorylation, stability, interactions with desmoglein and E-cadherin. *Cell Motil Cytoskel*: 258–272.

Paterson AD, Parton RG, Ferguson C, Stow JL, Yap AS. 2003. Characterization of E-cadherin endocytosis in isolated MCF-7 and chinese hamster ovary cells: The initial fate of unbound E-cadherin. *J Biol Chem* **278:** 21050–21057.

Pauken CM, Capco DG. 1999. Regulation of cell adhesion during embryonic compaction of mammalian embryos: Roles for PKC and β-catenin. *Mol Reprod Dev* **54:** 135–144.

Penn EJ, Burdett IDJ, Hobson C, Magee AI, Rees DA. 1987. Structure and assembly of desmosome junctions: Biosynthesis and turnover of the major desmosome components of Madin-Darby canine kidney cells in low calcium medium. *J Cell Biol* **105:** 2327–2334.

Perez-Moreno M, Fuchs E. 2006. Catenins: Keeping cells from getting their signals crossed. *Dev Cell* **11:** 601–612.

Perret E, Leung A, Feracci H, Evans E. 2004. Trans-bonded pairs of E-cadherin exhibit a remarkable hierarchy of mechanical strengths. *Proc Natl Acad Sci U S A* **101:** 16472–16477.

Pokutta S, Weis WI. 2007. Structure and mechanism of cadherins and catenins in cell-cell contacts. *Annu Rev Cell Dev Biol* **23:** 237–261.

Pokutta S, Drees F, Yamada S, Nelson WJ, Weis WI. 2008. Biochemical and structural analysis of α-catenin in cell-cell contacts. *Biochem Soc Trans* **36:** 141–147.

Raich WB, Agbunag C, Hardin J. 1999. Rapid epithelial-sheet sealing in the *Caenorhabditis elegans* embryo requires cadherin-dependent filopodial priming. *Curr Biol* **9:** 1139–1146.

Riethmacher D, Brinkmann V, Birchmeier C. 1995. A targeted mutation in the mouse E-cadherin gene results in defective preimplantation development. *Proc Natl Acad Sci* **92:** 855–859.

Ruiz P, Brinkmann V, Ledermann B, Behrend M, Grund C, Thalhammer C, Vogel F, Birchmeier C, Gunthert U, Franke WW, et al. 1996. Targeted mutation of plakoglobin in mice reveals essential functions of desmosomes in the embryonic heart. *J Cell Biol* **135:** 215–225.

Sahai E, Marshall CJ. 2002. ROCK and Dia have opposing effects on adherens junctions downstream of Rho. *Nat Cell Biol* **4:** 408–415.

Sakuntabhai A, Ruiz-Perez V, Carter S, Jacobsen N, Burge S, Monk S, Smith M, Munro CS, O'Donovan M, Craddock N, et al. 1999. Mutations in ATP2A2, encoding a Ca2+ pump, cause Darier disease [see comments]. *Nat Genet* **21:** 271–277.

Sato M, Aoyama Y, Kitajima Y. 2000. Assembly pathway of desmoglein 3 to desmosomes and its perturbation by pemphigus vulgaris-IgG in cultured keratinocytes, as revealed by time- lapsed labeling immunoelectron microscopy [In Process Citation]. *Lab Invest* **80:** 1583–1592.

Schneider SQ, Finnerty JR, Martindale MQ. 2003. Protein evolution: Structure-function relationships of the oncogene β-catenin in the evolution of multicellular animals. *J Exp Zoolog B Mol Dev Evol* **295:** 25–44.

Sefton M, Johnson MH, Clayton L. 1992. Synthesis and phosphorylation of uvomorulin during mouse early development. *Development* **115:** 313–318.

Shapiro L, Weis WI. 2009. Structure and biochemistry of cadherins and catenins. *Cold Spring Harb Perspect Biol* **1:** a003053.

Shapiro L, Fannon AM, Kwong PD, Thompson A, Lehmann MS, Grubel G, Legrand J-F, Als-Nielson J, Colman DR, Hendrickson WA. 1995. Structural basis of cell-cell adhesion by cadherins. *Nature* **374:** 327–337.

Shaw RM, Fay AJ, Puthenveedu MA, von Zastrow M, Jan YN, Jan LY. 2007. Microtubule plus-end-tracking proteins target gap junctions directly from the cell interior to adherens junctions. *Cell* **128:** 547–560.

Shen Y, Hirsch DS, Sasiela CA, Wu WJ. 2008. Cdc42 regulates E-cadherin ubiquitination and degradation through an epidermal growth factor receptor to Src-mediated pathway. *J Biol Chem* **283:** 5127–5137.

Sheu H-M, Kitajima Y, Yaoita H. 1989. Involvement of protein kinase C in translocation of desmoplakins from cytosol to plasma membrane during desmosome formation in human squamous cell carcinoma cells grown in low to normal calcium concentration. *Exp Cell Res* **185:** 176–190.

Shore EM, Nelson WJ. 1991. Biosynthesis of the cell adhesion molecule uvomorulin (E-cadherin) in Madin-Darby Canine Kidney epithelial cells. *J Biol Chem* **266:** 19672–19680.

Stappenbeck TS, Lamb JA, Corcoran CM, Green KJ. 1994. Phosphorylation of the desmoplakin COOH terminus negatively regulates its interaction with keratin intermediate filament networks. *J Biol Chem* **269:** 29351–29354.

Stehbens SJ, Paterson AD, Crampton MS, Shewan AM, Ferguson C, Akhmanova A, Parton RG, Yap AS. 2006. Dynamic microtubules regulate the local concentration of E-cadherin at cell-cell contacts. *J Cell Sci* **119:** 1801–1811.

Stehbens SJ, Akhmanova A, Yap AS. 2009. Microtubules and cadherins: A neglected partnership. *Front Biosci* **14:** 3159–3167.

Stemmler MP. 2008. Cadherins in development and cancer. *Mol Biosyst* **4:** 835–850.

Suyama K, Shapiro I, Guttman M, Hazan RB. 2002. A signaling pathway leading to metastasis is controlled by N-cadherin and the FGF receptor. *Cancer Cell* **2:** 301–314.

Syed SE, Trinnaman B, Martin S, Major S, Hutchinson J, Magee AI. 2002. Molecular interactions between desmosomal cadherins. *Biochem J* **362:** 317–327.

Tachibana K, Nakanishi H, Mandai K, Ozaki K, Ikeda W, Yamamoto Y, Nagafuchi A, Tsukita S, Takai Y. 2000. Two cell adhesion molecules, nectin and cadherin, interact through their cytoplasmic domain-associated proteins. *J Cell Biol* **150:** 1161–1176.

Tai CY, Mysore SP, Chiu C, Schuman EM. 2007. Activity-regulated N-cadherin endocytosis. *Neuron* **54:** 771–785.

Taniguchi T, Miyazaki M, Miyashita Y, Arima T, Ozawa M. 2005. Identification of regions of α-catenin required for desmosome organization in epithelial cells. *Int J Mol Med* **16:** 1003–1008.

Teng J, Rai T, Tanaka Y, Takei Y, Nakata T, Hirasawa M, Kulkarni AB, Hirokawa N. 2005. The KIF3 motor transports N-cadherin and organizes the developing neuroepithelium. *Nat Cell Biol* **7:** 474–482.

Tinkle CL, Pasolli HA, Stokes N, Fuchs E. 2008. New insights into cadherin function in epidermal sheet formation and maintenance of tissue integrity. *Proc Natl Acad Sci U S A* **105:** 15405–15410.

Torres M, Stoykova A, Huber O, Chowdhury K, Bonaldo P, Mansouri A, Butz S, Kemler R, Gruss P. 1997. An α-E-catenin gene trap mutation defines its function in preimplantation development. *Proc Natl Acad Sci U S A* **94:** 901–906.

Troyanovsky RB, Sokolov E, Troyanovsky SM. 2003. Adhesive and lateral E-cadherin dimers are mediated by the same interface. *Mol Cell Biol* **23:** 7965–7972.

Troyanovsky RB, Sokolov EP, Troyanovsky SM. 2006. Endocytosis of cadherin from intracellular junctions is the driving force for cadherin adhesive dimer disassembly. *Mol Biol Cell* **17:** 3484–3493.

Troyanovsky RB, Laur O, Troyanovsky SM. 2007. Stable and unstable cadherin dimers: Mechanisms of formation and roles in cell adhesion. *Mol Biol Cell* **18:** 4343–4352.

Tselepis C, Chidgey M, North A, Garrod D. 1998. Desmosomal adhesion inhibits invasive behavior. *Proc Natl Acad Sci* **95:** 8064–8069.

Tunggal JA, Helfrich I, Schmitz A, Schwarz H, Gunzel D, Fromm M, Kemler R, Krieg T, Niessen CM. 2005. E-cadherin is essential for in vivo epidermal barrier function by regulating tight junctions. *Embo J* **24:** 1146–1156.

Vaezi A, Bauer C, Vasioukhin V, Fuchs E. 2002. Actin cable dynamics and Rho/Rock orchestrate a polarized cytoskeletal architecture in the early steps of assembling a stratified epithelium. *Dev Cell* **3:** 367–381.

Vasioukhin V, Bauer C, Yin M, Fuchs E. 2000. Directed actin polymerization is the driving force for epithelial cell-cell adhesion. *Cell* **100:** 209–219.

Vasioukhin V, Bauer C, Degenstein L, Wise B, Fuchs E. 2001a. Hyperproliferation and defects in epithelial polarity upon conditional ablation of α-catenin in skin. *Cell* **104:** 605–617.

Vasioukhin V, Bowers E, Bauer C, Degenstein L, Fuchs E. 2001b. Desmoplakin is essential in epidermal sheet formation. *Nat Cell Biol* **3:** 1076–1085.

Vestweber D, Gossler A, Boller K, Kemler R. 1987. Expression and distribution of cell adhesion molecule uvomorulin in mouse preimplantation embryos. *Dev Biol* **124:** 451–456.

Wahl JK 3rd. 2005. A role for plakophilin-1 in the initiation of desmosome assembly. *J Cell Biochem* **96:** 390–403.

Wallis S, Lloyd S, Wise I, Ireland G, Fleming TP, Garrod D. 2000. The α isoform of protein kinase C is involved in signaling the response of desmosomes to wounding in cultured epithelial cells. *Mol Biol Cell* **11:** 1077–1092.

Wang Q, Chen XW, Margolis B. 2007. PALS1 regulates E-cadherin trafficking in mammalian epithelial cells. *Mol Biol Cell* **18:** 874–885.

Warren S, Nelson WJ. 1987. Nonmitogenic morphoregulatory action of pp60src on multicellular epithelial structures. *Mol Cell Biol* **7:** 1326–1337.

Waschke J, Bruggeman P, Baumgartner W, Zillikens D, Drenckhahn D. 2005. Pemphigus foliaceus IgG causes dissociation of desmoglein 1-containing junctions without blocking desmoglein 1 transinteraction. *J Clin Invest* **115:** 3157–3165.

Waschke J, Spindler V, Bruggeman P, Zillikens D, Schmidt G, Drenckhahn D. 2006. Inhibition of ρA activity causes pemphigus skin blistering. *J Cell Biol* **175:** 721–727.

Watabe M, Nagafuchi A, Tsukita S, Takeichi M. 1994. Induction of polarized cell-cell association and retardation of growth by activation of the E-cadherin-catenin adhesion system in a dispersed carcinoma line. *J Cell Biol* **127:** 247–256.

Waterman-Storer CM, Salmon WC, Salmon ED. 2000. Feedback interactions between cell-cell adherens junctions and cytoskeletal dynamics in newt lung epithelial cells [In Process Citation]. *Mol Biol Cell* **11:** 2471–2483.

Watt FM, Mattey DL, Garrod DR. 1984. Calcium-induced reorganization of desmosomal components in cultured human keratinocytes. *J Cell Biol* **99:** 2211–2215.

Wilanowski T, Caddy J, Ting SB, Hislop NR, Cerruti L, Auden A, Zhao LL, Asquith S, Ellis S, Sinclair R, et al. 2008. Perturbed desmosomal cadherin expression in grainy head-like 1-null mice. *Embo J* **27:** 886–897.

Wildenberg GA, Dohn MR, Carnahan RH, Davis MA, Lobdell NA, Settleman J, Reynolds AB. 2006. p120-catenin and p190RhoGAP regulate cell-cell adhesion by coordinating antagonism between Rac and Rho. *Cell* **127:** 1027–1039.

Windoffer R, Borchert-Stuhltrager M, Leube RE. 2002. Desmosomes: Interconnected calcium-dependent structures of remarkable stability with significant integral membrane protein turnover. *J Cell Sci* **115:** 1717–1732.

Wollner DA, Krzeminski KA, Nelson WJ. 1992. Remodeling the cell surface distribution of membrane proteins during the development of epithelial cell polarity. *J Cell Biol* **116:** 889–899.

Xiao K, Allison DF, Buckley KM, Kottke MD, Vincent PA, Faundez V, Kowalczyk AP. 2003. Cellular levels of p120 catenin function as a set point for cadherin expression levels in microvascular endothelial cells. *J Cell Biol* **163:** 535–545.

Xiao K, Garner J, Buckley KM, Vincent PA, Chiasson CM, Dejana E, Faundez V, Kowalczyk AP. 2005. p120-Catenin regulates clathrin-dependent endocytosis of VE-cadherin. *Mol Biol Cell* **16:** 5141–5151.

Yamada S, Nelson WJ. 2007. Localized zones of Rho and Rac activities drive initiation and expansion of epithelial cell-cell adhesion. *J Cell Biol* **178:** 517–527.

Yamada S, Pokutta S, Drees F, Weis WI, Nelson WJ. 2005. Deconstructing the cadherin-catenin-actin complex. *Cell* **123:** 889–901.

Yamamura R, Nishimura N, Nakatsuji H, Arase S, Sasaki T. 2008. The interaction of JRAB/MICAL-L2 with Rab8 and Rab13 coordinates the assembly of tight junctions and adherens junctions. *Mol Biol Cell* **19:** 971–983.

Yap AS, Crampton MS, Hardin J. 2007. Making and breaking contacts: The cellular biology of cadherin regulation. *Curr Opin Cell Biol* **19:** 508–514.

Yokouchi M, Saleh MA, Kuroda K, Hachiya T, Stanley JR, Amagai M, Ishii K. 2009. Pathogenic epitopes of autoantibodies in pemphigus reside in the amino-terminal adhesive region of desmogleins which are unmasked by proteolytic processing of prosequence. *J Invest Dermatol* doi: 10.1038/jid.2009.61.

Yonemura S, Itoh M, Nagafuchi A, Tsukita S. 1995. Cell-to-cell adherens junction formation and actin filament organization: Similarities and differences between non-polarized fibroblasts and polarized epithelial cells. *J Cell Sci* **108:** 127–142.

Zandy NL, Playford M, Pendergast AM. 2007. Abl tyrosine kinases regulate cell cell adhesion through Rho GTPases. *Proc Natl Acad Sci* **104:** 17686–17691.

Zhang J, Betson M, Erasmus J, Zeikos K, Bailly M, Cramer LP, Braga VM. 2005. Actin at cell-cell junctions is composed of two dynamic and functional populations. *J Cell Sci* **118:** 5549–5562.

Zhang Y, Sivasankar S, Nelson WJ, Chu S. 2009. Resolving cadherin interactions and binding cooperativity at the single-molecule level. *Proc Natl Acad Sci* **106:** 109–114.

Adhesive and Signaling Functions of Cadherins and Catenins in Vertebrate Development

Ewa Stepniak[1], Glenn L. Radice[2], and Valeri Vasioukhin[1,3]

[1]Division of Human Biology, Fred Hutchinson Cancer Research Center, Seattle, Washington 98109

[2]Center for Translational Medicine, Department of Medicine, Jefferson Medical College, Philadelphia, Pennsylvania 19107

[3]Department of Pathology and Institute for Stem Cell and Regenerative Medicine, University of Washington, Seattle, Washington 98195

Correspondence: vvasiouk@fhcrc.org

Properly regulated intercellular adhesion is critical for normal development of all metazoan organisms. Adherens junctions play an especially prominent role in development because they link the adhesive function of cadherin–catenin protein complexes to the dynamic forces of the actin cytoskeleton, which helps to orchestrate a spatially confined and very dynamic assembly of intercellular connections. Intriguingly, in addition to maintaining intercellular adhesion, cadherin–catenin proteins are linked to several major developmental signaling pathways crucial for normal morphogenesis. In this article we will highlight the key genetic studies that uncovered the role of cadherin–catenin proteins in vertebrate development and discuss the potential role of these proteins as molecular biosensors of external cellular microenvironment that may spatially confine signaling molecules and polarity cues to orchestrate cellular behavior throughout the complex process of normal morphogenesis.

Development of any multicellular organism is impossible without a dynamic and properly regulated intercellular adhesion. Adhesive contacts between cells provide a physical anchoring system that is necessary to form highly organized tissues, and these contacts are essential for effective intercellular communication that ensures the homeostasis and survival of the entire organism. A number of unique developmental processes, including such early events as embryonic compaction and first cell fate specification, as well as later tissue morphogenesis and organogenesis, rely on a dynamic balance between cellular adhesion and migration. Cadherin–catenin protein complexes, which constitute the core of a specialized subtype of cellular adhesion structures termed adherens junctions (AJs), play a particularly important role during these processes. Apart from maintaining adhesive contacts at the cell–cell junctions, they are actively involved in epithelial-to-mesenchymal and mesenchymal-to-epithelial transitions, which are crucial to sustain the tissue plasticity

 doi: 10.1101/cshperspect.a002949
Cite this article as *Cold Spring Harb Perspect Biol* 2009;1:a002949

during development. Most importantly, the components of cadherin–catenin complexes are tightly linked to several major signaling networks controlling cell division, differentiation, and apoptosis and this feature is crucial for the broad roles of the AJs throughout the vertebrate development (see Cavey and Lecuit 2009).

This article will focus on the role of cadherin–catenin proteins in regulating the signaling events critical for vertebrate development. Altering the expression pattern of particular cadherin–catenin complex components in the developing embryo often leads to major developmental defects, which reflect their role in both signaling and mechanical adhesion. In this article, we will highlight crucial findings suggesting that cadherin–catenin complexes provide not only the structural integrity of the tissue, but may also serve as biosensors of the external cellular microenvironment that modulate cellular behavior and make individual cells work together to ensure the fitness of the entire organism.

STRUCTURAL AND FUNCTIONAL FEATURES OF CADHERIN–CATENIN COMPLEXES

Core Components and their Functional Connection to the Actin Cytoskeleton

AJs are a specialized type of cellular adhesion structures formed by complex and highly dynamic interactions between two families of proteins: cadherins and catenins. The main building blocks of AJs involve classical cadherins, with E(epithelial)-cadherin being a prototypic member of this family, and several closely related members of the armadillo repeat protein superfamily: p120-, β- and γ-catenin (plakoglobin), as well as structurally unrelated and lacking armadillo domains α-catenin (Niessen 2007; Hartsock and Nelson 2008) (see also Meng and Takeichi 2009). Other well-known members of the classical cadherin family include N(neural)-, P(placental)-, VE-(vascular-endothelial), R(retinal)-, and K(kidney)-cadherins (Hulpiau and van Roy 2009). Similarly, depending on the tissue type, three α-catenin genes may be involved in formation of AJs: αE(epithelial)-, αN(neural)-, and αT(testis)-catenins (Nagafuchi et al. 1991; Hirano et al. 1992; Claverie et al. 1993; Janssens et al. 2001). The assembly of AJs is triggered by homophilic binding of the extracellular parts of cadherin molecules on the neighboring cells and a binding of their cytoplasmic tails to p120-catenin and either of two closely related proteins, β-catenin or plakoglobin (Fig. 1). While p120-catenin acts to stabilize cadherins at the cell surface (Reynolds and Carnahan 2004), β-catenin provides a link to α-catenin (Aberle et al. 1994), which in turn has the ability to provide a functional link to the actin cytoskeleton, thus promoting AJ protein clustering and stabilization of cellular adhesion (Hirano et al. 1992). Because binding of α-catenin to actin filaments and β-catenin is mutually exclusive, it is unlikely that it can directly bridge AJs with the actin cytoskeleton (Drees et al. 2005; Yamada et al. 2005). Instead, it can link them indirectly, through interaction with another actin-binding protein, eplin (Abe and Takeichi 2008). Moreover, through interaction with formins, α-catenin can regulate junctional actin polymerization (Vasioukhin et al. 2000; Kobielak et al. 2004). Besides α-catenin, p120-catenin has also a pivotal role in the actin cytoskeleton dynamics at the AJs by regulation of Rho-family small GTPases (Reynolds 2007). The ability of AJs' core components to reorganize the actin cytoskeleton makes the assembly of cadherin–catenin adhesion complexes a highly dynamic process, which allows spatial reorganization of cells during normal development.

A Crosstalk between the Cadherin–Catenin Complexes and Major Developmental Signaling Pathways

Besides their structural role in stabilizing adhesive contacts between the neighboring cells and directing actin cytoskeleton reorganization, components of the cadherin–catenin complex are tightly linked to several key signal

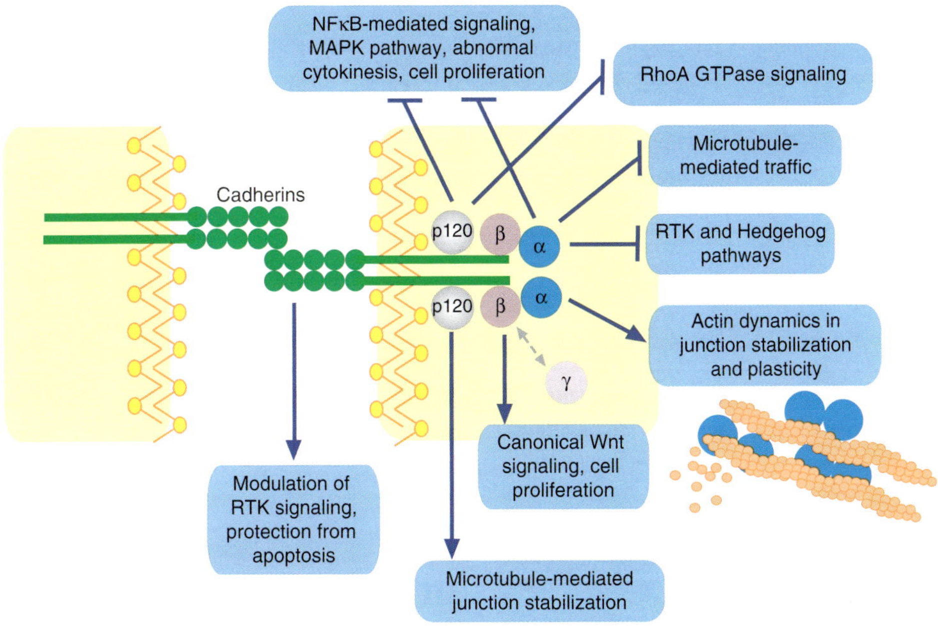

Figure 1. Cadherin–catenin proteins and their connections to developmental cell signaling pathways. Diagram showing major proteins involved in formation of cadherin, α-, β-, γ-, and p120-catenin complexes at the plasma membranes of two juxtaposed cells. Although cadherins are implicated in regulation of receptor-type tyrosine kinases (RTKs) and protection from apoptosis, catenins have multiple overlapping and distinct functions. Both p120-catenin and α-catenin regulate NFκB and MAPK signaling pathways, cytokinesis, and cell proliferation. However, p120-catenin has a unique function in the control of RhoA signaling and microtubule cytoskeleton organization, whereas α-catenin is implicated in regulation of RTKs, Hedgehog signaling, microtubule-mediated traffic, and functional coupling between the junctions and the actin cytoskeleton. β-catenin is a principal player in the canonical Wnt signaling pathway. Because RTKs, Hedgehog, and Wnt signaling mechanisms are heavily implicated in regulation of normal development, intriguing connections between cadherin–catenin receptors and these pathways provides potential tools for integrating the efforts of individual cells into a smoothly orchestrated developmental process.

transduction networks (Fig. 1). For example, β-catenin is a central player in the canonical Wnt signaling pathway, where it translocates to the nucleus and functions as a transcriptional cofactor (Clevers 2006; Grigoryan et al. 2008). One potential determinant of β-catenin's availability to participate in either the adhesion or transcriptional complexes may lay in its phosphorylation status. Indeed, β-catenin can be phosphorylated by a variety of serine/threonine and tyrosine kinases and this can profoundly change the ability of β-catenin to interact with cadherin and α-catenin, as well as its signaling function (Piedra et al. 2003; Brembeck et al. 2004; van Buul et al. 2005; Brembeck et al. 2006; Coluccia et al. 2006; Zinser et al. 2006; Coluccia et al. 2007; Daugherty and Gottardi 2007; Kajiguchi et al. 2007; Rhee et al. 2007). Overall, it appears that phosphorylation-mediated destabilization of cadherin–catenin complexes often results in activation of β-catenin signaling.

In addition to β-catenin, other members of the cadherin–catenin adhesion complex also play important roles in regulation of developmental cellular signaling networks. For example, receptor-type tyrosine kinase (RTK) pathways play a critical role in metazoan development. Cadherin–catenin complexes bind to and regulate the activity of a variety of RTKs (Erez et al. 2005). Interestingly, this interaction is often functionally complex. Although at steady state levels E-cadherin inhibits RTK signaling (Qian et al. 2004), formation of AJs can induce transient activation of RTKs (Pece and Gutkind 2000). Moreover, the outcome of functional interaction also depends on the type of involved cadherin and RTK molecules. For example, N-cadherin is generally a positive regulator of FGF receptor signaling (Suyama et al. 2002).

Both, p120- and α-catenins have been recently implicated in the control of NFκB (nuclear factor κB) signaling pathway, which is involved in cell stress and survival and often hyperactivated in cancer (Kobielak and Fuchs 2006; Perez-Moreno et al. 2006). Although exact mechanisms responsible for the connection between catenins and NFκB signaling are not well understood, p120-catenin's role in this context has been linked to the control of RhoA-dependent signaling (Perez-Moreno et al. 2006; Perez-Moreno et al. 2008). Likewise, studies on the role of α-catenin in epidermal development revealed an interesting link between this protein and Ras-MAPK pathway (Vasioukhin et al. 2001). Further studies using mice lacking α-catenin in developing neural progenitors uncovered the role of this protein in the negative regulation of the Hedgehog signaling pathway, one of the major developmental pathways in vertebrates (Lien et al. 2006b).

Overall, a significant amount of information has been accumulated, which indicates that cadherin–catenin complexes are not only responsible for mechanical intercellular adhesion, but they also provide a dynamic link to major developmental signaling networks. These features may explain why cadherin–catenin proteins are among the key regulators of principal vertebrate developmental events, which will be discussed in detail in the following section.

CADHERINS AND CATENINS IN VERTEBRATE DEVELOPMENT

Loss-of-function experiments have been very informative in uncovering the role and significance of specific members of cadherin–catenin complexes in vertebrate development. This is usually accomplished by the analyses of the mutant embryos, which are generated by targeted gene mutations, transgenic expression of dominant-negative constructs or injection of morpholino antisense oligos, interfering with translation of specific mRNAs. These experiments revealed that many core components of the AJs are essential for embryonic development (Table 1). Thus, conditional and tissue-specific gene knockout experiments were used to analyze the function of essential cadherin–catenin proteins at the later developmental time points. Altogether, these experiments revealed a highly complex picture concerning the function of these proteins in the developing embryo (Fig. 2).

Early Embryogenesis

During early embryonic development, significant amounts of E-cadherin and αE-catenin are provided maternally and mediate blastomere adhesion during embryonic compaction (Ohsugi et al. 1996; De Vries et al. 2004). Later, E-cadherin produced by the embryo becomes necessary for normal development, as zygotic E-cadherin-null embryos fail to form a trophectodermal epithelium and are unable to generate blastocyst cavity (Larue et al. 1994; Ohsugi et al. 1997). A unique role of E-cadherin in trophectoderm development was recently shown by gene replacement experiments (Kan et al. 2007). In this approach, N-cadherin expressed from the E-cadherin locus was unable to rescue the lethality of E-cadherin-null embryos, as it was insufficient to drive the formation of functional trophectoderm (Kan et al. 2007). Remarkably, ES cells generated from these mutants were able to form an epithelium, indicating that despite the fact that N-cadherin is capable of supporting epithelial morphogenesis, E-cadherin has a very unique function in trophectoderm formation. Interestingly, the phenotype of zygotic αE-catenin-null embryos has striking similarities to embryos missing E-cadherin. The gene-trap mutants missing the actin-binding domain of αE-catenin show disruption of trophoblast epithelium and the development is consequently blocked at the blastocyst stage (Torres et al. 1997). This striking similarity

Table 1. Phenotypes of zygotic and conditional mutations of cadherin–catenin genes in mice

Gene	Tissue/Time	Cre[a]	Phenotype	Ref.
E-cadherin (CDH1)	Maternal contribution to oocyte	ZP3-Cre	Normal development via zygotic expression	(De Vries et al. 2004)
	Zygote		Lethal at E4; defects in trophectoderm	(Larue et al. 1994; Riethmacher et al. 1995)
	Zygote (gene replacement, N-cadherin into E-cadherin locus)		Lethal; insufficient to rescue trophectoderm defects	(Kan et al. 2007)
	Epidermis/E11.5	K14-Cre	Hair follicle degeneration, progressive hyperplasia	(Tinkle et al. 2004)
	Epidermis/E11.5	K14-Cre	Lethal at P0; failure of tight junctions formation	(Tunggal et al. 2005)
	Epidermis/P15	Krox20-Cre	Loss of hair follicles; abnormal differentiation	(Young et al. 2003)
	Liver/E15	Alfp-Cre	No phenotype	(Battle et al. 2006)
	Mammary gland	MMTV-Cre	Impaired differentiation and apoptosis of alveolar epithelial cells	(Boussadia et al. 2002)
N-cadherin (CDH2)	Zygote		Lethal at E10; loss of cardiomyocyte adhesion	(Radice et al. 1997b)
	Zygote+transgene rescue		Lethal at E12; rescue of heart defects by both N-cadherin or E-cadherin expression	(Luo et al. 2001)
	Chimeric embryos/N-cad null ES cells		Exclusion and segregation of mutant cells from wild-type cells	(Kostetskii et al. 2001)
	Endothelial cells/E9.5-E11.5	Tie2-Cre	Lethal at mid-gestation; vascular morphogenesis defects	(Luo and Radice 2005)
	Cerebral cortex/E10.5	D6-Cre	Complete disorganization of cerebral cortex	(Kadowaki et al. 2007)
	Neural Crest	Wnt1-Cre	Lethal; aberrant remodeling of cardiac outflow tract	(Luo et al. 2006)
	Epicardium	Wnt1-Cre	Lethal; disruption of epicardial–myocardial cell interactions thinning of the myocardial wall	(Luo et al. 2006)
P-cadherin (CDH3)	Zygote		Precocious differentiation of the mammary gland	(Radice et al. 1997a)
R-cadherin (CDH4)	Zygote		Live and fertile; dilated kidney proximal tubules	(Dahl et al. 2002)
VE-cadherin (CDH5)	Zygote		Lethal at E9.5; severe vascular defects	(Carmeliet et al. 1999; Gory-Faure et al. 1999)
K-cadherin (CDH6)	Zygote		Live and fertile; delayed mesenchymal to epithelial conversion during kidney development	(Mah et al. 2000)

Continued

Table 1. *Continued*

Gene	Tissue/Time	Cre[a]	Phenotype	Ref.
Cadherin-8 (CDH8)	Zygote		Live and fertile; defective coupling between cold-sensitive sensory neurons and their target spinal cord neurons	(Suzuki et al. 2007)
OB-cadherin (CDH11)	Zygote		Live and fertile; increased long term neuronal potentiation, reduced fear- and anxiety-related responses, reduced bone density, hypoplastic synovial lining and resistance to arthritis	(Horikawa et al. 1999; Manabe et al. 2000; Kawaguchi et al. 2001; Lee et al. 2007)
M-cadherin (CDH15)	Zygote		Live and fertile; no developmental defects	(Hollnagel et al. 2002)
BILL-cadherin (CDH17)	Zygote		Abnormal immune B cell development	(Ohnishi et al. 2005)
Otocadherin (CDH23)	Zygote		Deafness and vestibular defects because of degeneration of the organ of Corti and other inner ear structures	(Wilson et al. 2001; Noben-Trauth et al. 2003)
E-cadherin+P-cadherin knockdown	Epidermis/E11.5	K14-Cre	Lethal; severe epidermal adhesion defects, apoptosis	(Tinkle et al. 2008)
αE-catenin	Zygote		Lethal at E4; defects in trophectoderm	(Torres et al. 1997)
	Epidermis/E11.5	K14-Cre	Lethal at P0; loss of adhesion, hyperproliferation	(Vasioukhin et al. 2001)
	CNS/E10.5	Nestin-Cre	Lethal at P14-21; brain dysplasia, hyperproliferation	(Lien et al. 2006b)
	Mammary epithelium/adult	MMTV-Cre	Abnormal differentiation, increased apoptosis	(Nemade et al. 2004)
	Myocardium/E9	MLC2v-Cre	Defects in intercalated discs, cardiomyopathy	(Sheikh et al. 2006)
αN-catenin	Zygote		Cerebellar and hippocampal lamination defects, axonal migration abnormalities	(Park et al. 2002a; Uemura and Takeichi 2006)
β-catenin	Zygote		Lethal at E8; failure of primitive streak formation	(Haegel et al. 1995)
	Primitive streak; presomitic mesoderm/E7.5	T-Cre (Brachyury-Cre)	Mesoderm formation defects and abnormal segmentation	(Aulehla et al. 2008; Dunty et al. 2008)
	CNS/E8.5	Nes8-Cre	Dorsal-ventral fate shift before neurogenesis	(Backman et al. 2005)
	CNS, neural crest progenitors/E9	Wnt1-Cre	Loss of cerebellum, failure of craniofacial development, cardiac outflow tract defects	(Brault et al. 2001) (Kioussi et al. 2002)
		Wnt1-Cre	Impaired migration of neural crest cells, loss of melanocytes and dorsal root ganglia	(Hari et al. 2002)
	CNS/E9.5	FoxG1-Cre	Lack of the entire forebrain and anterior facial structure	(Junghans et al. 2005)

CNS/E10	Brn4-Cre	Premature cell cycle withdrawal of neural progenitors	(Zechner et al. 2003)
Apical ectodermal ridge/E10	Brn4-Cre	Limb malformation	(Soshnikova et al. 2003)
CNS/E10.5	D6-Cre	Impaired radial migration, decreased proliferation	(Machon et al. 2003)
Embryonic endoderm/E6.25	K19-Cre	Failure of node formation & multiple hearts	(Lickert et al. 2002)
Head and limb mesenchyme/ E9 E10.5	Prx1-Cre	Limb truncation, overall failure to develop bone	(Hill et al. 2005; Hill et al. 2006)
Limb ectoderm/E9.5	Msx2-Cre	Loss of apical ectodermal ridge	(Barrow et al. 2003)
Epidermis/E11.5	K14-Cre	Failure of follicle morphogenesis	(Huelsken et al. 2001)
Vascular endothelium/ E8.5	Tie2-Cre	Lethal E11.5-13; vascular fragility, impaired development of the heart septum; impaired endocardial cushion formation	(Cattelino et al. 2003; Liebner et al. 2004)
Lung epithelium/E13.5	SP-C-Cre	Impaired development of peripheral lung	(Mucenski et al. 2003)
mesenchymal lineages/ E12.5-13.5	Dermo1-Cre	Synovial joint fusions; abnormal differentiation of osteoblasts and chondrocytes; increased chondrogenesis and ectopic cartilage formation	(Day et al. 2005)
		Impaired lung growth and loss of lung muscle progenitors	(De Langhe et al. 2008; Yin et al. 2008)
Kidney epithelium (Wolffian ducts)/E11.5	Hoxb7-Cre	Premature differentiation, kidney aplasia or hypoplasia	(Bridgewater et al. 2008; Marose et al. 2008)
Kidney (renal vesicle progenitor cells)/E12.5	Six2-GFP-Cre	Reduced nephron formation; reduced kidney size and branching	(Park et al. 2007)
Heart (myocardial precursors)/E8.5	Nkx2.5-Cre	Lethal at E12.5, impaired cardiac development, smaller right ventricles	(Kwon et al. 2007)
Heart (myocardial progenitors and early pharyngeal endoderm)/E8	Islet1-Cre	Lethal E13; dilated outflow tract; smaller right ventricles; pharyngeal arch defects	(Lin et al. 2007)
Hear (proepicardum)/ E9.5	GATA5-Cre	Lethal around E15.5; impaired cardiac growth and coronary artery formation	(Zamora et al. 2007)
Heart (myocardial precursors)/E9.5	SM22-Cre	Lethal E10.5-11.5; loss of right heart	(Cohen et al. 2007)
Heart (second heart field cells) /E-8 -E9.5	MesP1-Cre	Disrupted cardiac looping; shortened heart outflow tract	(Klaus et al. 2007)
	Mef2c-Cre	Heart outflow tract defects; right ventricle defects	(Ai et al. 2007)
Pancreatic epithelium/E8.5	Pdx1-Cre	Acute edematous pancreatitis	(Dessimoz et al. 2005)

Continued

Table 1. *Continued*

Gene	Tissue/Time	Cre[a]	Phenotype	Ref.
γ-catenin/plakoglobin	Zygote		Lethal at E12-E17; cardiac rupture, skin blistering, absence of desmosomes in heart, but not in epithelial tissues	(Bierkamp et al. 1996; Ruiz et al. 1996)
p120-catenin	Epidermis/E11.5	K14-Cre	Epidermal hyperplasia and neoplasia and chronic inflammation	(Perez-Moreno et al. 2006; Perez-Moreno et al. 2008)
	Salivary gland epithelium/E14	MMTV-Cre	High grade intraepithelial neoplasia	(Davis and Reynolds 2006)
	Dorsal forebrain/E9	Emx1-Cre	Reduced spine and synapse densities along dendrites	(Elia et al. 2006)

Note, detailed information on gain-of-function and loss-of-function β-*catenin* mutant mice can be found in (Grigoryan et al. 2008).
E, embryonic day.
[a]Cre is a site-specific recombinase used for gene deletion.

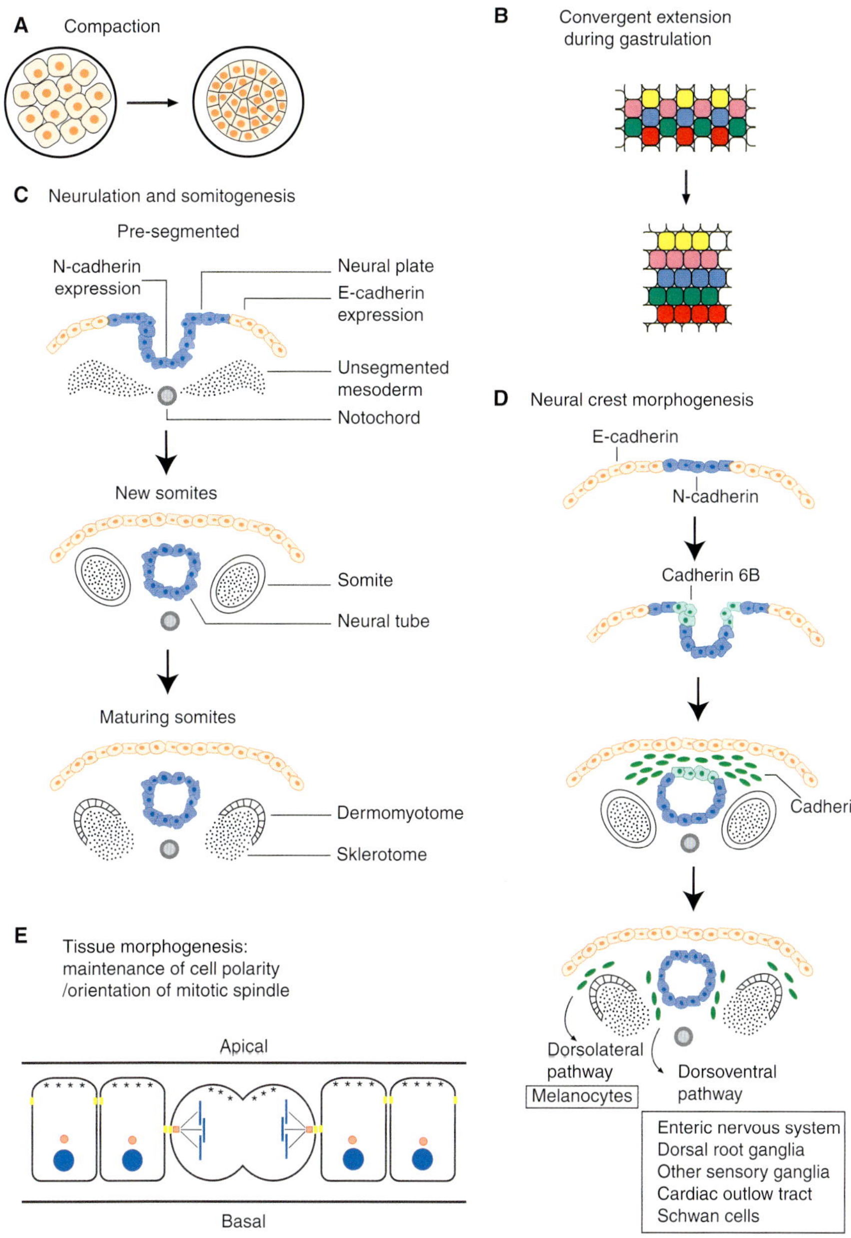

Figure 2. Cadherin–catenin complexes and major developmental events. (*A*) Functional cadherin–catenin adhesion is required for embryonic compaction during early preimplantation development, as well as multiple later morphogenetic events involving mesenchymal-to-epithelial transition. (*B*) Cadherin–catenin system and p120-catenin-mediated planar cell polarity signaling are required for embryonic gastrulation and "convergent extension," which results in lengthening of the embryonic body axis. (*C*) Cadherin–catenin adhesion is involved in tissue segregation and neural tube formation (neurulation), as well as formation of somites. (*D*) Down-regulation of cadherin–catenin adhesion (epithelial-to-mesenchymal transition) is necessary for emigration of neural crest cells from the neural tube and their subsequent migration. A reverse process of mesenchymal-to-epithelial transition is required for neural crest cell-derived structure formation, when neural crest cells reach their target locations. (*E*) Cadherin–catenin complexes are necessary to orient mitotic spindles of dividing cells to maintain proper tissue morphology.

indicates that αE-catenin is an indispensable component of the AJ complex. Although early embryonic phenotypes of E-cadherin- and αE-catenin-deficient mutants are adhesion-dependent, they clearly contrast with the phenotypes of β-catenin-null embryos, which seem to be adhesion-independent and rather result from β-catenin's signaling roles at later stages of development during embryonic gastrulation (Haegel et al. 1995; Huelsken et al. 2000). This is likely because of a compensatory effect of plakoglobin, and this hypothesis is supported by increased localization of plakoglobin to the lateral membrane of trophectodermal epithelial cells in *β-catenin*$^{-/-}$ embryos (Haegel et al. 1995).

Gastrulation, Embryo Patterning, and Somitogenesis

During gastrulation, the morphology of the newly formed embryo is dramatically restructured to form three embryonic germ layers (endoderm, ectoderm, and mesoderm) that will give rise to all tissues and organs in the developing organism. At the onset of gastrulation, cells of the epiblast migrate through the primitive streak to position the embryonic germ layers, and these morphogenetic movements are facilitated by the process of epithelial–mesenchymal transition (EMT), in which AJs play a fundamental role. During murine gastrulation, E-cadherin is down-regulated in the primitive streak as cells undergo embryonic EMT and this down-regulation is necessary for proper gastrulation and mesoderm cell fate specification in the mouse embryo (Burdsal et al. 1993; Ciruna and Rossant 2001).

Down-regulation of E-cadherin function is accomplished by multiple mechanisms including transcriptional repression via fibroblast growth factor/fibroblast growth factor receptor 1 (FGF/FGFR1)-dependent induction of *Snail* (Ciruna and Rossant 2001), down-regulation of E-cadherin protein levels by p38 mitogen-activated protein kinase (MAPK) (Zohn et al. 2006), and disruption of cadherin–catenin complexes by EPB41L5, a band 4.1 superfamily protein that interacts with p120-catenin and prevents its association with cadherin (Hirano et al. 2008). An additional pathway involving controlled internalization of cadherin was recently discovered in *Xenopus* (Ogata et al. 2007). TGF-β signaling, which is critical for gastrulation, induces the expression of transmembrane protein fibronectin leucine-rich repeat transmembrane 3 (FLRT3) and a small GTPase Rnd1. FLRT3 and Rnd1 interact physically and modulate cell adhesion during gastrulation by controlling dynamin-dependent endocytosis of epithelial cadherin (Ogata et al. 2007). Interestingly, mesodermal cells formed during gastrulation down-regulate E-cadherin and increase expression of N-cadherin, which is maintained by platelet-derived growth factor (PDGF) signaling and is required for normal mesodermal cell migration (Yang et al. 2008). Thus, gastrulation is a remarkable example of a naturally occurring cadherin-switch phenomenon, in which expression of one cadherin (E-cadherin) is replaced by another (N-cadherin), and both of them are required for normal developmental process, although they regulate different aspects of gastrulation.

In addition to EMT, the gastrulating embryo undergoes a significant extension of the body axis through the process known as convergent extension. Convergent extension involves highly organized and coordinated cell movements toward the midline, resulting in cell intercalation and rearrangement of their original positions (Fig. 2B). Morpholino-mediated knock down of p120-catenin or ARVCF in *Xenopus* revealed a critical role of these proteins in the regulation of convergent extension (Paulson et al. 1999; Fang et al. 2004). Subsequent mechanistic analyses indicated that signaling involving activation of Rac, inhibition of RhoA, and modulation of a repressive activity of transcription factor *Kaiso* on *Wnt11* expression were responsible for the phenotypes (Fang et al. 2004; Kim et al. 2004). Intriguingly, Wnt11 is a critical upstream regulator and RhoA is an important downstream target of the planar cell polarity pathway (see McNeill et al. 2009), which is known to be necessary for convergent extension. Thus, these experiments identified p120-catenin as an essential regulator of planar cell polarity and revealed at least two independent mechanisms responsible for this function.

Loss-of-function and gain-of-function experiments showed the requirement for β-catenin in proper patterning of various embryonic structures, tissues, and organs throughout embryonic development (Grigoryan et al. 2008). Although loss of zygotic β-catenin allows normal blastocyst development, the mutant embryos die shortly thereafter because of major defects in gastrulation, which is characterized by defects in the anterior–posterior axis formation and complete failure of mesoderm and head structure development (Haegel et al. 1995; Huelsken et al. 2000). Remarkably, these defects are not caused by the loss of cell–cell adhesion, which is maintained by plakoglobin, but because of deficiency in canonical Wnt signaling, which plays a critical role in regulation of gastrulation, and many other aspects of early embryonic development (see Cadigan and Peifer 2009; Heuberger and Birchmeier 2009).

In addition to major defects in gastrulation, loss of β-catenin also severely impairs somitogenesis. Conditional ablation of β-catenin in the visceral endoderm results in embryos that gastrulate, but are unable to form the node and display the failure of posterior axis elongation, somite formation, and endodermal cell fate specification (Lickert et al. 2002). Remarkably, these mutants show ectopic expression of Bmp2 in precardiac mesoderm and form multiple hearts. Here, as in many other early embryonic β-catenin loss-of-function studies, the authors found no defects in intercellular adhesion because of compensatory role of plakoglobin (Lickert et al. 2002). Furthermore, deletion of β-catenin in mesoderm precursors in the primitive streak blocks paraxial mesoderm formation and results in embryonic axis truncation (Aulehla et al. 2008; Dunty et al. 2008). Interestingly, β-catenin-mediated signaling is used again during embryonic dorsal–ventral axis specification. In *Xenopus*, canonical Wnt signaling is activated in dorsal part of the embryo to establish the dorsal axis (Tao et al. 2005). Because of the ancient genome duplication, Zebra fish genome contains two β-catenin genes. Although β-catenin-2 is essential for dorsal axis specification early in development, both β-catenin-1 and -2 are required for posteriorizing signaling required for normal embryonic patterning (Bellipanni et al. 2006). In addition to body patterning, Wnt/β-catenin signaling is also critical for the normal limb morphogenesis. Apical ectodermal ridge (AER) is an ectodermal structure overlying and inducing the development of the limb bud during vertebrate morphogenesis. β-catenin-mediated Wnt signaling lies upstream of the Bmp pathway in establishment of the AER and dorso-ventral limb patterning (Barrow et al. 2003; Soshnikova et al. 2003; Hill et al. 2006).

In summary, loss-of-function and gain-of-function experiments revealed critical roles for E-cadherin, N-cadherin, p120, and β-catenins in gastrulation and early embryonic patterning. Although most of the evidence so far points toward adhesive roles for cadherins, these experiments revealed very exciting signaling roles for p120-catenin in planar cell polarity-mediated convergent extension and for β-catenin in a variety of critical developmental decisions during early embryogenesis.

Early Heart and Vascular System Development

The primitive heart tube undergoes a series of morphological changes including looping into an S-shaped structure that is critical for proper alignment of the cardiac outflow tract. N-cadherin plays a critical role in multiple cardiac cell lineages and at different times during cardiac morphogenesis. Germline deletion of N-cadherin results in mid-gestation lethality associated with multiple developmental abnormalities including severe cardiovascular defects (Radice et al. 1997b). A dramatic cell adhesion defect was observed in the developing myocardium of N-cadherin mutant embryos resulting in a severely deformed heart tube. Interestingly, expression of either N-cadherin or E-cadherin under the control of a muscle-specific promoter was able to rescue the cardiac phenotype, demonstrating that these cadherins are interchangeable during cardiac looping morphogenesis (Luo et al. 2001). Chimeric mouse embryos generated from N-cadherin-null embryonic stem cells showed that N-cadherin-mediated adhesion is critical for maintaining cell–cell interactions in tissues undergoing active cellular rearrangements and increased mechanical stress (e.g., primitive heart) (Kostetskii et al. 2001). N-cadherin-deficient and wild-type cardiomyocytes initially intermix in the chimeras; however, the adhesion is short lived as the mutant cells are excluded from the myocardial wall during the transition from a cuboidal cell morphology to a flattened, tightly associated myocardial cell layer (Kostetskii et al. 2001). In addition to mediating cell adhesion, N-cadherin was shown to be involved in the migration of cardiomyocytes toward the endocardium in a process called trabeculation, which results in expansion of the cardiac chambers (Ong et al. 1998). N-cadherin is also involved in heterotypic cell interactions in the developing heart as loss of N-cadherin from the epicardium disrupts epicardial–myocardial cell–cell interactions resulting in a thinned ventricular myocardium (Luo et al. 2006).

Conditional deletion of N-cadherin in endothelial cells revealed an unexpected role for this protein in the developing vascular system (Luo and Radice 2005). Interestingly, the mid-gestation embryonic phenotype of the endothelial-specific N-cadherin knockout embryo bears similarities to the VE-cadherin-null phenotype. Both N-cadherin- and VE-cadherin-deficient endothelial cells are unable to form a normal vascular plexus leading to impaired vasculogenesis. VE-cadherin expression was perturbed in the endothelial-specific N-cadherin knockout embryo. In contrast, N-cadherin expression was not affected in VE-cadherin-deficient endothelial cells (Carmeliet et al. 1999). This is the only known example in which one cadherin subtype controls the expression of another, which may reflect complex and dynamic cell–cell rearrangements involved in vascular permeability. In addition to the similarities of endothelial phenotypes in N-cadherin and VE-cadherin mutant embryos, both knockouts show a myocardial cell adhesion defect indicating cadherin-mediated signaling between the endocardium and myocardium (Carmeliet et al. 1999; Luo and Radice 2005).

VE-cadherin itself is critical for a developing vascular system, where it has both mechanical and signaling functions (Carmeliet et al. 1999). Endothelial cells missing VE-cadherin fail to support remodeling and maturation of developing vasculature. In addition to the adhesion defect, these cells also display a prominent decrease in vascular endothelial growth factor receptor-2-phosphatidylinositol 3 kinase (VEGFR-2-PI3K) signaling and subsequent apoptosis (Carmeliet et al. 1999). VE-cadherin also uses a similar PI3K-Akt pathway to up-regulate the expression of claudin-5, which is required for the formation of tight junctions between endothelial cells to prevent vascular permeability (Taddei et al. 2008). Remarkably, although VE-cadherin is necessary for positive regulation of VEGFR-2-PI3K signaling and cell survival, it has an opposite effect on VEGFR-2-MAPK signaling and cell proliferation (Grazia Lampugnani et al. 2003). Subsequent analysis revealed a critical role of VE-cadherin in preventing the internalization of activated VEGFR-2 and stimulation of VEGFR-2-MAPK signaling from the endosomal compartment (Lampugnani et al. 2006). In addition to regulation of VEGFR-2, VE-cadherin also binds to all the components of the TGF-β receptor complex and promotes TGF-β signaling (Rudini et al. 2008). Because TGF-β is a major developmental signaling pathway, it is likely that a functional connection between VE-cadherin and TGF-β signaling plays an important role in vertebrate morphogenesis.

Plakoglobin is a close relative of β-catenin, but in contrast to the signaling function of β-catenin, plakoglobin loss-of-function experiments revealed primarily a structural role for this protein. Plakoglobin is unique in its ability to interact with both classical cadherins in AJs and desmosomal cadherins in desmosomes, which are distinct cell–cell adhesion structures linked to the intermediate cytoskeleton. Mice with zygotic disruption of plakoglobin die in utero showing defects in cardiac structural integrity and skin blistering (Bierkamp et al. 1996; Ruiz et al. 1996; Bierkamp et al. 1999). Interestingly, both heart and skin defects were caused by abnormal desmosome function, indicating that β-catenin cannot functionally substitute for plakoglobin in desmosomes. Therefore, the primary role of plakoglobin is in desmosome function and it is not essential for the function of AJs.

Neural Crest Development

Both adhesion and signaling activities of cadherin and catenin proteins play a pivotal role in neural crest development. Neural crest cells undergo EMT and migrate from the developing dorsal neural tube ultimately differentiating and forming a wide variety of adult structures such as parts of peripheral nervous system, melanocytes and some craniofacial structures (Fig. 2D). Initial EMT and emigration of neural crest cells from the neural tube requires down-regulation of N-cadherin (Nakagawa and Takeichi 1998), which is achieved by BMP4-induced cleavage of N-cadherin via ADAM10 (a disintegrin and metalloproteinase) (Shoval et al. 2007). Interestingly, this cleavage is not only required to down-regulate the adhesive function of N-cadherin, but it also generates cytoplasmic N-cadherin fragment that translocates to the nucleus and stimulates β-catenin-mediated transcription. Both adhesive and signaling activities are likely to participate in neural crest emigration. Deletion of N-cadherin from murine neural crest cells using Wnt1-Cre results in aberrant remodeling of the cardiac outflow tract and embryonic lethality (Luo et al. 2006). Cardiac neural crest cells normally elongate and align with their neighbors eventually aggregating at the midline to divide the outflow tract into aortic and pulmonary channels. In Wnt1-Cre-driven N-cadherin mutants, outflow tract cells are rounded and less condensed compared with wild-type cells (Luo et al. 2006). An adhesive function of N-cadherin at the neural crest cell migration target site was also shown for sympathetic ganglia (Kasemeier-Kulesa et al. 2006).

β-catenin plays a pivotal role in neural crest specification. Ablation of β-catenin in mouse embryos using Wnt1-*Cre* results in the failure of craniofacial development and loss of melanocyte and sensory neural cell lineages (Brault et al. 2001; Hari et al. 2002). This is reminiscent of the phenotype observed in conventional Wnt1 knockout mice (McMahon and Bradley 1990), indicating that β-catenin signaling function is required for these developmental processes.

P120-catenin has a major impact on migration of neural crest cells. Knockdown of p120-catenin in developing *Xenopus* embryos affects the migration of neural crest cells into the branchial arches, and causes defects in eye formation as well as several craniofacial cartilage structures (Ciesiolka et al. 2004). Importantly, defects induced by p120 depletion can be rescued by expression of either dominant-negative Rac or LIM kinase, indicating the involvement of p120-catenin in Rho GTPases signaling during development (Ciesiolka et al. 2004). Therefore, similar to its role in gastrulation, p120-catenin impacts neural crest morphogenesis by regulating Rho-family GTPases and the actin cytoskeleton.

Central Nervous System (CNS) Development

During formation of the neural plate, the precursor of the vertebrate neural system, E-cadherin expression is replaced by N-cadherin, which continues to be expressed at high levels in all neural tissues throughout the development (Fig. 2C). Knockout of N-cadherin in the developing mouse embryo results in severe malformation of the neural tube (Radice et al. 1997b). Concurrent heart defects and early embryonic death complicated the analysis of N-cadherin-null mice; however, loss-of-function N-cadherin experiments in zebrafish revealed impaired neuroectodermal cell adhesion and severely compromised cell movements during neurulation, which ultimately affected neuronal positioning and axon pathfinding (Lele et al. 2002). Interestingly, loss of N-cadherin in zebra fish destabilized membrane-associated β-catenin and caused increased mitoses in the dorsal midbrain and hindbrain, suggesting an additional signaling role for N-cadherin in this context (Lele et al. 2002). A CNS-specific role of mouse N-cadherin has been analyzed recently using a conditional gene knockout approach (Kadowaki et al. 2007). Deletion of N-cadherin in the developing cerebral cortex results in a massive cortical disorganization and scattering of mitotic cells throughout the cortex (Fig. 3) (Kadowaki et al. 2007). Future studies will likely analyze the potential changes in cell proliferation and cell death in these mutant animals.

Both αE- and αN-catenins are expressed in the developing mammalian brain; however, their expression patterns are different. αE-catenin is predominantly expressed in neural progenitors and αN-catenin in differentiated neurons. Conditional deletion of αE-catenin in neural progenitors resulted in loss of cell adhesion and polarity, which caused severe brain disorganization (Fig. 3) (Lien et al. 2006b). In addition to these mechanical defects, mutant brains also displayed shortening of the cell cycle and decreased apoptosis, which caused an abnormal increase of total brain cell number. Analyses of potential mechanisms identified abnormal activation of Hedgehog (Hh) signaling in hyperplastic dorsal brain structures of αE-catenin-null brains (Lien et al. 2006b). Because Hh signaling and cell proliferation were impacted only in dorsal, but not in the ventral part of αE-catenin-null brains, it is likely that this intriguing connection between αE-catenin and Hh signaling is modulated by other pathways, whose identity and mechanism of action will have to be discovered in future research.

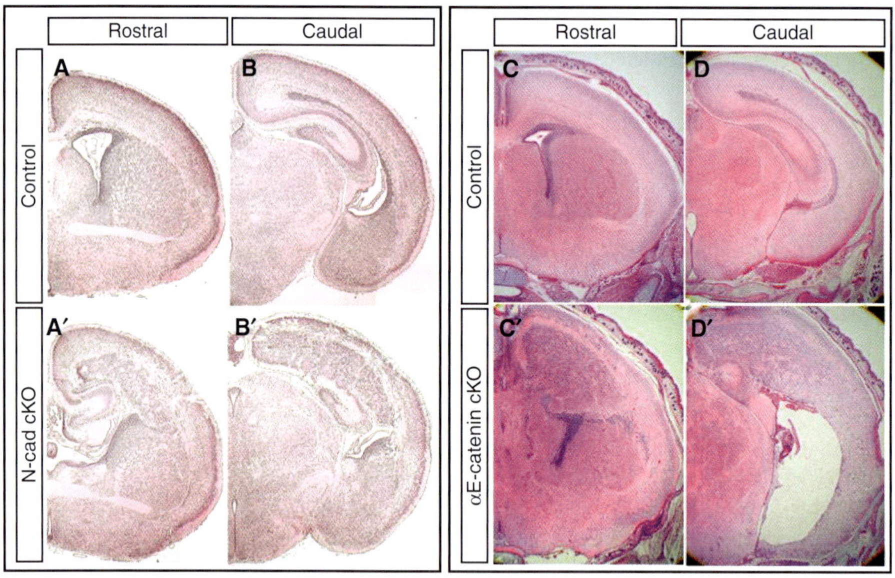

Figure 3. Cadherin–catenin complexes in embryonic brain development. (*A–B′*) Cortical disorganization in mice with embryonic brain-specific knockout of N-cadherin (N-cad cKO). Hematoxylin and eosin staining of coronal brain sections from one-day-old control (*A*, *B*) and D6-Cre N-cadherin cKO pups (A′, B′). (Reprinted with permission from Kadowaki et al. 2007, copyright 2007 Elsevier.) (*C–D′*) Brain disorganization and hyperplasia in mice with embryonic brain-specific knockout of αE-catenin. Hematoxylin and eosin staining of coronal brain sections from new-born control (*C*, *D*) and Nestin-Cre αE-catenin cKO pups (*C′*, *D′*). (Reprinted, with permission, from Lien et al. 2006b, copyright 2006 American Association for the Advancement in Science.)

Although αE-catenin is essential for early embryonic development, zygotic αN-catenin-null mice survive into adulthood; however, they display a prominent neurological phenotype. These animals show ataxia and their brains display underdevelopment and disorganization of the cerebellum (Park et al. 2002a,b).

Both adhesive and signaling activities of β-catenin play critical roles in the developing CNS. Conditional ablation of β-catenin in the murine forebrain early in development results in cell fate changes and ventralization of telencephalon (Backman et al. 2005). β-catenin is critical for normal cell–cell adhesion in neuroepithelial cells and conditional deletion of β-catenin in developing forebrain results in breakdown of neuroepithelial structures, an increase in apoptosis and loss of the entire forebrain and anterior facial structures (Junghans et al. 2005). Later in brain development, β-catenin is necessary for self-renewal and maintenance of neuronal progenitors (Machon et al. 2003; Zechner et al. 2003).

Epidermal Development

Functions of epithelial cadherin–catenin proteins were extensively studied in developing mouse epidermis. Loss-of-function mutants of cadherins, p120-, β-, and αE-catenins in skin epidermis offer an unprecedented opportunity for side-by-side comparison of the role of different members of cadherin–catenin complexes in the same tissue. These murine models taught us that cadherin–catenin proteins are required not only for cell–cell adhesion, but also for cellular proliferation and differentiation. Although cadherins seem to be mainly involved in the maintenance of the overall mechanical epidermal integrity and protection from apoptosis, catenins play a dual role; on the one hand they act as cadherin stabilizers during AJs formation, on the other hand they function as regulators of cellular signaling pathways (Fig. 4).

The principal cadherin responsible for AJ formation in epithelial cells is E-cadherin. Intriguingly, however, deletion of E-cadherin in embryonic skin epidermis did not result in a significant impairment of AJs, most likely because of compensatory up-regulation of P-cadherin (Young et al. 2003; Tinkle et al. 2004; Tunggal et al. 2005). The potential compensatory mechanism between E- and P-cadherin in the developing skin has recently been addressed by a study in which the conditional knockout approach was elegantly combined with transgenic shRNA technology to simultaneously ablate E-cadherin and knockdown P-cadherin ($E\text{-}cad^{-/-}/P\text{-}cad^{KD}$ mice) in the epidermal compartment (Tinkle et al. 2008). This resulted in the generation of mutant mice missing all major cadherins in skin epidermis and provided a first look at the epithelial tissue without classical cadherins. $E\text{-}cad^{-/-}/P\text{-}cad^{KD}$ epidermis displays loss of AJs, prominent disruption of intercellular adhesion and increase in apoptotic cell death (Tinkle et al. 2008). The cell–cell adhesion defects in $E\text{-}cad^{-/-}/P\text{-}cad^{KD}$ epidermis were similar to the defects in αE-catenin-deficient epidermis (Vasioukhin et al. 2001). Remarkably, however, in addition to the mechanical defects, the αE-catenin-deficient epidermis also showed an increase in cell proliferation and abnormal activation of Ras-MAPK and NFκB signaling pathways (Vasioukhin et al. 2001; Kobielak and Fuchs 2006). The increase in cell proliferation and activation of MAPK and NFκB pathways were also observed in p120-catenin-deficient epidermis (Perez-Moreno et al. 2006; Perez-Moreno et al. 2008). Moreover, when wild-type keratinocytes are transplanted to the backs of *nude* mice, they form normal skin epidermis. In contrast, transplanted αE- and p120-catenin-deficient keratinocytes display disorganization, inflammation and grow in a pattern resembling squamous cell carcinoma. Interestingly, inhibition of NFκB signaling in these transplants resulted in loss of αE-catenin-null cells and a decrease in proliferation of p120-catenin-null cells, indicating an important role of inflammatory pathways in the growth of these mutant cells in vivo (Kobielak and Fuchs 2006; Perez-Moreno et al. 2006).

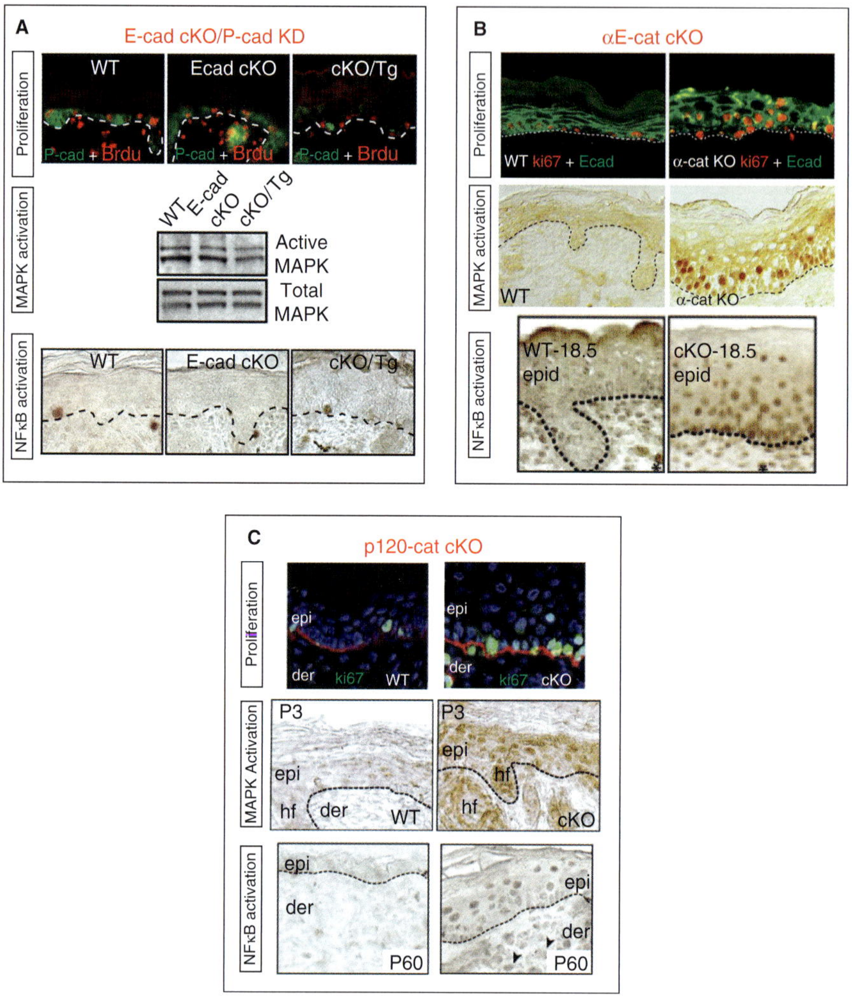

Figure 4. Overlapping and distinct functions of cadherin–catenin proteins in embryonic skin development. (*A*) Disruption of epidermal integrity, but normal proliferation, MAPK and NFκB signaling pathways in mice with embryonic epidermis-specific deficiency for E-cadherin (E-cad cKO), or both E- and P-cadherins (cKO/Tg). Proliferation was measured by BrdU incorporation (red). P-cadherin staining is shown in green channel. MAPK activity was measured by immunoblotting of total protein extracts with indicated antibodies. NFκB activity was revealed by staining with anti-phospho-NFκB antibodies. (Reprinted with permission from Tinkle et al. 2008, copyright 2008 National Academy of Sciences.) (*B*) Increase in proliferation, MAPK and NFκB signaling in mice with embryonic epidermis-specific deletion of αE-catenin (αE-cat cKO). Proliferation was measured by staining with anti-Ki67 antibodies (red). E-cadherin staining is shown in green channel. MAPK activity was measured by staining with anti-phospho-Erk1/2 antibodies. NFκB activity was revealed as described in (*A*). (Proliferation and MAPK panels reprinted from Vasioukhin et al. 2001, copyright 2001 Elsevier.) (NFκB panel reprinted with permission from Kobielak and Fuchs 2006, copyright 2006 National Academy of Sciences.) (*C*) Increase in proliferation, MAPK and NFκB signaling in mice with embryonic epidermis-specific deficiency for p120-catenin (p120-cat cKO). Proliferation was measured by staining with anti-Ki67 antibodies (green). β4-integrin staining is shown in red channel. MAPK and NFκB activities were revealed as described in (*B*). (Reprinted with permission from Perez-Moreno et al. 2006, copyright 2006 Elsevier.)

Because β-catenin is a prominent member of cadherin–catenin complexes, it would be logical to hypothesize that β-catenin becomes hyperactivated in αE- and p120-catenin-deficient cells and this may be responsible for hyperplasia in αE- and p120-catenin-null skin epidermis. However, loss of αE-catenin in skin epidermis did not affect β-catenin-dependent signaling (Vasioukhin et al. 2001), and this is in agreement with a recent study on the role of αE-catenin in the developing mammalian brain, which also found no changes in β-catenin transcriptional activity in αE-catenin-null neural progenitors (Lien et al. 2008). Therefore, functions of αE-catenin in regulation of cell proliferation is likely to be β-catenin-independent.

In contrast to p120- and αE-catenin, loss of β-catenin in the developing skin did not affect the integrity of the AJs, because of potential compensation by plakoglobin, but it resulted in the failure of proper hair follicle stem cell differentiation and loss of hair follicles (Huelsken et al. 2001). Although changes in cell proliferation, NFκB, and MAPK signaling in β-catenin-null epidermis have not been reported, β-catenin was required for maintenance of skin cancer stem cells, indicating that certain level of β-catenin activity is probably necessary for skin tumorigenesis (Malanchi et al. 2008). Moreover, epidermis-specific overexpression of stabilized, degradation resistant β-catenin is tumorigenic and results in hair follicle tumor formation (Gat et al. 1998). Therefore, it appears that αE-catenin, p120-catenin, and β-catenin have opposite functions in the regulation of epidermal cell proliferation. Although αE- and p120-catenins suppress in vivo cell proliferation, β-catenin promotes uncontrolled cell accumulation and causes cancer.

CONCLUSION AND PERSPECTIVES

Owing to extensive loss- and gain-of-function genetic analyses, it is now very clear that cadherin–catenin complexes play pivotal roles in multiple aspects of vertebrate development. These genetic studies accumulated significant evidence that besides their essential role in mechanical adhesion, cadherin–catenin proteins regulate signaling pathways important for such fundamental processes as cell division, migration, differentiation, and apoptosis. Intriguing connections between cadherin–catenin proteins and major developmental signaling networks led to a hypothesis that cells may use cadherin–catenin complexes as biosensors, which may provide them with information about the immediate external cellular microenvironment (Lien et al. 2006a). Indeed, from the individual cell point of view, neighbor-mediated clustering of cadherin–catenin complexes at a specific position of the plasma membrane provides information about the spatial localization of its immediate neighbors. This information may then be used to orient intracellular structures and maintain cellular polarity (Fig. 5). Thus, AJs may provide a vital link between cell polarity and three-dimensional polarity of the tissue, and help cells to exchange information and coordinate their activities as they build organs and tissues during development. Although this hypothesis is very attractive and we now know the major signaling networks affected by cadherin–catenin proteins, there are major gaps in our understanding of how cadherin–catenin complexes can transmit information from outside to inside the cell. Some of the possible scenarios include competition for protein availability between the junctional and cytoplasmic or nuclear pools; for example, competition between AJs and nucleus for β-catenin (Sehgal et al. 1997; Simcha et al. 1998; Gottardi et al. 2001; Onder et al. 2008). An alternative mechanism may involve a direct interaction and regulation of multiple growth factor receptors by cadherin–catenin complexes (Takahashi and Suzuki 1996; Qian et al. 2004; Perrais et al. 2007). Many questions remain unanswered. Is the function of the cadherin–catenin complex in adhesion really connected to its function in cell signaling, or do these proteins have multiple independent functions? How exactly is adhesion information converted into signaling information? How are connections between AJs and signaling

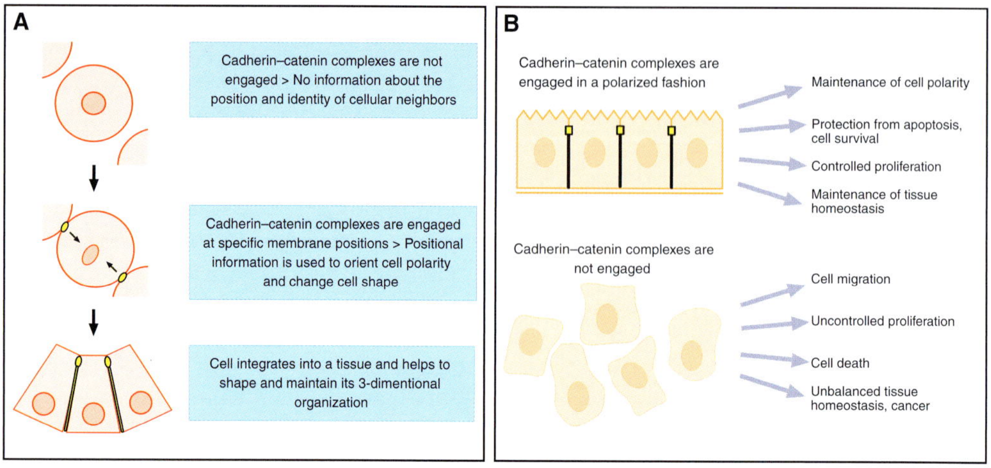

Figure 5. Hypothetical role of cadherin–catenin complexes as biosensors of external cellular environment. (*A*) Engagement and clustering of cadherin–catenin complexes at a specific position of the inner plasma membrane provides cell with information about the spatial position of its neighbors. This information may be translated into changes in apical–basal and planar cell polarity axes to integrate the cell with its neighbors and form 3-dimentional tissue. (*B*) Engaged and polarized cadherin–catenin complexes regulate cell polarity, survival and proliferation to control cell behavior and physically and functionally integrate cells into a tissue. Failure to form and cluster cadherin–catenin complexes in a wound or during tissue morphogenesis cause major cell-type-specific changes that may include migration and proliferation (to repair the wound), or cell death (to prevent an ectopic tissue formation).

pathways modulated by embryonic tissue- and region-specific microenvironment? Future research should help to answer these important questions.

ACKNOWLEDGMENTS

This work was supported by National Institute of Health (NIH) grants CA098161 (VV) and HL081569 (GLR).

REFERENCES

Abe K, Takeichi M. 2008. EPLIN mediates linkage of the cadherin catenin complex to F-actin and stabilizes the circumferential actin belt. *Proc Natl Acad Sci* **105:** 13–19.

Aberle H, Butz S, Stappert J, Weissig H, Kemler R, Hoschuetzky H. 1994. Assembly of the cadherin–catenin complex in vitro with recombinant proteins. *J Cell Sci* **107:** 3655–3663.

Ai D, Fu X, Wang J, Lu MF, Chen L, Baldini A, Klein WH, Martin JF. 2007. Canonical Wnt signaling functions in second heart field to promote right ventricular growth. *Proc Natl Acad Sci* **104:** 9319–9324.

Aulehla A, Wiegraebe W, Baubet V, Wahl MB, Deng C, Taketo M, Lewandoski M, Pourquie O. 2008. A β-catenin gradient links the clock and wavefront systems in mouse embryo segmentation. *Nat Cell Biol* **10:** 186–193.

Backman M, Machon O, Mygland L, van den Bout CJ, Zhong W, Taketo MM, Krauss S. 2005. Effects of canonical Wnt signaling on dorso-ventral specification of the mouse telencephalon. *Dev Biol* **279:** 155–168.

Barrow JR, Thomas KR, Boussadia-Zahui O, Moore R, Kemler R, Capecchi MR, McMahon AP. 2003. Ectodermal Wnt3/β-catenin signaling is required for the establishment and maintenance of the apical ectodermal ridge. *Genes Dev* **17:** 394–409.

Battle MA, Konopka G, Parviz F, Gaggl AL, Yang C, Sladek FM, Duncan SA. 2006. Hepatocyte nuclear factor 4α orchestrates expression of cell adhesion proteins during the epithelial transformation of the developing liver. *Proc Natl Acad Sci* **103:** 8419–8424.

Bellipanni G, Varga M, Maegawa S, Imai Y, Kelly C, Myers AP, Chu F, Talbot WS, Weinberg ES. 2006. Essential and opposing roles of zebrafish {β}-catenins in the formation of dorsal axial structures and neurectoderm. *Development* **133:** 1299–1309.

Bierkamp C, McLaughlin KJ, Schwarz H, Huber O, Kemler R. 1996. Embryonic heart and skin defects in mice lacking plakoglobin. *Dev Biol* **180:** 780–785.

Bierkamp C, Schwarz H, Huber O, Kemler R. 1999. Desmosomal localization of β-catenin in the skin of plakoglobin null-mutant mice. *Development* **126:** 371–381.

Boussadia O, Kutsch S, Hierholzer A, Delmas V, Kemler R. 2002. E-cadherin is a survival factor for the lactating mouse mammary gland. *Mech Dev* **115:** 53–62.

Brault V, Moore R, Kutsch S, Ishibashi M, Rowitch DH, McMahon AP, Sommer L, Boussadia O, Kemler R. 2001. Inactivation of the β-catenin gene by Wnt1-Cre-mediated deletion results in dramatic brain malformation and failure of craniofacial development. *Development* **128:** 1253–1264.

Brembeck FH, Schwarz-Romond T, Bakkers J, Wilhelm S, Hammerschmidt M, Birchmeier W. 2004. Essential role of BCL9-2 in the switch between β-catenin's adhesive and transcriptional functions. *Genes Dev* **18:** 2225–2230.

Brembeck FH, Rosario M, Birchmeier W. 2006. Balancing cell adhesion and Wnt signaling, the key role of β-catenin. *Curr Opin Genet Dev* **16:** 51–59.

Bridgewater D, Cox B, Cain J, Lau A, Athaide V, Gill PS, Kuure S, Sainio K, Rosenblum ND. 2008. Canonical WNT/β-catenin signaling is required for ureteric branching. *Dev Biol* **317:** 83–94.

Burdsal CA, Damsky CH, Pedersen RA. 1993. The role of E-cadherin and integrins in mesoderm differentiation and migration at the mammalian primitive streak. *Development* **118:** 829–844.

Cadigan KM, Peifer M. 2009. Wnt signaling from development to disease: Insights from model systems. *Cold Spring Harb Perspect Biol* **1:** a002881.

Carmeliet P, Lampugnani MG, Moons L, Breviario F, Compernolle V, Bono F, Balconi G, Spagnuolo R, Oostuyse B, Dewerchin M, et al. 1999. Targeted deficiency or cytosolic truncation of the VE-cadherin gene in mice impairs VEGF-mediated endothelial survival and angiogenesis. *Cell* **98:** 147–157.

Cattelino A, Liebner S, Gallini R, Zanetti A, Balconi G, Corsi A, Bianco P, Wolburg H, Moore R, Oreda B, et al. 2003. The conditional inactivation of the β-catenin gene in endothelial cells causes a defective vascular pattern and increased vascular fragility. *J Cell Biol* **162:** 1111–1122.

Ciesiolka M, Delvaeye M, Van Imschoot G, Verschuere V, McCrea P, van Roy F, Vleminckx K. 2004. p120 catenin is required for morphogenetic movements involved in the formation of the eyes and the craniofacial skeleton in *Xenopus*. *J Cell Sci* **117:** 4325–4339.

Cavey M, Lecuit T. 2009. Molecular bases of cell-cell junctions stability and dynamics. *Cold Spring Harb Perspect Biol* **1:** a002998.

Ciruna B, Rossant J. 2001. FGF signaling regulates mesoderm cell fate specification and morphogenetic movement at the primitive streak. *Dev Cell* **1:** 37–49.

Claverie JM, Hardelin JP, Legouis R, Levilliers J, Bougueleret L, Mattei MG, Petit C. 1993. Characterization and chromosomal assignment of a human cDNA encoding a protein related to the murine 102-kDa cadherin-associated protein (α-catenin). *Genomics* **15:** 13–20.

Clevers H. 2006. Wnt/β-catenin signaling in development and disease. *Cell* **127:** 469–480.

Cohen ED, Wang Z, Lepore JJ, Lu MM, Taketo MM, Epstein DJ, Morrisey EE. 2007. Wnt/β-catenin signaling promotes expansion of Isl-1-positive cardiac progenitor cells through regulation of FGF signaling. *The Journal of clinical investigation* **117:** 1794–1804.

Coluccia AM, Benati D, Dekhil H, De Filippo A, Lan C, Gambacorti-Passerini C. 2006. SKI-606 decreases growth and motility of colorectal cancer cells by preventing pp60(c-Src)-dependent tyrosine phosphorylation of β-catenin and its nuclear signaling. *Cancer Res* **66:** 2279–2286.

Coluccia AM, Vacca A, Dunach M, Mologni L, Redaelli S, Bustos VH, Benati D, Pinna LA, Gambacorti-Passerini C. 2007. Bcr-Abl stabilizes β-catenin in chronic myeloid leukemia through its tyrosine phosphorylation. *Embo J* **26:** 1456–1466.

Dahl U, Sjodin A, Larue L, Radice GL, Cajander S, Takeichi M, Kemler R, Semb H. 2002. Genetic dissection of cadherin function during nephrogenesis. *Mol Cell Biol* **22:** 1474–1487.

Daugherty RL, Gottardi CJ. 2007. Phospho-regulation of β-catenin adhesion and signaling functions. *Physiology (Bethesda)* **22:** 303–309.

Davis MA, Reynolds AB. 2006. Blocked acinar development, E-cadherin reduction, and intraepithelial neoplasia upon ablation of p120-catenin in the mouse salivary gland. *Dev Cell* **10:** 21–31.

Day TF, Guo X, Garrett-Beal L, Yang Y. 2005. Wnt/β-catenin signaling in mesenchymal progenitors controls osteoblast and chondrocyte differentiation during vertebrate skeletogenesis. *Dev Cell* **8:** 739–750.

De Langhe SP, Carraro G, Tefft D, Li C, Xu X, Chai Y, Minoo P, Hajihosseini MK, Drouin J, Kaartinen V, et al. 2008. Formation and differentiation of multiple mesenchymal lineages during lung development is regulated by β-catenin signaling. *PLoS ONE* **3:** e1516.

De Vries WN, Evsikov AV, Haac BE, Fancher KS, Holbrook AE, Kemler R, Solter D, Knowles BB. 2004. Maternal β-catenin and E-cadherin in mouse development. *Development* **131:** 4435–4445.

Dessimoz J, Bonnard C, Huelsken J, Grapin-Botton A. 2005. Pancreas-specific deletion of β-catenin reveals Wnt-dependent and Wnt-independent functions during development. *Curr Biol* **15:** 1677–1683.

Drees F, Pokutta S, Yamada S, Nelson WJ, Weis WI. 2005. α-catenin is a molecular switch that binds E-cadherin-β-catenin and regulates actin-filament assembly. *Cell* **123:** 903–915.

Dunty WC Jr, Biris KK, Chalamalasetty RB, Taketo MM, Lewandoski M, Yamaguchi TP. 2008. Wnt3a/β-catenin signaling controls posterior body development by coordinating mesoderm formation and segmentation. *Development* **135:** 85–94.

Elia LP, Yamamoto M, Zang K, Reichardt LF. 2006. p120 catenin regulates dendritic spine and synapse development through Rho-family GTPases and cadherins. *Neuron* **51:** 43–56.

Erez N, Bershadsky A, Geiger B. 2005. Signaling from adherens-type junctions. *European journal of cell biology* **84:** 235–244.

Fang X, Ji H, Kim SW, Park JI, Vaught TG, Anastasiadis PZ, Ciesiolka M, McCrea PD. 2004. Vertebrate development requires ARVCF and p120 catenins and their interplay with RhoA and Rac. *J Cell Biol* **165:** 87–98.

Gat U, DasGupta R, Degenstein L, Fuchs E. 1998. De Novo hair follicle morphogenesis and hair tumors in mice expressing a truncated β-catenin in skin. *Cell* **95:** 605–614.

Gory-Faure S, Prandini MH, Pointu H, Roullot V, Pignot-Paintrand I, Vernet M, Huber P. 1999. Role of vascular endothelial-cadherin in vascular morphogenesis. *Development* **126:** 2093–2102.

Gottardi CJ, Wong E, Gumbiner BM. 2001. E-cadherin suppresses cellular transformation by inhibiting β-catenin signaling in an adhesion-independent manner. *J Cell Biol* **153:** 1049–1060.

Grazia Lampugnani M, Zanetti A, Corada M, Takahashi T, Balconi G, Breviario F, Orsenigo F, Cattelino A, Kemler R, Daniel TO, et al. 2003. Contact inhibition of VEGF-induced proliferation requires vascular endothelial cadherin, β-catenin, and the phosphatase DEP-1/CD148. *J Cell Biol* **161:** 793–804.

Grigoryan T, Wend P, Klaus A, Birchmeier W. 2008. Deciphering the function of canonical Wnt signals in development and disease: Conditional loss- and gain-of-function mutations of β-catenin in mice. *Genes Dev* **22:** 2308–2341.

Haegel H, Larue L, Ohsugi M, Fedorov L, Herrenknecht K, Kemler R. 1995. Lack of β-catenin affects mouse development at gastrulation. *Development* **121:** 3529–3537.

Hari L, Brault V, Kleber M, Lee HY, Ille F, Leimeroth R, Paratore C, Suter U, Kemler R, Sommer L. 2002. Lineage-specific requirements of β-catenin in neural crest development. *J Cell Biol* **159:** 867–880.

Hartsock A, Nelson WJ. 2008. Adherens and tight junctions: Structure, function and connections to the actin cytoskeleton. *Biochim Biophys Acta* **1778:** 660–669.

Heuberger J, Birchmeier W. 2009. Interplay of cadherin-mediated cell adhesion and canonical Wnt signaling. *Cold Spring Harb Perspect Biol* **1:** a002915.

Hill TP, Spater D, Taketo MM, Birchmeier W, Hartmann C. 2005. Canonical Wnt/β-catenin signaling prevents osteoblasts from differentiating into chondrocytes. *Dev Cell* **8:** 727–738.

Hill TP, Taketo MM, Birchmeier W, Hartmann C. 2006. Multiple roles of mesenchymal {β}-catenin during murine limb patterning. *Development* **133:** 1219–1229.

Hirano S, Kimoto N, Shimoyama Y, Hirohashi S, Takeichi M. 1992. Identification of a neural α-catenin as a key regulator of cadherin function and multicellular organization. *Cell* **70:** 293–301.

Hirano M, Hashimoto S, Yonemura S, Sabe H, Aizawa S. 2008. EPB41L5 functions to post-transcriptionally regulate cadherin and integrin during epithelial-mesenchymal transition. *J Cell Biol* **182:** 1217–1230.

Hollnagel A, Grund C, Franke WW, Arnold HH. 2002. The cell adhesion molecule M-cadherin is not essential for muscle development and regeneration. *Mol Cell Biol* **22:** 4760–4770.

Horikawa K, Radice G, Takeichi M, Chisaka O. 1999. Adhesive subdivisions intrinsic to the epithelial somites. *Dev Biol* **215:** 182–189.

Huelsken J, Vogel R, Brinkmann V, Erdmann B, Birchmeier C, Birchmeier W. 2000. Requirement for β-catenin in anterior-posterior axis formation in mice. *J Cell Biol* **148:** 567–578.

Huelsken J, Vogel R, Erdmann B, Cotsarelis G, Birchmeier W. 2001. β-Catenin controls hair follicle morphogenesis and stem cell differentiation in the skin. *Cell* **105:** 533–545.

Hulpiau P, van Roy F. 2009. Molecular evolution of the cadherin superfamily. *Int J Biochem Cell Biol* **41:** 349–369.

Janssens B, Goossens S, Staes K, Gilbert B, van Hengel J, Colpaert C, Bruyneel E, Mareel M, van Roy F. 2001. αT-catenin: A novel tissue-specific β-catenin-binding protein mediating strong cell-cell adhesion. *J Cell Sci* **114:** 3177–3188.

Junghans D, Hack I, Frotscher M, Taylor V, Kemler R. 2005. β-catenin-mediated cell-adhesion is vital for embryonic forebrain development. *Dev Dyn* **233:** 528–539.

Kadowaki M, Nakamura S, Machon O, Krauss S, Radice GL, Takeichi M. 2007. N-cadherin mediates cortical organization in the mouse brain. *Dev Biol* **304:** 22–33.

Kajiguchi T, Chung EJ, Lee S, Stine A, Kiyoi H, Naoe T, Levis MJ, Neckers L, Trepel JB. 2007. FLT3 regulates β-catenin tyrosine phosphorylation, nuclear localization, and transcriptional activity in acute myeloid leukemia cells. *Leukemia* **21:** 2476–2484.

Kan NG, Stemmler MP, Junghans D, Kanzler B, de Vries WN, Dominis M, Kemler R. 2007. Gene replacement reveals a specific role for E-cadherin in the formation of a functional trophectoderm. *Development* **134:** 31–41.

Kasemeier-Kulesa JC, Bradley R, Pasquale EB, Lefcort F, Kulesa PM. 2006. Eph/ephrins and N-cadherin coordinate to control the pattern of sympathetic ganglia. *Development* **133:** 4839–4847.

Kawaguchi J, Azuma Y, Hoshi K, Kii I, Takeshita S, Ohta T, Ozawa H, Takeichi M, Chisaka O, Kudo A. 2001. Targeted disruption of cadherin-11 leads to a reduction in bone density in calvaria and long bone metaphyses. *J Bone Miner Res* **16:** 1265–1271.

Kim SW, Park JI, Spring CM, Sater AK, Ji H, Otchere AA, Daniel JM, McCrea PD. 2004. Non-canonical Wnt signals are modulated by the Kaiso transcriptional repressor and p120-catenin. *Nat Cell Biol* **6:** 1212–1220.

Kioussi C, Briata P, Baek SH, Rose DW, Hamblet NS, Herman T, Ohgi KA, Lin C, Gleiberman A, Wang J, et al. 2002. Identification of a Wnt/Dvl/β-Catenin –> Pitx2 pathway mediating cell-type-specific proliferation during development. *Cell* **111:** 673–685.

Klaus A, Saga Y, Taketo MM, Tzahor E, Birchmeier W. 2007. Distinct roles of Wnt/β-catenin and Bmp signaling during early cardiogenesis. *Proc Natl Acad Sci* **104:** 18531–18536.

Kobielak A, Fuchs E. 2006. Links between α-catenin, NF-kappaB, and squamous cell carcinoma in skin. *Proc Natl Acad Sci* **103:** 2322–2327.

Kobielak A, Pasolli HA, Fuchs E. 2004. Mammalian formin-1 participates in adherens junctions and polymerization of linear actin cables. *Nat Cell Biol* **6:** 21–30.

Kostetskii I, Moore R, Kemler R, Radice GL. 2001. Differential adhesion leads to segregation and exclusion of N-cadherin-deficient cells in chimeric embryos. *Dev Biol* **234:** 72–79.

Kwon C, Arnold J, Hsiao EC, Taketo MM, Conklin BR, Srivastava D. 2007. Canonical Wnt signaling is a positive regulator of mammalian cardiac progenitors. *Proc Natl Acad Sci* **104:** 10894–10899.

Lampugnani MG, Orsenigo F, Gagliani MC, Tacchetti C, Dejana E. 2006. Vascular endothelial cadherin controls

VEGFR-2 internalization and signaling from intracellular compartments. *J Cell Biol* **174:** 593–604.

Larue L, Ohsugi M, Hirchenhain J, Kemler R. 1994. E-cadherin null mutant embryos fail to form a trophectoderm epithelium. *Proc Natl Acad Sci* **91:** 8263–8267.

Lee DM, Kiener HP, Agarwal SK, Noss EH, Watts GF, Chisaka O, Takeichi M, Brenner MB. 2007. Cadherin-11 in synovial lining formation and pathology in arthritis. *Science* **315:** 1006–1010.

Lele Z, Folchert A, Concha M, Rauch GJ, Geisler R, Rosa F, Wilson SW, Hammerschmidt M, Bally-Cuif L. 2002. parachute/n-cadherin is required for morphogenesis and maintained integrity of the zebrafish neural tube. *Development* **129:** 3281–3294.

Lickert H, Kutsch S, Kanzler B, Tamai Y, Taketo MM, Kemler R. 2002. Formation of multiple hearts in mice following deletion of β-catenin in the embryonic endoderm. *Dev Cell* **3:** 171–181.

Liebner S, Cattelino A, Gallini R, Rudini N, Iurlaro M, Piccolo S, Dejana E. 2004. β-catenin is required for endothelial-mesenchymal transformation during heart cushion development in the mouse. *J Cell Biol* **166:** 359–367.

Lien WH, Klezovitch O, Vasioukhin V. 2006a. Cadherin–catenin proteins in vertebrate development. *Curr Opin Cell Biol* **18:** 499–506.

Lien WH, Klezovitch O, Fernandez TE, Delrow J, Vasioukhin V. 2006b. αE-catenin controls cerebral cortical size by regulating the hedgehog signaling pathway. *Science* **311:** 1609–1612.

Lien WH, Klezovitch O, Null M, Vasioukhin V. 2008. αE-catenin is not a significant regulator of β-catenin signaling in the developing mammalian brain. *J Cell Sci* **121:** 1357–1362.

Lin L, Cui L, Zhou W, Dufort D, Zhang X, Cai CL, Bu L, Yang L, Martin J, Kemler R, et al. 2007. β-catenin directly regulates Islet1 expression in cardiovascular progenitors and is required for multiple aspects of cardiogenesis. *Proc Natl Acad Sci* **104:** 9313–9318.

Luo Y, Radice GL. 2005. N-cadherin acts upstream of VE-cadherin in controlling vascular morphogenesis. *J Cell Biol* **169:** 29–34.

Luo Y, Ferreira-Cornwell M, Baldwin H, Kostetskii I, Lenox J, Lieberman M, Radice G. 2001. Rescuing the N-cadherin knockout by cardiac-specific expression of N- or E-cadherin. *Development* **128:** 459–469.

Luo Y, High FA, Epstein JA, Radice GL. 2006. N-cadherin is required for neural crest remodeling of the cardiac outflow tract. *Dev Biol* **299:** 517–528.

Machon O, van den Bout CJ, Backman M, Kemler R, Krauss S. 2003. Role of β-catenin in the developing cortical and hippocampal neuroepithelium. *Neuroscience* **122:** 129–143.

Mah SP, Saueressig H, Goulding M, Kintner C, Dressler GR. 2000. Kidney development in cadherin-6 mutants: Delayed mesenchyme-to-epithelial conversion and loss of nephrons. *Dev Biol* **223:** 38–53.

Malanchi I, Peinado H, Kassen D, Hussenet T, Metzger D, Chambon P, Huber M, Hohl D, Cano A, Birchmeier W, et al. 2008. Cutaneous cancer stem cell maintenance is dependent on β-catenin signalling. *Nature* **452:** 650–653.

Manabe T, Togashi H, Uchida N, Suzuki SC, Hayakawa Y, Yamamoto M, Yoda H, Miyakawa T, Takeichi M, Chisaka O. 2000. Loss of cadherin-11 adhesion receptor enhances plastic changes in hippocampal synapses and modifies behavioral responses. *Mol Cell Neurosci* **15:** 534–546.

Marose TD, Merkel CE, McMahon AP, Carroll TJ. 2008. β-catenin is necessary to keep cells of ureteric bud/Wolffian duct epithelium in a precursor state. *Dev Biol* **314:** 112–126.

McMahon AP, Bradley A. 1990. The Wnt-1 (int-1) proto-oncogene is required for development of a large region of the mouse brain. *Cell* **62:** 1073–1085.

McNeill H. 2009. Planar cell polarity: Keeping hairs straight is not so simple. *Cold Spring Harb Perspect Biol* **2:** a003376.

Meng W, Takeichi M. 2009. Adherens junction: Molecular architecture and regulation. *Cold Spring Harb Perspect Biol* **1:** a002899.

Mucenski ML, Wert SE, Nation JM, Loudy DE, Huelsken J, Birchmeier W, Morrisey EE, Whitsett JA. 2003. β-Catenin is required for specification of proximal/distal cell fate during lung morphogenesis. *J Biol Chem* **278:** 40231–40238.

Nagafuchi A, Takeichi M, Tsukita S. 1991. The 102 kd cadherin-associated protein: Similarity to vinculin and posttranscriptional regulation of expression. *Cell* **65:** 849–857.

Nakagawa S, Takeichi M. 1998. Neural crest emigration from the neural tube depends on regulated cadherin expression. *Development* **125:** 2963–2971.

Nemade RV, Bierie B, Nozawa M, Bry C, Smith GH, Vasioukhin V, Fuchs E, Hennighausen L. 2004. Biogenesis and function of mouse mammary epithelium depends on the presence of functional α-catenin. *Mech Dev* **121:** 91–99.

Niessen CM. 2007. Tight junctions/adherens junctions: Basic structure and function. *J Invest Dermatol* **127:** 2525–2532.

Noben-Trauth K, Zheng QY, Johnson KR. 2003. Association of cadherin 23 with polygenic inheritance and genetic modification of sensorineural hearing loss. *Nat Genet* **35:** 21–23.

Ogata S, Morokuma J, Hayata T, Kolle G, Niehrs C, Ueno N, Cho KW. 2007. TGF-β signaling-mediated morphogenesis: Modulation of cell adhesion via cadherin endocytosis. *Genes Dev* **21:** 1817–1831.

Ohnishi K, Melchers F, Shimizu T. 2005. Lymphocyte-expressed BILL-cadherin/cadherin-17 contributes to the development of B cells at two stages. *Eur J Immunol* **35:** 957–963.

Ohsugi M, Hwang SY, Butz S, Knowles BB, Solter D, Kemler R. 1996. Expression and cell membrane localization of catenins during mouse preimplantation development. *Dev Dyn* **206:** 391–402.

Ohsugi M, Larue L, Schwarz H, Kemler R. 1997. Cell-junctional and cytoskeletal organization in mouse blastocysts lacking E-cadherin. *Dev Biol* **185:** 261–271.

Onder TT, Gupta PB, Mani SA, Yang J, Lander ES, Weinberg RA. 2008. Loss of E-cadherin promotes metastasis via multiple downstream transcriptional pathways. *Cancer Res* **68:** 3645–3654.

Ong LL, Kim N, Mima T, Cohen-Gould L, Mikawa T. 1998. Trabecular myocytes of the embryonic heart require N-cadherin for migratory unit identity. *Dev Biol* **193:** 1–9.

Park C, Falls W, Finger JH, Longo-Guess CM, Ackerman SL. 2002a. Deletion in Catna2, encoding α N-catenin, causes cerebellar and hippocampal lamination defects and impaired startle modulation. *Nat Genet* **31:** 279–284.

Park C, Finger JH, Cooper JA, Ackerman SL. 2002b. The cerebellar deficient folia (cdf) gene acts intrinsically in Purkinje cell migrations. *Genesis* **32:** 32–41.

Park JS, Valerius MT, McMahon AP. 2007. Wnt/β-catenin signaling regulates nephron induction during mouse kidney development. *Development* **134:** 2533–2539.

Paulson AF, Fang X, Ji H, Reynolds AB, McCrea PD. 1999. Misexpression of the catenin p120(ctn)1A perturbs *Xenopus* gastrulation but does not elicit Wnt-directed axis specification. *Dev Biol* **207:** 350–363.

Pece S, Gutkind JS. 2000. Signaling from E-cadherins to the MAPK pathway by the recruitment and activation of epidermal growth factor receptors upon cell-cell contact formation. *J Biol Chem* **275:** 41227–41233.

Perez-Moreno M, Davis MA, Wong E, Pasolli HA, Reynolds AB, Fuchs E. 2006. p120-catenin mediates inflammatory responses in the skin. *Cell* **124:** 631–644.

Perez-Moreno M, Song W, Pasolli HA, Williams SE, Fuchs E. 2008. Loss of p120 catenin and links to mitotic alterations, inflammation, and skin cancer. *Proc Natl Acad Sci* **105:** 15399–15404.

Perrais M, Chen X, Perez-Moreno M, Gumbiner BM. 2007. E-Cadherin Homophilic Ligation Inhibits Cell Growth and Epidermal Growth Factor Receptor Signaling Independently of Other Cell Interactions. *Mol Biol Cell* **18:** 2013–2025.

Piedra J, Miravet S, Castano J, Palmer HG, Heisterkamp N, Garcia de Herreros A, Dunach M. 2003. p120 Catenin-associated Fer and Fyn tyrosine kinases regulate β-catenin Tyr-142 phosphorylation and β-catenin-α-catenin Interaction. *Mol Cell Biol* **23:** 2287–2297.

Qian X, Karpova T, Sheppard AM, McNally J, Lowy DR. 2004. E-cadherin-mediated adhesion inhibits ligand-dependent activation of diverse receptor tyrosine kinases. *Embo J* **23:** 1739–1748.

Radice GL, Ferreira-Cornwell MC, Robinson SD, Rayburn H, Chodosh LA, Takeichi M, Hynes RO. 1997a. Precocious mammary gland development in P-cadherin-deficient mice. *J Cell Biol* **139:** 1025–1032.

Radice GL, Rayburn H, Matsunami H, Knudsen KA, Takeichi M, Hynes RO. 1997b. Developmental defects in mouse embryos lacking N-cadherin. *Dev Biol* **181:** 64–78.

Reynolds AB. 2007. p120-catenin: Past and present. *Biochim Biophys Acta* **1773:** 2–7.

Reynolds AB, Carnahan RH. 2004. Regulation of cadherin stability and turnover by p120ctn: Implications in disease and cancer. *Semin Cell Dev Biol* **15:** 657–663.

Rhee J, Buchan T, Zukerberg L, Lilien J, Balsamo J. 2007. Cables links Robo-bound Abl kinase to N-cadherin-bound β-catenin to mediate Slit-induced modulation of adhesion and transcription. *Nat Cell Biol* **9:** 883–892.

Riethmacher D, Brinkmann V, Birchmeier C. 1995. A targeted mutation in the mouse E-cadherin gene results in defective preimplantation development. *Proc Natl Acad Sci* **92:** 855–859.

Rudini N, Felici A, Giampietro C, Lampugnani M, Corada M, Swirsding K, Garre M, Liebner S, Letarte M, ten Dijke P, et al. 2008. VE-cadherin is a critical endothelial regulator of TGF-β signalling. *Embo J* **27:** 993–1004.

Ruiz P, Brinkmann V, Ledermann B, Behrend M, Grund C, Thalhammer C, Vogel F, Birchmeier C, Gunthert U, Franke WW, et al. 1996. Targeted mutation of plakoglobin in mice reveals essential functions of desmosomes in the embryonic heart. *J Cell Biol* **135:** 215–225.

Sehgal RN, Gumbiner BM, Reichardt LF. 1997. Antagonism of cell adhesion by an α-catenin mutant, and of the Wnt-signaling pathway by α-catenin in *Xenopus* embryos. *J Cell Biol* **139:** 1033–1046.

Sheikh F, Chen Y, Liang X, Hirschy A, Stenbit AE, Gu Y, Dalton ND, Yajima T, Lu Y, Knowlton KU, et al. 2006. α-E-catenin inactivation disrupts the cardiomyocyte adherens junction, resulting in cardiomyopathy and susceptibility to wall rupture. *Circulation* **114:** 1046–1055.

Shoval I, Ludwig A, Kalcheim C. 2007. Antagonistic roles of full-length N-cadherin and its soluble BMP cleavage product in neural crest delamination. *Development* **134:** 491–501.

Simcha I, Shtutman M, Salomon D, Zhurinsky J, Sadot E, Geiger B, Ben-Ze'ev A. 1998. Differential nuclear translocation and transactivation potential of β-catenin and plakoglobin. *J Cell Biol* **141:** 1433–1448.

Soshnikova N, Zechner D, Huelsken J, Mishina Y, Behringer RR, Taketo MM, Crenshaw EB3rd, Birchmeier W. 2003. Genetic interaction between Wnt/β-catenin and BMP receptor signaling during formation of the AER and the dorsal-ventral axis in the limb. *Genes Dev* **17:** 1963–1968.

Suyama K, Shapiro I, Guttman M, Hazan RB. 2002. A signaling pathway leading to metastasis is controlled by N-cadherin and the FGF receptor. *Cancer Cell* **2:** 301–314.

Suzuki SC, Furue H, Koga K, Jiang N, Nohmi M, Shimazaki Y, Katoh-Fukui Y, Yokoyama M, Yoshimura M, Takeichi M. 2007. Cadherin-8 is required for the first relay synapses to receive functional inputs from primary sensory afferents for cold sensation. *J Neurosci* **27:** 3466–3476.

Taddei A, Giampietro C, Conti A, Orsenigo F, Breviario F, Pirazzoli V, Potente M, Daly C, Dimmeler S, Dejana E. 2008. Endothelial adherens junctions control tight junctions by VE-cadherin-mediated upregulation of claudin-5. *Nat Cell Biol* **10:** 923–934.

Takahashi K, Suzuki K. 1996. Density-dependent inhibition of growth involves prevention of EGF receptor activation by E-cadherin-mediated cell-cell adhesion. *Exp Cell Res* **226:** 214–222.

Tao Q, Yokota C, Puck H, Kofron M, Birsoy B, Yan D, Asashima M, Wylie CC, Lin X, Heasman J. 2005. Maternal wnt11 activates the canonical wnt signaling pathway required for axis formation in *Xenopus* embryos. *Cell* **120:** 857–871.

Tinkle CL, Lechler T, Pasolli HA, Fuchs E. 2004. Conditional targeting of E-cadherin in skin: Insights into hyperproliferative and degenerative responses. *Proc Natl Acad Sci* **101:** 552–557.

Tinkle CL, Pasoli HA, Stokes N, Fuchs E. 2008. New insights into cadherin function in epidermal sheet formation and maintenance of tissue integrity. *Proc Natl Acad Sci* **105:** 15405–15410.

Torres M, Stoykova A, Huber O, Chowdhury K, Bonaldo P, Mansouri A, Butz S, Kemler R, Gruss P. 1997. An α-E-catenin gene trap mutation defines its function in preimplantation development. *Proc Natl Acad Sci* **94:** 901–906.

Tunggal JA, Helfrich I, Schmitz A, Schwarz H, Gunzel D, Fromm M, Kemler R, Krieg T, Niessen CM. 2005. E-cadherin is essential for in vivo epidermal barrier function by regulating tight junctions. *Embo J* **24:** 1146–1156.

Uemura M, Takeichi M. 2006. α N-catenin deficiency causes defects in axon migration and nuclear organization in restricted regions of the mouse brain. *Dev Dyn* **235:** 2559–2566.

van Buul JD, Anthony EC, Fernandez-Borja M, Burridge K, Hordijk PL. 2005. Proline-rich tyrosine kinase 2 (Pyk2) mediates vascular endothelial-cadherin-based cell-cell adhesion by regulating β-catenin tyrosine phosphorylation. *J Biol Chem* **280:** 21129–21136.

Vasioukhin V, Bauer C, Yin M, Fuchs E. 2000. Directed actin polymerization is the driving force for epithelial cell-cell adhesion. *Cell* **100:** 209–219.

Vasioukhin V, Bauer C, Degenstein L, Wise B, Fuchs E. 2001. Hyperproliferation and defects in epithelial polarity upon conditional ablation of α-catenin in skin. *Cell* **104:** 605–617.

Wilson SM, Householder DB, Coppola V, Tessarollo L, Fritzsch B, Lee EC, Goss D, Carlson GA, Copeland NG, Jenkins NA. 2001. Mutations in Cdh23 cause nonsyndromic hearing loss in waltzer mice. *Genomics* **74:** 228–233.

Yamada S, Pokutta S, Drees F, Weis WI, Nelson WJ. 2005. Deconstructing the cadherin–catenin-actin complex. *Cell* **123:** 889–901.

Yang X, Chrisman H, Weijer CJ. 2008. PDGF signalling controls the migration of mesoderm cells during chick gastrulation by regulating N-cadherin expression. *Development* **135:** 3521–3530.

Yin Y, White AC, Huh SH, Hilton MJ, Kanazawa H, Long F, Ornitz DM. 2008. An FGF-WNT gene regulatory network controls lung mesenchyme development. *Dev Biol* **319:** 426–436.

Young P, Boussadia O, Halfter H, Grose R, Berger P, Leone DP, Robenek H, Charnay P, Kemler R, Suter U. 2003. E-cadherin controls adherens junctions in the epidermis and the renewal of hair follicles. *Embo J* **22:** 5723–5733.

Zamora M, Manner J, Ruiz-Lozano P. 2007. Epicardium-derived progenitor cells require β-catenin for coronary artery formation. *Proc Natl Acad Sci* **104:** 18109–18114.

Zechner D, Fujita Y, Hulsken J, Muller T, Walther I, Taketo MM, Crenshaw EB3rd, Birchmeier W, Birchmeier C. 2003. β-Catenin signals regulate cell growth and the balance between progenitor cell expansion and differentiation in the nervous system. *Dev Biol* **258:** 406–418.

Zinser GM, Leonis MA, Toney K, Pathrose P, Thobe M, Kader SA, Peace BE, Beauman SR, Collins MH, Waltz SE. 2006. Mammary-specific Ron receptor overexpression induces highly metastatic mammary tumors associated with β-catenin activation. *Cancer Res* **66:** 11967–11974.

Zohn IE, Li Y, Skolnik EY, Anderson KV, Han J, Niswander L. 2006. p38 and a p38-interacting protein are critical for downregulation of E-cadherin during mouse gastrulation. *Cell* **125:** 957–969.

Interplay of Cadherin-mediated Cell Adhesion and Canonical Wnt Signaling

Julian Heuberger and Walter Birchmeier

Max Delbrück Center for Molecular Medicine, 13125 Berlin, Germany

Correspondence: wbirch@mdc-berlin.de

The epithelial–mesenchymal transition is essential in both embryonic development and the progression of carcinomas. Wnt signaling and cadherin-mediated adhesion have been implicated in both processes; clarifying their role will depend on linking them to rearrangements of cellular structure and behavior. β-Catenin is an essential molecule both in cadherin-mediated cell adhesion and in canonical Wnt signaling. Numerous experiments have shown that the loss of cadherin-mediated cell adhesion can promote β-catenin release and signaling; this is accomplished by proteases, protein kinases and other molecules. Cadherin loss can also signal to several other regulatory pathways. Additionally, many target genes of Wnt signaling influence cadherin adhesion. The most conspicuous of these Wnt target genes encode the transcription factors Twist and Slug, which directly inhibit the E-cadherin gene promoter. Other Wnt/β-catenin target genes encode metalloproteases or the cell adhesion molecule L1, which favor the degradation of E-cadherin. These factors provide a mechanism whereby cadherin loss and increased Wnt signaling induce epithelial–mesenchymal transition in both carcinomas and development.

In both normal embryonic development and carcinomas, signals trigger epithelial–mesenchymal transitions (EMT). These biological programs convert polarized, rather immobile epithelial cells to fibroblastoid, highly motile mesenchymal cells (Fig. 1). EMT occurs over and over during embryonic development; for instance, it is crucial in the formation of mesoderm during gastrulation, which produces a three-layered embryo, and formation of the neural crest, the heart, and the craniofacial system (reviewed by Thiery 2002). EMT also occurs in carcinomas, where it promotes invasion and metastasis (reviewed by Thiery 2002; Yang and Weinberg 2008). A thorough understanding of these processes will require mechanistic explanations of how the signaling pathways implicated in EMTs trigger the rearrangements of cellular architecture and changes in behavior that permit it to happen. They should also explain how changes in healthy signaling lead to inappropriate EMTs in carcinomas. After three decades of work, an integrated picture of the connections between these processes is finally coming into focus. The present review centers on the role of cadherin-mediated cell adhesion and its interplay with Wnt/β-catenin signaling,

 doi: 10.1101/cshperspect.a002915
Cite this article as *Cold Spring Harb Perspect Biol* 2009;1:a002915

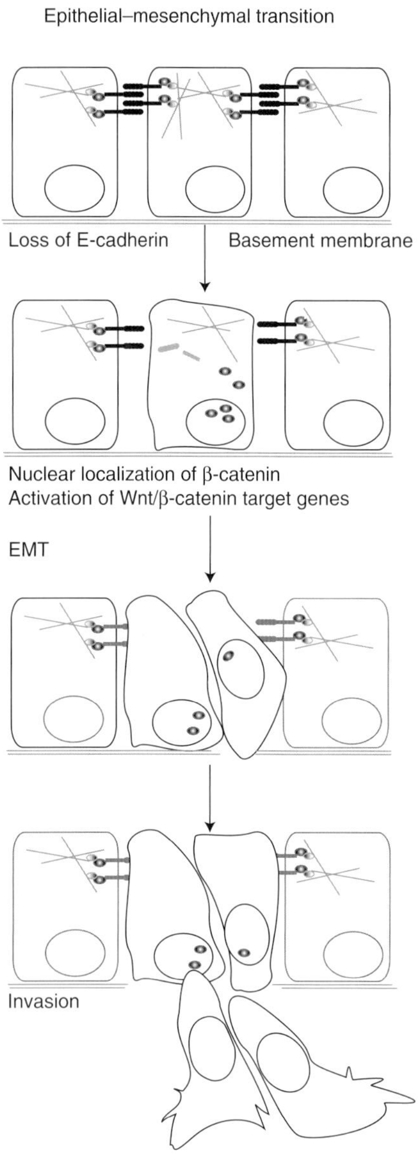

Figure 1. Epithelial–mesenchymal transition. EMT can be induced by the loss of E-cadherin and/or by the activation of canonical Wnt signaling. Epithelial cells become more motile and receive a mesenchymal character. The mesenchymal-like cells can pass the basement membrane and become invasive.

which are both crucial in the process of EMT (see also Nelson and Nusse 2004; Gavert and Ben-Ze'ev 2007).

In the late 1960s to early 1980s, Hay and collaborators recognized EMT as a distinct biological process in studies with embryos and three-dimensional cell cultures (Trelstad et al. 1967; Greenburg and Hay 1982). Soon afterwards, our group showed that Madin-Darby canine kidney (MDCK) epithelial cells are converted into fibroblastoid cells through the disruption of E-cadherin-mediated cell adhesion with specific antibodies (Imhof et al. 1983; Behrens et al. 1985). Stoker and Gherardi showed that MDCK epithelial cells could be converted into motile fibroblastoid cells by the addition of conditioned medium that contained scatter factor/hepatocyte growth factor, HGF/SF (Stoker et al. 1987; Weidner et al. 1991). Further work showed that MDCK epithelial cells acquired invasive properties, i.e., invaded collagen gels and embryonal heart tissue, following disturbed E-cadherin-mediated cell adhesion (Behrens et al. 1989), or by the addition of HGF/SF (Weidner et al. 1990; Weidner et al. 1991).

The process of EMT is reversible: Invasive human carcinoma cells that had lost E-cadherin expression and epithelial polarity could be made noninvasive by transfection with the E-cadherin cDNA (Frixen et al. 1991; Vleminckx et al. 1991). It could also be shown that poorly differentiated human breast and stomach carcinomas carried inactivating mutations in the E-cadherin gene (Becker et al. 1994; Berx et al. 1995), and inherited germline mutations of E-cadherin were found in early-onset, diffuse-type gastric cancer in human families (Guilford et al. 1998). The loss of E-cadherin expression coincided with the transition from well-differentiated adenoma to invasive carcinoma in a transgenic mouse model of pancreatic carcinogenesis (Perl et al. 1998). In other cases of human carcinomas, down-regulation of E-cadherin occurred by hypermethylation of the E-cadherin gene promoter (Hennig et al. 1995; Strathdee 2002). Beug and collaborators showed that activation of the c-Fos proto-oncogene in mammary gland epithelial cells also induced EMT, e.g., down-regulated E-cadherin, up-regulated several mesenchymal genes, and promoted invasion of the cells into collagen gels (Reichmann et al. 1992). Remarkably, overexpression of stabilized β-catenin also scattered MDCK epithelial cells in tissue culture (Barth et al. 1997). In the last decade, groups have identified several more signaling pathways and specific transcription factors that induce EMT in development and in the invasion and metastasis of carcinomas. These include:

- TGFβ signaling (Oft et al. 1996; Mercado-Pimentel and Runyan 2007; Nguyen and Massague 2007).
- Wnt signaling (Kinzler and Vogelstein 1996; Huelsken et al. 2000; Fodde et al. 2001).
- Tyrosine kinase signaling (the ras pathway; Behrens et al. 1993; Oft et al. 1996; Meiners et al. 1998; Dietrich et al. 1999; Muller et al. 1999; Khoury et al. 2001).
- The transcription factors Snail, Slug, ZEB, E47, and Twist, which suppress E-cadherin expression (Batlle et al. 2000; Cano et al. 2000; Comijn et al. 2001; Peinado et al. 2004b; Yang et al. 2004).
- Tiam1/Rac1 signaling (Habets et al. 1994; Collard et al. 1996).
- Goosecoid and FOXC2 (Hartwell et al. 2006; Mani et al. 2007).

The discovery that cytoplasmic catenins are essential in cadherin-mediated cell adhesion was made in the late 1980s and early 1990s. Kemler's group reported that the conserved cytoplasmic carboxyl terminus of the cell adhesion molecule E-cadherin binds three proteins with mol.wts. of 102, 88, and 80 kDa, i.e., α-catenin, β-catenin, and plakoglobin (γ-catenin) (Vestweber and Kemler 1984; Ozawa et al. 1989; Nagafuchi and Takeichi 1989). Plakoglobin has been molecularly cloned by Franke and collaborators (Cowin et al. 1986), α-catenin by Takeichi and Tsukita and collaborators (Nagafuchi et al. 1991), and β-catenin by Gumbiner and collaborators (McCrea et al. 1991; Butz et al. 1992). β-Catenin forms a direct link between cadherins and α-catenin, whereas α-catenin also talks with the actin cytoskeleton (Rimm et al. 1995; Drees et al. 2005; Yamada et al. 2005; Abe and Takeichi 2008). This link is essential for strong cadherin-mediated cell adhesion (Nagafuchi and Takeichi 1988; reviewed by Takeichi 1991; Kemler 1993; Gates and Peifer 2005; see Meng and Takeichi 2009 and Shapiro and Weis 2009).

The Wnt/β-catenin signaling pathway has a crucial role in the embryonic development of all animal species, in the regeneration of tissues in adult organisms and in numerous other processes (reviewed by Cadigan and Nusse 1997; Clevers 2006; Klaus and Birchmeier 2008). Mutations or deregulated expression of components of the canonical Wnt pathway can induce disease, most importantly cancer (Bienz and Clevers 2000; Polakis 2000; Moon et al. 2004). Through the Wnt pathway, signals are exchanged between neighboring cells and tissues: Wnt proteins that are secreted from one type of cell interact with surface receptors of neighboring cells. There the signals are passed through the cytoplasm to the nucleus, where gene regulation is modulated. This process controls proliferation, survival, cell migration, differentiation, and patterning in the receiving cells and tissues.

The protein β-catenin plays a critical role in canonical Wnt signaling. Noncanonical Wnt signaling, which does not involve β-catenin, is not discussed here (Veeman et al. 2003). It had been known since the late 1980s that β-catenin (armadillo in *Drosophila*) is involved in Wnt signaling, but the way this molecule moves from the cytoplasm to the nucleus and interacts with the transcription machinery was unknown (Wieschaus and Riggleman 1987; Riggleman et al. 1990; Peifer et al. 1992). In 1995, an interaction was found between β-catenin and the lymphocyte enhancer binding factor 1 (LEF1). This observation precipitated feverish work, which firmly established that LEF1, and its closest relatives, T cell factors (TCF), are activated by signaling through the Wnt pathway (Behrens et al. 1996; Molenaar et al. 1996; Huber et al. 1996; Riese et al. 1997). LEF1 and TCFs are high mobility group (HMG) proteins and were discovered as DNA-binding proteins that bind to specific sequences in lymphoid enhancers (Travis et al. 1991; van de Wetering et al. 1991; Waterman et al. 1991).

Canonical Wnt signaling works in the following fashion (for further detailed discussion of the Wnt pathway, see Cadigan and Pfeifer 2009; McCrea et al. 2009): In the absence of Wnt ligands, cytoplasmic β-catenin is recruited into a destruction complex, in which it interacts with APC and the axins, and is amino-terminally phosphorylated by axin-bound casein kinase 1α (CK1α) and GSK3β (Fig. 2A). Following phosphorylation, β-catenin is targeted for proteasome-dependent degradation involving an interaction with β-TrCP (β-transducin repeat-containing protein), a component of the E3 ubiquitin ligase complex. Therefore, in the absence of Wnt, cytoplasmic β-catenin levels remain low, and the transcription factors LEF1 and TCF interact with Grouchos in the nucleus to repress Wnt-specific target genes (Munemitsu et al. 1995; Aberle et al. 1997; Behrens et al. 1998; Roose et al. 1998; Liu et al. 2002). In the presence of Wnt ligands, LRP5–LRP6 surface receptors are phosphorylated by CK1γ and GSK3β (and possibly further kinases), and Dishevelled is recruited to the plasma membrane, where it interacts with Frizzled receptors and polymerizes with other Dishevelled molecules (Fig. 2B) (Davidson et al. 2005; Bilic et al. 2007; Schwarz-Romond et al. 2007). The inactivation of the destruction complex allows the cytoplasmic stabilization and translocation of β-catenin to the nucleus. Here, β-catenin forms a transcriptionally active complex with LEF1 and TCF transcription factors by displacing Grouchos and interacting with other coactivators (reviewed by Eastman and Grosschedl 1999; Stadeli et al. 2006; Klaus and Birchmeier 2008). This rounds out the picture of how Wnt's effects on genes are modulated by the activation and repression of β-catenin. What remains to be explained is how the different sets of target genes exert their influence on cell architecture and behavior, which is a major theme of this review.

CROSSTALK BETWEEN CADHERIN-MEDIATED CELL ADHESION AND CANONICAL Wnt SIGNALING THROUGH β-CATENIN

Wnt Signaling and Cadherins Compete for β-Catenin

The concept that canonical Wnt signaling and cadherin-mediated cell adhesion depend on the same pool of β-catenin is based on genetic and overexpression experiments in embryos and cultured cells. Heasman et al. showed in 1994 that overexpression of cadherins in *Xenopus* embryos inhibited dorsal axis formation, which is a clear function of canonical Wnt signaling (Heasman et al. 1994; see also Fagotto et al. 1996; Torres et al. 1996). Peifer and collaborators showed that armadillo (β-catenin) mutant embryos of *Drosophila*, which harbor only one E-cadherin allele, showed a less severe segment polarity phenotype than embryos with two cadherin alleles (Cox et al. 1996). Segment polarity is controlled by Wnt signaling (Nüsslein-Volhard et al. 1980; Rijsewijk et al. 1987). Similarly, cadherin overexpression mimicked the wingless (Wnt) phenotype in *Drosophila* embryos (Sanson et al. 1996). These data from the mid 1990s are thus consistent with a model in which there is crosstalk between β-catenin in two different compartments, the adhesion complex at the plasma membrane and a signaling complex in the nucleus.

Shortly after these findings in model organisms, Geiger and Ben-Ze'ev and collaborators showed an interplay between cadherin-mediated cell adhesion and canonical Wnt signaling in cell culture experiments. In colon cancer cells, expression of N-cadherin or an interleukin receptor-cadherin hybrid (in which the β-catenin binding region of N-cadherin was maintained) triggered a relocation of β-catenin from the nucleus to the plasma membrane and inhibited LEF1-mediated transcription (Sadot et al. 1998; Shtutman et al. 1999; Gottardi et al. 2001; Stockinger et al. 2001). Inducible expression of the Fos proto-oncogene in mammary gland epithelial cells resulted in the loss of

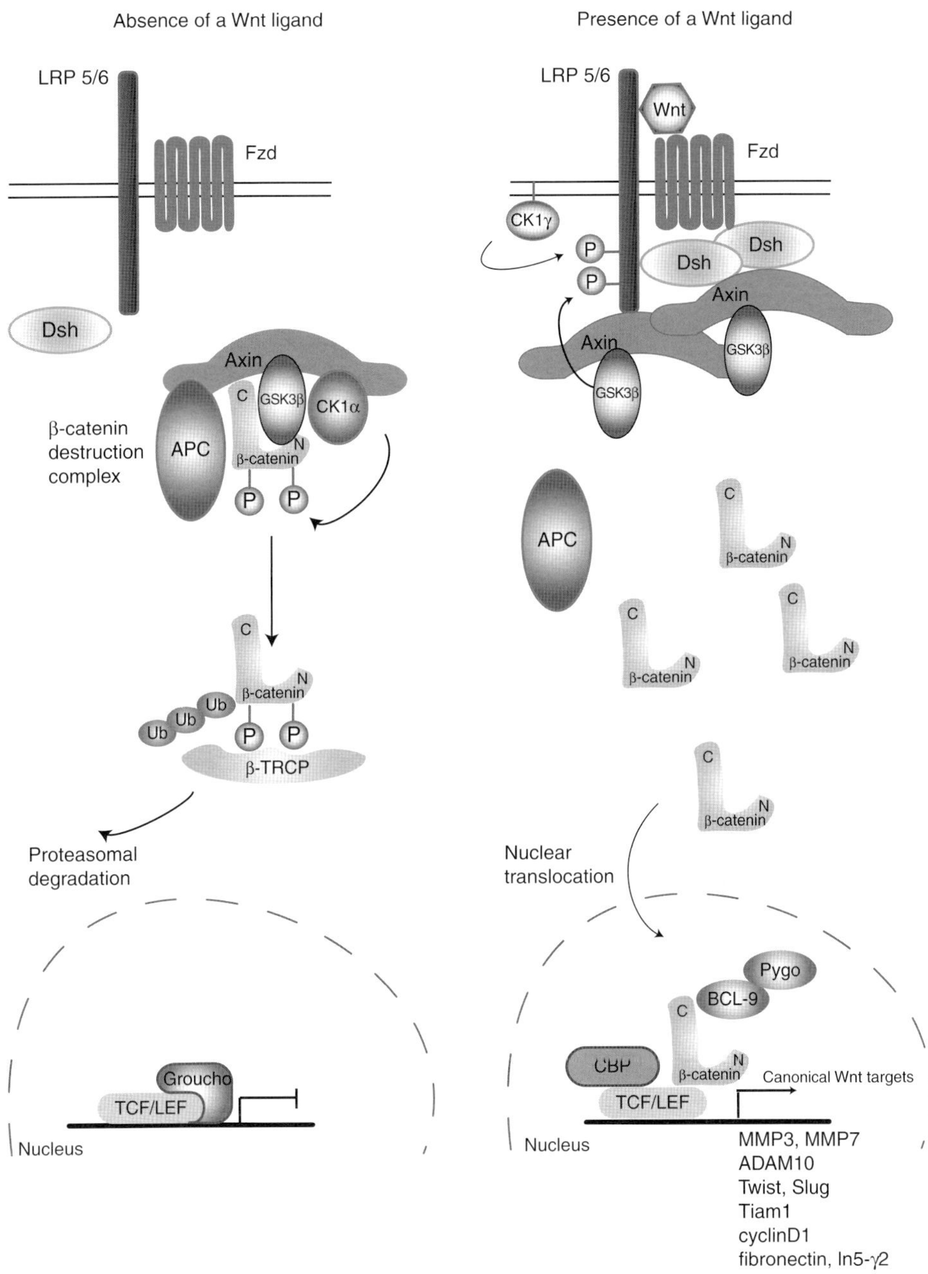

Figure 2. Canonical Wnt pathway. (*A*) In the absence of a Wnt ligand, β-catenin is sequentially phosphorylated in the destruction complex of axin and APC by the kinases CK1α and GSK3β, which results in ubiquitination by β-TRCP and, finally, in the degradation of β-catenin in the proteasom. (*B*) In the presence of a Wnt ligand, a Fzd/LRP5/6 complex is formed and is bound by Dishevelled, leading to the phosphorylation of the LRP5/6 receptor by GSK3β, CK1γ and other kinases. Phosphorylated LRP5/6 recruits axin to the plasma membrane, which leads to decay of the destruction complex, inhibited phosphorylation, and degradation of β catenin. Stabilized β-catenin is then translocated to the nucleus, a process that can involve BCl9-2. β-Catenin interacts with TCF/LEF, and the recruitment of cofactors like BCL9, Pygopus, or CBP as well as the replacement of Groucho leads to the activation of target genes.

E-cadherin and cell polarization, the colocalization of β-catenin with LEF1 in the nucleus, and increased Wnt/β-catenin signaling (Eger et al. 2000). Moreover, the absence of E-cadherin in E-cadherin -/- embryonic stem (ES) cells led to an accumulation of β-catenin with LEF1 in the nucleus and activation of a Wnt reporter (Orsulic et al. 1999). This could be antagonized by expression of E-cadherin.

Behrens and collaborators have recently shown that siRNA-mediated knockdown of E-cadherin augments β-catenin-dependent transcription in colon cancer cells in which the Wnt pathway is active. On the other hand, the same procedure has no effect in nontransformed keratinocytes that do not display Wnt signaling (Kuphal and Behrens 2006). These data indicate that the mere loss of E-cadherin does not activate Wnt signaling—except in cases in which the β-catenin degradation machinery is compromised. These results are consistent with data from breast cell cancer lines showing that the absence of E-cadherin alone does not result in activation of Wnt signaling (van de Wetering et al. 2001). Nor does a loss of E-cadherin function in Rip1Tag2 transgenic mice contribute to Wnt/β-catenin signaling (Herzig et al. 2007). Weinberg and collaborators recently showed that in ras-transformed mammary gland cells (HMLER), shRNA down-regulation of E-cadherin results in translocation of β-catenin from cell–cell junctions to the cytoplasm and nucleus (Onder et al. 2008). This type of β-catenin was nonphosphorylated and thus was not targeted for ubiquitination and degradation. However, in this system, the loss of E-cadherin affected numerous other signaling pathways that have been implicated in metastasis formation.

Gottardi and Gumbiner have performed precise studies to determine what controls β-catenin targeting to cadherin adhesion or to TCF transcriptional complexes (Gottardi and Gumbiner 2004). They showed that Wnt signaling generates a monomeric, intramolecularly folded-back form of β-catenin that binds TCF but not cadherins. In contrast, the cadherin-binding form of β-catenin builds a dimer with α-catenin. X-ray crystallographic studies have shown that cadherin-binding involves all 12 armadillo repeats of β-catenin, whereas TCF-binding requires only the central eight repeats (Graham et al. 2000; Huber and Weis 2001). Thus it is possible that the carboxyl terminus of Wnt-produced β-catenin folds back over armadillo repeats, affecting binding to cadherin but not TCF. The selective binding of β-catenin induced by Wnt could also involve post-translational modifications or the activation of further proteins. Overall, these data suggest that β-catenin's selectivity between adhesion and transcription are not always coupled; in other words, they might be regulated independently.

BCL9 protein, the product of a human proto-oncogene, also acts in the switch between cadherin cell adhesion and β-catenin signaling. This story has been worked out through work on BCL9 and its ortholog legless, a *Drosophila* segment polarity gene. Legless, which was isolated in 2002 by the group of Konrad Basler, is required for Wnt signaling in the fly (Kramps et al. 2002). It acts by binding directly to β-catenin. Human BCL9 was discovered in a B-cell lymphoma because of a translocation to the immunoglobulin locus, which caused BCL9 overexpression in the tumors (Willis et al. 1998). Remarkably, human BCL9 could rescue the segment polarity phenotype of the legless mutation (Kramps et al. 2002), indicating functional identity. Vertebrates have a second homolog BCL9-2 (Brembeck et al. 2004; Adachi et al. 2004), which also binds to β-catenin like BCL9. BCL9-2 promotes nuclear location of β-catenin, increased β-catenin signaling, and triggers EMT in vertebrate cells and Zebra fish embryos (Brembeck et al. 2004). BCL9-2 cannot colocalize with the E-cadherin/β-catenin/α-catenin complex at the plasma membrane, but following tyrosine phosphorylation of β-catenin, it is translocated to the nucleus and promotes β-catenin signaling. Thus BCL9 proteins may act in the switch between cadherin-mediated cell adhesion and Wnt signaling (Brembeck et al. 2004; Sampietro et al. 2006; Hoffmans and Basler 2007; de la Roche et al. 2008).

Proteolysis of Cadherins Affects Wnt Signaling

Competition for β-catenin is one mechanism by which cells can modulate Wnt signals. Another is the cleavage of E- and N-cadherin by proteases, which can lead to the release of β-catenin and to up-regulated β-catenin signaling. This occurs in normal development but also in processes such as wound healing, Ca^{2+}-influx, and apoptosis. Cadherins are cleaved intracellularly by proteases like caspase 3 or presenilin, or extracellularily by ADAM10 (Ito et al. 1999; Steinhusen et al. 2001; Marambaud et al. 2002; Maretzky et al. 2005; Shoval et al. 2007). Intracellular cleavage of cadherins releases nearly the entire cytoplasmic domain, whereas extracellular cleavage releases the adhesion domain (Fig. 3). ADAM10 belongs to the family of disintegrins and metalloproteases (ADAM), type I transmembrane proteins that combine cell–cell adhesion and proteinase activity (reviewed by Primakoff and Myles 2000; Seals and Courtneidge 2003). ADAM10 was recently identified as a β-catenin/TFC target gene (Gavert et al. 2007). The group of Saftig has shown that cleavage of E- and N-cadherin by ADAM10 in keratinocytes and neuronal cells, respectively, reduced cell adhesion, increased cell migration and led to translocation of β-catenin to the nucleus. Translocated β-catenin activated Wnt/β-catenin target genes like c-Myc and cyclinD1 (Maretzky et al. 2005; Reiss et al. 2005). These are clear examples of an immediate crosstalk between a loss of cadherin-mediated cell adhesion and Wnt/β-catenin signaling.

Presenilin1 (PS1) is an integral membrane protein and a component of the γ-secretase complex, which mediates an ε-cleavage of type I membrane proteins such as the amyloid precursor protein (APP). A mutant form of PS1 (PS1 FAD, familial Alzheimer's disease) is known to process APP, leading to increased production of the β-amyloid peptide (Aβ42), which is responsible for Alzheimer's disease (reviewed by Marambaud and Robakis 2005; Selkoe and Wolfe 2007). Mechanistically, PS1 acts in several ways to influence cell adhesion

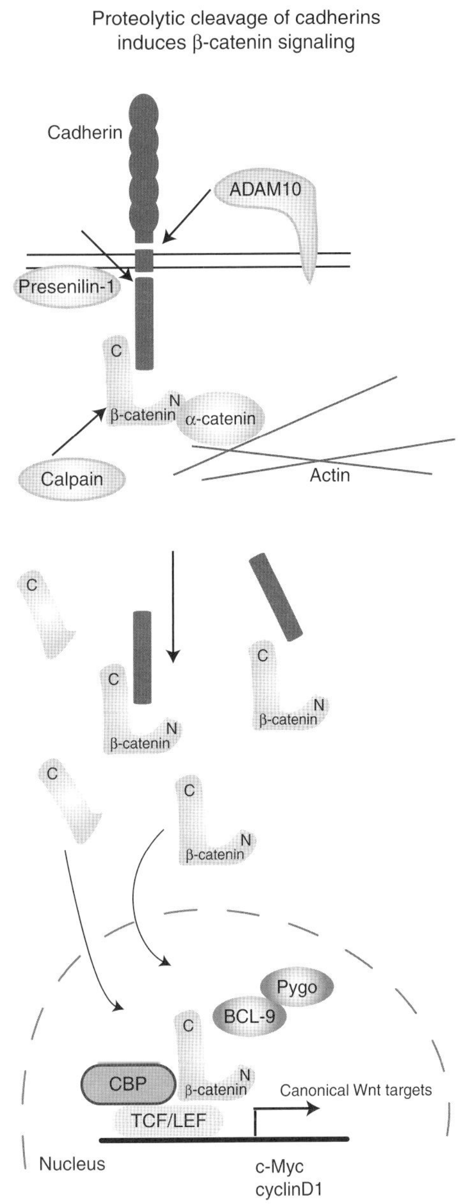

Figure 3. Proteolytic cleavage of cadherin can release β-catenin and induce canonical Wnt signaling. Under certain cellular conditions, cadherins can be cleaved by proteases like ADAM10 and Presenilin-1. This cleavage leads to loosening of cell adhesion, but also to the release of β-catenin, which is translocated into the nucleus and as a result, activates Wnt/β-catenin target genes. Another protease, calpain, is interfering with the adhesion complex by the amino-terminal cleavage of β-catenin. β-Catenin cleaved by calpain is stabilized and can enter the nucleus and activate target genes.

and transcription. PS1 can bind to E-cadherin and β-catenin and promotes their association to the cytoskeleton, and the overexpression of PS1 results in enhanced cadherin-mediated cell adhesion (Georgakopoulos et al. 1999; Baki et al. 2001). Under imbalanced calcium conditions or apoptosis, PS1 does not stabilize cell–cell adhesion but cleaves E-cadherin, which leads to the disassembly of adherens junctions, to the release of a cytoplasmic E-cadherin fragment, and to an increase of soluble β-catenin (Marambaud et al. 2002). It is possible that the release of β-catenin from adherens junctions caused by PS-1 modulates gene expression. It has been shown that the released N-cadherin fragments generated by PS1 favor nuclear localization of β-catenin and promote β-catenin signaling (Uemura et al. 2006). The released fragments of E-cadherin also translocate to the nucleus and affect transcriptional activity (Ferber et al. 2008). In contrast, PS1 can facilitate GSK3β-mediated phosphorylation of β-catenin, thereby favoring the degradation of β-catenin and negatively regulating β-catenin/LEF signaling (Zhang et al. 1998; Kang et al. 1999; Soriano et al. 2001; Kang et al. 2002). This latter type of cross-talk between PS-1 and β-catenin signaling has also been shown by genetic means; tumors with elevated β-catenin signaling developed in the skin of mice with PS-1 deficiency (Xia et al. 2001; Kang et al. 2002). Taken together, this indicates that PS-1 modulates junctional signaling (1) by cleaving cadherins, leading to increased β-catenin signaling and alterations in gene activity, and (2) by stabilizing cell-adhesion and promoting the degradation of β-catenin, thereby decreasing β-catenin mediated transcription.

Recently it has been proposed that the NMDA (*N*-methyl-D-aspartate) receptor in synapses links cadherin-mediated adhesion and β-catenin signaling (Abe and Takeichi 2007). The NMDA receptor is a glutamate-gated ion channel, which signals through the influx of Ca^{2+} and induces long-term synaptic plasticity in the nervous system (Carroll and Zukin 2002). N-cadherin has an important function in this process (Arikkath and Reichardt 2008). Triggering the NMDA receptor in neuronal cells activates the proteases ADAM10 and PS-1, which are able to process cadherins (Marambaud et al. 2003; Reiss et al. 2005). Another protease, calpain, has been shown to be critically involved in synaptic plasticity (Staubli et al. 1988). Abe and Takeichi have now shown that the NMDA-dependent activation of calpain cleaves β-catenin at its amino terminus (Fig. 3). As a result, β-catenin is stabilized, accumulates in the nucleus, and induces TCF/LEF1-dependent gene expression (Abe and Takeichi 2007). If it proves to be true that calpain cleaves β-catenin bound to cadherin and that this is essential in synaptic processes, this would establish a direct link between neuronal activity, adhesion between cells at the synapse, and β-catenin signaling.

Phosphorylation and the Control of β-Catenin in Signaling and Adhesion

Yet another link between adhesion and Wnt signaling can be found in the phosphorylation of components of the cadherin adhesion system, of the β-catenin degradation complex, or of the TCF transcription system, which regulate the strength of interactions. The action of various serine/threonine and tyrosine kinases therefore influences the function of β-catenin in adhesion and in signaling (reviewed by Daugherty and Gottardi 2007). For instance, the carboxy-terminal phosphorylation of cadherins can strengthen or weaken interactions with β-catenin, depending on the sites of phosphorylation (Choi et al. 2006; Qi et al. 2006). Tyrosine 654 phosphorylation of β-catenin reduces cadherin binding and adhesion function (Fig. 4) (Roura et al. 1999). The phosphorylation of β-catenin at tyrosine 489 or 142 strengthens Wnt signaling (Brembeck et al. 2004; Rhee et al. 2007). The phosphorylation of axin or APC, respectively, on serine or threonine residues can strengthen or weaken interactions with β-catenin, and the phosphorylation of TCFs also influences their binding to β-catenin (Rubinfeld et al. 1996; Willert et al. 1999; Lee et al. 2001; Ha et al. 2004).

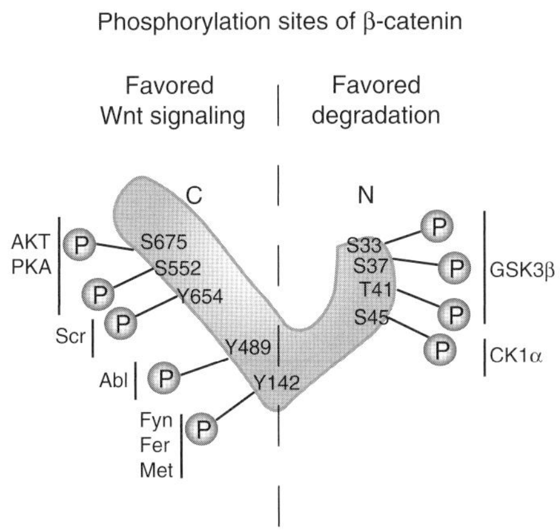

Figure 4. Phosphorylation sites of β-catenin. β-Catenin can be sequentially phosphorylated by different kinases, either promoting the degradation or the signaling activity of β-catenin. Phosphorylation at the amino terminus favors degradation, whereas phosphorylation in the armadillo domain changes the adhesiveness to cadherins and promotes nuclear localization.

Tyrosine kinases and tyrosine phosphatases have been shown to regulate the switch between cadherin-mediated cell adhesion and β-catenin signaling (Warren and Nelson 1987; Behrens et al. 1993; Muller et al. 1999; Roura et al. 1999; reviewed by Daugherty and Gottardi 2007). Insulin-like growth factor II leads to the internalization of E-cadherin/β-catenin and to E-cadherin degradation (Morali et al. 2001). β-Catenin then becomes nuclear, and β-catenin/TCF target genes are activated. HGF/SF action on Met can result in tyrosine 142 phosphorylation of β-catenin and induction of EMT in epithelial cells and axonal growth and branching of neuronal cells (Brembeck et al. 2004; David et al. 2008). A dominant-negative TCF4 can inhibit neural branching. The tyrosine kinase Met is itself a Wnt/β-catenin target gene (Boon et al. 2002). FGF and FGF receptor signaling also affects E-cadherin/β-catenin interaction and β-catenin/TCF signaling (Pai et al. 2008). Hunter and collaborators have shown that EGF receptor signaling induces EMT and caveolin-mediated endocytosis of E-cadherin (Lu et al. 2003). As a result, β-catenin localizes to the nucleus and activates Wnt/β-catenin target genes. EGF treatment also down-regulates caveolin-1, which enhances the transcriptional activity of β-catenin. Caveolin-1 negatively affects canonical Wnt signaling through the recruitment of β-catenin to caveolar membranes (Galbiati et al. 2000). E-cadherin can also be internalized by a different mechanism: Tyrosine phosphorylation of E-cadherin by Src kinase induces the recruitment of the E3 ubiquitin ligase Hakai and favors cell scattering (Fujita et al. 2002). Hakai ubiquitinates E-cadherin and mediates the clathrin-dependent internalization and lysosomal degradation of E-cadherin (Bonazzi et al. 2008; Shen et al. 2008). It is not yet clear whether the action of Hakai influences β-catenin signaling. Recently, the tyrosine phosphatases PTPRK and PCP-2 have been shown to inhibit β-catenin signaling and increase E-cadherin-dependent adhesion (Yan et al. 2006; Novellino et al. 2008). The cumulative evidence suggests that tyrosine phosphorylation-dependent destabilization of cadherin/β-catenin complexes and tyrosine-phosphorylation-dependent activation of β-catenin can both activate signaling and induce EMT.

A novel type of interplay between the loss of N-cadherin-mediated adhesion and transcriptional activity following tyrosine-phosphorylation of β-catenin has recently been discovered in Slit-activated neural retina cells (Rhee et al. 2007). Slit is a secreted axon guidance cue that binds and activates the Robo receptor, and this leads to the inactivation of N-cadherin (Brose et al. 1999). Balsamo and collaborators have shown that a population of N-cadherin-associated β-catenin, which is phosphorylated on tyrosine 489 in response to Slit-activation of the Robo-associated tyrosine kinase abl, is mobilized to the nucleus and becomes transcriptionally active (Rhee et al. 2007). This provides an example in which the down-regulation of cadherin-mediated adhesion in neuronal cells is coordinated with long-term effects on transcription.

The L1 adhesion system has recently also been shown to affect cadherin-mediated cell adhesion and β-catenin signaling (Shtutman et al. 2006). L1 is a transmembrane adhesion molecule with extracellular immunoglobulin-like domains and fibronectin type III repeats. It is up-regulated by Wnt/β-catenin signaling, it is expressed exclusively at the invasive front of colon cancer tissue, and its expression is correlated with tumor progression and metastasis (see also section B; Haspel and Grumet 2003; Gavert et al. 2005; Gavert et al. 2007). It has now been shown that L1 disrupts the formation of adherens junctions by releasing E-cadherin, that it promotes EMT-like transition, and that it induces β-catenin signals (Shtutman et al. 2006).

The Catenin p120 Relieves Kaiso-Mediated Repression of Wnt Target Genes

The turnover of cadherins is another crucial element of cell adhesion. It is critically dependent on the catenin p120, which binds to the highly conserved juxtamembrane domain of cadherins and increases their stability at the cell membrane. Depletion of p120 leads to an immediate turnover of E-cadherin and to a weakening of cell–cell adhesion (reviewed by Reynolds and Roczniak-Ferguson 2004). McCrea and collaborators have shown that p120 also influences the Wnt pathway through an interaction with the transcription factor Kaiso (Kim et al. 2004; Park et al. 2005; see also McCrea et al. 2009). Kaiso belongs to the BTB/POZ (zinc finger and broad-complex, tramtrack and bric-a-brac/poxvirus) protein family. The zinc-finger of Kaiso binds both CpG methylated sequences and a defined promoter consensus motif (Daniel et al. 2002; Kim et al. 2004). The association of Kaiso to promoters has been shown to repress Wnt target genes in *Xenopus* embryos and in mammalian cell lines by interfering with the binding of β-catenin to TCF/LEF (Kim et al. 2004; Park et al. 2005; Spring et al. 2005). It was recently suggested that Kaiso function may be mediated by its interaction with the DNA binding domain of TCF, leading to the displacement of TCF from promoters (Ruzov et al. 2009). The interaction of p120 with Kaiso influences the repressor function of Kaiso either by sequestrating Kaiso in the cytoplasm or by replacing it from the promoter (Park et al. 2005). It is likely that p120 affects the nuclear shuttle of Kaiso, since the NLS of p120 is crucial to the inhibition of Kaiso-mediated transcriptional repression (Kelly et al. 2004a; Kelly et al. 2004b). A functional interplay between p120, Kaiso, and Wnt target genes has also been observed in human cancer (reviewed by van Roy and McCrea 2005). The Wnt pathway also converges with the p120/Kaiso pathway further upstream, since p120 is stabilized by Frodo, a Dsh interacting partner. The interaction of Dsh with Frodo leads to stabilization of p120, which in turn enhances the displacement of Kaiso from target genes (Park et al. 2006). This means that the cell adhesion-modulating catenin p120 enables a further cross talk of cell junctions and the Wnt pathway in a β-catenin-independent fashion.

Adhesion and Wnt/β-Catenin Signaling in Cardiomyopathy

Mutations in genes that encode desmosomal proteins, including desmosomal cadherins, are the cause of arrhythmogenic right ventricular cardiomyopathies (ARVC) in humans. In this disease, muscle tissue of the right ventricle of the heart is replaced by fibrofatty tissue, resulting in arrhythmias and sudden death (reviewed by Sen-Chowdhry et al. 2004; MacRae et al. 2006). The mutated genes include all types of components of desmosomes: The cadherins desmocollins and desmogleins, the catenins plakoglobin and plakophilin2, and the linker protein to intermediate filaments, desmoplakin (MacRae et al. 2006; see also Heuser et al. 2006). Genetic experiments in mice have corroborated the essential role of desmosomal proteins in heart structure and function (see for instance Ruiz et al. 1996; Grossmann et al. 2004). Previously it had been firmly believed that the primary cause of ARVC is the weakening of desmosomal adhesion between cardiomyocytes in the diseased heart. However, the work of Garcia-Gras et al. has introduced a new twist in concepts: The cardiac-specific ablation of desmoplakin in mice is sufficient to cause nuclear translocation of plakoglobin (γ-catenin), which then inhibits Wnt/β-catenin signaling (Zhurinsky et al. 2000; Garcia-Gras et al. 2006). It is known that Wnt/β-catenin signals supports cardiomyogenesis, whereas inhibition promotes adipogenesis (Ross et al. 2000). This could explain the reduced production of cardiomyocytes and the increased production of adipocytes, which is observed in ARVC. These data nicely show how a switch from cadherin-mediated cell-adhesion to modulated Wnt signaling might result in the complex properties of a human disease.

General Remarks

The studies described above show how a loss of E-cadherin-mediated cell adhesion induces β-catenin release and signaling. These functions cause biological responses such as EMT, invasion and metastasis (Fig. 1). In other cases, loss of E-cadherin induces EMT without promoting β-catenin mobilization, but it affects other signaling pathways (see also Perrais et al. 2007; Onder et al. 2008). A compromised β-catenin degradation machinery appears to be the prerequisite for β-catenin signaling, which follows the loss of cadherins. β-Catenin occupies different functional niches in the cells, and the flow between these niches and the subsequent effects on cell adhesion and signaling are regulated by further events such as phosphorylation. Taken together, the data show intricate networks of both interacting and separate pathways that control cell adhesion and signaling, and their contribution to EMT needs to be unraveled in each case.

CANONICAL Wnt SIGNALING AFFECTS CADHERIN-MEDIATED CELL-ADHESION

So far, our focus has mainly been on events within the cytoplasm that affect the cell by modulating β-catenin's functions in cell adhesion and its availability as a transcription factor in combination with other molecules. This section summarizes the effects of Wnt signaling at the gene level, showing its impact on cell–cell adhesion through the regulation of the expression of diverse cell adhesion-modulating molecules. For a frequently updated overview of Wnt target genes see Roel Nusse's webpage (www.stanford.edu/~rnusse/wntwindow.html).

Some Wnt target genes encode for components of cell junctions such as L1-CAM (Gavert et al. 2005), Nr-CAM (Conacci-Sorrell et al. 2002) and connexin 43 (van der Heyden et al. 1998). E-cadherin is negatively regulated by Wnt signaling (Huber et al. 1996; Jamora et al. 2003). Other Wnt targets encode transcription factors that negatively control the expression of cadherins like Twist (Howe et al. 2003) and Slug (Vallin et al. 2001; Conacci-Sorrell et al. 2003). Additionally, Wnt targets also encode enzymes that are involved in regulating the stability of cell adhesions, such as Tiam1 (Malliri et al. 2006), Matrilysin (Brabletz et al. 1999; Crawford et al. 1999) and Stromelysin (Prieve and Moon 2003). Wnt/β-catenin signaling also directly controls the expression of the mesenchymal genes fibronectin and laminin-5γ2 (Gradl et al. 1999; Hlubek et al. 2004). Wnt target genes that affect cell motility also include fascin (Vignjevic et al. 2007), EphB/Ephrin (Batlle et al. 2005), CD44 (Wielenga et al. 1999), and S100/A4-metastasin (Stein et al. 2006). Overall, Wnt signaling appears to modulate target genes that cause a switch from primarily epithelial to more mesenchymal cells (Fig. 5).

Wnt Targeted Transcription Factors Regulate Cadherin Expression—Twist, Snail and Slug

Twist is a bHLH transcription factor that was first identified in *Drosophila* as an inducer of mesoderm formation (Nüsslein-Volhard et al. 1984; Simpson 1983; Thisse et al. 1988). Twist is also expressed in mesodermal cells of vertebrates (Wolf et al. 1991; Fuchtbauer 1995; Wang et al. 1997). Wnt1 regulates Twist expression in *Drosophila* and murine cells (Bate and Rushton 1993; Howe et al. 2003), and Wnt/β-catenin signaling regulates Twist expression in migrating cells in *Xenopus* (Borchers et al. 2001). Gene ablation experiments in the mouse show that Twist is

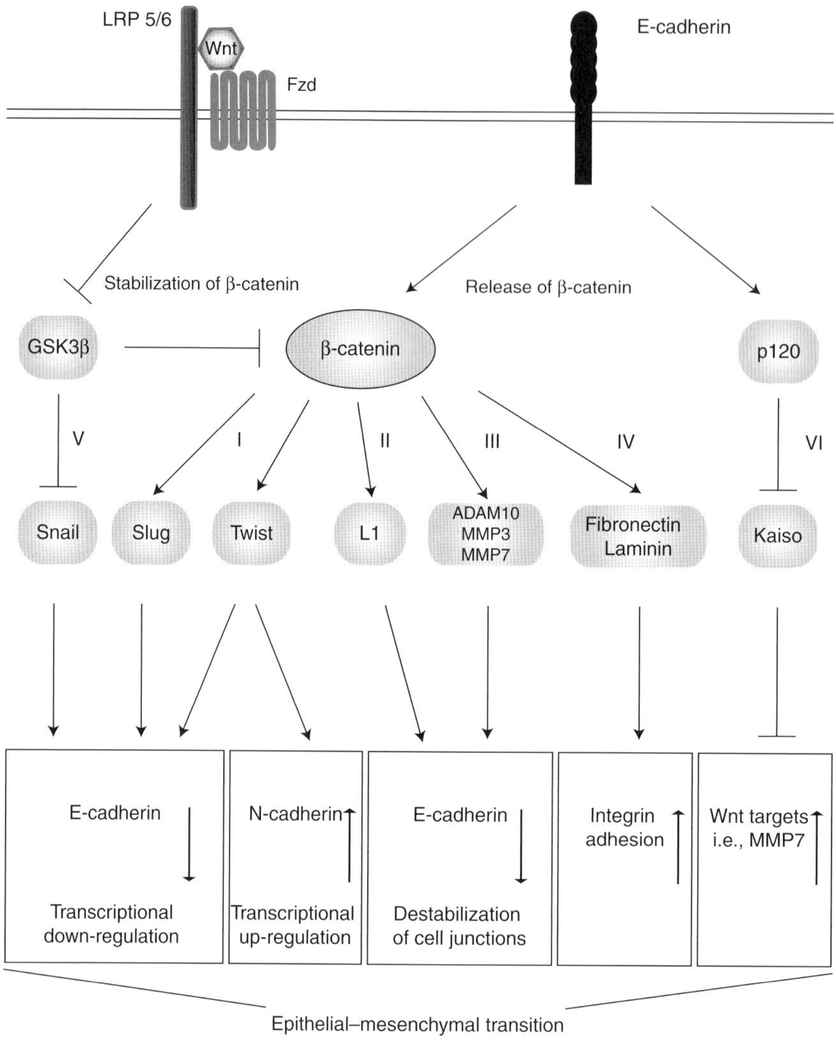

Figure 5. Canonical Wnt signaling modulates cell adhesions. Activated Wnt signaling leads to stabilization and nuclear localization of β-catenin, which induces the expression of Wnt target genes: I) The transcription factors Slug and Twist both repress the expression of E-cadherin, and Twist activates the expression of N-cadherin. II) The cell adhesion molecule L1 interferes with E-cadherin-mediated cell adhesion. III) The proteases MMP3 and MMP7 modulate the extracellular matrix, and MMP3, MMP7, and ADAM10 cleave E-cadherin. IV) The cell matrix molecules fibronectin and laminin favor cell-substrate adhesion. V) Wnt signaling inactivates GSK3β, leading to stabilization of Snail, which also inhibits the expression of E-cadherin. VI) p120 relieves the repressor Kaiso from Wnt target genes.

important for cell migration and cell localization during cranial neural tube morphogenesis (Chen and Behringer 1995; Soo et al. 2002). Twist down-regulates the expression of E-cadherin by binding to E-box elements of the promotor (Ip et al. 1992; Rohwedel et al. 1995; Lee et al. 1997; Yin et al. 1997; Kophengnavong et al. 2000), and it induces EMT and invasion and metastasis of mammary gland cells (Yang et al. 2004). In contrast, Twist is also involved in the up-regulation of N-cadherin (Oda et al. 1998; Yang et al. 2004; Alexander et al. 2006). N-cadherin can act as a proinvasive adhesion molecule that is produced in carcinomas following a so-called cadherin switch (Wheelock and Johnson 2003; Hazan et al. 2004). In spite of its opposing regulatory functions on the promoters of E- and N-cadherin, both of these functions of Twist promote EMT. The same type of duality has been observed in tumors and cancer cell lines (Hsu et al. 1996; Hazan et al. 1997; Sandig et al. 1997; Tomita et al. 2000; Rosivatz et al. 2002; Gravdal et al. 2007). Because transfections of E-cadherin cannot fully correct EMT in mammary gland cells, it is suggested that Twist affects additional EMT-inducing genes (Yang et al. 2004). These data reveal a mechanism by which Wnt signaling can promote EMT, tumor progression, and metastasis via the regulation of Twist.

Like Twist, the transcription repressors Snail and Slug are involved in the regulation of EMT (Nieto et al. 1994; Sefton et al. 1998; Batlle et al. 2000; Cano et al. 2000; Veltmaat et al. 2000; ten Berge et al. 2008). Snail was first identified in *Drosophila* (Nüsslein-Volhard et al. 1984; Grau et al. 1984) and its homolog Slug in the chick (Nieto et al. 1994). Both Snail and Slug bind to the E-box motive of the E-cadherin promoter through a carboxy-terminal zink-finger domain and suppress the expression of E-cadherin (Batlle et al. 2000; Cano et al. 2000; Hajra et al. 2002; Bolos et al. 2003). This is mediated by the amino-terminal SNAG domain of Snail, which recruits a histone deacetylase (Batlle et al. 2000; Peinado et al. 2004a). Slug is not able to repress the expression of E-cadherin in rat bladder carcinoma cells, but disrupts desmosomes (Savagner et al. 1997).

Snail and Slug not only play a role in cell adhesion, but also their functions are affected by the Wnt pathway. The Slug promoter harbors a LEF binding site and is responsive to β-catenin/TCF signaling, which leads to the expression of Slug and in turn to the down-regulation of E-cadherin (Vallin et al. 2001; Conacci-Sorrell et al. 2003). The Snail promoter is not directly responsive to β-catenin/TCF signaling (Conacci-Sorrell et al. 2003), but Wnt signaling leads to increased levels of Snail by inhibiting GSK3β. Blocking GSK3β activity increases the transcriptional level of Snail (Bachelder et al. 2005) and also increases its stability; Snail is a substrate of GSK3β, and phosphorylation marks Snail for degradation (Zhou et al. 2004; Yook et al. 2005). Additionally, Snail interacts directly with β-catenin and enhances Wnt target gene expression, and is believed to stimulate the Wnt pathway in a positive feedback loop (Stemmer et al. 2008). These data indicate that Slug and Snail are important regulators of cadherin-mediated cell adhesion and that the Wnt pathway affects cell–cell adhesion via regulation of these transcriptional regulators (Fig. 5).

Two other aspects of these transcription factors need to be mentioned here. First, Snail, Slug, and Twist are regulated by many signaling pathways, of which Wnt is only one. It has been shown that the cellular context determines which pathways crosstalk to Snail, Slug, and Twist and regulate the expression of E-cadherin (reviewed by Nieto 2002; Huber et al. 2005). Next to the E-boxes, the E-cadherin promoter harbors other regulatory elements, such as cis-regulatory elements (Stemmler et al. 2005), and can be regulated by other transcription factors like δEF1/ZEB1, Sip-1/ZEB2, and E12/E47 (Grooteclaes and Frisch 2000; Comijn et al. 2001; Perez-Moreno et al. 2001).

Second, EMT plays a crucial role in gastrulation, a process in which the regulation of E-cadherin has an important function (Burdsal et al. 1993). During gastrulation, epiblast cells form the primitive streak from which endoderm and mesoderm are generated. The importance of canonical Wnt signaling can be seen from the fact that Wnt3, β-catenin, and LRP5/LRP6-deficient mouse epiblasts fail to form the primitive streak (Liu et al. 1999; Huelsken et al. 2000; Kelly et al. 2004c). The down-regulation of E-cadherin in mouse gastrulation is not controlled by Twist, as Twist is not expressed in the primitive streak (Chen and Behringer 1995). However, Snail is expressed in the primitive streak in the mouse (Smith et al. 1992), and ablation of Snail leads to the continued expression of E-cadherin transcripts in the mesoderm. It also causes defects in the generated mesoderm cell layer (Carver et al. 2001). FGF signaling has an important role during mouse gastrulation, and a deficiency in

FGFR1 prevents the expression of Snail, in turn disturbing the down-regulation of E-cadherin and EMT (Deng et al. 1994; Yamaguchi et al. 1994; Ciruna et al. 1997; Deng et al. 1997; Ciruna and Rossant 2001). So the down-regulation of E-cadherin does not seem to be controlled by the Wnt pathway in mouse gastrulation, but further data indicate a cross talk of Wnt and FGF signaling. In FGFR1-deficient epiblasts, the expression of the Wnt target Brachyury is negatively affected, but lowering E-cadherin levels rescues Brachyury expression in the primitive streak (Ciruna and Rossant 2001). Therefore, increased E-cadherin levels are thought to sequester β-catenin and negatively affect canonical Wnt signaling.

Wnt Signaling Induces Expression of Proteases that Decrease Cell–Cell Adhesion

EMT and metastasis depend on modifications of the extracellular matrix and cell–cell adhesion through the cleavage of E-cadherin. This is achieved by proteases, which are regulated by Wnt signaling. A direct canonical Wnt target is the metalloproteinase Matrilysin (MMP7; Crawford et al. 1999). Stromelysin-1 (MMP3) expression is regulated by Wnt5a in a β-catenin-independent manner, indicating that Stromelysin-1 expression is controlled by the noncanonical Wnt pathway (Prieve and Moon 2003). Both Stromelysin-1 and Matrilysin degrade components of the extracellular matrix and are up-regulated in colorectal cancers, and MMP7 has been shown to promote metastasis (reviewed by Wagenaar-Miller et al. 2004). Both proteases also cleave E-cadherin, which results in the shedding of the E-cadherin ectodomain and causes cells to become motile and invasive (Fig. 5) (Lochter et al. 1997; Noe et al. 2001).

The Wnt Target Gene Tiam1 and IQGAP1 Modulate Cell–Cell Adhesion

Wnt has additional effects on cell–cell adhesion through its regulation of target genes. Tiam1 (T-cell lymphoma invasion and metastasis 1) is a nucleotide exchange factor that selectively activates Rac1 (Michiels et al. 1995) and is required for cadherin-based cell adhesions and suppression of epithelial motility (Hordijk et al. 1997; Malliri et al. 2004). Tiam1 modulates the dynamics of cell–cell adhesion by regulating tight junction assembly via activation of the Par complex (reviewed by Mertens et al. 2006) and by stabilizing the β-catenin/E-cadherin complex (Fukata et al. 1999). Tiam1 is a Wnt target gene, as shown by the fact that its expression can be induced by Wnt1 and correlates with active Wnt signaling in human colon adenoma, adenomatous polyps of Min/+ mice, and in the crypts of the small intestine (Liu et al. 2005; Malliri et al. 2006; Minard et al. 2006). Min/+ mice deficient in Tiam1 develop a reduced number of intestinal tumors and an impaired formation of mammary tumors. The developing tumors are however more aggressive and invade the submucosa in larger fractions (Malliri et al. 2006). These data indicate that Wnt signaling interferes with Tiam1 activity, which regulates cell growth, cell migration, and adherens junction formation.

A further mechanism is seen in the case of IQGAP1 (IQ motif containing GTPase-activating protein 1), which is involved in cytoskeletal dynamics and other cellular functions (reviewed by Briggs and Sacks 2003; Brandt and Grosse 2007). IQGAP1 regulates the dynamics of E-cadherin-based adherens junctions (Izumi et al. 2004) and interferes with β-catenin–α-catenin complex formation by binding the amino terminus of β-catenin, leading to reduced E-cadherin-mediated cell adhesion (Kuroda et al. 1998). It has recently been shown that IQGAP1 is up-regulated in colorectal carcinomas and promotes the invasiveness of colon cancer cells (Hayashi et al. 2009). GTP-bound Rac1 inhibits the interaction between IQGAP1 and β-catenin and stabilizes cell–cell adhesion (Fukata et al. 1999). As IQGAP1 acts downstream of Rac1, one likely possibility is that Tiam1/Rac1 positively regulates cell adhesion by inhibiting IQGAP1. Therefore, it is plausible that Wnt signaling affects cell adhesion via Tiam1/Rac1, which balances the activity of IQGAP1. Alongside its function in cell adhesion, IQGAP1 influences Wnt signaling, because in cells with stabilized β-catenin, IQGAP1 enhances the transcriptional coactivator function of β-catenin (Briggs et al. 2002). These results are an example of a cross-talk of canonical Wnt and Rac signaling including Tiam1 and IQGAP1 in the regulation of cell–cell adhesion.

General Remarks

The work described in this section shows several ways in which canonical Wnt signaling affects cadherin-mediated cell adhesion: (1) it down-regulates E-cadherin expression via the transcription factors Twist and Slug, (2) it up-regulates adhesion molecules that favor cell motility, such as N-cadherin and L1, and (3) it induces proteases and other EMT promoters. Wnt signaling can therefore induce a cadherin switch and weaken cell–cell adhesion (Fig. 5). Wnt/β-catenin signaling is up-regulated in many developmental processes and by mutations in the progression of tumors (reviewed by Clevers 2006; Grigoryan et al. 2008). Overall, cadherin loss and Wnt/β-catenin signaling can thus cooperate to promote EMT, invasion and metastasis.

Until recently there have been considerable gaps in our ability to link signals to the architectural and behavioral changes in cells which lead to EMT and metastases. But as this review shows, the Wnt pathway and the regulation of cell adhesion are strongly linked by a number of complementary mechanisms. This suggests a model that integrates a number of fundamental processes that underlie development and disease.

ACKNOWLEDGMENTS

We thank Russ Hodge (Berlin) for helpful discussions and improvements on the text. Our work was funded by the Deutsche Forschungsgemeinschaft (DFG), the Mildred-Scheel-Stiftung für Krebsforschung (Deutsche Krebshilfe), the European Union, and the Federal Ministry of Research and Technology of Germany (BMFT).

REFERENCES

Abe K, Takeichi M. 2007. NMDA-receptor activation induces calpain-mediated β-catenin cleavages for triggering gene expression. *Neuron* **53:** 387–397.

Abe K, Takeichi M. 2008. EPLIN mediates linkage of the cadherin catenin complex to F-actin and stabilizes the circumferential actin belt. *Proc Natl Acad Sci* **105:** 13–19.

Aberle H, Bauer A, Stappert J, Kispert A, Kemler R. 1997. β-catenin is a target for the ubiquitin-proteasome pathway. *EMBO J* **16:** 3797–3804.

Adachi S, Jigami T, Yasui T, Nakano T, Ohwada S, Omori Y, Sugano S, Ohkawara B, Shibuya H, Nakamura T, et al. 2004. Role of a BCL9-related β-catenin-binding protein, B9L, in tumorigenesis induced by aberrant activation of Wnt signaling. *Cancer Res* **64:** 8496–8501.

Alexander NR, Tran NL, Rekapally H, Summers CE, Glackin C, Heimark RL. 2006. N-cadherin gene expression in prostate carcinoma is modulated by integrin-dependent nuclear translocation of Twist1. *Cancer Res* **66:** 3365–3369.

Arikkath J, Reichardt LF. 2008. Cadherins and catenins at synapses: Roles in synaptogenesis and synaptic plasticity. *Trends Neurosci* **31:** 487–494.

Bachelder RE, Yoon SO, Franci C, de Herreros AG, Mercurio AM. 2005. Glycogen synthase kinase-3 is an endogenous inhibitor of Snail transcription: Implications for the epithelial-mesenchymal transition. *J Cell Biol* **168:** 29–33.

Baki L, Marambaud P, Efthimiopoulos S, Georgakopoulos A, Wen P, Cui W, Shioi J, Koo E, Ozawa M, Friedrich VL Jr, et al. 2001. Presenilin-1 binds cytoplasmic epithelial cadherin, inhibits cadherin/p120 association, and regulates stability and function of the cadherin/catenin adhesion complex. *Proc Natl Acad Sci* **98:** 2381–2386.

Barth AI, Pollack AL, Altschuler Y, Mostov KE, Nelson WJ. 1997. NH2-terminal deletion of β-catenin results in stable colocalization of mutant β-catenin with adenomatous polyposis coli protein and altered MDCK cell adhesion. *J Cell Biol* **136:** 693–706.

Bate M, Rushton E. 1993. Myogenesis and muscle patterning in *Drosophila. Comptes rendus de l'Academie des sciences* **316:** 1047–1061.

Batlle E, Sancho E, Franci C, Dominguez D, Monfar M, Baulida J, Garcia De Herreros A. 2000. The transcription factor snail is a repressor of E-cadherin gene expression in epithelial tumour cells. *Nature Cell Biol* **2:** 84–89.

Batlle E, Bacani J, Begthel H, Jonkheer S, Gregorieff A, van de Born M, Malats N, Sancho E, Boon E, Pawson T, et al. 2005. EphB receptor activity suppresses colorectal cancer progression. *Nature* **435:** 1126–1130.

Becker KF, Atkinson MJ, Reich U, Becker I, Nekarda H, Siewert JR, Hofler H. 1994. E-cadherin gene mutations provide clues to diffuse type gastric carcinomas. *Cancer Res* **54:** 3845–3852.

Behrens J, Birchmeier W, Goodman SL, Imhof BA. 1985. Dissociation of Madin-Darby canine kidney epithelial cells by the monoclonal antibody anti-arc-1: Mechanistic aspects and identification of the antigen as a component related to uvomorulin. *J Cell Biol* **101:** 1307–1315.

Behrens J, von Kries JP, Kuhl M, Bruhn L, Wedlich D, Grosschedl R, Birchmeier W. 1996. Functional interaction of β-catenin with the transcription factor LEF-1. *Nature* **382:** 638–642.

Behrens J, Mareel MM, Van Roy FM, Birchmeier W. 1989. Dissecting tumor cell invasion: Epithelial cells acquire invasive properties after the loss of uvomorulin-mediated cell-cell adhesion. *J Cell Biol* **108:** 2435–2447.

Behrens J, Jerchow BA, Wurtele M, Grimm J, Asbrand C, Wirtz R, Kuhl M, Wedlich D, Birchmeier W. 1998. Functional interaction of an axin homolog, conductin, with β-catenin, APC, and GSK3β. *Science* **280:** 596–599.

Behrens J, Vakaet L, Friis R, Winterhager E, Van Roy F, Mareel MM, Birchmeier W. 1993. Loss of epithelial differentiation and gain of invasiveness correlates with tyrosine phosphorylation of the E-cadherin/β-catenin complex in cells transformed with a temperature-sensitive v-SRC gene. *J Cell Biol* **120:** 757–766.

Berx G, Cleton-Jansen AM, Nollet F, de Leeuw WJ, van de Vijver M, Cornelisse C, van Roy F. 1995. E-cadherin is a tumour/invasion suppressor gene mutated in human lobular breast cancers. *The EMBO J* **14:** 6107–6115.

Bienz M, Clevers H. 2000. Linking colorectal cancer to Wnt signaling. *Cell* **103:** 311–320.

Bilic J, Huang YL, Davidson G, Zimmermann T, Cruciat CM, Bienz M, Niehrs C. 2007. Wnt induces LRP6 signalosomes and promotes dishevelled-dependent LRP6 phosphorylation. *Science* **316:** 1619–1622.

Bolos V, Peinado H, Perez-Moreno MA, Fraga MF, Esteller M, Cano A. 2003. The transcription factor Slug represses E-cadherin expression and induces epithelial to mesenchymal transitions: A comparison with Snail and E47 repressors. *J Cell Sci* **116:** 499–511.

Bonazzi M, Veiga E, Cerda JP, Cossart P. 2008. Successive post-translational modifications of E-cadherin are required for InlA-mediated internalisation of Listeria monocytogenes. *Cell Microbiol* **10:** 2208–2222.

Boon EM, van der Neut R, van de Wetering M, Clevers H, Pals ST. 2002. Wnt signaling regulates expression of the receptor tyrosine kinase met in colorectal cancer. *Cancer Res* **62:** 5126–5128.

Borchers A, David R, Wedlich D. 2001. *Xenopus* cadherin-11 restrains cranial neural crest migration and influences neural crest specification. *Development* **128:** 3049–3060.

Brabletz T, Jung A, Dag S, Hlubek F, Kirchner T. 1999. β-catenin regulates the expression of the matrix metalloproteinase-7 in human colorectal cancer. *Am Am J Pathol* **155:** 1033–1038.

Brandt DT, Grosse R. 2007. Get to grips: Steering local actin dynamics with IQGAPs. *EMBO Rep* **8:** 1019–1023.

Brembeck FH, Schwarz-Romond T, Bakkers J, Wilhelm S, Hammerschmidt M, Birchmeier W. 2004. Essential role of BCL9-2 in the switch between β-catenin's adhesive and transcriptional functions. *Genes Develop* **18:** 2225–2230.

Briggs MW, Sacks DB. 2003. IQGAP proteins are integral components of cytoskeletal regulation. *EMBO Rep* **4:** 571–574.

Briggs MW, Li Z, Sacks DB. 2002. IQGAP1-mediated stimulation of transcriptional co-activation by β-catenin is modulated by calmodulin. *J Biolog Chem* **277:** 7453–7465.

Brose K, Bland KS, Wang KH, Arnott D, Henzel W, Goodman CS, Tessier-Lavigne M, Kidd T. 1999. Slit proteins bind Robo receptors and have an evolutionarily conserved role in repulsive axon guidance. *Cell* **96:** 795–806.

Burdsal CA, Damsky CH, Pedersen RA. 1993. The role of E-cadherin and integrins in mesoderm differentiation and migration at the mammalian primitive streak. *Development* **118:** 829–844.

Butz S, Stappert J, Weissig H, Kemler R. 1992. Plakoglobin and β-catenin: Distinct but closely related. *Science* **257:** 1142–1144.

Cadigan KM, Nusse R. 1997. Wnt signaling: A common theme in animal development. *Genes Develop* **11:** 3286–3305.

Cadigan KM, Pfeifer M. 2009. Wnt signaling from development to disease: insights from model systems. *Cold Spring Harb Perspect Biol* **1:** a002881.

Cano A, Perez-Moreno MA, Rodrigo I, Locascio A, Blanco MJ, del Barrio MG, Portillo F, Nieto MA. 2000. The transcription factor snail controls epithelial-mesenchymal transitions by repressing E-cadherin expression. *Nat Cell Biol* **2:** 76–83.

Carroll RC, Zukin RS. 2002. NMDA-receptor trafficking and targeting: Implications for synaptic transmission and plasticity. *Trends Neurosci* **25:** 571–577.

Carver EA, Jiang R, Lan Y, Oram KF, Gridley T. 2001. The mouse snail gene encodes a key regulator of the epithelial-mesenchymal transition. *Mol Cell Biol* **21:** 8184–8188.

Chen ZF, Behringer RR. 1995. twist is required in head mesenchyme for cranial neural tube morphogenesis. *Genes Develop* **9:** 686–699.

Choi HJ, Huber AH, Weis WI. 2006. Thermodynamics of β-catenin-ligand interactions: The roles of the N- and C-terminal tails in modulating binding affinity. *J Biolog Chem* **281:** 1027–1038.

Ciruna B, Rossant J. 2001. FGF signaling regulates mesoderm cell fate specification and morphogenetic movement at the primitive streak. *Develop Cell* **1:** 37–49.

Ciruna BG, Schwartz L, Harpal K, Yamaguchi TP, Rossant J. 1997. Chimeric analysis of fibroblast growth factor receptor-1 (Fgfr1) function: A role for FGFR1 in morphogenetic movement through the primitive streak. *Development* **124:** 2829–2841.

Clevers H. 2006. Wnt/β-catenin signaling in development and disease. *Cell* **127:** 469–480.

Collard JG, Habets GG, Michiels F, Stam J, van der Kammen RA, van Leeuwen F. 1996. Role of Tiam 1 in Rac-mediated signal transduction pathways. *Current Topics Microbiol Immunol* **213:** 253–265.

Comijn J, Berx G, Vermassen P, Verschueren K, van Grunsven L, Bruyneel E, Mareel M, Huylebroeck D, van Roy F. 2001. The two-handed E box binding zinc finger protein SIP1 downregulates E-cadherin and induces invasion. *Mol Cell* **7:** 1267–1278.

Conacci-Sorrell ME, Ben-Yedidia T, Shtutman M, Feinstein E, Einat P, Ben-Ze'ev A. 2002. Nr-CAM is a target gene of the β-catenin/LEF-1 pathway in melanoma and colon cancer and its expression enhances motility and confers tumorigenesis. *Genes Develop* **16:** 2058–2072.

Conacci-Sorrell M, Simcha I, Ben-Yedidia T, Blechman J, Savagner P, Ben-Ze'ev A. 2003. Autoregulation of E-cadherin expression by cadherin-cadherin interactions: The roles of β-catenin signaling, Slug, and MAPK. *J Cell Biol* **163:** 847–857.

Cowin P, Kapprell HP, Franke WW, Tamkun J, Hynes RO. 1986. Plakoglobin: A protein common to different kinds of intercellular adhering junctions. *Cell* **46:** 1063–1073.

Cox RT, Kirkpatrick C, Peifer M. 1996. Armadillo is required for adherens junction assembly, cell polarity, and morphogenesis during *Drosophila* embryogenesis. *J Cell Biol* **134:** 133–148.

Crawford HC, Fingleton BM, Rudolph-Owen LA, Goss KJ, Rubinfeld B, Polakis P, Matrisian LM. 1999. The metalloproteinase matrilysin is a target of β-catenin transactivation in intestinal tumors. *Oncogene* **18:** 2883–2891.

Daniel JM, Spring CM, Crawford HC, Reynolds AB, Baig A. 2002. The p120(ctn)-binding partner Kaiso is a bi-modal DNA-binding protein that recognizes both a sequence-specific consensus and methylated CpG dinucleotides. *Nucleic Acids Res* **30:** 2911–2919.

Daugherty RL, Gottardi CJ. 2007. Phospho-regulation of β-catenin adhesion and signaling functions. *Physiology* **22:** 303–309.

David MD, Yeramian A, Dunach M, Llovera M, Canti C, de Herreros AG, Comella JX, Herreros J. 2008. Signalling by neurotrophins and hepatocyte growth factor regulates axon morphogenesis by differential β-catenin phosphorylation. *J Cell Sci* **121:** 2718–2730.

Davidson G, Wu W, Shen J, Bilic J, Fenger U, Stannek P, Glinka A, Niehrs C. 2005. Casein kinase 1 γ couples Wnt receptor activation to cytoplasmic signal transduction. *Nature* **438:** 867–872.

de la Roche M, Worm J, Bienz M. 2008. The function of BCL9 in Wnt/β-catenin signaling and colorectal cancer cells. *BMC Cancer* **8:** 199.

Deng CX, Wynshaw-Boris A, Shen MM, Daugherty C, Ornitz DM, Leder P. 1994. Murine FGFR-1 is required for early postimplantation growth and axial organization. *Genes Develop* **8:** 3045–3057.

Deng C, Bedford M, Li C, Xu X, Yang X, Dunmore J, Leder P. 1997. Fibroblast growth factor receptor-1 (FGFR-1) is essential for normal neural tube and limb development. *Develop Biol* **185:** 42–54.

Dietrich S, Abou-Rebyeh F, Brohmann H, Bladt F, Sonnenberg-Riethmacher E, Yamaai T, Lumsden A, Brand-Saberi B, Birchmeier C. 1999. The role of SF/HGF and c-Met in the development of skeletal muscle. *Development* **126:** 1621–1629.

Drees F, Pokutta S, Yamada S, Nelson WJ, Weis WI. 2005. α-catenin is a molecular switch that binds E-cadherin-β-catenin and regulates actin-filament assembly. *Cell* **123:** 903–915.

Eastman Q, Grosschedl R. 1999. Regulation of LEF-1/TCF transcription factors by Wnt and other signals. *Current Opinion Cell Biol* **11:** 233–240.

Eger A, Stockinger A, Schaffhauser B, Beug H, Foisner R. 2000. Epithelial mesenchymal transition by c-Fos estrogen receptor activation involves nuclear translocation of β-catenin and upregulation of β-catenin/lymphoid enhancer binding factor-1 transcriptional activity. *J Cell Biol* **148:** 173–188.

Fagotto F, Funayama N, Gluck U, Gumbiner BM. 1996. Binding to cadherins antagonizes the signaling activity of β-catenin during axis formation in *Xenopus*. *J Cell Biol* **132:** 1105–1114.

Ferber EC, Kajita M, Wadlow A, Tobiansky L, Niessen C, Ariga H, Daniel J, Fujita Y. 2008. A role for the cleaved cytoplasmic domain of E-cadherin in the nucleus. *J Biol Chem* **283:** 12691–12700.

Fodde R, Smits R, Clevers H. 2001. APC, signal transduction and genetic instability in colorectal cancer. *Nat Rev Cancer* **1:** 55–67.

Frixen UH, Behrens J, Sachs M, Eberle G, Voss B, Warda A, Lochner D, Birchmeier W. 1991. E-cadherin-mediated cell-cell adhesion prevents invasiveness of human carcinoma cells. *J Cell Biol* **113:** 173–185.

Fuchtbauer EM. 1995. Expression of M-twist during postimplantation development of the mouse. *Dev Dyn* **204:** 316–322.

Fujita Y, Krause G, Scheffner M, Zechner D, Leddy HE, Behrens J, Sommer T, Birchmeier W. 2002. Hakai, a c-Cbl-like protein, ubiquitinates and induces endocytosis of the E-cadherin complex. *Nat Cell Biol* **4:** 222–231.

Fukata M, Kuroda S, Nakagawa M, Kawajiri A, Itoh N, Shoji I, Matsuura Y, Yonehara S, Fujisawa H, Kikuchi A, et al. 1999. Cdc42 and Rac1 regulate the interaction of IQGAP1 with β-catenin. *J Biol Chem* **274:** 26044–26050.

Galbiati F, Volonte D, Brown AM, Weinstein DE, Ben-Ze'ev A, Pestell RG, Lisanti MP. 2000. Caveolin-1 expression inhibits Wnt/β-catenin/Lef-1 signaling by recruiting β-catenin to caveolae membrane domains. *J Biol Chem* **275:** 23368–23377.

Garcia-Gras E, Lombardi R, Giocondo MJ, Willerson JT, Schneider MD, Khoury DS, Marian AJ. 2006. Suppression of canonical Wnt/β-catenin signaling by nuclear plakoglobin recapitulates phenotype of arrhythmogenic right ventricular cardiomyopathy. *J Clin Invest* **116:** 2012–2021.

Gates J, Peifer M. 2005. Can 1000 reviews be wrong? Actin, α-Catenin, and adherens junctions. *Cell* **123:** 769–772.

Gavert N, Ben-Ze'ev A. 2007. β-Catenin signaling in biological control and cancer. *J Cell Biochem* **102:** 820–828.

Gavert N, Conacci-Sorrell M, Gast D, Schneider A, Altevogt P, Brabletz T, Ben-Ze'ev A. 2005. L1, a novel target of β-catenin signaling, transforms cells and is expressed at the invasive front of colon cancers. *J Cell Biol* **168:** 633–642.

Gavert N, Sheffer M, Raveh S, Spaderna S, Shtutman M, Brabletz T, Barany F, Paty P, Notterman D, Domany E, et al. 2007. Expression of L1-CAM and ADAM10 in human colon cancer cells induces metastasis. *Cancer Res* **67:** 7703–7712.

Georgakopoulos A, Marambaud P, Efthimiopoulos S, Shioi J, Cui W, Li HC, Schutte M, Gordon R, Holstein GR, Martinelli G, et al. 1999. Presenilin-1 forms complexes with the cadherin/catenin cell-cell adhesion system and is recruited to intercellular and synaptic contacts. *Mol cell* **4:** 893–902.

Gottardi CJ, Gumbiner BM. 2004. Distinct molecular forms of β-catenin are targeted to adhesive or transcriptional complexes. *J Cell Biol* **167:** 339–349.

Gottardi CJ, Wong E, Gumbiner BM. 2001. E-cadherin suppresses cellular transformation by inhibiting β-catenin signaling in an adhesion-independent manner. *J Cell Biol* **153:** 1049–1060.

Gradl D, Kuhl M, Wedlich D. 1999. The Wnt/Wg signal transducer β-catenin controls fibronectin expression. *Mol Cell Biol* **19:** 5576–5587.

Graham TA, Weaver C, Mao F, Kimelman D, Xu W. 2000. Crystal structure of a β-catenin/Tcf complex. *Cell* **103:** 885–896.

Grau Y, Carteret C, Simpson P. 1984. Mutations and chromosomal rearrangements affecting the expression of snail, a gene involved in embryonic patterning in *Drosophila melanogaster. Genetics* **108:** 347–360.

Gravdal K, Halvorsen OJ, Haukaas SA, Akslen LA. 2007. A switch from E-cadherin to N-cadherin expression indicates epithelial to mesenchymal transition and is of strong and independent importance for the progress of prostate cancer. *Clin Cancer Res* **13:** 7003–7011.

Greenburg G, Hay ED. 1982. Epithelia suspended in collagen gels can lose polarity and express characteristics of migrating mesenchymal cells. *J Cell Biol* **95:** 333–339.

Grigoryan T, Wend P, Klaus A, Birchmeier W. 2008. Deciphering the function of canonical Wnt signals in development and disease: Conditional loss- and gain-of-function mutations of β-catenin in mice. *Genes Develop* **22:** 2308–2341.

Grooteclaes ML, Frisch SM. 2000. Evidence for a function of CtBP in epithelial gene regulation and anoikis. *Oncogene* **19:** 3823–3828.

Grossmann KS, Grund C, Huelsken J, Behrend M, Erdmann B, Franke WW, Birchmeier W. 2004. Requirement of plakophilin 2 for heart morphogenesis and cardiac junction formation. *J Cell Biol* **167:** 149–160.

Guilford P, Hopkins J, Harraway J, McLeod M, McLeod N, Harawira P, Taite H, Scoular R, Miller A, Reeve AE. 1998. E-cadherin germline mutations in familial gastric cancer. *Nature* **392:** 402–405.

Ha NC, Tonozuka T, Stamos JL, Choi HJ, Weis WI. 2004. Mechanism of phosphorylation-dependent binding of APC to β-catenin and its role in β-catenin degradation. *Mol Cell* **15:** 511–521.

Habets GG, Scholtes EH, Zuydgeest D, van der Kammen RA, Stam JC, Berns A, Collard JG. 1994. Identification of an invasion-inducing gene, Tiam-1, that encodes a protein with homology to GDP-GTP exchangers for Rho-like proteins. *Cell* **77:** 537–549.

Hajra KM, Chen DY, Fearon ER. 2002. The SLUG zinc-finger protein represses E-cadherin in breast cancer. *Cancer Res* **62:** 1613–1618.

Hartwell KA, Muir B, Reinhardt F, Carpenter AE, Sgroi DC, Weinberg RA. 2006. The Spemann organizer gene, Goosecoid, promotes tumor metastasis. *Proc Natl Acad Sci* **103:** 18969–18974.

Haspel J, Grumet M. 2003. The L1CAM extracellular region: A multi-domain protein with modular and cooperative binding modes. *Front Biosci* **8:** s1210–1225.

Hayashi H, Nabeshima K, Aoki M, Hamasaki M, Enatsu S, Yamauchi Y, Yamashita Y, Iwasaki H. 2009. Overexpression of IQGAP1 in advanced colorectal cancer correlates with poor prognosis. Dysregulation of E-cadherin-based cell-to-cell adhesion. *Int J Cancer* (in press).

Hazan RB, Kang L, Whooley BP, Borgen PI. 1997. N-cadherin promotes adhesion between invasive breast cancer cells and the stroma. *Cell Adhes Commun* **4:** 399–411.

Hazan RB, Qiao R, Keren R, Badano I, Suyama K. 2004. Cadherin switch in tumor progression. *Ann NY Acad Sci* **1014:** 155–163.

Heasman J, Crawford A, Goldstone K, Garner-Hamrick P, Gumbiner B, McCrea P, Kintner C, Noro CY, Wylie C. 1994. Overexpression of cadherins and underexpression of β-catenin inhibit dorsal mesoderm induction in early *Xenopus* embryos. *Cell* **79:** 791–803.

Hennig G, Behrens J, Truss M, Frisch S, Reichmann E, Birchmeier W. 1995. Progression of carcinoma cells is associated with alterations in chromatin structure and factor binding at the E-cadherin promoter in vivo. *Oncogene* **11:** 475–484.

Herzig M, Savarese F, Novatchkova M, Semb H, Christofori G. 2007. Tumor progression induced by the loss of E-cadherin independent of β-catenin/Tcf-mediated Wnt signaling. *Oncogene* **26:** 2290–2298.

Heuser A, Plovie ER, Ellinor PT, Grossmann KS, Shin JT, Wichter T, Basson CT, Lerman BB, Sasse-Klaassen S, Thierfelder L, et al. 2006. Mutant desmocollin-2 causes arrhythmogenic right ventricular cardiomyopathy. *Am J Human Gen* **79:** 1081–1088.

Hlubek F, Spaderna S, Jung A, Kirchner T, Brabletz T. 2004. β-catenin activates a coordinated expression of the proinvasive factors laminin-5 γ2 chain and MT1-MMP in colorectal carcinomas. *Int J Cancer* **108:** 321–326.

Hoffmans R, Basler K. 2007. BCL9-2 binds Arm/β-catenin in a Tyr142-independent manner and requires Pygopus for its function in Wg/Wnt signaling. *Mech Develop* **124:** 59–67.

Hordijk PL, ten Klooster JP, van der Kammen RA, Michiels F, Oomen LC, Collard JG. 1997. Inhibition of invasion of epithelial cells by Tiam1-Rac signaling. *Science* **278:** 1464–1466.

Howe LR, Watanabe O, Leonard J, Brown AM. 2003. Twist is up-regulated in response to Wnt1 and inhibits mouse mammary cell differentiation. *Cancer Res* **63:** 1906–1913.

Hsu MY, Wheelock MJ, Johnson KR, Herlyn M. 1996. Shifts in cadherin profiles between human normal melanocytes and melanomas. *J Invest Dermatol Symp* **1:** 188–194.

Huber AH, Weis WI. 2001. The structure of the β-catenin/E-cadherin complex and the molecular basis of diverse ligand recognition by β-catenin. *Cell* **105:** 391–402.

Huber O, Korn R, McLaughlin J, Ohsugi M, Herrmann BG, Kemler R. 1996. Nuclear localization of β-catenin by interaction with transcription factor LEF-1. *Mech Develop* **59:** 3–10.

Huber MA, Kraut N, Beug H. 2005. Molecular requirements for epithelial-mesenchymal transition during tumor progression. *Current Opinion Cell Biol* **17:** 548–558.

Huelsken J, Vogel R, Brinkmann V, Erdmann B, Birchmeier C, Birchmeier W. 2000. Requirement for β-catenin in anterior-posterior axis formation in mice. *J Cell Biol* **148:** 567–578.

Imhof BA, Vollmers HP, Goodman SL, Birchmeier W. 1983. Cell-cell interaction and polarity of epithelial cells: Specific perturbation using a monoclonal antibody. *Cell* **35:** 667–675.

Ip YT, Park RE, Kosman D, Bier E, Levine M. 1992. The dorsal gradient morphogen regulates stripes of rhomboid expression in the presumptive neuroectoderm of the *Drosophila* embryo. *Genes Develop* **6:** 1728–1739.

Ito K, Okamoto I, Araki N, Kawano Y, Nakao M, Fujiyama S, Tomita K, Mimori T, Saya H. 1999. Calcium influx triggers the sequential proteolysis of extracellular and cytoplasmic domains of E-cadherin, leading to loss of β-catenin from cell-cell contacts. *Oncogene* **18:** 7080–7090.

Izumi G, Sakisaka T, Baba T, Tanaka S, Morimoto K, Takai Y. 2004. Endocytosis of E-cadherin regulated by Rac and Cdc42 small G proteins through IQGAP1 and actin filaments. *J Cell Biol* **166:** 237–248.

Jamora C, DasGupta R, Kocieniewski P, Fuchs E. 2003. Links between signal transduction, transcription and adhesion in epithelial bud development. *Nature* **422:** 317–322.

Kang DE, Soriano S, Frosch MP, Collins T, Naruse S, Sisodia SS, Leibowitz G, Levine F, Koo EH. 1999. Presenilin 1 facilitates the constitutive turnover of β-catenin: Differential activity of Alzheimer's disease-linked PS1 mutants in the β-catenin-signaling pathway. *J Neurosci* **19:** 4229–4237.

Kang DE, Soriano S, Xia X, Eberhart CG, De Strooper B, Zheng H, Koo EH. 2002. Presenilin couples the paired phosphorylation of β-catenin independent of axin: Implications for β-catenin activation in tumorigenesis. *Cell* **110:** 751–762.

Kelly KF, Otchere AA, Graham M, Daniel JM. 2004a. Nuclear import of the BTB/POZ transcriptional regulator Kaiso. *J Cell Sci* **117:** 6143–6152.

Kelly KF, Spring CM, Otchere AA, Daniel JM. 2004b. NLS-dependent nuclear localization of p120ctn is necessary to relieve Kaiso-mediated transcriptional repression. *J Cell Sci* **117:** 2675–2686.

Kelly OG, Pinson KI, Skarnes WC. 2004c. The Wnt co-receptors Lrp5 and Lrp6 are essential for gastrulation in mice. *Development* **131:** 2803–2815.

Kemler R. 1993. From cadherins to catenins: Cytoplasmic protein interactions and regulation of cell adhesion. *Trends Genet* **9:** 317–321.

Khoury H, Dankort DL, Sadekova S, Naujokas MA, Muller WJ, Park M. 2001. Distinct tyrosine autophosphorylation sites mediate induction of epithelial mesenchymal like transition by an activated ErbB-2/Neu receptor. *Oncogene* **20:** 788–799.

Kim SW, Park JI, Spring CM, Sater AK, Ji H, Otchere AA, Daniel JM, McCrea PD. 2004. Non-canonical Wnt signals are modulated by the Kaiso transcriptional repressor and p120-catenin. *Nat Cell Biol* **6:** 1212–1220.

Kinzler KW, Vogelstein B. 1996. Lessons from hereditary colorectal cancer. *Cell* **87:** 159–170.

Klaus A, Birchmeier W. 2008. Wnt signalling and its impact on development and cancer. *Nat Rev Cancer* **8:** 387–398.

Kophengnavong T, Michnowicz JE, Blackwell TK. 2000. Establishment of distinct MyoD, E2A, and twist DNA binding specificities by different basic region-DNA conformations. Mol Cell Biol **20:** 261–272.

Kramps T, Peter O, Brunner E, Nellen D, Froesch B, Chatterjee S, Murone M, Zullig S, Basler K. 2002. Wnt/wingless signaling requires BCL9/legless-mediated recruitment of pygopus to the nuclear β-catenin-TCF complex. *Cell* **109:** 47–60.

Kuphal F, Behrens J. 2006. E-cadherin modulates Wnt-dependent transcription in colorectal cancer cells but does not alter Wnt-independent gene expression in fibroblasts. *Exp Cell Res* **312:** 457–467.

Kuroda S, Fukata M, Nakagawa M, Fujii K, Nakamura T, Ookubo T, Izawa I, Nagase T, Nomura N, Tani H, et al. 1998. Role of IQGAP1, a target of the small GTPases Cdc42 and Rac1, in regulation of E-cadherin- mediated cell-cell adhesion. *Science* **281:** 832–835.

Lee YM, Park T, Schulz RA, Kim Y. 1997. Twist-mediated activation of the NK-4 homeobox gene in the visceral mesoderm of *Drosophila* requires two distinct clusters of E-box regulatory elements. *J Biol Chem* **272:** 17531–17541.

Lee E, Salic A, Kirschner MW. 2001. Physiological regulation of [β]-catenin stability by Tcf3 and CK1epsilon. *J Cell Biol* **154:** 983–993.

Liu P, Wakamiya M, Shea MJ, Albrecht U, Behringer RR, Bradley A. 1999. Requirement for Wnt3 in vertebrate axis formation. *Nat Gen* **22:** 361–365.

Liu C, Li Y, Semenov M, Han C, Baeg GH, Tan Y, Zhang Z, Lin X, He X. 2002. Control of β-catenin phosphorylation/degradation by a dual-kinase mechanism. *Cell* **108:** 837–847.

Liu L, Wu DH, Ding YQ. 2005. Tiam1 gene expression and its significance in colorectal carcinoma. *World J Gastroenterol* **11:** 705–707.

Lochter A, Galosy S, Muschler J, Freedman N, Werb Z, Bissell MJ. 1997. Matrix metalloproteinase stromelysin-1 triggers a cascade of molecular alterations that leads to stable epithelial-to-mesenchymal conversion and a premalignant phenotype in mammary epithelial cells. *J Cell Biol* **139:** 1861–1872.

Lu Z, Ghosh S, Wang Z, Hunter T. 2003. Downregulation of caveolin-1 function by EGF leads to the loss of E-cadherin, increased transcriptional activity of β-catenin, and enhanced tumor cell invasion. *Cancer cell* **4:** 499–515.

MacRae CA, Birchmeier W, Thierfelder L. 2006. Arrhythmogenic right ventricular cardiomyopathy: Moving toward mechanism. *J Clin Invest* **116:** 1825–1828.

Malliri A, Rygiel TP, van der Kammen RA, Song JY, Engers R, Hurlstone AF, Clevers H, Collard JG. 2006. The rac activator Tiam1 is a Wnt-responsive gene that modifies intestinal tumor development. *J Biol Chem* **281:** 543–548.

Malliri A, van Es S, Huveneers S, Collard JG. 2004. The Rac exchange factor Tiam1 is required for the establishment and maintenance of cadherin-based adhesions. *J Biol Chem* **279:** 30092–30098.

Mani SA, Yang J, Brooks M, Schwaninger G, Zhou A, Miura N, Kutok JL, Hartwell K, Richardson AL, Weinberg RA. 2007. Mesenchyme Forkhead 1 (FOXC2) plays a key role in metastasis and is associated with aggressive basal-like breast cancers. *Proc Natl Acad Sci* **104:** 10069–10074.

Marambaud P, Robakis NK. 2005. Genetic and molecular aspects of Alzheimer's disease shed light on new mechanisms of transcriptional regulation. *Genes, Brain, Behavior* **4:** 134–146.

Marambaud P, Shioi J, Serban G, Georgakopoulos A, Sarner S, Nagy V, Baki L, Wen P, Efthimiopoulos S, Shao Z, et al. 2002. A presenilin-1/γ-secretase cleavage releases the E-cadherin intracellular domain and regulates disassembly of adherens junctions. *EMBO J* **21:** 1948–1956.

Marambaud P, Wen PH, Dutt A, Shioi J, Takashima A, Siman R, Robakis NK. 2003. A CBP binding transcriptional repressor produced by the PS1/epsilon-cleavage of N-cadherin is inhibited by PS1 FAD mutations. *Cell* **114:** 635–645.

Maretzky T, Reiss K, Ludwig A, Buchholz J, Scholz F, Proksch E, de Strooper B, Hartmann D, Saftig P. 2005. ADAM10 mediates E-cadherin shedding and regulates epithelial cell-cell adhesion, migration, and β-catenin translocation. *Proc Natl Acad Sci* **102:** 9182–9187.

McCrea PD, Turck CW, Gumbiner B. 1991. A homolog of the armadillo protein in *Drosophila* (plakoglobin) associated with E-cadherin. *Science* **254:** 1359–1361.

McCrea PD, Gu D, Balda M. 2009. Junctional music that the nucleus hears. *Cold Spring Harb Perspect Biol* **1:** a002923.

Meiners S, Brinkmann V, Naundorf H, Birchmeier W. 1998. Role of morphogenetic factors in metastasis of mammary carcinoma cells. *Oncogene* **16:** 9–20.

Meng W, Takeichi M. 2009. Adherens junction: Molecular architecture and regulation. *Cold Spring Harb Perspect Biol* **1:** a002899.

Mercado-Pimentel ME, Runyan RB. 2007. Multiple transforming growth factor-β isoforms and receptors function during epithelial-mesenchymal cell transformation in the embryonic heart. *Cells, Tissues, Organs* **185:** 146–156.

Mertens AE, Pegtel DM, Collard JG. 2006. Tiam1 takes PARt in cell polarity. *Trends Cell Biol* **16:** 308–316.

Michiels F, Habets GG, Stam JC, van der Kammen RA, Collard JG. 1995. A role for Rac in Tiam1-induced membrane ruffling and invasion. *Nature* **375:** 338–340.

Minard ME, Ellis LM, Gallick GE. 2006. Tiam1 regulates cell adhesion, migration and apoptosis in colon tumor cells. *Clin Exp Metas* **23:** 301–313.

Molenaar M, van de Wetering M, Oosterwegel M, Peterson-Maduro J, Godsave S, Korinek V, Roose J, Destree O, Clevers H. 1996. XTcf-3 transcription factor mediates β-catenin-induced axis formation in *Xenopus* embryos. *Cell* **86:** 391–399.

Moon RT, Kohn AD, De Ferrari GV, Kaykas A. 2004. WNT and β-catenin signalling: Diseases and therapies. *Nat Rev Genet* **5:** 691–701.

Morali OG, Delmas V, Moore R, Jeanney C, Thiery JP, Larue L. 2001. IGF-II induces rapid β-catenin relocation to the nucleus during epithelium to mesenchyme transition. *Oncogene* **20:** 4942–4950.

Muller T, Choidas A, Reichmann E, Ullrich A. 1999. Phosphorylation and free pool of β-catenin are regulated by tyrosine kinases and tyrosine phosphatases during epithelial cell migration. *J Biol Chem* **274:** 10173–10183.

Munemitsu S, Albert I, Souza B, Rubinfeld B, Polakis P. 1995. Regulation of intracellular β-catenin levels by the adenomatous polyposis coli (APC) tumor-suppressor protein. *Proc Natl Acad Sci* **92:** 3046–3050.

Nagafuchi A, Takeichi M. 1988. Cell binding function of E-cadherin is regulated by the cytoplasmic domain. *EMBO J* **7:** 3679–3684.

Nagafuchi A, Takeichi M. 1989. Transmembrane control of cadherin-mediated cell adhesion: A 94 kDa protein functionally associated with a specific region of the cytoplasmic domain of E-cadherin. *Cell Reg* **1:** 37–44.

Nagafuchi A, Takeichi M, Tsukita S. 1991. The 102 kd cadherin-associated protein: Similarity to vinculin and posttranscriptional regulation of expression. *Cell* **65:** 849–857.

Nelson WJ, Nusse R. 2004. Convergence of Wnt, β-catenin, and cadherin pathways. *Science* **303:** 1483–1487.

Nguyen DX, Massague J. 2007. Genetic determinants of cancer metastasis. *Nat Rev Genet* **8:** 341–352.

Nieto MA. 2002. The snail superfamily of zinc-finger transcription factors. *Nat Rev* **3:** 155–166.

Nieto MA, Sargent MG, Wilkinson DG, Cooke J. 1994. Control of cell behavior during vertebrate development by Slug, a zinc finger gene. *Science* **264:** 835–839.

Noe V, Fingleton B, Jacobs K, Crawford HC, Vermeulen S, Steelant W, Bruyneel E, Matrisian LM, Mareel M. 2001. Release of an invasion promoter E-cadherin fragment by matrilysin and stromelysin-1. *J Cell Sci* **114:** 111–118.

Novellino L, De Filippo A, Deho P, Perrone F, Pilotti S, Parmiani G, Castelli C. 2008. PTPRK negatively regulates transcriptional activity of wild type and mutated oncogenic β-catenin and affects membrane distribution of β-catenin/E-cadherin complexes in cancer cells. *Cell Signal* **20:** 872–883.

Nüsslein-Volhard C, Lohs-Schardin M, Sander K, Cremer C. 1980. A dorso-ventral shift of embryonic primordia in a new maternal-effect mutant of *Drosophila*. *Nature* **283:** 474–476.

Nüsslein-Volhard C, Wieschaus E, Kluding H. 1984. Mutations affecting the pattern of the larval cuticle in *Drosophila* melanogaster. I. Zygotic loci on the second chromosome. *Roux's Arch Develop Biol* **193:** 267–282.

Oda H, Tsukita S, Takeichi M. 1998. Dynamic behavior of the cadherin-based cell-cell adhesion system during *Drosophila* gastrulation. *Develop Biol* **203:** 435–450.

Oft M, Peli J, Rudaz C, Schwarz H, Beug H, Reichmann E. 1996. TGF-β1 and Ha-Ras collaborate in modulating the phenotypic plasticity and invasiveness of epithelial tumor cells. *Genes Develop* **10:** 2462–2477.

Onder TT, Gupta PB, Mani SA, Yang J, Lander ES, Weinberg RA. 2008. Loss of E-cadherin promotes metastasis via multiple downstream transcriptional pathways. *Cancer Res* **68:** 3645–3654.

Orsulic S, Huber O, Aberle H, Arnold S, Kemler R. 1999. E-cadherin binding prevents β-catenin nuclear localization and β-catenin/LEF-1-mediated transactivation. *J Cell Sci* **112:** 1237–1245.

Ozawa M, Baribault H, Kemler R. 1989. The cytoplasmic domain of the cell adhesion molecule uvomorulin associates with three independent proteins structurally related in different species. *EMBO J* **8:** 1711–1717.

Pai R, Dunlap D, Qing J, Mohtashemi I, Hotzel K, French DM. 2008. Inhibition of fibroblast growth factor 19 reduces tumor growth by modulating β-catenin signaling. *Cancer Res* **68:** 5086–5095.

Park JI, Kim SW, Lyons JP, Ji H, Nguyen TT, Cho K, Barton MC, Deroo T, Vleminckx K, Moon RT, et al. 2005. Kaiso/p120-catenin and TCF/β-catenin complexes coordinately regulate canonical Wnt gene targets. *Develop Cell* **8:** 843–854.

Park JI, Ji H, Jun S, Gu D, Hikasa H, Li L, Sokol SY, McCrea PD. 2006. Frodo links Dishevelled to the p120-catenin/Kaiso pathway: Distinct catenin subfamilies promote Wnt signals. *Develop Cell* **11:** 683–695.

Peifer M, McCrea PD, Green KJ, Wieschaus E, Gumbiner BM. 1992. The vertebrate adhesive junction proteins β-catenin and plakoglobin and the *Drosophila* segment polarity gene armadillo form a multigene family with similar properties. *J Cell Biol* **118:** 681–691.

Peinado H, Ballestar E, Esteller M, Cano A. 2004a. Snail mediates E-cadherin repression by the recruitment of the Sin3A/histone deacetylase 1 (HDAC1)/HDAC2 complex. *Mol Cell Biol* **24:** 306–319.

Peinado H, Marin F, Cubillo E, Stark HJ, Fusenig N, Nieto MA, Cano A. 2004b. Snail and E47 repressors of E-cadherin induce distinct invasive and angiogenic properties in vivo. *J Cell Sci* **117:** 2827–2839.

Perez-Moreno MA, Locascio A, Rodrigo I, Dhondt G, Portillo F, Nieto MA, Cano A. 2001. A new role for E12/E47 in the repression of E-cadherin expression and epithelial-mesenchymal transitions. *J Biol Chem* **276:** 27424–27431.

Perl AK, Wilgenbus P, Dahl U, Semb H, Christofori G. 1998. A causal role for E-cadherin in the transition from adenoma to carcinoma. *Nature* **392:** 190–193.

Perrais M, Chen X, Perez-Moreno M, Gumbiner BM. 2007. E-cadherin homophilic ligation inhibits cell growth and epidermal growth factor receptor signaling independently of other cell interactions. *Mol Biol Cell* **18:** 2013–2025.

Polakis P. 2000. Wnt signaling and cancer. *Genes Develop* **14:** 1837–1851.

Prieve MG, Moon RT. 2003. Stromelysin-1 and mesothelin are differentially regulated by Wnt-5a and Wnt-1 in C57 mg mouse mammary epithelial cells. *BMC Develop Biol* **3:** 2.

Primakoff P, Myles DG. 2000. The ADAM gene family: Surface proteins with adhesion and protease activity. *Trends Genet* **16:** 83–87.

Qi J, Wang J, Romanyuk O, Siu CH. 2006. Involvement of Src family kinases in N-cadherin phosphorylation and β-catenin dissociation during transendothelial migration of melanoma cells. *Mol Biol Cell* **17:** 1261–1272.

Reichmann E, Schwarz H, Deiner EM, Leitner I, Eilers M, Berger J, Busslinger M, Beug H. 1992. Activation of an inducible c-FosER fusion protein causes loss of epithelial polarity and triggers epithelial-fibroblastoid cell conversion. *Cell* **71:** 1103–1116.

Reiss K, Maretzky T, Ludwig A, Tousseyn T, de Strooper B, Hartmann D, Saftig P. 2005. ADAM10 cleavage of N-cadherin and regulation of cell-cell adhesion and β-catenin nuclear signalling. *EMBO J* **24:** 742–752.

Reynolds AB, Roczniak-Ferguson A. 2004. Emerging roles for p120-catenin in cell adhesion and cancer. *Oncogene* **23:** 7947–7956.

Rhee J, Buchan T, Zukerberg L, Lilien J, Balsamo J. 2007. Cables links Robo-bound Abl kinase to N-cadherin-bound β-catenin to mediate Slit-induced modulation of adhesion and transcription. *Nat Cell Biol* **9:** 883–892.

Riese J, Yu X, Munnerlyn A, Eresh S, Hsu SC, Grosschedl R, Bienz M. 1997. LEF-1, a nuclear factor coordinating signaling inputs from wingless and decapentaplegic. *Cell* **88:** 777–787.

Riggleman B, Schedl P, Wieschaus E. 1990. Spatial expression of the *Drosophila* segment polarity gene armadillo is posttranscriptionally regulated by wingless. *Cell* **63:** 549–560.

Rijsewijk F, Schuermann M, Wagenaar E, Parren P, Weigel D, Nusse R. 1987. The *Drosophila* homolog of the mouse mammary oncogene int-1 is identical to the segment polarity gene wingless. *Cell* **50:** 649–657.

Rimm DL, Koslov ER, Kebriaei P, Cianci CD, Morrow JS. 1995. α 1(E)-catenin is an actin-binding and -bundling protein mediating the attachment of F-actin to the membrane adhesion complex. *Proc Natl Acad Sci* **92:** 8813–8817.

Rohwedel J, Horak V, Hebrok M, Fuchtbauer EM, Wobus AM. 1995. M-twist expression inhibits mouse embryonic stem cell-derived myogenic differentiation in vitro. *Exp Cell Res* **220:** 92–100.

Roose J, Molenaar M, Peterson J, Hurenkamp J, Brantjes H, Moerer P, van de Wetering M, Destree O, Clevers H. 1998. The *Xenopus* Wnt effector XTcf-3 interacts with Groucho-related transcriptional repressors. *Nature* **395:** 608–612.

Rosivatz E, Becker I, Specht K, Fricke E, Luber B, Busch R, Hofler H, Becker KF. 2002. Differential expression of the epithelial-mesenchymal transition regulators snail, SIP1, and twist in gastric cancer. *Am J Pathol* **161:** 1881–1891.

Ross SE, Hemati N, Longo KA, Bennett CN, Lucas PC, Erickson RL, MacDougald OA. 2000. Inhibition of adipogenesis by Wnt signaling. *Science* **289:** 950–953.

Roura S, Miravet S, Piedra J, Garcia de Herreros A, Dunach M. 1999. Regulation of E-cadherin/Catenin association by tyrosine phosphorylation. *J Biol Chem* **274:** 36734–36740.

Rubinfeld B, Albert I, Porfiri E, Fiol C, Munemitsu S, Polakis P. 1996. Binding of GSK3β to the APC-β-catenin complex and regulation of complex assembly. *Science* **272:** 1023–1026.

Ruiz P, Brinkmann V, Ledermann B, Behrend M, Grund C, Thalhammer C, Vogel F, Birchmeier C, Gunthert U, Franke WW, et al. 1996. Targeted mutation of plakoglobin in mice reveals essential functions of desmosomes in the embryonic heart. *J Cell Biol* **135:** 215–225.

Ruzov A, Hackett JA, Prokhortchouk A, Reddington JP, Madej MJ, Dunican DS, Prokhortchouk E, Pennings S, Meehan RR. 2009. The interaction of xKaiso with xTcf3: A revised model for integration of epigenetic and Wnt signalling pathways. *Development* **136:** 723–727.

Sadot E, Simcha I, Shtutman M, Ben-Ze'ev A, Geiger B. 1998. Inhibition of β-catenin-mediated transactivation by cadherin derivatives. *Proc Natl Acad Sci* **95:** 15339–15344.

Sampietro J, Dahlberg CL, Cho US, Hinds TR, Kimelman D, Xu W. 2006. Crystal structure of a β-catenin/BCL9/Tcf4 complex. *Mol Cell* **24:** 293–300.

Sandig M, Voura EB, Kalnins VI, Siu CH. 1997. Role of cadherins in the transendothelial migration of melanoma cells in culture. *Cell Motil Cytoskel* **38:** 351–364.

Sanson B, White P, Vincent JP. 1996. Uncoupling cadherin-based adhesion from wingless signalling in *Drosophila. Nature* **383:** 627–630.

Savagner P, Yamada KM, Thiery JP. 1997. The zinc-finger protein slug causes desmosome dissociation, an initial and necessary step for growth factor-induced epithelial-mesenchymal transition. *J Cell Biol* **137:** 1403–1419.

Schwarz-Romond T, Fiedler M, Shibata N, Butler PJ, Kikuchi A, Higuchi Y, Bienz M. 2007. The DIX domain of Dishevelled confers Wnt signaling by dynamic polymerization. *Nat Struct Mol Bioll* **14:** 484–492.

Seals DF, Courtneidge SA. 2003. The ADAMs family of metalloproteases: Multidomain proteins with multiple functions. *Genes Develop* **17:** 7–30.

Sefton M, Sanchez S, Nieto MA. 1998. Conserved and divergent roles for members of the Snail family of transcription factors in the chick and mouse embryo. *Development* **125:** 3111–3121.

Selkoe DJ, Wolfe MS. 2007. Presenilin: Running with scissors in the membrane. *Cell* **131:** 215–221.

Sen-Chowdhry S, Lowe MD, Sporton SC, McKenna WJ. 2004. Arrhythmogenic right ventricular cardiomyopathy: Clinical presentation, diagnosis, and management. *Am J Med Am J* **117:** 685–695.

Shapiro L, Weis WI. 2009. Structure and biochemistry of cadherins and catenins. *Cold Spring Harb Perspect Biol* **1:** a003053.

Shen Y, Hirsch DS, Sasiela CA, Wu WJ. 2008. Cdc42 regulates E-cadherin ubiquitination and degradation through an epidermal growth factor receptor to Src-mediated pathway. *J Biol Chem* **283:** 5127–5137.

Shoval I, Ludwig A, Kalcheim C. 2007. Antagonistic roles of full-length N-cadherin and its soluble BMP cleavage product in neural crest delamination. *Development* **134:** 491–501.

Shtutman M, Zhurinsky J, Simcha I, Albanese C, D'Amico M, Pestell R, Ben-Ze'ev A. 1999. The cyclin D1 gene is a target of the β-catenin/LEF-1 pathway. *Proc Natl Acad Sci* **96:** 5522–5527.

Shtutman M, Levina E, Ohouo P, Baig M, Roninson IB. 2006. Cell adhesion molecule L1 disrupts E-cadherin-containing adherens junctions and increases scattering and motility of MCF7 breast carcinoma cells. *Cancer Res* **66:** 11370–11380.

Simpson P. 1983. Maternal-Zygotic Gene Interactions during Formation of the Dorsoventral Pattern in *Drosophila* Embryos. *Genetics* **105:** 615–632.

Smith DE, Franco del Amo F, Gridley T. 1992. Isolation of Sna, a mouse gene homologous to the *Drosophila* genes snail and escargot: Its expression pattern suggests multiple roles during postimplantation development. *Development* **116:** 1033–1039.

Soo K, O'Rourke MP, Khoo PL, Steiner KA, Wong N, Behringer RR, Tam PP. 2002. Twist function is required for the morphogenesis of the cephalic neural tube and the differentiation of the cranial neural crest cells in the mouse embryo. *Develop Biol* **247:** 251–270.

Soriano S, Kang DE, Fu M, Pestell R, Chevallier N, Zheng H, Koo EH. 2001. Presenilin 1 negatively regulates β-catenin/T cell factor/lymphoid enhancer factor-1 signaling independently of β-amyloid precursor protein and notch processing. *J Cell Biol* **152:** 785–794.

Spring CM, Kelly KF, O'Kelly I, Graham M, Crawford HC, Daniel JM. 2005. The catenin p120ctn inhibits Kaiso-mediated transcriptional repression of the β-catenin/TCF target gene matrilysin. *Exp Cell Res* **305:** 253–265.

Stadeli R, Hoffmans R, Basler K. 2006. Transcription under the control of nuclear Arm/β-catenin. *Curr Biol* **16:** R378–385.

Staubli U, Larson J, Thibault O, Baudry M, Lynch G. 1988. Chronic administration of a thiol-proteinase inhibitor blocks long-term potentiation of synaptic responses. *Brain Res* **444:** 153–158.

Stein U, Arlt F, Walther W, Smith J, Waldman T, Harris ED, Mertins SD, Heizmann CW, Allard D, Birchmeier W, et al. 2006. The metastasis-associated gene S100A4 is a novel target of β-catenin/T-cell factor signaling in colon cancer. *Gastroenterology* **131:** 1486–1500.

Steinhusen U, Weiske J, Badock V, Tauber R, Bommert K, Huber O. 2001. Cleavage and shedding of E-cadherin after induction of apoptosis. *J Biol Chem* **276:** 4972–4980.

Stemmer V, de Craene B, Berx G, Behrens J. 2008. Snail promotes Wnt target gene expression and interacts with β-catenin. *Oncogene* **27:** 5075–5080.

Stemmler MP, Hecht A, Kemler R. 2005. E-cadherin intron 2 contains cis-regulatory elements essential for gene expression. *Development* **132:** 965–976.

Stockinger A, Eger A, Wolf J, Beug H, Foisner R. 2001. E-cadherin regulates cell growth by modulating proliferation-dependent β-catenin transcriptional activity. *J Cell Biol* **154:** 1185–1196.

Stoker M, Gherardi E, Perryman M, Gray J. 1987. Scatter factor is a fibroblast-derived modulator of epithelial cell mobility. *Nature* **327:** 239–242.

Strathdee G. 2002. Epigenetic versus genetic alterations in the inactivation of E-cadherin. *Sem Cancer Biol* **12:** 373–379.

Takeichi M. 1991. Cadherin cell adhesion receptors as a morphogenetic regulator. *Science* **251:** 1451–1455.

ten Berge D, Koole W, Fuerer C, Fish M, Eroglu E, Nusse R. 2008. Wnt signaling mediates self-organization and axis formation in embryoid bodies. *Cell Stem cell* **3:** 508–518.

Thiery JP. 2002. Epithelial-mesenchymal transitions in tumour progression. *Nat Rev Cancer* **2:** 442–454.

Thisse B, Stoetzel C, Gorostiza-Thisse C, Perrin-Schmitt F. 1988. Sequence of the twist gene and nuclear localization of its protein in endomesodermal cells of early *Drosophila* embryos. *EMBO J* **7:** 2175–2183.

Tomita K, van Bokhoven A, van Leenders GJ, Ruijter ET, Jansen CF, Bussemakers MJ, Schalken JA. 2000. Cadherin switching in human prostate cancer progression. *Cancer Res* **60:** 3650–3654.

Torres MA, Yang-Snyder JA, Purcell SM, DeMarais AA, McGrew LL, Moon RT. 1996. Activities of the Wnt-1 class of secreted signaling factors are antagonized by the Wnt-5A class and by a dominant negative cadherin in early *Xenopus* development. *J Cell Biol* **133:** 1123–1137.

Travis A, Amsterdam A, Belanger C, Grosschedl R. 1991. LEF-1, a gene encoding a lymphoid-specific protein with an HMG domain, regulates T-cell receptor α enhancer function [corrected]. *Genes Develop* **5:** 880–894.

Trelstad RL, Hay ED, Revel JD. 1967. Cell contact during early morphogenesis in the chick embryo. *Develop Biol* **16:** 78–106.

Uemura K, Kihara T, Kuzuya A, Okawa K, Nishimoto T, Bito H, Ninomiya H, Sugimoto H, Kinoshita A, Shimohama S. 2006. Activity-dependent regulation of β-catenin via epsilon-cleavage of N-cadherin. *Biochem Biophys Res Comm* **345:** 951–958.

Vallin J, Thuret R, Giacomello E, Faraldo MM, Thiery JP, Broders F. 2001. Cloning and characterization of three *Xenopus* slug promoters reveal direct regulation by Lef/β-catenin signaling. *J Biol Chem* **276:** 30350–30358.

van de Wetering M, Oosterwegel M, Dooijes D, Clevers H. 1991. Identification and cloning of TCF-1, a T lymphocyte-specific transcription factor containing a sequence-specific HMG box. *EMBO J* **10:** 123–132.

van de Wetering M, Barker N, Harkes IC, van der Heyden M, Dijk NJ, Hollestelle A, Klijn JG, Clevers H, Schutte M. 2001. Mutant E-cadherin breast cancer cells do not display constitutive Wnt signaling. *Cancer Res* **61:** 278–284.

van der Heyden MA, Rook MB, Hermans MM, Rijksen G, Boonstra J, Defize LH, Destree OH. 1998. Identification of connexin43 as a functional target for Wnt signalling. *J Cell Sci* **111:** 1741–1749.

van Roy FM, McCrea PD. 2005. A role for Kaiso-p120ctn complexes in cancer? *Nat Rev Cancer* **5:** 956–964.

Veeman MT, Axelrod JD, Moon RT. 2003. A second canon. Functions and mechanisms of β-catenin-independent Wnt signaling. *Develop Cell* **5:** 367–377.

Veltmaat JM, Orelio CC, Ward-Van Oostwaard D, Van Rooijen MA, Mummery CL, Defize LH. 2000. Snail is an immediate early target gene of parathyroid hormone related peptide signaling in parietal endoderm formation. *Int J Develop Biol* **44:** 297–307.

Vestweber D, Kemler R. 1984. Some structural and functional aspects of the cell adhesion molecule uvomorulin. *Cell Different* **15:** 269–273.

Vignjevic D, Schoumacher M, Gavert N, Janssen KP, Jih G, Lae M, Louvard D, Ben-Ze'ev A, Robine S. 2007. Fascin, a novel target of β-catenin-TCF signaling, is expressed at the invasive front of human colon cancer. *Cancer Res* **67:** 6844–6853.

Vleminckx K, Vakaet L Jr, Mareel M, Fiers W, van Roy F. 1991. Genetic manipulation of E-cadherin expression by epithelial tumor cells reveals an invasion suppressor role. *Cell* **66:** 107–119.

Wagenaar-Miller RA, Gorden L, Matrisian LM. 2004. Matrix metalloproteinases in colorectal cancer: Is it worth talking about? *Cancer Metast Rev* **23:** 119–135.

Wang SM, Coljee VW, Pignolo RJ, Rotenberg MO, Cristofalo VJ, Sierra F. 1997. Cloning of the human twist gene: Its expression is retained in adult mesodermally-derived tissues. *Gene* **187:** 83–92.

Warren SL, Nelson WJ. 1987. Nonmitogenic morphoregulatory action of pp60v-src on multicellular epithelial structures. *Mol Cell Biol* **7:** 1326–1337.

Waterman ML, Fischer WH, Jones KA. 1991. A thymus-specific member of the HMG protein family regulates the human T cell receptor C α enhancer. *Genes Develop* **5:** 656–669.

Weidner KM, Behrens J, Vandekerckhove J, Birchmeier W. 1990. Scatter factor: Molecular characteristics and effect on the invasiveness of epithelial cells. *J Cell Biol* **111:** 2097–2108.

Weidner KM, Arakaki N, Hartmann G, Vandekerckhove J, Weingart S, Rieder H, Fonatsch C, Tsubouchi H, Hishida T, Daikuhara Y, et al. 1991. Evidence for the identity of human scatter factor and human hepatocyte growth factor. *Proc Natl Acad Sci* **88:** 7001–7005.

Wheelock MJ, Johnson KR. 2003. Cadherins as modulators of cellular phenotype. *Ann Rev Cell Develop Biol* **19:** 207–235.

Wielenga VJ, Smits R, Korinek V, Smit L, Kielman M, Fodde R, Clevers H, Pals ST. 1999. Expression of CD44 in Apc and Tcf mutant mice implies regulation by the WNT pathway. *Am J Pathol* **154:** 515–523.

Wieschaus E, Riggleman R. 1987. Autonomous requirements for the segment polarity gene armadillo during *Drosophila* embryogenesis. *Cell* **49:** 177–184.

Willert K, Shibamoto S, Nusse R. 1999. Wnt-induced dephosphorylation of axin releases β-catenin from the axin complex. *Genes Develop* **13:** 1768–1773.

Willis TG, Zalcberg IR, Coignet LJ, Wlodarska I, Stul M, Jadayel DM, Bastard C, Treleaven JG, Catovsky D, Silva ML, et al. 1998. Molecular cloning of translocation t(1;14)(q21;q32) defines a novel gene (BCL9) at chromosome 1q21. *Blood* **91:** 1873–1881.

Wolf C, Thisse C, Stoetzel C, Thisse B, Gerlinger P, Perrin-Schmitt F. 1991. The M-twist gene of Mus is expressed in subsets of mesodermal cells and is closely related to the *Xenopus* X-twi and the *Drosophila* twist genes. *Develop Biol* **143:** 363–373.

Xia X, Qian S, Soriano S, Wu Y, Fletcher AM, Wang XJ, Koo EH, Wu X, Zheng H. 2001. Loss of presenilin 1 is associated with enhanced β-catenin signaling and skin tumorigenesis. *Proc Natl Acad Sci* **98:** 10863–10868.

Yamada S, Pokutta S, Drees F, Weis WI, Nelson WJ. 2005. Deconstructing the cadherin-catenin-actin complex. *Cell* **123:** 889–901.

Yamaguchi TP, Harpal K, Henkemeyer M, Rossant J. 1994. fgfr-1 is required for embryonic growth and mesodermal patterning during mouse gastrulation. *Genes Develop* **8:** 3032–3044.

Yan HX, Yang W, Zhang R, Chen L, Tang L, Zhai B, Liu SQ, Cao HF, Man XB, Wu HP, et al. 2006. Protein-tyrosine phosphatase PCP-2 inhibits β-catenin signaling and increases E-cadherin-dependent cell adhesion. *J Biol Chem* **281:** 15423–15433.

Yang J, Weinberg RA. 2008. Epithelial-mesenchymal transition: At the crossroads of development and tumor metastasis. *Develop Cell* **14:** 818–829.

Yang J, Mani SA, Donaher JL, Ramaswamy S, Itzykson RA, Come C, Savagner P, Gitelman I, Richardson A, Weinberg RA. 2004. Twist, a master regulator of morphogenesis, plays an essential role in tumor metastasis. *Cell* **117:** 927–939.

Yin Z, Xu XL, Frasch M. 1997. Regulation of the twist target gene tinman by modular cis-regulatory elements during early mesoderm development. *Development* **124:** 4971–4982.

Yook JI, Li XY, Ota I, Fearon ER, Weiss SJ. 2005. Wnt-dependent regulation of the E-cadherin repressor snail. *J Biol Chem* **280:** 11740–11748.

Zhang Z, Hartmann H, Do VM, Abramowski D, Sturchler-Pierrat C, Staufenbiel M, Sommer B, van de Wetering M, Clevers H, Saftig P, et al. 1998. Destabilization of β-catenin by mutations in presenilin-1 potentiates neuronal apoptosis. *Nature* **395:** 698–702.

Zhou BP, Deng J, Xia W, Xu J, Li YM, Gunduz M, Hung MC. 2004. Dual regulation of Snail by GSK-3β-mediated phosphorylation in control of epithelial-mesenchymal transition. *Nat Cell Biol* **6:** 931–940.

Zhurinsky J, Shtutman M, Ben-Ze'ev A. 2000. Plakoglobin and β-catenin: Protein interactions, regulation and biological roles. *J Cell Sci* **113:** 3127–3139.

Wnt Signaling from Development to Disease: Insights from Model Systems

Ken M. Cadigan[1] **and Mark Peifer**[2]

[1]Department of Molecular, Cellular, and Developmental Biology, University of Michigan, Ann Arbor, Michigan 48109-1048

[2]Department of Biology and Lineberger Comprehensive Cancer Center, University of North Carolina at Chapel Hill, Chapel Hill, North Carolina 27599-3280

Correspondence: peifer@unc.edu

One of the early surprises in the study of cell adhesion was the discovery that β-catenin plays dual roles, serving as an essential component of cadherin-based cell–cell adherens junctions and also serving as the key regulated effector of the Wnt signaling pathway. Here, we review our current model of Wnt signaling and discuss how recent work using model organisms has advanced our understanding of the roles Wnt signaling plays in both normal development and in disease. These data help flesh out the mechanisms of signaling from the membrane to the nucleus, revealing new protein players and providing novel information about known components of the pathway.

Modern biomedical science is a partnership between scientists studying basic cell and developmental processes in model systems and clinicians exploring the basis of human disease. Few fields exemplify this better than Wnt signaling, born 22 years ago with the realization that the oncogene *int1* and the *Drosophila* developmental patterning gene *wingless (wg)* are homologs (Cabrera et al. 1987; Rijsewijk et al. 1987). Additional connections further fueled research. *Drosophila* Armadillo (Arm), a component of the Wg pathway, is the homolog of the cell junction proteins β-catenin (βcat) and plakoglobin (McCrea et al. 1991; Peifer et al. 1992; Peifer and Wieschaus 1990) joining Wnt signaling and cadherin-based cell adhesion, a connection we still do not fully understand (see Heuberger and Birchmeier 2009). Adenomatous polyposis coli (APC), the tumor suppressor mutated in most colon cancers, binds βcat and is a key regulator of Wnt signaling (Rubinfeld et al. 1993; Su et al. 1993), putting the Wnt field even more squarely in the center of cancer research. Here, we outline recent advances in understanding Wnt signaling, casting new light on these critical regulators of development, homeostasis, and disease.

THE CURRENT MODEL OF WNT SIGNALING

We first outline the reigning model for Wnt signaling, focusing on canonical signaling involving βcat (for reviews of alternate Wnt

 doi: 10.1101/cshperspect.a002881
Cite this article as *Cold Spring Harb Perspect Biol* 2009;1:a002881

pathways, see e.g., Fanto and McNeill 2004; Kohn and Moon 2005; Semenov et al. 2007; van Amerongen et al. 2008; Veeman et al. 2003). The key regulated Wnt effector is βcat. In cells not receiving signal, βcat accumulates in adherens junctions, but outside junctions βcat levels are low, because of its short half-life (Fig. 1, left). Free βcat is bound by the multiprotein "destruction complex" (Clevers 2006), in which the tumor suppressors APC and Axin present βcat to the kinases glycogen synthase kinase 3 (GSK3) and casein kinase I (CKI), facilitating sequential phosphorylation of sites in βcat's amino terminus. An SCF-class E3-ubiquitin ligase containing the F-box protein Slimb/βTrCP recognizes correctly phosphorylated βcat, and targets it for polyubiqitination and proteasomal destruction. Meanwhile, TCF/LEF proteins bound to Groucho-family corepressors keep Wnt target genes tightly off (Arce et al. 2006).

Wnts bind a two-part receptor: a seven-transmembrane Frizzled (Fz) and LRP5/6 (Fig. 1, right; Clevers 2006). Both are required for canonical signaling. Ligand binding triggers phosphorylation of LRP5/6's cytoplasmic tail, creating an Axin-binding site. Axin recruitment inactivates the destruction complex, in a process requiring Disheveled (Dvl). This stabilizes βcat, and it enters nuclei. βcat displaces Groucho from TCF, nucleating formation of a multiprotein activator complex including Pygopus and Legless/Bcl9, activating Wnt target genes (Arce et al. 2006). The last 5 years have tested this proposed model, revealing new mechanistic insights and further complexity. In the following section, we examine different steps in Wnt signaling in turn and describe these new insights.

PREPARING FOR DEPARTURE—POSTTRANSLATIONAL WNT MODIFICATIONS

After signal sequence cleavage and translocation into the endoplasmic reticulum (ER), Wnts are

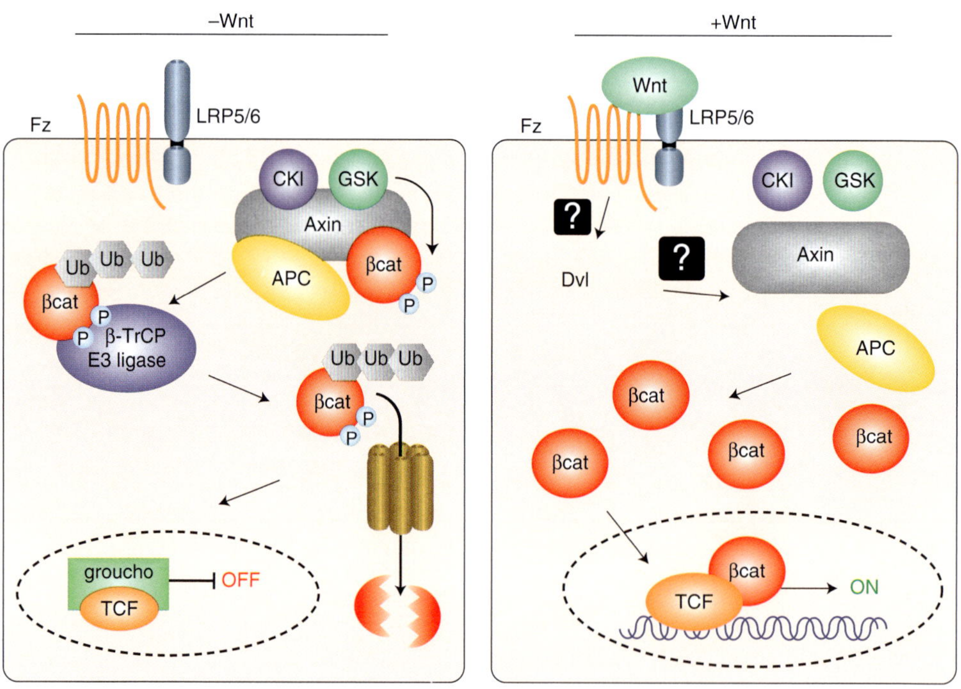

Figure 1. A simplified current view of Wnt signaling. See text for details.

transported through the endomembrane system to the cell surface and undergo several modifications. Wnts undergo N-linked glycosylation (Burrus and McMahon 1995; Kadowaki et al. 1996; Komekado et al. 2007; Kurayoshi et al. 2007; Mason et al. 1992; Smolich et al. 1993; Tanaka et al. 2002). Several Wnts also are palmitoylated at the first conserved cysteine (Galli et al. 2007; Kadowaki et al. 1996; Komekado et al. 2007; Willert et al. 2003). In addition, Wnt3a is modified with palmitoleic acid at a conserved serine (Takada et al. 2006). These acylations likely cause the notoriously hydrophobic nature of secreted Wnts (Willert et al. 2003).

Posttranslational lipidation of mammalian Wnts is clearly important for function. Mutating the conserved cysteine of Wnt1, Wnt3a, or Wnt5a prevents palmitoylation in cell culture. These mutant Wnts are secreted but have little or no signaling activity (Galli et al. 2007; Komekado et al. 2007; Kurayoshi et al. 2007; Willert et al. 2003), and unpalmitoylated Wnts cannot bind Fz receptors (Komekado et al. 2007; Kurayoshi et al. 2007). Mutating the conserved serine in Wnt3a prevents palmitoleic acid addition and blocks secretion (Takada et al. 2006). These studies suggest a model in which palmitoleic acid-modification is required for secretion, and palmitate for Fz binding.

Research on *Drosophila* Wg confirms the importance of acylation, but results differ from those in mammals. For example, mutating the conserved serine in Wg to which palmitoleic acid is added in mouse Wnt3a does not prevent secretion, though it reduces signaling activity (Franch-Marro et al. 2008a). Wg is palmitoylated (Willert et al. 2003), and mutating this amino-terminal cysteine results in secreted but inactive Wg in cell culture (Franch-Marro et al. 2008a). However, the same mutant Wg accumulates in the ER in fly tissues (Franch-Marro et al. 2008a), and mutating this cysteine in the endogenous *wg* gene prevents secretion (Nusse 2003). These data fit well with the *porcupine (porc)* phenotype, a strong loss of Wg signaling (van den Heuvel et al. 1993). Porc is an ER-localized integral membrane O-acyl transferase (Kadowaki et al. 1996) required for Wg palmitoylation (Zhai et al., 2004), and for Wg ER exit (Tanaka et al. 2002). Vertebrate Porc also promotes Wnt lipidation and is required for signaling (Galli et al. 2007). This suggests Porc acylates Wnts, though this has not yet been shown.

Apparent differences in palmitoylation's importance in Wnt signaling may reflect differences in levels of Wnt expression. In mammalian and fly cell culture, overexpression may overcome lack of palmitoylation. Consistent with this, whereas Porc is required for endogenous Wg signaling, it is not required when Wg is overexpressed (Noordermeer et al. 1994). In exploring relationships between Wnt glycosylation and lipidation, conflicting results were also obtained in culture versus in flies. Site-directed mutagenesis of Wnt3a and Wnt5a in cell culture suggested glycosylation precedes acylation and is required for ER export (Komekado et al. 2007; Kurayoshi et al. 2007). However, *porc* mutant fly embryos lacking acylation have defects in Wg glycosylation (Tanaka et al. 2002), suggesting acylation precedes glycosylation. While species or Wnt-specific differences may explain these discrepancies, care must be taken in interpreting experiments in which Wnts are expressed at nonphysiological levels.

GETTING OUT AND ABOUT—WNT SECRETION AND TRANSPORT

For several Wnts, reaching the cell surface also requires Wntless (Wls) and Retromer. Wls (also known as Evi or Sprinter in flies and MIG-14 in *Caenorhabditis elegans*) is an integral membrane protein found in the Golgi, plasma membrane, and endosomes (Banziger et al. 2006; Belenkaya et al. 2008; Franch-Marro et al. 2008b; Port et al. 2008; Yang et al. 2008). Retromer is a multiprotein complex that shuttles cargo from endosomes to the trans-Golgi (Seaman 2005). Several recent reviews covered these proteins in detail (Bartscherer and Boutros 2008; Ching and Nusse 2006; Eaton 2008; Hardin and King 2008). In short, the data suggest Wls is a "Wnt chaperone," guiding Wnts from the Golgi to the cell surface.

Retromer mutants block Wnt secretion because Wls is missorted to lysosomes (Belenkaya et al. 2008; Franch-Marro et al. 2008b; Pan et al. 2008a; Port et al. 2008; Yang et al. 2008). Although this secretion mechanism is highly conserved, it is not universal. *Drosophila* WntD is secreted independently of lipidation, Porc, and Wls (Ching et al. 2008).

On secretion, Wnts move away from producing cells to influence neighbors (Cadigan 2002; Capdevila and Izpisua Belmonte 2001; Cayuso and Marti 2005; Strigini and Cohen 1999). One superb model for this is the *Drosophila* larval wing imaginal disc, a flat columnar epithelium. Wg is expressed along the dorsoventral (D/V) boundary and moves in both directions, forming a morphogen gradient and regulating short- and long-range targets in a concentration-dependent manner (Cadigan 2002; Strigini and Cohen 1999). Wg can directly act up to 20 cell diameters from its synthesis site (Zecca et al. 1996).

Several factors influence Wg movement. Glypicans are heparin sulfate proteoglycans anchored to cell membranes via a glycerol phosphatidylinositol (GPI) linkage (Blair 2005). Two glypicans, Dally and Dally-like (Dly), influence Wg signaling in wing discs (Fig. 2). Dally promotes signaling and is suggested to facilitate Wg movement or act as a coreceptor (Franch-Marro et al. 2005; Han et al. 2005). Interestingly, Dly has different functions than Dally. Loss of Dly increases expression of short-range targets but decreases long-range Wg signaling (Franch-Marro et al. 2005; Han et al. 2005; Kirkpatrick et al. 2004; Kreuger et al. 2004). Consistent with this, *dly* mutants have reduced spread of Wg from its synthesis site (Han et al. 2005; Marois et al. 2006). A recent study suggests Dly mediates transcytosis of apically secreted

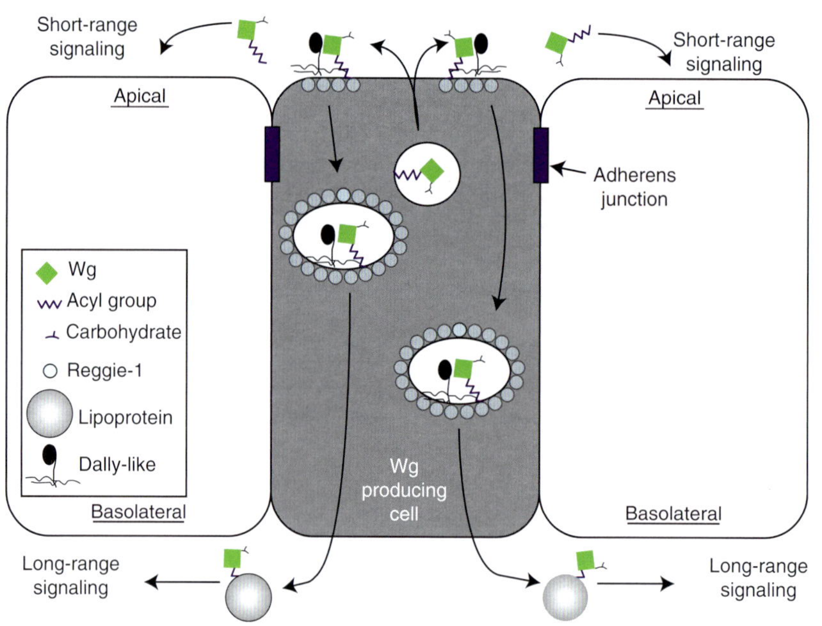

Figure 2. Model depicting Wg transcytosis and how this process could influence short- and long-range Wg signaling in the wing imaginal disc. Wg is secreted apically from producing cells, where association with extracellular glypican Dally-like directs it into endocytic vesicles. These Reggie-1-associated vesicles undergo transcytosis and fuse with the plasma membrane to release Wg on the basolateral side of the epithelial sheet. There, it can undergo long-range diffusion and signaling by associating with lipoproteins. See text for further explanation.

Wg, transporting it to the basolateral compartment (Gallet et al. 2008) where it then is presumed to diffuse to activate long-range targets.

Membrane microdomains also play roles in Wg secretion/diffusion (Katanaev et al. 2008). Reggie-1 (Flotillin-2) is a cytoplasmic component of noncaveolin lipid microdomains (Babuke and Tikkanen 2007; Langhorst et al. 2005). Reggie-1 loss strongly reduces long-range Wg targets but has no effect on short-range targets (Katanaev et al. 2008). Conversely, Reggie-1 overexpression inhibits short-range and activates long-range targets, even when Reggie-1 expression is restricted to Wg-expressing cells. As Reggie-1 is implicated in vesicular trafficking (Babuke and Tikkanen 2007; Langhorst et al. 2008) and Wg can associate with lipid rafts (Zhai et al. 2004), these data suggest Reggie-1 works with Dly to promote Wg transcytosis and subsequent long-range diffusion (Fig. 2).

After transcytosis, how does Wg travel across wing discs? Several proteins participate. Wg can associate with lipoprotein particles, which are required for long-range signaling (Panakova et al. 2005). Endocytosis is proposed to be less efficient basolaterally, facilitating Wg diffusion (Marois et al. 2006). The Wg receptors Fz, Fz-2, and Arrow can promote Wg degradation (Han et al. 2005; Piddini et al. 2005), while the secreted hydrolase Notum can inhibit Wg signaling by modifying Dally and Dly (Giraldez et al. 2002; Han et al. 2005; Kirkpatrick et al. 2004; Kreuger et al. 2004).

It is not clear whether wing imaginal discs provide a general paradigm for Wnt transport. Even in the fly embryonic epidermis, where Wg signaling regulates cell fates (DiNardo et al. 1994), things may be different. Wg is internalized by secreting cells and recycled (Pfeiffer et al. 2002), consistent with transcytosis. However, while Dly is required for Wg signaling in this tissue (Franch-Marro et al. 2005), the inhibitory role evident in wing discs is not observed. Distinguishing general from tissue-specific mechanisms for transport of Wg and other Wnts remains an important goal.

ACROSS THE PLASMA MEMBRANE—ASSEMBLING WNT SIGNALOSOMES

Wnt/βcat signaling is generally mediated by two families of cell surface proteins: Fz family serpentine receptors, and lipoprotein receptor-related proteins (LRP) mammalian LRP5 and LRP6 and fly Arrow (Arr) (Cadigan and Liu 2006; Gordon and Nusse 2006; He et al. 2004). Forced association of Fz and LRP/Arr activates Wnt/βcat signaling (Cong et al. 2004; Holmen et al. 2005; Tolwinski et al. 2003), consistent with reports that Wnt promotes Fz-LRP6 association in vitro (Tamai et al. 2000).

Wnt signaling also promotes phosphorylation of PPPSPxS motifs in LRP6's cytoplasmic tail by GSK3 and CKI (Tamai et al. 2004; Davidson et al. 2005; Zeng et al. 2005). LRP6 contains five PPPSPxP motifs and systematic mutagenesis revealed that all five contribute to signaling activity (MacDonald et al. 2008). Phosphorylated LRP6 recruits Axin to the plasma membrane, presumably inactivating the Axin-APC βcat-destruction complex (Fig. 3) (Tamai et al. 2004; Zeng et al. 2005).

Wnt stimulation also induces LRP6 oligomerization into a large ribosome-size structure, termed the LRP6 signalosome (Fig. 3) (Bilic et al. 2007). Wnt induces phospho-LRP6 aggregates at the cell surface that colocalize with Dvls (Bilic et al. 2007), cytosolic proteins containing PDZ and DIX domains that are required for Wnt/βcat signaling (Wallingford and Habas 2005; Wharton 2003). Intriguingly, Dvl forms microscopic puncta in cells (Roberts et al. 2007; Wharton 2003) that are recruited to the plasma membrane by Wnt signaling by interaction with Fzs (Cong et al. 2004; Wong et al. 2003). Dvl puncta are dynamic and form through interactions between two portions of Dvl DIX domains (Schwarz-Romond et al. 2007a). Dvl dimerization is not sufficient to activate Wnt/βcat signaling; instead, higher order oligomers are required (Schwarz-Romond et al. 2007a). Although Dvl could not be detected in LRP6 signalosomes (Bilic et al. 2007), this may be because of its dynamic properties and instability in vitro.

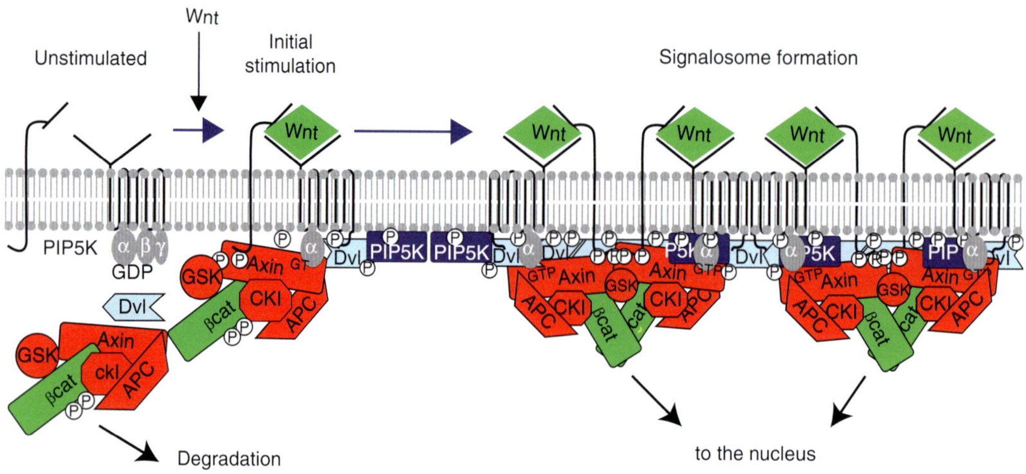

Figure 3. Model for the Wnt-induced formation of LRP5/6 signalosomes. In the absence of Wnt (at the *left*), LRP and Fz receptors are unassociated in the plasma membrane. Some data suggest that Fz may be coupled to trimeric G proteins, with the Gαo subunit in the inactive GDP-bound form. PIPKIIα (not shown for simplicity) and PIP5KI are membrane associated but inactive. The destruction complex (Axin, APC, GSK, and CKI) phosphorylates βcat, tagging the protein for ubiquitination and proteosomal degradation (see text for further explanation). Initially, Wnt promotes association of LRP5/6 and Fz receptor, leading to phosphorylation of the cytoplasmic tail of LRP5/6 by CKI and GSK3 and recruitment of the destruction complex by binding to Axin. In addition, Gαo is converted to its active GTP-bound form. One potential target of Gαo-GTP is Dvl, which is recruited to the receptor complex, where it can bind and activate PIP5KI, causing an increase in PIP2. All of these modifications appear to reinforce each other in a complex positive feedback loop, leading to a large complex of Wnt, LRP5/6, Fz, Dvl, and degradation complex at the cell surface. GSK activity is inhibited by interaction with the LRP phosphorylated tail, resulting in the accumulation of hypophosphorylated βcat, which can then translocate to the nucleus. See text for more details.

However, imaging studies in living cells support a model in which Wnt induces a large plasma membrane complex of phospho-LRP, Fz, and Dvl.

Axin is found in LRP signalosomes (Bilic et al. 2007), consistent with phospho-LRP binding Axin (Tamai et al. 2004; Zeng et al. 2005). GSK3 is also in signalosomes (Bilic et al. 2007) and Axin is required for LRP phosphorylation (Zeng et al. 2008). This suggests signalosome formation is not strictly linear. Rather, LRP phosphorylation initially recruits Axin, bringing additional GSK3 to promote more LRP phosphorylation. This positive feedback loop may be important in signalosome formation (Fig. 3); consistent with this, Dvl puncta can recruit Axin and CKI (Schwarz-Romond et al. 2007b). In *Drosophila*, biochemical evidence for Arrow signalosomes has not yet been reported. Analysis of chimeric Arrow receptors argues for distinct initiation and amplification steps in Wg signaling (Baig-Lewis et al. 2007), which could be consistent with the vertebrate model (Fig. 3).

A recent report added to the complexity of LRP6 signalosomes (Pan et al. 2008b). siRNA screening revealed that phosphatidylinositol 4-kinase type IIα (PI4KIIα) and phosphatidylinositol-4-phosphate 5-kinase type I (PIP5KI) are required for Wnt-induced LRP-phosphorylation and βcat accumulation in culture, and Wnt/βcat signaling in *Xenopus* (Pan et al. 2008b). PI4KII and PIP5KI produce phosphatidylinositol (4,5)bis-phosphate (PIP2) (Doughman et al. 2003), and delivery of PIP2 to cells stimulated Wnt signaling, whereas PIP2 depletion reduced LRP-phosphorylation by Wnt (Pan et al. 2008b). Wnt stimulation also increased PIP2 formation. This was Dvl-dependent, and recombinant Dvl activates

PIPKI in vitro, suggesting a new role for Dvl. PIP2 is required for LRP6 signalosome formation, and PIP2 is enriched in fractions containing signalosomes (Pan et al. 2008b). This provides strong evidence for PI kinases and PIP2 in Wnt receptor function (Fig. 3).

If Dvl binds and activates PIP5KI, how is Dvl activated by Wnt signaling? One possibility is through Fz-mediated G protein signaling. Studies in both mammalian cell culture (Liu et al. 2005) and *Drosophila* (Katanaev et al. 2005) suggest Gαo is required for Wnt/βcat signaling. This requirement is upstream of Dvl (Katanaev et al. 2005), suggesting a model in which information is passed from Fz to Gαo to Dvl (Fig. 3).

How do LRP signalosomes promote βcat accumulation? Wnt stimulation or LRP overexpression decrease Axin protein levels in several systems (Kofron et al. 2007; Mao et al. 2001; Tolwinski et al. 2003), which should compromise the destruction complex. LRP-mediated Axin down-regulation and inhibition of βcat degradation can be recapitulated in vitro, but LRP still stabilizes βcat when endogenous Axin is replaced with a nondegradable version (Cselenyi et al. 2008). LRP's phosphorylated tail can directly inhibit GSK3 activity (Cselenyi et al. 2008; Piao et al. 2008), which may contribute to destruction complex inhibition. Consistent with this, dephosphorylated βcat and APC are recruited to the plasma membrane on Wnt signaling (Hendriksen et al. 2008). Interestingly, dephosphorylated βcat generated by LRP6 activation is much more potent (molecule for molecule) at activating Wnt target genes than overexpressed nonphosphorylatable βcat (Hendriksen et al. 2008); thus, additional βcat activation events may occur on receptor activation.

LRP signalosomes also colocalize with Caveolin-1, a caveolae marker (Bilic et al. 2007). Functional connections between Wnt/βcat signaling and Caveolin-1 were reported, suggesting that caveolae-induced endocytosis of LRP6 is required for signaling (Yamamoto et al. 2006). However, caveolin-1, caveolin-2, and caveolin-3 knockouts are all viable, as are caveolin-1 caveolin-3 double knockouts (Le Lay and Kurzchalia 2005), suggesting that they do not play essential roles. In fact, caveolin-1 mutants were reported to have increased Wnt/βcat signaling (Sotgia et al. 2005). *Drosophila* lacks recognizable Caveolins (Le Lay and Kurzchalia 2005), suggesting that caveolae are not essential for Wg signaling.

IF IT IS IN THE TEXTBOOK, WHY ARE YOU STILL STUDYING IT?

The canonical model for Wnt regulation was an important advance, and is widely accepted and incorporated into cell biology textbooks. However, this overstates our knowledge of how things actually work: Many key questions remain. Given that Axin, which binds numerous partners, is likely the scaffold for destruction-complex assembly, APC's mechanistic role remains mysterious. Second, it is not clear where the destruction complex resides, or whether its localization changes on Wnt signaling to modulate inactivation. Third, the detailed biochemical interactions modulating βcat phosphorylation and release to the E3 ubiquitin-ligase remain to be elucidated. Finally, the multiple APC and Axin family members in many animals raise questions about differential function or redundancy.

With the realization that Axin is the scaffold for destruction-complex assembly, the search began for other APC mechanisms of action. One model is that APC helps localize the destruction complex to the correct location (Fig. 4A). There has been substantial difficulty in localizing endogenous Axin, though if overexpressed it forms large cytoplasmic puncta. GSK3 and CKI both accumulate throughout cells, reflecting their many substrates. Thus, the clearest picture came from studying APC. Both mammalian APC (Näthke et al. 1996) and *Drosophila* APC2 (McCartney et al. 1999; Yu et al. 1999) are enriched at the cortex of epithelial cells, suggesting this may be the normal location of the destruction complex. Consistent with this, missense mutations in fly *APC2* exhibit a strong correlation between loss of cortical protein localization and loss of function in Wnt regulation (McCartney et al. 2006).

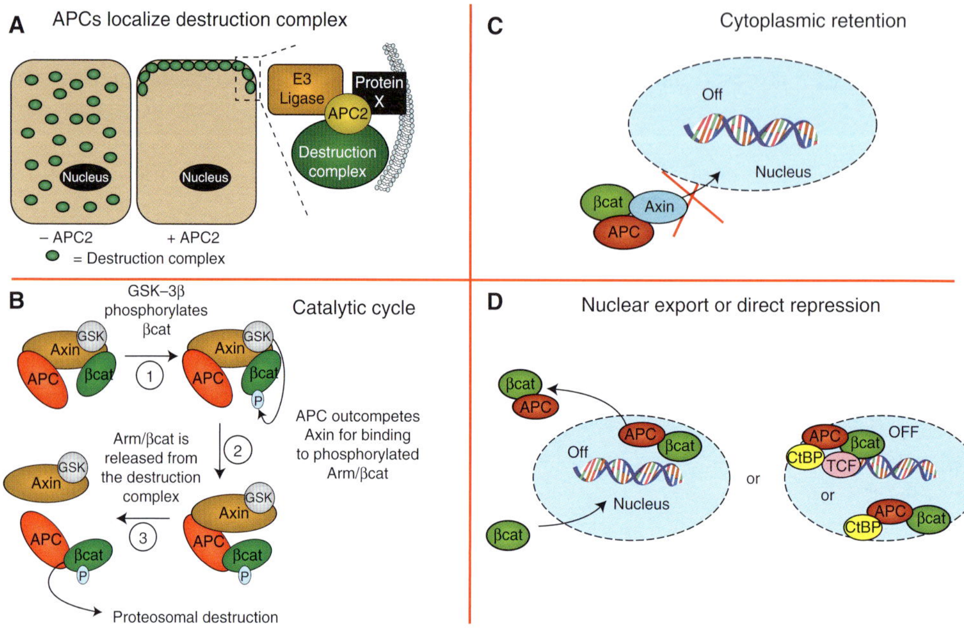

Figure 4. Four proposed models for the role of APC in the destruction complex. (*A*) APC binds the destruction complex and unknown protein X, recruiting the destruction complex to the apical cell cortex and connecting it to the E3 ubiquitin ligase. (*B*) The catalytic cycle model in which βcat is sequentially bound to Axin and then APC before transfer to the E3 ligase. (*C*) Cytoplasmic retention model. Axin and APC bind cytoplasmic βcat and tether it there. (*D*) APC in nucleus. APC exports βcat from nucleus or acts directly at promotors in transcriptional repression. See text for more details.

The apical cortex is attractive, as it brings APC into proximity of at least a subset of Fz receptors (Wu et al. 2004).

However, other data is less consistent with this model. Although fly APC1 and APC2 are redundant in many tissues (Ahmed et al. 2002; Akong et al. 2002a), their intracellular localizations are quite distinct. APC1 localizes to axons in neurons, and to centrosomes and microtubules when overexpressed (Akong et al. 2002a; Akong et al. 2002b; Hayashi et al. 1997). APC1 and APC2 have similar differential localization in male germline stem cells (Yamashita et al. 2003). How do we explain this? Possibilities include: (1) Each APC localizes to the other location at lower but still functional levels, (2) The destruction complex can function at several locations, or (3) APC does not localize the destruction complex. Overexpressed Axin recruits APC to cytoplasmic puncta (e.g., Faux et al. 2008), perhaps consistent with the latter conclusion, but colocalization of endogenous APC and Axin is difficult to detect. It is also important to remember that APC proteins have Wnt independent cytoskeletal functions (Näthke 2006), and thus many of the sites where APC proteins localize, including their predominant localization sites, may reflect these cytoskeletal functions rather than the localization of the destruction complex.

Strikingly, in *C. elegans*, localization of Wnt signaling proteins during Wnt-mediated asymmetric divisions suggests a different way of wiring the pathway (Mizumoto and Sawa 2007; Takeshita and Sawa 2005). Apr-1 (APC), Pry-1 (Axin), and Dvl homologs all localize to the cortex, and do so asymmetrically. Surprisingly, Wrm-1/βcat acts at the cortex to inhibit Wnt signaling, a striking reversal of canonical signaling, whereas cortical Apr-1 mediates nuclear "export" of Wrm-1 (these data might also be explained by cytoplasmic Wrm-1 retention). Thus, natural selection can reshape even this highly conserved pathway; similar asymmetries may occur in other animals (Schneider and Bowerman 2007).

REGULATING βCAT DESTRUCTION IS A COMPLEX TOPIC

Other models for APC function arose from structural/biochemical studies. The Weis and Xu laboratories examined affinities of βcat for both APC and Axin (Choi et al. 2006; Ha et al. 2004; Liu et al. 2006; Xing et al. 2003). The results were intriguing. Under basal conditions, βcat's affinity for Axin is higher than that for APC. However, if APC is phosphorylated by GSK3 within the βcat-binding 20-amino acid (aa) repeats, APC has a higher affinity for βcat than Axin. This led Kimelman and Xu (2006) to propose that the destruction complex goes through a cycle of structural rearrangements (Fig. 4B). It assembles with Axin bound to βcat, because of its higher affinity. βcat and APC are both phosphorylated by GSK3, triggering transfer of βcat to APC. This is suggested to facilitate βcat transfer to the E3 ubiquitin ligase, with presumed APC dephosphorylation by PP2A resetting the system. This is an intriguing model, which can be tested by site-directed mutations in APC.

In contrast, Weis' laboratory interpreted their data differently. They directly showed that Axin can be a scaffold for substrate phosphorylation (Ha et al. 2004). However, phosphorylated APC bound to βcat is not accessible to dephosphorylation (albeit by PP1, not PP2), inconsistent with the catalytic cycle model. They suggest that altered APC affinity for βcat allows it to accommodate drastically different βcat levels in the presence or absence of Wnt signal.

Recent data support a role for APC in facilitating transfer of phospho-βcat to the E3 ubiquitin ligase (Su et al. 2008). In a cell-free system, extracts from *APC* mutant colon cancer cells can phosphorylate but not ubiquitinate βcat, consistent with Axin playing the primary scaffolding role. Adding tagged APC restores ubiquitination. Added APC also restores the ability of phospho-βcat to coIP with βTrCP. Finally, APC can protect phospho-βcat from dephosphorylation. These results are quite interesting, and suggest transferring βcat from the destruction complex to the E3 ligase is an important APC role.

DECONSTRUCTING THE DESTRUCTION COMPLEX

Full mechanistic understanding requires dissecting the moving parts of the destruction machine. Several laboratories did so. Wehrli's laboratory examined Axin's mechanism of action by site-directed mutagenesis and analysis in vivo in *Drosophila* (Peterson-Nedry et al. 2008). Their results were quite surprising. Substantial Axin function was retained by proteins lacking binding sites for APC (the RGS domain) or for βcat. Even Axin lacking the GSK3 binding site retained some function. These data suggest that the destruction complex is stabilized by multiple interactions, with individual protein contacts somewhat dispensable. Strikingly, mutants lacking the PP2A binding site or Dix domain had phenotypes suggesting that these mutant proteins could not be turned off by Wnt signaling, consistent with proposed roles for PP2A and Dvl in inactivating the destruction complex. Finally, an Axin mutant lacking the βcat binding site was complemented in trans by one lacking the APC binding site, suggesting that the functional destruction complex contains multiple copies of Axin.

APCs are also complex mosaics (Näthke 2006; Polakis 2007). All share core domains that together are sufficient for Wnt regulation: Arm repeats, 15- and 20-aa repeats that bind βcat, and SAMP repeats that bind Axin. Many APCs also carry carboxy-terminal extensions, allowing interactions with microtubules, the microtubule plus-end binding protein EB1, and other partners, presumably modulating cytoskeletal functions. Colon tumors carry truncated APC proteins lacking a subset of βcat binding sites and all the Axin-binding SAMP repeats (Polakis 1995). In mice, a truncated APC lacking the cytoskeletal-interacting carboxyl terminus but retaining one SAMP motif still regulates Wnt signaling (it is homozygous viable) (Smits et al. 1999). In contrast, loss of 65 additional amino acids abrogated Wnt signaling. This complemented earlier work in cultured colon cancer cells, which also suggested key roles for 20-aa and SAMP repeats (Munemitsu et al. 1995; Rubinfeld et al. 1997).

All APCs share amino-terminal Arm repeats (Peifer et al. 1994), known binding sites for several protein partners (Näthke 2006). In *Drosophila* APC2, the Arm repeats play key roles in Wnt signaling (McCartney et al. 2006; Hamada and Bienz 2002; McCartney et al. 1999). Both the Arm repeats and the region including the 20-aa and SAMP repeats are important for APC2's cortical localization (McCartney et al. 2006), supporting the hypothesis that APC2 acts at the cortex in Wnt regulation and suggesting the existence of an unidentified cortical partner.

Two *APC2* alleles precisely mimic truncations in human tumors (Polakis 2007), allowing testing of hypotheses concerning roles of truncated proteins. Some suggested that they have dominant-negative effects on Wnt signaling or chromosome segregation, and it remained unclear whether they are null for Wnt regulation. Phenotypic comparisons with the null allele revealed that truncated APCs are severely impaired but not null for Wnt signaling. They do not have dominant-negative effects on Wnt signaling, although they can have dominant-negative effects on cytoskeletal events (McCartney et al. 2006). These data support the "just-right" hypothesis (Albuquerque et al. 2002), suggesting that selection during development of colon polyps favors mutations that reduce but do not eliminate APC regulation, producing just the right level of Wnt signal.

Behrens' laboratory explored possible roles of another conserved sequence they term the CID (Kohler et al. 2009), using a cultured cell assay pioneered by Polakis (Munemitsu et al. 1995; Rubinfeld et al. 1997). To their surprise, truncated APC lacking all SAMP repeats substantially rescued βcat regulation, in contrast to what is seen in mice (Smits et al. 1999). Instead, they saw a sharp drop in rescuing ability when they removed the CID, a sequence conserved in both insect and both mammalian APCs. As the authors point out, these data are somewhat paradoxical, because some colon tumors carry truncated APCs retaining CID. They find that different colon cancer cell lines differ in their ability to be rescued by SAMP-less fragments, suggesting that differences at other loci are important. These data also reinforce the idea that multiple protein interactions are likely important to assemble a functional destruction complex.

APC IN AND OUT OF NUCLEI

Although regulating βcat stability is critical for Wnt signaling, some data suggest additional roles for APC and Axin in Wnt regulation. One early, plausible suggestion was that the destruction complex is also a cytoplasmic anchor for βcat, reducing activation of Wnt target genes (Fig. 4C) (Tolwinski and Wieschaus 2001). Wieschaus' laboratory found that Wnt signaling alters Axin stability, thus regulating its cytoplasmic anchoring function (Tolwinski et al. 2003). Others provided evidence for an anchoring role for APCs (Ahmed et al. 2002; Akong et al. 2002a; Krieghoff et al. 2006). It seems likely that cytoplasmic retention plays an important modulatory role, working together with targeted destruction.

Others suggested APC acts in other cellular compartments. Nuclear localization and export signals on APC led to the suggestion that it regulates βcat nuclear export (Fig. 4D) (reviewed in Brocardo and Henderson 2008). Two papers support an even more direct role for APCs in nuclei (Fig. 4D). Hamada and Bienz (Hamada and Bienz 2004) identified interactions between APC and the transcriptional repressor (carboxy-terminal binding protein [CtBP]) by proteomics. They argue that nuclear CtBP, by binding APC and thus indirectly recruiting βcat, is a nuclear βcat sink (see Fig. 5), reducing TCF association and thus reducing Wnt-regulated transcription. Even more direct action is envisioned by Jones' laboratory (Sierra et al. 2006). APC can be ChIPed at the Wnt target gene c-*myc*.

Strikingly, it binds there with βTrCP and CtBP. They suggest that transient APC recruitment leads to long-term repression, and argue that carboxy-terminally truncated APCs in colon tumors, which no longer interact with CtBP, are impaired in this. These possible direct roles for APC in repressing βcat-driven transcription are intriguing. However, activating βcat mutants, lacking single phosphorylation sites and thus not targeted for destruction, can cause colorectal and other cancers (Polakis 2007), suggesting that regulating βcat stability is APC's primary role, with other roles likely to be modulatory.

DOUBLE THE FUN

Questions about APC function are further complicated by the two family members in mammals (APC and APC2) and *Drosophila* (APC1 and APC2). All four regulate βcat stability, so their shared core domains must be sufficient for this. Fly APC1 negatively regulates Wnt signaling in the eye (Ahmed et al. 1998), whereas fly APC2, which is more broadly expressed, regulates Wnt signaling in the embryonic epidermis (McCartney et al. 1999). However, in many tissues, they are functionally redundant (Ahmed et al. 2002; Akong et al. 2002a). Further, even in tissues in which one has the predominant role, removing the other enhances Wnt activation (Akong et al. 2002a; Benchabane et al. 2008). This argues that even low levels of APC, near the detection threshold, confer residual function. Redundancy among mammalian APCs could explain why *APC* mutants are found only in colorectal tumors and not other tumors in which Wnt activation is implicated (Polakis 2007); however, the lack of published mouse *APC2* mutants precludes this analysis.

Exploring functions of the two fly APCs led to further insights (Takacs et al. 2008). In screening for mutations suppressing the *APC1* eye phenotype, Ahmed's laboratory obtained *APC2* alleles. This was surprising, because in other contexts the two APCs act redundantly, and suggests that APCs play positive as well as negative roles in Wnt signaling. Their data further suggest that APC can promote Axin turnover, likely in response to Wnt signals, and suggest that this requires the Arm repeats, while the carboxy-terminal region is dispensable. These striking findings remind us that feedback regulation is likely to play a role in signaling. The mechanism for targeting Axin for proteolysis remains to be determined. The HECT domain E3 ubiquitin-ligase Edd binds to APC, but stabilizes rather than destabilizes Axin and APC (Ohshima et al. 2007) suggesting even more complexity in feedback regulation.

There are also two mammalian Axins, with presumed functional overlap. Axin loss leads to early embryonic lethality (Zeng et al. 1997). Axin2/Conductin mutants are viable with craniofacial defects in mice (Yu et al. 2005) and human patients (Lammi et al. 2004). Chia and Constantini (2005) tested functional equivalence by knocking *Axin2* into the *Axin* locus; this resulted in a viable, normal mouse. Flies have only a single Axin, which is essential for Wnt regulation (Hamada et al. 1999; Willert et al. 1999). In contrast, *C. elegans* also has two Axins. Both diverged dramatically from mammalian and fly Axins, retaining only recognizable RGS and Dix domains (Korswagen et al. 2002; Oosterveen et al. 2007). The two worm Axins functionally overlap; *axl-1* mutations enhance many *pry-1* phenotypes, while Axl-1 has phenotypes in some tissues unaffected by Pry-1 loss. The reduced sequence similarity and altered protein interactions call into question the universality of mechanisms of regulating βcat.

AND IF THAT WAS NOT COMPLEX ENOUGH, LET US ADD NEW PLAYERS

Recent work identified several other potential Wnt regulators. One is Wtx/Amer, originally identified as a gene mutated in renal Wilms tumors (Rivera et al. 2007). Other Wilms tumors have activating mutations in βcat (Koesters et al. 1999), suggesting that Wnt signaling plays a role in pathogenesis. Two recent papers independently connected Wtx to Wnt signaling. Moon's laboratory used proteomics to identify new partners of βcat, APC, and

Axin (Major et al. 2007). Among them was Wtx, which bound all three TAP-tagged baits. Wtx can also interact directly with βTrCP, suggesting that it might help bridge the destruction complex and E3 ligase. In vivo assays in *Xenopus* supported a role in negative Wnt regulation. However, changes in βcat stability and Wnt pathway activation were milder than those caused by loss of core components of the destruction machinery.

In parallel, Behrens' laboratory identified Wtx (they refer to it as Amer1) in a two-hybrid screen for proteins interacting with APC's Arm repeats (Grohmann et al. 2007). Wtx can recruit APC to the plasma membrane, perhaps by binding PIP2, and Wtx positively regulates APC stability. These two datasets are distinct but consistent—stabilizing APC would promote βcat destruction. However, much more remains to be learned about Wtx function. Wtx has a second vertebrate paralog, which might explain the relatively modest effects of knockdown in *Xenopus* and the presence of Wtx mutations in Wilms' but not other tumors (Yoo et al. 2008). Surprisingly, however, there are no Wtx homologs in flies or worms, suggesting that it is not a universal part of the pathway.

Another possible vertebrate-specific Wnt regulator is the cytoskeletal protein MCAF, which cross-links actin, intermediate filaments, and microtubules. *MACF1* mutant mice are embryonic lethal (unlike its paralog BPAG, which has postnatal neural defects) (Chen et al. 2006). Although *MACF* embryos do not die until E11.5, they arrest at E7.5, and lack mesoderm and the primitive streak. The phenotype is similar to that of Wnt3a or LRP 5 + 6 knockouts, though marker analysis in the *MACF* study is relatively limited. MACF1 can coIP with Axin, APC, βcat, and GSK3, suggesting that it associates with the destruction complex, and that *MACF1* siRNA blunts Wnt3-induced transcription and reduces βcat levels. MACF1 also coIPs with LRP5/6, suggesting a possible role in destruction complex inactivation. Consistent with this, MACF1 knockdown reduces Axin recruitment to the membrane after Wnt treatment. It will be interesting to further examine phenotypic similarities and differences between *MACF1* and other key Wnt pathway proteins. The lack of reported Wnt phenotypes of mutations in Shortstop, the fly homolog, suggest that MACF1's role in Wnt signaling may also be vertebrate-specific.

Bejsovec's laboratory discovered another novel negative regulator, Rac-GTPase-activating-protein RacGAP50C, in screening for modifiers of *Drosophila wg* mutants (Jones and Bejsovec 2005). It may act in synergy with Naked cuticle (Nkd). RacGAP50C has a known role in cytokinesis (Gregory et al. 2008; Zavortink et al. 2005), but its role in Wnt signaling is independent of this, and, surprisingly, independent of GAP activity. The mechanism by which RacGAP50C acts, and whether its mammalian homolog regulates Wnt signaling remain to be determined.

Nkd and its mammalian homologs also may differ in their importance in flies and mammals. Both bind Dvl (e.g., Rousset et al. 2001; Wharton et al. 2001), but their function is unknown. Fly Nkd is a key negative regulator (Zeng et al. 2000), although signaling is not activated to quite the same degree as is seen on complete destruction complex inactivation. Overexpressing zebrafish Nkd homologs suppresses both canonical and noncanonical Wnt signaling (Van Raay et al. 2007). Surprisingly, however, *Nkd1* and *Nkd2* single mutant mice and even *Nkd1 Nkd2* double mutants are viable and relatively normal (Zhang et al. 2007). Double mutants have subtle cranial abnormalities like those of *Axin2*. Thus, Nkd is a relatively minor modulator of mammalian Wnt signaling.

INTO THE NUCLEUS—TARGET GENE REGULATION BY βCAT/ARM

On entry into the nucleus, stabilized βcat/Arm transcriptionally regulates Wnt target genes. However, βcat/Arm cannot bind DNA. Rather, it is a coregulator, binding to transcription factors. Members of the TCF family of high mobility group (HMG) DNA-binding proteins are responsible for much of this regulation. Flies and worms have only one TCF (TCF/Pangolin and POP-1, respectively). Mammals have four TCFs (TCF1, Lef-1, TCF-3, and TCF-4). In addition to TCFs, several other transcription factors can recruit βcat to target genes, including mouse PitX1 (Kioussi et al. 2002), *Xenopus* SOX17 (Sinner et al. 2004), and *C. elegans* FOXO-1 (Essers et al. 2005). Here, we focus on the mechanism by which βcat/Arm regulates transcriptional activity of TCFs.

TCFs are thought to be bimodal regulators of Wnt targets (Fig. 5). In the absence of signal, they act with corepressors to keep Wnt target genes silenced. βcat/Arm binding to TCF antagonizes this repression and recruits additional coactivators, inducing target gene expression. Thus, TCF is a transcriptional switch, with βcat/Arm converting repression to activation (reviewed in Parker et al. 2007; Stadeli et al. 2006; Willert and Jones 2006).

TCFs recognize specific DNA sequences through their HMG domains (Laudet et al. 1993). Several studies of different TCFs largely agree that CCTTTGAT is a high affinity site for the HMG domain (Barolo 2006). However,

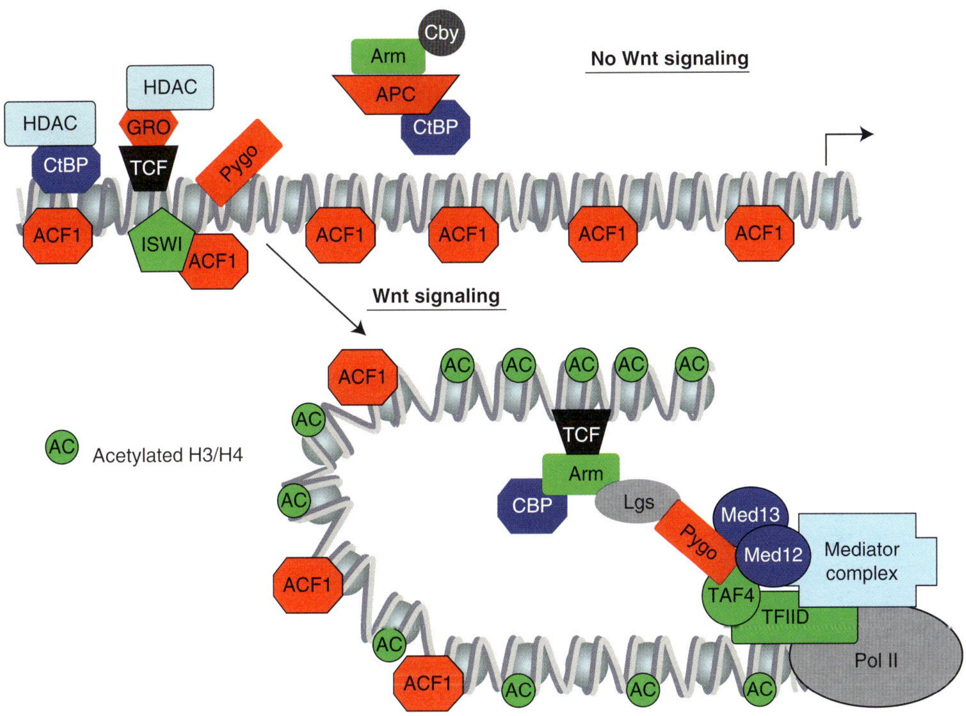

Figure 5. Model for target gene regulation by Wg signaling in *Drosophila*. The top part of the figure outlines some of the mechanisms by which target genes are repressed in the absence of Wg signaling. TCF recruits Gro through direct binding, while CtBP is recruited to WREs in an TCF-independent manner. Both corepressors may act by recruiting HDACs. In addition, the ISWI-ACF1 is recruited to the WRE in a TCF-independent way. Although ISWI remains near the WRE, ACF1 spreads across the loci. Low levels of nuclear Armadillo (Arm) are diverted away from TCF by an APC-CtBP complex or Cby. When high levels of Arm accumulate in the nucleus after Wg signaling, Arm binds to TCF, displacing Gro and recruiting additional coactivators. For simplicity, only two are shown in the bottom part of the figure. The histone acetyltransferase CBP is recruited by the carboxy-terminal Arm transactivation domain and causes a widespread increase in histone acetylation at Wg target loci. ISWI and ACF1 are bound less efficiently to acetylated histone chromatin. The amino-terminal transactivation domain of Arm recruits Lgs and Pygo. Pygo then contacts the Med12 and Med13 subunits of the mediator complex, as well as the TAF4 subunit of the TFIID basal transcription complex. Thus, Pygo acts as an adaptor, connecting the TCF activation complex with the basal transcription machinery to activate target gene expression.

TCF can also bind motifs differing from this consensus (Hallikas et al. 2006). In addition, several Wnt response elements (WREs) in flies have functional TCF-binding sites varying significantly from the consensus (Chang et al. 2008a; Lee and Frasch 2000; Riese et al. 1997). Given the degeneracy of DNA recognition by TCF, potential binding sites are found with high frequency throughout the genome. This suggests that additional sequence information may exist to specify which TCF sites can act as WREs.

For some TCFs, an additional DNA-binding domain was recently shown to provide greater DNA-binding specificity. Some mammalian TCF isoforms (TCF-1E and TCF-4E) contain an additional motif carboxy-terminal to the HMG domain, allowing them to bind an extended DNA sequence (Atcha et al. 2007). This 30-residue motif—the C-clamp—is highly conserved in invertebrate TCFs (Atcha et al. 2007). The extended sequence it binds (RCCG; R=A or G) resembles the sequence of "TCF helper sites" (GCCGCCR), recently discovered in several fly WREs, where they are as essential for Wnt responsiveness as classic TCF-binding sites (Chang et al. 2008b). These WREs require the TCF C-clamp for activation, and in vitro DNA binding studies suggest fly TCF binds in a bipartite manner, with the HMG domain binding the classic site and the C-clamp binding the Helper site (Chang et al. 2008b). This mechanism appears to be essential in flies (where all major TCF isoforms contain a C-clamp) and raises the question of how vertebrate TCFs lacking a C-clamp efficiently locate their targets.

In the absence of Wnt stimulation, TCF is thought to silence target gene expression by recruiting corepressors (Fig. 5). TCFs can bind TLE/Groucho/Grg family corepressors (Cavallo et al. 1998; Daniels and Weis 2005; Roose et al. 1998). Flies only contain one family member (Groucho), which clearly contributes to silencing in the absence of Wnt signaling (Cavallo et al. 1998; Fang et al. 2006; Mieszczanek et al. 2008). TLE and βcat compete for binding to TCF (Daniels and Weis 2005). Consistent with this, TLE1 can occupy Wnt target gene chromatin in a reciprocal manner as βcat (Sierra et al. 2006). Like TLEs, mouse myeloid translocation gene related-1 (Mtgr1) was recently reported to bind to TCF4 and contribute to target gene silencing (Moore et al. 2008).

In addition to this corepressor mechanism, other factors contribute to silencing Wnt target genes in flies by acting in parallel to TCF (Fig. 5). CtBP is required for silencing Wnt targets, and is at WREs in a pattern similar to fly TCF (Fang et al. 2006). However, depleting TCF by RNAi had no effect on CtBP occupancy and CtBP appears to repress target gene expression in parallel to TCF/Groucho (Fang et al. 2006). The DNA-dependent ATPase ISWI, part of the ACF chromatin remodeling complex, is required for repression of Wnt targets in flies (Liu et al. 2008). Like CtBP, it is found at WREs and its localization is TCF-independent (Liu et al. 2008). Presumably, ISWI silences Wnt targets by repositioning nucleosomes at WREs. Intriguingly, although ISWI is localized to WREs, the ACF1 subunit of ACF is broadly distributed across Wnt target loci (Liu et al. 2008) (Fig. 5).

Repression of Wnt targets is relieved when sufficient βcat enters nuclei and binds TCF. TCF's amino terminus directly binds βcat's Arm repeats in vitro (Daniels and Weis 2002; Graham et al. 2000), but several other factors regulate this association in cells. Chibby and ICAT bind to βcat's carboxy-terminal region, preventing TCF-βcat interaction (Tago et al. 2000; Takemaru et al. 2003). Sox 9 can also compete with TCFs for βcat binding (Akiyama et al. 2004), and a complex of APC and CtBP can bind nuclear βcat and divert it from TCF binding (Hamada and Bienz 2004). These proteins act as TCF-βcat "buffers," ensuring the proper threshold of βcat is achieved before it complexes with TCFs (Fig. 5).

In addition to proteins antagonizing TCF–βcat interactions, recent reports identified factors required to stabilize the complex on target gene chromatin. The related transducin beta-like proteins TBL1 and TBLR1 are required to stabilize TCF-βcat on target genes in mammalian and fly cell culture. They appear to do so through direct interactions with both TCF and βcat (Li and Wang 2008). In mammalian cells and zebrafish embryos, nuclear Dvl and c-Jun form a complex with TCF and βcat to promote target gene expression (Gan et al. 2008). These studies suggest that TCF-βcat interaction on its own is not sufficient under physiological conditions, and needs additional protein-protein contacts to remain on target chromatin long enough to activate gene expression.

After recruitment to TCFs at target gene loci, βcat/Arm acts as a platform for recruiting additional coactivators (Fig. 5). Many βcat-binding factors are implicated in activating Wnt targets (reviewed in Parker et al. 2007; Stadeli et al. 2006; Willert and Jones 2006). Many, such as the histone acetyltransferases CBP and p300 (Hecht et al. 2000; Li et al. 2007b; Sun et al. 2000; Takemaru and Moon 2000) and Parafibromin/Hyrax (Mosimann et al. 2006), bind the last two Arm repeats and carboxyl terminus of βcat and promote target gene activation. In addition, the amino-terminal portion of βcat/Arm recruits Legless (Lgl) in flies (Hoffmans and Basler 2004; Kramps et al. 2002) and BCL9 or BCL9-2 in vertebrates (Brembeck et al. 2004; Hoffmans and Basler 2007; Kramps et al. 2002; Sustmann et al. 2008). Lgl/BCL9 in turn recruits the PHD-finger protein Pygopus (Pygo) to the activation complex (Kramps et al. 2002; Stadeli and Basler 2005). These studies suggest that βcat/Arm's amino and carboxyl termini are both required to recruit distinct coactivators for activation of Wnt targets.

Intensive study of fly Pygo offers a glimpse of the complexities of transcriptional activation. In addition to Pygo's PHD domain, which binds Lgs, Pygo also contains a conserved amino-terminal homology domain (NHD). This is proposed to promote Wnt target activation by binding the mediator complex subunits Med12 and Med13 (Fig. 5) (Carrera et al. 2008), as well as the TFIID complex subunit TAF4 (Wright and Tjian 2009). In addition, Pygo's PHD domain can bind histone H3 methylated at lysine 4 (H3K4me). This is dependent on Lgs and is important for Pygo function in vivo (Fiedler et al. 2008). In addition to a direct role in connecting the TCF-Arm-Lgs complex to Mediator and basal transcription machinery, Pygo and Lgs are also reported to regulate Arm nuclear translocation (Townsley et al. 2004). Furthermore, Pygo is found at WREs in the absence of signaling (de la Roche and Bienz 2007) and functions as an antirepressor, counteracting Groucho (Mieszczanek et al. 2008). Thus, Pygo acts at multiple levels to promote activation of Wnt target genes in flies.

Given the several roles it plays in Wnt signaling, it is perhaps not surprising that Pygo is required for Wnt signaling throughout fly development (Belenkaya et al. 2002; Kramps et al. 2002; Parker et al. 2002; Thompson et al. 2002). However, even though Pygo's biochemical properties are conserved in mammalian Pygo1 and Pygo2 (Fiedler et al. 2008; Kramps et al. 2002), *pygo1*, *pygo2* or double knockouts in mice have a surprisingly mild decrease in Wnt signaling (Li et al. 2007a; Schwab et al. 2007; Song et al. 2007). Perhaps this is because mouse BCL9 contains its own transactivation domain (Sustmann et al. 2008). As the field becomes more sophisticated in understanding how Wnt target genes are activated, discrepancies between different phyla may become more commonplace.

Although extensive protein–protein interactions are clearly critical for TCF-βcat to activate target gene expression, chromatin modifications also play a key role. Several studies in mammals showed an increase in acetylation of the histone H3 and H4 N-termini at WREs in response to pathway activation (Fig. 5) (Feng et al. 2003; Kioussi et al. 2002; Sierra et al. 2006), consistent with correlation between histone acetylation and gene activation (Grewal and Moazed 2003; Robyr et al. 2002). Increased H3 and H4 acetylation was also observed in fly cells, but here the increase was observed over the entire target gene, up to 30 kB away from the WREs (Parker et al. 2008). This increased histone acetylation still occurred when transcription was blocked, but was dependent on CBP. Interestingly, CBP histone acetyltransferase was restricted to the WRE (Parker et al. 2008). The purpose of widespread histone acetylation is not clear, though it appears to be needed to displace ACF1, which is broadly distributed across silenced Wnt targets (Liu et al. 2008).

Finally, TCF-βcat can also directly repress transcription (Hoverter and Waterman 2008); targets include *decapentaplegic* in fly imaginal discs (Theisen et al. 2007), *E-cadherin* in mouse keratinocytes (Jamora et al. 2003), and *p16INK4a* in melanomas (Delmas et al. 2007). In these cases, TCF acts through traditional sites, but TCF-Arm repression of *Ugt36Bc* in fly hemocytes occurs through highly divergent sites (Blauwkamp et al. 2008). Converting these divergent sites to classic ones results in Wnt signaling activating this WRE, arguing that DNA allosterically regulates the TCF-Arm complex in a profound manner. The prevalence of Wnt-mediated direct repression relative to the more commonly recognized transcriptional activation is one of many important questions requiring further study.

CONCLUSIONS

The past 5 years provided dramatic new insights into the mechanisms of Wnt signaling and its roles in development and disease. However, they also raised many new questions that promise to make the next 5 years equally exciting. For example, the roles of lipid modifications and the mechanisms of Wnt transport remain to be clarified. Signaling by Fz and LRP/Arrow, including possible roles for G-proteins, the function of Dvl, and the mechanisms by which the destruction complex is inactivated, are active areas of research. Interesting questions remain about the function of APC in the destruction complex, the location at which βcat regulation occurs, and the nature of the catalytic cycle. Within the nucleus, hot topics include how TCF selects sites from the entire genome, how TCF complexes mediate both activation and repression, and which transcriptional partners are general and which are target gene or tissue specific. Despite the size of the Wnt community, we will have our hands full addressing these issues.

ACKNOWLEDGMENTS

We apologize to colleagues whose work was not cited because of space constraints. We thank David Roberts for comments and help with Figures 1 and 3. Work on Wnt signaling in the Peifer laboratory is supported by National Institutes of Health (NIH) RO1 GM67236 and in the Cadigan laboratory by NIH RO1 GM82994.

REFERENCES

Ahmed Y, Hayashi S, Levine A, Wieschaus E. 1998. Regulation of Armadillo by a *Drosophila* APC inhibits neuronal apoptosis during retinal development. *Cell* **93:** 1171–1182.

Ahmed Y, Nouri A, Wieschaus E. 2002. Drosophila Apc1 and Apc2 regulate Wingless transduction throughout development. *Development* **129:** 1751–1762.

Akiyama H, Lyons JP, Mori-Akiyama Y, Yang X, Zhang R, Zhang Z, Deng JM, Taketo MM, Nakamura T, Behringer RR, et al. 2004. Interactions between Sox9 and β-catenin control chondrocyte differentiation. *Genes Dev* **18:** 1072–1087.

Akong K, Grevengoed E, Price M, McCartney B, Hayden M, DeNofrio J, Peifer M. 2002a. Drosophila APC2 and APC1 play overlapping roles in wingless signaling in the embryo and imaginal discs. *Dev Biol* **250:** 91–100.

Akong K, McCartney B, Peifer M. 2002b. Drosophila APC2 and APC1 have overlapping roles in the larval brain despite their distinct intracellular localizations. *Dev Biol* **250:** 71–90.

Albuquerque C, Breukel C, van der Luijt R, Fidalgo P, Lage P, Slors FJ, Leitao CN, Fodde R, Smits R. 2002. The 'just-right' signaling model: APC somatic mutations are selected based on a specific level of activation of the β-catenin signaling cascade. *Hum Mol Genet* **11:** 1549–1560.

Arce L, Yokoyama NN, Waterman ML. 2006. Diversity of LEF/TCF action in development and disease. *Oncogene* **25:** 7492–7504.

Atcha FA, Syed A, Wu B, Hoverter NP, Yokoyama NN, Ting JH, Munguia JE, Mangalam HJ, Marsh JL, Waterman ML. 2007. A unique DNA binding domain converts T-cell factors into strong Wnt effectors. *Mol Cell Biol* **27:** 8352–8363.

Babuke T, Tikkanen R. 2007. Dissecting the molecular function of reggie/flotillin proteins. *Eur J Cell Biol* **86:** 525–532.

Baig-Lewis S, Peterson-Nedry W, Wehrli M. 2007. Wingless/Wnt signal transduction requires distinct initiation and amplification steps that both depend on Arrow/LRP. *Dev Biol* **306:** 94–111.

Banziger C, Soldini D, Schutt C, Zipperlen P, Hausmann G, Basler K. 2006. Wntless, a conserved membrane protein dedicated to the secretion of Wnt proteins from signaling cells. *Cell* **125:** 509–522.

Barolo S. 2006. Transgenic Wnt/TCF pathway reporters: All you need is Lef? *Oncogene* **25:** 7505–7511.

Bartscherer K, Boutros M. 2008. Regulation of Wnt protein secretion and its role in gradient formation. *EMBO Rep* **9:** 977–982.

Belenkaya TY, Han C, Standley HJ, Lin X, Houston DW, Heasman J, Lin X. 2002. pygopus Encodes a nuclear protein essential for wingless/Wnt signaling. *Development* **129:** 4089–4101.

Belenkaya TY, Wu Y, Tang X, Zhou B, Cheng L, Sharma YV, Yan D, Selva EM, Lin X. 2008. The retromer complex influences Wnt secretion by recycling wntless from endosomes to the trans-Golgi network. *Dev Cell* **14:** 120–131.

Benchabane H, Hughes EG, Takacs CM, Baird JR, Ahmed Y. 2008. Adenomatous polyposis coli is present near the minimal level required for accurate graded responses to the Wingless morphogen. *Development* **135:** 963–971.

Bilic J, Huang YL, Davidson G, Zimmermann T, Cruciat CM, Bienz M, Niehrs C. 2007. Wnt induces LRP6 signalosomes and promotes dishevelled-dependent LRP6 phosphorylation. *Science* **316:** 1619–1622.

Blair SS. 2005. Cell signaling: Wingless and glypicans together again. *Curr Biol* **15:** R92–94.

Blauwkamp TA, Chang MV, Cadigan KM. 2008. Novel TCF-binding sites specify transcriptional repression by Wnt signalling. *Embo J* **27:** 1436–1446.

Brembeck FH, Schwarz-Romond T, Bakkers J, Wilhelm S, Hammerschmidt M, Birchmeier W. 2004. Essential role of BCL9–2 in the switch between β-catenin's adhesive and transcriptional functions. *Genes Dev* **18:** 2225–2230.

Brocardo M, Henderson BR. 2008. APC shuttling to the membrane, nucleus and beyond. *Trends Cell Biol* **18:** 587–596.

Burrus LW, McMahon AP. 1995. Biochemical analysis of murine Wnt proteins reveals both shared and distinct properties. *Exp Cell Res* **220:** 363–373.

Cabrera CV, Alonso MC, Johnston P, Phillips RG, Lawrence PA. 1987. Phenocopies induced with antisense RNA identify the *wingless* gene. *Cell* **50:** 659–663.

Cadigan KM. 2002. Regulating morphogen gradients in the Drosophila wing. *Semin Cell Dev Biol* **13:** 83–90.

Cadigan KM, Liu YI. 2006. Wnt signaling: Complexity at the surface. *J Cell Sci* **119:** 395–402.

Capdevila J, Izpisua Belmonte JC. 2001. Patterning mechanisms controlling vertebrate limb development. *Annu Rev Cell Dev Biol* **17:** 87–132.

Carrera I, Janody F, Leeds N, Duveau F, Treisman JE. 2008. Pygopus activates Wingless target gene transcription through the mediator complex subunits Med12 and Med13. *Proc Natl Acad Sci* **105:** 6644–6649.

Cavallo RA, Cox RT, Moline MM, Roose J, Polevoy GA, Clevers H, Peifer M, Bejsovec A. 1998. Drosophila Tcf and Groucho interact to repress Wingless signalling activity. *Nature* **395:** 604–608.

Cayuso J, Marti E. 2005. Morphogens in motion: Growth control of the neural tube. *J Neurobiol* **64:** 376–387.

Chang JL, Chang MV, Barolo S, Cadigan KM. 2008a. Regulation of the feedback antagonist naked cuticle by Wingless signaling. *Dev Biol* **321:** 446–454.

Chang MV, Chang JL, Gangopadhyay A, Shearer A, Cadigan KM. 2008b. Activation of wingless targets requires bipartite recognition of DNA by TCF. *Curr Biol* **18:** 1877–1881.

Chen HJ, Lin CM, Lin CS, Perez-Olle R, Leung CL, Liem RK. 2006. The role of microtubule actin cross-linking factor 1 (MACF1) in the Wnt signaling pathway. *Genes Dev* **20:** 1933–1945.

Chia IV, Costantini F. 2005. Mouse axin and axin2/conductin proteins are functionally equivalent in vivo. *Mol Cell Biol* **25:** 4371–4376.

Ching W, Nusse R. 2006. A dedicated Wnt secretion factor. *Cell* **125:** 432–433.

Ching W, Hang HC, Nusse R. 2008. Lipid-independent secretion of a Drosophila Wnt protein. *J Biol Chem* **283:** 17092–17098.

Choi HJ, Huber AH, Weis WI. 2006. Thermodynamics of β-catenin-ligand interactions: The roles of the N-and C-terminal tails in modulating binding affinity. *J Biol Chem* **281:** 1027–1038.

Clevers H. 2006. Wnt/β-catenin signaling in development and disease. *Cell* **127:** 469–480.

Cong F, Schweizer L, Varmus H. 2004. Wnt signals across the plasma membrane to activate the β-catenin pathway by forming oligomers containing its receptors, Frizzled and LRP. *Development* **131:** 5103–5115.

Cselenyi CS, Jernigan KK, Tahinci E, Thorne CA, Lee LA, Lee E. 2008. LRP6 transduces a canonical Wnt signal independently of Axin degradation by inhibiting GSK3's phosphorylation of β-catenin. *Proc Natl Acad Sci* **105:** 8032–8037.

Daniels DL, Weis WI. 2002. ICAT inhibits β-catenin binding to Tcf/Lef-family transcription factors and the general coactivator p300 using independent structural modules. *Mol Cell* **10:** 573–584.

Daniels DL, Weis WI. 2005. β-Catenin directly displaces Groucho/TLE repressors from Tcf/Lef in Wnt-mediated transcription activation. *Nat Struct Mol Biol* **12:** 364–371.

Davidson G, Wu W, Shen J, Bilic J, Fenger U, Stannek P, Glinka A, Niehrs C. 2005. Casein kinase 1γ couples Wnt receptor activation to cytoplasmic signal transduction. *Nature* **438:** 867–872.

de la Roche M, Bienz M. 2007. Wingless-independent association of Pygopus with dTCF target genes. *Curr Biol* **17:** 556–561.

Delmas V, Beermann F, Martinozzi S, Carreira S, Ackermann J, Kumasaka M, Denat L, Goodall J, Luciani F, Viros A, et al. 2007. β-Catenin induces immortalization of melanocytes by suppressing p16INK4a expression and cooperates with N-Ras in melanoma development. *Genes Dev* **21:** 2923–2935.

DiNardo S, Heemskerk J, Dougan S, O'Farrell PH. 1994. The making of a maggot: Patterning the Drosophila embryonic epidermis. *Curr Opin Genet Dev* **4:** 529–534.

Doughman RL, Firestone AJ, Anderson RA. 2003. Phosphatidylinositol phosphate kinases put PI4,5P(2) in its place. *J Membr Biol* **194:** 77–89.

Eaton S. 2008. Retromer retrieves wntless. *Dev Cell* **14:** 4–6.

Essers MA, de Vries-Smits LM, Barker N, Polderman PE, Burgering BM, Korswagen HC. 2005. Functional interaction between β-catenin and FOXO in oxidative stress signaling. *Science* **308:** 1181–1184.

Fang M, Li J, Blauwkamp T, Bhambhani C, Campbell N, Cadigan KM. 2006. C-terminal-binding protein directly activates and represses Wnt transcriptional targets in Drosophila. *Embo J* **25:** 2735–2745.

Fanto M, McNeill H. 2004. Planar polarity from flies to vertebrates. *J Cell Sci* **117:** 527–533.

Faux MC, Coates JL, Catimel B, Cody S, Clayton AH, Layton MJ, Burgess AW. 2008. Recruitment of adenomatous polyposis coli and β-catenin to axin-puncta. *Oncogene* **27:** 5808–5820.

Feng Y, Lee N, Fearon ER. 2003. TIP49 regulates β-catenin-mediated neoplastic transformation and T-cell factor target gene induction via effects on chromatin remodeling. *Cancer Res* **63:** 8726–8734.

Fiedler M, Sanchez-Barrena MJ, Nekrasov M, Mieszczanek J, Rybin V, Muller J, Evans P, Bienz M. 2008. Decoding of methylated histone H3 tail by the Pygo-BCL9 Wnt signaling complex. *Mol Cell* **30:** 507–518.

Franch-Marro X, Marchand O, Piddini E, Ricardo S, Alexandre C, Vincent JP. 2005. Glypicans shunt the Wingless signal between local signalling and further transport. *Development* **132:** 659–666.

Franch-Marro X, Wendler F, Griffith J, Maurice MM, Vincent JP. 2008a. In vivo role of lipid adducts on Wingless. *J Cell Sci* **121:** 1587–1592.

Franch-Marro X, Wendler F, Guidato S, Griffith J, Baena-Lopez A, Itasaki N, Maurice MM, Vincent JP. 2008b. Wingless secretion requires endosome-to-Golgi retrieval of Wntless/Evi/Sprinter by the retromer complex. *Nat Cell Biol* **10:** 170–177.

Gallet A, Staccini-Lavenant L, Therond PP. 2008. Cellular trafficking of the glypican Dally-like is required for full-strength Hedgehog signaling and wingless transcytosis. *Dev Cell* **14:** 712–725.

Galli LM, Barnes TL, Secrest SS, Kadowaki T, Burrus LW. 2007. Porcupine-mediated lipid-modification regulates the activity and distribution of Wnt proteins in the chick neural tube. *Development* **134:** 3339–3348.

Gan XQ, Wang JY, Xi Y, Wu ZL, Li YP, Li L. 2008. Nuclear Dvl, c-Jun, β-catenin, and TCF form a complex leading to stabilization of β-catenin-TCF interaction. *J Cell Biol* **180:** 1087–1100.

Giraldez AJ, Copley RR, Cohen SM. 2002. HSPG modification by the secreted enzyme Notum shapes the Wingless morphogen gradient. *Dev Cell* **2:** 667–676.

Gordon MD, Nusse R. 2006. Wnt signaling: Multiple pathways, multiple receptors, and multiple transcription factors. *J Biol Chem* **281:** 22429–22433.

Graham TA, Weaver C, Mao F, Kimelman D, Xu W. 2000. Crystal structure of a β-catenin/Tcf complex. *Cell* **103:** 885–896.

Gregory SL, Ebrahimi S, Milverton J, Jones WM, Bejsovec A, Saint R. 2008. Cell division requires a direct link between microtubule-bound RacGAP and Anillin in the contractile ring. *Curr Biol* **18:** 25–29.

Grewal SI, Moazed D. 2003. Heterochromatin and epigenetic control of gene expression. *Science* **301:** 798–802.

Grohmann A, Tanneberger K, Alzner A, Schneikert J, Behrens J. 2007. AMER1 regulates the distribution of the tumor suppressor APC between microtubules and the plasma membrane. *J Cell Sci* **120:** 3738–3747.

Ha NC, Tonozuka T, Stamos JL, Choi HJ, Weis WI. 2004. Mechanism of phosphorylation-dependent binding of APC to β-catenin and its role in β-catenin degradation. *Mol Cell* **15:** 511–521.

Hallikas O, Palin K, Sinjushina N, Rautiainen R, Partanen J, Ukkonen E, Taipale J. 2006. Genome-wide prediction of mammalian enhancers based on analysis of transcription-factor binding affinity. *Cell* **124:** 47–59.

Hamada F, Bienz M. 2002. A Drosophila APC tumour suppressor homologue functions in cellular adhesion. *Nat Cell Biol* **4:** 208–213.

Hamada F, Bienz M. 2004. The APC tumor suppressor binds to C-terminal binding protein to divert nuclear β-catenin from TCF. *Dev Cell* **7:** 677–685.

Hamada F, Tomoyasu Y, Takatsu Y, Nakamura M, Nagai S, Suzuki A, Fujita F, Shibuya H, Toyoshima K, Ueno N, et al. 1999. Negative regulation of Wingless signaling by D-axin, a Drosophila homolog of axin. *Science* **283:** 1739–1742.

Han C, Yan D, Belenkaya TY, Lin X. 2005. Drosophila glypicans Dally and Dally-like shape the extracellular Wingless morphogen gradient in the wing disc. *Development* **132:** 667–679.

Hardin J, King RS. 2008. The long and the short of Wnt signaling in *C. elegans*. *Curr Opin Genet Dev* **18:** 362–367.

Hayashi S, Rubinfeld B, Souza B, Polakis P, Wieschaus E, Levine A. 1997. A *Drosophila* homolog of the tumor suppressor gene adenomatous polyposis coli down-regulates ß-catenin but its zygotic expression is not essential for the regulation of Armadillo. *Proc Natl Acad Sci* **94:** 242–247.

He X, Semenov M, Tamai K, Zeng X. 2004. LDL receptor-related proteins 5 and 6 in Wnt/β-catenin signaling: Arrows point the way. *Development* **131:** 1663–1677.

Hecht A, Vleminckx K, Stemmler MP, van Roy F, Kemler R. 2000. The p300/CBP acetyltransferases function as transcriptional coactivators of β-catenin in vertebrates. *Embo J* **19:** 1839–1850.

Hendriksen J, Jansen M, Brown CM, van der Velde H, van Ham M, Galjart N, Offerhaus GJ, Fagotto F, Fornerod M. 2008. Plasma membrane recruitment of dephosphorylated β-catenin upon activation of the Wnt pathway. *J Cell Sci* **121:** 1793–1802.

Heuberger J, Birchmeier W. 2009. Interplay of cadherin-mediated cell adhesion and canonical Wnt signaling. *Cold Spring Harb Perspect Biol* **2:** a002915.

Hoffmans R, Basler K. 2004. Identification and in vivo role of the Armadillo-Legless interaction. *Development* **131:** 4393–4400.

Hoffmans R, Basler K. 2007. BCL9–2 binds Arm/β-catenin in a Tyr142-independent manner and requires Pygopus for its function in Wg/Wnt signaling. *Mech Dev* **124:** 59–67.

Holmen SL, Robertson SA, Zylstra CR, Williams BO. 2005. Wnt-independent activation of β-catenin mediated by a Dkk1-Fz5 fusion protein. *Biochem Biophys Res Commun* **328:** 533–539.

Hoverter NP, Waterman ML. 2008. AWnt-fall for gene regulation: Repression. *Sci Signal* **1:** pe43.

Jamora C, DasGupta R, Kocieniewski P, Fuchs E. 2003. Links between signal transduction, transcription and adhesion in epithelial bud development. *Nature* **422:** 317–322.

Jones WM, Bejsovec A. 2005. RacGap50C negatively regulates wingless pathway activity during Drosophila embryonic development. *Genetics* **169:** 2075–2086.

Kadowaki T, Wilder E, Klingensmith J, Zachary K, Perrimon N. 1996. The segment polarity gene porcupine encodes a putative multitransmembrane protein involved in Wingless processing. *Genes Dev* **10:** 3116–3128.

Katanaev VL, Ponzielli R, Semeriva M, Tomlinson A. 2005. Trimeric G protein-dependent frizzled signaling in Drosophila. *Cell* **120:** 111–122.

Katanaev VL, Solis GP, Hausmann G, Buestorf S, Katanayeva N, Schrock Y, Stuermer CA, Basler K. 2008. Reggie-1/flotillin-2 promotes secretion of the long-range signalling forms of Wingless and Hedgehog in Drosophila. *Embo J* **27:** 509–521.

Kimelman D, Xu W. 2006. β-Catenin destruction complex: Insights and questions from a structural perspective. *Oncogene* **25:** 7482–7491.

Kioussi C, Briata P, Baek SH, Rose DW, Hamblet NS, Herman T, Ohgi KA, Lin C, Gleiberman A, Wang J, et al. 2002. Identification of a Wnt/Dvl/β-Catenin –> Pitx2 pathway mediating cell-type-specific proliferation during development. *Cell* **111:** 673–685.

Kirkpatrick CA, Dimitroff BD, Rawson JM, Selleck SB. 2004. Spatial regulation of Wingless morphogen distribution and signaling by Dally-like protein. *Dev Cell* **7:** 513–523.

Koesters R, Ridder R, Kopp-Schneider A, Betts D, Adams V, Niggli F, Briner J, von Knebel Doeberitz M. 1999. Mutational activation of the β-catenin proto-oncogene is a common event in the development of Wilms' tumors. *Cancer Res* **59:** 3880–3882.

Kofron M, Birsoy B, Houston D, Tao Q, Wylie C, Heasman J. 2007. Wnt11/β-catenin signaling in both oocytes and early embryos acts through LRP6-mediated regulation of axin. *Development* **134:** 503–513.

Kohler EM, Chandra SH, Behrens J, Schneikert J. 2009. β-Catenin degradation mediated by the CID domain of APC provides a model for the selection of APC mutations in colorectal, desmoid and duodenal tumours. *Hum Mol Genet* **18:** 213–226.

Kohn AD, Moon RT. 2005. Wnt and calcium signaling: β-catenin-independent pathways. *Cell Calcium* **38:** 439–446.

Komekado H, Yamamoto H, Chiba T, Kikuchi A. 2007. Glycosylation and palmitoylation of Wnt-3a are coupled to produce an active form of Wnt-3a. *Genes Cells* **12:** 521–534.

Korswagen HC, Coudreuse DY, Betist MC, van de Water S, Zivkovic D, Clevers HC. 2002. The Axin-like protein PRY-1 is a negative regulator of a canonical Wnt pathway in C. elegans. *Genes Dev* **16:** 1291–1302.

Kramps T, Peter O, Brunner E, Nellen D, Froesch B, Chatterjee S, Murone M, Zullig S, Basler K. 2002. Wnt/wingless signaling requires BCL9/legless-mediated recruitment of pygopus to the nuclear β-catenin-TCF complex. *Cell* **109.** 47–60.

Kreuger J, Perez L, Giraldez AJ, Cohen SM. 2004. Opposing activities of Dally-like glypican at high and low levels of Wingless morphogen activity. *Dev Cell* **7:** 503–512.

Krieghoff E, Behrens J, Mayr B. 2006. Nucleo-cytoplasmic distribution of β-catenin is regulated by retention. *J Cell Sci* **119:** 1453–1463.

Kurayoshi M, Yamamoto H, Izumi S, Kikuchi A. 2007. Post-translational palmitoylation and glycosylation of Wnt-5a are necessary for its signalling. *Biochem J* **402:** 515–523.

Lammi L, Arte S, Somer M, Jarvinen H, Lahermo P, Thesleff I, Pirinen S, Nieminen P. 2004. Mutations in AXIN2 cause familial tooth agenesis and predispose to colorectal cancer. *Am J Hum Genet* **74:** 1043–1050.

Langhorst MF, Reuter A, Stuermer CA. 2005. Scaffolding microdomains and beyond: The function of reggie/flotillin proteins. *Cell Mol Life Sci* **62:** 2228–2240.

Langhorst MF, Reuter A, Jaeger FA, Wippich FM, Luxenhofer G, Plattner H, Stuermer CA. 2008. Trafficking of the microdomain scaffolding protein reggie-1/flotillin-2. *Eur J Cell Biol* **87:** 211–226.

Laudet V, Stehelin D, Clevers H. 1993. Ancestry and diversity of the HMG box superfamily. *Nucleic Acids Res* **21:** 2493–2501.

Le Lay S, Kurzchalia TV. 2005. Getting rid of caveolins: Phenotypes of caveolin-deficient animals. *Biochim Biophys Acta* **1746:** 322–333.

Lee HH, Frasch M. 2000. Wingless effects mesoderm patterning and ectoderm segmentation events via induction of its downstream target sloppy paired. *Development* **127:** 5497–5508.

Li J, Wang CY. 2008. TBL1-TBLR1 and β-catenin recruit each other to Wnt target-gene promoter for transcription activation and oncogenesis. *Nat Cell Biol* **10:** 160–169.

Li B, Rheaume C, Teng A, Bilanchone V, Munguia JE, Hu M, Jessen S, Piccolo S, Waterman ML, Dai X. 2007a. Developmental phenotypes and reduced Wnt signaling in mice deficient for pygopus 2. *Genesis* **45:** 318–325.

Li J, Sutter C, Parker DS, Blauwkamp T, Fang M, Cadigan KM. 2007b. CBP/p300 are bimodal regulators of Wnt signaling. *Embo J* **26:** 2284–2294.

Liu X, Rubin JS, Kimmel AR. 2005. Rapid, Wnt-induced changes in GSK3β associations that regulate β-catenin stabilization are mediated by Gα proteins. *Curr Biol* **15:** 1989–1997.

Liu J, Xing Y, Hinds TR, Zheng J, Xu W. 2006. The third 20 amino acid repeat is the tightest binding site of APC for β-catenin. *J Mol Biol* **360:** 133–144.

Liu YI, Chang MV, Li HE, Barolo S, Chang JL, Blauwkamp TA, Cadigan KM. 2008. The chromatin remodelers ISWI and ACF1 directly repress Wingless transcriptional targets. *Dev Biol* **323:** 41–52.

MacDonald BT, Yokota C, Tamai K, Zeng X, He X. 2008. Wnt signal amplification via activity, cooperativity, and regulation of multiple intracellular PPPSP motifs in the Wnt co-receptor LRP6. *J Biol Chem* **283:** 16115–16123.

Major MB, Camp ND, Berndt JD, Yi X, Goldenberg SJ, Hubbert C, Biechele TL, Gingras AC, Zheng N, Maccoss MJ, et al. 2007. Wilms tumor suppressor WTX negatively regulates WNT/β-catenin signaling. *Science* **316:** 1043–1046.

Mao J, Wang J, Liu B, Pan W, Farr GH3rd, Flynn C, Yuan H, Takada S, Kimelman D, Li L, Wu D. 2001. Low-density lipoprotein receptor-related protein-5 binds to Axin and regulates the canonical Wnt signaling pathway. *Mol Cell* **7:** 801–809.

Marois E, Mahmoud A, Eaton S. 2006. The endocytic pathway and formation of the Wingless morphogen gradient. *Development* **133:** 307–317.

Mason JO, Kitajewski J, Varmus HE. 1992. Mutational analysis of mouse Wnt-1 identifies two temperature-sensitive alleles and attributes of Wnt-1 protein essential for transformation of a mammary cell line. *Mol Biol Cell* **3:** 521–533.

McCartney BM, Dierick HA, Kirkpatrick C, Moline MM, Baas A, Peifer M, Bejsovec A. 1999. Drosophila APC2 is a cytoskeletally-associated protein that regulates Wingless signaling in the embryonic epidermis. *J Cell Biol* **146:** 1303–1318.

McCartney BM, Price MH, Webb RL, Hayden MA, Holot LM, Zhou M, Bejsovec A, Peifer M. 2006. Testing hypotheses for the functions of APC family proteins using null and truncation alleles in Drosophila. *Development* **133:** 2407–2418.

McCrea P, Turck C, Gumbiner B. 1991. A homolog of *Drosophila armadillo* (plakoglobin) associated with the cytoplasmic tail of E-cadherin. *Science* **254:** 1359–1361.

Mieszczanek J, de la Roche M, Bienz M. 2008. A role of Pygopus as an anti-repressor in facilitating Wnt-dependent transcription. *Proc Natl Acad Sci* **105:** 19324–19329.

Mizumoto K, Sawa H. 2007. Cortical β-catenin and APC regulate asymmetric nuclear β-catenin localization during asymmetric cell division in *C. elegans*. *Dev Cell* **12:** 287–299.

Moore AC, Amann JM, Williams CS, Tahinci E, Farmer TE, Martinez JA, Yang G, Luce KS, Lee E, Hiebert SW. 2008. Myeloid translocation gene family members associate with T-cell factors (TCFs) and influence TCF-dependent transcription. *Mol Cell Biol* **28:** 977–987.

Mosimann C, Hausmann G, Basler K. 2006. Parafibromin/Hyrax activates Wnt/Wg target gene transcription by direct association with β-catenin/Armadillo. *Cell* **125:** 327–341.

Munemitsu S, Souza B, Albert I, Rubinfeld B, Polakis P. 1995. Regulation of intracellular ß-catenin levels by the adenomatous polyposis coli (APC) tumor suppressor protein. *Proc Natl Acad Sci* **92:** 3046–3050.

Näthke I. 2006. Cytoskeleton out of the cupboard: Colon cancer and cytoskeletal changes induced by loss of APC. *Nat Rev Cancer* **6:** 967–974.

Näthke IS, Adams CL, Polakis P, Sellin JH, Nelson WJ. 1996. The Adenomatous Polyposis Coli (APC) tumor suppressor protein localizes to plasma membrane sites involved in active cell migration. *J Cell Biol* **134:** 165–180.

Noordermeer J, Klingensmith J, Perrimon N, Nusse R. 1994. dishevelled and armadillo act in the wingless signalling pathway in Drosophila. *Nature* **367:** 80–83.

Nusse R. 2003. Wnts and Hedgehogs: Lipid-modified proteins and similarities in signaling mechanisms at the cell surface. *Development* **130:** 5297–5305.

Ohshima R, Ohta T, Wu W, Koike A, Iwatani T, Henderson M, Watts CK, Otsubo T. 2007. Putative tumor suppressor EDD interacts with and up-regulates APC. *Genes Cells* **12:** 1339–1345.

Oosterveen T, Coudreuse DY, Yang PT, Fraser E, Bergsma J, Dale TC, Korswagen HC. 2007. Two functionally distinct Axin-like proteins regulate canonical Wnt signaling in *C. elegans*. *Dev Biol* **308:** 438–448.

Pan CL, Baum PD, Gu M, Jorgensen EM, Clark SG, Garriga G. 2008a. C. elegans AP-2 and retromer control Wnt signaling by regulating mig-14/Wntless. *Dev Cell* **14:** 132–139.

Pan W, Choi SC, Wang H, Qin Y, Volpicelli-Daley L, Swan L, Lucast L, Khoo C, Zhang X, Li L, et al. 2008b. Wnt3a-mediated formation of phosphatidylinositol 4,5-bisphosphate regulates LRP6 phosphorylation. *Science* **321:** 1350–1353.

Panakova D, Sprong H, Marois E, Thiele C, Eaton S. 2005. Lipoprotein particles are required for Hedgehog and Wingless signalling. *Nature* **435:** 58–65.

Parker DS, Jemison J, Cadigan KM. 2002. Pygopus, a nuclear PHD-finger protein required for Wingless signaling in Drosophila. *Development* **129:** 2565–2576.

Parker DS, Blauwkamp T, Cadigan KM. 2007. Wnt/β-catenin-mediated transcriptional regulation. In *Wnt Signaling in Embryonic Development*, (ed. S. Sokol) Advances in Developmental Biology, Wassarman PM (ed), Vol. 17, pp. 1–61. San Diego: Elsevier.

Parker DS, Ni YY, Chang JL, Li J, Cadigan KM. 2008. Wingless signaling induces widespread chromatin remodeling of target loci. *Mol Cell Biol* **28:** 1815–1828.

Peifer M, Wieschaus E. 1990. The segment polarity gene *armadillo* encodes a functionally modular protein that is the *Drosophila* homolog of human plakoglobin. *Cell* **63:** 1167–1178.

Peifer M, Berg S, Reynolds AB. 1994. A repeating amino acid motif shared by proteins with diverse cellular roles. *Cell* **76:** 789–791.

Peifer M, McCrea PD, Green KJ, Wieschaus E, Gumbiner BM. 1992. The vertebrate adhesive junction proteins β-catenin and plakoglobin and the *Drosophila* segment polarity gene *armadillo* form a multigene family with similar properties. *J Cell Biol* **118:** 681–691.

Peterson-Nedry W, Erdeniz N, Kremer S, Yu J, Baig-Lewis S, Wehrli M. 2008. Unexpectedly robust assembly of the Axin destruction complex regulates Wnt/Wg signaling in Drosophila as revealed by analysis in vivo. *Dev Biol* **320:** 226–241.

Pfeiffer S, Ricardo S, Manneville JB, Alexandre C, Vincent JP. 2002. Producing cells retain and recycle Wingless in Drosophila embryos. *Curr Biol* **12:** 957–962.

Piao S, Lee SH, Kim H, Yum S, Stamos JL, Xu Y, Lee SJ, Lee J, Oh S, Han JK, et al. 2008. Direct inhibition of GSK3β by the phosphorylated cytoplasmic domain of LRP6 in Wnt/β-catenin signaling. *PLoS ONE* **3:** e4046.

Piddini E, Marshall F, Dubois L, Hirst E, Vincent JP. 2005. Arrow (LRP6) and Frizzled2 cooperate to degrade Wingless in Drosophila imaginal discs. *Development* **132:** 5479–5489.

Polakis P. 1995. Mutations in the APC gene and their implications for protein structure and function. *Curr Opin Genet Dev* **5:** 66–71.

Polakis P. 2007. The many ways of Wnt in cancer. *Curr Opin Genet Dev* **17:** 45–51.

Port F, Kuster M, Herr P, Furger E, Banziger C, Hausmann G, Basler K. 2008. Wingless secretion promotes and requires retromer-dependent cycling of Wntless. *Nat Cell Biol* **10:** 178–185.

Riese J, Yu X, Munnerlyn A, Eresh S, Hsu SC, Grosschedl R, Bienz M. 1997. LEF-1, a nuclear factor coordinating signaling inputs from wingless and decapentaplegic. *Cell* **88:** 777–787.

Rijsewijk F, Schuermann M, Wagenaar E, Parren P, Weigel D, Nusse R. 1987. The *Drosophila* homologue of the mouse mammary oncogene *int*-1 is identical to the segment polarity gene *wingless*. *Cell* **50:** 647–657.

Rivera MN, Kim WJ, Wells J, Driscoll DR, Brannigan BW, Han M, Kim JC, Feinberg AP, Gerald WL, Vargas SO, et al. 2007. An X chromosome gene, WTX, is commonly inactivated in Wilms tumor. *Science* **315:** 642–645.

Roberts DM, Slep KC, Peifer M. 2007. It takes more than two to tango: Dishevelled polymerization and Wnt signaling. *Nat Struct Mol Biol* **14:** 463–465.

Robyr D, Suka Y, Xenarios I, Kurdistani SK, Wang A, Suka N, Grunstein M. 2002. Microarray deacetylation maps determine genome-wide functions for yeast histone deacetylases. *Cell* **109:** 437–446.

Roose J, Molenaar M, Peterson J, Hurenkamp J, Brantjes H, Moerer P, van de Wetering M, Destree O, Clevers H. 1998. The *Xenopus* Wnt effector XTcf-3 interacts with Groucho-related transcriptional repressors. *Nature* **395:** 608–612.

Rousset R, Mack JA, Wharton KAJr, Axelrod JD, Cadigan KM, Fish MP, Nusse R, Scott MP. 2001. naked cuticle targets dishevelled to antagonize Wnt signal transduction. *Genes Dev* **15:** 658–671.

Rubinfeld B, Souza B, Albert I, Muller O, Chamberlain SH, Masiarz FR, Munemitsu S, Polakis P. 1993. The APC gene product associates with ß-catenin. *Science* **262:** 1731–1734.

Rubinfeld B, Albert I, Porfiri E, Munemitsu S, Polakis P. 1997. Loss of ß-Catenin Regulation by the APC Tumor Suppressor Protein Correlates with Loss of Structure Due to Common Somatic Mutations of the Gene. *Cancer Res* **57:** 4624–4630.

Schneider SQ, Bowerman B. 2007. β-Catenin asymmetries after all animal/vegetal- oriented cell divisions in *Platynereis dumerilii* embryos mediate binary cell-fate specification. *Dev Cell* **13:** 73–86.

Schwab KR, Patterson LT, Hartman HA, Song N, Lang RA, Lin X, Potter SS. 2007. Pygo1 and Pygo2 roles in Wnt signaling in mammalian kidney development. *BMC Biol* **5:** 15.

Schwarz-Romond T, Fiedler M, Shibata N, Butler PJ, Kikuchi A, Higuchi Y, Bienz M. 2007a. The DIX domain of Dishevelled confers Wnt signaling by dynamic polymerization. *Nat Struct Mol Biol* **14:** 484–492.

Schwarz-Romond T, Metcalfe C, Bienz M. 2007b. Dynamic recruitment of axin by Dishevelled protein assemblies. *J Cell Sci* **120:** 2402–2412.

Seaman MN. 2005. Recycle your receptors with retromer. *Trends Cell Biol* **15:** 68–75.

Semenov MV, Habas R, Macdonald BT, He X. 2007. SnapShot: Noncanonical Wnt signaling pathways. *Cell* **131:** 1378.

Sierra J, Yoshida T, Joazeiro CA, Jones KA. 2006. The APC tumor suppressor counteracts β-catenin activation and H3K4 methylation at Wnt target genes. *Genes Dev* **20:** 586–600.

Sinner D, Rankin S, Lee M, Zorn AM. 2004. Sox17 and β-catenin cooperate to regulate the transcription of endodermal genes. *Development* **131:** 3069–3080.

Smits R, Kielman MF, Breukel C, Zurcher C, Neufeld K, Jagmohan-Changur S, Hofland N, van Dijk J, White R, Edelmann W, et al. 1999. Apc1638T: A mouse model delineating critical domains of the adenomatous polyposis coli protein involved in tumorigenesis and development. *Genes Dev* **13:** 1309–1321.

Smolich BD, McMahon JA, McMahon AP, Papkoff J. 1993. Wnt family proteins are secreted and associated with the cell surface. *Mol Biol Cell* **4:** 1267–1275.

Song N, Schwab KR, Patterson LT, Yamaguchi T, Lin X, Potter SS, Lang RA. 2007. pygopus 2 has a crucial, Wnt pathway-independent function in lens induction. *Development* **134:** 1873–1885.

Sotgia F, Williams TM, Cohen AW, Minetti C, Pestell RG, Lisanti MP. 2005. Caveolin-1-deficient mice have an increased mammary stem cell population with upregulation of Wnt/β-catenin signaling. *Cell Cycle* **4:** 1808–1816.

Stadeli R, Basler K. 2005. Dissecting nuclear Wingless signalling: Recruitment of the transcriptional co-activator Pygopus by a chain of adaptor proteins. *Mech Dev* **122:** 1171–1182.

Stadeli R, Hoffmans R, Basler K. 2006. Transcription under the control of nuclear Arm/β-catenin. *Curr Biol* **16:** R378–385.

Strigini M, Cohen SM. 1999. Formation of morphogen gradients in the Drosophila wing. *Semin Cell Dev Biol* **10:** 335–344.

Su L-K, Vogelstein B, Kinzler KW. 1993. The APC tumor suppressor protein associates with catenins. *Science* **262:** 1734–1737.

Su Y, Fu C, Ishikawa S, Stella A, Kojima M, Shitoh K, Schreiber EM, Day BW, Liu B. 2008. APC is essential for targeting phosphorylated β catenin to the SCFβ-TrCP ubiquitin ligase. *Mol Cell* **32:** 652–661.

Sun Y, Kolligs FT, Hottiger MO, Mosavin R, Fearon ER, Nabel GJ. 2000. Regulation of β-catenin transformation by the p300 transcriptional coactivator. *Proc Natl Acad Sci* **97:** 12613–12618.

Sustmann C, Flach H, Ebert H, Eastman Q, Grosschedl R. 2008. Cell-type-specific function of BCL9 involves a transcriptional activation domain that synergizes with β-catenin. *Mol Cell Biol* **28:** 3526–3537.

Tago K, Nakamura T, Nishita M, Hyodo J, Nagai S, Murata Y, Adachi S, Ohwada S, Morishita Y, Shibuya H, Akiyama T. 2000. Inhibition of Wnt signaling by ICAT, a novel β-catenin-interacting protein. *Genes Dev* **14:** 1741–1749.

Takacs CM, Baird JR, Hughes EG, Kent SS, Benchabane H, Paik R, Ahmed Y. 2008. Dual positive and negative regulation of wingless signaling by adenomatous polyposis coli. *Science* **319:** 333–336.

Takada R, Satomi Y, Kurata T, Ueno N, Norioka S, Kondoh H, Takao T, Takada S. 2006. Monounsaturated fatty acid modification of Wnt protein: Its role in Wnt secretion. *Dev Cell* **11:** 791–801.

Takemaru KI, Moon RT. 2000. The transcriptional coactivator CBP interacts with β-catenin to activate gene expression. *J Cell Biol* **149:** 249–254.

Takemaru K, Yamaguchi S, Lee YS, Zhang Y, Carthew RW, Moon RT. 2003. Chibby, a nuclear β-catenin-associated antagonist of the Wnt/Wingless pathway. *Nature* **422:** 905–909.

Takeshita H, Sawa H. 2005. Asymmetric cortical and nuclear localizations of WRM-1/β-catenin during asymmetric cell division in *C. elegans*. *Genes Dev* **19:** 1743–1748.

Tamai K, Semenov M, Kato Y, Spokony R, Liu C, Katsuyama Y, Hess F, Saint-Jeannet JP, He X. 2000. LDL-receptor-related proteins in Wnt signal transduction. *Nature* **407:** 530–535.

Tamai K, Zeng X, Liu C, Zhang X, Harada Y, Chang Z, He X. 2004. A mechanism for Wnt coreceptor activation. *Mol Cell* **13:** 149–156.

Tanaka K, Kitagawa Y, Kadowaki T. 2002. Drosophila segment polarity gene product porcupine stimulates the posttranslational N-glycosylation of wingless in the endoplasmic reticulum. *J Biol Chem* **277:** 12816–12823.

Theisen H, Syed A, Nguyen BT, Lukacsovich T, Purcell J, Srivastava GP, Iron D, Gaudenz K, Nie Q, Wan FY, et al. 2007. Wingless directly represses DPP morphogen expression via an armadillo/TCF/Brinker complex. *PLoS ONE* **2:** e142.

Thompson B, Townsley F, Rosin-Arbesfeld R, Musisi H, Bienz M. 2002. A new nuclear component of the Wnt signalling pathway. *Nat Cell Biol* **4:** 367–373.

Tolwinski NS, Wieschaus E. 2001. Armadillo nuclear import is regulated by cytoplasmic anchor Axin and nuclear anchor dTCF/Pan. *Development* **128:** 2107–2117.

Tolwinski NS, Wehrli M, Rives A, Erdeniz N, DiNardo S, Wieschaus E. 2003. Wg/Wnt signal can be transmitted through arrow/LRP5,6 and Axin independently of Zw3/Gsk3β activity. *Dev Cell* **4:** 407–418.

Townsley FM, Cliffe A, Bienz M. 2004. Pygopus and Legless target Armadillo/β-catenin to the nucleus to enable its transcriptional co-activator function. *Nat Cell Biol* **6:** 626–633.

van Amerongen R, Mikels A, Nusse R. 2008. Alternative wnt signaling is initiated by distinct receptors. *Sci Signal* **1:** pre9.

van den Heuvel M, Harryman-Samos C, Klingensmith J, Perrimon N, Nusse R. 1993. Mutations in the segment polarity genes wingless and porcupine impair secretion of the wingless protein. *Embo J* **12:** 5293–5302.

Van Raay TJ, Coffey RJ, Solnica-Krezel L. 2007. Zebrafish Naked1 and Naked2 antagonize both canonical and non-canonical Wnt signaling. *Dev Biol* **309:** 151–168.

Veeman MT, Axelrod JD, Moon RT. 2003. A second canon. Functions and mechanisms of β-catenin-independent Wnt signaling. *Dev Cell* **5:** 367–377.

Wallingford JB, Habas R. 2005. The developmental biology of Dishevelled: An enigmatic protein governing cell fate and cell polarity. *Development* **132:** 4421–4436.

Wharton KAJr. 2003. Runnin' with the Dvl: Proteins that associate with Dsh/Dvl and their significance to Wnt signal transduction. *Dev Biol* **253:** 1–17.

Wharton KAJr, Zimmermann G, Rousset R, Scott MP. 2001. Vertebrate proteins related to *Drosophila* naked cuticle bind dishevelled and antagonize wnt signaling. *Dev Biol* **234:** 93–106.

Willert K, Jones KA. 2006. Wnt signaling: Is the party in the nucleus? *Genes Dev* **20:** 1394–1404.

Willert K, Logan CY, Arora A, Fish M, Nusse R. 1999. A Drosophila Axin homolog, Daxin, inhibits Wnt signaling. *Development* **126:** 4165–4173.

Willert K, Brown JD, Danenberg E, Duncan AW, Weissman IL, Reya T, Yates JR3rd, Nusse R. 2003. Wnt proteins are lipid-modified and can act as stem cell growth factors. *Nature* **423:** 448–452.

Wong HC, Bourdelas A, Krauss A, Lee HJ, Shao Y, Wu D, Mlodzik M, Shi DL, Zheng J. 2003. Direct binding of the PDZ domain of Dishevelled to a conserved internal sequence in the C-terminal region of Frizzled. *Mol Cell* **12:** 1251–1260.

Wright KJ, Tjian R. 2009. Wnt signaling targets ETO coactivation domain of TAF4/TFIID in vivo. *Proc Natl Acad Sci* **106:** 55–60.

Wu J, Klein TJ, Mlodzik M. 2004. Subcellular localization of frizzled receptors, mediated by their cytoplasmic tails, regulates signaling pathway specificity. *PLoS Biol* **2:** E158.

Xing Y, Clements WK, Kimelman D, Xu W. 2003. Crystal structure of a β-catenin/axin complex suggests a mechanism for the β-catenin destruction complex. *Genes Dev* **17:** 2753–2764.

Yamamoto H, Komekado H, Kikuchi A. 2006. Caveolin is necessary for Wnt-3a-dependent internalization of LRP6 and accumulation of β-catenin. *Dev Cell* **11:** 213–223.

Yamashita YM, Jones DL, Fuller MT. 2003. Orientation of Asymmetric Stem Cell Division by the APC Tumor Suppressor and Centrosome. *Science* **301:** 1547–1550.

Yang PT, Lorenowicz MJ, Silhankova M, Coudreuse DY, Betist MC, Korswagen HC. 2008. Wnt signaling requires retromer-dependent recycling of MIG-14/Wntless in Wnt-producing cells. *Dev Cell* **14:** 140–147.

Yoo NJ, Kim S, Lee SH. 2008. Mutational analysis of WTX gene in Wnt/β-catenin pathway in gastric, colorectal, and hepatocellular carcinomas. *Dig Dis Sci* doi: 10.1007/s10620-008-0458-9.

Yu X, Waltzer L, Bienz M. 1999. A new Drosophila APC homologue associated with adhesive zones of epithelial cells. *Nat Cell Biol* **1:** 144–151.

Yu HM, Jerchow B, Sheu TJ, Liu B, Costantini F, Puzas JE, Birchmeier W, Hsu W. 2005. The role of Axin2 in calvarial morphogenesis and craniosynostosis. *Development* **132:** 1995–2005.

Zavortink M, Contreras N, Addy T, Bejsovec A, Saint R. 2005. Tum/RacGAP50C provides a critical link between anaphase microtubules and the assembly of the contractile ring in *Drosophila melanogaster*. *J Cell Sci* **118:** 5381–5392.

Zecca M, Basler K, Struhl G. 1996. Direct and long-range action of a wingless morphogen gradient. *Cell* **87:** 833–844.

Zeng L, Fagatto F, Zhang T, Hsu W, Vasicek TJ, Perry WL, Lee JJ, Tilghman SM, Gumbiner BM, Constantini F. 1997. The mouse *Fused* locus encodes axin, an inhibitor of the Wnt signaling pathway that regulates embryonic axis formation. *Cell* **90:** 181–192.

Zeng W, Wharton KAJr, Mack JA, Wang K, Gadbaw M, Suyama K, Klein PS, Scott MP. 2000. naked cuticle encodes an inducible antagonist of Wnt signalling. *Nature* **403:** 789–795.

Zeng X, Tamai K, Doble B, Li S, Huang H, Habas R, Okamura H, Woodgett J, He X. 2005. A dual-kinase mechanism for Wnt co-receptor phosphorylation and activation. *Nature* **438:** 873–877.

Zeng X, Huang H, Tamai K, Zhang X, Harada Y, Yokota C, Almeida K, Wang J, Doble B, Woodgett J, et al. 2008. Initiation of Wnt signaling: Control of Wnt coreceptor Lrp6 phosphorylation/activation via frizzled, dishevelled and axin functions. *Development* **135:** 367–375.

Zhai L, Chaturvedi D, Cumberledge S. 2004. Drosophila wnt-1 undergoes a hydrophobic modification and is targeted to lipid rafts, a process that requires porcupine. *J Biol Chem* **279:** 33220–33227.

Zhang S, Cagatay T, Amanai M, Zhang M, Kline J, Castrillon DH, Ashfaq R, Oz OK, Wharton KAJr. 2007. Viable mice with compound mutations in the Wnt/Dvl pathway antagonists nkd1 and nkd2. *Mol Cell Biol* **27:** 4454–4464.

Cadherin-mediated Intercellular Adhesion and Signaling Cascades Involving Small GTPases

Takashi Watanabe[1,2], Kazuhide Sato[1], and Kozo Kaibuchi[1,3]

[1]Department of Cell Pharmacology, Nagoya University Graduate School of Medicine, 65 Tsurumai, Showa, Nagoya 466-8550, Japan

[2]Institute for Advanced Research, Nagoya University, Furo, Showa, Nagoya 464-8601, Japan

[3]JST, CREST, 4-1-8 Honcho, Kawaguchi 332-0012, Japan

Correspondence: kaibuchi@med.nagoya-u.ac.jp

Epithelia form physical barriers that separate the internal milieu of the body from its external environment. The biogenesis of functional epithelia requires the precise coordination of many cellular processes. One of the key events in epithelial biogenesis is the establishment of cadherin-dependent cell–cell contacts, which initiate morphological changes and the formation of other adhesive structures. Cadherin-mediated adhesions generate intracellular signals that control cytoskeletal reorganization, polarity, and vesicle trafficking. Among such signaling pathways, those involving small GTPases play critical roles in epithelial biogenesis. Assembly of E-cadherin activates several small GTPases and, in turn, the activated small GTPases control the effects of E-cadherin-mediated adhesions on epithelial biogenesis. Here, we focus on small GTPase signaling at E-cadherin-mediated epithelial junctions.

Cell–cell adhesions are involved in a diverse range of physiological processes, including morphological changes during tissue development, cell scattering, wound healing, and synaptogenesis (Adams and Nelson 1998; Gumbiner 2000; Halbleib and Nelson 2006; Takeichi 1995; Tepass et al. 2000). In epithelial cells, cell–cell adhesions are classified into three kinds of adhesions: adherens junction, tight junction, and desmosome (for more details, see Meng and Takeichi 2009, Furuse 2009, and Delva et al. 2009, respectively). A key event in epithelial polarization and biogenesis is the establishment of cadherin-dependent cell–cell contacts. Cadherins belong to a large family of adhesion molecules that require Ca^{2+} for their homophilic interactions (Adams and Nelson 1998; Blanpain and Fuchs 2009; Gumbiner 2000; Hartsock and Nelson 2008; Takeichi 1995; Tepass et al. 2000). Cadherins form transinteraction on the surface of neighboring cells (for details, see Shapiro and Weis 2009). For the development of strong and rigid adhesions, cadherins are clustered concomitantly with changes in the organization of the actin cytoskeleton (Tsukita et al. 1992). Classical cadherins are required, but not sufficient, to initiate cell–cell contacts, and other adhesion protein complexes subsequently assemble (for details, see Green et al. 2009).

Cite this article as *Cold Spring Harb Perspect Biol* 2009;1:a003020

These complexes include the tight junction, which controls paracellular permeability, and desmosomes, which support the structural continuum of epithelial cells. A fundamental problem is to understand how these diverse cellular processes are regulated and coordinated. Intracellular signals, generated when cells attach with one another, mediate these complicated processes.

Several signaling pathways upstream or downstream of cadherin-mediated cell–cell adhesions have been identified (Perez-Moreno et al. 2003) (see also McCrea et al. 2009). Among these pathways, small GTPases including the Rho and Ras family GTPases play critical roles in epithelial biogenesis and have been studied extensively. Many key morphological and functional changes are induced when these small GTPases act at epithelial junctions, where they mediate an interplay between cell–cell adhesion molecules and fundamental cellular processes including cytoskeletal activity, polarity, and vesicle trafficking. In addition to these small GTPases, Ca^{2+} signaling and phosphorylation of cadherin complexes also play pivotal roles in the formation and maintenance of cadherin-mediated adhesions. Here, we focus on signaling pathways involving the small GTPases in E-cadherin-mediated cell–cell adhesions. Other signaling pathways are described in recent reviews (Braga 2002; Fukata and Kaibuchi 2001; Goldstein and Macara 2007; McLachlan et al. 2007; Tsukita et al. 2008; Yap and Kovacs 2003; see also McCrea et al. 2009).

MODE OF ACTION OF THE SMALL GTPases

Small GTPases have GDP/GTP binding and GTPase activity. They cycle between a GTP-bound active state and a GDP-bound inactive state, thus functioning as molecular switches in cells (Fig. 1). The nucleotide state of the small GTPases is generally controlled by three

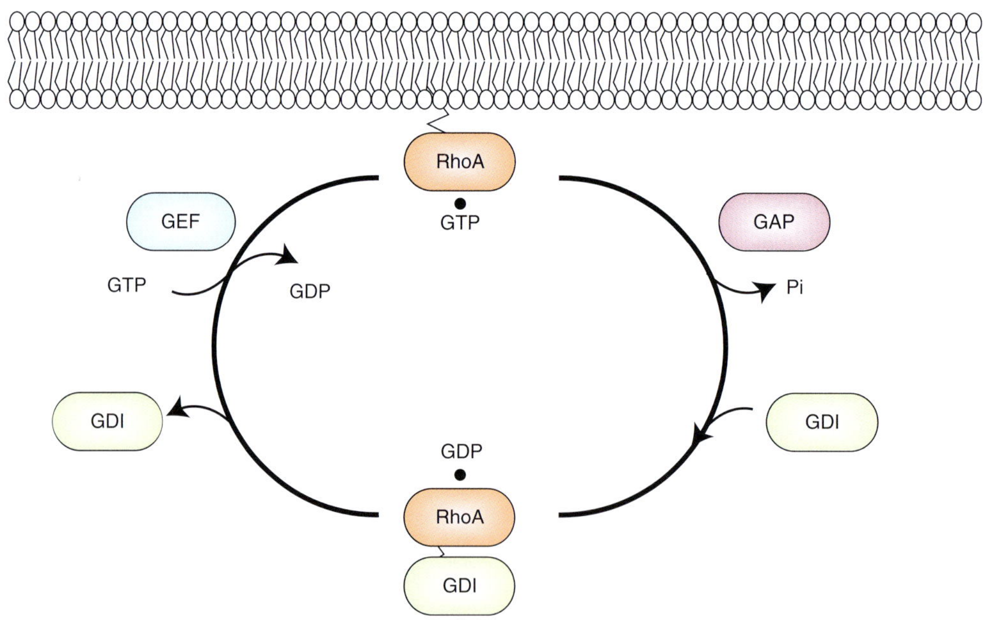

Figure 1. Regulation of the small GTPases. In this figure, RhoA is depicted. In resting cells, Rho exists mostly in the GDP-bound form (GDP· Rho) and in complexes with Rho GDIs in the cytosol. On stimulation with extracellular signals, Rho is likely to be dissociated from Rho GDIs and targeted to specific membranes by its carboxy-terminal prenyl group. At the membrane, specific GEFs for Rho are activated: GDP· Rho is then converted to GTP· Rho. GTP· Rho interacts with its specific effectors and exerts its functions. GAPs enhance the GTPase activity of Rho and reconvert Rho to its inactive GDP-bound form. Rho GDI can then form a complex with GDP· Rho and extract it from the membrane back into the cytosol.

classes of key regulators: Guanine nucleotide exchange factors (GEFs), which promote the exchange of GDP for GTP; GDP dissociation inhibitors (GDIs), which interact with GDP-bound small GTPases, inhibit the exchange of GDP for GTP, and sequester the small GTPases into the cytosol (note that a GDI for the Ras family has not been identified); and GTPase-activating proteins (GAPs), which enhance the intrinsic GTPase activity of small GTPases. These regulators ensure that activation and inactivation of small GTPases is tightly regulated both spatially and temporally in order to generate specific and localized effects (Gulli and Peter 2001; Jaffe and Hall 2005; Kaibuchi et al. 1999; Van Aelst and D'souza-Schorey 1997). The modes of action of small GTPases have been elucidated by the identification and characterization of specific effectors. Such effector molecules interact with small GTPases only in their GTP-bound state to transmit signals downstream and exert physiological functions (Gulli and Peter 2001; Jaffe and Hall 2005; Kaibuchi et al. 1999; Van Aelst and D'souza-Schorey 1997).

The Rho family GTPases are believed to shuttle between the cytosol and specific membrane sites after extracellular stimulation (Fleming et al. 1996; Kranenburg et al. 1997). In epithelial cells, Rac1 and Cdc42, which are members of the Rho family GTPases, localize to sites of cell–cell junctions (Jou and Nelson 1998; Kuroda et al. 1998; Nakagawa et al. 2001; Takaishi et al. 1997). In Madin-Darby canine kidney II (MDCKII) cells, Rac1 is colocalized with E-cadherin at sites of cell–cell contact and is translocated to the cytosol during disruption of cell–cell adhesion by Ca^{2+} chelation with EGTA (Nakagawa et al. 2001). Furthermore, studies using green fluorescent protein (GFP) show that an anti-E-cadherin antibody or the functional extracellular domain of E-cadherin fused to Fc region can recruit Rac1-GFP immediately at contact sites (Kovacs et al. 2002; Nakagawa et al. 2001; Niessen and Gumbiner 2002; Perez et al. 2008). Rac1 is also colocalized to E-cadherin-containing vesicles in keratinocytes (Akhtar and Hotchin 2001). Cdc42 is found at sites of cell–cell contact as well as the Golgi apparatus (Kroschewski et al. 1999), where it is thought to regulate the establishment of epithelial polarity during secretory and endocytic vesicle transport. On the other hand, RhoA, another member of the Rho family GTPases, localizes diffusely in the cytosol (Takaishi et al. 1997). Rap1, a member of the Ras family GTPases, accumulates at sites of cell–cell contact in an activity-independent manner. Similar to the case of Rac1, E-cadherin-Fc can recruit Rap1 to the plasma membrane (Hogan et al. 2004).

Several regulators of small GTPases accumulate at the site of cell–cell contacts through PDZ (PSD-95, Dlg, and ZO-1) domain-mediated interactions. β-catenin possesses a PDZ-binding consensus sequence at its carboxyl terminus and can recruit several PDZ domain-containing scaffolding and signaling molecules (Dobrosotskaya and James 2000; Ide et al. 1999; Kawajiri et al. 2000; Perego et al. 2000). PDZ domain-containing GEFs for small GTPases have also been shown to localize to intercellular adhesion sites. These GEFs include Tiam1 (Michiels et al. 1995), STEF (Hoshino et al. 1999), PDZ-RhoGEF (Fukuhara et al. 1999), and LARG (Kourlas et al. 2000; Taya et al. 2001). This localization of GEFs may account for the activation of the Rho family GTPases at intercellular adhesion sites (see the following).

The function of small GTPases is investigated with the use of three kinds of tools: point mutants of small GTPases to serve as constitutively active and dominant negative forms, affinity precipitation with a specific effector to monitor the level of activated small GTPases, and FRET (fluorescence resonance energy transfer)-based imaging techniques to show the spatiotemporal activation states of small GTPases. These biosensors provide valuable insights because small GTPases are activated separately in space and time within a single epithelial cell. Recent RNAi-based methods and the availability of knockout mice provide additional approaches to examine effects of the loss of function of small GTPases. These tools have increased our understanding of the roles of small GTPases in epithelial as well as nonepithelial cells, such as fibroblasts and neuronal cells (Braga and Yap 2005; Fukata and Kaibuchi 2001; Kiyokawa et al. 2006; Pertz and Hahn 2004; Van Aelst and Symons 2002).

DE NOVO CADHERIN-MEDIATED ADHESIONS CONTROL THE ACTIVITY OF SMALL GTPases

Like stimulation by growth factors, the engagement of integrins with fibronectin induces the activation of small GTPases (Price et al. 1998). Analogies with integrin-mediated cell–substratum adhesions lead to the speculation that cadherins might also function as receptors that transduce signals to small GTPases (outside-in-signal) (Table 1). Indeed, E-cadherin-mediated cell–cell interactions result in the rapid activation of Cdc42 in MCF-7 epithelial cells (Kim et al. 2000). Homophilic E-cadherin ligations recruit Rac1 at adhesion sites and induce its activation in MDCKII cells, and basal levels of active Rac1 are still maintained even when cells become confluent (Betson et al. 2002; Braga et al. 1997; Nakagawa et al. 2001; Noren et al. 2001). These findings raise the question of how cadherin-mediated cell–cell adhesions modulate the activity of Rho family GTPases. It has been reported that E-cadherin-mediated cell–cell adhesions stimulate phosphatidylinositol 3-kinase (PI3K) activity in MDCKII cells (Pece et al. 1999). Moreover, PI3K has been shown to interact with E-cadherin (Pece et al. 1999) and β-catenin (Espada et al. 1999). Because PI3K is thought to function upstream of Rac1 (Kotani et al. 1994), these observations indicate the possible involvement of PI3K in E-cadherin-dependent Rac1 activation. Indeed, wortmannin—an inhibitor of PI3K—inhibits E-cadherin-induced Rac1 activation but does not affect the localization of Rac1 or E-cadherin (Nakagawa et al. 2001).

Activation of Rac1 through PI3K by E-cadherin-mediated cell–cell adhesions seems to require at least two steps (Fig. 2): (1) Rac1 recruitment to sites of cell–cell contacts, and (2) Rac1 activation by a GEF that responds to PI3K products. Consistently, Rac1 rapidly accumulates at the sites of E-cadherin engagement (1–3 min), and this accumulation is independent of Rac1 GTP binding/hydrolysis (Perez et al. 2008). Considering that Tiam1 is localized at sites of cell–cell contacts and functions downstream of PI3K, Tiam1 appears to act as a Rac1 GEF that functions downstream of E-cadherin engagement. Another possible mechanism for the regulation of Rho family GTPases involves p120ctn, which binds directly to the juxtamembrane domain of E-cadherin. p120ctn likely activates Rac1 and Cdc42 and inhibits RhoA. Vav2, a RhoGEF, directly binds to p120ctn, which may account for the ability of p120ctn to activate Rac1 and Cdc42 (Noren et al. 2000). Because Rac1 antagonizes RhoA through a mechanism that involves p190 RhoGAP, it is possible that the inhibition of RhoA by p120ctn occurs indirectly through the activation of Rac1 (Bustos et al. 2008; Nimnual et al. 2003). However, it remains controversial how p120ctn regulates Rho family GTPases (Anastasiadis 2007; Reynolds 2007; Wildenberg et al. 2006).

RhoA is also activated during calcium-induced keratinocyte differentiation (Calautti et al. 2002). Localization of Rac1 and RhoA activity using FRET biosensors shows

Table 1. Activation of small GTPases by cadherin

Small GTPase	E-cadherin	N-cadherin	VE-cadherin	R-cadherin	C-cadherin
Cdc42	+ (MCF-7)	–	+	–/+	–
Rac1	+	–	+	+	+
RhoA	+	+	–	–	–
Rap1	+	ND	+	ND	ND

Activation of small GTPases by various cadherins. Increased levels of activated Cdc42, Rac1, RhoA, and Rap1 are assessed by biochemical or FRET-based methods. (ND) not determined; (+) activated; (–) not activated.

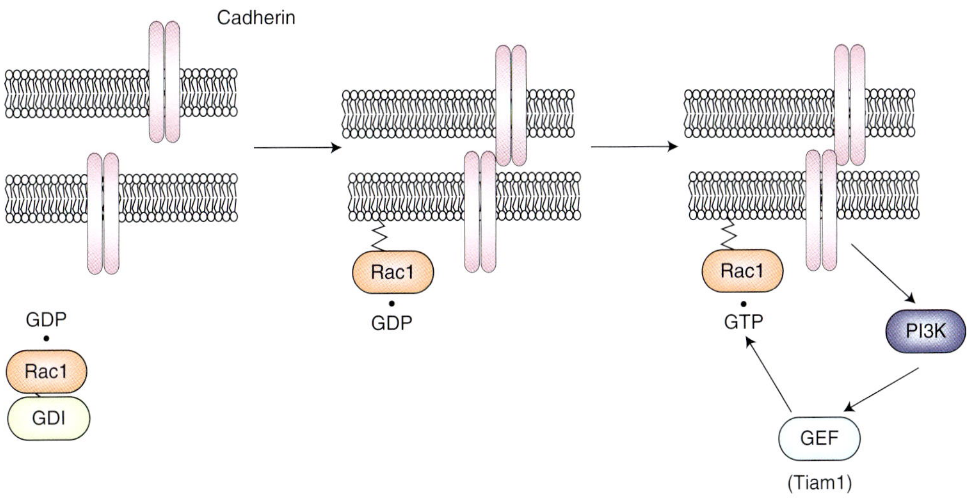

Figure 2. Mode of activation of Rac1 by the formation of E-cadherin-mediated cell–cell adhesions. Before the establishment of E-cadherin-mediated cell–cell adhesions, GDP· Rac1 is sequestered in the cytosol by Rho GDI. When cadherin-mediated homophilic interactions occur, GDP· Rac1 is dissociated from Rho GDI by an unknown mechanism and is targeted to the plasma membrane. GDP· Rac1 is converted to GTP· Rac1 through the action of a GEF (e.g., Tiam1) downstream of PI3k. Activated Rac1 then positively regulates E-cadherin-mediated cell–cell adhesions.

spatiotemporal activation during the initiation and expansion of epithelial cell–cell junctions (Yamada and Nelson 2007). The zones of Rac1 and lamellipodia activity as well as those of RhoA and actomyosin contractility are restricted to the periphery of the contacting membranes and together drive the initiation, expansion, and completion of cell–cell adhesions. These observations are consistent with the proposal that Rac1/Cdc42 and RhoA exert mutually antagonistic effects at some point during contact initiation, maturation, and reorganization, as in the case of the formation of focal adhesions (Rottner et al. 1999) and neurite extension (Kozma et al. 1997). It has also been shown that Tiam1 or constitutively active Rac1 (Rac1V12) can down-regulate RhoA activity and revert the mesenchymal phenotype of RasV12-transformed MDCKII cells to an epithelial phenotype (Sander et al. 1999). Thus, E-cadherin-mediated adhesions activate Rac1 and Cdc42, and active Rac1 down-regulates RhoA spatially, possibly through certain types of Rho GAPs, such as p190 RhoGAP (Bustos et al. 2008; Nimnual et al. 2003; Wildenberg et al. 2006).

Rap1 is activated by many extracellular stimuli and is strongly implicated in the control of integrin-mediated cell adhesions. Recent evidence indicates that Rap1 also plays key roles in the formation of cadherin-based cell–cell adhesions (Fig. 3) (Kooistra et al. 2007; Retta et al. 2006). Rap1 is activated by homophilic E-cadherin ligation (Hogan et al. 2004). This Rap1 activation is required for subsequent Cdc42 activation. Interestingly, Rap1 GEFs such as PDZ-GEF1 and C3G are linked to cadherin directly and indirectly, respectively, and contribute to Rap1 activation (Hogan et al. 2004; Kawajiri et al. 2000; Sakurai et al. 2006). Hogan and colleagues (2004) found that C3G binds to the cadherin cytoplasmic tail, providing a potential mechanism for E-cadherin to locally activate Rap1. Furthermore, C3G has been implicated in activating Rap1 downstream of the junctional protein nectin. In this case, C3G requires activation of Src after nectin clustering. It should be noted that nectin-based adhesions sequentially activate Rap1, Cdc42, and Rac1 (Fukuhara et al. 2004; Fukuyama et al. 2005; Kawakatsu et al. 2005; Kawakatsu et al. 2002).

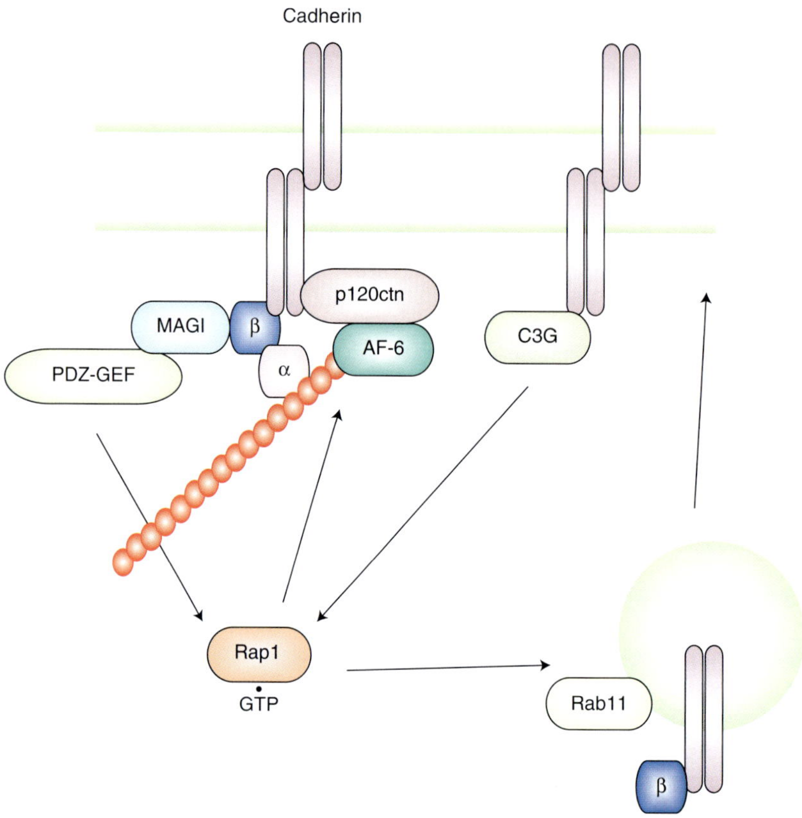

Figure 3. Model for the mode of action of Rap1 at cell–cell junctions. At initial cell–cell contacts, C3G bound to E-cadherin activates Rap1. Alternatively, PDZ-GEF linked to E-cadherin through MAGI and β-catenin can activate Rap1. Rap1 is also activated during E-cadherin endocytosis and locates to the E-cadherin/Rab11-containing vesicles. This population is likely involved in E-cadherin trafficking.

Additionally, cAMP analogues can influence cadherin-mediated adhesions in a Rap1-dependent manner, suggesting that a cAMP-responsible Rap1GEF, EPAC, is involved in signaling from cadherins (Price et al. 2004; Wittchen et al. 2005).

Thus, E-cadherin-mediated cell–cell contacts activate small GTPases including Rac1, Cdc42, and Rap1. However, it remains elusive how E-cadherin spatially and temporally regulates these small GTPases.

SMALL GTPases REGULATE ESTABLISHED ADHERENS JUNCTIONS

When small GTPases are activated by E-cadherin-mediated cell–cell contacts, they control existing E-cadherin-mediated adhesions (inside-out-signal) (Fig. 4). Braga and colleagues (Braga et al. 1997) found that the Rho family GTPases affect the formation of cell–cell junctions. When dominant–negative Rac1 (Rac1N17) or C3 botulinum toxin (an inhibitor of RhoA by ADP-ribosylation) is microinjected into keratinocytes, cadherin accumulation is inhibited at sites of cell–cell contacts. Subsequent studies from Takaishi and colleagues (1997) revealed that overexpression of constitutively active Rac1 (Rac1V12) in MDCKII cells promotes basal accumulation of E-cadherin, β-catenin, and actin filaments at sites of cell–cell contacts, whereas overexpression of Rac1N17 reduces their accumulation. Moreover, Cdc42, as well as Rac1, is

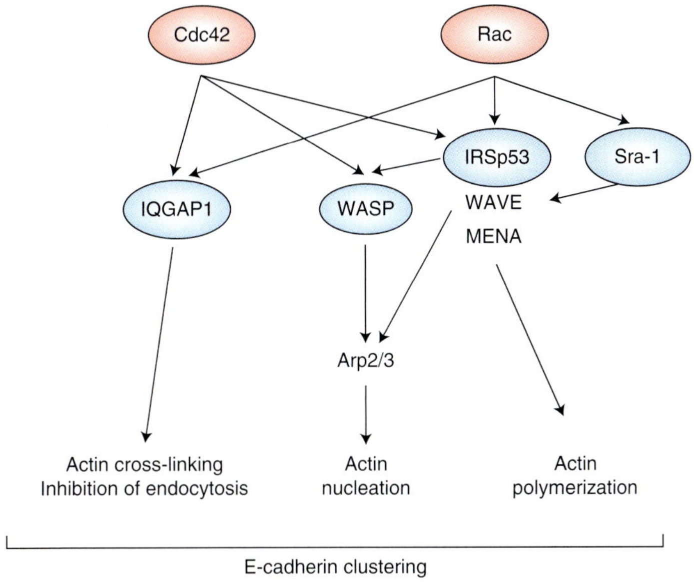

Figure 4. Possible mode of action of activated Rac1 and Cdc42 for E-cadherin-mediated adhesions. Activated Cdc42 induces actin nucleation through WASP and the Arp2/3 complex. Acting through WAVEs, Rac1 enhances actin polymerization locally at sites of E-cadherin-mediated adhesions. This results in the local remodeling of actin filaments near the Rac1/Cdc42-activated membrane. In addition, IQGAP1, an effector of Rac1 and Cdc42, positively regulates E-cadherin-dependent intercellular adhesions through cross-linking actin filaments as well as inhibiting of E-cadherin endocytosis.

required for E-cadherin-mediated cell–cell adhesions in MDCKII cells.

Those studies indicate that Rho family GTPases control cadherin-mediated cell–cell adhesions. More direct evidence was obtained using a quantitative, cell-dissociation assay for E-cadherin activity. These experiments sought to determine whether Rho family GTPases regulate cadherin-mediated cell–cell adhesions by acting on either the cadherin–catenin complex or the actin cytoskeleton and other components. Two stable cell lines derived from mouse L fibroblasts, which lack cadherin, are used: (1) EL cells express wild-type E-cadherin; (2) nEαCL cells express an E-cadherin mutant in which the distal β-catenin binding is replaced by the carboxy-terminal domain of α-catenin, thereby preventing remodeling of the cadherin–catenin complex. Although both cell lines adhere to each other in an E-cadherin-dependent manner, cell–cell adhesions in nEaCL cells do not require the distal β-catenin-binding domain of E-cadherin, β-catenin, or the amino terminus of α-catenin. The cadherin–catenin complex in nEαCL cells is not remodeled (Nagafuchi et al. 1994). The expression of Rac1N17 or dominant–negative Cdc42 (Cdc42N17) in EL cells markedly reduces E-cadherin activity, whereas expression of Rac1N17 or Cdc42N17 in nEαCL cells hardly reduces mutant E-cadherin activity (Fukata et al. 1999). This observation indicates that Rac1 and Cdc42 regulate E-cadherin activity through the cadherin–catenin complex. On the other hand, the expression of dominant–negative RhoA (RhoAN19) slightly reduces E-cadherin activity in both EL cells and nEαCL cells, suggesting that RhoA affects E-cadherin-mediated adhesive activity (presumably through the actin cytoskeleton or other components) but does not affect the cadherin–catenin complex.

How Rho family GTPases control E-cadherin-mediated intercellular adhesions remains an interesting question. One of the explanations involves the coordination of downstream effectors (Fig. 4). IQGAP1, an effector of Rac1 and Cdc42, has been implicated in E-cadherin-mediated intercellular adhesions. IQGAP1 has actin-cross-linking activity that is enhanced by activated Rac1/Cdc42. IQGAP1 localizes to sites of cell–cell contact in epithelial cells and can associate with β-catenin in vivo and in vitro (Kuroda et al. 1998). When overexpressed in EL cells, IQGAP1 interacts with β-catenin and induces the dissociation of α-catenin from β-catenin. This, in turn, weakens E-cadherin-mediated intercellular adhesions. When the amount of activated Rac1 increases, Rac1 interacts with IQGAP1, thereby stabilizing actin filaments. Under these conditions, IQGAP1 does not bind to β-catenin and cannot dissociate α-catenin from the cadherin–catenin complex, leading to strong adhesions. In contrast, when the amount of inactivated Rac1 increases, IQGAP1 is freed from Rac1 and interacts with β-catenin to dissociate α-catenin from the cadherin–catenin complex, resulting in weak adhesions (Noritake et al. 2005). Because IQGAP1 has anti-GTPase activity (Hart et al. 1996), it is possible that IQGAP1 controls the amount of GTP-bound Rac1 at sites of cell–cell contacts and leads to stable adhesions (Noritake et al. 2005). Furthermore, it has been shown that the inhibition of transinteraction E-cadherin endocytosis is mediated by the reorganization of the actin cytoskeleton by the IQGAP1-Rac/Cdc42 complex (Izumi et al. 2004). Thus, IQGAP1 behaves as a positive and negative regulator of cell–cell adhesion downstream of Rac1.

The actin polymerization machinery is another mediator between Rac1 activation and E-cadherin-mediated cell–cell adhesion (Fig. 4). Recruitment and activation of Rac1 and Cdc42 by homophilic E-cadherin ligations can induce actin polymerization through the Arp2/3 complex and WAVEs or WASP. Together with other actin-binding proteins such as cortactin, polymerized actin filaments reinforce cadherin-mediated adhesions (Gates and Peifer 2005; Scott and Yap 2006; Vasioukhin and Fuchs 2001). Significantly, inhibition of Rac1 signaling blocks actin assembly at sites of cell–cell adhesion (Braga et al. 1997; Takaishi et al. 1997). It is interesting to note that Rac1 also functions in the clustering of integrin receptors, suggesting that it could play a similar role in the clustering of E-cadherin. However, the effect of constitutively active Rac1 on adherens junctions depends on the cell type. In keratinocytes, constitutively active Rac1 (Rac1L61) causes the disassembly of adherens junctions (Braga et al. 2000), which is opposite to what is observed in MDCKII cells (see previous). The molecular mechanisms underlying these cell type-dependent effects need to be clarified.

Downstream signaling pathways that mediate the effects of RhoA on adherens junctions remain largely unknown. In keratinocytes, adherens junction formation has been shown to depend on the activity of the tyrosine kinase Fyn (Calautti et al. 2002). Constitutively active RhoA (RhoAV14) stimulates Fyn-mediated tyrosine phosphorylation of catenins and cell–cell adhesions, indicating that Fyn and possibly other Src family kinases can function downstream of RhoA in the establishment of adherens junctions. As described previously, during the establishment of cell–cell adhesions, RhoA is activated at the most distal edges of the expanding cell–cell contacts along with phosphorylated myosin II (Yamada and Nelson 2007). Rho-kinase, an effector of RhoA, directly phosphorylates myosin light chain and myosin phosphatase target subunit-1 and inhibits myosin phosphatase activity (Kaibuchi et al. 1999). This results in an increase in myosin light chain phosphorylation and subsequent actomyosin contraction (Kaibuchi et al. 1999). E-cadherin-mediated activation of RhoA is likely involved in the expansion of cell–cell adhesions through Rho-kinase.

Activated Rho family GTPases affect the surface level of E-cadherin expression, presumably by regulating endocytic transport (Mosesson et al. 2008). Constitutively active Rac1 (Rac1L61) and RhoA (RhoAL63) inhibit clathrin-dependent endocytosis in fibroblasts (Lamaze et al. 1996). This raises the possibility that Rho family GTPases could regulate cadherin-mediated cell–cell adhesions by controlling cadherin transport. Indeed, Rac1V12 and RhoAV14 inhibit HGF- and TPA-induced endocytosis of cadherin in MDCKII cells (Kamei et al. 1999). However, microinjection of Rac1L61 in keratinocytes causes E-cadherin to relocalize in large intracellular vesicles via a clathrin-independent pathway and disrupts cell–cell contacts. This discrepancy may be because of differences in cell types or levels of Rac1 activity (Braga et al. 2000; Braga et al. 1999). Regardless, the molecular mechanisms by which Rac1 and RhoA regulate the endocytosis of E-cadherin remains to be clarified.

The first evidence indicating that Rap1 is required for cell–cell junctions came from genetic studies in *Drosophila melanogaster* (Knox and Brown 2002). In MDCKII cells, Rap1 activity is required not only for Cdc42 activation during cell–cell junction formation but also for maintaining E-cadherin-mediated cell–cell adhesions (Fukuyama et al. 2005; Hogan et al. 2004). "Fast cycling" constitutively active Cdc42 (Cdc42L28) can rescue the effects of Rap1GAP on cell–cell junction formation. Furthermore, as described previously, activation of Rap1 by nectins is required for the subsequent activation of Cdc42 and Rac1. Thus, one of the functions of Rap1 activation may be the recruitment of Rac1 and Cdc42 GEFs to the site of initial cell–cell contacts to provide a link with the actin cytoskeleton. In addition, Rap1 may also be involved in the recruitment of junctional proteins. One protein heavily involved in this process is afadin/AF6, an effector of Rap1. In vitro studies show that afadin/AF6 in the presence of Rap1 inhibits the endocytosis of E-cadherins that are not engaged in homophilic transinteractions (Hoshino et al. 2005). Furthermore, strong activation of Rap1 occurs on adherens junction disassembly triggered by E-cadherin internalization and trafficking in the endocytic pathway (Retta et al. 2006). Such internalized E-cadherin is colocalized with Rap1 at the perinuclear Rab11-positive recycling endosome compartment, and Rap1 associates with a subset of E-cadherin–catenin complexes lacking p120ctn. Although the mode of action of Rap1 at the cell–cell adhesions remains largely unknown, Rap1 appears to act as a critical regulator of E-cadherin-mediated intercellular adhesion.

Recent evidence indicates that the Rab and Arf family of small GTPases are critical regulators that control the surface level of E-cadherin. Even in confluent cells, E-cadherin undergoes continuous endocytosis and exocytosis. The majority of studies implicates a clathrin-dependent route of E-cadherin internalization (Kimura et al. 2006), whereas some suggest the existence of non-clathrin-dependent pathways (e.g., caveola-mediated endocytosis) (Lu et al. 2003) and an EGF-induced macropinocytosis pathway (Bryant et al. 2007). Important mechanisms controlling E-cadherin endocytosis involve binding of p120ctn to the juxtamembrane domain of E-cadherin, and the tyrosine phosphorylation of E-cadherin by receptor tyrosine kinases and Src family kinases (Mellman and Nelson 2008; Mosesson et al. 2008).

The intracellular fate of internalized E-cadherin (i.e., sorting for degradation or recycling) is crucial, as it eliminates existing junctions or re-deploys E-cadherin to new junctions. Along with constitutive clathrin-mediated endocytosis, cadherin internalization is selectively induced by growth factors. The recycling endosome serves as a major location of cadherin for sorting back to the plasma membrane. Arf6, an endosomal GTP-binding protein, regulates cadherin endocytosis in response to EGF. Rab4 and Rab11 on the recycling endosomes also participate in E-cadherin endocytosis. Activated Rab4 interacts within the trans-Golgi network, whereas Rab11 seems to regulate not only the delivery of nascent E-cadherin from the Golgi to the basolateral surface but also the ability of internalized E-cadherin to activate Rap1 while residing in recycling endosomes (Balzac et al. 2005). Mechanisms that regulate the surface level of E-cadherin expression have begun to be uncovered. It will be worthwhile for future studies to reveal the functional cross talk among small GTPases in this process.

CONCLUDING REMARKS

Small GTPases have emerged as key players mediating the effects of signals, from cadherin engagement to downstream cellular machines that control the organization of the actin cytoskeleton, contractility, and vesicle trafficking. The identification and characterization of effectors for small GTPases have unraveled the action of small GTPases, especially those with relevance to the actin cytoskeleton, in E-cadherin-mediated intercellular adhesions. It is worthwhile in the future to examine in more detail the molecular mechanisms that control E-cadherin trafficking in the maintenance of intercellular adhesions. In contrast, much less is known about the signaling mechanisms that translate cadherin-mediated cues into spatio-temporal regulation of the small GTPases. It is likely, however, that the concerted action of GEFs and GAPs is critical for the precisely timing and localized activation of small GTPases.

REFERENCES

Adams CL, Nelson WJ. 1998. Cytomechanics of cadherin-mediated cell–cell adhesion. *Current Opinion Cell Biol* **10:** 572–577.

Akhtar N, Hotchin NA. 2001. RAC1 regulates adherens junctions through endocytosis of E-cadherin. *Mol Biol Cell* **12:** 847–862.

Anastasiadis PZ. 2007. p120-ctn: A nexus for contextual signaling via Rho GTPases. *Biochim Biophys Acta* **1773:** 34–46.

Balzac F, Avolio M, Degani S, Kaverina I, Torti M, Silengo L, Small JV, Retta SF. 2005. E-cadherin endocytosis regulates the activity of Rap1: A traffic light GTPase at the crossroads between cadherin and integrin function. *J Cell Sci* **118:** 4765–4783.

Betson M, Lozano E, Zhang J, Braga VM. 2002. Rac activation upon cell–cell contact formation is dependent on signaling from the epidermal growth factor receptor. *J Biol Chem* **277:** 36962–36969.

Blanpain C, Fuchs E. 2009. Epidermal homeostasis: A balancing act of stem cells in the skin. *Nat Rev Mol Cell Biol* **10:** 207–217.

Braga VM. 2002. Cell–cell adhesion and signalling. *Current Opinion Cell Biol* **14:** 546–556.

Braga VM, Yap AS. 2005. The challenges of abundance: Epithelial junctions and small GTPase signalling. *Current Opinion Cell Biol* **17:** 466–474.

Braga VM, Betson M, Li X, Lamarche-Vane N. 2000. Activation of the small GTPase Rac is sufficient to disrupt cadherin-dependent cell–cell adhesion in normal human keratinocytes. *Mol Biol Cell* **11:** 3703–3721.

Braga VM, Del Maschio A, Machesky L, Dejana E. 1999. Regulation of cadherin function by Rho and Rac: Modulation by junction maturation and cellular context. *Mol Biol Cell* **10:** 9–22.

Braga VM, Machesky LM, Hall A, Hotchin NA. 1997. The small GTPases Rho and Rac are required for the establishment of cadherin-dependent cell–cell contacts. *J Cell Biol* **137:** 1421–1431.

Bryant DM, Kerr MC, Hammond LA, Joseph SR, Mostov KE, Teasdale RD, Stow JL. 2007. EGF induces macropinocytosis and SNX1-modulated recycling of E-cadherin. *J Cell Sci* **120:** 1818–1828.

Bustos RI, Forget MA, Settleman JE, Hansen SH. 2008. Coordination of Rho and Rac GTPase function via p190B RhoGAP. *Current Biol* **18:** 1606–1611.

Calautti E, Grossi M, Mammucari C, Aoyama Y, Pirro M, Ono Y, Li J, Dotto GP. 2002. Fyn tyrosine kinase is a downstream mediator of Rho/PRK2 function in keratinocyte cell–cell adhesion. *J Cell Biol* **156:** 137–148.

Delva E, Tucker DK, Kowalczyk AP. 2009. The desmosome. *Cold Spring Harb Perspect Biol* **1:** a002543.

Dobrosotskaya IY, James GL. 2000. MAGI-1 interacts with β-catenin and is associated with cell–cell adhesion structures. *Biochem Biophys Res Comm* **270:** 903–909.

Espada J, Perez-Moreno M, Braga VM, Rodriguez-Viciana P, Cano A. 1999. H-Ras activation promotes cytoplasmic accumulation and phosphoinositide 3-OH kinase association of β-catenin in epidermal keratinocytes. *J Cell Biol* **146:** 967–980.

Fleming IN, Elliott CM, Exton JH. 1996. Differential translocation of rho family GTPases by lysophosphatidic acid, endothelin-1, and platelet-derived growth factor. *J Biol Chem* **271:** 33067–33073.

Fukata M, Kaibuchi K. 2001. Rho-family GTPases in cadherin-mediated cell–cell adhesion. *Nat Rev Mol Cell Biol* **2:** 887–897.

Fukata M, Kuroda S, Nakagawa M, Kawajiri A, Itoh N, Shoji I, Matsuura Y, Yonehara S, Fujisawa H, Kikuchi A, et al. 1999. Cdc42 and Rac1 regulate the interaction of IQGAP1 with β-catenin. *J Biol Chem* **274:** 26044–26050.

Fukuhara S, Murga C, Zohar M, Igishi T, Gutkind JS. 1999. A novel PDZ domain containing guanine nucleotide exchange factor links heterotrimeric G proteins to Rho. *J Biol Chem* **274:** 5868–5879.

Fukuhara T, Shimizu K, Kawakatsu T, Fukuyama T, Minami Y, Honda T, Hoshino T, Yamada T, Ogita H, Okada M, et al. 2004. Activation of Cdc42 by trans interactions of the cell adhesion molecules nectins through c-Src and Cdc42-GEF FRG. *Journal of Cell Biology* **166:** 393–405.

Fukuyama T, Ogita H, Kawakatsu T, Fukuhara T, Yamada T, Sato T, Shimizu K, Nakamura T, Matsuda M, Takai Y. 2005. Involvement of the c-Src-Crk-C3G-Rap1 signaling in the nectin-induced activation of Cdc42 and formation of adherens junctions. *J Biol Chem* **280:** 815–825.

Furuse M. 2009. Molecular basis of the core structure of tight junctions. *Cold Spring Harb Perspect Biol* **2:** a002907.

Gates J, Peifer M. 2005. Can 1000 reviews be wrong? Actin, α-Catenin, and adherens junctions. *Cell* **123:** 769–772.

Goldstein B, Macara IG. 2007. The PAR proteins: Fundamental players in animal cell polarization. *Develop Cell* **13:** 609–622.

Green KJ, Getsios S, Troyanovsky S, Godsel LM. 2009. Intercellular junction assembly, dynamics and homeostasis. *Cold Spring Harb Perspect Biol* **2:** a000125.

Gulli MP, Peter M. 2001. Temporal and spatial regulation of Rho-type guanine-nucleotide exchange factors: The yeast perspective. *Genes Develop* **15:** 365–379.

Gumbiner BM. 2000. Regulation of cadherin adhesive activity. *J Cell Biol* **148:** 399–404.

Halbleib JM, Nelson WJ. 2006. Cadherins in development: Cell adhesion, sorting, and tissue morphogenesis. *Genes Develop* **20:** 3199–3214.

Hart MJ, Callow MG, Souza B, Polakis P. 1996. IQGAP1, a calmodulin-binding protein with a rasGAP-related domain, is a potential effector for cdc42Hs. EMBO *J* **15:** 2997–3005.

Hartsock A, Nelson WJ. 2008. Adherens and tight junctions: Structure, function and connections to the actin cytoskeleton. *Biochim Biophys Acta* **1778:** 660–669.

Hogan C, Serpente N, Cogram P, Hosking CR, Bialucha CU, Feller SM, Braga VM, Birchmeier W, Fujita Y. 2004. Rap1 regulates the formation of E-cadherin-based cell–cell contacts. *Mol Cell Biol* **24:** 6690–6700.

Hoshino T, Sakisaka T, Baba T, Yamada T, Kimura T, Takai Y. 2005. Regulation of E-cadherin endocytosis by nectin through afadin, Rap1, and p120ctn. *J Biol Chem* **280:** 24095–24103.

Hoshino M, Sone M, Fukata M, Kuroda S, Kaibuchi K, Nabeshima Y, Hama C. 1999. Identification of the stef gene that encodes a novel guanine nucleotide exchange factor specific for Rac1. *J Biol Chem* **274:** 17837–17844.

Ide N, Hata Y, Deguchi M, Hirao K, Yao I, Takai Y. 1999. Interaction of S-SCAM with neural plakophilin-related Armadillo-repeat protein/delta-catenin. *Biochem Biophys Res Comm* **256:** 456–461.

Izumi G, Sakisaka T, Baba T, Tanaka S, Morimoto K, Takai Y. 2004. Endocytosis of E-cadherin regulated by Rac and Cdc42 small G proteins through IQGAP1 and actin filaments. *J Cell Biol* **166:** 237–248.

Jaffe AB, Hall A. 2005. Rho GTPases: Biochemistry and biology. *Ann Rev Cell Develop Biol* **21:** 247–269.

Jou TS, Nelson WJ. 1998. Effects of regulated expression of mutant RhoA and Rac1 small GTPases on the development of epithelial (MDCK) cell polarity. *J Cell Biol* **142:** 85–100.

Kaibuchi K, Kuroda S, Amano M. 1999. Regulation of the cytoskeleton and cell adhesion by the Rho family GTPases in mammalian cells. *Ann Rev Biochem* **68:** 459–486.

Kamei T, Matozaki T, Sakisaka T, Kodama A, Yokoyama S, Peng YF, Nakano K, Takaishi K, Takai Y. 1999. Coendocytosis of cadherin and c-Met coupled to disruption of cell–cell adhesion in MDCK cells–regulation by Rho, Rac and Rab small G proteins. *Oncogene* **18:** 6776–6784.

Kawajiri A, Itoh N, Fukata M, Nakagawa M, Yamaga M, Iwamatsu A, Kaibuchi K. 2000. Identification of a novel β-catenin-interacting protein. *Biochem Biophys Res Commun* **273:** 712–717.

Kawakatsu T, Ogita H, Fukuhara T, Fukuyama T, Minami Y, Shimizu K, Takai Y. 2005. Vav2 as a Rac-GDP/GTP exchange factor responsible for the nectin-induced, c-Src- and Cdc42-mediated activation of Rac. *J Biol Chem* **280:** 4940–4947.

Kawakatsu T, Shimizu K, Honda T, Fukuhara T, Hoshino T, Takai Y. 2002. Trans-interactions of nectins induce formation of filopodia and Lamellipodia through the respective activation of Cdc42 and Rac small G proteins. *J Biol Chem* **277:** 50749–50755.

Kim SH, Li Z, Sacks DB. 2000. E-cadherin-mediated cell–cell attachment activates Cdc42. *J Biol Chem* **275:** 36999–37005.

Kimura T, Sakisaka T, Baba T, Yamada T, Takai Y. 2006. Involvement of the Ras-Ras-activated Rab5 guanine nucleotide exchange factor RIN2-Rab5 pathway in the hepatocyte growth factor-induced endocytosis of E-cadherin. *J Biol Chem* **281:** 10598–10609.

Kiyokawa E, Hara S, Nakamura T, Matsuda M. 2006. Fluorescence (Forster) resonance energy transfer imaging of oncogene activity in living cells. *Cancer Sci* **97:** 8–15.

Knox AL, Brown NH. 2002. Rap1 GTPase regulation of adherens junction positioning and cell adhesion. *Science* **295:** 1285–1288.

Kooistra MR, Dube N, Bos JL. 2007. Rap1: A key regulator in cell–cell junction formation. *J Cell Sci* **120:** 17–22.

Kotani K, Yonezawa K, Hara K, Ueda H, Kitamura Y, Sakaue H, Ando A, Chavanieu A, Calas B, Grigorescu F, et al. 1994. Involvement of phosphoinositide 3-kinase in insulin- or IGF-1-induced membrane ruffling. *EMBO J* **13:** 2313–2321.

Kourlas PJ, Strout MP, Becknell B, Veronese ML, Croce CM, Theil KS, Krahe R, Ruutu T, Knuutila S, Bloomfield CD, et al. 2000. Identification of a gene at 11q23 encoding a guanine nucleotide exchange factor: Evidence for its fusion with MLL in acute myeloid leukemia. *Proc Natl Acad Sci* **97:** 2145–2150.

Kovacs EM, Ali RG, McCormack AJ, Yap AS. 2002. E-cadherin homophilic ligation directly signals through Rac and phosphatidylinositol 3-kinase to regulate adhesive contacts. *J Biol Chem* **277:** 6708–6718.

Kozma R, Sarner S, Ahmed S, Lim L. 1997. Rho family GTPases and neuronal growth cone remodelling: Relationship between increased complexity induced by Cdc42Hs, Rac1, and acetylcholine and collapse induced by RhoA and lysophosphatidic acid. *Mol Cell Biol* **17:** 1201–1211.

Kranenburg O, Poland M, Gebbink M, Oomen L, Moolenaar WH. 1997. Dissociation of LPA-induced cytoskeletal contraction from stress fiber formation by differential localization of RhoA. *J Cell Sci* **110:** 2417–2427.

Kroschewski R, Hall A, Mellman I. 1999. Cdc42 controls secretory and endocytic transport to the basolateral plasma membrane of MDCK cells. *Nat Cell Biol* **1:** 8–13.

Kuroda S, Fukata M, Nakagawa M, Fujii K, Nakamura T, Ookubo T, Izawa I, Nagase T, Nomura N, Tani H, et al. 1998. Role of IQGAP1, a target of the small GTPases Cdc42 and Rac1, in regulation of E-cadherin-mediated cell–cell adhesion. *Science* **281:** 832–835.

Lamaze C, Chuang TH, Terlecky LJ, Bokoch GM, Schmid SL. 1996. Regulation of receptor-mediated endocytosis by Rho and Rac. *Nature* **382:** 177–179.

Lu Z, Ghosh S, Wang Z, Hunter T. 2003. Downregulation of caveolin-1 function by EGF leads to the loss of E-cadherin, increased transcriptional activity of β-catenin, and enhanced tumor cell invasion. *Cancer Cell* **4:** 499–515.

McCrea PD, Gu D, Balda M. 2009. Junctional music that the nucleus hears: Cell-cell junction signaling and the modulation of gene activity. *Cold Spring Harb Perspect Biol* **1:** a002923.

McLachlan RW, Kraemer A, Helwani FM, Kovacs EM, Yap AS. 2007. E-cadherin adhesion activates c-Src signaling at cell–cell contacts. *Mol Biol Cell* **18:** 3214–3223.

Mellman I, Nelson WJ. 2008. Coordinated protein sorting, targeting and distribution in polarized cells. *Nat Rev Mol Cell Biol* **9:** 833–845.

Meng W, Takeichi M. 2009. Adherens junction: Molecular architecture and regulation. *Cold Spring Harb Perspect Biol* **1:** a002899.

Michiels F, Habets GG, Stam JC, van der Kammen RA, Collard JG. 1995. A role for Rac in Tiam1-induced membrane ruffling and invasion. *Nature* **375:** 338–340.

Mosesson Y, Mills GB, Yarden Y. 2008. Derailed endocytosis: An emerging feature of cancer. *Nat Rev Cancer* **8:** 835–850.

Nagafuchi A, Ishihara S, Tsukita S. 1994. The roles of catenins in the cadherin-mediated cell adhesion: Functional analysis of E-cadherin-α catenin fusion molecules. *J Cell Biol* **127:** 235–245.

Nakagawa M, Fukata M, Yamaga M, Itoh N, Kaibuchi K. 2001. Recruitment and activation of Rac1 by the formation of E-cadherin-mediated cell–cell adhesion sites. *J Cell Sci* **114:** 1829–1838.

Niessen CM, Gumbiner BM. 2002. Cadherin-mediated cell sorting not determined by binding or adhesion specificity. *J Cell Biol* **156:** 389–399.

Nimnual AS, Taylor LJ, Bar-Sagi D. 2003. Redox-dependent downregulation of Rho by Rac. *Nat Cell Biol* **5:** 236–241.

Noren NK, Liu BP, Burridge K, Kreft B. 2000. p120 catenin regulates the actin cytoskeleton via Rho family GTPases. *J Cell Biol* **150:** 567–580.

Noren NK, Niessen CM, Gumbiner BM, Burridge K. 2001. Cadherin engagement regulates Rho family GTPases. *J Biol Chem* **276:** 33305–33308.

Noritake J, Watanabe T, Sato K, Wang S, Kaibuchi K. 2005. IQGAP1: A key regulator of adhesion and migration. *J Cell Sci* **118:** 2085–2092.

Pece S, Chiariello M, Murga C, Gutkind JS. 1999. Activation of the protein kinase Akt/PKB by the formation of E-cadherin-mediated cell–cell junctions. Evidence for the association of phosphatidylinositol 3-kinase with the E-cadherin adhesion complex. *J Biol Chem* **274:** 19347–19351.

Perego C, Vanoni C, Massari S, Longhi R, Pietrini G. 2000. Mammalian LIN-7 PDZ proteins associate with β-catenin at the cell- cell junctions of epithelia and neurons. *EMBO J* **19:** 3978–3989.

Perez TD, Tamada M, Sheetz MP, Nelson WJ. 2008. Immediate-early signaling induced by E-cadherin engagement and adhesion. *J Biol Chem* **283:** 5014–5022.

Perez-Moreno M, Jamora C, Fuchs E. 2003. Sticky business: Orchestrating cellular signals at adherens junctions. *Cell* **112:** 535–548.

Pertz O, Hahn KM. 2004. Designing biosensors for Rho family proteins–deciphering the dynamics of Rho family GTPase activation in living cells. *J Cell Sci* **117:** 1313–1318.

Price LS, Hajdo-Milasinovic A, Zhao J, Zwartkruis FJ, Collard JG, Bos JL. 2004. Rap1 regulates E-cadherin-mediated cell–cell adhesion. *J Biol Chem* **279:** 35127–35132.

Price LS, Leng J, Schwartz MA, Bokoch GM. 1998. Activation of Rac and Cdc42 by integrins mediates cell spreading. *Mol Biol Cell* **9:** 1863–1871.

Retta SF, Balzac F, Avolio M. 2006. Rap1: A turnabout for the crosstalk between cadherins and integrins. *Eur J Cell Biol* **85:** 283–293.

Reynolds AB. 2007. p120-catenin: Past and present. *Biochim Biophys Acta* **1773:** 2–7.

Rottner K, Hall A, Small JV. 1999. Interplay between Rac and Rho in the control of substrate contact dynamics. *Current Biol* **9:** 640–648.

Sakurai A, Fukuhara S, Yamagishi A, Sako K, Kamioka Y, Masuda M, Nakaoka Y, Mochizuki N. 2006. MAGI-1 is required for Rap1 activation upon cell–cell contact and for enhancement of vascular endothelial cadherin-mediated cell adhesion. *Mol Biol Cell* **17:** 966–976.

Sander EE, ten Klooster JP, van Delft S, van der Kammen RA, Collard JG. 1999. Rac downregulates Rho activity: Reciprocal balance between both GTPases determines cellular morphology and migratory behavior. *J Cell Biol* **147:** 1009–1022.

Scott JA, Yap AS. 2006. Cinderella no longer: α-catenin steps out of cadherin's shadow. *J Cell Sci* **119:** 4599–4605.

Shapiro L, Weis WI. 2009. Structure and biochemistry of cadherins and catenins. *Cold Spring Harb Perspect Biol* **1:** a003053.

Takaishi K, Sasaki T, Kotani H, Nishioka H, Takai Y. 1997. Regulation of cell–cell adhesion by rac and rho small G proteins in MDCK cells. *J Cell Biol* **139:** 1047–1059.

Takeichi M. 1995. Morphogenetic roles of classic cadherins. *Current Opinion Cell Biol* **7:** 619–627.

Taya S, Inagaki N, Sengiku H, Makino H, Iwamatsu A, Urakawa I, Nagao K, Kataoka S, Kaibuchi K. 2001. Direct interaction of insulin-like growth factor-1 receptor with leukemia-associated RhoGEF. *J Cell Biol* **155:** 809–820.

Tepass U, Truong K, Godt D, Ikura M, Peifer M. 2000. Cadherins in embryonic and neural morphogenesis. *Nat Rev Mol Cell Biol* **1:** 91–100.

Tsukita S, Tsukita S, Nagafuchi A, Yonemura S. 1992. Molecular linkage between cadherins and actin filaments in cell–cell adherens junctions. *Current Opinion Cell Biol* **4:** 834–839.

Tsukita S, Yamazaki Y, Katsuno T, Tamura A. 2008. Tight junction-based epithelial microenvironment and cell proliferation. *Oncogene* **27:** 6930–6938.

Van Aelst L, D'souza-Schorey C. 1997. Rho GTPases and signaling networks. *Genes Develop* **11:** 2295–2322.

Van Aelst L, Symons M. 2002. Role of Rho family GTPases in epithelial morphogenesis. *Genes Develop* **16:** 1032–1054.

Vasioukhin V, Fuchs E. 2001. Actin dynamics and cell–cell adhesion in epithelia. *Current Opinion Cell Biol* **13:** 76–84.

Wildenberg GA, Dohn MR, Carnahan RH, Davis MA, Lobdell NA, Settleman J, Reynolds AB. 2006. p120-catenin and p190RhoGAP regulate cell–cell adhesion by coordinating antagonism between Rac and Rho. *Cell* **127:** 1027–1039.

Wittchen ES, Worthylake RA, Kelly P, Casey PJ, Quilliam LA, Burridge K. 2005. Rap1 GTPase inhibits leukocyte transmigration by promoting endothelial barrier function. *J Biol Chem* **280:** 11675–11682.

Yamada S, Nelson WJ. 2007. Localized zones of Rho and Rac activities drive initiation and expansion of epithelial cell–cell adhesion. *J Cell Biol* **178:** 517–527.

Yap AS, Kovacs EM. 2003. Direct cadherin-activated cell signaling: A view from the plasma membrane. *J Cell Biol* **160:** 11–16.

Junctional Music that the Nucleus Hears: Cell–Cell Contact Signaling and the Modulation of Gene Activity

Pierre D. McCrea[1], Dongmin Gu[1], and Maria S. Balda[2]

[1]Department of Biochemistry and Molecular Biology, University of Texas MD Anderson Cancer Center, Program in Genes and Development, University of Texas Graduate School of Biomedical Sciences, Houston, Texas 77030

[2]Division of Cell Biology, UCL Institute of Ophthalmology, University College London, London EC1V 9EL, United Kingdom

Correspondence: pdmccrea@mdanderson.org, m.balda@ucl.ac.uk

Cell–cell junctions continue to capture the interest of cell and developmental biologists, with an emerging area being the molecular means by which junctional signals relate to gene activity in the nucleus. Although complexities often arise in determining the direct versus indirect nature of such signal transduction, it is clear that such pathways are essential for the function of tissues and that alterations may contribute to many pathological outcomes. This review assesses a variety of cell–cell junction-to-nuclear signaling pathways, and outlines interesting areas for further study.

The evolution of multicellular life forms has to a significant extent involved refinements of each cell's capacity to sense the state of its directly contacting neighbors. This exchange of information often occurs within tissues, with the result that gene activity in the nucleus is altered or maintained accordingly. In this article, we focus on how signals arise at cell–cell junctions and are transduced to the nucleus; we do not include discussion of mechanical/cytoskeletal signals influencing nuclear decisions, and the reader is directed to a recent review of this topic (Ingber 2008).

An issue that arises when addressing cell–cell junction(s), referred to as CCJ(s), -to-nuclear signals, is that homotypic or heterotypic junctional proteins responsible for conferring adhesive activity are often in a much larger complex of proteins. These interactions may be either in *cis* (interacting within the plasma membrane of the cell) or *trans* orientations (interacting through ectodomain contacts extended between cells). Most of these transmembrane proteins are likely to have the potential to contribute to downstream signaling events, and many may associate with one another only under specific physiological conditions. For example, certain receptor tyrosine kinases (RTKs) associate with particular cadherins, and when associated are relevant to that cadherin's functions (Wheelock and Johnson 2003; Andl and Rustgi 2005). In this

 doi: 10.1101/cshperspect.a002923
Cite this article as *Cold Spring Harb Perspect Biol* 2009;1:a002923

article, we discuss relationships such as these in the context of CCJ-nuclear signaling. A topic not represented here is the CCJ signaling of immune surveillance cells, for example, pathways activated following leukocyte–endothelia contact. This area is of great basic and biomedical interest, but is addressed elsewhere (Dustin 2007).

We focus on signaling by a select number of junction types, including adherens, desmosomal, and tight junctions, and to a lesser extent, gap junctions. Details of the structure and function of each of these junctions are presented in other articles (see Meng and Takeichi 2009, Delva et al. 2009, Furuse 2009, and Goodenough and Paul 2009, respectively). These junctions are often represented in textbooks as distinct entities in the context of epithelial tissues, but their structures and how they respond to or generate signaling cues vary according to cellular context. Select components within these junctions may be shared, for example between desmosomal, adherens, and tight junctions, and in some instances, intimate physical proximities are likely to advance these junctions' functional interrelation. Further, different cell types show less common junctional organizations (Straub et al. 2003; Wuchter et al. 2007), such that the total spectrum of CCJ signals is likely to be impressive, and far beyond what is currently known or understood. Given the interdependence of cell neighbors in forming and maintaining cell groupings, high diversity and sophistication arose in complex organisms, both in CCJ structures themselves and their associated nuclear signaling pathways. Compared with the knowledge accumulated over the past two decades on cell–extracellular matrix signaling via integrins (Abram and Lowell 2009), we know less about signals initiated from forming or mature cell–cell contacts in epithelial, neural, or endothelial tissues. Thus, as the field moves forward, there is the potential to achieve a deepened understanding of how the cell–extracellular matrix and cell–cell adhesion systems are coupled in a signaling context, and how they collectively relate to the adhesion, motility, and differentiation of cells and tissues.

ADHERENS AND DESMOSOMAL JUNCTIONS

Adherens junctions are comprised of a number of transmembrane proteins, including the classic cadherins (see Meng and Takeichi 2009) and nectins, which are members of the distinct Ig (immunoglobulin) family of adhesion proteins/receptors (Fig. 1) (Takai et al. 2008a; Takai et al. 2008b). In some cases, cadherins and nectins associate through intracellular proteins, including p120-catenin bound to the juxtamembrane domain of cadherin and afadin bound to nectin (Ogita and Takai 2008). RTKs are also present at adherens junctions, although unlike cadherins and nectins, they are not viewed as adhesion proteins, but rather as signaling entities instructing, responding, or working in parallel with other signal generators within the cadherin/nectin macromolecular complex.

Desmosomal junctions contain additional members of the cadherin superfamily, represented in the transmembrane polypeptides of the desmocollin and desmoglein subfamilies (Fig. 2) (Green and Simpson 2007; Holthofer et al. 2007; Garrod and Chidgey 2008) (see also Delva et al. 2009). Desmosomal cadherins bind catenins that are different from those found in complex with classic cadherins at adherens junctions (Hatzfeld 2007). There are, however, catenins such as plakoglobin that are found at more than one junction (Zhurinsky et al. 2000). This may occur more often than realized as β-catenin, which is generally thought to bind only cadherins at adherens junctions, is also found at desmosomes under certain conditions (Bierkamp et al. 1999). In addition, p120-catenin has been identified at desmosomal junctions (Kanno et al. 2008), and ZO-1, generally found at tight junctions, associates with adherens junctions during their assembly (Itoh et al. 1993; Muller et al. 2005). Finally, p0071-catenin was first characterized in association with desmosomal cadherins but also localizes to adherens junction (Hatzfeld et al. 2003; Calkins et al. 2003). Such examples of catenins being shared across junctional types may reflect a means to

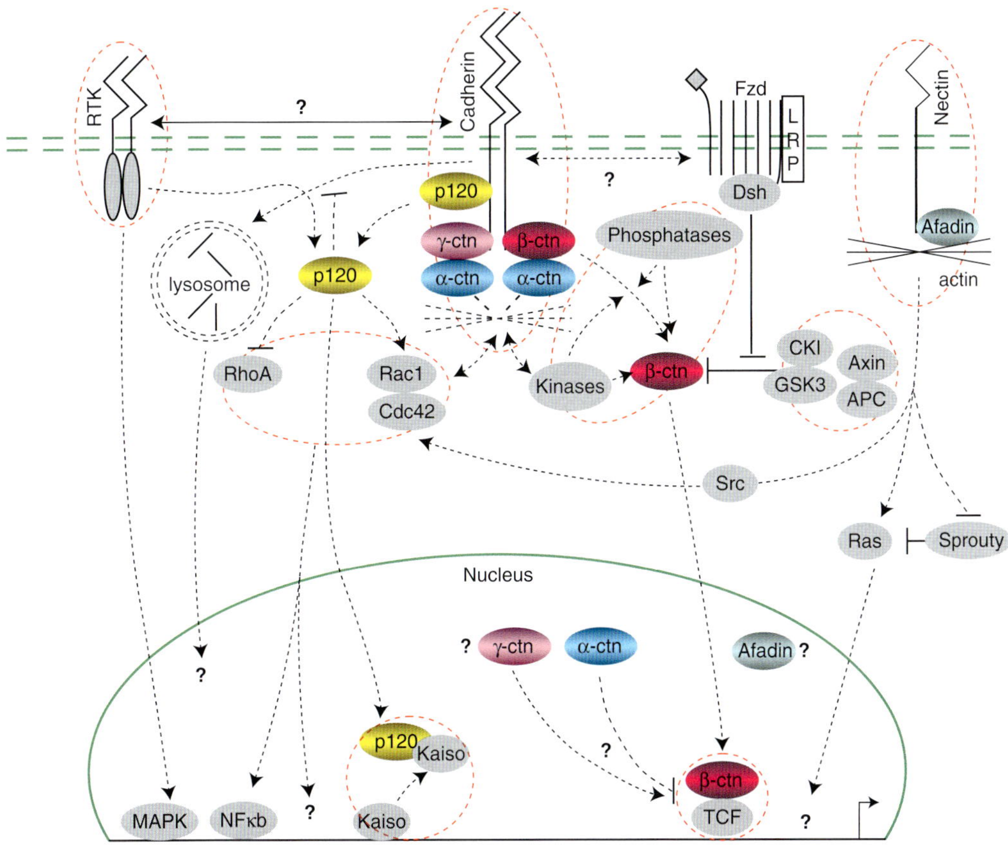

Figure 1. Adherens junction signaling to the nucleus. Proteins binding to the intracellular regions of cadherins include catenins, with β-catenin further associating with α-catenin and perhaps indirectly/dynamically with actin microfilaments. The structurally related p120-catenin subfamily is proposed to modulate cadherin lateral clustering, cadherin endocytosis, and activities of Rho-family GTPases. β-catenin acts in various ways within the nucleus, including as a transcriptional coactivator in conjunction with LEF/TCF, whereas p120-catenin relieves Kaiso-mediated gene repression. Other catenins have also been observed in the nucleus, where they have putative or shared roles in transcriptional regulation. Signaling from the nectin/afadin complex to the nucleus may occur indirectly through Ras or Rho, or more directly through afadin's potential association with chromatin or transcriptional modulators. Some evidence has indicated that receptor tyrosine kinases/growth factor receptors, or the frizzled/LRP complex, modulate signals initiated from the adherens junction and vice versa. Dotted circles highlight select signaling entities where research has been focused over the past decade.

coordinately regulate distinct cell–cell contacts, to enhance cross talk between junctions, or to provide additive junctional signals directed to other cellular compartments such as the nucleus.

How signals cross the membrane following *trans*-cadherin ectodomain contact is not well understood. The relevance of cadherin *cis*-dimerization to subsequent *trans* interactions is proposed in some but not all models (Brieher et al. 1996; Yap et al. 1997; Yap et al. 1998; Patel et al. 2003; Yap and Kovacs 2003; Troyanovsky 2005; Troyanovsky et al. 2007; Nelson 2008), with higher order clustering possibly correlating with raised levels of interacting phosphatases, kinases, or small GTPases (see also Watenabe et al. 2009). This may be similar to models of RTK clustering and activation in which cytoplasmic kinase domains come in closer proximity to initiate signaling.

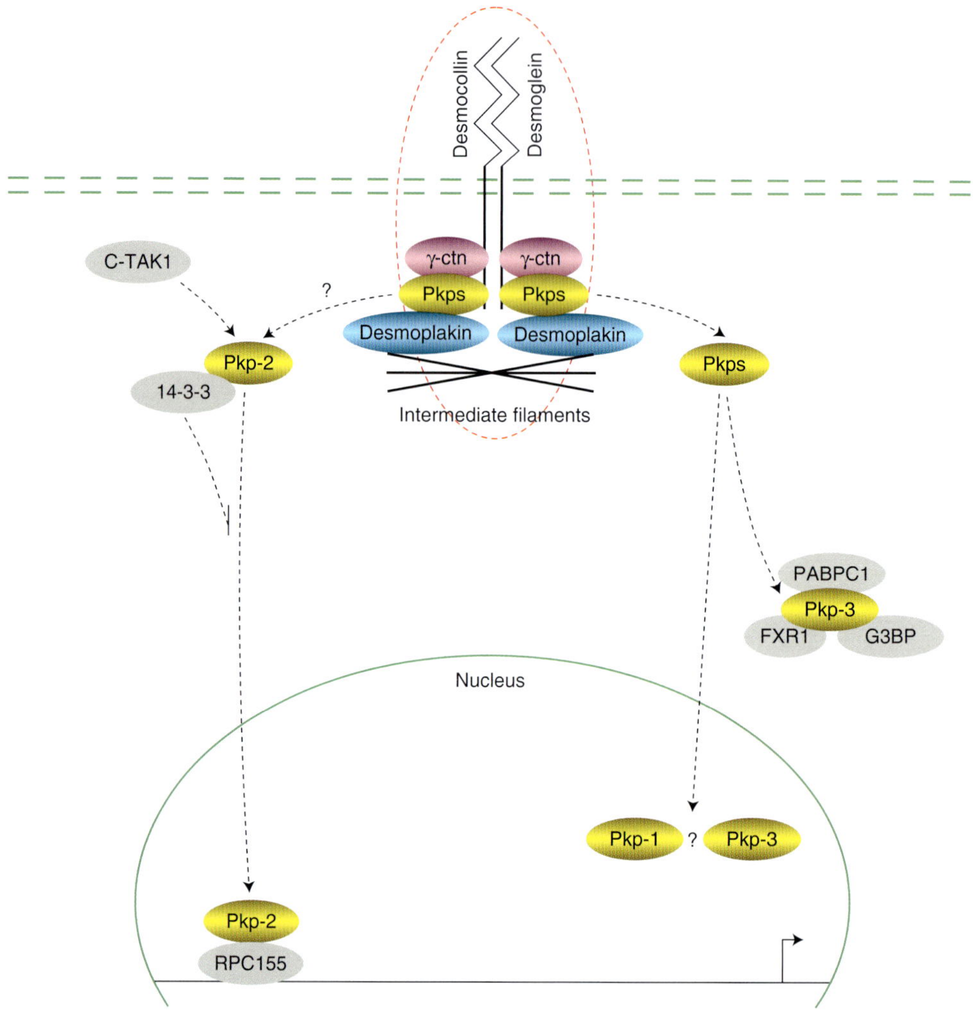

Figure 2. Desmosome junction signaling to the nucleus. Desmosomes are adhesive plaques containing the transmembrane cadherin superfamily members desmocollin and desmoglein, as well as intracellular binding partners such as plakoglobin (γ-catenin), plakophilins (Pkps), and desmoplakin. A number of desmosomal catenins have been localized to the nucleus although the functional relevance remains to be elucidated. Plakophilin-2, for example, has been found to reside within the PolIII complex and to associate with RPC155 and the transcription factor TFIIIB.

Following ectodomain engagement, transmembrane signaling might occur through displacement of adhesion molecules relative to the plasma membrane, perhaps exposing or removing an intracellular binding site, or displacements relative to other transmembrane polypeptides, analogous to that occurring between the α- and β-integrin subunits on binding extracellular matrix.

Our focus here is on the nuclear signaling pathways that are initiated at cell–cell junctions. From a conceptual perspective, it seems that evolution would favor signals passing in both directions, thereby informing nuclear as well as cell–cell contact decision-making in a reciprocal manner. This would include still incompletely understood concepts such as contact inhibition, where chromosomal replication/cell division slows with increased cell–cell association or, conversely, the lessening or remodeling of cell–cell contacts in cells initiating gene programs that favor a migratory phase or epithelial-to-mesenchymal transitions.

CATENIN SIGNALING

A signaling trajectory that might at first seem obvious between cell–cell junctions and downstream nuclear events is that involving β-catenin. β-catenin binds the distal (carboxy-terminal) domain of cadherin cytoplasmic tails via β-catenin's central Armadillo domain (Huber and Weis 2001; Xu and Kimelman 2007). Concomitantly, β-catenin binds to α-catenin via its amino-terminal domain (Pokutta and Weis 2000). A recent model proposes that the cadherin–β-catenin complex may locally enrich α-catenin (Yamada et al. 2005; Drees et al. 2005; Weis and Nelson 2006), which when released forms homodimers capable of lowering Arp 2/3 activity (Yamada and Nelson 2007). This is then expected to reduce actin branching and cellular protrusive activity within the microenvironment, facilitating contact formation and stability.

Another major role of β-catenin is in the canonical Wnt signaling pathway (Clevers 2006; Gavert and Ben-Ze'ev 2007; Huang and He 2008; Barker 2008; Mosimann et al. 2009) (see also Heuberger and Birchmeier 2009; Cadigan and Peifer 2009). Following Wnt-ligand association with the plasma membrane receptor–coreceptor complex of Frizzled–Lrp, a signaling pool of β-catenin is formed that is not bound to cadherin. This process involves Wnt-ligand-directed inactivation of an intracellular β-catenin destruction complex (involving components such as GSK3β, CK1α, Axin, and APC), that in the absence of Wnt ligand would otherwise basally suppress the Wnt pathway (see Cadigan and Peifer 2009 for details). Once stabilized, a number of poorly understood events allow β-catenin to enter the nucleus to activate Wnt target genes important in many developmental and pathological contexts (Suh and Gumbiner 2003; Gottardi and Gumbiner 2004; Mosimann et al. 2009). Activation occurs on β-catenin's association with TCF/LEF transcription factors resident at TCF/LEF consensus binding sites within Wnt target gene promoters.

It was envisaged that β-catenin released from the cadherin complex, perhaps in response to specific phosphorylation events (Daniel and Reynolds 1997; Potter et al. 2005; Sallee et al. 2006; Alema and Salvatore 2007), might also enter a signaling pool and activate Wnt target genes (Perez-Moreno et al. 2003; Nelson and Nusse 2004; Bienz 2005; Perez-Moreno and Fuchs 2006). However, unless Wnt signals are already active, it remains reasonable to ask how β-catenin released from cadherin escapes the destruction complex. Indeed, in a number of contexts, it appears that cadherin loss alone may not be sufficient to activate β-catenin signaling (Caca et al. 1999; van de Wetering et al. 2001; Herzig et al. 2007), even as cadherin loss increases Wnt signals activated by other means (Gottardi et al. 2001; Kuphal and Behrens 2006).

Evidence has accumulated over the past decade that β-catenin released from cadherin is associated with increased Wnt pathway activity in some experimental contexts, such as in response to RTK or Src activity (Nelson and Nusse 2004; Bienz 2005; Lilien and Balsamo 2005; Brembeck et al. 2006; Gavert and Ben-Ze'ev 2007; McLachlan and Yap 2007). Src is capable of phosphorylating β-catenin at tyrosine 654, and reducing its association with cadherin (Roura et al. 1999). Most studies have been correlative, with the dissociation of β-catenin from cadherin occurring concurrent with increased transcriptional activity from endogenous Wnt gene targets or luciferase reporters. However, the kinases examined may well have produced effects beyond β-catenin release from cadherin. For example, activation of Akt and/or inhibition of GSK3β could result in β-catenin stabilization and modulation of the Wnt endogenous genes or reporter constructs by an alternative upstream path (Larue and Bellacosa 2005).

Just as kinases bind and act on the cadherin–catenin complex, so too do a variety of phosphatases that dephosphorylate both cadherins and catenins (Lilien et al. 2002; Sallee et al. 2006; McLachlan and Yap 2007). The phosphatases involved are numerous and include membrane-spanning receptor-type phosphatases such as DEP-1, PTPmu, LAR, and VE-PTP, as well as cytoplasmic phosphatases including PTP1B and Shp-2. In addition to the opposing impact of kinases and phosphatases on catenin signaling, a further means of promoting β-catenin's displacement from the larger cadherin–catenin complex toward a nuclear signaling pool involves the β-catenin-binding protein BCL9-2, a homolog of the human protooncogene product BCL9 (Brembeck et al. 2004; Brembeck et al. 2006). In response to RTK activation, direct or indirect β-catenin phosphorylation at tyrosine 142 occurs (e.g., via the RTK Met versus the cytoplasmic tyrosine kinases fer or fyn), decreasing β-catenin's association with α-catenin in favor of BCL9, and promoting β-catenin nuclear entry and activity. Such examples have brought attention to the likely interrelation of adherens junction states, inclusive of secondary protein modifications, with transcriptional outcomes (see also Heuberger and Birchmeier 2009).

Because the Wnt pathway impinges directly or indirectly on multiple classes of genes and produces various cellular responses (proliferation, differentiation, stem-cell maintenance), the impact of increased β-catenin signaling cannot be predicted without knowing the molecular details of the tissue or larger developmental programs involved (Clevers 2006; Gavert and Ben-Ze'ev 2007; Huang and He 2008; Barker 2008; Mosimann et al. 2009). However, in the case of RTK effects on the cadherin–catenin complex in particular, assessments have generally been made of endogenous β-catenin gene targets such as cyclin-D1 and c-myc, whose increased expression in many cases correlates with increased cell-cycle progression under normal and pathological conditions. Even though a large number of factors are contributory, decreased E-cadherin levels or function in many carcinomas is correlated with increased proliferation. Thus, the field will continue to be interested in the biological roles of catenin nuclear signals originating from the cadherin–catenin macromolecular complex in future studies of normal or diseased cells and tissues.

OTHER CATENINS

Cadherins bind a number of other catenins distinct from β-catenin that may be involved in CCJ signaling to the nucleus. One of these is plakoglobin (γ-catenin), which being very similar in structure to β-catenin, competes with β-catenin for binding to the membrane-distal region of cadherin tails, and can associate with α-catenin (Zhurinsky et al. 2000). In common with β-catenin, signaling pools of plakoglobin have been found to respond to signals initiated by Wnt ligands (thereby escaping destruction), although plakoglobin's nuclear role has been characterized as either activating, repressive, or relatively neutral in nature probably because of context issues (Merriam et al. 1997; Kodama et al. 1999; Kolligs et al. 2000; Williams et al. 2000; Charpentier et al. 2000; Shtutman et al. 2002; Maeda et al. 2004; Garcia-Gras et al. 2006; Dusek et al. 2007; Shimizu et al. 2008; Martin et al. 2009). Although less studied than β-catenin, it is possible that plakoglobin once dissociated from the cadherin complex might also represent a signaling entity arising from cell–cell junctions. Interestingly, plakoglobin is further present in association with the desmoglein and desmocollin cadherin family members (Garrod and Chidgey 2008), the principal transmembrane proteins conferring adhesive activity at desmosomal junctions. Thus, a signaling pool of plakoglobin could conceivably arise from either of these cell–cell junctions, or perhaps coordinately from both.

Loss of desmosome-associated desmoplakin in mouse cardiomyocytes was associated with increased levels of plakoglobin in the nucleus and a reduction in β-catenin-mediated Wnt signaling. This led to fibroadipocytic replacement of cardiac myocytes that mimics the human genetic condition arrhythmogenic right ventricular dysplasia/cardiomyopathy (ARVC) (Garcia-Gras et al. 2006). Additional evidence showed that plakoglobin entered the nucleus to modulate some Wnt/β-catenin target genes (Maeda et al. 2004), and that it might modulate cellular Src activity (Yin et al. 2005). Correlative links have also been made between misdirected desmocollin (3a and 3b) expression in transgenic mice and altered keratinocyte differentiation, with altered β-catenin stability and, thereby, Wnt activity being a possible basis of the phenotype (Hardman et al. 2005). Thus, desmosome-to-nuclear communication occurs in pathologic/experimental contexts, and presumably also under physiologic conditions.

α-Catenin, which is structurally unrelated to the other catenins (Kobielak and Fuchs 2004; Pokutta et al. 2008), binds to β-catenin within the cadherin–catenin complex, and independently modulates, and associates with the actin cytoskeleton (Yamada et al. 2005; Drees et al. 2005). Phosphorylation of α-catenin at tyrosine 148 increases its affinity for β-catenin (Burks and Agazie 2006), whereas β-catenin phosphorylation at tyrosine 142 lowers its affinity for α-catenin (Piedra et al. 2003). Just as the absence or presence of these secondary modifications represent points that might assist β-catenin release into a signaling pool, so might α-catenin be released in response to upstream kinases/phosphatases to modulate cytoskeletal or perhaps nuclear outcomes. α-Catenin has been observed in the nucleus (Giannini et al. 2004; El-Bahrawy et al. 2002), and reports suggest it may modulate β-catenin nuclear signaling (Giannini et al. 2004). Although the mechanisms are uncertain, genetic ablation of α-catenin alters NFκB, HH, MAPK, and Wnt signaling pathways (Kobielak and Fuchs 2006; Lien et al. 2006; Vasioukhin et al. 2001; Merdek et al. 2004; Hwang et al. 2005). Further studies may reveal activities of α-catenin that strengthen its position as a nuclear signaling entity, whereas additional work will be needed to determine if α-catenin originating from the cadherin–catenin complex contributes to CCJ-nuclear signaling.

p120 Subfamily

The p120-catenin subfamily includes p120, and the ARVCF-, δ-, and p0071-catenins (Hatzfeld 2005; Reynolds 2007; McCrea and Park 2007). Each binds weakly to the membrane proximal tail region of classic cadherins, and some members, such as p120, also interact with desmosomal cadherins (desmoglein 3) (Kanno et al. 2008). Interestingly, p120 subfamily members have a prominent role in determining the rate of cadherin endocytosis (Kowalczyk and Reynolds 2004; Xiao et al. 2007), and thus in determining the state of cadherin-mediated contacts.

p120 subfamily members directly or indirectly modulate small GTPases such as Rac1 and RhoA (Anastasiadis et al. 2000; Noren et al. 2001; Yap and Kovacs 2003; Wolf et al. 2006; Anastasiadis 2007) (see also Watanabe et al. 2009). This activity functionally distinguishes p120 family members from β-catenin and plakoglobin. An open question is the extent to which p120 members modulate small GTPases in the immediate molecular vicinity of the cadherin complex (Braga 2002). Although cadherin association appears to lower their capacity to modulate Rac1 or RhoA, p120 subfamily members may cycle on-and-off cadherins to modulate local GTPases and thereby cytoskeletal structures. Once fully dissociated from cadherin, p120 subfamily members could diffuse elsewhere to activate small GTPases whose downstream impact may include the modulation of gene activity via effects on cell-cycle progression (Villalonga and Ridley 2006; Wolf et al. 2006).

The most established outcome of p120 entry into the nucleus is loss of gene repression conferred by Kaiso, a POZ/zinc-finger family member. This outcome is conceptually analogous to β-catenin relieving TCF/LEF mediated repression, although the underlying mechanism involving p120 is different (Kelly et al. 2004; Daniel 2007). Although results relating to Kaiso's functions in Wnt signaling and TCF/LEF differ (Ruzov et al. 2009a; Ruzov et al. 2009b; Iioka et al. 2009), other evidence suggests that p120-catenin assists in de-repressing select Wnt/β-catenin target genes whose promoters have both TCF/LEF and Kaiso binding sites (Kim et al. 2004; Park et al. 2005; Spring et al. 2005; van Roy and McCrea 2005; Park et al. 2006). In addition to Kaiso, recent work revealed that the transcriptional repressor Glis2 associates with p120 (Hosking et al. 2007). Glis2 and Src promote p120's localization to the nucleus, and functional studies suggest that Glis2 participates in neuronal differentiation. Thus, as for β-catenin, we should consider the possibility that p120 has a nuclear function on its dissociation from cadherin. A recent report documented the nuclear- and p120-catenin-dependent entry of a cadherin cyto-domain fragment generated by presenilin, with consequent effects on Kaiso-mediated gene repression (Ferber et al. 2008). Recent results further indicate that, in common with β-catenin and plakoglobin, p120 subfamily members are stabilized in response to upstream Wnt signals (Park et al. 2006; JY Hong and PDM, unpublished results). This suggests an integrated network of catenin downstream effects, including those involving gene regulation. Such integrated responses could arise on Wnt-ligand signaling (Wnt-ligand–Frizzled–LRP complex formation), or in response to local kinase or other activities at CCJs. More work is required to address the validity of this latter possibility, as the graphic dissociation of multiple catenins into signaling pools is not often reported in response to stimuli.

It is intriguing to consider that activities of the cadherin and Wnt pathways may be more closely linked than presently recognized. For example, in addition to reports of Wnt pathway activation having effects on cadherin function (Bradley et al. 1993; Hinck et al. 1994b; Hinck et al. 1994a; Yanagawa et al. 1997; Shariatmadari et al. 2005; Ulrich et al. 2005; Wodarz et al. 2006), there are indications of physical associations between Wnt and cadherin/protocadherin pathway components (Medina et al. 2004; Qin et al. 2005; Hay et al. 2009), suggesting that cadherins are likely to modulate canonical (β-catenin mediated) or noncanonical (alternative) Wnt pathways. As cadherin-mediated adhesion is required for neighboring cells to engage in multiple ligand–receptor interactions, for example those occurring at varied cell synapses or contacts, cadherins contribute to quite indirect yet essential effects on downstream/nuclear signaling. Although not addressed in this article in detail, it is also important to note that the cadherin superfamily is diverse in structure and thus function, with the unifying element being the presence of cadherin-like repeats within the extracellular domain (Hulpiau and van Roy 2009). Some members appear to act more as signaling receptors (or ligands), as opposed to combined adhesion/signaling receptors. Interesting examples include the Dachsous and FAT cadherins (see also McNeill et al. 2009), the putative ligand and receptor of the newly characterized Hippo pathway, which shows a central and conserved role in tissue growth and size control in both vertebrates and invertebrates (Zhao et al. 2008). A further interesting subgroup of cadherins covered elsewhere is the protocadherins (Morishita and Yagi 2007). Protocadherins are expressed in multiple cell types but most prominently in neural tissues, and appear to regulate a number of signaling pathways relevant to cytoskeletal functions or gene activity (Medina et al. 2004; Unterseher et al. 2004; Yang et al. 2005; Chen and Gumbiner 2006; Wang et al. 2008).

Plakophilin Subfamily

A third class of catenins are the plakophilins (PKPs) (Hatzfeld 2007). Plakophilins contain a central Armadillo domain, similar to the β-catenin and p120-catenin subfamilies. In the case of plakophilins, however, the amino-terminal domains mediate interactions with desmogleins or desmocollins. A number of plakophilins have been localized to the nucleus (Mertens et al. 1996; Schmidt et al. 1997). For example, PKP2 associates with desmosomal cadherins and components of the PolIII transcriptional complex, such as RPC155 and the transcription factor TFIIIB (Mertens et al. 2001), and has been found to influence β-catenin-mediated Wnt signaling (Chen et al. 2002). The intracellular localization of PKP2 is modulated through phosphorylation by the serine kinase Cdc25C-associated kinase 1, which promotes the association of PKP2 with 14-3-3 proteins and localizes PKP2 to junctions rather than the nucleus (Muller et al. 2003). Another PKP family member, PKP1b, is mostly localized to the nucleus, whereas the alternatively spliced isoform PKP1a is mostly at junctions, suggesting that RNA processing of PKP family members affects PKP nuclear functions (Schmidt et al. 1997). Thus, plakophilins appear to function at both the desmosome and nucleus, although it remains unknown if plakophilins enter a signaling pool as a consequence of dissociating from desmosomal cadherins (*bona fide* junctional–nuclear signaling).

Another Perspective Relating to Junctional–Nuclear Catenin Signaling

As presented above, catenins may transduce signals to the nucleus on their dissociation from cell–cell junctions. An alternative view, not incompatible with the first, is that cell–cell junctions and nuclear components compete for catenin binding. Thus, when the level of cadherin is relatively high, catenins may be sequestered away from nuclear complexes (Choi et al. 2006). In experimental contexts, this has been shown to occur for β-catenin in which downstream Wnt pathway activity (e.g., transcription from endogenous Wnt gene targets or luciferase reporters) is suppressed upon cadherin cyto-domain sequestration of β-catenin (Heasman et al. 1994; Sanson et al. 1996; Orsulic et al. 1999). In such a case, the degree of catenin nuclear signaling might be modulated by the expression of cadherin, or through changes in cadherin–catenin affinity states or physical access. For example, in epithelial-to-mesenchymal transitions, the transcriptional repressors Snail and Slug repress E-cadherin expression (Peinado et al. 2007; Alves et al. 2009). If another cadherin does not sequester the population of catenin that would become available, and this population escapes metabolic destruction, effects on gene expression or small GTPase activity may result. Whether increased junctional–nuclear signaling results from catenin dissociation from cadherins (initial control at the protein level) and/or from increased catenin signaling pools that arise because of reduced cadherin expression (initial control at the transcriptional level), nuclear decisions are sensitive to the state of cadherin–catenin cell–cell contacts.

Receptor Tyrosine Kinases (RTKs)/Growth Factor Receptors

Cadherins and Ig-family adhesion receptors such as nectins (see the following) have been linked to the functions of growth factor receptors/RTKs (Comoglio et al. 2003). Examples include the roles of N-cadherin in neurite outgrowth (Williams et al. 1994; Sanchez-Heras et al. 2006) and of the Ig-family member N-CAM in tumor cell adhesion (Cavallaro et al. 2001). In the context of neural cells, N-caherin is thought to facilitate FGF receptor dimerization and signaling in a manner that is independent of FGF ligand (Doherty et al. 2000; Skaper et al. 2001). This relationship also appears to be independent of N-cadherin adhesive activity (Utton et al. 2001) and to involve N-cadherin and FGF receptor clustering (Williams et al. 2002). Work in epithelial and endothelial cells has indicated physical and functional interactions between N-cadherin or another classic cadherin, with the FGF receptor (Nieman et al. 1999; Kim et al. 2000; Suyama et al. 2002; Erez et al. 2004). N-cadherin–FGF-receptor interactions appear to involve interactions through their ecto-domains (Kim et al. 2000; Suyama et al. 2002; Sanchez-Heras et al. 2006), although in some contexts this association is difficult to detect at the biochemical level (Kim et al. 2005). Enhanced MAPK-ERK signaling correlates with N-cadherin-mediated cell surface retention and clustering of the FGF receptor, as well as an increase in FGF ligand responsiveness (Suyama et al. 2002). Regardless of the specific molecular underpinnings, N-cadherin's interaction with RTKs is likely relevant to its functional association with more dynamic/motile cell states in physiologic and pathologic contexts (Wheelock et al. 2008), and shows the role of cadherin signaling both in the context of localized phenomena such as cytoskeletal rearrangement and proliferative signaling by gene activity.

Cell–cell contact, presumably mediated by cadherins, has been shown to reduce cell proliferation promoted by the EGF ligand (Qian et al. 2004). E-cadherin has been found in complex with the EGF, FGF, and cErbB2 receptors (Ochiai et al. 1994; Fedor-Chaiken et al. 2003; Qian et al. 2004; Bryant et al. 2005). E-cadherin ligation reduces Stat 5b proliferative signaling downstream of the EGF receptor (Perrais et al. 2007), and FGF-induced MAPK signaling and receptor internalization (Bryant et al. 2005). In keratinocytes and mammary epithelial cells, however, E-cadherin was found to activate the EGF receptor (Fedor-Chaiken et al. 2003), which on cell–cell contact occurred even in the absence of EGF ligand (Pece and Gutkind 2000). In endothelial cells, VE-cadherin associates with, and reduces VEGF receptor activity, an effect that may occur with the cell-density dependent phosphatase PTP1 (Grazia Lampugnani et al. 2003). This VE-cadherin–RTK interaction appears to occur indirectly via cytoplasmic domain interactions, requiring β-catenin as reported also for the E-cadherin–EGF-receptor interaction (Hoschuetzky et al. 1994). Analogous to some cadherin–RTK contexts, VE-cadherin limits VEGF-receptor internalization, thereby lowering PLCγ/MAPK downstream signaling in response to growth signals (Lampugnani et al. 2006). This is similar to the effects of the E-cadherin–FGF-receptor complex, but not the N-cadherin–FGF-receptor complex. Several possibilities have been proposed for how cadherins might alter RTK (or other receptor) signaling, such as lateral receptor clustering to facilitate RTK auto-activation, alteration of ligand–RTK interactions, or modulation of RTK endocytosis.

In summary, although cadherin and Ig-family adhesion protein interactions with RTKs (as well as phosphatases) are incompletely understood, they represent a link between adhesive junction components and signaling pathways that modulate cell growth and a host of other processes. Further, although some of these relationships may not be dependent on cell–cell adhesion per se, others appear adhesion- or contact-responsive, such as E-cadherin-mediated growth arrest via effects on EGF-receptor interactions (Perrais et al. 2007). Interestingly, the FGF receptor appears to associate more with nonjunctional cadherins or Ig-family (N-CAM) adhesion proteins in cell lines derived from a number of sources (e.g., epithelial, myoblastic, and fibroblastic) (Sanchez-Heras et al. 2006). Further, N-cadherin and the PDGF receptor associate to promote cell movements at the leading edge of cells, with their cytoplasmic domain interactions occurring indirectly via β-catenin and NHERF (Theisen et al. 2007). These considerations indicate that much remains to be learned about how adhesive-protein–RTK signaling is engineered within cells, while already suggesting an intimate cell-surface interplay that ultimately modulates gene activity.

Small GTPases

We will discuss briefly junction-mediated effects on small GTPases (see by Watanabe et al). Cadherin-mediated adhesion has been shown to modulate Rac and less directly Rho and Cdc42, with these GTPases reciprocally contributing to determining cell–cell junction states (Anastasiadis 2007; Yap and Kovacs 2003; Braga and Yap 2005; Yap et al. 2007). Junctional effects on small GTPases (or the reverse) are complex, as differences are apparent as a function of cell type and context. Most of the effects studied have been within the junctional area itself, or cytoplasmic effects that largely relate to cytoskeletal structure–function.

Although GTPase effects are thought to be compartmentalized, including a proportion of those acting at cell–cell contact sites, it is likely that some GTPase signals contribute directly to longer-range signaling to the nucleus. Future studies might address, for example, the impact of junctionally activated (or inhibited) small GTPases on cell cycle progression. In such cases, it is likely that p120-catenin subfamily members (Anastasiadis 2007), or other molecular players such as PI3 kinase or Akt (Yap and Kovacs 2003; Yap et al. 2007) (see the following), would act as intermediates. Importantly, Rho, Rac, and Cdc42 are known to play roles in cell cycle progression (Coleman et al. 2004; Villalonga and Ridley 2006), and each is responsive to cadherin-mediated cell–cell contact in context-dependent manners (Yap and Kovacs 2003; Burridge and Wennerberg 2004; Yap et al. 2007; Anastasiadis 2007). Thus, as we learn more from future studies, small GTPases are likely to become recognized players in CCJ-nuclear signaling.

PI3 Kinase and Akt

In addition to small GTPase signals arising from CCJs, other signaling mediators such as PI3K and Akt have attracted attention (Larue and Bellacosa 2005; Giehl and Menke 2008). For example, the survival of melanoma cells is enhanced by N-cadherin through an Akt-dependent mechanism (Li et al. 2001), whereas reduced N- or VE-cadherin function in endothelial, granulose, or hepatocellular carcinoma cells results in apoptosis (Peluso et al. 1996; Makrigiannakis et al. 1999; Erez et al. 2004; Gwak et al. 2006; Jiang et al. 2007). E-cadherin also promotes cell survival, inclusive of cancer contexts (St Croix and Kerbel 1997; Kantak and Kramer 1998; Kang et al. 2007). As junctional molecules such as cadherins allow for multiple ligand–receptor interactions, it is not possible in many cases to conclude that cell survival effects are mediated as a direct consequence of cadherin cytoplasmic domain interactions. However, the involvement of cadherin–catenin complexes in producing intracellular signals suggest that direct cadherin effects are contributory and perhaps even predominant in some instances. In this regard, strong evidence has accumulated that PI3 kinase and Rac activation results following cadherin–cadherin *trans* interactions, and that the cadherin cytoplasmic domain is involved (Kovacs et al. 2002; Goodwin et al. 2003). Future work will be required to determine the relative impact of such signals locally, relative to those that might quickly reach the nucleus to modulate gene programs.

NECTIN BASED SIGNALING

Thus far, our discussion has largely taken place from a cadherin-centric perspective. However, it is clear that other adhesive transmembrane polypeptides, such as the Ig-family member N-CAM, also modulate signaling pathways reaching the nucleus (Williams et al. 1994; Choi et al. 2001; Little et al. 2001). Although we will not discuss N-CAM further, we will summarize evidence that the nectin Ig-family of adhesion receptors have important roles in junction formation in epithelia, fibroblasts, and neurons (Takai and Nakanishi 2003; Takai et al. 2008a; Sakisaka et al. 2007). Heterophilic *trans* interactions between nectin1 and nectin3 occur at synaptic junctions (Mizoguchi et al. 2002; Togashi et al. 2006), whereas homophilic N-cadherin *trans* interactions are from both sides of the synapse. Nectins bind several cytoplasmic proteins including afadin, which binds actin, and the cell polarity protein Par3. Indirectly, nectins associate with α-catenin, annexin II, IQGAP1, and ZO1. This latter ZO1 interaction is relevant to nectin's roles in tight junction formation (Ooshio et al. 2007). With respect to signaling, nectin3 associates in *cis* with the integrin α(v)β(3), which appears important for peripheral cytoskeletal organization and activity, and formation of adherens and tight junctions (Sakamoto et al. 2006; Ozaki et al. 2007; Sakamoto et al. 2008). PKC, FAK, and c-Src participate in nectin function downstream of α(v)β(3) (Ozaki et al. 2007; Sakamoto et al. 2008). The fact that nectins interact with components of both the adherens junction (cadherins through α-catenin–afadin associations), tight junctions (ZO1), and α(v)β(3) integrin, appears to provide an example of junctional cross talk. Nectin3 has also been found in *cis* association with the PDGF, but not the

EGF receptor, raising the question whether nectin modulates RTK signaling or vice-versa, and if this nectin complex might also include cadherin or $\alpha(v)\beta(3)$ integrin (unpublished in Sakisaka et al. 2007). A number of studies have further established functional links between nectins and the activation of the small G proteins Rap1, Cdc42, and Rac through c-Src (Ogita and Takai 2006; Kawakatsu et al. 2005).

It has been proposed that distinct nectin-like (Necl) transmembrane molecules link nectin pathways directly with the nucleus. Much of the focus has been on Necl5, which appears to contribute to contact-mediated inhibition of cell growth, and is capable of binding in *trans* to nectin3 in confluent cell culture. In nonconfluent cells, however, Necl5 associates in *cis* with the PDGF receptor, and $\alpha(v)\beta(3)$ integrin. These binary associations appear to influence Rap1 and Rac signals involved in cell movements, and Ras signals that induce cell proliferation (H. Amano et al. unpublished; Sakisaka et al. 2007; Minami et al. 2007; Kakunaga et al. 2004). Once cells reach higher densities, Necl5 is thought to engage in heterotypic interactions with nectin3, reducing Ras-mediated proliferative signaling and enhancing signals that strengthen junctions through cadherin and cytoskeletal reorganization. Necl5 is also associated with Sprouty, a negative regulator of growth factor signaling through inhibition of Ras (Kajita et al. 2007). In subconfluent cells, Necl5 appears to block Sprouty inhibition of Ras, favoring a proliferative cell state. In confluent cells, Necl5 is endocytosed more rapidly, and this reduction in Necl5 levels is proposed to allow Sprouty to inhibit Ras, leading to slowed cell proliferation. Finally, recent work indicates that the nectin–afadin complex contributes to PDGF-induced cell survival through the PI3K-Akt signaling pathway (Kanzaki et al. 2008). Cadherins have also been linked to cell survival signaling (Peluso et al. 1996; Kantak and Kramer 1998; St Croix and Kerbel 1997; Makrigiannakis et al. 1999; Gwak et al. 2006; Jiang et al. 2007; Wheelock et al. 2008), and as above noted, associate with RTKs. Prosurvival effects arise from Akt and β-catenin signaling in the case of N-cadherin (Li et al. 2001), and Akt signaling upon E-cadherin association with ErbB4 (Kang et al. 2007). Given the previously noted associations that have been resolved between cadherins and nectins, it may well be that their respective downstream signals are coordinated or interdependent.

GAP JUNCTIONS

Among CCJs, gap junctions represent a unique nonadhesive connection formed from the docking of two hemichannels (or connexon, a hexameric cluster of various connexin isoforms), allowing direct intercellular communications between adjacent cells (Fig. 3) (see Goodenough and Paul 2009). Such an exchange of chemical or electrical signals has been implicated in a broad spectrum of biological events, including metabolic and ion homeostasis, synchronization of muscle contraction, and cell proliferation, survival, differentiation, and migration (Wong et al. 2008; Harris 2007; Mese et al. 2007).

Unlike many classical outside-in signaling pathways, in which ligands bind receptors to trigger events such as phosphorylation or oligomerization and ultimately alter gene expression, gap junctions or hemichannels function as gate keepers to permit molecular diffusion in their opened conformation and prevent such diffusion in closed states (Harris 2007). Entering signaling entities include calcium and other critical metabolites or second messengers that may alter gene expression through known (e.g., CxRE dependent) or unknown mechanisms (Rodriguez-Sinovas et al. 2007).

A number of connexins appear to localize to the nucleus. For example, on immunostaining, connexin 43 has been observed in the nuclei of magnocellular neurons (Eisner and Colombo 2002), whereas its carboxy-terminal fragment localizes to cardiomyocyte and HeLa cell nuclei and correlates with decreased cell proliferation (Dang et al. 2003). Further experimentation is needed, however, to determine if physiologic gap junction–nuclear signaling occurs as a function of connexin

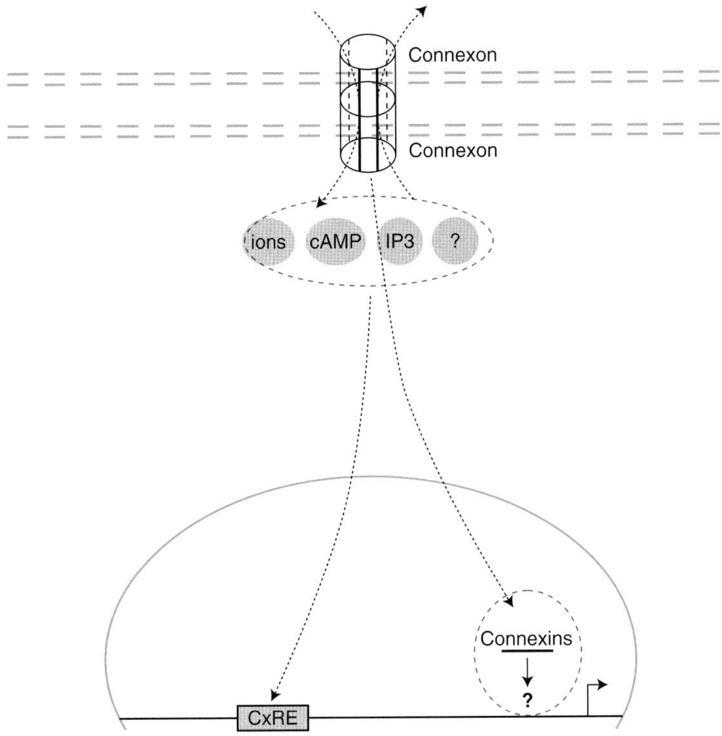

Figure 3. Gap junction signaling to the nucleus. Gap junctions are nonadhesive channels that allow regulated intercellular transmission of electrical or small molecule signals. Connexin 43, a gap junction structural component, has been observed in the nuclei of certain cells and associated with decreased cell proliferation.

or connexin-fragment entry to the nucleus, or through other proteins originally associated with plasma membrane-associated gap junctions.

TIGHT JUNCTIONS

Tight junctions (TJs) are areas of close contact between neighboring plasma membranes, where intramembrane strands at the contact sites can be visualized by freeze-fracture electron microscopy (Tsukita et al. 2001; see Furuse 2009). Tight junctions act as barriers that show both gating and fencing functions: Gating determines the paracellular diffusion of molecules according to size and charge, whereas fencing restricts lipid diffusion between the apical and basolateral intramembrane domains (Balda and Matter 2008; Cereijido et al. 2008; Paris et al. 2008; see Anderson and Van Itallie 2009). Certain proteins of the epithelial TJ show cellular localizations other than the TJ. These proteins localize to both nuclei and TJ, and tend to be involved in the regulation of cell proliferation, gene expression, and cell differentiation (Fig. 4) (Balda and Matter 2003; Matter and Balda 2007). Here, we discuss TJ proteins having both junctional and nuclear localizations, and summarize what is known about their functions, regulation, and expression.

Six TJ-associated transmembrane proteins have been identified: occludin, claudins, tricellulin, JAMs (Junction Adhesion Molecules), CRB-3 (a human homolog of *Drosophila* Crumbs), and Bves (blood vessel/epicardial substance) (Lemmers et al. 2004; Osler et al. 2005; Furuse and Tsukita 2006; Wang and Margolis 2007; Chiba et al. 2008). It remains unclear how each of these proteins interacts with components of the TJ cytoplasmic plaque

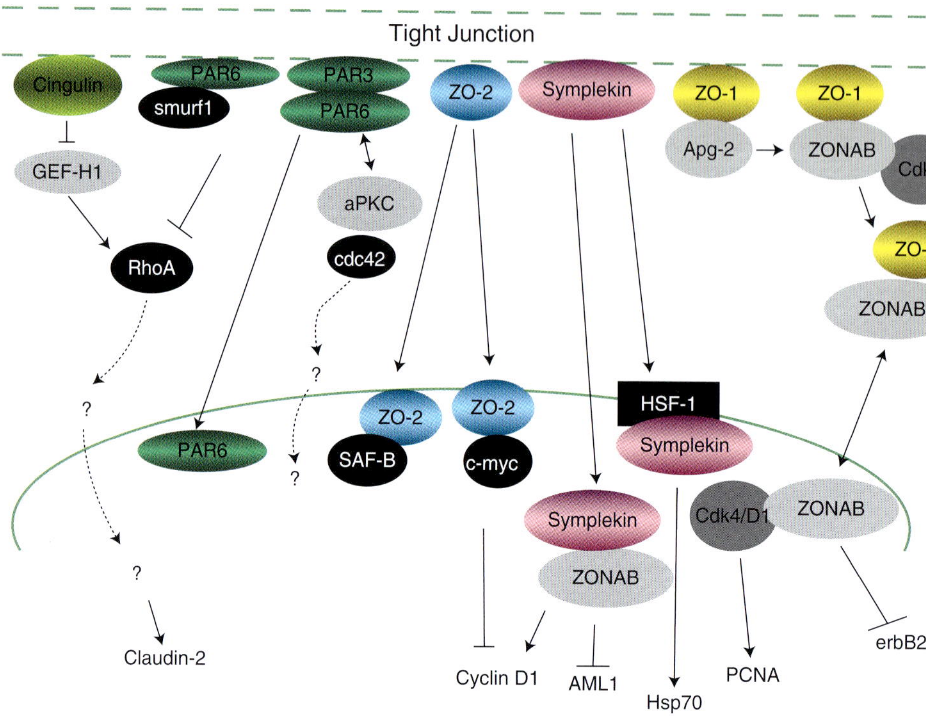

Figure 4. Tight junction signaling to the nucleus. Displayed are pathways associated with tight junctions and suggested or showed to regulate gene expression. Arrows refer to positive functional interactions that could be because of either indirect positive effects or direct physical interactions. Conversely, lines ending with a "T" are negative functional interactions reflecting indirect effects or direct physical interactions. Although most of the junctional proteins depicted associate with junctional membrane proteins, many of the transmembrane entities relevant to the indicated pathways are not known and therefore not indicated.

and how these interactions result in a protein network that controls paracellular permeability, gene expression, junctional dynamics, proliferation, and polarity. The TJ cytoplasmic plaque is formed by multiple adaptor and scaffolding proteins (e.g., ZO-1/2/3, PATJ, Pals1, PAR-3, and PAR-6), different types of signaling components such as GTP-binding proteins, protein kinases, and phosphatases, as well as transcriptional and post-transcriptional regulators (Matter and Balda 2007; Paris et al. 2008). Several proteins have dual localizations at TJs and the nucleus, including ZONAB, cdk4, symplekin, ZO-2, PAR-6, huASH1, and ubinuclein. Most of these proteins have been linked to cell proliferation, gene expression, and/or cell differentiation, and interact with and regulate transcription factors.

ZO-1 AND ZONAB IN CELL PROLIFERATION AND GENE EXPRESSION

ZO-1, the first TJ protein identified (Stevenson et al. 1986), contains distinct protein–protein interaction domains including three PDZ, an SH3 domain, and a domain homologous to yeast guanylate kinase (GUK domain) (Willott et al. 1993). ZO-1 belongs to the same protein family as DlgA (Discs large A), a *Drosophila* tumor suppressor (Tsukita et al. 1993; Woods and Bryant 1993; Fanning et al. 2007). ZO-1 is present not only in epithelial TJs, but also in fibroblasts, neurons, and astrocytes (Howarth et al. 1992), suggesting it has more general functions in cells that do not have tight junctions. Although ZO-1 has been observed in the nucleus of certain cells and conditions, some of the data is conflicting (Gottardi et al. 1996;

Balda and Matter 2000; Reichert et al. 2000; Glaunsinger et al. 2001; Matter and Balda 2003; Traweger et al. 2003). Moreover, the known signaling functions of ZO-1 do not require a nuclear pool of ZO-1, but involve interactions that occur in the cytoplasm and lead to the cytoplasmic sequestration of the dual localization protein ZONAB.

ZONAB is a Y-box transcription factor that belongs to a small family of multifunctional regulators of transcription and post-transcriptional processes (Gallia et al. 2000; Kohno et al. 2003). The Y-box factor family includes DbpA/ZONAB, DbpB/YB-1, and DbpC/contrin (Sakura et al. 1988; Lloberas et al. 1995). ZONAB controls expression of the growth factor coreceptor erbB2, cell cycle regulators such as cyclin D1 and PCNA, and regulates epithelial morphogenesis (Balda and Matter 2000; Balda et al. 2003; Sourisseau et al. 2006). Knockout experiments in mice have shown that mice deficient in DbpA/ZONAB (also called MSY3 in mice) develop normally. However, functional redundancy is indicated as DbpA/ZONAB and DbpB/YB-1 double knockout mice die earlier than animals knocked out for DbpB/YB-1 alone (Lu et al. 2006).

ZONAB binds to the SH3 domain of ZO-1, which contributes to ZONAB's regulation (Balda and Matter 2000; Balda et al. 2003). In MDCK cells, ZO-1 and ZONAB expression are inversely regulated by cell density, and results from different experimental approaches indicate that ZO-1 suppresses proliferation by inhibiting ZONAB's nuclear accumulation (Balda and Matter 2000; Balda et al. 2003). Indeed, the SH3 domain of ZO-1 that binds to ZONAB is necessary and sufficient to sequester ZONAB in the cytoplasm and reduce proliferation (Balda et al. 2003). ZONAB associates with the cell division kinase (CDK) 4, which also has a TJ-nuclear localization; thus, reduction of nuclear ZONAB levels results in reduced nuclear CDK4. Hence, the cell-density dependent accumulation of ZO-1 at TJs works to inhibit the proliferation-promoting ZONAB and CDK4 pathways.

The relative expression levels of ZONAB and ZO-1 are tightly regulated. ZO-1 is expressed at low levels in sparse cultures of MDCK cells, and its levels rises as a function of increased cell density (Balda and Matter 2000). As ZO-1 interacts with many junctional proteins and its half-life increases with greater cell–cell contact, its up-regulation seems to involve junctional stabilization (Gumbiner et al. 1991). Transcriptional mechanisms also affect ZO-1 expression. ZO-1 levels are decreased during corneal wound repair (Cao et al. 2002), in response to β–catenin (Mann et al. 1999), on experimental cell transformation and in a variety of cancers (Miettinen et al. 1994; Chen et al. 2000; Tian and Phillips 2002; Eger et al. 2004; Harten et al. 2009). Analysis of the ZO-1 promoter revealed response elements targeted during EMT (Venkov et al. 2007), as well as by signaling networks activated in neoplastic, metabolic, and viral diseases (Chen et al. 2008). In contrast to ZO-1, ZONAB/DbpA expression increases in contexts that favor proliferation, such as transcriptional regulation by E2F1, and thus is a prognostic marker for hepatocellular carcinomas (Yasen et al. 2005).

ZONAB activity is also regulated by other proteins that control its interaction with ZO-1. One such protein is Apg-2, a heat shock protein that like ZONAB binds the SH3 domain of ZO-1 (Tsapara et al. 2006). Apg-2 is a member of the Hsp110 family of chaperones in the cytosol, the nucleus, and to a lesser extent at the cell–cell junctions in MDCK cells (Tsapara et al. 2006). In response to heat shock, Apg-2 is recruited to junctions and binds ZO-1, leading to ZONAB dissociation, nuclear localization, and activation. Apg-2 depletion retards TJ assembly in calcium switch experiments using two-dimensional cultures, and affects the development of well-organized polarized cysts in three-dimensional cultures (Aijaz et al. 2007), similar to the effects of ZO-1 depletion (Aijaz et al. 2006; Sourisseau et al. 2006). However, whether this effect is due solely to Apg-2's relationship with ZO-1 and ZONAB is not clear. Nevertheless, a functional interaction of these components is further suggested by Apg-2's over-expression in hepatocellular carcinomas

(Gotoh et al. 2004), where ZONAB/DbpA is a prognostic marker (Yasen et al. 2005), and ZO-1 is down-regulated (Orban et al. 2008). Thus, it seems that liver cancer correlates with increased ZONAB activity either because of increased expression of ZONAB itself or of Apg-2, or because of down-regulation of ZO-1.

SYMPLEKIN IN mRNA PROCESSING AND CELL PROLIFERATION

Symplekin, a nuclear protein that can associate with TJs, has been linked to the machinery involved in 3′-end processing of pre-mRNA and polyadenylation (Keon et al. 1996; Takagaki and Manley 2000; Hofmann et al. 2002). In stressed cells, symplekin interacts with HSF-1 (heat shock inducible transcription factor one) to regulate hsp70 mRNA polyadenylation (Xing et al. 2004). In *Xenopus*, symplekin and an unusual poly(A) polymerase, GLD-2, are required for cytoplasmic polyadenylation (Barnard et al. 2004). Furthermore, symplekin and the polyadenylation factor CPSF-73 are targets of the small ubiquitin-like modifier SUMO (Vethantham et al. 2007). Thus, symplekin has a role in mRNA processing.

Whether symplekin's junctional localization reflects a regulatory role of cell–cell adhesion in polyadenylation is not known. As symplekin appears to bind transcription factors, it may localize to sites where those interacting transcription factors are found. This possibility is supported by the finding that symplekin interacts with ZONAB. Symplekin depletion by RNA interference reduces ZONAB nuclear accumulation, transcriptional activity, cell proliferation and expression of cyclin D1 in a ZONAB-dependent manner (Kavanagh et al. 2006). Furthermore, symplekin cooperates with ZONAB to negatively regulate intestinal goblet cell differentiation, acting via repression of the AML1 transcription factor (Buchert et al. 2009). As ZONAB is a Y-box factor, a family of multifunctional proteins that can interact with DNA as well as RNA (Kohno et al. 2003), it is tempting to speculate that the interaction between symplekin and ZONAB is not only functionally important for transcription but also for RNA processing, stability, and/or localization.

ZO-2 IN CELL PROLIFERATION AND GENE EXPRESSION

ZO-2 has the same domain structure as ZO-1 and was identified because of its interaction with ZO-1 (Gumbiner et al. 1991; Jesaitis and Goodenough 1994). ZO-2 shuttles from the cell periphery to the nucleus (Islas et al. 2002; Traweger et al. 2003; Kausalya et al. 2004), interacts with the nuclear ribonucleoprotein scaffold attachment factor-B (SAF-B) (Traweger et al. 2003) and several transcription factors (Betanzos et al. 2004; Huerta et al. 2007). ZO-2 inhibits cell proliferation by affecting transcription, translation, and degradation of cyclin D1 (Huerta et al. 2007; Tapia et al. 2009), coincident with observations that its expression is reduced in certain tumors (Chlenski et al. 2000; Fink et al. 2006; Paschoud et al. 2007). ZO-2 binds and inactivates oncogenic viral proteins (Glaunsinger et al. 2000; Lee et al. 2000; Glaunsinger et al. 2001), and its degradation is induced by HIV type 1 gp120 (Nakamuta et al. 2008). Conversely, in some epithelial and endothelial cells, ZO-2 has been reported to increase proliferation and regulate the expression of M2 pyruvate kinase (Traweger et al. 2008).

In mice, ZO-2 deficiency is embryonic lethal; embryos show decreased proliferation at embryonic day 6.5, and increased apoptosis at embryonic day 7.5, indicating an arrest in early gastrulation (Xu et al. 2008). Depletion of ZO-2 protein by RNA interference in two-cell embryos delays blastocoel cavity formation with normal cell proliferation and morphogenesis. In contrast, ZO-1 knockdown, or combined ZO-1 and ZO-2 knockdown, produces a more severe inhibition of blastocoel formation, indicating distinct but possibly overlapping roles for ZO proteins in blastocyst morphogenesis (Sheth et al. 2008). Further studies are needed to address these contradictions in ZO-2's role in cell proliferation, possibly because of different experimental systems/contexts, and to identify genes regulated by ZO-2 in association with SAF-B or other transcription factors.

HuASH1 AND UBINUCLEIN IN CHROMATIN REMODELING AND GENE EXPRESSION

HuASH1, the human homolog of *Drosophila* ASH1 (absent, small, or homeotic discs 1), belongs to the trithorax group of transcription factors. It was found to colocalize with TJ markers by immunofluorescence (Nakamura et al. 2000). HuASH1 functions as a histone methyltransferase, an enzymatic activity involved in chromatin remodeling and gene expression (Beisel et al. 2002; Gregory et al. 2007), and interacts with HDAC1 repression complexes (Tanaka et al. 2008). It remains unknown how huASH1 is recruited to intercellular junctions, or the functional relevance of its junctional association in chromatin remodeling and gene expression.

Ubinuclein belongs to a novel family of histone chaperones conserved throughout eukaryotes (Balaji et al. 2009), and has been found to interact with viral transcription factors in the nucleus and ZO-1 at epithelial TJs (Aho et al. 2008). When overexpressed, ubinuclein prevents MDCK cells from going through cytokinesis (Aho et al. 2008). Whether and how ZO-1 might regulate the function and localization of ubinuclein, or the specifics of ubinuclein's interactions with other proteins in the nucleus, remain open questions.

TIGHT JUNCTION-ASSOCIATED REGULATION OF SMALL GTPases

TJs also participate in the regulation of small GTPases (see also Watanabe et al. 2009). RhoA activation affects actin organization and stimulates epithelial cell proliferation, gene expression, and differentiation (Jaffe and Hall 2002; Sahai and Marshall 2002; Nelson 2008). RhoA is down-regulated when epithelial cells reach confluence, resulting in inhibition of signaling pathways that stimulate proliferation. GEF-H1/Lfc, a guanine nucleotide exchange factor specific for RhoA, is involved in the regulation of paracellular permeability and cell proliferation (Benais-Pont et al. 2003; Aijaz et al. 2005). GEF-H1/Lfc directly interacts with cingulin, a junctional adaptor, resulting in its inhibition and thus reduced RhoA activity. In MDCK cells with reduced cingulin levels, increased claudin-2 expression is RhoA-dependent, suggesting it might be mediated by GEF-H1 signaling (Guillemot and Citi 2006).

JACOP/paracingulin, a TJ plaque protein with sequence similarity to cingulin (Ohnishi et al. 2004), interacts with GEF-H1 as well as the Rac GEF Tiam1, thereby influencing junction assembly (Guillemot et al. 2008). Tiam1 and Cdc42 have been linked to the assembly of the TJ-associated, evolutionarily conserved cell polarity signaling pathway that includes the PAR3/PAR6/aPKC complex, and is required for the formation of distinct tight and adherens junctions (Ebnet et al. 2004; Assemat et al. 2008; Iden and Collard 2008; Macara and Mili 2008).

PAR6 itself is a protein with multiple interaction partners and also shows dual localization. In different experimental systems, PAR6 was shown to interact with small GTPases such as cdc42, Rac2, Rin, Rit, and Wrch-1 (Johansson et al. 2000; Hoshino et al. 2005; Macara and Mili 2008; Brady et al. 2009), as well as p190RhoGAP (Zhang and Macara 2008). PAR6 is also part of the TGF-β-activated pathway leading to ubiquitination and degradation of RhoA (Wang and Margolis 2007). PAR6 localizes to nuclear speckles and colocalizes with Tax, a transcriptional activator of the human T-cell leukemia virus type 1 long terminal repeat (Cline and Nelson 2007). Although its nuclear functions are not clear, PAR6 does not seem to be directly involved in transcription but likely instead plays an architectural role.

TIGHT JUNCTION CROSS TALK WITH GROWTH FACTORS AND EXTRACELLULAR MATRIX PATHWAYS

The relationship of TJs with growth factor and matrix signaling is an emerging area of research. Occludin regulates localization of the TGF-β type I receptor and thus plays a role in TGF-β-dependent dissolution of TJs during epithelial-to-mesenchymal transitions (Bose and Wrana 2006). FGF-2 fails to induce angiogenesis in JAM-A-deficient mice (Cooke et al. 2006), suggesting a role of JAM-A in FGF-2 signaling, which might be mediated by the PAR3/PAR6/aPKC complex (Ebnet et al. 2004). Additionally, EGF signaling seems to affect the expression and localization of several TJ proteins that regulate TJ assembly and barrier functions (Van Itallie et al. 1995; Basuroy et al. 2006; Wang et al. 2006; Flores-Benitez et al. 2007; Singh et al. 2007), suggesting that EGF signaling could have an indirect role in regulating TJ-nuclear signaling by modulating the activities of the PAR3/PAR6/aPKC and/or ZO-1/ZONAB pathways.

It is generally accepted that integrin signaling pathways regulate cell migration, junctional complex stability, and cell–cell interactions including those taking place at TJs (Ojakian et al. 2001). There are also examples linking TJ-ssociated proteins with integrins. For example, claudin-11 forms a complex with OAP-1, a tetraspan family protein, and with β1 integrin to regulate proliferation and migration of oligodendrocytes (Tiwari-Woodruff et al. 2001). JAMs regulate cell migration through effects on β(1) integrin or, depending on the cellular context, on α(v)β(3) integrin (Mandell et al. 2005; Naik and Naik 2006; Mandicourt et al. 2007). These results suggest that transmembrane TJ proteins may influence nuclear events through their effects on integrins. It is tempting to speculate that such relationships exist because down-regulation of ZO-1 or manipulation of ZO-1 interactions with ZONAB produce mild phenotypes in monolayers but inhibit cyst formation in three dimensional (collagen/matrigel) cultures (Umeda et al. 2004; Aijaz et al. 2006; McNeil et al. 2006; Sourisseau et al. 2006; Aijaz et al. 2007). Future studies will need to address the specific molecular mechanisms involved; for example, if/how the TJ affects gene expression profiles relevant to extracellular matrix adhesion and signaling.

ADHERENS AND TIGHT JUNCTION CROSS TALK

In revising earlier ideas in which junctions may have been viewed as rather distinct functional entities, it seems likely that in many cases shared functions occur including those involving signals directed to the nucleus. For example, the p120- and ARVCF-catenins, traditionally viewed as components of adherens junctions, also interact with ZO-1 and ZO-2 (Kausalya et al. 2004). In polarized epithelial MDCK cells, ARVCF and ZO-1 partially colocalize in the vicinity of the apical adhesion complex. Disruption of cell–cell adhesion releases ARVCF from the plasma membrane, with an increased ARVCF fraction localizing to the nucleus where it binds ZO-2 (Kausalya et al. 2004). Thus, even as the nuclear roles of ARVCF are not known, the interaction of TJ and adherens junction proteins may allow cooperation between junctions in regulating gene expression.

Cyclin D1 is an example of a gene target responsive to both TJ and adherens junction function, and which is up-regulated in many tumors (Coqueret 2002; Fu et al. 2004). Several signaling pathways that regulate cell–cell and cell–extracellular matrix adhesion have an impact on cyclin D1 expression (Shtutman et al. 1999; Tetsu and McCormick 1999; D'Amico et al. 2000; Amanatullah et al. 2001; Zhao et al. 2001; Balda and Matter 2003; Zhao et al. 2003). The identification of other genes regulated by both adherens and tight junctions should help us understand downstream cross talk occurring between adherens and tight junctions, as well as appreciate how these pathways interact with those activated by growth factors or the extracellular matrix (Balda and Matter 2003).

Junctional cross talk involving nuclear signals is further reflected in roles of the adherens junction in transcriptional regulation of TJ gene products. In endothelia, for example, the level of VE-cadherin affects that of the TJ protein claudin-5. This effect appears to occur through Akt, which phosphorylates the forkhead box factor FoxO1, and regulates the

TCF4–β-catenin transcriptional repressor complex (Taddei et al. 2008). It has been suggested that potential TCF/LEF binding sites present in the promoter of claudin-3 may account for its down-regulation in response to reduced β-catenin activity (Liebner et al. 2008). In the future, it will be interesting to further examine whether signaling pathways downstream of one junction have an impact on the expression of gene products comprising another junction.

CONCLUDING REMARKS

Although it is clear that CCJ-nuclear signaling takes place in cells and tissues, it is also apparent that our present knowledge is not satisfactory in understanding the larger roles of CCJs in transcriptional modulation. Although we are able to list a number of intriguing examples for each junction type, we are far from having a coherent picture of the networks involved at the plasma membrane, cytoplasmic, or nuclear levels. An inherent difficulty in identifying the network and its details is that downstream signals originating from cell–cell contacts are both direct and indirect, and are likely to be concurrent. For example, signals may be blocked or promoted by a component being junctionally sequestered or released from the adhesion complex to the nucleus (direct) or, alternatively, mediated through associated signaling molecules such as an RTKs, or other ligand–receptor or adaptor/signaling pairs dependent on the presence of the junction (each indirect).

One approach that has been helpful is the use of engineered contacts bearing purified adhesion receptors (coated beads, etc.). Although this is not physiologic, it assure that the physical interaction presented in *trans* is defined and, if desired, singular. Such methods, combined with the use of reagents to report pathway activation in real time, can also be useful in working out the temporal sequence of pathway activation following contact. In time, it may be that single molecule tracking methods will be used to more firmly establish the fate of individual molecules putatively traveling between junctions and the nucleus. This could assist, for example, in ascertaining if a specific catenin molecule dissociating from a cadherin complex is actually capable of reaching transcriptional regulatory regions in the nucleus. Although much work remains, we can appreciate the findings that have already been made, as they tell us of two fascinating worlds that listen to one another, one at the cell surface where cells meet and touch, and the other internal where transcriptional and proliferative/differentiation outcomes are further processed.

ACKNOWLEDGMENTS

This article is dedicated to Margaret Jean (Peggy) Wheelock, PhD (1945–2009). We thank W.J. Nelson for his critical reading, helpful comments, and patience during the review process. The laboratory of PDM is supported through funding from NIH (RO1GM52112), Texas ARP (003657-0008-2006), and UT MDACC NCI Core Grants (CA-16672). The laboratory of M.S.B. is supported through funding from the Wellcome Trust, MRC, BBSRC, AICR, and Fight for Sight.

REFERENCES

Abram CL, Lowell CA. 2009. The ins and outs of leukocyte integrin signaling. *Annu Rev Immunol* **27:** 339–362.

Aho S, Lupo J, Coly PA, Sabine A, Castellazzi M, Morand P, Sergeant A, Manet E, Boyer V, Gruffat H. 2008. Characterization of the ubinuclein protein as a new member of the nuclear and adhesion complex components (NaCos). *Biol Cell* **101:** 319–334.

Aijaz S, D'Atri F, Citi S, Balda MS, Matter K. 2005. Binding of GEF-H1 to the tight junction-associated adaptor cingulin results in inhibition of Rho signaling and G1/S phase transition. *Dev Cell* **8:** 777–786.

Aijaz S, Balda MS, Matter K. 2006. Tight junctions: Molecular architecture and function. *Int Rev Cytol* **248:** 261–298.

Aijaz S, Sanchez-Heras E, Balda MS, Matter K. 2007. Regulation of tight junction assembly and epithelial morphogenesis by the heat shock protein Apg-2. *BMC Cell Biol* **8:** 49.

Alema S, Salvatore AM. 2007. 120 catenin and phosphorylation: Mechanisms and traits of an unresolved issue. *Biochim Biophys Acta* **1773:** 47–58.

Alves CC, Carneiro F, Hoefler H, Becker KF. 2009. Role of the epithelial-mesenchymal transition regulator Slug in primary human cancers. *Front Biosci* **14:** 3035–3050.

Amanatullah DF, Zafonte BT, Albanese C, Fu M, Messiers C, Hassell J, Pestell RG. 2001. Ras regulation of cyclin D1 promoter. *Methods Enzymol* **333:** 116–127.

Anastasiadis PZ. 2007. 120-ctn: A nexus for contextual signaling via Rho GTPases. *Biochim Biophys Acta* **1773:** 34–46.

Anastasiadis PZ, Moon SY, Thoreson MA, Mariner DJ, Crawford HC, Zheng Y, Reynolds AB. 2000. Inhibition of RhoA by p120 catenin. *Nat Cell Biol* **2:** 637–644.

Anderson JM, Van Itallie CM. 2009. Physiology and function of the tight junction. *Cold Spring Harb Perspect Biol* **1:** a002584.

Andl CD, Rustgi AK. 2005. No one-way street: Cross-talk between e-cadherin and receptor tyrosine kinase (RTK) signaling: A mechanism to regulate RTK activity. *Cancer Biol Ther* **4:** 28–31.

Assemat E, Bazellieres E, Pallesi-Pocachard E, Le Bivic A, Massey-Harroche D. 2008. Polarity complex proteins. *Biochim Biophys Acta* **1778:** 614–630.

Balaji S, Iyer LM, Aravind L. 2009. HPC2 and ubinuclein define a novel family of histone chaperones conserved throughout eukaryotes. *Mol Biosyst* **5:** 269–275.

Balda MS, Matter K. 2000. The tight junction protein ZO-1 and an interacting transcription factor regulate ErbB-2 expression. *Embo J* **19:** 2024–2033.

Balda MS, Matter K. 2003. Epithelial cell adhesion and the regulation of gene expression. *Trends Cell Biol* **13:** 310–318.

Balda MS, Matter K. 2008. Tight junctions at a glance. *J Cell Sci* **121:** 3677–3682.

Balda MS, Garrett MD, Matter K. 2003. The ZO-1-associated Y-box factor ZONAB regulates epithelial cell proliferation and cell density. *J Cell Biol* **160:** 423–432.

Barker N. 2008. The canonical Wnt/β-catenin signalling pathway. *Methods Mol Biol* **468:** 5–15.

Barnard DC, Ryan K, Manley JL, Richter JD. 2004. Symplekin and xGLD-2 are required for CPEB-mediated cytoplasmic polyadenylation. *Cell* **119:** 641–651.

Basuroy S, Seth A, Elias B, Naren AP, Rao R. 2006. MAPK interacts with occludin and mediates EGF-induced prevention of tight junction disruption by hydrogen peroxide. *Biochem J* **393:** 69–77.

Beisel C, Imhof A, Greene J, Kremmer E, Sauer F. 2002. Histone methylation by the *Drosophila* epigenetic transcriptional regulator Ash1. *Nature* **419:** 857–862.

Benais-Pont G, Punn A, Flores-Maldonado C, Eckert J, Raposo G, Fleming TP, Cereijido M, Balda MS, Matter K. 2003. Identification of a tight junction-associated guanine nucleotide exchange factor that activates Rho and regulates paracellular permeability. *J Cell Biol* **160:** 729–740.

Betanzos A, Huerta M, Lopez-Bayghen E, Azuara E, Amerena J, Gonzalez-Mariscal L. 2004. The tight junction protein ZO-2 associates with Jun, Fos and C/EBP transcription factors in epithelial cells. *Exp Cell Res* **292:** 51–66.

Bienz M. 2005. β-Catenin: A pivot between cell adhesion and Wnt signalling. *Curr Biol* **15:** R64–67.

Bierkamp C, Schwarz H, Huber O, Kemler R. 1999. Desmosomal localization of β-catenin in the skin of plakoglobin null-mutant mice. *Development* **126:** 371–381.

Bose R, Wrana JL. 2006. Regulation of Par6 by extracellular signals. *Curr Opin Cell Biol* **18:** 206–212.

Bradley RS, Cowin P, Brown AM. 1993. Expression of Wnt-1 in PC12 cells results in modulation of plakoglobin and E-cadherin and increased cellular adhesion. *J Cell Biol* **123:** 1857–1865.

Brady DC, Alan JK, Madigan JP, Fanning AS, Cox AD. 2009. The transforming Rho family GTPase Wrch-1 disrupts epithelial cell tight junctions and epithelial morphogenesis. *Mol Cell Biol* **29:** 1035–1049.

Braga VM. 2002. Cell–cell adhesion and signalling. *Curr Opin Cell Biol* **14:** 546–556.

Braga VM, Yap AS. 2005. The challenges of abundance: Epithelial junctions and small GTPase signalling. *Curr Opin Cell Biol* **17:** 466–474.

Brembeck FH, Rosario M, Birchmeier W. 2006. Balancing cell adhesion and Wnt signaling, the key role of β-catenin. *Curr Opin Genet Dev* **16:** 51–59.

Brembeck FH, Schwarz-Romond T, Bakkers J, Wilhelm S, Hammerschmidt M, Birchmeier W. 2004. Essential role of BCL9-2 in the switch between β-catenin's adhesive and transcriptional functions. *Genes Dev* **18:** 2225–2230.

Brieher WM, Yap AS, Gumbiner BM. 1996. Lateral dimerization is required for the homophilic binding activity of C-cadherin. *J Cell Biol* **135:** 487–496.

Bryant DM, Wylie FG, Stow JL. 2005. Regulation of endocytosis, nuclear translocation, and signaling of fibroblast growth factor receptor 1 by E-cadherin. *Mol Biol Cell* **16:** 14–23.

Buchert M, Darido C, Lagerqvist E, Sedello A, Cazevieille C, Buchholz F, Pannequin J, Joubert D, Hollande F. 2009. The symplekin/ZONAB complex inhibits intestinal cell differentiation via the repression of AML1/Runx1. *Gastroenterology* doi: 10.1053/j.gastro.2009.03.037.

Burks J, Agazie YM. 2006. Modulation of α-catenin Tyr phosphorylation by SHP2 positively effects cell transformation induced by the constitutively active FGFR3. *Oncogene* **25:** 7166–7179.

Burridge K, Wennerberg K. 2004. Rho and Rac take center stage. *Cell* **116:** 167–179.

Caca K, Kolligs FT, Ji X, Hayes M, Qian J, Yahanda A, Rimm DL, Costa J, Fearon ER. 1999. β- and γ-catenin mutations, but not E-cadherin inactivation, underlie T-cell factor/lymphoid enhancer factor transcriptional deregulation in gastric and pancreatic cancer. *Cell Growth Differ* **10:** 369–376.

Cadigan KM, Peifer M. 2009. Wnt signaling from development to disease: Insights from model systems. *Cold Spring Harb Perspect Biol* **1:** a002881.

Calkins CC, Hoepner BL, Law CM, Novak MR, Setzer SV, Hatzfeld M, Kowalczyk AP. 2003. The Armadillo family protein p0071 is a VE-cadherin- and desmoplakin-binding protein. *J Biol Chem* **278:** 1774–1783.

Cao Z, Wu HK, Bruce A, Wollenberg K, Panjwani N. 2002. Detection of differentially expressed genes in healing mouse corneas, using cDNA microarrays. *Invest Ophthalmol Vis Sci* **43:** 2897–2904.

Cavallaro U, Niedermeyer J, Fuxa M, Christofori G. 2001. N-CAM modulates tumour-cell adhesion to matrix by inducing FGF-receptor signalling. *Nat Cell Biol* **3:** 650–657.

Cereijido M, Contreras RG, Shoshani L, Flores-Benitez D, Larre I. 2008. Tight junction and polarity interaction in the transporting epithelial phenotype. *Biochim Biophys Acta* **1778:** 770–793.

Charpentier E, Lavker RM, Acquista E, Cowin P. 2000. Plakoglobin suppresses epithelial proliferation and hair growth in vivo. *J Cell Biol* **149:** 503–520.

Chen X, Gumbiner BM. 2006. Paraxial protocadherin mediates cell sorting and tissue morphogenesis by regulating C-cadherin adhesion activity. *J Cell Biol* **174:** 301–313.

Chen J, Xiao L, Rao JN, Zou T, Liu L, Bellavance E, Gorospe M, Wang JY. 2008. JunD represses transcription and translation of the tight junction protein zona occludens-1 modulating intestinal epithelial barrier function. *Mol Biol Cell* **19:** 3701–3712.

Chen X, Bonne S, Hatzfeld M, van Roy F, Green KJ. 2002. Protein binding and functional characterization of plakophilin 2. Evidence for its diverse roles in desmosomes and β-catenin signaling. *J Biol Chem* **277:** 10512–10522.

Chen Y, Lu Q, Schneeberger EE, Goodenough DA. 2000. Restoration of tight junction structure and barrier function by down-regulation of the mitogen-activated protein kinase pathway in ras-transformed Madin-Darby canine kidney cells. *Mol Biol Cell* **11:** 849–862.

Chiba H, Osanai M, Murata M, Kojima T, Sawada N. 2008. Transmembrane proteins of tight junctions. *Biochim Biophys Acta* **1778:** 588–600.

Chlenski A, Ketels KV, Korovaitseva GI, Talamonti MS, Oyasu R, Scarpelli DG. 2000. Organization and expression of the human zo-2 gene (tjp-2) in normal and neoplastic tissues. *Biochim Biophys Acta* **1493:** 319–324.

Choi HJ, Huber AH, Weis WI. 2006. Thermodynamics of β-catenin-ligand interactions: The roles of the N- and C-terminal tails in modulating binding affinity. *J Biol Chem* **281:** 1027–1038.

Choi J, Krushel LA, Crossin KL. 2001. NF-kappaB activation by N-CAM and cytokines in astrocytes is regulated by multiple protein kinases and redox modulation. *Glia* **33:** 45–56.

Clevers H. 2006. Wnt/β-catenin signaling in development and disease. *Cell* **127:** 469–480.

Cline EG, Nelson WJ. 2007. Characterization of mammalian Par 6 as a dual-location protein. *Mol Cell Biol* **27:** 4431–4443.

Coleman ML, Marshall CJ, Olson MF. 2004. RAS and RHO GTPases in G1-phase cell-cycle regulation. *Nat Rev Mol Cell Biol* **5:** 355–366.

Comoglio PM, Boccaccio C, Trusolino L. 2003. Interactions between growth factor receptors and adhesion molecules: Breaking the rules. *Curr Opin Cell Biol* **15:** 565–571.

Cooke VG, Naik MU, Naik UP. 2006. Fibroblast growth factor-2 failed to induce angiogenesis in junctional adhesion molecule-A-deficient mice. *Arterioscler Thromb Vasc Biol* **26:** 2005–2011.

Coqueret O. 2002. Linking cyclins to transcriptional control. *Gene* **299:** 35–55.

D'Amico M, Hulit J, Amanatullah DF, Zafonte BT, Albanese C, Bouzahzah B, Fu M, Augenlicht LH, Donehower LA, Takemaru KI, et al. 2000. The integrin-linked kinase regulates the cyclin D1 gene through glycogen synthase kinase 3β and cAMP-responsive element-binding protein-dependent pathways. *J Biol Chem* **275:** 32649–32657.

Dang X, Doble BW, Kardami E. 2003. The carboxy-tail of connexin-43 localizes to the nucleus and inhibits cell growth. *Mol Cell Biochem* **242:** 35–38.

Daniel JM. 2007. Dancing in and out of the nucleus: p120(ctn) and the transcription factor Kaiso. *Biochim Biophys Acta* **1773:** 59–68.

Daniel JM, Reynolds AB. 1997. Tyrosine phosphorylation and cadherin/catenin function. *Bioessays* **19:** 883–891.

Delva E, Tucker DK, Kowalczyk AP. 2009. The Desmosome. *Cold Spring Harb Perspect Biol* **1:** a002543.

Doherty P, Williams G, Williams EJ. 2000. CAMs and axonal growth: A critical evaluation of the role of calcium and the MAPK cascade. *Mol Cell Neurosci* **16:** 283–295.

Drees F, Pokutta S, Yamada S, Nelson WJ, Weis WI. 2005. α-catenin is a molecular switch that binds E-cadherin-β-catenin and regulates actin-filament assembly. *Cell* **123:** 903–915.

Dusek RL, Godsel LM, Chen F, Strohecker AM, Getsios S, Harmon R, Muller EJ, Caldelari R, Cryns VL, Green KJ. 2007. Plakoglobin deficiency protects keratinocytes from apoptosis. *J Invest Dermatol* **127:** 792–801.

Dustin ML. 2007. Cell adhesion molecules and actin cytoskeleton at immune synapses and kinapses. *Curr Opin Cell Biol* **19:** 529–533.

Ebnet K, Suzuki A, Ohno S, Vestweber D. 2004. Junctional adhesion molecules (JAMs): More molecules with dual functions? *J Cell Sci* **117:** 19–29.

Eger A, Stockinger A, Park J, Langkopf E, Mikula M, Gotzmann J, Mikulits W, Beug H, Foisner R. 2004. β-Catenin and TGFβ signalling cooperate to maintain a mesenchymal phenotype after FosER-induced epithelial to mesenchymal transition. *Oncogene* **23:** 2672–2680.

Eisner I, Colombo JA. 2002. Detection of a novel pattern of connexin 43 immunoreactivity responsive to dehydration in rat hypothalamic magnocellular nuclei. *Exp Neurol* **177:** 321–325.

El-Bahrawy M, Talbot I, Poulsom R, Alison M. 2002. Variable nuclear localization of α-catenin in colorectal carcinoma. *Lab Invest* **82:** 1167–1174.

Erez N, Zamir E, Gour BJ, Blaschuk OW, Geiger B. 2004. Induction of apoptosis in cultured endothelial cells by a cadherin antagonist peptide: Involvement of fibroblast growth factor receptor-mediated signalling. *Exp Cell Res* **294:** 366–378.

Fanning AS, Little BP, Rahner C, Utepbergenov D, Walther Z, Anderson JM. 2007. The unique-5 and -6 motifs of ZO-1 regulate tight junction strand localization and scaffolding properties. *Mol Biol Cell* **18:** 721–731.

Fedor-Chaiken M, Hein PW, Stewart JC, Brackenbury R, Kinch MS. 2003. E-cadherin binding modulates EGF receptor activation. *Cell Commun Adhes* **10:** 105–118.

Ferber EC, Kajita M, Wadlow A, Tobiansky L, Niessen C, Ariga H, Daniel J, Fujita Y. 2008. A role for the cleaved cytoplasmic domain of E-cadherin in the nucleus. *J Biol Chem* **283:** 12691–12700.

Fink C, Weigel R, Hembes T, Lauke-Wettwer H, Kliesch S, Bergmann M, Brehm RH. 2006. Altered expression of ZO-1 and ZO-2 in Sertoli cells and loss of blood-testis barrier integrity in testicular carcinoma in situ. *Neoplasia* **8:** 1019–1027.

Flores-Benitez D, Ruiz-Cabrera A, Flores-Maldonado C, Shoshani L, Cereijido M, Contreras RG. 2007. Control of tight junctional sealing: Role of epidermal growth factor. *Am J Physiol Renal Physiol* **292:** F828–836.

Fu M, Wang C, Li Z, Sakamaki T, Pestell RG. 2004. Minireview: Cyclin D1: Normal and abnormal functions. *Endocrinology* **145:** 5439–5447.

Furuse M. 2009. Molecular basis of the core structure of tight junctions. *Cold Spring Harb Perspect Biol* **2:** a002907.

Furuse M, Tsukita S. 2006. Claudins in occluding junctions of humans and flies. *Trends Cell Biol* **16:** 181–188.

Gallia GL, Johnson EM, Khalili K. 2000. Purα: A multifunctional single-stranded DNA- and RNA-binding protein. *Nucleic Acids Res* **28:** 3197–3205.

Garcia-Gras E, Lombardi R, Giocondo MJ, Willerson JT, Schneider MD, Khoury DS, Marian AJ. 2006. Suppression of canonical Wnt/β-catenin signaling by nuclear plakoglobin recapitulates phenotype of arrhythmogenic right ventricular cardiomyopathy. *J Clin Invest* **116:** 2012–2021.

Garrod D, Chidgey M. 2008. Desmosome structure, composition and function. *Biochim Biophys Acta* **1778:** 572–587.

Gavert N, Ben-Ze'ev A. 2007. β-Catenin signaling in biological control and cancer. *J Cell Biochem* **102:** 820–828.

Giannini A, Mazor M, Orme M, Vivanco M, Waxman J, Kypta R. 2004. Nuclear export of α-catenin: Overlap between nuclear export signal sequences and the β-catenin binding site. *Exp Cell Res* **295:** 150–160.

Giehl K, Menke A. 2008. Microenvironmental regulation of E-cadherin-mediated adherens junctions. *Front Biosci* **13:** 3975–3985.

Glaunsinger BA, Lee SS, Thomas M, Banks L, Javier R. 2000. Interactions of the PDZ-protein MAGI-1 with adenovirus E4-ORF1 and high-risk papillomavirus E6 oncoproteins. *Oncogene* **19:** 5270–5280.

Glaunsinger BA, Weiss RS, Lee SS, Javier R. 2001. Link of the unique oncogenic properties of adenovirus type 9 E4-ORF1 to a select interaction with the candidate tumor suppressor protein ZO-2. *EMBO J* **20:** 5578–5586.

Goodenough DA, Paul DL. 2009. Gap Junctions. *Cold Spring Harb Perspect Biol* **1:** a002576.

Goodwin M, Kovacs EM, Thoreson MA, Reynolds AB, Yap AS. 2003. Minimal mutation of the cytoplasmic tail inhibits the ability of E-cadherin to activate Rac but not phosphatidylinositol 3-kinase: Direct evidence of a role for cadherin-activated Rac signaling in adhesion and contact formation. *J Biol Chem* **278:** 20533–20539.

Gotoh K, Nonoguchi K, Higashitsuji H, Kaneko Y, Sakurai T, Sumitomo Y, Itoh K, Subjeck JR, Fujita J. 2004. Apg-2 has a chaperone-like activity similar to Hsp110 and is over-expressed in hepatocellular carcinomas. *FEBS Lett* **560:** 19–24.

Gottardi CJ, Gumbiner BM. 2004. Distinct molecular forms of β-catenin are targeted to adhesive or transcriptional complexes. *J Cell Biol* **167:** 339–349.

Gottardi CJ, Wong E, Gumbiner BM. 2001. E-cadherin suppresses cellular transformation by inhibiting β-catenin signaling in an adhesion-independent manner. *J Cell Biol* **153:** 1049–1060.

Gottardi CJ, Arpin M, Fanning AS, Louvard D. 1996. The junction-associated protein, zonula occludens-1, localizes to the nucleus before the maturation and during the remodeling of cell-cell contacts. *Proc Natl Acad Sci* **93:** 10779–10784.

Grazia Lampugnani M, Zanetti A, Corada M, Takahashi T, Balconi G, Breviario F, Orsenigo F, Cattelino A, Kemler R, Daniel TO, et al. 2003. Contact inhibition of VEGF-induced proliferation requires vascular endothelial cadherin, β-catenin, and the phosphatase DEP-1/CD148. *J Cell Biol* **161:** 793–804.

Green KJ, Simpson CL. 2007. Desmosomes: New perspectives on a classic. *J Invest Dermatol* **127:** 2499–2515.

Gregory GD, Vakoc CR, Rozovskaia T, Zheng X, Patel S, Nakamura T, Canaani E, Blobel GA. 2007. Mammalian ASH1L is a histone methyltransferase that occupies the transcribed region of active genes. *Mol Cell Biol* **27:** 8466–8479.

Guillemot L, Citi S. 2006. Cingulin regulates claudin-2 expression and cell proliferation through the small GTPase RhoA. *Mol Biol Cell* **17:** 3569–3577.

Guillemot L, Paschoud S, Jond L, Foglia A, Citi S. 2008. Paracingulin regulates the activity of Rac1 and RhoA GTPases by recruiting Tiam1 and GEF-H1 to epithelial junctions. *Mol Biol Cell* **19:** 4442–4453.

Gumbiner B, Lowenkopf T, Apatira D. 1991. Identification of a 160-kDa polypeptide that binds to the tight junction protein ZO-1. *Proc Natl Acad Sci* **88:** 3460–3464.

Gwak GY, Yoon JH, Yu SJ, Park SC, Jang JJ, Lee KB, Lee SH, Lee SM, Shin CS, Suh KS, Lee HS. 2006. Anti-apoptotic N-cadherin signaling and its prognostic implication in human hepatocellular carcinomas. *Oncol Rep* **15:** 1117–1123.

Hardman MJ, Liu K, Avilion AA, Merritt A, Brennan K, Garrod DR, Byrne C. 2005. Desmosomal cadherin misexpression alters β-catenin stability and epidermal differentiation. *Mol Cell Biol* **25:** 969–978.

Harris AL. 2007. Connexin channel permeability to cytoplasmic molecules. *Prog Biophys Mol Biol* **94:** 120–143.

Harten SK, Shukla D, Barod R, Hergovich A, Balda MS, Matter K, Esteban MA, Maxwell PH. 2009. Regulation of renal epithelial tight junctions by the von Hippel-Lindau tumor suppressor gene involves occludin and claudin 1 and is independent of E-cadherin. *Mol Biol Cell* **20:** 1089–1101.

Hatzfeld M. 2005. The p120 family of cell adhesion molecules. *Eur J Cell Biol* **84:** 205–214.

Hatzfeld M. 2007. Plakophilins: Multifunctional proteins or just regulators of desmosomal adhesion? *Biochim Biophys Acta* **1773:** 69–77.

Hatzfeld M, Green KJ, Sauter H. 2003. Targeting of 0071 to desmosomes and adherens junctions is mediated by different protein domains. *J Cell Sci* **116:** 1219–1233.

Hay E, Laplantine E, Geoffroy V, Frain M, Kohler T, Muller R, Marie PJ. 2009. N-cadherin interacts with axin and LRP5 to negatively regulate Wnt/β-catenin signaling, osteoblast function, and bone formation. *Mol Cell Biol* **29:** 953–964.

Heasman J, Crawford A, Goldstone K, Garner-Hamrick P, Gumbiner B, McCrea P, Kintner C, Noro CY, Wylie C. 1994. Overexpression of cadherins and underexpression of β-catenin inhibit dorsal mesoderm induction in early Xenopus embryos. *Cell* **79:** 791–803.

Herzig M, Savarese F, Novatchkova M, Semb H, Christofori G. 2007. Tumor progression induced by the loss of E-cadherin independent of β-catenin/Tcf-mediated Wnt signaling. *Oncogene* **26:** 2290–2298.

Heuberger J, Birchmeier W. 2009. Interplay of cadherin-mediated cell adhesion and canonical Wnt signaling. *Cold Spring Harb Perspect Biol* **2:** a002915.

Hinck L, Nelson WJ, Papkoff J. 1994b. Wnt-1 modulates cell-cell adhesion in mammalian cells by stabilizing β-catenin binding to the cell adhesion protein cadherin. *J Cell Biol* **124:** 729–741.

Hinck L, Nathke IS, Papkoff J, Nelson WJ. 1994a. β-catenin: A common target for the regulation of cell adhesion by Wnt-1 and Src signaling pathways. *Trends Biochem Sci* **19:** 538–542.

Hofmann I, Schnolzer M, Kaufmann I, Franke WW. 2002. Symplekin, a Constitutive Protein of Karyo- and Cytoplasmic Particles Involved in mRNA Biogenesis in Xenopus laevis Oocytes. *Mol Biol Cell* **13:** 1665–1676.

Holthofer B, Windoffer R, Troyanovsky S, Leube RE. 2007. Structure and function of desmosomes. *Int Rev Cytol* **264:** 65–163.

Hoschuetzky H, Aberle H, Kemler R. 1994. β-catenin mediates the interaction of the cadherin-catenin complex with epidermal growth factor receptor. *J Cell Biol* **127:** 1375–1380.

Hoshino M, Yoshimori T, Nakamura S. 2005. Small GTPase proteins Rin and Rit Bind to PAR6 GTP-dependently and regulate cell transformation. *J Biol Chem* **280:** 22868–22874.

Hosking CR, Ulloa F, Hogan C, Ferber EC, Figueroa A, Gevaert K, Birchmeier W, Briscoe J, Fujita Y. 2007. The transcriptional repressor Glis2 is a novel binding partner for pp120 catenin. *Mol Biol Cell* **18:** 1918–1927.

Howarth AG, Hughes MR, Stevenson BR. 1992. Detection of the tight junction-associated protein ZO-1 in astrocytes and other nonepithelilal cell types. *Am J Physiol* **262:** C461–C469.

Huang H, He X. 2008. Wnt/β-catenin signaling: New (and old) players and new insights. *Curr Opin Cell Biol* **20:** 119–125.

Huber AH, Weis WI. 2001. The structure of the β-catenin/E-cadherin complex and the molecular basis of diverse ligand recognition by β-catenin. *Cell* **105:** 391–402.

Huerta M, Munoz R, Tapia R, Soto-Reyes E, Ramirez L, Recillas-Targa F, Gonzalez-Mariscal L, Lopez-Bayghen E. 2007. Cyclin D1 Is Transcriptionally Down-Regulated by ZO-2 via an E Box and the Transcription Factor c-Myc. *Mol Biol Cell* **18:** 4826–4836.

Hulpiau P, van Roy F. 2009. Molecular evolution of the cadherin superfamily. *Int J Biochem Cell Biol* **41:** 349–369.

Hwang SG, Yu SS, Ryu JH, Jeon HB, Yoo YJ, Eom SH, Chun JS. 2005. Regulation of β-catenin signaling and maintenance of chondrocyte differentiation by ubiquitin-independent proteasomal degradation of α-catenin. *J Biol Chem* **280:** 12758–12765.

Iden S, Collard JG. 2008. Crosstalk between small GTPases and polarity proteins in cell polarization. *Nat Rev Mol Cell Biol* **9:** 846–859.

Iioka H, Doerner SK, Tamai K. 2009. Kaiso is a bimodal modulator for Wnt/β-catenin signaling. *FEBS Lett* **583:** 627–632.

Ingber DE. 2008. Tensegrity-based mechanosensing from macro to micro. *Prog Biophys Mol Biol* **97:** 163–179.

Islas S, Vega J, Ponce L, Gonzalez-Mariscal L. 2002. Nuclear Localization of the Tight Junction Protein ZO-2 in Epithelial Cells. *Exp Cell Res* **274:** 138–148.

Itoh M, Nagafuchi A, Yonemura S, Kitani-Yasuda T, Tsukita S, Tsukita S. 1993. The 220-kD protein colocalizing with cadherins in nonepithelial cells is identical to ZO-1, a tight junction-associated protein in epithelial cells: cDNA cloning and immunoelectron microscopy. *J Cell Biol* **121:** 491–502.

Jaffe AB, Hall A. 2002. Rho GTPases in transformation and metastasis. *Adv Cancer Res* **84:** 57–80.

Jesaitis LA, Goodenough DA. 1994. Molecular characterization and tissue distribution of ZO-2, a tight junction protein homologous to ZO-1 and Drosophila tumor suppressor gene dlg-A. *J Cell Biol* **124:** 949–961.

Jiang C, Pecha J, Hoshino I, Ankrapp D, Xiao H. 2007. TIP30 mutant derived from hepatocellular carcinoma specimens promotes growth of HepG2 cells through up-regulation of N-cadherin. *Cancer Res* **67:** 3574–3582.

Johansson A, Driessens M, Aspenstrom P. 2000. The mammalian homologue of the Caenorhabditis elegans polarity protein PAR-6 is a binding partner for the Rho GTPases Cdc42 and Rac1. *J Cell Sci* **113:** 3267–3275.

Kajita M, Ikeda W, Tamaru Y, Takai Y. 2007. Regulation of platelet-derived growth factor-induced Ras signaling by poliovirus receptor Necl-5 and negative growth regulator Sprouty2. *Genes Cells* **12:** 345–357.

Kakunaga S, Ikeda W, Shingai T, Fujito T, Yamada A, Minami Y, Imai T, Takai Y. 2004. Enhancement of serum- and platelet-derived growth factor-induced cell proliferation by Necl-5/Tage4/poliovirus receptor/CD155 through the Ras-Raf-MEK-ERK signaling. *J Biol Chem* **279:** 36419–36425.

Kang HG, Jenabi JM, Zhang J, Keshelava N, Shimada H, May WA, Ng T, Reynolds CP, Triche TJ, Sorensen PH. 2007. E-cadherin cell-cell adhesion in ewing tumor cells mediates suppression of anoikis through activation of the ErbB4 tyrosine kinase. *Cancer Res* **67:** 3094–3105.

Kanno M, Isa Y, Aoyama Y, Yamamoto Y, Nagai M, Ozawa M, Kitajima Y. 2008. P120-catenin is a novel desmoglein 3 interacting partner: Identification of the p120-catenin association site of desmoglein 3. *Exp Cell Res* **314:** 1683–1692.

Kantak SS, Kramer RH. 1998. E-cadherin regulates anchorage-independent growth and survival in oral squamous cell carcinoma cells. *J Biol Chem* **273:** 16953–16961.

Kanzaki N, Ogita H, Komura H, Ozaki M, Sakamoto Y, Majima T, Ijuin T, Takenawa T, Takai Y. 2008. Involvement of the nectin-afadin complex in PDGF-induced cell survival. *J Cell Sci* **121:** 2008–2017.

Kausalya PJ, Phua DC, Hunziker W. 2004. Association of ARVCF with zonula occludens (ZO)-1 and ZO-2: Binding to PDZ-domain proteins and cell-cell adhesion regulate plasma membrane and nuclear localization of ARVCF. *Mol Biol Cell* **15:** 5503–5515.

Kavanagh E, Buchert M, Tsapara A, Choquet A, Balda MS, Hollande F, Matter K. 2006. Functional interaction between the ZO-1-interacting transcription factor ZONAB/DbpA and the RNA processing factor symplekin. *J Cell Sci* **119:** 5098–5105.

Kawakatsu T, Ogita H, Fukuhara T, Fukuyama T, Minami Y, Shimizu K, Takai Y. 2005. Vav2 as a Rac-GDP/GTP exchange factor responsible for the nectin-induced, c-Src- and Cdc42-mediated activation of Rac. *J Biol Chem* **280:** 4940–4947.

Kelly KF, Spring CM, Otchere AA, Daniel JM. 2004. NLS-dependent nuclear localization of p120ctn is necessary to relieve Kaiso-mediated transcriptional repression. *J Cell Sci* **117:** 2675–2686.

Keon BH, Schäfer S, Kuhn C, Grund C, Franke WW. 1996. Symplekin, a novel type of tight junction plaque protein. *J Cell Biol* **134:** 1003–1018.

Kim YJ, Johnson KR, Wheelock MJ. 2005. N-cadherin-mediated cell motility requires cis dimers. *Cell Commun Adhes* **12:** 23–39.

Kim JB, Islam S, Kim YJ, Prudoff RS, Sass KM, Wheelock MJ, Johnson KR. 2000. N-Cadherin extracellular repeat 4 mediates epithelial to mesenchymal transition and increased motility. *J Cell Biol* **151:** 1193–1206.

Kim SW, Park JI, Spring CM, Sater AK, Ji H, Otchere AA, Daniel JM, McCrea PD. 2004. Non-canonical Wnt signals are modulated by the Kaiso transcriptional repressor and p120-catenin. *Nat Cell Biol* **6:** 1212–1220.

Kobielak A, Fuchs E. 2004. α-catenin: At the junction of intercellular adhesion and actin dynamics. *Nat Rev Mol Cell Biol* **5:** 614–625.

Kobielak A, Fuchs E. 2006. Links between α-catenin, NF-kappaB, and squamous cell carcinoma in skin. *Proc Natl Acad Sci* **103:** 2322–2327.

Kodama S, Ikeda S, Asahara T, Kishida M, Kikuchi A. 1999. Axin directly interacts with plakoglobin and regulates its stability. *J Biol Chem* **274:** 27682–27688.

Kohno K, Izumi H, Uchiumi T, Ashizuka M, Kuwano M. 2003. The pleiotropic functions of the Y-box-binding protein, YB-1. *Bioessays* **25:** 691–698.

Kolligs FT, Kolligs B, Hajra KM, Hu G, Tani M, Cho KR, Fearon ER. 2000. γ-catenin is regulated by the APC tumor suppressor and its oncogenic activity is distinct from that of β-catenin. *Genes Dev* **14:** 1319–1331.

Kovacs EM, Ali RG, McCormack AJ, Yap AS. 2002. E-cadherin homophilic ligation directly signals through Rac and phosphatidylinositol 3-kinase to regulate adhesive contacts. *J Biol Chem* **277:** 6708–6718.

Kowalczyk AP, Reynolds AB. 2004. Protecting your tail: Regulation of cadherin degradation by p120-catenin. *Curr Opin Cell Biol* **16:** 522–527.

Kuphal F, Behrens J. 2006. E-cadherin modulates Wnt-dependent transcription in colorectal cancer cells but does not alter Wnt-independent gene expression in fibroblasts. *Exp Cell Res* **312:** 457–467.

Lampugnani MG, Orsenigo F, Gagliani MC, Tacchetti C, Dejana E. 2006. Vascular endothelial cadherin controls VEGFR-2 internalization and signaling from intracellular compartments. *J Cell Biol* **174:** 593–604.

Larue L, Bellacosa A. 2005. Epithelial-mesenchymal transition in development and cancer: Role of phosphatidylinositol 3′ kinase/AKT pathways. *Oncogene* **24:** 7443–7454.

Lee SS, Glaunsinger B, Mantovani F, Banks L, Javier RT. 2000. Multi-PDZ domain protein MUPP1 is a cellular target for both adenovirus E4-ORF1 and high-risk papillomavirus type 18 E6 oncoproteins. *J Virol* **74:** 9680–9693.

Lemmers C, Michel D, Lane-Guermonprez L, Delgrossi MH, Medina E, Arsanto JP, Le Bivic A. 2004. CRB3 binds directly to Par6 and regulates the morphogenesis of the tight junctions in mammalian epithelial cells. *Mol Biol Cell* **15:** 1324–1333.

Li G, Satyamoorthy K, Herlyn M. 2001. N-cadherin-mediated intercellular interactions promote survival and migration of melanoma cells. *Cancer Res* **61:** 3819–3825.

Liebner S, Corada M, Bangsow T, Babbage J, Taddei A, Czupalla CJ, Reis M, Felici A, Wolburg H, Fruttiger M, et al. 2008. Wnt/β-catenin signaling controls development of the blood-brain barrier. *J Cell Biol* **183:** 409–417.

Lien WH, Klezovitch O, Fernandez TE, Delrow J, Vasioukhin V. 2006. αE-catenin controls cerebral cortical size by regulating the hedgehog signaling pathway. *Science* **311:** 1609–1612.

Lilien J, Balsamo J. 2005. The regulation of cadherin-mediated adhesion by tyrosine phosphorylation/dephosphorylation of β-catenin. *Curr Opin Cell Biol* **17:** 459–465.

Lilien J, Balsamo J, Arregui C, Xu G. 2002. Turn-off, drop-out: Functional state switching of cadherins. *Dev Dyn* **224:** 18–29.

Little EB, Crossin KL, Krushel LA, Edelman GM, Cunningham BA. 2001. A short segment within the cytoplasmic domain of the neural cell adhesion molecule (N-CAM) is essential for N-CAM-induced NF-kappa B activity in astrocytes. *Proc Natl Acad Sci* **98:** 2238–2243.

Lloberas J, Maki RA, Celada A. 1995. Repression of major histocompatibility complex I-A β gene expression by dbpA and dbpB (mYB-1) proteins. *Mol Cell Biol* **15:** 5092–5099.

Lu ZH, Books JT, Ley TJ. 2006. Cold shock domain family members YB-1 and MSY4 share essential functions during murine embryogenesis. *Mol Cell Biol* **26:** 8410–8417.

Macara IG, Mili S. 2008. Polarity and differential inheritance–universal attributes of life? *Cell* **135:** 801–812.

Maeda O, Usami N, Kondo M, Takahashi M, Goto H, Shimokata K, Kusugami K, Sekido Y. 2004. Plakoglobin (γ-catenin) has TCF/LEF family-dependent transcriptional activity in β-catenin-deficient cell line. *Oncogene* **23:** 964–972.

Makrigiannakis A, Coukos G, Christofidou-Solomidou M, Gour BJ, Radice GL, Blaschuk O, Coutifaris C. 1999. N-cadherin-mediated human granulosa cell adhesion prevents apoptosis: A role in follicular atresia and luteolysis? *Am J Pathol* **154:** 1391–1406.

Mandell KJ, Babbin BA, Nusrat A, Parkos CA. 2005. Junctional adhesion molecule 1 regulates epithelial cell morphology through effects on β1 integrins and Rap1 activity. *J Biol Chem* **280:** 11665–11674.

Mandicourt G, Iden S, Ebnet K, Aurrand-Lions M, Imhof BA. 2007. JAM-C regulates tight junctions and integrin-mediated cell adhesion and migration. *J Biol Chem* **282:** 1830–1837.

Mann B, Gelos M, Siedow A, Hanski ML, Gratchev A, Ilyas M, Bodmer WF, Moyer MP, Riecken EO, Buhr HJ, et al. 1999. Target genes of β-catenin-T cell-factor/lymphoid-enhancer-factor signaling in human colorectal carcinomas. *Proc Natl Acad Sci* **96:** 1603–1608.

Martin ED, Moriarty MA, Byrnes L, Grealy M. 2009. Plakoglobin has both structural and signalling roles in zebrafish development. *Dev Biol* **327:** 83–96.

Matter K, Balda MS. 2003. Signalling to and from tight junctions. *Nat Rev Mol Cell Biol* **4:** 225–236.

Matter K, Balda MS. 2007. Epithelial tight junctions, gene expression and nucleo-junctional interplay. *J Cell Sci* **120:** 1505–1511.

McCrea PD, Park JI. 2007. Developmental functions of the P120-catenin sub-family. *Biochim Biophys Acta* **1773:** 17–33.

McLachlan RW, Yap AS. 2007. Not so simple: The complexity of phosphotyrosine signaling at cadherin adhesive contacts. *J Mol Med* **85:** 545–554.

McNeil E, Capaldo CT, Macara IG. 2006. Zonula occludens-1 function in the assembly of tight junctions in Madin-Darby canine kidney epithelial cells. *Mol Biol Cell* **17:** 1922–1932.

McNeill H. 2009. Planar cell polarity: keeping hairs straight is not so simple. *Cold Spring Harb Perspect Biol* **2:** a003376.

Medina A, Swain RK, Kuerner KM, Steinbeisser H. 2004. Xenopus paraxial protocadherin has signaling functions and is involved in tissue separation. *Embo J* **23:** 3249–3258.

Meng W, Takeichi M. 2009. Adherens junction: molecular architecture and regulation. *Cold Spring Harb Perspect Biol* **1:** a002899.

Merdek KD, Nguyen NT, Toksoz D. 2004. Distinct activities of the α-catenin family, α-catulin and α-catenin, on β-catenin-mediated signaling. *Mol Cell Biol* **24:** 2410–2422.

Merriam JM, Rubenstein AB, Klymkowsky MW. 1997. Cytoplasmically anchored plakoglobin induces a WNT-like phenotype in Xenopus. *Dev Biol* **185:** 67–81.

Mertens C, Kuhn C, Franke WW. 1996. Plakophilins 2a and 2b: Constitutive proteins of dual location in the karyoplasm and the desmosomal plaque. *J Cell Biol* **135:** 1009–1025.

Mertens C, Hofmann I, Wang Z, Teichmann M, Sepehri Chong S, Schnolzer M, Franke WW. 2001. Nuclear particles containing RNA polymerase III complexes associated with the junctional plaque protein plakophilin 2. *Proc Natl Acad Sci* **98:** 7795–7800.

Mese G, Richard G, White TW. 2007. Gap junctions: Basic structure and function. *J Invest Dermatol* **127:** 2516–2524.

Miettinen PJ, Ebner R, Lopez AR, Derynck R. 1994. TGF-β induced transdifferentiation of mammary epithelial cells to mesenchymal cells: Involvement of type I receptors. *J Cell Biol* **127:** 2021–2036.

Minami Y, Ikeda W, Kajita M, Fujito T, Amano H, Tamaru Y, Kuramitsu K, Sakamoto Y, Monden M, Takai Y. 2007. Necl-5/poliovirus receptor interacts in cis with integrin αVβ3 and regulates its clustering and focal complex formation. *J Biol Chem* **282:** 18481–18496.

Mizoguchi A, Nakanishi H, Kimura K, Matsubara K, Ozaki-Kuroda K, Katata T, Honda T, Kiyohara Y, Heo K, Higashi M, et al. 2002. Nectin: An adhesion molecule involved in formation of synapses. *J Cell Biol* **156:** 555–565.

Morishita H, Yagi T. 2007. Protocadherin family: Diversity, structure, and function. *Curr Opin Cell Biol* **19:** 584–592.

Mosimann C, Hausmann G, Basler K. 2009. β-catenin hits chromatin: Regulation of Wnt target gene activation. *Nat Rev Mol Cell Biol* **10:** 276–286.

Muller SL, Portwich M, Schmidt A, Utepbergenov DI, Huber O, Blasig IE, Krause G. 2005. The tight junction protein occludin and the adherens junction protein α-catenin share a common interaction mechanism with ZO-1. *J Biol Chem* **280:** 3747–3756.

Muller J, Ritt DA, Copeland TD, Morrison DK. 2003. Functional analysis of C-TAK1 substrate binding and identification of PKP2 as a new C-TAK1 substrate. *Embo J* **22:** 4431–4442.

Naik MU, Naik UP. 2006. Junctional adhesion molecule-A-induced endothelial cell migration on vitronectin is integrin α v β 3 specific. *J Cell Sci* **119:** 490–499.

Nakamura T, Blechman J, Tada S, Rozovskaia T, Itoyama T, Bullrich F, Mazo A, Croce CM, Geiger B, Canaani E. 2000. huASH1 protein, a putative transcription factor encoded by a human homologue of the Drosophila ash1 gene, localizes to both nuclei and cell-cell tight junctions. *Proc Natl Acad Sci* **97:** 7284–7289.

Nakamuta S, Endo H, Higashi Y, Kousaka A, Yamada H, Yano M, Kido H. 2008. Human immunodeficiency virus type 1 gp120-mediated disruption of tight junction proteins by induction of proteasome-mediated degradation of zonula occludens-1 and -2 in human brain microvascular endothelial cells. *J Neurovirol* **14:** 186–195.

Nelson WJ. 2008. Regulation of cell-cell adhesion by the cadherin-catenin complex. *Biochem Soc Trans* **36:** 149–155.

Nelson WJ, Nusse R. 2004. Convergence of Wnt, β-catenin, and cadherin pathways. *Science* **303:** 1483–1487.

Nieman MT, Prudoff RS, Johnson KR, Wheelock MJ. 1999. N-cadherin promotes motility in human breast cancer cells regardless of their E-cadherin expression. *J Cell Biol* **147:** 631–644.

Noren NK, Niessen CM, Gumbiner BM, Burridge K. 2001. Cadherin engagement regulates Rho family GTPases. *J Biol Chem* **276:** 33305–33308.

Ochiai A, Akimoto S, Kanai Y, Shibata T, Oyama T, Hirohashi S. 1994. c-erbB-2 gene product associates with catenins in human cancer cells. *Biochem Biophys Res Commun* **205:** 73–78.

Ogita H, Takai Y. 2006. Activation of Rap1, Cdc42, and rac by nectin adhesion system. *Methods Enzymol* **406:** 415–424.

Ogita H, Takai Y. 2008. Cross-talk among integrin, cadherin, and growth factor receptor: Roles of nectin and nectin-like molecule. *Int Rev Cytol* **265:** 1–54.

Ohnishi H, Nakahara T, Furuse K, Sasaki H, Tsukita S, Furuse M. 2004. JACOP, a novel plaque protein localizing at the apical junctional complex with sequence similarity to cingulin. *J Biol Chem* **279:** 46014–46022.

Ojakian GK, Ratcliffe DR, Schwimmer R. 2001. Integrin regulation of cell-cell adhesion during epithelial tubule formation. *J Cell Sci* **114:** 941–952.

Ooshio T, Fujita N, Yamada A, Sato T, Kitagawa Y, Okamoto R, Nakata S, Miki A, Irie K, Takai Y. 2007. Cooperative roles of Par-3 and afadin in the formation of adherens and tight junctions. *J Cell Sci* **120:** 2352–2365.

Orban E, Szabo E, Lotz G, Kupcsulik P, Paska C, Schaff Z, Kiss A. 2008. Different expression of occludin and ZO-1 in primary and metastatic liver tumors. *Pathol Oncol Res* **14:** 299–306.

Orsulic S, Huber O, Aberle H, Arnold S, Kemler R. 1999. E-cadherin binding prevents β-catenin nuclear localization and β-catenin/LEF-1-mediated transactivation. *J Cell Sci* **112:** 1237–1245.

Osler ME, Chang MS, Bader DM. 2005. Bves modulates epithelial integrity through an interaction at the tight junction. *J Cell Sci* **118:** 4667–4678.

Ozaki M, Ogita H, Takai Y. 2007. Involvement of integrin-induced activation of protein kinase C in the formation of adherens junctions. *Genes Cells* **12:** 651–662.

Paris L, Tonutti L, Vannini C, Bazzoni G. 2008. Structural organization of the tight junctions. *Biochim Biophys Acta* **1778:** 646–659.

Park JI, Ji H, Jun S, Gu D, Hikasa H, Li L, Sokol SY, McCrea PD. 2006. Frodo links Dishevelled to the p120-catenin/Kaiso pathway: Distinct catenin subfamilies promote Wnt signals. *Dev Cell* **11:** 683–695.

Park JI, Kim SW, Lyons JP, Ji H, Nguyen TT, Cho K, Barton MC, Deroo T, Vleminckx K, Moon RT, et al. 2005. Kaiso/p120-catenin and TCF/β-catenin complexes coordinately regulate canonical Wnt gene targets. *Dev Cell* **8:** 843–854.

Paschoud S, Bongiovanni M, Pache JC, Citi S. 2007. Claudin-1 and claudin-5 expression patterns differentiate lung squamous cell carcinomas from adenocarcinomas. *Mod Pathol* **20:** 947–954.

Patel SD, Chen CP, Bahna F, Honig B, Shapiro L. 2003. Cadherin-mediated cell-cell adhesion: Sticking together as a family. *Curr Opin Struct Biol* **13:** 690–698.

Pece S, Gutkind JS. 2000. Signaling from E-cadherins to the MAPK pathway by the recruitment and activation of epidermal growth factor receptors upon cell-cell contact formation. *J Biol Chem* **275:** 41227–41233.

Peinado H, Olmeda D, Cano A. 2007. Snail, Zeb and bHLH factors in tumour progression: An alliance against the epithelial phenotype? *Nat Rev Cancer* **7:** 415–428.

Peluso JJ, Pappalardo A, Trolice MP. 1996. N-cadherin-mediated cell contact inhibits granulosa cell apoptosis in a progesterone-independent manner. *Endocrinology* **137:** 1196–1203.

Perez-Moreno M, Fuchs E. 2006. Catenins: keeping cells from getting their signals crossed. *Dev Cell* **11:** 601–612.

Perez-Moreno M, Jamora C, Fuchs E. 2003. Sticky business: Orchestrating cellular signals at adherens junctions. *Cell* **112:** 535–548.

Perrais M, Chen X, Perez-Moreno M, Gumbiner BM. 2007. E-cadherin homophilic ligation inhibits cell growth and epidermal growth factor receptor signaling independently of other cell interactions. *Mol Biol Cell* **18:** 2013–2025.

Piedra J, Miravet S, Castano J, Palmer HG, Heisterkamp N, Garcia de Herreros A, Dunach M. 2003. p120 Catenin-associated Fer and Fyn tyrosine kinases regulate β-catenin Tyr-142 phosphorylation and β-catenin-α-catenin Interaction. *Mol Cell Biol* **23:** 2287–2297.

Pokutta S, Weis WI. 2000. Structure of the dimerization and β-catenin-binding region of α-catenin. *Mol Cell* **5:** 533–543.

Pokutta S, Drees F, Yamada S, Nelson WJ, Weis WI. 2008. Biochemical and structural analysis of α-catenin in cell-cell contacts. *Biochem Soc Trans* **36:** 141–147.

Potter MD, Barbero S, Cheresh DA. 2005. Tyrosine phosphorylation of VE-cadherin prevents binding of pp120- and β-catenin and maintains the cellular mesenchymal state. *J Biol Chem* **280:** 31906–31912.

Qian X, Karpova T, Sheppard AM, McNally J, Lowy DR. 2004. E-cadherin-mediated adhesion inhibits ligand-dependent activation of diverse receptor tyrosine kinases. *Embo J* **23:** 1739–1748.

Qin Y, Capaldo C, Gumbiner BM, Macara IG. 2005. The mammalian Scribble polarity protein regulates epithelial cell adhesion and migration through E-cadherin. *J Cell Biol* **171:** 1061–1071.

Reichert M, Muller T, Hunziker W. 2000. The PDZ domains of zonula occludens-1 induce an epithelial to mesenchymal transition of Madin-Darby canine kidney I cells. Evidence for a role of β-catenin/Tcf/Lef signaling. *J Biol Chem* **275:** 9492–9500.

Reynolds AB. 2007. 120-catenin: Past and present. *Biochim Biophys Acta* **1773:** 2–7.

Rodriguez-Sinovas A, Cabestrero A, Lopez D, Torre I, Morente M, Abellan A, Miro E, Ruiz-Meana M, Garcia-Dorado D. 2007. The modulatory effects of connexin 43 on cell death/survival beyond cell coupling. *Prog Biophys Mol Biol* **94:** 219–232.

Roura S, Miravet S, Piedra J, Garcia de Herreros A, Dunach M. 1999. Regulation of E-cadherin/Catenin association by tyrosine phosphorylation. *J Biol Chem* **274:** 36734–36740.

Ruzov A, Hackett JA, Prokhortchouk A, Reddington JP, Madej MJ, Dunican DS, Prokhortchouk E, Pennings S, Meehan RR. 2009a. The interaction of xKaiso with xTcf3: A revised model for integration of epigenetic and Wnt signalling pathways. *Development* **136:** 723–727.

Ruzov A, Savitskaya E, Hackett JA, Reddington JP, Prokhortchouk A, Madej MJ, Chekanov N, Li M, Dunican DS, Prokhortchouk E, et al. 2009b. The non-methylated DNA-binding function of Kaiso is not required in early Xenopus laevis development. *Development* **136:** 729–738.

Sahai E, Marshall CJ. 2002. RHO-GTPases and cancer. *Nat Rev Cancer* **2:** 133–142.

Sakamoto Y, Ogita H, Hirota T, Kawakatsu T, Fukuyama T, Yasumi M, Kanzaki N, Ozaki M, Takai Y. 2006. Interaction of integrin α(v)β3 with nectin. Implication in cross-talk between cell-matrix and cell-cell junctions. *J Biol Chem* **281:** 19631–19644.

Sakamoto Y, Ogita H, Komura H, Takai Y. 2008. Involvement of nectin in inactivation of integrin α(v)β(3) after the establishment of cell-cell adhesion. *J Biol Chem* **283:** 496–505.

Sakisaka T, Ikeda W, Ogita H, Fujita N, Takai Y. 2007. The roles of nectins in cell adhesions: Cooperation with other cell adhesion molecules and growth factor receptors. *Curr Opin Cell Biol* **19:** 593–602.

Sakura H, Maekawa T, Imamoto F, Yasuda K, Ishii S. 1988. Two human genes isolated by a novel method encode DNA-binding proteins containing a common region of homology. *Gene* **73:** 499–507.

Sallee JL, Wittchen ES, Burridge K. 2006. Regulation of cell adhesion by protein-tyrosine phosphatases: II. Cell-cell adhesion. *J Biol Chem* **281:** 16189–16192.

Sanchez-Heras E, Howell FV, Williams G, Doherty P. 2006. The fibroblast growth factor receptor acid box is essential for interactions with N-cadherin and all of the major isoforms of neural cell adhesion molecule. *J Biol Chem* **281:** 35208–35216.

Sanson B, White P, Vincent JP. 1996. Uncoupling cadherin-based adhesion from wingless signalling in Drosophila. *Nature* **383:** 627–630.

Schmidt A, Langbein L, Rode M, Pratzel S, Zimbelmann R, Franke WW. 1997. Plakophilins 1a and 1b: Widespread nuclear proteins recruited in specific epithelial cells as desmosomal plaque components. *Cell Tissue Res* **290:** 481–499.

Shariatmadari M, Peyronnet J, Papachristou P, Horn Z, Sousa KM, Arenas E, Ringstedt T. 2005. Increased Wnt levels in the neural tube impair the function of adherens junctions during neurulation. *Mol Cell Neurosci* **30:** 437–451.

Sheth B, Nowak RL, Anderson R, Kwong WY, Papenbrock T, Fleming TP. 2008. Tight junction protein ZO-2 expression and relative function of ZO-1 and ZO-2 during mouse blastocyst formation. *Exp Cell Res* **314:** 3356–3368.

Shimizu M, Fukunaga Y, Ikenouchi J, Nagafuchi A. 2008. Defining the roles of β-catenin and plakoglobin in LEF/T-cell factor-dependent transcription using β-catenin/plakoglobin-null F9 cells. *Mol Cell Biol* **28:** 825–835.

Shtutman M, Zhurinsky J, Oren M, Levina E, Ben-Ze'ev A. 2002. PML is a target gene of β-catenin and plakoglobin, and coactivates β-catenin-mediated transcription. *Cancer Res* **62:** 5947–5954.

Shtutman M, Zhurinsky J, Simcha I, Albanese C, D'Amico M, Pestell R, Ben-Ze'ev A. 1999. The cyclin D1 gene is a target of the β-catenin/LEF-1 pathway. *Proc Natl Acad Sci* **96:** 5522–5527.

Singh AB, Sugimoto K, Dhawan P, Harris RC. 2007. Juxtacrine activation of EGFR regulates claudin expression and increases transepithelial resistance. *Am J Physiol Cell Physiol* **293:** C1660–1668.

Skaper SD, Moore SE, Walsh FS. 2001. Cell signalling cascades regulating neuronal growth-promoting and inhibitory cues. *Prog Neurobiol* **65:** 593–608.

Sourisseau T, Georgiadis A, Tsapara A, Ali RR, Pestell R, Matter K, Balda MS. 2006. Regulation of PCNA and cyclin D1 expression and epithelial morphogenesis by the ZO-1-regulated transcription factor ZONAB/DbpA. *Mol Cell Biol* **26:** 2387–2398.

Spring CM, Kelly KF, O'Kelly I, Graham M, Crawford HC, Daniel JM. 2005. The catenin p120ctn inhibits Kaiso-mediated transcriptional repression of the β-catenin/TCF target gene matrilysin. *Exp Cell Res* **305:** 253–265.

St Croix B, Kerbel RS. 1997. Cell adhesion and drug resistance in cancer. *Curr Opin Oncol* **9:** 549–556.

Stevenson BR, Siliciano JD, Mooseker MS, Goodenough DA. 1986. Identification of ZO-1: A high molecular weight polypeptide associated with the tight junction (zonula occludens) in a variety of epithelia. *J Cell Biol* **103:** 755–766.

Straub BK, Boda J, Kuhn C, Schnoelzer M, Korf U, Kempf T, Spring H, Hatzfeld M, Franke WW. 2003. A novel cell-cell junction system: The cortex adhaerens mosaic of lens fiber cells. *J Cell Sci* **116:** 4985–4995.

Suh EK, Gumbiner BM. 2003. Translocation of β-catenin into the nucleus independent of interactions with FG-rich nucleoporins. *Exp Cell Res* **290:** 447–456.

Suyama K, Shapiro I, Guttman M, Hazan RB. 2002. A signaling pathway leading to metastasis is controlled by N-cadherin and the FGF receptor. *Cancer Cell* **2:** 301–314.

Taddei A, Giampietro C, Conti A, Orsenigo F, Breviario F, Pirazzoli V, Potente M, Daly C, Dimmeler S, Dejana E. 2008. Endothelial adherens junctions control tight junctions by VE-cadherin-mediated upregulation of claudin-5. *Nat Cell Biol* **10:** 923–934.

Takagaki Y, Manley JL. 2000. Complex protein interactions within the human polyadenylation machinery identify a novel component. *Mol Cell Biol* **20:** 1515–1525.

Takai Y, Ikeda W, Ogita H, Rikitake Y. 2008a. The immunoglobulin-like cell adhesion molecule nectin and its associated protein afadin. *Annu Rev Cell Dev Biol* **24:** 309–342.

Takai Y, Miyoshi J, Ikeda W, Ogita H. 2008b. Nectins and nectin-like molecules: Roles in contact inhibition of cell movement and proliferation. *Nat Rev Mol Cell Biol* **9:** 603–615.

Takai Y, Nakanishi H. 2003. Nectin and afadin: Novel organizers of intercellular junctions. *J Cell Sci* **116:** 17–27.

Tanaka Y, Nakayama Y, Taniguchi M, Kioussis D. 2008. Regulation of early T cell development by the PHD finger of histone lysine methyltransferase ASH1. *Biochem Biophys Res Commun* **365:** 589–594.

Tapia R, Huerta M, Islas S, Avila-Flores A, Lopez-Bayghen E, Weiske J, Huber O, Gonzalez-Mariscal L. 2009. Zona occludens-2 inhibits cyclin D1 expression and cell proliferation and exhibits changes in localization along the cell cycle. *Mol Biol Cell* **20:** 1102–1117.

Tetsu O, McCormick F. 1999. β-catenin regulates expression of cyclin D1 in colon carcinoma cells. *Nature* **398:** 422–426.

Theisen CS, Wahl JK 3rd, Johnson KR, Wheelock MJ. 2007. NHERF links the N-cadherin/catenin complex to the platelet-derived growth factor receptor to modulate the actin cytoskeleton and regulate cell motility. *Mol Biol Cell* **18:** 1220–1232.

Tian YC, Phillips AO. 2002. Interaction between the transforming growth factor-β type II receptor/Smad pathway and β-catenin during transforming growth factor-β1-mediated adherens junction disassembly. *Am J Pathol* **160:** 1619–1628.

Tiwari-Woodruff SK, Buznikov AG, Vu TQ, Micevych PE, Chen K, Kornblum HI, Bronstein JM. 2001. OSP/claudin-11 forms a complex with a novel member of the tetraspanin super family and β1 integrin and regulates proliferation and migration of oligodendrocytes. *J Cell Biol* **153:** 295–305.

Togashi H, Miyoshi J, Honda T, Sakisaka T, Takai Y, Takeichi M. 2006. Interneurite affinity is regulated by heterophilic nectin interactions in concert with the cadherin machinery. *J Cell Biol* **174:** 141–151.

Traweger A, Fuchs R, Krizbai IA, Weiger TM, Bauer HC, Bauer H. 2003. The Tight Junction Protein ZO-2 Localizes to the Nucleus and Interacts with the Heterogeneous Nuclear Ribonucleoprotein Scaffold Attachment Factor-B. *J Biol Chem* **278:** 2692–2700.

Traweger A, Lehner C, Farkas A, Krizbai IA, Tempfer H, Klement E, Guenther B, Bauer HC, Bauer H. 2008. Nuclear Zonula occludens-2 alters gene expression and junctional stability in epithelial and endothelial cells. *Differentiation* **76:** 99–106.

Troyanovsky S. 2005. Cadherin dimers in cell-cell adhesion. *Eur J Cell Biol* **84:** 225–233.

Troyanovsky RB, Laur O, Troyanovsky SM. 2007. Stable and unstable cadherin dimers: Mechanisms of formation and roles in cell adhesion. *Mol Biol Cell* **18:** 4343–4352.

Tsapara A, Matter K, Balda MS. 2006. The heat-shock protein Apg-2 binds to the tight junction protein ZO-1 and regulates transcriptional activity of ZONAB. *Mol Biol Cell* **17:** 1322–1330.

Tsukita S, Furuse M, Itoh M. 2001. Multifunctional strands in tight junctions. *Nat Rev Mol Cell Biol* **2:** 286–293.

Tsukita S, Itoh M, Nagafuchi A, Yonemura S, Tsukita S. 1993. Submembranous junctional plaque proteins include potential tumor suppressor molecules. *J Cell Biol* **123:** 1049–1053.

Ulrich F, Krieg M, Schotz EM, Link V, Castanon I, Schnabel V, Taubenberger A, Mueller D, Puech PH, Heisenberg CP. 2005. Wnt11 functions in gastrulation by controlling cell cohesion through Rab5c and E-cadherin. *Dev Cell* **9:** 555–564.

Umeda K, Matsui T, Nakayama M, Furuse K, Sasaki H, Furuse M, Tsukita S. 2004. Establishment and characterization of cultured epithelial cells lacking expression of ZO-1. *J Biol Chem* **279:** 44785–44794.

Unterseher F, Hefele JA, Giehl K, De Robertis EM, Wedlich D, Schambony A. 2004. Paraxial protocadherin coordinates cell polarity during convergent extension via Rho A and JNK. *Embo J* **23:** 3259–3269.

Utton MA, Eickholt B, Howell FV, Wallis J, Doherty P. 2001. Soluble N-cadherin stimulates fibroblast growth factor receptor dependent neurite outgrowth and N-cadherin and the fibroblast growth factor receptor co-cluster in cells. *J Neurochem* **76:** 1421–1430.

van de Wetering M, Barker N, Harkes IC, van der Heyden M, Dijk NJ, Hollestelle A, Klijn JG, Clevers H, Schutte M. 2001. Mutant E-cadherin breast cancer cells do not display constitutive Wnt signaling. *Cancer Res* **61:** 278–284.

Van Itallie CM, Balda MS, Anderson JM. 1995. Epidermal growth factor induces tyrosine phosphorylation and reorganization of the tight junction protein ZO-1 in A431 cells. *J Cell Sci* **108:** 1735–1742.

van Roy FM, McCrea PD. 2005. A role for Kaiso-p120ctn complexes in cancer? *Nat Rev Cancer* **5:** 956–964.

Vasioukhin V, Bauer C, Degenstein L, Wise B, Fuchs E. 2001. Hyperproliferation and defects in epithelial polarity upon conditional ablation of α-catenin in skin. *Cell* **104:** 605–617.

Venkov CD, Link AJ, Jennings JL, Plieth D, Inoue T, Nagai K, Xu C, Dimitrova YN, Rauscher FJ, Neilson EG. 2007. A proximal activator of transcription in epithelial-mesenchymal transition. *J Clin Invest* **117:** 482–491.

Vethantham V, Rao N, Manley JL. 2007. Sumoylation modulates the assembly and activity of the pre-mRNA 3′ processing complex. *Mol Cell Biol* **27:** 8848–8858.

Villalonga P, Ridley AJ. 2006. Rho GTPases and cell cycle control. *Growth Factors* **24:** 159–164.

Wang Q, Margolis B. 2007. Apical junctional complexes and cell polarity. *Kidney Int* **72:** 1448–1458.

Wang Y, Du D, Fang L, Yang G, Zhang C, Zeng R, Ullrich A, Lottspeich F, Chen Z. 2006. Tyrosine phosphorylated Par3 regulates epithelial tight junction assembly promoted by EGFR signaling. *Embo J* **25:** 5058–5070.

Wang Y, Janicki P, Koster I, Berger CD, Wenzl C, Grosshans J, Steinbeisser H. 2008. Xenopus Paraxial Protocadherin regulates morphogenesis by antagonizing Sprouty. *Genes Dev* **22:** 878–883.

Watanabe T, Sato K, Kaibuchi K. 2009. Cadherin-mediated intercellular adhesions and signaling cascades involving in small GTPases. *Cold Spring Harb Perspect Biol* **1:** a003020.

Weis WI, Nelson WJ. 2006. Re-solving the cadherin-catenin-actin conundrum. *J Biol Chem* **281:** 35593–35597.

Wheelock MJ, Johnson KR. 2003. Cadherin-mediated cellular signaling. *Curr Opin Cell Biol* **15:** 509–514.

Wheelock MJ, Shintani Y, Maeda M, Fukumoto Y, Johnson KR. 2008. Cadherin switching. *J Cell Sci* **121:** 727–735.

Williams G, Williams EJ, Doherty P. 2002. Dimeric versions of two short N-cadherin binding motifs (HAVDI and INPISG) function as N-cadherin agonists. *J Biol Chem* **277:** 4361–4367.

Williams BO, Barish GD, Klymkowsky MW, Varmus HE. 2000. A comparative evaluation of β-catenin and plakoglobin signaling activity. *Oncogene* **19:** 5720–5728.

Williams EJ, Furness J, Walsh FS, Doherty P. 1994. Activation of the FGF receptor underlies neurite outgrowth stimulated by L1, N-CAM, and N-cadherin. *Neuron* **13:** 583–594.

Willott E, Balda MS, Fanning AS, Jameson B, Van Itallie C, Anderson JM. 1993. The tight junction protein ZO-1 is homologous to the Drosophila discs-large tumor suppressor protein of septate junctions. *Proc Natl Acad Sci* **90:** 7834–7838.

Wodarz A, Stewart DB, Nelson WJ, Nusse R. 2006. Wingless signaling modulates cadherin-mediated cell adhesion in Drosophila imaginal disc cells. *J Cell Sci* **119:** 2425–2434.

Wolf A, Keil R, Gotzl O, Mun A, Schwarze K, Lederer M, Huttelmaier S, Hatzfeld M. 2006. The armadillo protein p0071 regulates Rho signalling during cytokinesis. *Nat Cell Biol* **8:** 1432–1440.

Wong RC, Pera MF, Pebay A. 2008. Role of gap junctions in embryonic and somatic stem cells. *Stem Cell Rev* **4:** 283–292.

Woods DF, Bryant PJ. 1993. ZO-1, DlgA and PSD-95/SAP90: Homologous proteins in tight, septate and synaptic junctions. *Mech Dev* **44:** 85–89.

Wuchter P, Boda-Heggemann J, Straub BK, Grund C, Kuhn C, Krause U, Seckinger A, Peitsch WK, Spring H, Ho AD, Franke WW. 2007. Processus and recessus adhaerentes: Giant adherens cell junction systems connect and attract human mesenchymal stem cells. *Cell Tissue Res* **328:** 499–514.

Xiao K, Oas RG, Chiasson CM, Kowalczyk AP. 2007. Role of p120-catenin in cadherin trafficking. *Biochim Biophys Acta* **1773:** 8–16.

Xing H, Mayhew CN, Cullen KE, Park-Sarge OK, Sarge KD. 2004. HSF1 modulation of Hsp70 mRNA polyadenylation via interaction with symplekin. *J Biol Chem* **279:** 10551–10555.

Xu J, Kausalya PJ, Phua DC, Ali SM, Hossain Z, Hunziker W. 2008. Early embryonic lethality of mice lacking ZO-2, but Not ZO-3, reveals critical and nonredundant roles for individual zonula occludens proteins in mammalian development. *Mol Cell Biol* **28:** 1669–1678.

Xu W, Kimelman D. 2007. Mechanistic insights from structural studies of β-catenin and its binding partners. *J Cell Sci* **120:** 3337–3344.

Yamada S, Nelson WJ. 2007. Localized zones of Rho and Rac activities drive initiation and expansion of epithelial cell-cell adhesion. *J Cell Biol* **178:** 517–527.

Yamada S, Pokutta S, Drees F, Weis WI, Nelson WJ. 2005. Deconstructing the cadherin-catenin-actin complex. *Cell* **123:** 889–901.

Yanagawa S, Lee JS, Haruna T, Oda H, Uemura T, Takeichi M, Ishimoto A. 1997. Accumulation of Armadillo induced by Wingless, Dishevelled, and dominant-negative Zeste-White 3 leads to elevated DE-cadherin in Drosophila clone 8 wing disc cells. *J Biol Chem* **272:** 25243–25251.

Yang X, Chen MW, Terry S, Vacherot F, Chopin DK, Bemis DL, Kitajewski J, Benson MC, Guo Y, Buttyan R. 2005. A human- and male-specific protocadherin that acts through the wnt signaling pathway to induce neuroendocrine transdifferentiation of prostate cancer cells. *Cancer Res* **65:** 5263–5271.

Yap AS, Kovacs EM. 2003. Direct cadherin-activated cell signaling: A view from the plasma membrane. *J Cell Biol* **160:** 11–16.

Yap AS, Brieher WM, Gumbiner BM. 1997. Molecular and functional analysis of cadherin-based adherens junctions. *Annu Rev Cell Dev Biol* **13:** 119–146.

Yap AS, Crampton MS, Hardin J. 2007. Making and breaking contacts: The cellular biology of cadherin regulation. *Curr Opin Cell Biol* **19:** 508–514.

Yap AS, Niessen CM, Gumbiner BM. 1998. The juxtamembrane region of the cadherin cytoplasmic tail supports lateral clustering, adhesive strengthening, and interaction with p120ctn. *J Cell Biol* **141:** 779–789.

Yasen M, Kajino K, Kano S, Tobita H, Yamamoto J, Uchiumi T, Kon S, Maeda M, Obulhasim G, Arii S, et al. 2005. The up-regulation of Y-box binding proteins (DNA binding protein A and Y-box binding protein-1) as prognostic markers of hepatocellular carcinoma. *Clin Cancer Res* **11:** 7354–7361.

Yin T, Getsios S, Caldelari R, Kowalczyk AP, Muller EJ, Jones JC, Green KJ. 2005. Plakoglobin suppresses keratinocyte motility through both cell-cell adhesion-dependent and -independent mechanisms. *Proc Natl Acad Sci* **102:** 5420–5425.

Zhang H, Macara IG. 2008. The PAR-6 polarity protein regulates dendritic spine morphogenesis through p190 RhoGAP and the Rho GTPase. *Dev Cell* **14:** 216–226.

Zhao B, Lei QY, Guan KL. 2008. The Hippo-YAP pathway: New connections between regulation of organ size and cancer. *Curr Opin Cell Biol* **20:** 638–646.

Zhao J, Pestell R, Guan JL. 2001. Transcriptional activation of cyclin D1 promoter by FAK contributes to cell cycle progression. *Mol Biol Cell* **12:** 4066–4077.

Zhao J, Bian ZC, Yee K, Chen BP, Chien S, Guan JL. 2003. Identification of transcription factor KLF8 as a downstream target of focal adhesion kinase in its regulation of cyclin D1 and cell cycle progression. *Mol Cell* **11:** 1503–1515.

Zhurinsky J, Shtutman M, Ben-Ze'ev A. 2000. Plakoglobin and β-catenin: Protein interactions, regulation and biological roles. *J Cell Sci* **113:** 3127–3139.

Cell Adhesion, the Backbone of the Synapse: "Vertebrate" and "Invertebrate" Perspectives

Nikolaos Giagtzoglou[1], Cindy V. Ly[2], and Hugo J. Bellen[1,2,3,4]

[1]Howard Hughes Medical Institute, Baylor College of Medicine, Houston, Texas 77030

[2]Department of Neuroscience, Baylor College of Medicine, Houston, Texas 77030

[3]Department of Molecular and Human Genetics, Baylor College of Medicine, Houston, Texas 77030

[4]Program in Developmental Biology, Baylor College of Medicine, Houston, Texas 77030

Correspondence: giagtzog@bcm.tmc.edu, hbellen@bcm.tmc.edu

Synapses are asymmetric intercellular junctions that mediate neuronal communication. The number, type, and connectivity patterns of synapses determine the formation, maintenance, and function of neural circuitries. The complexity and specificity of synaptogenesis relies upon modulation of adhesive properties, which regulate contact initiation, synapse formation, maturation, and functional plasticity. Disruption of adhesion may result in structural and functional imbalance that may lead to neurodevelopmental diseases, such as autism, or neurodegeneration, such as Alzheimer's disease. Therefore, understanding the roles of different adhesion protein families in synapse formation is crucial for unraveling the biology of neuronal circuit formation, as well as the pathogenesis of some brain disorders. The present review summarizes some of the knowledge that has been acquired in vertebrate and invertebrate genetic model organisms.

Synapses are asymmetric, intercellular junctions that are the basic structural units of neuronal transmission. The correct development of synaptic specializations and the establishment of appropriate connectivity patterns are crucial for the assembly of functional neuronal circuits. Improper synapse formation and function may cause neurodevelopmental disorders, such as mental retardation (MsR) and autism spectrum disorders (ASD) (McAllister 2007; Sudhof 2008), and likely play a role in neurodegenerative disorders, such as Alzheimer's disease (AD) (Haass and Selkoe 2007).

At chemical synapses (reviewed in Sudhof 2004; Zhai and Bellen 2004; Waites et al. 2005; McAllister 2007; Jin and Garner 2008), the presynaptic compartment contains synaptic vesicles (SV), organized in functionally distinct subcellular pools. A subset of SVs docks to the presynaptic membrane around protein-dense release sites, named active zones (AZ). Upon the arrival of an action potential at the terminal, the docked and "primed" SVs fuse with the plasma membrane and release neurotransmitter molecules into the synaptic cleft. Depending on the type of synapse (i.e., excitatory vs. inhibitory synapses), neurotransmitters ultimately

 doi: 10.1101/cshperspect.a003079
Cite this article as *Cold Spring Harb Perspect Biol* 2009;1:a003079

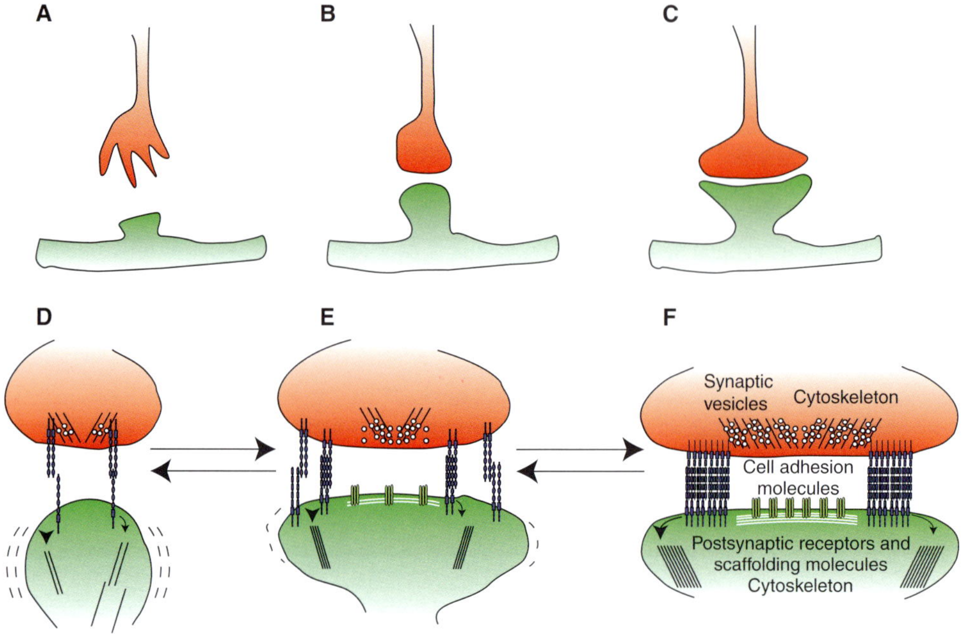

Figure 1. (*A–C*) Different stages of synapse formation. (*A*) Target selection, (*B*) Synapse assembly, (*C*) Synapse maturation and stabilization. (*D–F*) The role of cell adhesion molecules in synapse formation is exemplified by the paradigm of N-cadherin and catenins in regulation of the morphology and strength of dendritic spine heads. (*D*) At an early stage the dendritic spines are elongated from motile structures "seeking" their synaptic partners. (*E*) The contacts between the presynaptic and postsynaptic compartments are stabilized by recruitment of additional cell adhesion molecules. Adhesional interactions activate downstream pathways that remodel the cytoskeleton and organize pre- and postsynaptic apparatuses. (*F*) Cell adhesion complexes, stabilized by increased synaptic activity, promote the expansion of the dendritic spine head and the maturation/stabilization of the synapse. Retraction and expansion is dependent on synaptic plasticity.

activate an appropriate set of postsynaptic receptors that are accurately apposed to the AZ.

Synapse formation occurs in several steps (Fig. 1) (reviewed in Eaton and Davis 2003; Goda and Davis 2003; Waites et al. 2005; Garner et al. 2006; Gerrow and El-Husseini 2006; McAllister 2007). Spatiotemporal signals guide axons through heterogeneous cellular environments to contact appropriate postsynaptic targets. At their destination, axonal growth cones initiate synaptogenesis through adhesive interactions with target cells. In the mammalian central nervous system (CNS), immature postsynaptic dendritic spines initially protrude as thin, actin-rich filopodia on the surface of dendrites. Similarly, at the *Drosophila* neuromuscular junction (NMJ), myopodia develop from the muscles (Ritzenthaler et al. 2000). The stabilization of intercellular contacts and their elaboration into mature, functional synapses involves cytoskeletal arrangements and recruitment of pre- and postsynaptic components to contact sites in spines and boutons. Conversely, retraction of contacts results in synaptic elimination. Both stabilization and retraction sculpt a functional neuronal circuitry.

In addition to the plastic nature of synapse formation, the vast heterogeneity of synapses (in terms of target selection, morphology, and type of neurotransmitter released) greatly enhances the complexity of synaptogenesis (reviewed in Craig and Boudin 2001; Craig et al. 2006; Gerrow and El-Husseini 2006). The complexity and specificity of synaptogenesis relies upon the modulation of adhesion between the pre- and postsynaptic components (reviewed in Craig et al. 2006; Gerrow and El-Husseini 2006; Piechotta et al. 2006; Dalva et al. 2007; Shapiro et al. 2007; Yamada and Nelson 2007; Gottmann 2008). Cell adhesive interactions enable cell–cell recognition via extracellular domains and also mediate intracellular signaling cascades that affect synapse morphology and organize scaffolding complexes. Thus, cell adhesion molecules (CAMs) coordinate multiple synaptogenic steps.

However, in vitro and in vivo studies of vertebrate CAMs are often at odds with each other. Indeed, there are no examples of mutants for synaptic CAMs that exhibit prominent defects in synapse formation. This apparent "resilience" of synapses is probably caused by functional redundancy or compensatory effects among different CAMs (Piechotta et al. 2006). Hence, studies using simpler organisms less riddled by redundancy, such as *Caenorhabditis elegans* and *Drosophila*, have aided in our understanding of the role that these molecules play in organizing synapses.

In this survey, we discuss the roles of the best characterized CAM families of proteins involved in synaptogenesis. Our focus is to highlight the complex principles that govern the molecular basis of synapse formation and function from a comparative perspective. We will present results from cell culture studies as well as in vivo analyses in vertebrate systems and refer to invertebrate studies, mainly performed in *Drosophila* and *C. elegans*, when they have provided important insights into the role of particular CAM protein families. However, we do not discuss secreted factors, for which we refer the reader to numerous excellent reviews (as for example Washbourne et al. 2004; Salinas 2005; Piechotta et al. 2006; Shapiro et al. 2006; Dalva 2007; Yamada and Nelson 2007; Biederer and Stagi 2008; Salinas and Zou 2008).

THE NEUROLIGINS–NEUREXINS: SUFFICIENCY VERSUS NECESSITY

The Neuroligin (Nlg)–Neurexin (Nrx) proteins play a major role in synapse formation and function (reviewed in Dean and Dresbach 2006; Dalva et al. 2007; Sudhof 2008). Neuroligins are type I membrane proteins consisting of an extracellular, catalytically inactive acetylcholinesterase domain that mediates binding to the following: neurexins, an *O*-glycosylation motif, a transmembrane (TM) domain, a cytoplasmic PDZ (post synaptic density protein [PSD95]), *Drosophila* disc large tumor suppressor (Dlg), and a zonula occludens-1 protein (zo-1) domain (Fig. 2) (Craig and Kang 2007; Sudhof 2008). There are four neuroligin genes in vertebrates, but alternative splicing increases the complexity of this protein family (Ichtchenko et al. 1996; Bolliger et al. 2001). Neuroligins, which are enriched postsynaptically, interact with presynaptic neurexins (Song et al. 1999). Three neurexin genes have been reported in mammals (Nrx 1–3), each with dual promoters that enable expression of two distinct protein isoforms, a long α- and a short β-isoform. The latter short isoforms bind to neuroligins (Tabuchi and Sudhof 2002; Sudhof 2008), neurexins contain adhesive, extracellular laminin, nectin, sex-hormone binding globulin (LNS) domains prone to extensive alternative splicing (Fig. 2) (Rudenko et al. 1999, 2001). Though found presynaptically at both excitatory and inhibitory synapses (Ushkaryov et al. 1992; Ullrich et al. 1995; Dean et al. 2003; Graf et al. 2004), neurexins have also been reported to be present on postsynaptic cells (Kattenstroth et al. 2004; Taniguchi et al. 2007).

Neuroligins induce presynaptic differentiation of contacting neurites when expressed in heterologous nonneuronal cells (Scheiffele et al. 2000) and regulate synapse number and spine density in cell culture (Chih et al. 2005). In addition, recruitment of PSD-95 to the Neuroligin PDZ domain clusters NMDA receptors and other postsynaptic proteins (Fig. 2) (Chih et al. 2004). Neurexin overexpression in heterophilic cells induces postsynaptic differentiation of contacting neurites in vitro (Graf et al. 2004; Nam and Chen 2005; Chih et al. 2006). Neurexin signaling induces presynaptic specialization and is linked with the synaptic release machinery, voltage gated calcium channel (VGCC) clustering, and actin cytoskeletal changes (Hata et al. 1996; Butz et al. 1998; Biederer and Sudhof 2000; Atasoy et al. 2007).

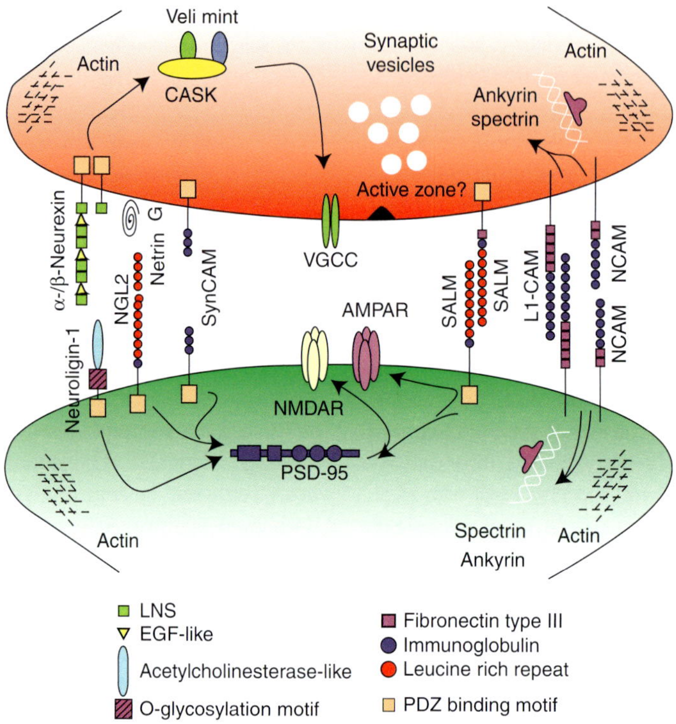

Figure 2. Diagram of Nrx –Nlg, LRR, and Ig mediated adhesion at excitatory synapses. Neurexins participate in the organization of presynaptic apparatuses and are linked to the synaptic vesicle cycle through interaction with CASK/Mint/Veli. Neuroligins interact with the postsynaptic PDZ scaffolding molecule PSD-95, which organizes the postsynaptic receptors (mainly NMDAR). Neurexins and neuroligins also participate at inhibitory synapses. Therefore, they are important for the balance of excitatory and inhibitory neurotransmission. Postsynaptic NGL2 binds to presynaptic GPI anchored Netrin G. NGL2 binds to the postsynaptic PDZ scaffolding molecule PSD95 and recruits NMDAR, but not AMPAR, in coculture experiments. SALM family members regulate the differentiation of neurites, organize postsynaptic sites by interacting with PSD-95, and recruit mainly NMDAR, and to a lesser extend AMPAR, in coculture experiments. Although no effects of SALMs on presynaptic organization are known, SALMs form distinct homophilic and heterophilic interactions suggesting adhesive roles on both sides of synapses. Ig molecules participate in both the structural and the functional organization of the synapse. SynCAM forms preferentially heterophilic interactions and induces synapse formation in coculture experiments. Presynaptically, it participates in the functional organization of the terminal through the interaction with CASK/Mint/Veli complexes. Postsynaptically, it interacts with the scaffolding molecule PSD-95. NCAM and L1-CAM form homophilic complexes and regulate cytoskeleton dynamics pre- and postsynaptically through interactions with spectrin and ankyrin.

In summary, the neurexin–neuroligin interactions are sufficient in cell culture assays to induce synapse formation, suggesting that they play a critical role in synapse formation.

The synaptogenic activities of Nlg and Nrx depend on their physical interaction (Scheiffele et al. 2000). These heterophilic associations regulate both excitatory and inhibitory synaptogenesis. Their ability to define different synapse types likely involves cell-type specific localization, as well as particular neuroligin–neurexin combinations (Boucard et al. 2005; Chih et al. 2006). For example, Nlg-1 and -4 exclusively localize to excitatory synapses, whereas Nlg-2 is found only at inhibitory synapses, suggesting distinct roles at these terminals (Graf et al. 2004; Song et al. 1999; Varoqueaux et al. 2004). Nlg-3 can be found at both types of synapses (Budreck and Scheiffele 2007). Interestingly, an insert in splice site B of Nlg-1 disrupts binding to Nrx-1β and favors its localization to excitatory synapses. On the other end, Nlg-1 isoforms without the insert in splice site B bind to both α- and β-neurexins (Boucard et al. 2005; Chih et al. 2006). Hence, subtype and splice code variants of neurexin and neuroligin may dictate their binding specificity and roles in functional differentiation during synapse formation (Dalva et al. 2007).

The aforementioned data indicate that the Nrx–Nlg axis is sufficient to induce synapse formation in cell culture. However, in vivo studies paint a different picture. Removal of Nlg 1–3 or α-Nrx in mice does not support a singular inductive role for these proteins (Missler et al. 2003; Varoqueaux et al. 2006). Nlg triple knockout mice die soon after birth from respiratory failure, likely because of a subtle synaptic imbalance of excitatory and inhibitory synapses in the brainstem, although synapse numbers in most areas of the brain, including respiratory centers, are intact (Varoqueaux et al. 2006). Similarly, α-Nrx triple knockout mice exhibit neonatal lethality and reduced synaptic transmission at both excitatory and inhibitory synapses (Missler et al. 2003), though synapse number and morphology are also unaffected (Dudanova et al. 2007). Electrophysiological studies, however, indicate a requirement for a presynaptic function of the neurexins and a role in regulating the activity of synaptic Ca^{2+} channels (Missler 2003). Hence, the in vivo analysis indicates that the Nrx–Nlg proteins are sufficient in vitro but not necessary in vivo for synapse formation. Redundancy with other CAM proteins may underlie the apparent discrepancy between in vitro and in vivo studies. Alternatively, the Nrx–Nlg may be required in synapse stabilization at a later stage.

Drosophila contains a single neurexin (*dnrx*) and a single neuroligin gene homologous to the vertebrate counterparts (Tabuchi and Sudhof 2002; Li et al. 2007; Zeng et al. 2007). A neurexin IV has been reported in *Drosophila*, but this protein is clearly less homologous to Nrx I and was later shown to be more similar to the vertebrate Caspr protein (Bhat et al. 2001). Both NrxIV and Caspr were shown to play a role in septate junction formation and axon insulation in flies and mice (Baumgartner et al. 1996; Bhat et al. 2001). dNrx (α-neurexin) is expressed in neurons and localizes to active zones at the NMJ (Li et al. 2007). *dnrx1* controls synapse growth, differentiation, and function, in agreement with a synaptogenic role for neurexin proteins. Overexpression of *dnrx* leads to synaptic overgrowth. Conversely, loss of *dnrx* leads to a reduction of the number of synaptic boutons at the NMJ and misalignment of the pre- and postsynaptic compartments. The SV protein synaptotagmin and the AZ protein Bruchpilot are mislocalized in *dnrx* mutant axons, indicating a role in the recruitment, localization, or transport of key synaptic components (Li et al. 2007). Future studies in *Drosophila* should shed a better insight into how the neuroligin–neurexin system regulates synapse development. An important step will be to determine whether the *Drosophila Nlg* gene, which reportedly exhibits the same spatiotemporal pattern of expression as dNrx, also affects synapse growth and function (Li et al. 2007).

In summary, neurexins and neuroligins are not absolutely required to form synapses, but they clearly play a role in the formation of the ultrastructural morphological features of synapses in flies and affect synapse function in vertebrates in vivo. These observations also emphasize the importance of carrying out in vivo experiments as they reveal a higher order of complexity than the in vitro cell culture data.

LEUCINE RICH REPEAT (LRR) PROTEINS

The leucine rich repeat (LRR) motifs are 20–29 amino acid (aa)-long protein interaction sequences, present in extracellular domains of a wide range of transmembrane as well as cytosolic proteins. Many LRR proteins regulate neurite outgrowth and migration (reviewed in Ko and Kim 2007), but two specific members of this family have been implicated in synaptogenesis: the netrin-G ligand (NGL) and the synaptic-like adhesion molecule (SALM) (Fig. 2).

The NGL family is composed of three type I transmembrane members, NGL 1–3. The NGL proteins contain nine LRR extracellular interaction domains, immunoglobulin (Ig) domains, a TM domain, and a PDZ motif, which interacts with PSD-95 (Fig. 2) (reviewed in Ko and Kim 2007). Extracellularly, the NGLs associate with netrin-G1 or netrin-G2, a family of netrin-like cell adhesion molecules, with a great number of spatiotemporally regulated splice variants, mainly expressed in the brain. NGL1 interacts with netrin-G1 (and possibly with other receptors as well) to regulate the growth of thalamocortical axons (Lin et al. 2003).

Interestingly, the NGL/netrin-G system shares key functional features with the Nlg-Nrx axis. Overexpression of NGL in nonneuronal cells in vitro induces the differentiation of presynapses in contacting axons of cocultured neurons, although overexpression of netrin-G does not have the reciprocal capacity of inducing postsynaptic differentiation, as observed for Nrx (Kim et al. 2006). The possibility that netrin-G might not be the only ligand for NGLs may account for the unidirectional synaptogenic activity. For example, NGL3 does not interact with any of the netrin-G ligands (Kim et al. 2006). Furthermore, putative isoform specific interactions increase both the complexity and redundancy of adhesion. A role in synapse formation is further supported by the elevated expression levels of NGL in the brain postnatally and the ability of NGL-2 to induce clustering of NMDA receptors at excitatory synapses (Fig. 2) (Kim et al. 2006).

Interestingly, in *C. elegans*, a molecule distantly related to netrin-G, called netrin (a ligand for netrin receptors), is secreted by glial cells and plays a dual role in a pair of specific interneurons (Colon-Ramos et al. 2007). In this experimental system, netrin guides the axon of the postsynaptic neuron to the glial site of secretion, but unexpectedly, also orchestrates the differentiation of presynaptic boutons.

The SALM family of adhesion molecules contains five members in vertebrates. SALMs contain six LRR domains, an Ig domain, and a fibronectin III domain, followed by a transmembrane region and carboxy-terminal PDZ binding motif that interacts with PSD-95 (Fig. 2) (Ko et al. 2006; Ko and Kim 2007). The SALMs are mainly expressed in the brain. SALM-1 overexpression promotes excitatory synapse maturation and NMDA receptor clustering (Wang et al. 2006). SALM-2 regulates the number of excitatory synapses in cocultured neurons and associates with AMPA receptors more than with NMDA receptors (Ko et al. 2006; Wang et al. 2006). SALMs form distinct homo- and heterophilic interactions, suggesting adhesive roles pre- and postsynaptically (Seabold et al. 2008). However, the ligands and the nature of the molecular interactions of different SALMS need to be further elucidated.

Interestingly, in *Drosophila*, an overexpression genetic screen was recently conducted to identify novel cell membrane and secreted molecules with a role in target selection by the motor neuron axons (Kurusu et al. 2008). Out of 55 genes that were found to affect muscle targeting of motor axons and/or NMJ arborization patterns, 16 belong to the LRR family, including the *tartan* and *capricious* genes. *Capricious* is involved in target selectivity of both motor neuron and photoreceptor axons (Shishido et al. 1998; Shinza-Kameda et al. 2006). Tartan and Capricious cooperate in axonal pathfinding and have interchangeable extracellular domains (Kurusu et al. 2008).

In summary, LRR proteins play an important role in various aspects of neuronal development. However, additional studies are necessary to identify their mode of action in synaptogenesis.

CONTROL OF SYNAPTOGENESIS BY MEMBERS OF THE IG SUPERFAMILY

The Ig superfamily is the largest, most diverse, and most ancient superfamily represented in the genomes of both vertebrate and invertebrate organisms (Vogel et al. 2003). Members of the Ig protein family are single-pass type I transmembrane proteins that contain variable numbers of globular, extracellular, cysteine looped, Ig or Ig-like domains that mediate homophilic or heterophilic interactions. They are typically followed by fibronectin III type repeats (Fig. 2) (reviewed in Walsh and Doherty 1997; Piechotta et al. 2006).

The Synaptic Cell Adhesion Molecule (SynCAM)

The family of synaptic cell adhesion molecules (SynCAM) proteins, SynCAM1–4 (Biederer 2006), are expressed in the nervous system, where they constitute a *trans*-synaptic adhesion system (Fig. 2). SynCAMs can undergo further alternative splicing and are often post-translationally modified through glycosylation (Fogel et al. 2007), thus increasing their complexity. The differential expression pattern and the heterophilic pattern of interactions among SynCAMs establish an adhesion code for neuronal circuitry formation (Biederer and Stagi 2008; Thomas et al. 2008).

SynCAM1 (also named nectin-like 2) localizes to both sides of the synaptic cleft. SynCAM1 can trigger the differentiation of functional presynaptic specializations in the axons of cocultured neurons (Fig. 2), similarly to Nlg (Biederer et al. 2002), although the induced synapses may be functionally distinct (Sara et al. 2005). Upon coexpression of SynCAM with glutamate receptors in HEK cells, glutamate-receptor-mediated currents can be detected (Biederer et al. 2002). The synaptogenic effect of SynCAM might be mediated through interaction of intracellular motifs (common among all members of SynCAM) with FERM (fourpoint-one, ezrin, radixin, and moesin) domains of cytoskeletal adaptors and PDZ domains of synaptic scaffolding proteins, such as CASK and syntenin (Fig. 2) (Biederer and Stagi 2008).

Despite the potent synaptogenic activity of SynCAM1 in vitro, SynCAM 1 knockout mutant mice are viable and fertile and synaptogenesis is not altered, possibly because of functional redundancy with other SynCAM proteins or compensation by other adhesion signaling complexes (Fujita et al. 2006; van der Weyden et al. 2006). Hence, redundancy is a recurrent theme to explain the lack of expected phenotypes when these proteins with potent in vitro functions are removed in vivo.

Neural Cell Adhesion Molecule (NCAM)

Neural cell adhesion molecule (NCAM) was one of the first membrane proteins implicated in cell adhesion between neurons. Three isoforms are produced from a single gene, one of which is glycosyl-phospahtidyl-inositol (GPI) linked to the membrane, whereas the other two are transmembrane molecules. NCAM regulates synapse formation, maturation, and function through homo- and heterophilic interactions (Fig. 2).

In vertebrates, NCAM localizes to developing synapses (Uryu et al. 1999) and its expression is activity-dependent (Schuster et al. 1998). Excitatory synapse density is reduced in NCAM-deficient cells when cocultured with wild-type neurons (Dityatev et al. 2000). Interestingly, synaptogenesis in this context depends on postsynaptic NCAM, its modification with polysialic acid (PSA) and activation of NMDA and fibroblast growth factor (FGF) receptors (Dityatev et al. 2004). However, synapse numbers from cultures of NCAM-deficient neurons alone are not different from wild-type cultures. Furthermore, mice lacking all three isoforms of NCAM are viable and fertile, although synaptic plasticity is impaired (Cremer et al. 1994; Muller et al. 1996; Cremer et al. 2000; Eckhardt et al. 2000; Bukalo et al. 2004). Synaptic transmission and vesicle dynamics at NCAM-deficient neuromuscular junctions are also altered (Polo-Parada et al. 2001). Hence, NCAM is essential for synapse function, but not synapse formation, possibly because of redundancy or compensation by other adhesion molecules. Alternatively, NCAM might only play a role in synaptic plasticity.

In *Drosophila*, fasciclin II (Fas II) is the only ortholog of vertebrate NCAM (reviewed in Packard et al. 2003; Kristiansen and Hortsch 2008). Similar to NCAM, there are three major protein isoforms of Fas II, two of which are transmembrane proteins and one that is GPI-anchored to the cell membrane. However, unlike the vertebrate homologs, none of the Fas II protein isoforms is modified by PSA. Fas II is expressed in numerous tissues, including the developing nervous system. Notably, the GPI-anchored isoform is expressed in nonneuronal cell types, especially glial cells where it provides an adhesive signal for guiding extending axons. The role of Fas II has been studied in the context of axonal pathfinding, synapse growth, and function at the NMJ and in the CNS.

Fas II is expressed in motor neuron axons and is especially enriched in their growth cones (Schuster et al. 1996a,b). Upon contact of the motor axon with the target muscle at the NMJ and initiation of synapse formation, Fas II becomes enriched pre- and postsynaptically. Overexpression of Fas II in the muscle affects target-selection and changes neuronal connectivity (Davis et al. 1997). However, in the absence of Fas II, synapses still form but eventually retract, suggesting that Fas II is not required for synapse induction, but rather stabilizes the nascent synapse (Schuster et al. 1996a,b). Lack of stability allows or induces dynamic structural changes during synapse growth. Indeed, reduced levels of Fas II allow synapse growth and sprouting in mutant flies (Schuster et al. 1996a,b). Interestingly, mutations that presynaptically increase synaptic activity by affecting cyclic AMP (cAMP) degradation result in reduction of FasII and expansion of the NMJ arborization pattern (Budnik et al. 1990; Zhong et al. 1992). Postsynaptically, Fas II localization at the NMJ is regulated through interaction with Dlg (the fly homolog of PSD-95) (Thomas et al. 1997). Ca^{2+}-calmodulin protein kinase II (CaMKII) phosphorylates Dlg in an activity-dependent manner, causing the dissociation of Dlg from the postsynaptic density, reducing the binding of Dlg with FasII. This results in declustering and removal of Fas II from the synapse, thereby allowing an expansion of the synapse (Koh et al. 1999).

In the CNS, Fas II is not required for initial synapse formation, similarly to the principles of NMJ development, but a lack of balance between the pre- and postsynaptic levels of Fas II causes a decrease in synaptic input to the motor neurons and a reduction of the number of synapses (Baines et al. 2002). Fas II regulates synaptic function, and this has been documented in learning and memory related studies. Fas II is expressed in the axons of the mushroom bodies, the learning centers of the fly brain, where it is involved in memory storage, and maintenance and retrieval of memory (Cheng et al. 2001).

L1-CAM

In vertebrates, L1-CAM, neurofascin, Nr-CAM, and CHL1 comprise the L1-type CAM protein group. L1-CAM interacts both homophilicaly and heterophilicaly. L1 is present in the CNS at the onset of differentiation. The highly conserved intracellular domains of most L1-type proteins contain ankyrin and FERM binding motifs linking it to spectrin and the actin cytoskeleton (Fig. 2) (Davis and Bennett 1994; Dickson et al. 2002; Cheng et al. 2005). However, L1 can also interact with microtubule stabilizing proteins such as doublecortin (Kizhatil et al. 2002).

L1-type proteins play a role in cell migration, neurite extension, axon pathfinding, myelination, fasciculation, synapse targeting, and long term potentiation (LTP) (Crossin and Krushel 2000; Gerrow and El-Husseini 2006; Piechotta et al. 2006; Maness and Schachner 2007). The L1-CAM gene is causally related to a variety of neurological disorders that are associated with mental retardation, hydrocephalus, and spasticity of the lower limbs (CRASH syndrome) (Fransen et al. 1995, 1997; Bateman et al. 1996; Fransen et al. 1997; Kenwrick and Doherty 1998; Kenwrick et al. 2000; De Angelis et al. 2002). Two L1-CAM knockout mice strains were independently generated (Dahme et al. 1997; Cohen et al. 1998). Dahme et al. (1997) reported that deletion of L1-CAM causes a high rate of embryonic lethality and compromises the postnatal survival of knockout mice, which also display a deficiency of peripheral sensory and motor innervation, a reduction of the corticospinal tract, an enlargement of the cerebral ventricles, and an impaired axon–Schwann cell association (Dahme et al. 1997). In other studies, the L1 knockout mice were viable but exhibited an axon guidance defect in the corticospinal tract, disturbed axonal targeting, altered dendritic morphology and orientation, and abnormalities of spatial learning and exploratory behavior (Cohen et al. 1998; Fransen et al. 1998; Demyanenko et al. 1999, 2001). An electrophysiological study of L1 mice has revealed a reduction of γ-aminobutyric acid (GABA)-ergic transmission in the CA1 region of the hippocampus. Morphologically, a corresponding decrease in the density of perisomatic symmetrical synapses has been observed and synaptic vesicles show a more diffuse distribution (Saghatelyan et al. 2004). In a hippocampus-specific conditional L1-CAM knockout mouse model, basal excitatory transmission is enhanced in the mutant neurons and behavioral deficits are only partly similar to those observed in constitutive $L1^{-/-}$ mice (Law et al. 2003). The above findings suggest an involvement of L1 in the formation and/or maintenance of inhibitory synapses and imply that L1 might be functioning along with other cell adhesion systems.

The only homolog of L1-CAM in the fly is neuroglian (Nrg) (Bieber et al. 1989; Walsh and Doherty 1997; Kristiansen and Hortsch 2008). Two transcripts are generated encoding a short and a long isoform (Hortsch et al. 1990; Hall and Bieber 1997). The two isoforms differ in their cytoplasmic domains and the short isoform is expressed in epidermis, muscle, trachea, and glia, whereas the long isoform is restricted to the nervous system (Hortsch et al. 1990). Nrg plays a role in axon growth in the peripheral nervous system during embryonic and postembryonic development. Null *nrg* mutations are lethal, and motor axons are stalled in embryos (Bieber et al. 1989; Hall and Bieber 1997; Martin et al. 2008). Neuronal expression of Nrg rescues the mutant phenotype suggesting that Nrg mediates heterophilic adhesion of the advancing axons with their substrate (Martin et al. 2008). In the adult ocellar sensory system, Nrg establishes two distinct axonal pathways, one originating from the bristle mechanoreceptors and the other originating from the ocellar pioneer neurons (Garcia-Alonso et al. 2000; Kristiansen et al. 2005). Especially in the ocellar pioneer neurons, Nrg functions redundantly with Fas II. Interestingly, it has been proposed that Nrg mediated activation of epidermal growth factor (EGF) and FGF receptors is needed for the pathfinding process (Garcia-Alonso et al. 2000).

It has been suggested that ankyrin-binding to L1-type CAMs provides a "master switch," so that L1-type adhesive proteins can function in different contexts: in an ankyrin-independent manner during neurite outgrowth and axonal pathfinding, or in ankyrin-dependent manner during synaptogenesis (Hortsch et al. 2009). Indeed, Nrg also functions along with the intracellular adaptor ankyrin to suppress axonal sprouting and dendritic branching in a subset of multidendritic neurons (Yamamoto et al. 2006) and control proper synapse formation and function in the giant fiber synapse by organizing the microtubules in the presynaptic terminals (Godenschwege et al. 2006). The significance of ankyrin in organizing the stability of presynaptic cytoskeleton is further supported by experiments at the fly NMJ, which show that in the absence of the neuronally expressed ankyrin 2, synapses retract, leaving behind "footprints" of postsynaptic specializations (Koch et al. 2008; Pielage et al. 2008). In addition, postsynaptic spectrin is necessary for the development and organization of both the presynaptic active zone and the opposing postsynaptic density (Pielage et al. 2006). It would be interesting to assess whether Nrg participates in these processes as well.

Synaptogenesis Abnormal (SYG) -1 and -2

In the *C. elegans* egg-laying circuitry, two transmembrane proteins, SYG-1 and SYG-2, control target recognition and synapse formation of the hermaphrodite specific motor neuron HSNL. SYG-2 is present on vulval epithelial cells and interacts with SYG-1, which is expressed on HSNL neuron (Shen and Bargmann 2003; Shen et al. 2004). Upon interaction with SYG-2, SYG-1 initiates the hierarchical assembly of presynaptic specializations in HSNL (Patel et al. 2006) by recruiting SYD-1 (a rhoGTPase activating protein factor) and

SYD-2 (liprin-α). Furthermore, the interaction of SYG-1 and SYG-2 enables SYG-1 to bind to SKR-1, an E3 ubiquitin ligase, and inhibit its activity, protecting the nascent synapses from proteasome mediated, aberrant elimination (Ding et al. 2007). *syg-2* mutants lack synapses at the normal location and instead exhibit ectopic synapses and a reduced number of active zones (Shen et al. 2004). In *syg-1* mutants, vesicles fail to accumulate at normal synaptic locations and form ectopic clusters instead (Shen and Bargmann 2003).

There are two SYG-1 homologs in *Drosophila*, irregular chiasm C-roughest (IrreC-Rst) and kin of IrreC also named dumbfounded (Kirre or Duf), and two SYG-2 homologs, sticks and stones (Sns) and Hibris (Chao and Shen 2008). SYG-1/IrreC/Kirre are homologous to Neph receptor and SYG-2/Hibris/Sns are homologues of nephrin ligands (Chao and Shen 2008). IrreC-Rst, Kirre, and Hibris are involved in patterning of the *Drosophila* eye and in synaptic wiring of the visual system (Ramos et al. 1993; Bao and Cagan 2005; Carthew 2005, 2007; Bazigou et al. 2007). Heterophilic interactions between IrreC-Rst and Kirre, which are expressed in muscle founder cells with Hibris and Sns, respectively (which are expressed on fusion competent myoblasts), are important for *Drosophila* myoblast fusion (Bour et al. 2000; Ruiz-Gomez et al. 2000; Artero et al. 2001; Dworak et al. 2001; Strunkelnberg et al. 2001; Dworak and Sink 2002). Neph/nephrin are instrumental in kidney function (Patrakka and Tryggvason 2007). However, their expression pattern in the nervous system and their ability to interact with CASK implies that the vertebrate homologs may be involved in synapse formation as well (Gerke et al. 2006).

THE INSTRUMENTAL ROLE OF LIPRIN-α AND LEUCOCYTE-COMMON ANTIGEN RELATED (LAR) PROTEINS IN SYNAPSE FORMATION AND ACTIVE ZONE MORPHOGENESIS

Liprin-α family proteins contain an amino-terminal coiled coil domain and a carboxy-terminal liprin homology domain, comprised of three sterile α motifs (SAMs) (reviewed in Stryker and Johnson 2007). Through these motifs, liprin-α proteins interact with leucocyte-common antigen related (LAR) proteins. In vertebrates, there are four liprin-α (liprin α 1–4) and three LAR (LAR, PTP-σ, and PTP-δ) paralogues. Members of the LAR family are transmembrane tyrosine phosphatases with highly conserved extracellular and cytoplasmic domains. Their extracellular region is structurally similar to cell adhesion molecules of the Ig family. Intracellularly, the phosphatase domain proximal to the membrane possesses the majority of the catalytic activity, whereas the distal one seems to interact with numerous downstream effectors, including β-catenin and members of the liprin-α protein family (reviewed in Stryker and Johnson 2007) (Fig. 3).

LAR and liprin-α regulate synapse formation and function (reviewed in Spangler and Hoogenraad 2007; Stryker and Johnson 2007). Presynaptically, mammalian liprin-α associates with two major synaptic complexes. One of the complexes is formed by liprin-α interacting through the coiled coil regions with the cytomatrix at the active zone-associated structural protein (CAST) and Rab3-interacting molecule (RIM) (Fig. 3) (Ohtsuka et al. 2002; Schoch et al. 2002; Ko et al. 2003). The other complex liprin-α associates with is the mammalian-LIN-seven protein (MALS)-CASK-Mint complex (Olsen et al. 2005). Postsynaptically, liprin-α interacts with glutamate receptor interacting protein 1 (GRIP1), an AMPA receptor binding protein (Wyszynski et al. 2002)(Fig. 3). The interaction of LAR and liprin-α regulates AMPA receptor clustering and dendritic spine morphology. Intriguingly, the phosphatase activity of LAR and its interactions with liprin-α and GRIP1 regulate the delivery of cadherin/β-catenin complexes during the development and maintenance of excitatory synapses (Dunah et al. 2005). Liprin-α is regulated by CAMKII. Overexpression of nondegradable mutant forms of liprin or dominant negative forms of LAR that cannot bind to liprin impair signaling

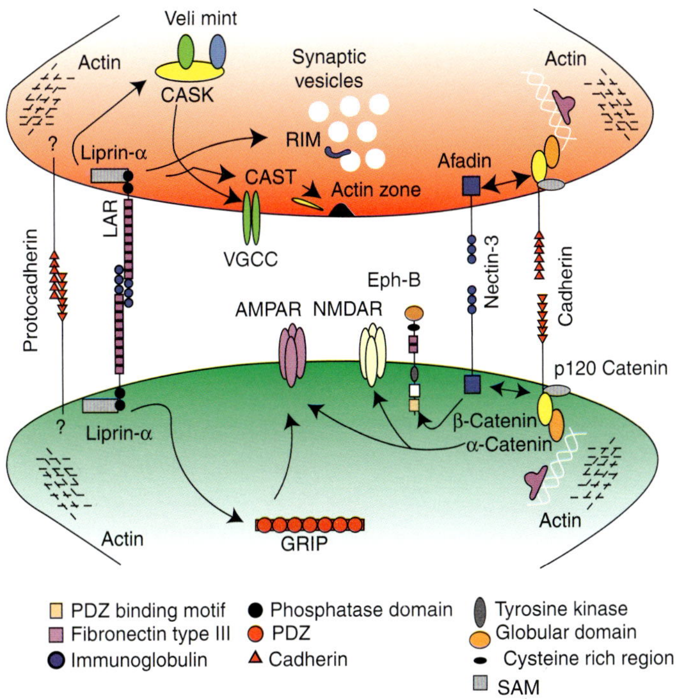

Figure 3. Diagram of LAR-liprin-α signaling and cadherin-nectin mediated adhesion at excitatory synapses. Liprin-α is the mediator of LAR signaling in both pre- and postsynaptic compartments. Thus, LAR/liprin-α complexes control neurotransmitter release and active zone formation through interactions with CAS/Mint/Veli, RIM, and ERC. LAR/liprin-α complexes regulate the postsynaptic organization of neurotransmitter receptors through interaction with GRIP, an AMPAR interacting protein. Cadherins are localized at *puncta adherentia*, the region flanking the active zone, along with nectins. Cadherins form homophilic complexes. The intracellular domain of cadherins interacts with catenins that ultimately link cadherins to the cytoskeleton. Cadherins are the physical link between synaptic activity and dendritic spine morphology (see text and Fig. 1C–F). Cadherins also control the organization of postsynaptic receptors. Nectins form heterophilic complexes and are able to recruit cadherins, but they can also interact with other adhesions systems, such as ephrins. Nectins are indirectly linked to the cytoskeleton through Afadin.

and lead to a reduction of dendritic branching and spine density (Hoogenraad et al. 2007).

LAR-liprin-α signaling has been studied in *C. elegans* and *Drosophila* (reviewed in Stryker and Johnson 2007). In *C. elegans*, SYD-2 (liprin-α) is required presynaptically for the formation of active zones and synaptic function (Zhen and Jin 1999). In addition, the localization of PTP-3A, one of the protein isoforms of LAR, is disrupted in *syd-2* mutants. Mutants lacking PTP-3A phenocopy *syd-2* mutants in terms of active zone organization (Zhen and Jin 1999). The precise localization at the synapse of PTP3A also depends on interactions of the extracellular Ig domains with nidogen, a sulfated glycoprotein, and animals lacking nidogen phenocopy *ptp-3a* and *syd-2* mutants (Ackley et al. 2005).

In *Drosophila*, fly LAR and liprin-α homologs are required in the visual system, along with N-cadherin, at an early developmental stage for unbundling the R1–R6 afferent photoreceptor growth cones, so that they can project properly into assembling laminal cartridges (Clandinin et al. 2001; Choe et al. 2006) (also see Fig. 4). Unlike N-cadherin, which is required both pre- and postsynaptically (see the following discussion), liprin-α and LAR are required only in the presynaptic terminals, where they may regulate N-cadherin

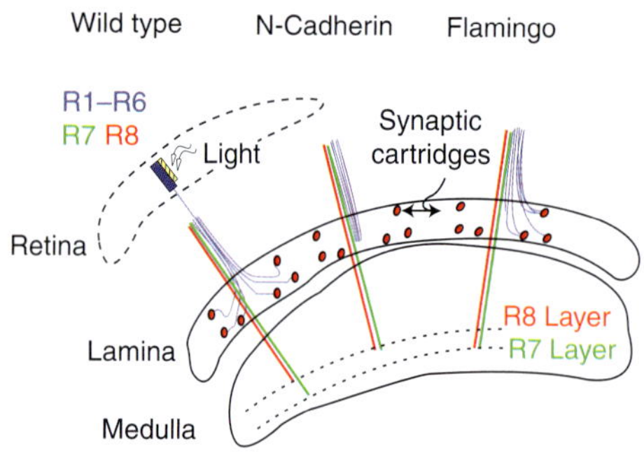

Figure 4. Synaptic wiring in the visual system of *Drosophila*. The compound eye of *Drosophila* is comprised of 750 ommatidia, each of which has eight photoreceptors, named R1–R8. Light stimulates the photoreceptors at the outer level of the retina, where the cell bodies lie. The signal is transmitted to the postsynaptic partners at the inner neuropils of the lamina and the medulla. R1–R6 photoreceptor axons from each ommatidium stop at the lamina, where they defasciculate and innervate nearby synaptic cartridges according to the principle of neural superposition. R7 and R8 photoreceptor axons continue to the medulla, where they terminate at distinct layers. In loss of function of N-cadherin, R1 and R6 fail to defasciculate, whereas R7 retracts to the wrong R8 specific layer. In loss of function of Flamingo cadherin, R1–R6 defasciculate but innervate abberrantly the synaptic cartridges, whereas the R8 photoreceptor axons terminate in the R7 layer (see text and also Mast et al. 2006; Ting and Lee 2007).

activity to establish a form of functional asymmetry (Clandinin et al. 2001; Choe et al. 2006). LAR and liprin-α are required for proper layer target selection of the R7 photoreceptor axons (Newsome et al. 2000; Maurel-Zaffran et al. 2001; Hofmeyer et al. 2006), where LAR and liprin-α form physical complexes. Although there is evidence of LAR independent liprin-α function in certain contexts (Hofmeyer et al. 2006), at the NMJ they control synapse growth and active zones formation together (Kaufmann et al. 2002). In the absence of LAR or liprin-α, synapses are smaller and less elaborate than in wild-type animals, whereas presynaptic overexpression of LAR is adequate to increase synaptic size in a liprin-α dependent manner. Interestingly, liprin-α is normally down-regulated by the ubiquitinating activity of anaphase promoting complex (APC), which, when absent, results in stabilization of liprin-α and overgrown synapses (van Roessel et al. 2004). The interaction between liprin-α and APC is conserved in vertebrates (Hoogenraad et al. 2007).

Another parallel among *Drosophila* and *C. elegans* is the interaction with glycoproteins. Two heparin sulfate proteoglycans, *syndecan* (sdc) and *dallylike* (dlp) regulate in opposite ways LAR activity. Sdc promotes LAR activity and synapse growth presynaptically, whereas Dlp inhibits it (Johnson et al. 2006). In the absence of *sdc*, synapse growth is impaired, similar to LAR and liprin-α mutants (Johnson et al. 2006). However, overexpression of Dlp postsynaptically inhibits LAR and results in decreased synaptic growth, whereas in parallel it increases the area of the active zones (Johnson et al. 2006), phenocopying in this manner LAR and liprin-α mutations (Kaufmann et al. 2002). Interestingly, in both mammalian cells and *Drosophila*, liprin-α interacts with KIF1A, a neuron specific motor transport protein, capable of interacting with synaptic proteins (Shin et al. 2003; Miller et al. 2005), indicating that LAR–liprin interactions regulate the recruitment of molecular complexes that mediate active zone and synapse formation.

CADHERIN SUPERFAMILY

Cadherins are involved in cell adhesion in numerous cell types in invertebrate and vertebrate organisms (reviewed in Shapiro et al. 2007; Takeichi 2007; Tai et al. 2008). Structurally, cadherins are single pass transmembrane molecules whose ectodomain contains a variable number of calcium binding cadherin repeats (Fig. 3). On the basis of their protein sequence, the domain structure, and the genomic organization, cadherins belong to two major branches, the cadherins and the cadherin related proteins, further subdivided into a total of six subfamilies (Nollet et al. 2000; Hulpiau and van Roy 2009). We mainly refer to the classical cadherins and protocadherins with respect to their role in the nervous system.

Nonclassical Cadherins: Protocadherins and Flamingo (Fmi)

Protocadherins (Pcdh) are considered the primordial type of classical cadherins (reviewed in Frank and Kemler 2002). Protocadherins bear a variable number of cadherin repeats (Fig. 3). Based on their genomic organization and protein sequence, protocadherins are divided into the clustered and the nonclustered protocadherins (Redies et al. 2005; Morishita and Yagi 2007). All protocadherins undergo alternative splicing, and some members of the protocadherin family have region-specific expression patterns in the developing brain (Kohmura et al. 1998; Carroll et al. 2001; Hirano et al. 2002; Wang et al. 2002b; Phillips et al. 2003), providing an adhesive code for the establishment of specific brain structures (Serafini 1999; Shapiro and Colman 1999; Redies 2000; Shapiro et al. 2007). Moreover, protocadherins are localized to synaptic junctions and may modulate synaptic structure and transmission properties (Fig. 3) (Phillips et al. 2003).

The role of protocadherins in synapse formation and maturation has been documented genetically in null mutants for the Pcdh-γ gene cluster. In *Pcdh-γ*$^{-/-}$ animals, there is an apoptotic degeneration of spinal interneurons, which leads to neonatal death (Wang et al. 2002a). Although prevention of cell death in mutants by removal of the proapoptotic Bax protein rescues the survival of the neurons, the density of synaptic puncta and of currents recorded from in vitro cultured neurons are still reduced, indicating a role of Pcdh-γ in synaptogenesis (Weiner et al. 2005). The lack of general effects of Pcdh-γ on the nervous system could be caused by compensation by Pcdh-α and Pcdh-β. Further genetic analysis of Pcdh-α and Pcdh-β mutants is needed.

CELSR1–3 belong to the Flamingo/CELSR group of cadherins and are characterized by seven transmembrane domains (Takeichi 2007). CELSR2 and 3 have been implicated in nervous system development. Knockdown of CELSR2 results in retraction of dendrites, suggesting that the CELSR2 mediated homophilic interactions are important for maintenance of dendritic branching (Shima et al. 2004). Ablation of CELSR3 results in axonal tract defects (Tissir et al. 2005). The fly protein Flamingo (Fmi) is a unique nonclassical cadherin with seven transmembrane domains, that regulates planar cell polarity in epithelial cells (Lu et al. 1999), dendritic tiling of peripheral neurons (Gao et al. 1999, 2000; Grueber 2002), and axon target selection in the visual system of *Drosophila* (Lee et al. 2003; Senti et al. 2003; Chen and Clandinin 2008). In the visual system, Fmi plays a role in both the lamina and the medulla neuropils (Fig. 4). In the medulla, Fmi is expressed in the afferent R7 and R8 photoreceptor axons, as well as in their target neurons. Fmi controls the layer target selection of R8 photoreceptor axons, which in the absence of Fmi retract to superficial layers (Lee et al. 2003; Senti et al. 2003). In addition, Fmi controls the tiling pattern of R8 axons, because mutations in Fmi cause a significant degree of overlap among the growth cones, as if the repulsion is impaired (Lee et al. 2003; Senti et al. 2003). In the lamina, Fmi functions strictly nonautonomously as a short-range, homophilic adhesive signal between specific R1–R6 photoreceptor cell growth cones, ensuring that the proper trajectory toward the assembling cartridges is followed (Chen and Clandinin 2008).

Classical Cadherins

Multiple classic cadherins exhibit distinct spatiotemporal patterns of expression (Suzuki et al. 1997; Inoue et al. 1998; Bekirov et al. 2002), suggesting that homophilic interactions among the extracellular cadherin repeats of different types of cadherins play a role in target recognition and synapse formation (Inoue et al. 1998; Miskevich et al. 1998; Wohrn et al. 1998; Ranscht 2000).

Classic cadherins and their associated partners, the catenins, localize at synapses pre- and postsynaptically (Fig. 3) (Yamagata et al. 1995, 2003; Fannon and Colman 1996; Uchida et al. 1996). In the mature synapse, the cadherin–catenin complexes are present in *puncta adherentia*, which are the regions flanking the active zone and are considered sites of mechanical adhesion (Uchida et al. 1996). N-cadherin is cotransported with active zone components to sites of synapse formation (Jontes et al. 2004) and colocalizes with other synaptic markers (Benson and Tanaka 1998). Interestingly, N-cadherin is retained mostly in excitatory and lost from inhibitory synapses (Benson and Tanaka 1998). The proximal region of the cytoplasmic domain of cadherins binds to p120-catenin family members, whereas the distal region binds to β-catenin (Fig. 3). β-catenins contain a PDZ domain that enables interactions with synaptic proteins (Arikkath and Reichardt 2008). The PDZ domain interacts with α-catenin, which interacts with cytoskeletal actin binding proteins, such as α-actinin or profilin (Fig. 3) (Pokutta and Weis 2007; Arikkath and Reichardt 2008). Clustering of cadherins may increase the critical concentration of α-catenin locally, which may favor the formation of stable cell contacts by suppressing the Arp2/3 mediated actin polymerization (Pokutta and Weis 2007). Notably, the extracellular cadherin repeats are also able to bind to calcium (Ca^{2+}), which rigidifies the molecular structure of cadherins, enhances their dimerization, and stabilizes them by rendering them resistant to proteolytic degradation for extended periods of time (Fig. 1) (Shapiro et al. 2007). Ca^{2+} also reverses the inhibitory effect of peptides or antibodies directed against the extracellular domain of N-cadherin (Tang et al. 1998). Thus, cadherins become trans-synaptic sensors of synaptic activity and regulators of both presynaptic and postsynaptic components. Indeed, synaptic activity stabilizes dimers of N-cadherin and renders them more resistant to proteolytic degradation (Tanaka et al. 2000). NMDA activation retains N-cadherin at the membrane by inhibiting its endocytosis (Tai et al. 2007). Activation of AMPA receptors on the other hand, leads to lateral expansion of the dendritic spine heads, during which N-cadherin is redistributed to facilitate enhanced adhesion (Fig. 1) (Okamura et al. 2004).

Morphological changes in dendritic spines reflect alterations in synaptic strength (synaptic plasticity). Loss of function of N-cadherin, β-catenin, or α-catenin results in a less stable spine structure, reduction in the accumulation of pre- and postsynaptic markers, and impaired synaptic function (Elia et al. 2006; Jungling et al. 2006; Tanabe et al. 2006; Togashi et al. 2006; Okuda et al. 2007; Saglietti et al. 2007). N-cadherin also regulates the trafficking of AMPA receptors (Nuriya and Huganir 2006; Saglietti et al. 2007) and postsynaptic loss of N-cadherin causes defects in neurotransmitter exocytosis by presynaptic wild-type neurons (Jungling et al. 2006), suggesting a role in retrograde signaling. Loss of function of p120 (pre or post) catenin leads to decreased spine density, fewer mature spines, and mislocalization of pre- and postsynaptic components (Elia et al. 2006). Conversely, overexpression of α-catenin increases dendritic spine density and stability (Abe et al. 2004).

Genetic analysis in *Drosophila* has greatly contributed to determining the function of N-cadherins in vivo. Loss of N-cadherin in *Drosophila* embryos causes the trajectories of longitudinal CNS axons and the guidance of growth cones to be affected (Iwai et al. 1997). The role of N-cadherins in synaptic "wiring" was more clearly shown by elegant studies in the visual (Fig. 4) and olfactory systems in *Drosophila*. N-cadherin is instrumental in the assembly of the visual neuronal circuitry, where it functions in the afferent axons of the R1–R6 photoreceptor sensory neurons as well as their target neurons in the lamina neuropil of the fly visual system (Prakash et al. 2005). Similarly, N-cadherin is crucial for the coordinated targeting of multiple neuronal types, such as R7 photoreceptor axons and L1–L5 lamina neurons, to the correct target layer in the medulla neuropil of the visual system (Lee et al. 2001; Nern et al. 2005, 2008; Ting et al. 2005). There are 12 isoforms of N-cadherin in the fly visual system, but the expression of a single isoform is able to rescue null mutations, suggesting functional redundancy (Ting et al. 2005). In the olfactory system, loss of N-cadherin does not affect the pathfinding of olfactory receptor neuronal axons, but it influences their ability to converge onto a single glomerulus and form synapses (Hummel and Zipursky 2004). Similarly, the dendritic projections of the N-cadherin mutant second-order neurons, which also innervate the olfactory lobes, expand on nonappropriate glomeruli (Zhu and Luo 2004). In both systems, N-cadherin probably provides the adhesive force that mediates and stabilizes the attraction between neurites. The functions of N-cadherins appear to be evolutionarily maintained. For example, misrouting of axon fibers has also been observed in the developing optic tract of N-cadherin mutated zebrafish (Masai et al. 2003). Neurite nonextension/retraction has also been observed in cadherin blocked horizontal cell dendrites in the chicken retina (Tanabe et al. 2006).

NECTINS-AFADIN

Nectins (nectins 1–4) (reviewed in Takai et al. 2008a,b) are Ig-like transmembrane molecules with a pleiotropic role in cell–cell adhesion, cell movement, proliferation, survival, differentiation, polarization, and the entry of viruses (Fig. 3). All nectin genes have two or three splicing variants and are ubiquitously expressed (Takai et al. 2008a,b). Nectins interact in *trans*, in a Ca^{2+}-independent manner through their extracellular regions with each other or with other Ig-like molecules (Takai et al. 2008a,b). They are anchored to the actin cytoskeleton, through the interaction of the carboxy-terminal of nectins with the PDZ domain of the actin-binding protein afadin, which, like nectins, are expressed ubiquitously (Takai et al. 2008a,b).

The nectin–afadin system resembles the cadherin–catenin system. Both systems colocalize at *puncta adherentia* junctions (Fig. 3) (Fannon and Colman 1996). Nectins recruit cadherins through the interaction with afadin, α-catenin, and their binding proteins (Fig. 3) (Tachibana et al. 2000; Pokutta et al. 2002; Asada et al. 2003; Honda et al. 2003, 2006; Ooshio et al. 2004), with which they cooperate in vitro to regulate neurite overextension (Togashi et al. 2006). Afadin also binds to other cell adhesion systems, such as Eph receptors (Hock et al. 1998; Buchert et al. 1999), indicating how multiple cell adhesion complexes may operate cooperatively in the formation of junctional complexes (Fig. 3).

At synapses between mossy fiber terminals and the dendrites of CA3 pyramidal cells in the hippocampus of the adult brain, nectin-1 and -3 localize at presynaptic and postsynaptic sides, respectively, whereas afadin is symmetrically present at both sides (Fig. 3) (Mizoguchi et al. 2002). When presynaptic nectin-1 is blocked and interaction with postsynaptic nectin-3 is lost, the formation of synapses in hippocampal cell culture is impaired (Mizoguchi et al. 2002). However, nectin-1 and nectin-3 mutant mice are viable and fertile (except for male nectin-3 knockout mice). The mutant mice display micropthalmia, impairments of the mossy fiber tract, aberrantly localized afadin and N-cadherin, and reduced number of *puncta adherentia* junctions at hippocampal synapses (Honda et al. 2006). In spite of these deficits, basic synaptic transmission and LTP at the mossy fiber synapse of nectin-1 knockout mice are identical to wild-type mice (Honda et al. 2006). These results suggest that there might be functional redundancy among different members of the family. However, disruption of the single afadin locus in mice, which consequently disrupts the function of all nectins (but also of other binding proteins) leads to embryonic lethality with pleiotropic defects, suggesting a widespread role of nectin–afadin in cell adhesion, migration, and differentiation (Ikeda et al. 1999).

In *Drosophila*, the nectin–afadin system could be paralleled by the echinoid–canoe complexes (Tepass and Harris 2007). Canoe is the fly homolog of afadin, which plays a pleiotropic role in morphogenesis. However, it is unknown if it serves a role in synapse formation and function. There is no clear homolog of nectins in *Drosophila*. The homophilic Ig-type adhesion protein Echinoid binds to canoe, as nectins bind to afadin, but there are many functional differences between echinoid and nectins (Tepass and Harris 2007).

CONCLUDING REMARKS: GENETICS OF CELL ADHESION AND NEUROPSYCHIATRIC DISEASES

Cell adhesion plays an important role in the maintenance and modulation of synaptogenic activity within neuronal circuitries. Synaptogenic cell adhesion molecules have been identified by a wide variety of methods, and their activity has been clearly assesed in in vitro synaptogenesis assays, where the complexity of trans-synaptic signaling is greatly simplified (Biederer and Scheiffele 2007; Sudhof 2008). However, "loss of function" in vivo studies often do not corroborate the in vitro results because of redundancy within a protein family and among different adhesive systems or because of subtle or different function. The intrinsic complexity and redundancy may also account for why so few components have been identified in forward genetic screens in flies and worms. Perhaps the lack of sensitivity of the screening assays in the past will be compensated for by current, more sophisticated tools. For example, screens for neuronal connectivity in the visual system of *Drosophila* (reviewed in Mast et al. 2006; Ting and Lee 2007) and synaptogenesis in *C. elegans* (Zhen and Jin 1999; Shen and Bargmann 2003) have illustrated the power of these approaches. Genetic screens can now be carried out in sensitized genetic backgrounds using GFP based-markers, such as GRASP (a split GFP method that relies upon reconstitution of GFP along synaptic partners) (Feinberg et al. 2008), and a combination of genetic methodologies and transcriptional profiling or computational methods (Jin and Garner 2008).

Synaptic function and structure may be the converging point of malfunction in neuropsychiatric disorders, such as mental retardation and autism (Geschwind and Levitt 2007; Geschwind 2008; Walsh et al. 2008; Kramer and van Bokhoven 2009). Subtle changes in the connectivity patterns of a subset of synapses or in their basal function (Geschwind and Levitt 2007; Krey and Dolmetsch 2007; Abrahams and Geschwind 2008; Geschwind 2008; Sudhof 2008) and plasticity (Ramocki and Zoghbi 2008) might alter dramatically the output of the neuronal circuitry. Interestingly, mutations in neurexin 1 and neuroligins 3 and 4 are associated with autism spectrum disorders and mental retardation (Sudhof 2008). The scaffolding molecule and interacting protein of neuroligin SHANK3 (ProSAP2) is also associated with autism (Durand et al. 2007). More than 140 different mutations in L1-CAM have been associated with neurological disorders, including mental retardation (Bateman et al. 1996; Kenwrick and Doherty 1998; Kenwrick et al. 2000; De Angelis et al. 2002).

The heritability of psychiatric disorders, such as bipolar disorder, schizophrenia, and autism, has been estimated to be significantly higher than that of cancer (Burmeister et al. 2008). However, the genetics of neuropsychiatric disorders are obscured by incomplete penetrance, variable expressivity, complex interactions among genetic and environmental factors, and imprecisely defined traits (Burmeister et al. 2008). Intriguingly, most of the genes associated with mental retardation are evolutionarily conserved in *Drosophila* (Inlow and Restifo 2004), which is regarded as a succesful model for human diseases (Bier 2005).

In conclusion, it is important to provide more candidate genes and a better, more detailed understanding of their in vivo role in model organisms in order to further unravel the complexity of synaptogenesis and its role in neuronal development, function, and disease.

ABBREVIATIONS

CASK	Ca^{2+}/calmodulin-dependent serine protein kinase
aa	amino acid
Alk	Anaplastic lymphoma kinase
AMPA	α-amino-3-hydroxy-5-methyl-4-isoxazole propionic acid (AMPA)
AMPAR	AMPA receptors
APC	Anaphase Promoting Complex
ASD	Autism Spectrum Disorders
BDNF	Brain Derived Neurotrophic Factor
BMP	Bone Morphogenetic Protein

Ca^{2+}	Calcium
CABAR	GABA receptors
CAM	Cell Adhesion Molecule
CAMKII	Ca2+/calmodulin dependent protein kinase II
CASK	Calcium/calmodulin-dependent serine protein kinase
CAST	cytomatrix at the active zone-associated structural protein
CNS	Central Nervous System
DGC	dystrophin glycoprotein complex
dNrx	*Drosophila* Neurexin
Eph	Erythropoietin-producing hepatocellular carcinoma
FAK	focal adhesion kinase
FERM	Four-point-one, Ezrin, Radixin Moesin
FGF	fibroblast growth factor
Fmi	Flamingo
GABA	γ-aminobutyric acid
GAD	glutamate decarboxylase
GEF	guanine exchange factor
GPI	glycosyl-phospahtidyl-inositol
GRB2	growth factor receptor-bound protein 2
GRIP	glutamate receptor interacting protein
HEK	Human Embryonic Kindey
Ig	Immunoglobulin
LAR	leucocyte-common antigen related
LNS	Laminin, Nectin, Sex-hormone binding globulin
LRR	Leucine Rich Repeat
MALS	Mammalian-LIN-seven protein
Mint	Munc-18 interacting protein
MR	Mental Retardation
Narp	Neuronal activity-regulated pentraxin
NCAM	Neural Cell Adhesion Molecule
NGL	Netrin-G Ligand
Nlg	Neuroligin
NMDA	N-methyl-D-aspartate
NMDAR	NMDA receptors
NMJ	Neuromuscular Junction
Nrx	Neurexin
N-terminal	aminoterminal
PAK	p21-activated kinase
PAK	P21-activated kinase
Pcdh	protocadherins
PDZ	post synaptic density protein (PSD95), *Drosophila* disc large tumor suppressor(Dlg), zonula occludens-1 protein (zo-1)
PSA	polysialic acid
PSD	PSD
PTP	Protein tyrosine phosphatase
RIM	Rab3-Interacting Molecule
SALM	Synaptic Adhesion Like Molecule
SAM	sterile α motifs
S-SCAM	synaptic scaffolding molecule
SV	Synaptic Vesicles
SYD-2	synaptogenesis defective
SynCAM	Synaptic Cell Adhesion Molecule
TEM	Transmission Electron Microscopy
TM	Transmembrane
Veli	vertebrate LIN-7

ACKNOWLEDGMENTS

We are grateful to Manish Jaiswal and Tomoko Ohyama for their insightful comments during the preparation of the manuscript. Nikolaos Giagtzoglou is supported by a long-term European Molecular Biology Organization postdoctoral fellowship and the Howard Hughes Medical Institute. Cindy V. Ly is supported by a Ruth L. Kirschstein National Research Service Award from the National Institute of Neurological Disorders and Stroke. Hugo J. Bellen is a Howard Hughes Medical Institute Investigator. N.G. and C.V.L. contributed equally to this work.

REFERENCES

Abe K, Chisaka O, Van Roy F, Takeichi M. 2004. Stability of dendritic spines and synaptic contacts is controlled by α N-catenin. *Nat Neurosci* **7:** 357–363.

Abrahams BS, Geschwind DH. 2008. Advances in autism genetics: On the threshold of a new neurobiology. *Nat Rev Genet* **9:** 341–355.

Ackley BD, Harrington RJ, Hudson ML, Williams L, Kenyon CJ, Chisholm AD, Jin Y. 2005. The two isoforms of the *Caenorhabditis elegans* leukocyte-common antigen related receptor tyrosine phosphatase PTP-3 function independently in axon guidance and synapse formation. *J Neurosci* **25:** 7517–7528.

Arikkath J, Reichardt LF. 2008. Cadherins and catenins at synapses: Roles in synaptogenesis and synaptic plasticity. *Trends Neurosci* **31:** 487–494.

Artero RD, Castanon I, Baylies MK. 2001. The immunoglobulin-like protein Hibris functions as a dose-dependent regulator of myoblast fusion and is differentially controlled by Ras and Notch signaling. *Development* **128:** 4251–4264.

Asada M, Irie K, Morimoto K, Yamada A, Ikeda W, Takeuchi M, Takai Y. 2003. ADIP, a novel Afadin- and α-actinin-binding protein localized at cell-cell adherens junctions. *J Biol Chem* **278:** 4103–4111.

Atasoy D, Schoch S, Ho A, Nadasy KA, Liu X, Zhang W, Mukherjee K, Nosyreva ED, Fernandez-Chacon R, Missler M, et al. 2007. Deletion of CASK in mice is lethal and impairs synaptic function. *Proc Natl Acad Sci* **104:** 2525–2530.

Baines RA, Seugnet L, Thompson A, Salvaterra PM, Bate M. 2002. Regulation of synaptic connectivity: Levels of Fasciclin II influence synaptic growth in the *Drosophila* CNS. *J Neurosci* **22:** 6587–6595.

Bao S, Cagan R. 2005. Preferential adhesion mediated by Hibris and Roughest regulates morphogenesis and patterning in the *Drosophila* eye. *Dev Cell* **8:** 925–935.

Bateman A, Jouet M, MacFarlane J, Du JS, Kenwrick S, Chothia C. 1996. Outline structure of the human L1 cell adhesion molecule and the sites where mutations cause neurological disorders. *Embo J* **15:** 6050–6059.

Baumgartner S, Littleton JT, Broadie K, Bhat MA, Harbecke R, Lengyel JA, Chiquet-Ehrismann R, Prokop A, Bellen HJ. 1996. A *Drosophila* neurexin is required for septate junction and blood-nerve barrier formation and function. *Cell* **87:** 1059–1068.

Bazigou E, Apitz H, Johansson J, Loren CE, Hirst EM, Chen PL, Palmer RH, Salecker I. 2007. Anterograde Jelly belly and Alk receptor tyrosine kinase signaling mediates retinal axon targeting in *Drosophila. Cell* **128:** 961–975.

Bekirov IH, Needleman LA, Zhang W, Benson DL. 2002. Identification and localization of multiple classic cadherins in developing rat limbic system. *Neuroscience* **115:** 213–227.

Benson DL, Tanaka H. 1998. N-cadherin redistribution during synaptogenesis in hippocampal neurons. *J Neurosci* **18:** 6892–6904.

Bhat MA, Rios JC, Lu Y, Garcia-Fresco GP, Ching W, St Martin M, Li J, Einheber S, Chesler M, Rosenbluth J, et al. 2001. Axon-glia interactions and the domain organization of myelinated axons requires neurexin IV/Caspr/Paranodin. *Neuron* **30:** 369–383.

Bieber AJ, Snow PM, Hortsch M, Patel NH, Jacobs JR, Traquina ZR, Schilling J, Goodman CS. 1989. *Drosophila* neuroglian: A member of the immunoglobulin superfamily with extensive homology to the vertebrate neural adhesion molecule L1. *Cell* **59:** 447–460.

Biederer T. 2006. Bioinformatic characterization of the SynCAM family of immunoglobulin-like domain-containing adhesion molecules. *Genomics* **87:** 139–150.

Biederer T, Scheiffele P. 2007. Mixed-culture assays for analyzing neuronal synapse formation. *Nat Protoc* **2:** 670–676.

Biederer T, Stagi M. 2008. Signaling by synaptogenic molecules. *Curr Opin Neurobiol* **18:** 261–269.

Biederer T, Sudhof TC. 2000. Mints as adaptors. Direct binding to neurexins and recruitment of munc18. *J Biol Chem* **275:** 39803–39806.

Biederer T, Sara Y, Mozhayeva M, Atasoy D, Liu X, Kavalali ET, Sudhof TC. 2002. SynCAM, a synaptic adhesion molecule that drives synapse assembly. *Science* **297:** 1525–1531.

Bier E. 2005. *Drosophila*, the golden bug, emerges as a tool for human genetics. *Nat Rev Genet* **6:** 9–23.

Bolliger MF, Frei K, Winterhalter KH, Gloor SM. 2001. Identification of a novel neuroligin in humans which binds to PSD-95 and has a widespread expression. *Biochem J* **356:** 581–588.

Boucard AA, Chubykin AA, Comoletti D, Taylor P, Sudhof TC. 2005. A splice code for trans-synaptic cell adhesion mediated by binding of neuroligin 1 to α- and β-neurexins. *Neuron* **48:** 229–236.

Bour BA, Chakravarti M, West JM, Abmayr SM. 2000. *Drosophila* SNS, a member of the immunoglobulin superfamily that is essential for myoblast fusion. *Genes & Dev* **14:** 1498–1511.

Buchert M, Schneider S, Meskenaite V, Adams MT, Canaani E, Baechi T, Moelling K, Hovens CM. 1999. The junction-associated protein AF-6 interacts and clusters with specific Eph receptor tyrosine kinases at specialized sites of cell-cell contact in the brain. *J Cell Biol* **144:** 361–371.

Budnik V, Zhong Y, Wu CF. 1990. Morphological plasticity of motor axons in *Drosophila* mutants with altered excitability. *J Neurosci* **10:** 3754–3768.

Budreck EC, Scheiffele P. 2007. Neuroligin-3 is a neuronal adhesion protein at GABAergic and glutamatergic synapses. *Eur J Neurosci* **26:** 1738–1748.

Bukalo O, Fentrop N, Lee AY, Salmen B, Law JW, Wotjak CT, Schweizer M, Dityatev A, Schachner M. 2004. Conditional ablation of the neural cell adhesion molecule reduces precision of spatial learning, long-term potentiation, and depression in the CA1 subfield of mouse hippocampus. *J Neurosci* **24:** 1565–1577.

Burmeister M, McInnis MG, Zollner S. 2008. Psychiatric genetics: Progress amid controversy. *Nat Rev Genet* **9:** 527–540.

Butz S, Okamoto M, Sudhof TC. 1998. A tripartite protein complex with the potential to couple synaptic vesicle exocytosis to cell adhesion in brain. *Cell* **94:** 773–782.

Carroll P, Gayet O, Feuillet C, Kallenbach S, de Bovis B, Dudley K, Alonso S. 2001. Juxtaposition of CNR protocadherins and reelin expression in the developing spinal cord. *Mol Cell Neurosci* **17:** 611–623.

Carthew RW. 2005. Adhesion proteins and the control of cell shape. *Curr Opin Genet Dev* **15:** 358–363.

Carthew RW. 2007. Pattern formation in the *Drosophila* eye. *Curr Opin Genet Dev* **17:** 309–313.

Chao DL, Shen K. 2008. Functional dissection of SYG-1 and SYG-2, cell adhesion molecules required for selective synaptogenesis in *C. elegans. Mol Cell Neurosci* **39:** 248–257.

Chen PL, Clandinin TR. 2008. The cadherin Flamingo mediates level-dependent interactions that guide photoreceptor target choice in *Drosophila. Neuron* **58:** 26–33.

Cheng Y, Endo K, Wu K, Rodan AR, Heberlein U, Davis RL. 2001. *Drosophila* fasciclinII is required for the formation of odor memories and for normal sensitivity to alcohol. *Cell* **105:** 757–768.

Cheng L, Itoh K, Lemmon V. 2005. L1-mediated branching is regulated by two ezrin-radixin-moesin (ERM)-binding sites, the RSLE region and a novel juxtamembrane ERM-binding region. *J Neurosci* **25:** 395–403.

Chih B, Afridi SK, Clark L, Scheiffele P. 2004. Disorder-associated mutations lead to functional inactivation of neuroligins. *Hum Mol Genet* **13:** 1471–1477.

Chih B, Engelman H, Scheiffele P. 2005. Control of excitatory and inhibitory synapse formation by neuroligins. *Science* **307:** 1324–1328.

Chih B, Gollan L, Scheiffele P. 2006. Alternative splicing controls selective trans-synaptic interactions of the neuroligin-neurexin complex. *Neuron* **51:** 171–178.

Choe KM, Prakash S, Bright A, Clandinin TR. 2006. Liprin-α is required for photoreceptor target selection in *Drosophila*. *Proc Natl Acad Sci* **103:** 11601–11606.

Clandinin TR, Lee CH, Herman T, Lee RC, Yang AY, Ovasapyan S, Zipursky SL. 2001. *Drosophila* LAR regulates R1-R6 and R7 target specificity in the visual system. *Neuron* **32:** 237–248.

Cohen NR, Taylor JS, Scott LB, Guillery RW, Soriano P, Furley AJ. 1998. Errors in corticospinal axon guidance in mice lacking the neural cell adhesion molecule L1. *Curr Biol* **8:** 26–33.

Colon-Ramos DA, Margeta MA, Shen K. 2007. Glia promote local synaptogenesis through UNC-6 (netrin) signaling in *C. elegans*. *Science* **318:** 103–106.

Craig AM, Boudin H. 2001. Molecular heterogeneity of central synapses: Afferent and target regulation. *Nat Neurosci* **4:** 569–578.

Craig AM, Kang Y. 2007. Neurexin-neuroligin signaling in synapse development. *Curr Opin Neurobiol* **17:** 43–52.

Craig AM, Graf ER, Linhoff MW. 2006. How to build a central synapse: Clues from cell culture. *Trends Neurosci* **29:** 8–20.

Cremer H, Lange R, Christoph A, Plomann M, Vopper G, Roes J, Brown R, Baldwin S, Kraemer P, Scheff S, et al. 1994. Inactivation of the N-CAM gene in mice results in size reduction of the olfactory bulb and deficits in spatial learning. *Nature* **367:** 455–459.

Cremer H, Chazal G, Lledo PM, Rougon G, Montaron MF, Mayo W, Le Moal M, Abrous DN. 2000. PSA-NCAM: An important regulator of hippocampal plasticity. *Int J Dev Neurosci* **18:** 213–220.

Crossin KL, Krushel LA. 2000. Cellular signaling by neural cell adhesion molecules of the immunoglobulin superfamily. *Dev Dyn* **218:** 260–279.

Dahme M, Bartsch U, Martini R, Anliker B, Schachner M, Mantei N. 1997. Disruption of the mouse L1 gene leads to malformations of the nervous system. *Nat Genet* **17:** 346–349.

Dalva MB. 2007. There's more than one way to skin a chimaerin. *Neuron* **55:** 681–684.

Dalva MB, McClelland AC, Kayser MS. 2007. Cell adhesion molecules: Signalling functions at the synapse. *Nat Rev Neurosci* **8:** 206–220.

Davis JQ, Bennett V. 1994. Ankyrin binding activity shared by the neurofascin/L1/NrCAM family of nervous system cell adhesion molecules. *J Biol Chem* **269:** 27163–27166.

Davis GW, Schuster CM, Goodman CS. 1997. Genetic analysis of the mechanisms controlling target selection: Target-derived Fasciclin II regulates the pattern of synapse formation. *Neuron* **19:** 561–573.

De Angelis E, Watkins A, Schafer M, Brummendorf T, Kenwrick S. 2002. Disease-associated mutations in L1 CAM interfere with ligand interactions and cell-surface expression. *Hum Mol Genet* **11:** 1–12.

Dean C, Dresbach T. 2006. Neuroligins and neurexins: Linking cell adhesion, synapse formation and cognitive function. *Trends Neurosci* **29:** 21–29.

Dean C, Scholl FG, Choih J, DeMaria S, Berger J, Isacoff E, Scheiffele P. 2003. Neurexin mediates the assembly of presynaptic terminals. *Nat Neurosci* **6:** 708–716.

Demyanenko GP, Tsai AY, Maness PF. 1999. Abnormalities in neuronal process extension, hippocampal development, and the ventricular system of L1 knockout mice. *J Neurosci* **19:** 4907–4920.

Demyanenko GP, Shibata Y, Maness PF. 2001. Altered distribution of dopaminergic neurons in the brain of L1 null mice. *Brain Res Dev Brain Res* **126:** 21–30.

Dickson TC, Mintz CD, Benson DL, Salton SR. 2002. Functional binding interaction identified between the axonal CAM L1 and members of the ERM family. *J Cell Biol* **157:** 1105–1112.

Ding M, Chao D, Wang G, Shen K. 2007. Spatial regulation of an E3 ubiquitin ligase directs selective synapse elimination. *Science* **317:** 947–951.

Dityatev A, Dityateva G, Schachner M. 2000. Synaptic strength as a function of post- versus presynaptic expression of the neural cell adhesion molecule NCAM. *Neuron* **26:** 207–217.

Dityatev A, Dityateva G, Sytnyk V, Delling M, Toni N, Nikonenko I, Muller D, Schachner M. 2004. Polysialylated neural cell adhesion molecule promotes remodeling and formation of hippocampal synapses. *J Neurosci* **24:** 9372–9382.

Dudanova I, Tabuchi K, Rohlmann A, Sudhof TC, Missler M. 2007. Deletion of α-neurexins does not cause a major impairment of axonal pathfinding or synapse formation. *J Comp Neurol* **502:** 261–274.

Dunah AW, Hueske E, Wyszynski M, Hoogenraad CC, Jaworski J, Pak DT, Simonetta A, Liu G, Sheng M. 2005. LAR receptor protein tyrosine phosphatases in the development and maintenance of excitatory synapses. *Nat Neurosci* **8:** 458–467.

Durand CM, Betancur C, Boeckers TM, Bockmann J, Chaste P, Fauchereau F, Nygren G, Rastam M, Gillberg IC, Anckarsater H, et al. 2007. Mutations in the gene encoding the synaptic scaffolding protein SHANK3 are associated with autism spectrum disorders. *Nat Genet* **39:** 25–27.

Dworak HA, Sink H. 2002. Myoblast fusion in *Drosophila*. *Bioessays* **24:** 591–601.

Dworak HA, Charles MA, Pellerano LB, Sink H. 2001. Characterization of *Drosophila* hibris, a gene related to human nephrin. *Development* **128:** 4265–4276.

Eaton BA, Davis GW. 2003. Synapse disassembly. *Genes & Dev* **17:** 2075–2082.

Eckhardt M, Bukalo O, Chazal G, Wang L, Goridis C, Schachner M, Gerardy-Schahn R, Cremer H, Dityatev A. 2000. Mice deficient in the polysialyltransferase ST8SiaIV/PST-1 allow discrimination of the roles of neural cell adhesion molecule protein and polysialic acid in neural development and synaptic plasticity. *J Neurosci* **20:** 5234–5244.

Elia LP, Yamamoto M, Zang K, Reichardt LF. 2006. p120 catenin regulates dendritic spine and synapse development through Rho-family GTPases and cadherins. *Neuron* **51:** 43–56.

Fannon AM, Colman DR. 1996. A model for central synaptic junctional complex formation based on the differential adhesive specificities of the cadherins. *Neuron* **17:** 423–434.

Feinberg EH, Vanhoven MK, Bendesky A, Wang G, Fetter RD, Shen K, Bargmann CI. 2008. GFP Reconstitution Across Synaptic Partners (GRASP) defines cell contacts and synapses in living nervous systems. *Neuron* **57:** 353–363.

Fogel AI, Akins MR, Krupp AJ, Stagi M, Stein V, Biederer T. 2007. SynCAMs organize synapses through heterophilic adhesion. *J Neurosci* **27:** 12516–12530.

Frank M, Kemler R. 2002. Protocadherins. *Curr Opin Cell Biol* **14:** 557–562.

Fransen E, Lemmon V, Van Camp G, Vits L, Coucke P, Willems PJ. 1995. CRASH syndrome: Clinical spectrum of corpus callosum hypoplasia, retardation, adducted thumbs, spastic paraparesis and hydrocephalus due to mutations in one single gene, L1. *Eur J Hum Genet* **3:** 273–284.

Fransen E, Van Camp G, Vits L, Willems PJ. 1997. L1-associated diseases: Clinical geneticists divide, molecular geneticists unite. *Hum Mol Genet* **6:** 1625–32.

Fransen E, Van Camp G, D'Hooge R, Vits L, Willems PJ. 1998. Genotype-phenotype correlation in L1 associated diseases. *J Med Genet* **35:** 399–404.

Fujita E, Kouroku Y, Ozeki S, Tanabe Y, Toyama Y, Maekawa M, Kojima N, Senoo H, Toshimori K, Momoi T. 2006. Oligo-astheno-teratozoospermia in mice lacking RA175/TSLC1/SynCAM/IGSF4A, a cell adhesion molecule in the immunoglobulin superfamily. *Mol Cell Biol* **26:** 718–726.

Gao FB, Brenman JE, Jan LY, Jan YN. 1999. Genes regulating dendritic outgrowth, branching, and routing in *Drosophila*. *Genes & Dev* **13:** 2549–2561.

Gao FB, Kohwi M, Brenman JE, Jan LY, Jan YN. 2000. Control of dendritic field formation in *Drosophila*: The roles of flamingo and competition between homologous neurons. *Neuron* **28:** 91–101.

Garcia-Alonso L, Romani S, Jimenez F. 2000. The EGF and FGF receptors mediate neuroglian function to control growth cone decisions during sensory axon guidance in *Drosophila*. *Neuron* **28:** 741–752.

Garner CC, Waites CL, Ziv NE. 2006. Synapse development: Still looking for the forest, still lost in the trees. *Cell Tissue Res* **326:** 249–262.

Gerke P, Benzing T, Hohne M, Kispert A, Frotscher M, Walz G, Kretz O. 2006. Neuronal expression and interaction with the synaptic protein CASK suggest a role for Neph1 and Neph2 in synaptogenesis. *J Comp Neurol* **498:** 466–475.

Gerrow K, El-Husseini A. 2006. Cell adhesion molecules at the synapse. *Front Biosci* **11:** 2400–2419.

Geschwind DH. 2008. Autism: Many genes, common pathways? *Cell* **135:** 391–395.

Geschwind DH, Levitt P. 2007. Autism spectrum disorders: Developmental disconnection syndromes. *Curr Opin Neurobiol* **17:** 103–111.

Goda Y, Davis GW. 2003. Mechanisms of synapse assembly and disassembly. *Neuron* **40:** 243–64.

Godenschwege TA, Kristiansen LV, Uthaman SB, Hortsch M, Murphey RK. 2006. A conserved role for *Drosophila* Neuroglian and human L1-CAM in central-synapse formation. *Curr Biol* **16:** 12–23.

Gottmann K. 2008. Transsynaptic modulation of the synaptic vesicle cycle by cell-adhesion molecules. *J Neurosci Res* **86:** 223–232.

Graf ER, Zhang X, Jin SX, Linhoff MW, Craig AM. 2004. Neurexins induce differentiation of GABA and glutamate postsynaptic specializations via neuroligins. *Cell* **119:** 1013–1026.

Grueber WB, Jan LY, Jan YN. 2002. Tiling of the *Drosophila* epidermis by multidendritic sensory neurons. *Development* **129:** 2867–2878.

Haass C, Selkoe DJ. 2007. Soluble protein oligomers in neurodegeneration: Lessons from the Alzheimer's amyloid β-peptide. *Nat Rev Mol Cell Biol* **8:** 101–112.

Hall SG, Bieber AJ. 1997. Mutations in the *Drosophila* neuroglian cell adhesion molecule affect motor neuron pathfinding and peripheral nervous system patterning. *J Neurobiol* **32:** 325–340.

Hata Y, Butz S, Sudhof TC. 1996. CASK: A novel dlg/PSD95 homolog with an N-terminal calmodulin-dependent protein kinase domain identified by interaction with neurexins. *J Neurosci* **16:** 2488–2494.

Hirano S, Wang X, Suzuki ST. 2002. Restricted expression of protocadherin 2A in the developing mouse brain. *Brain Res Mol Brain Res* **98:** 119–123.

Hock B, Bohme B, Karn T, Yamamoto T, Kaibuchi K, Holtrich U, Holland S, Pawson T, Rubsamen-Waigmann H, Strebhardt K. 1998. PDZ-domain-mediated interaction of the Eph-related receptor tyrosine kinase EphB3 and the ras-binding protein AF6 depends on the kinase activity of the receptor. *Proc Natl Acad Sci* **95:** 9779–9784.

Hofmeyer K, Maurel-Zaffran C, Sink H, Treisman JE. 2006. Liprin-α has LAR-independent functions in R7 photoreceptor axon targeting. *Proc Natl Acad Sci* **103:** 11595–11600.

Honda T, Shimizu K, Kawakatsu T, Fukuhara A, Irie K, Nakamura T, Matsuda M, Takai Y. 2003. Cdc42 and Rac small G proteins activated by trans-interactions of nectins are involved in activation of c-Jun N-terminal kinase, but not in association of nectins and cadherin to form adherens junctions, in fibroblasts. *Genes Cells* **8:** 481–491.

Honda T, Sakisaka T, Yamada T, Kumazawa N, Hoshino T, Kajita M, Kayahara T, Ishizaki H, Tanaka-Okamoto M, Mizoguchi A, et al. 2006. Involvement of nectins in the formation of puncta adherentia junctions and the mossy fiber trajectory in the mouse hippocampus. *Mol Cell Neurosci* **31:** 315–325.

Hoogenraad CC, Feliu-Mojer MI, Spangler SA, Milstein AD, Dunah AW, Hung AY, Sheng M. 2007. Liprinα1 degradation by calcium/calmodulin-dependent protein kinase II regulates LAR receptor tyrosine phosphatase distribution and dendrite development. *Dev Cell* **12:** 587–602.

Hortsch M, Bieber AJ, Patel NH, Goodman CS. 1990. Differential splicing generates a nervous system-specific form of *Drosophila* neuroglian. *Neuron* **4:** 697–709.

Hortsch M, Nagaraj K, Godenschwege TA. 2009. The interaction between L1-type proteins and ankyrins—a master switch for L1-type CAM function. *Cell Mol Biol Lett* **14:** 57–69.

Hulpiau P, van Roy F. 2009. Molecular evolution of the cadherin superfamily. *Int J Biochem Cell Biol* **41:** 349–369.

Hummel T, Zipursky SL. 2004. Afferent induction of olfactory glomeruli requires N-cadherin. *Neuron* **42:** 77–88.

Ichtchenko K, Nguyen T, Sudhof TC. 1996. Structures, alternative splicing, and neurexin binding of multiple neuroligins. *J Biol Chem* **271:** 2676–2682.

Ikeda W, Nakanishi H, Miyoshi J, Mandai K, Ishizaki H, Tanaka M, Togawa A, Takahashi K, Nishioka H, Yoshida H, et al. 1999. Afadin: A key molecule essential for structural organization of cell-cell junctions of polarized epithelia during embryogenesis. *J Cell Biol* **146:** 1117–1132.

Inlow JK, Restifo LL. 2004. Molecular and comparative genetics of mental retardation. *Genetics* **166:** 835–881.

Inoue T, Tanaka T, Suzuki SC, Takeichi M. 1998. Cadherin-6 in the developing mouse brain: Expression along restricted connection systems and synaptic localization suggest a potential role in neuronal circuitry. *Dev Dyn* **211:** 338–351.

Iwai Y, Usui T, Hirano S, Steward R, Takeichi M, Uemura T. 1997. Axon patterning requires DN-cadherin, a novel neuronal adhesion receptor, in the *Drosophila* embryonic CNS. *Neuron* **19:** 77–89.

Jin Y, Garner CC. 2008. Molecular mechanisms of presynaptic differentiation. *Annu Rev Cell Dev Biol* **24:** 237–262.

Johnson KG, Tenney AP, Ghose A, Duckworth AM, Higashi ME, Parfitt K, Marcu O, Heslip TR, Marsh JL, Schwarz TL, et al. 2006. The HSPGs Syndecan and Dallylike bind the receptor phosphatase LAR and exert distinct effects on synaptic development. *Neuron* **49:** 517–531.

Jontes JD, Emond MR, Smith SJ. 2004. In vivo trafficking and targeting of N-cadherin to nascent presynaptic terminals. *J Neurosci* **24:** 9027–9034.

Jungling K, Eulenburg V, Moore R, Kemler R, Lessmann V, Gottmann K. 2006. N-cadherin transsynaptically regulates short-term plasticity at glutamatergic synapses in embryonic stem cell-derived neurons. *J Neurosci* **26:** 6968–6978.

Kattenstroth G, Tantalaki E, Sudhof TC, Gottmann K, Missler M. 2004. Postsynaptic N-methyl-D-aspartate receptor function requires α-neurexins. *Proc Natl Acad Sci* **101:** 2607–2612.

Kaufmann N, DeProto J, Ranjan R, Wan H, Van Vactor D. 2002. *Drosophila* liprin-α and the receptor phosphatase Dlar control synapse morphogenesis. *Neuron* **34:** 27–38.

Kenwrick S, Doherty P. 1998. Neural cell adhesion molecule L1: Relating disease to function. *Bioessays* **20:** 668–675.

Kenwrick S, Watkins A, De Angelis E. 2000. Neural cell recognition molecule L1: Relating biological complexity to human disease mutations. *Hum Mol Genet* **9:** 879–886.

Kim S, Burette A, Chung HS, Kwon SK, Woo J, Lee HW, Kim K, Kim H, Weinberg RJ, Kim E. 2006. NGL family PSD-95-interacting adhesion molecules regulate excitatory synapse formation. *Nat Neurosci* **9:** 1294–1301.

Kizhatil K, Wu YX, Sen A, Bennett V. 2002. A new activity of doublecortin in recognition of the phospho-FIGQY tyrosine in the cytoplasmic domain of neurofascin. *J Neurosci* **22:** 7948–7958.

Ko J, Kim E. 2007. Leucine-rich repeat proteins of synapses. *J Neurosci Res* **85:** 2824–2832.

Ko J, Na M, Kim S, Lee JR, Kim E. 2003. Interaction of the ERC family of RIM-binding proteins with the liprin-α family of multidomain proteins. *J Biol Chem* **278:** 42377–42385.

Ko J, Kim S, Chung HS, Kim K, Han K, Kim H, Jun H, Kaang BK, Kim E. 2006. SALM synaptic cell adhesion-like molecules regulate the differentiation of excitatory synapses. *Neuron* **50:** 233–245.

Koch I, Schwarz H, Beuchle D, Goellner B, Langegger M, Aberle H. 2008. *Drosophila* ankyrin 2 is required for synaptic stability. *Neuron* **58:** 210–222.

Koh YH, Popova E, Thomas U, Griffith LC, Budnik V. 1999. Regulation of DLG localization at synapses by CaMKII-dependent phosphorylation. *Cell* **98:** 353–363.

Kohmura N, Senzaki K, Hamada S, Kai N, Yasuda R, Watanabe M, Ishii H, Yasuda M, Mishina M, Yagi T. 1998. Diversity revealed by a novel family of cadherins expressed in neurons at a synaptic complex. *Neuron* **20:** 1137–1151.

Kramer JM, van Bokhoven H. 2009. Genetic and epigenetic defects in mental retardation. *Int J Biochem Cell Biol* **41:** 96–107.

Krey JF, Dolmetsch RE. 2007. Molecular mechanisms of autism: A possible role for Ca2+ signaling. *Curr Opin Neurobiol* **17:** 112–119.

Kristiansen LV, Hortsch M. 2008. Fasciclin II: The NCAM Ortholog in *Drosophila* melanogaster. *Neurochem Res.* Doi: 10.1007/s11064-007-9566-9568.

Kristiansen LV, Velasquez E, Romani S, Baars S, Berezin V, Bock E, Hortsch M, Garcia-Alonso L. 2005. Genetic analysis of an overlapping functional requirement for L1- and NCAM-type proteins during sensory axon guidance in *Drosophila*. *Mol Cell Neurosci* **28:** 141–152.

Kurusu M, Cording A, Taniguchi M, Menon K, Suzuki E, Zinn K. 2008. A screen of cell-surface molecules identifies leucine-rich repeat proteins as key mediators of synaptic target selection. *Neuron* **59:** 972–985.

Law JW, Lee AY, Sun M, Nikonenko AG, Chung SK, Dityatev A, Schachner M, Morellini F. 2003. Decreased anxiety, altered place learning, and increased CA1 basal excitatory synaptic transmission in mice with conditional ablation of the neural cell adhesion molecule L1. *J Neurosci* **23:** 10419–10432.

Lee CH, Herman T, Clandinin TR, Lee R, Zipursky SL. 2001. N-cadherin regulates target specificity in the *Drosophila* visual system. *Neuron* **30:** 437–450.

Lee RC, Clandinin TR, Lee CH, Chen PL, Meinertzhagen IA, Zipursky SL. 2003. The protocadherin Flamingo is required for axon target selection in the *Drosophila* visual system. *Nat Neurosci* **6:** 557–563.

Li J, Ashley J, Budnik V, Bhat MA. 2007. Crucial role of *Drosophila* neurexin in proper active zone apposition to postsynaptic densities, synaptic growth, and synaptic transmission. *Neuron* **55:** 741–755.

Lin JC, Ho WH, Gurney A, Rosenthal A. 2003. The netrin-G1 ligand NGL-1 promotes the outgrowth of thalamocortical axons. *Nat Neurosci* **6:** 1270–1276.

Lu B, Usui T, Uemura T, Jan L, Jan YN. 1999. Flamingo controls the planar polarity of sensory bristles and asymmetric division of sensory organ precursors in *Drosophila*. *Curr Biol* **9:** 1247–1250.

Maness PF, Schachner M. 2007. Neural recognition molecules of the immunoglobulin superfamily: Signaling transducers of axon guidance and neuronal migration. *Nat Neurosci* **10:** 19–26.

Martin V, Mrkusich E, Steinel MC, Rice J, Merritt DJ, Whitington PM. 2008. The L1-type cell adhesion molecule Neuroglian is necessary for maintenance of sensory axon advance in the *Drosophila* embryo. *Neural Develop* **3:** 10.

Masai I, Lele Z, Yamaguchi M, Komori A, Nakata A, Nishiwaki Y, Wada H, Tanaka H, Nojima Y, Hammerschmidt M, et al. 2003. N-cadherin mediates retinal lamination, maintenance of forebrain compartments and patterning of retinal neurites. *Development* **130:** 2479–2494.

Mast JD, Prakash S, Chen PL, Clandinin TR. 2006. The mechanisms and molecules that connect photoreceptor axons to their targets in *Drosophila*. *Semin Cell Dev Biol* **17:** 42–49.

Maurel-Zaffran C, Suzuki T, Gahmon G, Treisman JE, Dickson BJ. 2001. Cell-autonomous and -nonautonomous functions of LAR in R7 photoreceptor axon targeting. *Neuron* **32:** 225–235.

McAllister AK. 2007. Dynamic aspects of CNS synapse formation. *Annu Rev Neurosci* **30:** 425–450.

Miller KE, DeProto J, Kaufmann N, Patel BN, Duckworth A, Van Vactor D. 2005. Direct observation demonstrates that Liprin-α is required for trafficking of synaptic vesicles. *Curr Biol* **15:** 684–689.

Miskevich F, Zhu Y, Ranscht B, Sanes JR. 1998. Expression of multiple cadherins and catenins in the chick optic tectum. *Mol Cell Neurosci* **12:** 240–255.

Missler M. 2003. Synaptic cell adhesion goes functional. *Trends Neurosci* **26:** 176–178.

Missler M, Zhang W, Rohlmann A, Kattenstroth G, Hammer RE, Gottmann K, Sudhof TC. 2003. α-neurexins couple Ca2+ channels to synaptic vesicle exocytosis. *Nature* **423:** 939–948.

Mizoguchi A, Nakanishi H, Kimura K, Matsubara K, Ozaki-Kuroda K, Katata T, Honda T, Kiyohara Y, Heo K, Higashi M, et al. 2002. Nectin: An adhesion molecule involved in formation of synapses. *J Cell Biol* **156:** 555–565.

Morishita H, Yagi T. 2007. Protocadherin family: Diversity, structure, and function. *Curr Opin Cell Biol* **19:** 584–592.

Muller D, Wang C, Skibo G, Toni N, Cremer H, Calaora V, Rougon G, Kiss JZ. 1996. PSA-NCAM is required for activity-induced synaptic plasticity. *Neuron* **17:** 413–422.

Nam CI, Chen L. 2005. Postsynaptic assembly induced by neurexin-neuroligin interaction and neurotransmitter. *Proc Natl Acad Sci* **102:** 6137–6142.

Nern A, Nguyen LV, Herman T, Prakash S, Clandinin TR, Zipursky SL. 2005. An isoform-specific allele of *Drosophila* N-cadherin disrupts a late step of R7 targeting. *Proc Natl Acad Sci* **102:** 12944–12949.

Nern A, Zhu Y, Zipursky SL. 2008. Local N-cadherin interactions mediate distinct steps in the targeting of lamina neurons. *Neuron* **58:** 34–41.

Newsome TP, Asling B, Dickson BJ. 2000. Analysis of *Drosophila* photoreceptor axon guidance in eye-specific mosaics. *Development* **127:** 851–860.

Nollet F, Kools P, van Roy F. 2000. Phylogenetic analysis of the cadherin superfamily allows identification of six major subfamilies besides several solitary members. *J Mol Biol* **299:** 551–572.

Nuriya M, Huganir RL. 2006. Regulation of AMPA receptor trafficking by N-cadherin. *J Neurochem* **97:** 652–661.

Ohtsuka T, Takao-Rikitsu E, Inoue E, Inoue M, Takeuchi M, Matsubara K, Deguchi-Tawarada M, Satoh K, Morimoto K, Nakanishi H, et al. 2002. Cast: A novel protein of the cytomatrix at the active zone of synapses that forms a ternary complex with RIM1 and munc13–1. *J Cell Biol* **158:** 577–590.

Okamura K, Tanaka H, Yagita Y, Saeki Y, Taguchi A, Hiraoka Y, Zeng LH, Colman DR, Miki N. 2004. Cadherin activity is required for activity-induced spine remodeling. *J Cell Biol* **167:** 961–972.

Okuda T, Yu LM, Cingolani LA, Kemler R, Goda Y. 2007. β-Catenin regulates excitatory postsynaptic strength at hippocampal synapses. *Proc Natl Acad Sci* **104:** 13479–13484.

Olsen O, Moore KA, Fukata M, Kazuta T, Trinidad JC, Kauer FW, Streuli M, Misawa H, Burlingame AL, Nicoll RA, et al. 2005. Neurotransmitter release regulated by a MALS-liprin-α presynaptic complex. *J Cell Biol* **170:** 1127–1134.

Ooshio T, Irie K, Morimoto K, Fukuhara A, Imai T, Takai Y. 2004. Involvement of LMO7 in the association of two cell-cell adhesion molecules, nectin and E-cadherin, through afadin and α-actinin in epithelial cells. *J Biol Chem* **279:** 31365–31373.

Packard M, Mathew D, Budnik V. 2003. FASt remodeling of synapses in *Drosophila*. *Curr Opin Neurobiol* **13:** 527–534.

Patel MR, Lehrman EK, Poon VY, Crump JG, Zhen M, Bargmann CI, Shen K. 2006. Hierarchical assembly of presynaptic components in defined *C. elegans* synapses. *Nat Neurosci* **9:** 1488–98.

Patrakka J, Tryggvason K. 2007. Nephrin–a unique structural and signaling protein of the kidney filter. *Trends Mol Med* **13:** 396–403.

Phillips GR, Tanaka H, Frank M, Elste A, Fidler L, Benson DL, Colman DR. 2003. γ-protocadherins are targeted to subsets of synapses and intracellular organelles in neurons. *J Neurosci* **23:** 5096–5104.

Piechotta K, Dudanova I, Missler M. 2006. The resilient synapse: Insights from genetic interference of synaptic cell adhesion molecules. *Cell Tissue Res* **326:** 617–642.

Pielage J, Fetter RD, Davis GW. 2006. A postsynaptic spectrin scaffold defines active zone size, spacing, and efficacy at the *Drosophila* neuromuscular junction. *J Cell Biol* **175:** 491–503.

Pielage J, Cheng L, Fetter RD, Carlton PM, Sedat JW, Davis GW. 2008. A presynaptic giant ankyrin stabilizes the NMJ through regulation of presynaptic microtubules and transsynaptic cell adhesion. *Neuron* **58:** 195–209.

Pokutta S, Weis WI. 2007. Structure and mechanism of cadherins and catenins in cell-cell contacts. *Annu Rev Cell Dev Biol* **23:** 237–261.

Pokutta S, Drees F, Takai Y, Nelson WJ, Weis WI. 2002. Biochemical and structural definition of the l-afadin- and actin-binding sites of α-catenin. *J Biol Chem* **277:** 18868–18874.

Polo-Parada L, Bose CM, Landmesser LT. 2001. Alterations in transmission, vesicle dynamics, and transmitter release machinery at NCAM-deficient neuromuscular junctions. *Neuron* **32:** 815–828.

Prakash S, Caldwell JC, Eberl DF, Clandinin TR. 2005. *Drosophila* N-cadherin mediates an attractive interaction between photoreceptor axons and their targets. *Nat Neurosci* **8:** 443–450.

Ramocki MB, Zoghbi HY. 2008. Failure of neuronal homeostasis results in common neuropsychiatric phenotypes. *Nature* **455:** 912–918.

Ramos RG, Igloi GL, Lichte B, Baumann U, Maier D, Schneider T, Brandstatter JH, Frohlich A, Fischbach KF. 1993. The irregular chiasm C-roughest locus of *Drosophila*, which affects axonal projections and programmed cell death, encodes a novel immunoglobulin-like protein. *Genes & Dev* **7:** 2533–2547.

Ranscht B. 2000. Cadherins: Molecular codes for axon guidance and synapse formation. *Int J Dev Neurosci* **18:** 643–651.

Redies C. 2000. Cadherins in the central nervous system. *Prog Neurobiol* **61:** 611–48.

Redies C, Vanhalst K, Roy F. 2005. δ-Protocadherins: Unique structures and functions. *Cell Mol Life Sci* **62:** 2840–2852.

Ritzenthaler S, Suzuki E, Chiba A. 2000. Postsynaptic filopodia in muscle cells interact with innervating motoneuron axons. *Nat Neurosci* **3:** 1012–1017.

Rudenko G, Nguyen T, Chelliah Y, Sudhof TC, Deisenhofer J. 1999. The structure of the ligand-binding domain of neurexin Iβ: Regulation of LNS domain function by alternative splicing. *Cell* **99:** 93–101.

Rudenko G, Hohenester E, Muller YA. 2001. LG/LNS domains: Multiple functions—one business end? *Trends Biochem Sci* **26:** 363–368.

Ruiz-Gomez M, Coutts N, Price A, Taylor MV, Bate M. 2000. *Drosophila* dumbfounded: A myoblast attractant essential for fusion. *Cell* **102:** 189–198.

Saghatelyan AK, Nikonenko AG, Sun M, Rolf B, Putthoff P, Kutsche M, Bartsch U, Dityatev A, Schachner M. 2004. Reduced GABAergic transmission and number of hippocampal perisomatic inhibitory synapses in juvenile mice deficient in the neural cell adhesion molecule L1. *Mol Cell Neurosci* **26:** 191–203.

Saglietti L, Dequidt C, Kamieniarz K, Rousset MC, Valnegri P, Thoumine O, Beretta F, Fagni L, Choquet D, Sala C, et al. 2007. Extracellular interactions between GluR2 and N-cadherin in spine regulation. *Neuron* **54:** 461–477.

Salinas PC. 2005. Signaling at the vertebrate synapse: New roles for embryonic morphogens? *J Neurobiol* **64:** 435–445.

Salinas PC, Zou Y. 2008. Wnt signaling in neural circuit assembly. *Annu Rev Neurosci* **31:** 339–358.

Sara Y, Biederer T, Atasoy D, Chubykin A, Mozhayeva MG, Sudhof TC, Kavalali ET. 2005. Selective capability of SynCAM and neuroligin for functional synapse assembly. *J Neurosci* **25:** 260–270.

Scheiffele P, Fan J, Choih J, Fetter R, Serafini T. 2000. Neuroligin expressed in nonneuronal cells triggers presynaptic development in contacting axons. *Cell* **101:** 657–669.

Schoch S, Castillo PE, Jo T, Mukherjee K, Geppert M, Wang Y, Schmitz F, Malenka RC, Sudhof TC. 2002. RIM1α forms a protein scaffold for regulating neurotransmitter release at the active zone. *Nature* **415:** 321–326.

Schuster CM, Davis GW, Fetter RD, Goodman CS. 1996a. Genetic dissection of structural and functional components of synaptic plasticity. I. Fasciclin II controls synaptic stabilization and growth. *Neuron* **17:** 641–654.

Schuster CM, Davis GW, Fetter RD, Goodman CS. 1996b. Genetic dissection of structural and functional components of synaptic plasticity. II. Fasciclin II controls presynaptic structural plasticity. *Neuron* **17:** 655–667.

Schuster T, Krug M, Hassan H, Schachner M. 1998. Increase in proportion of hippocampal spine synapses expressing neural cell adhesion molecule NCAM180 following long-term potentiation. *J Neurobiol* **37:** 359–372.

Seabold GK, Wang PY, Chang K, Wang CY, Wang YX, Petralia RS, Wenthold RJ. 2008. The SALM family of adhesion-like molecules forms heteromeric and homomeric complexes. *J Biol Chem* **283:** 8395–8405.

Senti KA, Usui T, Boucke K, Greber U, Uemura T, Dickson BJ. 2003. Flamingo regulates R8 axon-axon and axon-target interactions in the *Drosophila* visual system. *Curr Biol* **13:** 828–832.

Serafini T. 1999. Finding a partner in a crowd: Neuronal diversity and synaptogenesis. *Cell* **98:** 133–136.

Shapiro L, Colman DR. 1999. The diversity of cadherins and implications for a synaptic adhesive code in the CNS. *Neuron* **23:** 427–430.

Shapiro EM, Gonzalez-Perez O, Manuel Garcia-Verdugo J, Alvarez-Buylla A, Koretsky AP. 2006. Magnetic resonance imaging of the migration of neuronal precursors generated in the adult rodent brain. *Neuroimage* **32:** 1150–1157.

Shapiro L, Love J, Colman DR. 2007. Adhesion molecules in the nervous system: Structural insights into function and diversity. *Annu Rev Neurosci* **30:** 451–474.

Shen K, Bargmann CI. 2003. The immunoglobulin superfamily protein SYG-1 determines the location of specific synapses in *C. elegans*. *Cell* **112:** 619–630.

Shen K, Fetter RD, Bargmann CI. 2004. Synaptic specificity is generated by the synaptic guidepost protein SYG-2 and its receptor, SYG-1. *Cell* **116:** 869–881.

Shima Y, Kengaku M, Hirano T, Takeichi M, Uemura T. 2004. Regulation of dendritic maintenance and growth by a mammalian 7-pass transmembrane cadherin. *Dev Cell* **7:** 205–216.

Shin H, Wyszynski M, Huh KH, Valtschanoff JG, Lee JR, Ko J, Streuli M, Weinberg RJ, Sheng M, Kim E. 2003. Association of the kinesin motor KIF1A with the multimodular protein liprin-α. *J Biol Chem* **278:** 11393–11401.

Shinza-Kameda M, Takasu E, Sakurai K, Hayashi S, Nose A. 2006. Regulation of layer-specific targeting by reciprocal expression of a cell adhesion molecule, capricious. *Neuron* **49:** 205–13.

Shishido E, Takeichi M, Nose A. 1998. *Drosophila* synapse formation: Regulation by transmembrane protein with Leu-rich repeats, CAPRICIOUS. *Science* **280:** 2118–2121.

Song JY, Ichtchenko K, Sudhof TC, Brose N. 1999. Neuroligin 1 is a postsynaptic cell-adhesion molecule of excitatory synapses. *Proc Natl Acad Sci* **96:** 1100–1105.

Spangler SA, Hoogenraad CC. 2007. Liprin-α proteins: Scaffold molecules for synapse maturation. *Biochem Soc Trans* **35:** 1278–1282.

Strunkelnberg M, Bonengel B, Moda LM, Hertenstein A, de Couet HG, Ramos RG, Fischbach KF. 2001. rst and its paralogue kirre act redundantly during embryonic muscle development in *Drosophila*. *Development* **128:** 4229–4239.

Stryker E, Johnson KG. 2007. LAR, liprin α and the regulation of active zone morphogenesis. *J Cell Sci* **120:** 3723–3728.

Sudhof TC. 2004. The synaptic vesicle cycle. *Annu Rev Neurosci* **27:** 509–547.

Sudhof TC. 2008. Neuroligins and neurexins link synaptic function to cognitive disease. *Nature* **455:** 903–911.

Suzuki SC, Inoue T, Kimura Y, Tanaka T, Takeichi M. 1997. Neuronal circuits are subdivided by differential expression of type-II classic cadherins in postnatal mouse brains. *Mol Cell Neurosci* **9:** 433–447.

Tabuchi K, Sudhof TC. 2002. Structure and evolution of neurexin genes: Insight into the mechanism of alternative splicing. *Genomics* **79:** 849–859.

Tachibana K, Nakanishi H, Mandai K, Ozaki K, Ikeda W, Yamamoto Y, Nagafuchi A, Tsukita S, Takai Y. 2000. Two cell adhesion molecules, nectin and cadherin, interact through their cytoplasmic domain-associated proteins. *J Cell Biol* **150:** 1161–1176.

Tai CY, Mysore SP, Chiu C, Schuman EM. 2007. Activity-regulated N-cadherin endocytosis. *Neuron* **54:** 771–785.

Tai CY, Kim SA, Schuman EM. 2008. Cadherins and synaptic plasticity. *Curr Opin Cell Biol* **20:** 567–575.

Takai Y, Ikeda W, Ogita H, Rikitake Y. 2008a. The immunoglobulin-like cell adhesion molecule nectin and its associated protein afadin. *Annu Rev Cell Dev Biol* **24:** 309–342.

Takai Y, Miyoshi J, Ikeda W, Ogita H. 2008b. Nectins and nectin-like molecules: Roles in contact inhibition of cell movement and proliferation. *Nat Rev Mol Cell Biol* **9:** 603–615.

Takeichi M. 2007. The cadherin superfamily in neuronal connections and interactions. *Nat Rev Neurosci* **8:** 11–20.

Tanabe K, Takahashi Y, Sato Y, Kawakami K, Takeichi M, Nakagawa S. 2006. Cadherin is required for dendritic morphogenesis and synaptic terminal organization of retinal horizontal cells. *Development* **133:** 4085–4096.

Tanaka H, Shan W, Phillips GR, Arndt K, Bozdagi O, Shapiro L, Huntley GW, Benson DL, Colman DR. 2000. Molecular modification of N-cadherin in response to synaptic activity. *Neuron* **25:** 93–107.

Tang L, Hung CP, Schuman EM. 1998. A role for the cadherin family of cell adhesion molecules in hippocampal long-term potentiation. *Neuron* **20:** 1165–1175.

Taniguchi H, Gollan L, Scholl FG, Mahadomrongkul V, Dobler E, Limthong N, Peck M, Aoki C, Scheiffele P. 2007. Silencing of neuroligin function by postsynaptic neurexins. *J Neurosci* **27:** 2815–2824.

Tepass U, Harris KP. 2007. Adherens junctions in *Drosophila* retinal morphogenesis. *Trends Cell Biol* **17:** 26–35.

Thomas U, Kim E, Kuhlendahl S, Koh YH, Gundelfinger ED, Sheng M, Garner CC, Budnik V. 1997. Synaptic clustering of the cell adhesion molecule fasciclin II by discs-large and its role in the regulation of presynaptic structure. *Neuron* **19:** 787–799.

Thomas LA, Akins MR, Biederer T. 2008. Expression and adhesion profiles of SynCAM molecules indicate distinct neuronal functions. *J Comp Neurol* **510:** 47–67.

Ting CY, Lee CH. 2007. Visual circuit development in *Drosophila*. *Curr Opin Neurobiol* **17:** 65–72.

Ting CY, Yonekura S, Chung P, Hsu SN, Robertson HM, Chiba A, Lee CH. 2005. *Drosophila* N-cadherin functions in the first stage of the two-stage layer-selection process of R7 photoreceptor afferents. *Development* **132:** 953–963.

Tissir F, Bar I, Jossin Y, De Backer O, Goffinet AM. 2005. Protocadherin Celsr3 is crucial in axonal tract development. *Nat Neurosci* **8:** 451–457.

Togashi H, Miyoshi J, Honda T, Sakisaka T, Takai Y, Takeichi M. 2006. Interneurite affinity is regulated by heterophilic nectin interactions in concert with the cadherin machinery. *J Cell Biol* **174:** 141–151.

Uchida N, Honjo Y, Johnson KR, Wheelock MJ, Takeichi M. 1996. The catenin/cadherin adhesion system is localized in synaptic junctions bordering transmitter release zones. *J Cell Biol* **135:** 767–779.

Ullrich B, Ushkaryov YA, Sudhof TC. 1995. Cartography of neurexins: More than 1000 isoforms generated by alternative splicing and expressed in distinct subsets of neurons. *Neuron* **14:** 497–507.

Uryu K, Butler AK, Chesselet MF. 1999. Synaptogenesis and ultrastructural localization of the polysialylated neural cell adhesion molecule in the developing striatum. *J Comp Neurol* **405:** 216–232.

Ushkaryov YA, Petrenko AG, Geppert M, Sudhof TC. 1992. Neurexins: Synaptic cell surface proteins related to the α-latrotoxin receptor and laminin. *Science* **257:** 50–56.

van der Weyden L, Arends MJ, Chausiaux OE, Ellis PJ, Lange UC, Surani MA, Affara N, Murakami Y, Adams DJ, Bradley A. 2006. Loss of TSLC1 causes male infertility due to a defect at the spermatid stage of spermatogenesis. *Mol Cell Biol* **26:** 3595–3609.

van Roessel P, Elliott DA, Robinson IM, Prokop A, Brand AH. 2004. Independent regulation of synaptic size and activity by the anaphase-promoting complex. *Cell* **119:** 707–718.

Varoqueaux F, Jamain S, Brose N. 2004. Neuroligin 2 is exclusively localized to inhibitory synapses. *Eur J Cell Biol* **83:** 449–456.

Varoqueaux F, Aramuni G, Rawson RL, Mohrmann R, Missler M, Gottmann K, Zhang W, Sudhof TC, Brose N. 2006. Neuroligins determine synapse maturation and function. *Neuron* **51:** 741–754.

Vogel C, Teichmann SA, Chothia C. 2003. The immunoglobulin superfamily in *Drosophila* melanogaster and *Caenorhabditis elegans* and the evolution of complexity. *Development* **130:** 6317–6328.

Waites CL, Craig AM, Garner CC. 2005. Mechanisms of vertebrate synaptogenesis. *Annu Rev Neurosci* **28:** 251–274.

Walsh FS, Doherty P. 1997. Neural cell adhesion molecules of the immunoglobulin superfamily: Role in axon growth and guidance. *Annu Rev Cell Dev Biol* **13:** 425–456.

Walsh CA, Morrow EM, Rubenstein JL. 2008. Autism and brain development. *Cell* **135:** 396–400.

Wang X, Su H, Bradley A. 2002a. Molecular mechanisms governing Pcdh-γ gene expression: Evidence for a multiple promoter and cis-alternative splicing model. *Genes & Dev* **16:** 1890–1905.

Wang X, Weiner JA, Levi S, Craig AM, Bradley A, Sanes JR. 2002b. γ protocadherins are required for survival of spinal interneurons. *Neuron* **36:** 843–854.

Wang CY, Chang K, Petralia RS, Wang YX, Seabold GK, Wenthold RJ. 2006. A novel family of adhesion-like molecules that interacts with the NMDA receptor. *J Neurosci* **26:** 2174–2183.

Washbourne P, Dityatev A, Scheiffele P, Biederer T, Weiner JA, Christopherson KS, El-Husseini A. 2004. Cell adhesion molecules in synapse formation. *J Neurosci* **24:** 9244–9249.

Weiner JA, Wang X, Tapia JC, Sanes JR. 2005. γ protocadherins are required for synaptic development in the spinal cord. *Proc Natl Acad Sci* **102:** 8–14.

Wohrn JC, Puelles L, Nakagawa S, Takeichi M, Redies C. 1998. Cadherin expression in the retina and retinofugal pathways of the chicken embryo. *J Comp Neurol* **396:** 20–38.

Wyszynski M, Kim E, Dunah AW, Passafaro M, Valtschanoff JG, Serra-Pages C, Streuli M, Weinberg RJ, Sheng M. 2002. Interaction between GRIP and liprin-α/SYD2 is required for AMPA receptor targeting. *Neuron* **34:** 39–52.

Yamada S, Nelson WJ. 2007. Synapses: Sites of cell recognition, adhesion, and functional specification. *Annu Rev Biochem* **76:** 267–294.

Yamagata M, Herman JP, Sanes JR. 1995. Lamina-specific expression of adhesion molecules in developing chick optic tectum. *J Neurosci* **15:** 4556–4571.

Yamagata M, Sanes JR, Weiner JA. 2003. Synaptic adhesion molecules. *Curr Opin Cell Biol* **15:** 621–632.

Yamamoto M, Ueda R, Takahashi K, Saigo K, Uemura T. 2006. Control of axonal sprouting and dendrite branching by the Nrg-Ank complex at the neuron-glia interface. *Curr Biol* **16:** 1678–1683.

Zeng X, Sun M, Liu L, Chen F, Wei L, Xie W. 2007. Neurexin-1 is required for synapse formation and larvae associative learning in *Drosophila*. *FEBS Lett* **581:** 2509–2516.

Zhai RG, Bellen HJ. 2004. The architecture of the active zone in the presynaptic nerve terminal. *Physiology (Bethesda)* **19:** 262–270.

Zhen M, Jin Y. 1999. The liprin protein SYD-2 regulates the differentiation of presynaptic termini in *C. elegans*. *Nature* **401:** 371–375.

Zhong Y, Budnik V, Wu CF. 1992. Synaptic plasticity in *Drosophila* memory and hyperexcitable mutants: Role of cAMP cascade. *J Neurosci* **12:** 644–651.

Zhu H, Luo L. 2004. Diverse functions of N-cadherin in dendritic and axonal terminal arborization of olfactory projection neurons. *Neuron* **42:** 63–75.

Modular Design of Immunological Synapses and Kinapses

Michael L. Dustin

Program in Molecular Pathogenesis, Skirball Institute for Biomolecular Medicine, The Helen L and Martin S Kimmel Center for Biology and Medicine, New York University School of Medicine, New York 10016

Correspondence: michael.dustin@med.nyu.edu

The concept of an immunological synapse goes back to the early 1980s with the discovery of the relationship between T-cell antigen receptor mediated Ca^{2+} signaling, adhesion, and directed secretion. However, this concept did not gain traction until images were published starting in 1998 that revealed a specific molecular pattern in the interface between T cells and model antigen-presenting cells or supported planar bilayers. The dominant pattern, a ring of adhesion molecules surrounding a central cluster of antigen receptors, was observed in both model systems. Analysis of the origins of this pattern over the past 10 years has presented a solution for a difficult problem in lymphocyte biology—how a highly motile cell can suddenly stop when it encounters a signal delivered by just a few antigenic ligands on the surface of another cell without disabling the sensory machinery of the motile cell. The T lymphocyte actively assembles the immunological synapse pattern following a modular design with roots in actin–myosin-based motility.

The immune system provides an outstanding model for study of both dynamic and stable cell–cell adhesion (Dustin and Springer 1989; Lawrence and Springer 1991; Miller et al. 2002; Mempel et al. 2004). Early studies on molecules important for the function of cytotoxic lymphocytes, cells that kill virally infected cells and contribute to destruction of transplanted organs and tumors, identified two families of adhesion molecules (Davignon et al. 1981) (Fig. 1). Monoclonal antibodies that blocked the activity of cytotoxic lymphocytes identified an integrin, still widely referred to as LFA-1, and two immunoglobulin superfamily members, LFA-2 and LFA-3, now known more commonly as CD2 and CD58, respectively (Sanchez-Madrid et al. 1982). CD2 and CD58 were defined as a receptor ligand pair making it the first clearly defined heterophilic cell–cell adhesion system (Dustin et al. 1987a; Selvaraj et al. 1987). The identification of a ligand for LFA-1, ICAM-1, revealed that it too was a transmembrane member of the immunoglobulin superfamily providing a mechanism for involvement of integrins in cell–cell junction formation (Marlin and Springer 1987). These studies were contemporaneous with evidence for adhesion via direct ligand binding as a mechanism of matrix binding integrins, and homophilic interactions of NCAM and cadherins

 doi: 10.1101/cshperspect.a002873
Cite this article as *Cold Spring Harb Perspect Biol* 2009;1:a002873

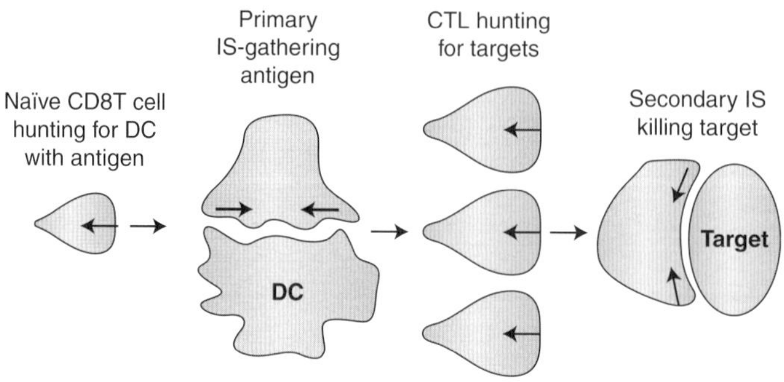

Figure 1. Cytotoxic T lymphocyte (CTL) life cycle. Naïve cells hunt for evidence of infection on the surface of self dendritic cells in lymphoid tissue. One antigen, if found on the primary immunological synapse, leads to activation, proliferation, and development of a CTL. This CTL exits the lymph node and uses the blood to reach a tissue. In the tissue, the CTL migrates to find its target and then uses a secondary synapse to kill the target.

(Rutishauser et al. 1982; Peyrieras et al. 1983; Gardner and Hynes 1985; Horwitz et al. 1985; Pytela et al. 1985; Wright and Meyer 1985; Cunningham et al. 1987; Nagafuchi et al. 1987). In some respects, the complexity of the role of oligosaccharides in NCAM-mediated adhesion and studying homophilic systems has resulted in clearer results coming from the heterophilic immune cell adhesion systems in the early to mid 1980s. It was speculated that the CD2–CD58 interaction might have evolved from an ancestral homophilic system like NCAM and subsequently it has been determined that CD58 is located in a rapidly evolving gene cluster including several important homophilic adhesion molecules on chromosome 1 (Wong et al. 1990). The number of receptor ligand interactions that are involved in immune cell interactions has grown significantly since these early studies, but the LFA-1/ICAM-1 and CD2 family interactions still appear to be major contributors in cell adhesion in many functional settings. In this article, I review the role of LFA-1/ICAM-1, CD2/CD58, and CD2 family homophilic adhesion molecules, like SLAM, in immune cell interactions. I describe the supported planar bilayer model in some detail because this has played an important role in the characterization of immune-cell adhesion systems, but also discuss recent studies using in vivo imaging that have also provided insight into the unique demands of in situ immune-cell interactions leading to specific molecular requirements.

DISCOVERY OF ADHESION RECEPTORS IN THE IMMUNE SYSTEM

An antigen is any chemical structure that can be recognized by the immune system (Lindenmann 1984). The adaptive immune system uses two types of cells to produce distinct types of antigen receptors. B cells recognize a wide variety of chemical structures from small molecules to conformation-sensitive determinants of biological macromolecules. B cells secrete copious amounts of these receptors, as antibodies, making them highly available biochemically such that their activity was discovered around 1900 and the molecule had been purified and sequenced by the 1960s. T cells were known to be important for protection against viruses and intracellular bacterial infections, and also provided help for B-cell responses, but the T-cell antigen receptor, which is not produced in an abundant secreted form, was more elusive. The focus of T-cell biology in the early 1980s was to discover this antigen receptor, which binds a ligand composed of antigen fragments bound to self major histocompatibility complex (MHC) molecules (Trombetta and Mellman 2005). Monoclonal antibodies were generated after immunization of mice with T cells and thousands were screened for ability to inhibit function one at a time. The Allison and Kappler and Marrack groups used monoclonal T cells as immunogens and testers to produce blocking antibodies to the T-cell receptor (McIntyre and Allison 1983; White et al. 1983). When polyclonal T cells were used, the major function-blocking antibodies were not against the antigen receptor, because a polyclonal population contains many different structurally distinct antigen receptors, but were against adhesion molecules that are invariant and required for killing. Sanchez-Madrid and Springer produced the most complete panel, which included antibodies to LFA-1, CD2, and CD58, mentioned above (Sanchez-Madrid et al. 1982).

By 1984, it was shown that these antibodies inhibited function by blocking adhesion (conjugate formation between CTLs and target cells), and evidence was generated by 1986 that CD2 and CD58 were a receptor ligand pair, as mentioned above (Krensky et al. 1984; Shaw et al. 1986). Like NCAM, CD58 exists in transmemebrane and glycolipid-anchored splice variants (Dustin et al. 1987b). In the rodent models, there is no CD58, but CD2 binds to the related glycolipid-anchored CD48 (Kato et al. 1992). Antibodies to LFA-1 ligands were not discovered in these screens because there are multiple ligands that all need to be blocked at the same time to prevent killing, but the mAb were only tested one at a time. The first LFA-1 ligand to be identified was the inflammation-regulated protein ICAM-1, which was discovered through generation of monoclonal antibodies to cells from children lacking LFA-1 and related β2 integrins (Dustin et al. 1986; Rothlein et al. 1986). ICAM-2 was then identified from an endothelial-cell library by expression cloning (Dustin and Springer 1988; Staunton et al. 1989). ICAM-3 is expressed on leukocytes, ICAM-4 is expressed on red blood cells, and ICAM-5 is expressed in the nervous system. CD2–CD58 and LFA-1–ICAM are the major cell–cell adhesion systems for human T cells.

REGULATION OF LYMPHOCYTE ADHESION

There are two major paradigms for regulation of lymphocyte adhesion that are directly related to function: extravasation and immunological synapse formation. In both cases, the adhesion molecules are subjected to significant although very different forces. To transit from the blood to tissues, lymphocytes must interact with the vessel wall under flow, arrest motion with the blood flow, and polarize and crawl between or through endothelial cells (Springer 1995). The cells then read tissue queues to drive cell migration. To communicate with antigen-presenting cells, the highly polarized migrating T cells must arrest migration and form a tight cell–cell junction for minutes to hours (Mempel et al. 2004). The specialized cell–cell junction that mediates this arrest and subsequent communication is referred to as an immunological synapse (Grakoui et al. 1999). The forces experienced by the extravasating cells are external forces of blood flow generating ~200 pN forces on molecular interactions that are tethering cells, whereas the forces experienced by cells migrating in tissues or forming an immunological synapse are on the order of 2 pN, the force transmitted by the cytoskeletal adapter talin (Jiang et al. 2003; Alon and Dustin 2007). Surprisingly, a single versatile adhesion molecule, LFA-1, dominates both processes.

LFA-1 functions as an adhesion molecule and is regulated by at least three mechanisms—changes in conformation driven by inside-out signaling, changes in organization secondary to increased lateral mobility, and force-dependent adhesion strengthening that is related to fluid shear. The LFA-1 ligand binding domain can exist in three conformations: a low affinity conformation with a Kd of approximately 1 mM, an intermediate affinity form with a Kd of approximately 10 μM, and a high affinity form with a Kd of approximately 100 nM (Shimaoka et al. 2003). The intermediate affinity form is the likely substrate for regulation of LFA-1 through changes in lateral mobility and clustering, and the high affinity form may be achieved when LFA-1 is subjected to specific forces following intermediate affinity binding. This ability to shift to a higher affinity in response to force may give LFA-1 the appearance of a catch bond and allow it to operate under conditions that would undermine a fixed affinity adhesion system. Application of relatively low forces applied parallel to the membrane plane or greater forces applied perpendicular to the membrane may induce this high-affinity form (Zhu et al. 2008). High-affinity LFA-1 interacts with ICAM-1 with a half-life of 30 s in solution, but likely has a much shorter half-life with application of forces from the cytoskeleton or from fluid shear. Thus, even when the high-affinity form is generated in physiological settings, the interactions may appear much more dynamic because the high-affinity form may only be generated under forces that will dramatically shorten the interaction compared with solution measurements. Keeping the half-life in the interaction in the subsecond range under all conditions, with strengthening of the interaction when under shear forces, is ideal for the situation in which the lymphocyte not only needs to arrest on the vessel wall, but also to migrate to endothelial cell junctions for extra-vasation.

LFA-1 displays regulated adhesive interactions in both extravasation and immunological synapse formation. Regulation of LFA-1-mediated adhesion was first documented using phorbol-ester stimulated aggregation of lymphoblasts, a simple in vitro model (Rothlein and Springer 1986). In retrospect, the original descriptions of cell dynamics in phorbol-ester stimulated lymphocyte aggregates are remarkably similar to recent description of B lymphoblasts in germinal centers of lymph nodes (Allen et al. 2007; Schwickert et al. 2007). The aggregation assay was used to identify ICAM-1 as a ligand for LFA-1, and it was also determined that phorbol esters stimulated aggregation without altering expression levels of LFA-1 and ICAM-1 (Dustin et al. 1986; Rothlein et al. 1986). Live cell adhesion to surfaces coated with purified LFA-1 or ICAM-1 revealed that T cell receptor (TCR) signaling stimulated changes in LFA-1 activity through signaling processes—described as inside-out signaling (Dustin and Springer 1989).

The concept of inside-out signaling was extended to the extravasation process, but using chemokines as triggers for inside-out signaling to LFA-1 and other integrins to induce arrest of leukocytes undergoing selectin-mediated rolling adhesion on the vessel wall (Lawrence and Springer 1991). During extravasation, it has been argued by Alon and colleagues that LFA-1 is subject to very rapid activation that may reflect a preformed complex between LFA-1 and chemokine receptors (Shamri et al. 2005).

The detailed mechanism by which inside-out signaling operates is not known. TCR cross-linking initiates a tyrosine kinase cascade that activates the Rac GTP exchange factor (GEF) Vav and the Rap 1 GEF C3G (Krawczyk et al. 2002; Nolz et al. 2008). Rap-1 activation of RIAM is thought be an important step in the recruitment of Vinclulin and Talin to nascent adhesion (Nolz et al. 2007). The activation of LFA-1 by inside-out signaling results in significant recruitment of the cytoskeletal adapter talin to the interface, consistent with Talin as an integrator (Tadokoro et al. 2003). Talin is recruited to areas of LFA-1/ICAM-1 interaction whether this adhesion is triggered by acute TCR signaling or by cytokine and cell-cycle mediated activation (Somersalo et al. 2004; Smith et al. 2005). Talin knockdown results in decreased LFA-1 mediated adhesion both in migrating lymphoblasts and in cells forming immunological synapses. Whether this also impacts adhesion under shear flow has not been determined.

Recently, it has been shown that the in vivo interaction of T cells and B cells, but not T cells and dendritic cells, depends on the small SH2 domain-only adapter protein SAP (Qi et al. 2008). SAP is mutated in X-linked immunodeficiency, which is characterized by loss of T-cell dependent antibody production. SAP interacts with a subfamily of homophilic adhesion molecules, including SLAM, NTB-A, Ly9, and CD84. These interactions appear to be important for stable adhesion between T and B cells, but it is not clear if this is because of adhesive energy provided by the homophilic interactions or if they mediate a signal via SAP that enhances stability of LFA-1 dependent adhesion. SAP interacts with the phosphorylated cytoplasmic domain of SLAM, NTB-A, Ly9, and CD84, and with the Src family kinase Fyn via an unconventional interaction with its SH3 domain (Latour et al. 2003). The immunological synapse formed by T and B cells in vivo is much more dynamic than the structure characterized in vitro in that the B cell remains highly motile and appears to drag the T cell (Okada et al. 2005). Thus, B cells do not appear to experience a stop signal in this interaction. It is not clear whether the essential role of SAP and the associated adhesion systems is based on direct stabilization of adhesion by these interactions or if signaling through this pathway helps to stabilize other adhesion mechanisms.

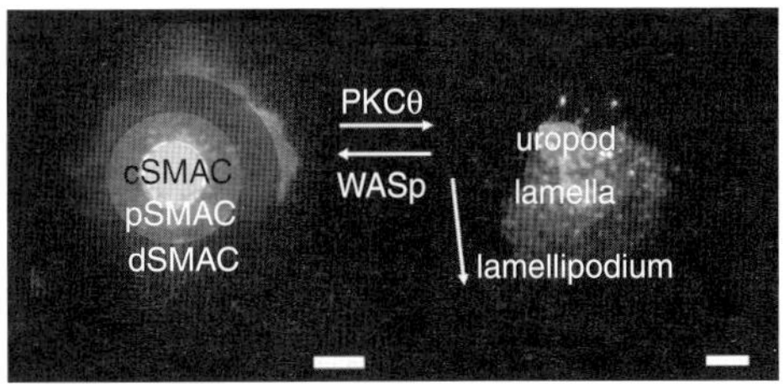

Figure 2. SMAC and kinapse zones. Immunological synapses may have a central cSMAC, an annular pSMAC, and an outer dSMAC. The cSMAC can be rich in TCR–peptide–MHC interaction. The pSMAC is enriched in LFA-1–ICAM-1 interactions. These structures are normally in the interface between two cells, but can be modeled using a live T cell interacting with a supported planar bilayer. A kinapse is a mobile junction with the arrow reflecting the direction of motion. The leading edge is the lamellipodium, the wide region is the lamella or focal zone, and the trailing edge is the uropod. These zones are shaded over images showing the pattern of T-cell receptors.

DEFINING *SYNAPSE*

Synapse is a term derived from Greek that means "to fasten onto": As chromosomes fasten to each other during meiosis, synaptic boutons fasten onto dendritic spines or T lymphocytes fasten onto antigen-presenting cells. The concrete definition proposed in essentially all of these contexts is a stable junction that allows for a functional outcome—recombination, relaying electrical impulses, or mediating immune cell communication or function. "Stable" is a relative term with times lasting from minutes in meiosis, to hours or days for immune cells, and months to years for neural synapses. Thus, the immune cell synapses are considered stable relative to the highly dynamic behavior of lymphocytes in secondary lymphoid tissues where they migrate at up to 30 µm/min (Miller et al. 2002). "Immunological synapse" was not used in a peer-reviewed publication until 1998, although it appeared in two speculative reviews published in 1984 and 1994 (Norcross 1984; Paul and Seder 1994). The observations that gave this concept traction in 1998 were images from Kupfer of the interface between T cells and antigen-presenting cells that were fixed at different time points, bound with fluorescently tagged antibodies, and then visualized by deconvolution microscopy (Monks et al. 1998). These images revealed a discrete central cluster of T-cell receptors surrounded by a ring of adhesion molecules on a supramolecular scale. Kupfer defined three supramolecular activation clusters (SMACs): central (cSMAC), perpheral (pSMAC), and distal (dSMAC) (Fig. 2) (Monks et al. 1998; Freiberg et al. 2002). The cSMAC contained TCR, peptide-MHC, and protein kinase C-θ. The pSMAC contained LFA-1, talin, and ICAM-1. The dSMAC appeared enriched in CD45 (Johnson et al. 2000; Freiberg et al. 2002), but this may be an optical artifact arising from two closely apposed plasma membranes in a sheet-like projection of lamellipodium, as we discuss in the following section. We observed similar segregation and organization of adhesion molecules in the interface between T cells and supported planar bilayers (Fig. 3) containing purified ligands from the antigen-presenting cells and further linked this to control of cell polarity (Dustin et al. 1998). We initially proposed a definition of immunological synapse as having a combination of a "specialized (adhesive) junction, cell polarization, and positional stability" (Dustin et al. 1998). In 2002, David Coleman and I considered four criteria: (1) cells remain individual; (2) the apposed membranes are held together by adhesion molecules; (3) the junction is stable

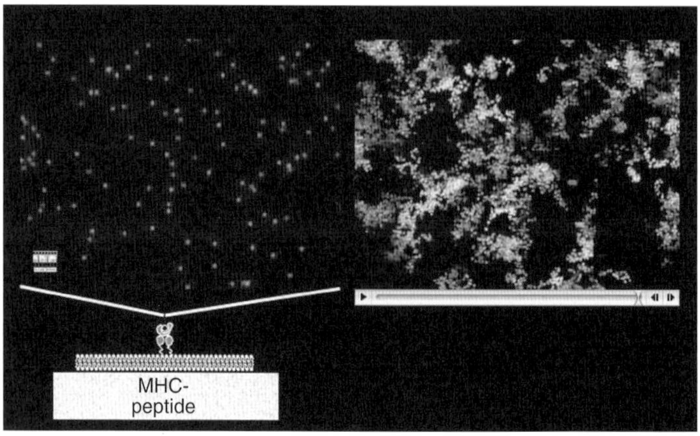

Figure 3. Supported planar bilayers. Supported planar bilayers have been used to model the immunological synapse. A remarkable property of planar bilayers is that the lipids, and anything linked to them, are laterally mobile. The *left* image shows single CD58 molecules in a bilayer labeled with 100 nm fluorescent latex spheres. The *left* image is based on running background subtraction with peak detection from a 30-s movie and illustrates the areas covered by single diffusing particles in the bilayer. Each particle covers roughly the area of an immunological synapse in 30 s.

either through adhesion clamp or cytoskeletal polarity; and (4) directed secretion in response to signals from microdomains (Dustin and Colman 2002). These criteria applied equally well to neural and immunological synapses. Immunological synapses are considered stable based on their remaining symmetric and within the same cell sized area for 20–30 minutes (Grakoui et al. 1999; Sims et al. 2007). The immunological synapse defined by our studies is very similar to the SMAC defined in Kupfer's 3D analysis of T-cell–APC conjugates such that the SMACs are considered components of the immunological synapse (Fig. 4).

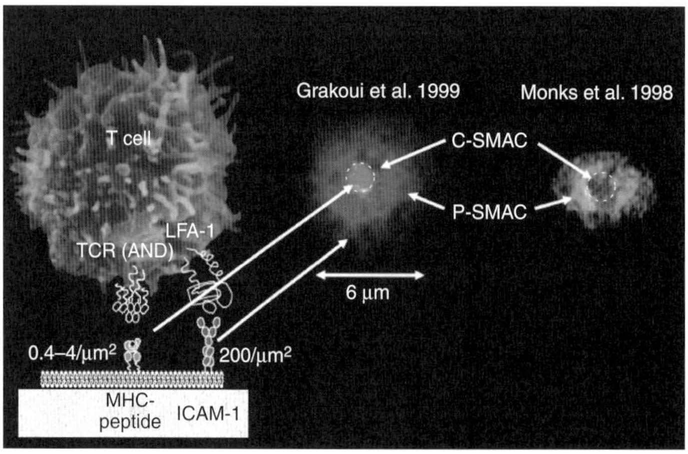

Figure 4. Reconstitution of the immunological synapse with a planar bilayer. The image on the far *right* is a slice from the interface between B cells and T cells showing the cSMAC and pSMAC. The schematic and image on the *left* are of the planar bilayer containing only ICAM-1 and peptide-MHC. The same pattern is formed, although the contact synapse area is larger on the planar surface. The dSMAC was not labeled in these images.

DEFINING *KINAPSE*

Parallel studies on T-cell interactions with dendritic cells in vitro observed highly dynamic interactions between T cells and dendritic cells or macrophages, particularly when the cells were embedded in collagen gels (Underhill et al. 1999; Gunzer et al. 2000). These observations raised questions about whether the more stable cell–cell interactions were physiological. In the past 6 years the use of intravital microscopy to peer directly into living lymphoid tissues of mice revealed that both modes of interaction are observed in vivo. As mentioned previously, T cells are highly motile in secondary lymphoid tissues with average speeds between 10 and 15 μm/min and peak speeds of over 30 μm/min. The mechanism of propulsion is not fully understood. The speed of motility is only weakly dependent on integrins like LFA-1 and is more dependent on G_i-protein coupled receptors that are sensitive to pertussis toxin. In vitro reconstitutions show that chemokines like CCL21, which bind to the receptor CCR7, can reconstitute strong T-cell polarization and rapid motility (Woolf et al. 2007). Thus, both polarity and sufficient traction can be generated by G protein coupled receptors (GPCRs). We have observed that this mode of adhesion involves very small contact areas with essentially no detectable area of contact observed by interference reflection microscopy. This suggests that the T cells are only contacting the surface through small projections that make contact below the diffraction limit (~200–300 nm). Including ICAM-1 on the substrate along with CCL21 allows the formation of measurable contact areas on the order of 50 μm^2 in size (Huang and Dustin, unpublished observations). It is known that the LFA-1–ICAM-1 interaction increases T-cell sensitivity to antigen by 10- to 100-fold and this may correspond to the increase in contact area formed by T cells and APC, which is on the order of 10- to 100-fold (Schmits et al. 1996; Bachmann et al. 1997). These mobile scanning contact areas are the foundation of the kinapse—a moving junction that allows T cells to scan self and foreign MHC-peptide complexes and costimulatory ligands. Although the synapse is symmetric in the plane of the contact area, the kinapse is asymmetric and this is critical to control stopping and going. The asymmetric kinapse has a leading lamellipodium, an intermediate lamella, and a trailing uropod (Fig. 2). Therefore the kinapse is an adhesion-dependent junction, like the synapse, but it is mobile rather than stable, and it is not an efficient platform for directed secretion.

ELEMENTS OF THE IMMUNE CELL CYTOSKELETON

Like other motile cell types, immune cells possess and use the three major filament systems—actin thin filaments, vimentin intermediate filaments, and tubulin-based microtubules. The role of microtubules and the actin cytoskeleton in organizing conjugate formation has been appreciated since the early 1980s based on studies with natural killer (NK) cells and T lymphocytes. The interest in the intermediate filament system has focused on the manner in which circulating lymphocytes survive blood flow by building a vimentin cage that needs to be rapidly disassembled in response to chemokine signals to permit adhesion to the vessel wall and extravasation (Brown et al. 2001).

The microtubule system in lymphocytes is characterized by an astral organization with a central hub or microtubule organizing center (MTOC) collecting all of the minus ends and the plus ends extending through the thin cytoplasm around the nucleus toward the opposite pole of the cell. Early studies from Berke and Singer laboratories showed that the microtubule organizing center of CTLs and NK cells, respectively, docks to the synapse during cell mediated killing (Geiger et al. 1982; Kupfer et al. 1983). Kupfer and Singer extended these findings during the 1980s and 1990s to helper T cells, and developed the hypothesis that the MTOC brings with it the Golgi apparatus and a means for directed secretion of cytokines by helper T lymphocytes (Kupfer et al. 1986). More recently, it has been shown that although the Golgi is focused at the synapse with the

MTOC, some proteins that transit the Golgi, like the cytokine IL-4, are released away from the synapse, whereas others like IFN-γ are released toward the synapse (Huse et al. 2006). The polarity of release of cytolytic granules, on the other hand, is independent of the Golgi positioning and requires the direct docking of the MTOC through which the microtubules recruit the granules to the membrane to allow efficient release of granule contents into the synaptic cleft (Stinchcombe et al. 2006).

The actin system of lymphocytes is dominated by microvilli in circulating lymphocytes that are remodeled into an amoeboid organization during chemokine-mediated transit from blood to tissues (Nijhara et al. 2004). The amoeboid pattern is characterized by dynamic actin at the leading edge and a myosin II enriched uropod at the trailing edge. The MTOC remains behind the nucleus in the uropod. In the early 1980s, Carpen and Saksela showed a distinct polarization of F-actin to the synapse between NK cells and target cells (Carpen et al. 1983). Subsequently, Kupfer and Singer showed that talin, a protein identified as linking the extracellular matrix receptor complexes to F-actin, was also concentrated in the synapse in an antigen-dependent manner (Kupfer and Singer 1989). The discovery that LFA-1 and the extracellular matrix families were both members of the same protein family, the integrins, solidified the appreciation of talin as a link between ICAM-1 on the antigen-presenting cell, LFA-1 and the T cells, and synaptic F-actin in the T-cell cytoskeleton. Because ICAM-1 is also cytoskeletally associated with ezrin/radixin/moesin (ERM) family proteins in the antigen presenting cells, this system "integrates" the antigen presenting cell and T cell surfaces and cytoskeleton into a single supramolecular system—later defined by Kupfer as the pSMAC.

T cells also use the ERM system of proteins, but the exact manner and functional significance of these interactions in T cells has been controversial. T cells express ezrin and moesin. ERM proteins are important in formation of microvilli and the dephosphorylation of ERM proteins mediated by chemokine signals in circulating lymphocytes allows spreading and extravasation (Brown et al. 2003; Nijhara et al. 2004). There is general agreement that moesin is excluded from the immunological synapse and moves either out of the synapse or to a distinct distal-pole complex where a number of signaling molecules are deposited during activation (Allenspach et al. 2001; Delon et al. 2001; Ilani et al. 2007; Shaffer et al. 2009). The general model is that moesin is constitutively active in linking F-actin to proteins like CD43. When T cells are activated, moesin is phosphorylated in a TCR and Vav-1 dependent manner and this causes its dissociation from membrane proteins. This dephosphorylation is partial and much moesin remains membrane associated outside the contact area. Ezrin dynamics are more complicated and it's agreed that it remains associated with the F-actin rich regions of the synapse, at least transiently (Allenspach et al. 2001; Ilani et al. 2007; Shaffer et al. 2009). Movement of CD43 to the uropod of migrating T cells, which is dependent on interaction with ERM proteins, assists disengagement of LFA-1 (Savage et al. 2002). Global dephosphorylation of ERM proteins in the cell also appears to "relax" the cortex of the cell and allow spreading using dynamic F-actin (Faure et al. 2004). It has also been proposed that ezrin, which has an immunotyrosine activation motif-like (ITAM-like) sequence and lingers in the actin rich areas of the synapse, may be important for activation of ZAP-70 in the immunological synapse (Ilani et al. 2007), although mouse T cells deficient in ezrin appeared to have normal ZAP-70 activation (Shaffer et al. 2009). In the later study, both ezrin and moesin contributed to T cell IL-2 production, although they displayed a different localization (Shaffer et al. 2009).

A highly dynamic F-actin network characterizes the lamellipodia of motile cells. Dynamic actin is composed of a mixture of branched F-actin nucleated by the Arp2/3 and linear actin bundles nucleated by formins. At least two pathways contribute to dynamic F-actin networks in T cells, the WAVE2 complex, which is controlled by Rac small G-proteins (Gomez et al. 2006; Nolz et al. 2006; Gomez et al. 2007), and WASp, which is controlled by Cdc42 (Snapper et al. 1998; Cannon et al. 2001; Dupre et al. 2002; Sasahara et al. 2002; Labno et al. 2003). T cells also express myosin IIA, which may contribute to the cortical tension of the F-actin network along with poorly characterized myosin I isoforms (Pierce et al. 2001; Jacobelli et al. 2004; Kim et al. 2006; Morin et al. 2008). Lamellipodia undergo periodic extensions mediated by actin polymerization and retraction mediated by myosin II dependent contraction (Giannone et al. 2004; Giannone et al. 2007). The propagation of these contractile oscillations in lamellipodia represents a universal dynamic pattern in many motile cell types, including lymphocytes (Dobereiner et al. 2006). This appears to be a basic module from which both kinapses and synapses can be initiated.

MODULAR ASSEMBLY OF KINAPSES AND SYNAPSES

The lamellipodium functions as a sensory structure in T lymphocytes. The leading lamellipodium is several-fold more sensitive to antigen than other parts of a polarized T cell (Negulescu et al. 1996; Wei et al. 1999). These studies focused on the motile T cells or kinapse configuration. Immunological synapse formation also appeared to involve the extension of lamellipodial structures over the surface of the antigen-presenting cell (Grakoui et al. 1999; Dustin and Cooper 2000; Bunnell et al. 2001; Tskvitaria-Fuller et al. 2003). In these studies, this phase of extension and contraction appeared to be a one-time event, taking place in the first minutes of interaction, which then set up a highly stable adhesive clamp in the cSMAC (Grakoui et al. 1999). Two lines of experimentation shattered this perception. First, Varma applied TIRFM imaging to track TCR in the immunological synapse formed on supported planar bilayers with laterally mobile ligands and observed that the stable synapse contained within it a continual evolution of small TCR clusters in the periphery that streamed toward the center to sustain signaling (Fig. 5) (Varma et al. 2006). These TCR microclusters recruited a number of signaling molecules that were not recruited to the clamplike cSMAC, in which signaling appeared to be terminated (Campi et al. 2005; Yokosuka et al. 2005). T-cell activation by immobilized anti-CD3 antibodies also incorporated dynamic elements, including single-molecule movement of Src family kinase Lck between signaling foci (Douglass and Vale 2005) and signaling adapter SLP-76–rich

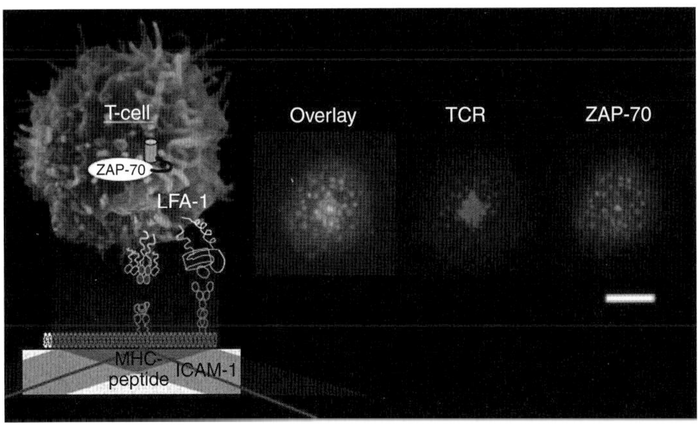

Figure 5. Snap shots of TCR microclusters in an immunological synapse. The images are frames from a movie showing the colocalization of TCR and the kinase ZAP-70 (tagged with GFP) as detected by Campi and Varma by TIRFM microscopy (illustrated schematically on the *left*).

vesicles that underwent microtubule-directed centripetal motion (Bunnell et al. 2002; Barda-Saad et al. 2005; Barr et al. 2006). TCR microcluster movement could be interrupted by inhibition of actin polymerization, but not by depolymerization of microtubules (Varma et al. 2006). Static analysis of proteins involved in dynamic actin polymerization revealed the periphery of the immunological synapse was rich in Arp2/3, cofilin, and F-actin consistent with sustained lamellipodia-like activity (Sims et al. 2007). Dynamic analysis of the outer edge of stable synapses also showed a pattern of circumferential waves of extension and contraction that were remarkably similar to lateral waves in fibroblasts and epithelial cells (Dobereiner et al. 2006; Sims et al. 2007). Speckle microscopy of GFP-actin in a model immunological synapse showed rapid centripetal flow that was moving 2.4-fold faster than the TCR microclusters (Kaizuka et al. 2007). Rapid binding and dissociation of TCR microclusters from the centripetally moving F-actin was also supported by the ability of microclusters to run through nanofabricated mazes (DeMond et al. 2008). Not only the TCR, but the LFA-1 ring is a highly dynamic system built from microclusters that undergo constant turnover (Kaizuka et al. 2007; Sims et al. 2007). The stability of the synapse seems not to be built by clamping adhesion per se, but by taking a highly dynamic motile system and balancing out its motile tendencies through symmetry. Thus, the highly sensitive lamellipodial module can be used in the motile kinapse and the stable synapse.

Amoeboid locomotion uses a leading lamellipodium, a force integrating lamella and a contractile uropod for substrate sensing, traction, and disengagement (Jay et al. 1995; Gupton et al. 2005; Giannone et al. 2007). The kinapse appears to be similarly constructed with relatively discrete zones defined by F-actin, talin, and myosin IIA enrichment (Sanchez-Madrid and del Pozo 1999; Smith et al. 2005; Morin et al. 2008). The synapse formed on laterally mobile substrates such as planar bilayers displays a well-organized radial lamellipodium or dSMAC, an annular lamella or pSMAC, and, in the relative position of the uropod, a relatively F-actin free cSMAC. Although each of these modules has similarities between the kinapse and synapse, the imposition of radial symmetry and the functional specialization for secretion require modifications to each module.

The radially symmetric pSMAC has two important functions: stopping motility and containing materials directly secreted into the cSMAC. Stopping motility is the first critical function. The symmetry of the pSMAC can be modulated by intrinsic and extrinsic factors. Intrinsic factors include anergy, antagonism, and PKC-θ activation. Self-tolerance is a critical function of the immune system. T cells can be made self tolerant through deletion of the clone bearing a certain TCR, in which case this TCR specificity is lost, or by making the clone of T cells nonresponsive under most conditions—a state called "anergy." Anergic T cells activate a gene expression profile that leads them to express an array of ubiquitin ligases that targets signaling machinery and destabilizes the synapse (Heissmeyer et al. 2004). Anergic T cells break symmetry of the pSMAC within 10 minutes of contact and do not reform them. The default process when TCR signaling is disrupted is for the cells to break symmetry and convert to a kinapse, as is also observed when antagonist peptide-MHC complexes kinetically poison TCR signaling (Sumen et al. 2004). Protein kinase C-θ is in the novel subgroup, meaning that it is activated by diacylglycerol, but is insensitive to Ca^{2+}. It is the only PKC isoform that is specifically recruited to the immunological synapse (Monks et al. 1997; Monks et al. 1998). PKC-θ deficient mice have a profound defect in T-cell activation because of deficits in NFκB activation via Carma-1 (Sun et al. 2000; Egawa et al. 2003; Lee et al. 2005). Time lapse imaging of T-cell activation using naïve T cells from these mice and with wild-type T cells treated with PKC-θ inhibitors revealed hyperstable immunological synapses (Sims et al. 2007). Analysis of Wiskott-Aldrich syndrome protein (WASp) deficient T cells revealed gross instability, which was converted to hyperstability by the PKC-θ inhibitor. In naïve T cells responding to antigens for the first time, the activity or PKC-θ induces symmetry breaking and kinapse formation, whereas WASp induced symmetry formation and synapse formation. This system is poised in wild-type T cells to induce cycles of synapse and kinapse formation with a periodicity of approximately 20 minutes (Sims et al. 2007). Our model is that PKC-θ undermines the integrity of the pSMAC actin network to break the annular structure open to generate a kinapse, whereas WASp contributes to actin polymerization positioned to repair the annular network and restore symmetry and the synapse. Extrinsic signals that can contribute to symmetry breaking are delivered by dominant chemokines like CCR7 and CXCR3 (Bromley et al. 2000; Molon et al. 2005), although this process has not been studied in detail. The concept that the pSMAC plays an important role in containing components that are released into the synapse has been tested directly by stabilizing pSMACs formed by $CD4^+$ cytolytic T cells with a PKC-θ inhibitor, which increased killing efficiency by threefold (Beal et al. 2008). Immune cells appear to have a unique ability among cellular systems to modulate lamellar symmetry to achieve fine control of migration and cell–cell communication.

The correspondence of the uropod and cSMAC is less clear because the cSMAC is neither rich in myosin II nor a site of de-adhesion. The cSMAC is a relatively F-actin free zone, but it is not clear what mechanisms enforce this. The paucity of F-actin is functionally important because this appears to promote TCR signal termination (TCR signaling is F-actin dependent) (Varma et al. 2006), exclusion of LFA-1 (its interaction with ICAM-1 is F-actin dependent) (Kaizuka et al. 2007), and the access of secretory vesicles to the plasma membrane for synaptically directed secretion (Stinchcombe et al. 2006). How TCR clusters are moved into this F-actin free zone is not clear, but cSMAC formation appears to involve a fusion of structures that are initially jostling about in the center of the synapse (Varma et al. 2006). An important function of the cSMAC may be to sort different components from incoming TCR microclusters into distinct compartments. To be fully activated, a T cell needs to integrate both TCR and costimulatory signals. One type of costimulatory signal is delivered when CD28 on the surface of T cells interacts with CD80 or CD86 on the surface of the antigen-presenting cells. The level of expression of CD80 and CD86 on dendritic cells reflects the level of innate activation of the dendritic cells and thus serves to alert the immune system to microbial invasion, foreign bodies, and tissue injury. The CD28–CD80 complex has the same extracellular dimensions as the TCR–peptide–MHC complex, such that these pairs can and do comingle in the same microclusters (Springer 1990). Although the variable affinity TCR–peptide–MHC interactions can become irreversible at the higher end of the affinity spectrum and need to be degraded, the fixed affinity CD28–CD80 interaction is tuned to be readily reversible and can be recycled. Recently, Yokosuka et al. have shown that stable TCR-peptide-MHC interactions are sorted from more dynamic CD28–CD80 interaction to form two compartments within the cSMAC (Yokosuka et al. 2008). The mechanism of this sorting process is currently not established, but my laboratory is currently working to test the hypothesis that endosomal sorting complexes required for transport (ESCRT) are required for this sorting and for cSMAC formation. Therefore, the cSMAC is likely the most unique module of the synapse configuration that has no counterpart in a kinapse. The combination of sorting, signaling, and secretion in this site suggest that this region combines several functional modules that have in common a need for relatively F-actin free access to the plasma membrane.

CONCLUSIONS

I have presented a reductionist view of this complex model system based largely on consideration of only a few adhesion systems used during T-cell function, which to some extent appears to provide a framework on which other molecules can be added. There are still many open issues in our understanding of the immunological synapse in the context of cell biology and immunology. Many of the findings I have described are based on surrogate antigen presenting cells and have simplified the system to better understand the T cell side of the synapse/kinapse equation, but it is clear that different types of antigen-presenting cells make profound contributions to the outcome and are far from passive (Boes et al. 2002; Al-Alwan et al. 2003; Brossard et al. 2005; Tseng et al. 2008). Improved methods for high-resolution imaging of T cells-APC conjugates will help to advance this cause (Oddos et al. 2008). Improvements in intravital microscopy and fluorescent proteins may also enable better descriptions of in vivo immune synapses and kinapses. Despite these limitations, the concept of symmetry and kinapse/synapse interconversion appears to be useful in vivo in which both behaviors are readily observed in protective responses and immunopathology (Sims et al. 2007; Skokos et al. 2007; Kim et al. 2009). A major goal of this work is to predict immune-cell behavior toward development of immunotherapies. It is likely that this depth of understanding will advance discoveries in cell biology.

REFERENCES

Al-Alwan MM, Liwski RS, Haeryfar SM, Baldridge WH, Hoskin DW, Rowden G, West KA. 2003. Cutting edge: Dendritic cell actin cytoskeletal polarization during immunological synapse formation is highly antigen-dependent. *J Immunol* **171:** 4479–4483.

Allen CD, Okada T, Tang HL, Cyster JG. 2007. Imaging of germinal center selection events during affinity maturation. *Science* **315:** 528–531.

Allenspach EJ, Cullinan P, Tong J, Tang Q, Tesciuba AG, Cannon JL, Takahashi SM, Morgan R, Burkhardt JK, Sperling AI. 2001. ERM-dependent movement of CD43 defines a novel protein complex distal to the immunological synapse. *Immunity* **15:** 739–750.

Alon R, Dustin ML. 2007. Force as a facilitator of integrin conformational changes during leukocyte arrest on blood vessels and antigen-presenting cells. *Immunity* **26:** 17–27.

Bachmann MF, McKall-Faienza K, Schmits R, Bouchard D, Beach J, Speiser DE, Mak TW, Ohashi PS. 1997. Distinct roles for LFA-1 and CD28 during activation of naive T cells: Adhesion versus costimulation. *Immunity* **7:** 549–557.

Barda-Saad M, Braiman A, Titerence R, Bunnell SC, Barr VA, Samelson LE. 2005. Dynamic molecular interactions linking the T cell antigen receptor to the actin cytoskeleton. *Nat Immunol* **6:** 80–89.

Barr VA, Balagopalan L, Barda-Saad M, Polishchuk R, Boukari H, Bunnell SC, Bernot KM, Toda Y, Nossal R, Samelson LE. 2006. T-cell antigen receptor-induced signaling complexes: Internalization via a cholesterol-dependent endocytic pathway. *Traffic* **7:** 1143–1162.

Beal AM, Anikeeva N, Varma R, Cameron TO, Norris PJ, Dustin ML, Sykulev Y. 2008. Protein Kinase C θ Regulates Stability of the Peripheral Adhesion Ring Junction and Contributes to the Sensitivity of Target Cell Lysis by CTL. *J Immunol* **181:** 4815–4824.

Boes M, Cerny J, Massol R, Op Den Brouw M, Kirchhausen T, Chen J, Ploegh HL. 2002. T-cell engagement of dendritic cells rapidly rearranges MHC class II transport. *Nature* **418:** 983–988.

Bromley SK, Peterson DA, Gunn MD, Dustin ML. 2000. Cutting edge: Hierarchy of chemokine receptor and TCR signals regulating T cell migration and proliferation. *J Immunol* **165:** 15–19.

Brossard C, Feuillet V, Schmitt A, Randriamampita C, Romao M, Raposo G, Trautmann A. 2005. Multifocal structure of the T cell - dendritic cell synapse. *Eur J Immunol* **35:** 1741–1753.

Brown MJ, Hallam JA, Colucci-Guyon E, Shaw S. 2001. Rigidity of circulating lymphocytes is primarily conferred by vimentin intermediate filaments. *J Immunol* **166:** 6640–6646.

Brown MJ, Nijhara R, Hallam JA, Gignac M, Yamada KM, Erlandsen SL, Delon J, Kruhlak M, Shaw S. 2003. Chemokine stimulation of human peripheral blood T lymphocytes induces rapid dephosphorylation of ERM proteins, which facilitates loss of microvilli and polarization. *Blood* **102:** 3890–3899.

Bunnell SC, Kapoor V, Trible RP, Zhang W, Samelson LE. 2001. Dynamic actin polymerization drives T cell receptor-induced spreading: a role for the signal transduction adaptor LAT. *Immunity* **14:** 315–329.

Bunnell SC, Hong DI, Kardon JR, Yamazaki T, McGlade CJ, Barr VA, Samelson LE. 2002. T cell receptor ligation induces the formation of dynamically regulated signaling assemblies. *J Cell Biol* **158:** 1263–1275.

Campi G, Varma R, Dustin ML. 2005. Actin and agonist MHC-peptide complex-dependent T cell receptor microclusters as scaffolds for signaling. *J Exp Med* **202:** 1031–1036.

Cannon JL, Labno CM, Bosco G, Seth A, McGavin MH, Siminovitch KA, Rosen MK, Burkhardt JK. 2001. Wasp recruitment to the T cell: APC contact site occurs independently of Cdc42 activation. *Immunity* **15:** 249–259.

Carpen O, Virtanen I, Lehto VP, Saksela E. 1983. Polarization of NK cell cytoskeleton upon conjugation with sensitive target cells. *J Immunol* **131:** 2695–2698.

Cunningham BA, Hemperly JJ, Murray BA, Prediger EA, Brackenbury R, Edelman GM. 1987. Neural cell adhesion molecule: Structure, immunoglobulin-like domains, cell surface modulation, and alternative RNA splicing. *Science* **236:** 799–806.

Davignon D, Martz E, Reynolds T, Kurzinger K, Springer TA. 1981. Lymphocyte function-associated antigen 1

(LFA-1): A surface antigen distinct from Lyt-2,3 that participates in T lymphocyte-mediated killing. *Proc Natl Acad Sci* **78:** 4535–4539.

Delon J, Kaibuchi K, Germain RN. 2001. Exclusion of CD43 from the immunological synapse is mediated by phosphorylation-regulated relocation of the cytoskeletal adaptor moesin. *Immunity* **15:** 691–701.

DeMond AL, Mossman KD, Starr T, Dustin ML, Groves JT. 2008. T cell receptor microcluster transport through molecular mazes reveals mechanism of translocation. *Biophys J* **94:** 3286–3292.

Dobereiner HG, Dubin-Thaler BJ, Hofman JM, Xenias HS, Sims TN, Giannone G, Dustin ML, Wiggins CH, Sheetz MP. 2006. Lateral membrane waves constitute a universal dynamic pattern of motile cells. *Phys Rev Lett* **97:** 038102.

Douglass AD, Vale RD. 2005. Single-Molecule Microscopy Reveals Plasma Membrane Microdomains Created by Protein-Protein Networks that Exclude or Trap Signaling Molecules in T Cells. *Cell* **121:** 937–950.

Dupre L, Aiuti A, Trifari S, Martino S, Saracco P, Bordignon C, Roncarolo MG. 2002. Wiskott-Aldrich syndrome protein regulates lipid raft dynamics during immunological synapse formation. *Immunity* **17:** 157–166.

Dustin ML, Springer TA. 1988. Lymphocyte function-associated antigen-1 (LFA-1) interaction with intercellular adhesion molecule-1 (ICAM-1) is one of at least three mechanisms for lymphocyte adhesion to cultured endothelial cells. *J Cell Biol* **107:** 321–331.

Dustin ML, Springer TA. 1989. T cell receptor cross-linking transiently stimulates adhesiveness through LFA-1. *Nature* **341:** 619–624.

Dustin ML, Colman DR. 2002. Neural and immunological synaptic relations. *Science* **298:** 785–789.

Dustin ML, Cooper JA. 2000. The immunological synapse and the actin cytoskeleton: Molecular hardware for T cell signaling. *Nat Immunol* **1:** 23–29.

Dustin ML, Rothlein R, Bhan AK, Dinarello CA, Springer TA. 1986. Induction by IL-1 and interferon, tissue distribution, biochemistry, and function of a natural adherence molecule (ICAM-1). *J Immunol* **137:** 245–254.

Dustin ML, Sanders ME, Shaw S, Springer TA. 1987a. Purified lymphocyte function-associated antigen 3 binds to CD2 and mediates T lymphocyte adhesion. *J Exp Med* **165:** 677–692.

Dustin ML, Selvaraj P, Mattaliano RJ, Springer TA. 1987b. Anchoring mechanisms for LFA-3 cell adhesion glycoprotein at membrane surface. *Nature* **329:** 846–848.

Dustin ML, Olszowy MW, Holdorf AD, Li J, Bromley S, Desai N, Widder P, Rosenberger F, van der Merwe PA, Allen PM, et al. 1998. A novel adapter protein orchestrates receptor patterning and cytoskeletal polarity in T cell contacts. *Cell* **94:** 667–677.

Egawa T, Albrecht B, Favier B, Sunshine MJ, Mirchandani K, O'Brien W, Thome M, Littman DR. 2003. Requirement for CARMA1 in antigen receptor-induced NF-κB activation and lymphocyte proliferation. *Curr Biol* **13:** 1252–1258.

Faure S, Salazar-Fontana LI, Semichon M, Tybulewicz VL, Bismuth G, Trautmann A, Germain RN, Delon J. 2004. ERM proteins regulate cytoskeleton relaxation promoting T cell-APC conjugation. *Nat Immunol* **5:** 272–279.

Freiberg BA, Kupfer H, Maslanik W, Delli J, Kappler J, Zaller DM, Kupfer A. 2002. Staging and resetting T cell activation in SMACs. *Nat Immunol* **3:** 911–917.

Gardner JM, Hynes RO. 1985. Interaction of fibronectin with its receptor on platelets. *Cell* **42:** 439–448.

Geiger B, Rosen D, Berke G. 1982. Spatial relationships of microtubule-organizing centers and the contact area of cytotoxic T lymphocytes and target cells. *J Cell Biol* **95:** 137–143.

Giannone G, Dubin-Thaler BJ, Dobereiner HG, Kieffer N, Bresnick AR, Sheetz MP. 2004. Periodic lamellipodial contractions correlate with rearward actin waves. *Cell* **116:** 431–443.

Giannone G, Dubin-Thaler BJ, Rossier O, Cai Y, Chaga O, Jiang G, Beaver W, Dobereiner HG, Freund Y, Borisy G, et al. 2007. Lamellipodial actin mechanically links Myosin activity with adhesion-site formation. *Cell* **128:** 561–575.

Gomez TS, McCarney SD, Carrizosa E, Labno CM, Comiskey EO, Nolz JC, Zhu P, Freedman BD, Clark MR, Rawlings DJ, et al. 2006. HS1 Functions as an Essential Actin-Regulatory Adaptor Protein at the Immune Synapse. *Immunity* **24:** 741–752.

Gomez TS, Kumar K, Medeiros RB, Shimizu Y, Leibson PJ, Billadeau DD. 2007. Formins regulate the actin-related protein 2/3 complex-independent polarization of the centrosome to the immunological synapse. *Immunity* **26:** 177–190.

Grakoui A, Bromley SK, Sumen C, Davis MM, Shaw AS, Allen PM, Dustin ML. 1999. The immunological synapse: A molecular machine controlling T cell activation. *Science* **285:** 221–227.

Gunzer M, Schafer A, Borgmann S, Grabbe S, Zanker KS, Brocker EB, Kampgen E, Friedl P. 2000. Antigen presentation in extracellular matrix: Interactions of T cells with dendritic cells are dynamic, short lived, and sequential. *Immunity* **13:** 323–332.

Gupton SL, Anderson KL, Kole TP, Fischer RS, Ponti A, Hitchcock-DeGregori SE, Danuser G, Fowler VM, Wirtz D, Hanein D, et al. 2005. Cell migration without a lamellipodium: Translation of actin dynamics into cell movement mediated by tropomyosin. *J Cell Biol* **168:** 619–631.

Heissmeyer V, Macian F, Im SH, Varma R, Feske S, Venuprasad K, Gu H, Liu YC, Dustin ML, Rao A. 2004. Calcineurin imposes T cell unresponsiveness through targeted proteolysis of signaling proteins. *Nat Immunol* **5:** 255–265.

Horwitz A, Duggan K, Greggs R, Decker C, Buck C. 1985. The cell substrate attachment (CSAT) antigen has properties of a receptor for laminin and fibronectin. *J Cell Biol* **101:** 2134–2144.

Huse M, Lillemeier BF, Kuhns MS, Chen DS, Davis MM. 2006. T cells use two directionally distinct pathways for cytokine secretion. *Nat Immunol* **7:** 247–255.

Ilani T, Khanna C, Zhou M, Veenstra TD, Bretscher A. 2007. Immune synapse formation requires ZAP-70 recruitment by ezrin and CD43 removal by moesin. *J Cell Biol* **179:** 733–746.

Jacobelli J, Chmura SA, Buxton DB, Davis MM, Krummel MF. 2004. A single class II myosin modulates T cell motility and stopping, but not synapse formation. *Nat Immunol* **5:** 531–538.

Jay PY, Pham PA, Wong SA, Elson EL. 1995. A mechanical function of myosin II in cell motility. *J Cell Sci* **108:** 387–393.

Jiang G, Giannone G, Critchley DR, Fukumoto E, Sheetz MP. 2003. Two-piconewton slip bond between fibronectin and the cytoskeleton depends on talin. *Nature* **424:** 334–337.

Johnson KG, Bromley SK, Dustin ML, Thomas ML. 2000. A supramolecular basis for CD45 tyrosine phosphatase regulation in sustained T cell activation. *Proc Natl Acad Sci* **97:** 10138–10143.

Kaizuka Y, Douglass AD, Varma R, Dustin ML, Vale RD. 2007. Mechanisms for segregating T cell receptor and adhesion molecules during immunological synapse formation in Jurkat T cells. *Proc Natl Acad Sci* **104:** 20296–20301.

Kato K, Koyanagi M, Okada H, Takanashi T, Wong YW, Williams AF, Okumura K, Yagita H. 1992. CD48 is a counter-receptor for mouse CD2 and is inolved in T cell activation. *Journal of Experimental Medicine* **176:** 1241–1249.

Kim SV, Mehal WZ, Dong X, Heinrich V, Pypaert M, Mellman I, Dembo M, Mooseker MS, Wu D, Flavell RA. 2006. Modulation of cell adhesion and motility in the immune system by Myo1f. *Science* **314:** 136–139.

Kim JV, Kang SS, Dustin ML, McGavern DB. 2009. Myelomonocytic cell recruitment causes fatal CNS vascular injury during acute viral meningitis. *Nature* **457:** 191–195.

Krawczyk C, Oliveira-dos-Santos A, Sasaki T, Griffiths E, Ohashi PS, Snapper S, Alt F, Penninger JM. 2002. Vav1 controls integrin clustering and MHC/peptide-specific cell adhesion to antigen-presenting cells. *Immunity* **16:** 331–343.

Krensky AM, Robbins E, Springer TA, Burakoff SJ. 1984. LFA-1, LFA-2, and LFA-3 antigens are involved in CTL-target conjugation. *J Immunol* **132:** 2180–2182.

Kupfer A, Dennert G, Singer SJ. 1983. Polarization of the Golgi apparatus and the microtubule-organizing center within cloned natural killer cells bound to their targets. *Proc Natl Acad Sci* **80:** 7224–7228.

Kupfer A, Swain SL, Janeway CA Jr, Singer SJ. 1986. The specific direct interaction of helper T cells and antigen-presenting B cells. *Proc Natl Acad Sci* **83:** 6080–6083.

Kupfer A, Singer SJ. 1989. The specific interaction of helper T cells and antigen-presenting B cells. IV. Membrane and cytoskeletal reorganizations in the bound T cell as a function of antigen dose. *J ExpMed* **170:** 1697–1713.

Labno CM, Lewis CM, You D, Leung DW, Takesono A, Kamberos N, Seth A, Finkelstein LD, Rosen MK, Schwartzberg PL, et al. 2003. Itk functions to control actin polymerization at the immune synapse through localized activation of Cdc42 and WASP. *Curr Biol* **13:** 1619–1624.

Latour S, Roncagalli R, Chen R, Bakinowski M, Shi X, Schwartzberg PL, Davidson D, Veillette A. 2003. Binding of SAP SH2 domain to FynT SH3 domain reveals a novel mechanism of receptor signalling in immune regulation. *Nat Cell Biol* **5:** 149–154.

Lawrence MB, Springer TA. 1991. Leukocytes roll on a selectin at physiologic flow rates: Distinction from and prerequisite for adhesion through integrins. *Cell* **65:** 859–873.

Lee KY, D'Acquisto F, Hayden MS, Shim JH, Ghosh S. 2005. PDK1 nucleates T cell receptor-induced signaling complex for NF-κB activation. *Science* **308:** 114–118.

Lindenmann J. 1984. Origin of the terms 'antibody' and 'antigen'. *Scand J Immunol* **19:** 281–285.

Marlin SD, Springer TA. 1987. Purified intercellular adhesion molecule-1 (ICAM-1) is a ligand for lymphocyte function-associated antigen 1 (LFA-1). *Cell* **51:** 813–819.

McIntyre BW, Allison JP. 1983. The mouse T cell receptor: Structural heterogeneity of molecules of normal T cells defined by xenoantiserum. *Cell* **34:** 739–746.

Mempel TR, Henrickson SE, Von Andrian UH. 2004. T-cell priming by dendritic cells in lymph nodes occurs in three distinct phases. *Nature* **427:** 154–159.

Miller MJ, Wei SH, Parker I, Cahalan MD. 2002. Two-photon imaging of lymphocyte motility and antigen response in intact lymph node. *Science* **296:** 1869–1873.

Molon B, Gri G, Bettella M, Gomez-Mouton C, Lanzavecchia A, Martinez AC, Manes S, Viola A. 2005. T cell costimulation by chemokine receptors. *Nat Immunol* **6:** 465–471.

Monks CR, Kupfer H, Tamir I, Barlow A, Kupfer A. 1997. Selective modulation of protein kinase C-θ during T-cell activation. *Nature* **385:** 83–86.

Monks CR, Freiberg BA, Kupfer H, Sciaky N, Kupfer A. 1998. Three-dimensional segregation of supramolecular activation clusters in T cells. *Nature* **395:** 82–86.

Morin NA, Oakes PW, Hyun YM, Lee D, Chin EY, King MR, Springer TA, Shimaoka M, Tang JX, Reichner JS, Kim M. 2008. Nonmuscle myosin heavy chain IIA mediates integrin LFA-1 de-adhesion during T lymphocyte migration. *J Exp Med* **205:** 195–205.

Nagafuchi A, Shirayoshi Y, Okazaki K, Yasuda K, Takeichi M. 1987. Transformation of cell adhesion properties by exogenously introduced E-cadherin cDNA. *Nature* **329:** 341–343.

Negulescu PA, Krasieva TB, Khan A, Kerschbaum HH, Cahalan MD. 1996. Polarity of T cell shape, motility, and sensitivity to antigen. *Immunity* **4:** 421–430.

Nijhara R, van Hennik PB, Gignac ML, Kruhlak MJ, Hordijk PL, Delon J, Shaw S. 2004. Rac1 mediates collapse of microvilli on chemokine-activated T lymphocytes. *J Immunol* **173:** 4985–4993.

Nolz JC, Gomez TS, Zhu P, Li S, Medeiros RB, Shimizu Y, Burkhardt JK, Freedman BD, Billadeau DD. 2006. The WAVE2 complex regulates actin cytoskeletal reorganization and CRAC-mediated calcium entry during T cell activation. *Curr Biol* **16:** 24–34.

Nolz JC, Medeiros RB, Mitchell JS, Zhu P, Freedman BD, Shimizu Y, Billadeau DD. 2007. WAVE2 regulates high-affinity integrin binding by recruiting vinculin and talin to the immunological synapse. *Mol Cell Biol* **27:** 5986–6000.

Nolz JC, Nacusi LP, Segovis CM, Medeiros RB, Mitchell JS, Shimizu Y, Billadeau DD. 2008. The WAVE2 complex regulates T cell receptor signaling to integrins via

Abl- and CrkL-C3G-mediated activation of Rap1. *J Cell Biol* **182:** 1231–1244.

Norcross MA. 1984. A synaptic basis for T-lymphocyte activation. *Ann Immunol* **135D:** 113–134.

Oddos S, Dunsby C, Purbhoo MA, Chauveau A, Owen DM, Neil MA, Davis DM, French PM. 2008. High-speed high-resolution imaging of intercellular immune synapses using optical tweezers. *Biophys J* **95:** L66–68.

Okada T, Miller MJ, Parker I, Krummel MF, Neighbors M, Hartley SB, O'Garra A, Cahalan MD, Cyster JG. 2005. Antigen-engaged B cells undergo chemotaxis toward the T zone and form motile conjugates with helper T cells. *PLoS Biol* **3:** e150.

Paul WE, Seder RA. 1994. Lymphocyte responses and cytokines. *Cell* **76:** 241–251.

Peyrieras N, Hyafil F, Louvard D, Ploegh HL, Jacob F. 1983. Uvomorulin: A nonintegral membrane protein of early mouse embryo. *Proc Natl Acad Sci* **80:** 6274–6277.

Pierce RA, Field ED, Mutis T, Golovina TN, Von Kap-Herr C, Wilke M, Pool J, Shabanowitz J, Pettenati MJ, Eisenlohr LC, et al. 2001. The HA-2 minor histocompatibility antigen is derived from a diallelic gene encoding a novel human class I myosin protein. *J Immunol* **167:** 3223–3230.

Pytela R, Pierschbacher MD, Ruoslahti E. 1985. A 125/115-kDa cell surface receptor specific for vitronectin interacts with the arginine-glycine-aspartic acid adhesion sequence derived from fibronectin. *Proc Natl Acad Sci* **82:** 5766–5770.

Qi H, Cannons JL, Klauschen F, Schwartzberg PL, Germain RN. 2008. SAP-controlled T-B cell interactions underlie germinal centre formation. *Nature* **455:** 764–769.

Rothlein R, Springer TA. 1986. The requirement for lymphocyte function-associated antigen 1 in homotypic leukocyte adhesion stimulated by phorbol ester. *J Exp Med* **163:** 1132–1149.

Rothlein R, Dustin ML, Marlin SD, Springer TA. 1986. A human intercellular adhesion molecule (ICAM-1) distinct from LFA-1. *J Immunol* **137:** 1270–1274.

Rutishauser U, Hoffman S, Edelman GM. 1982. Binding properties of a cell adhesion molecule from neural tissue. *Proc Natl Acad Sci* **79:** 685–689.

Sanchez-Madrid F, del Pozo MA. 1999. Leukocyte polarization in cell migration and immune interactions. *Embo J* **18:** 501–511.

Sanchez-Madrid F, Krensky AM, Ware CF, Robbins E, Strominger JL, Burakoff SJ, Springer TA. 1982. Three distinct antigens associated with human T-lymphocyte-mediated cytolysis: LFA-1, LFA-2, and LFA-3. *Proc Natl Acad Sci* **79:** 7489–7493.

Sasahara Y, Rachid R, Byrne MJ, de la Fuente MA, Abraham RT, Ramesh N, Geha RS. 2002. Mechanism of recruitment of WASP to the immunological synapse and of its activation following TCR ligation. *Mol Cell* **10:** 1269–1281.

Savage ND, Kimzey SL, Bromley SK, Johnson KG, Dustin ML, Green JM. 2002. Polar redistribution of the sialoglycoprotein CD43: Implications for T cell function. *J Immunol* **168:** 3740–3746.

Schmits R, Kundig TM, Baker DM, Shumaker G, Simard JJ, Duncan G, Wakeham A, Shahinian A, van der Heiden A, Bachmann MF, et al. 1996. LFA-1-deficient mice show normal CTL responses to virus but fail to reject immunogenic tumor. *J Exp Med* **183:** 1415–1426.

Schwickert TA, Lindquist RL, Shakhar G, Livshits G, Skokos D, Kosco-Vilbois MH, Dustin ML, Nussenzweig MC. 2007. In vivo imaging of germinal centres reveals a dynamic open structure. *Nature* **446:** 83–87.

Selvaraj P, Plunkett ML, Dustin M, Sanders ME, Shaw S, Springer TA. 1987. The T lymphocyte glycoprotein CD2 (LFA-2/T11/E-Rosette receptor) binds the cell surface ligand LFA-3. *Nature* **326:** 400–403.

Shaffer MH, Dupree RS, Zhu P, Saotome I, Schmidt RF, McClatchey AI, Freedman BD, Burkhardt JK. 2009. Ezrin and moesin function together to promote T cell activation. *J Immunol* **182:** 1021–1032.

Shamri R, Grabovsky V, Gauguet JM, Feigelson S, Manevich E, Kolanus W, Robinson MK, Staunton DE, von Andrian UH, Alon R. 2005. Lymphocyte arrest requires instantaneous induction of an extended LFA-1 conformation mediated by endothelium-bound chemokines. *Nat Immunol* **6:** 497–506.

Shaw S, Luce GE, Quinones R, Gress RE, Springer TA, Sanders ME. 1986. Two antigen-independent adhesion pathways used by human cytotoxic T-cell clones. *Nature* **323:** 262–264.

Shimaoka M, Xiao T, Liu JH, Yang Y, Dong Y, Jun CD, McCormack A, Zhang R, Joachimiak A, Takagi J, et al. 2003. Structures of the alpha L I domain and its complex with ICAM-1 reveal a shape-shifting pathway for integrin regulation. *Cell* **112:** 99–111.

Sims TN, Soos TJ, Xenias HS, Dubin-Thaler B, Hofman JM, Waite JC, Cameron TO, Thomas VK, Varma R, Wiggins CH, et al. 2007. Opposing Effects of PKC-θ and WASp on Symmetry Breaking and Relocation of the Immunological Synapse. *Cell* **129:** 773–785.

Skokos D, Shakhar G, Varma R, Waite JC, Cameron TO, Lindquist RL, Schwickert T, Nussenzweig MC, Dustin ML. 2007. Peptide-MHC potency governs dynamic interactions between T cells and dendritic cells in lymph nodes. *Nat Immunol* **8:** 835–844.

Smith A, Carrasco YR, Stanley P, Kieffer N, Batista FD, Hogg N. 2005. A talin-dependent LFA-1 focal zone is formed by rapidly migrating T lymphocytes. *J Cell Biol* **170:** 141–151.

Snapper SB, Rosen FS, Mizoguchi E, Cohen P, Khan W, Liu CH, Hagemann TL, Kwan SP, Ferrini R, Davidson L, et al. 1998. Wiskott-Aldrich syndrome protein-deficient mice reveal a role for WASP in T but not B cell activation. *Immunity* **9:** 81–91.

Somersalo K, Anikeeva N, Sims TN, Thomas VK, Strong RK, Spies T, Lebedeva T, Sykulev Y, Dustin ML. 2004. Cytotoxic T lymphocytes form an antigen-independent ring junction. *J Clin Invest* **113:** 49–57.

Springer TA. 1990. Adhesion receptors of the immune system. *Nature* **346:** 425–434.

Springer TA. 1995. Traffic signals on endothelium for lymphocyte recirculation and leukocyte emigration. *Annu Rev Physiol* **57:** 827–872.

Staunton DE, Dustin ML, Springer TA. 1989. Functional cloning of ICAM-2, a cell adhesion ligand for LFA-1 homologous to ICAM-1. *Nature* **339:** 61–64.

Stinchcombe JC, Majorovits E, Bossi G, Fuller S, Griffiths GM. 2006. Centrosome polarization delivers secretory granules to the immunological synapse. *Nature* **443:** 462–465.

Sumen C, Dustin ML, Davis MM. 2004. T cell receptor antagonism interferes with MHC clustering and integrin patterning during immunological synapse formation. *J Cell Biol* **166:** 579–590.

Sun Z, Arendt CW, Ellmeier W, Schaeffer EM, Sunshine MJ, Gandhi L, Annes J, Petrzilka D, Kupfer A, Schwartzberg PL, et al. 2000. PKC-θ is required for TCR-induced NF-κB activation in mature but not immature T lymphocytes. *Nature* **404:** 402–407.

Tadokoro S, Shattil SJ, Eto K, Tai V, Liddington RC, De Pereda JM, Ginsberg MH, Calderwood DA. 2003. Talin binding to integrin β tails: A final common step in integrin activation. *Science* **302:** 103–106.

Trombetta ES, Mellman I. 2005. Cell biology of antigen processing in vitro and in vivo. *Annu Rev Immunol* **23:** 975–1028.

Tseng SY, Waite JC, Liu M, Vardhana S, Dustin ML. 2008. T cell-dendritic cell immunological synapses contain TCR-dependent CD28-CD80 clusters that recruit protein kinase C-θ. *J Immunol* **181:** 4852–4863.

Tskvitaria-Fuller I, Rozelle AL, Yin HL, Wulfing C. 2003. Regulation of sustained actin dynamics by the TCR and costimulation as a mechanism of receptor localization. *J Immunol* **171:** 2287–2295.

Underhill DM, Bassetti M, Rudensky A, Aderem A. 1999. Dynamic interactions of macrophages with T cells during antigen presentation. *J Exp Med* **190:** 1909–1914.

Varma R, Campi G, Yokosuka T, Saito T, Dustin ML. 2006. T cell receptor-proximal signals are sustained in peripheral microclusters and terminated in the central supramolecular activation cluster. *Immunity* **25:** 117–127.

Wei X, Tromberg BJ, Cahalan MD. 1999. Mapping the sensitivity of T cells with an optical trap: Polarity and minimal number of receptors for Ca(2+) signaling. *Proc Natl Acad Sci* **96:** 8471–8476.

White J, Haskins KM, Marrack P, Kappler J. 1983. Use of I region-restricted, antigen-specific T cell hybridomas to produce idiotypically specific anti-receptor antibodies. *J Immunol* **130:** 1033–1037.

Wong YW, Williams AF, Kingsmore SF, Seldin MF. 1990. Structure, expression, and genetic linkage of the mouse BCM1 (OX45 or Blast-1) antigen. Evidence for genetic duplication giving rise to the BCM1 region on mouse chromosome 1 and the CD2/LFA3 region on mouse chromosome 3. *J Exp Med* **171:** 2115–2130.

Woolf E, Grigorova I, Sagiv A, Grabovsky V, Feigelson SW, Shulman Z, Hartmann T, Sixt M, Cyster JG, Alon R. 2007. Lymph node chemokines promote sustained T lymphocyte motility without triggering stable integrin adhesiveness in the absence of shear forces. *Nat Immunol* **8:** 1076–1085.

Wright SD, Meyer BC. 1985. Fibronectin receptor of human macrophages recognizes the sequence Arg-Gly-Asp-Ser. *J Exp Med* **162:** 762–767.

Yokosuka T, Sakata-Sogawa K, Kobayashi W, Hiroshima M, Hashimoto-Tane A, Tokunaga M, Dustin ML, Saito T. 2005. Newly generated T cell receptor microclusters initiate and sustain T cell activation by recruitment of Zap70 and SLP-76. *Nat Immunol* **6:** 1253–1262.

Yokosuka T, Kobayashi W, Sakata-Sogawa K, Takamatsu M, Hashimoto-Tane A, Dustin ML, Tokunaga M, Saito T. 2008. Spatiotemporal regulation of T cell costimulation by TCR-CD28 microclusters and protein kinase C-θ translocation. *Immunity* **29:** 589–601.

Zhu J, Luo BH, Xiao T, Zhang C, Nishida N, Springer TA. 2008. Structure of a complete integrin ectodomain in a physiologic resting state and activation and deactivation by applied forces. *Mol Cell* **32:** 849–861.

Membrane Domains Based on Ankyrin and Spectrin Associated with Cell–Cell Interactions

Vann Bennett and Jane Healy

Howard Hughes Medical Institute, and Departments of Cell Biology and Biochemistry, Duke University Medical Center, Durham, North Carolina 27710

Correspondence: v.bennett@cellbio.duke.edu

Nodes of Ranvier and axon initial segments of myelinated nerves, sites of cell–cell contact in early embryos and epithelial cells, and neuromuscular junctions of skeletal muscle all perform physiological functions that depend on clustering of functionally related but structurally diverse ion transporters and cell adhesion molecules within microdomains of the plasma membrane. These specialized cell surface domains appeared at different times in metazoan evolution, involve a variety of cell types, and are populated by distinct membrane-spanning proteins. Nevertheless, recent work has shown that these domains all share on their cytoplasmic surfaces a membrane skeleton comprised of members of the ankyrin and spectrin families. This review will summarize basic features of ankyrins and spectrins, and will discuss emerging evidence that these proteins are key players in a conserved mechanism responsible for assembly and maintenance of physiologically important domains on the surfaces of diverse cells.

Spectrins are flexible rods 0.2 microns in length with actin-binding sites at each end (Shotton et al. 1979; Bennett et al. 1982) (Fig. 1A). Spectrins are assembled from α and β subunits, each comprised primarily of multiple copies of a 106-amino acid repeat (Speicher and Marchesi 1984). In addition to the canonical 106-residue repeat, β spectrins also have a carboxy-terminal pleckstrin homology domain (Zhang et al. 1995; Macias et al. 1994) and tandem amino-terminal calponin homology domains (Bañuelos et al. 1998), whereas α spectrins contain an Src homology domain 3 (SH3) site (Musacchio et al. 1992), a calmodulin-binding site (Simonovic et al. 2006), and EF hands (Travé et al. 1995) (Fig. 1A). Spectrin α and β subunits are assembled antiparallel and side-to-side into heterodimers, which in turn are associated head-to-head to form tetramers (Clarke 1971; Shotton et al. 1979; Davis and Bennett 1983) (Fig. 1A). In human erythrocytes, in which spectrin was first characterized (Marchesi and Steers 1968; Clarke 1971), actin oligomers containing 10–14 monomers are each linked to five to six spectrin tetramers by accessory proteins to form a geodesic domelike structure that has been resolved by electron microscopy (Byers and Branton 1985). The principal proteins at the spectrin–actin junction are protein 4.1,

Cite this article as *Cold Spring Harb Perspect Biol* 2009;1:a003012

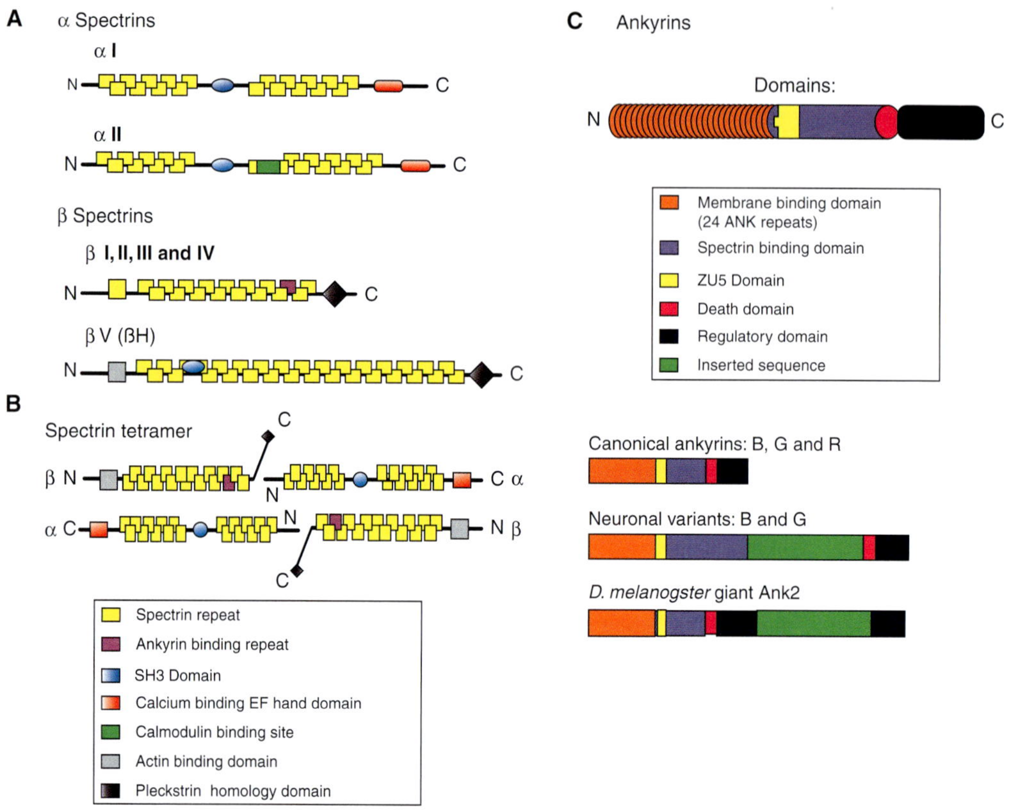

Figure 1. Domain structure and variants of spectrin and ankyrin proteins. (*A*) Molecular domains of spectrins: Two α spectrins and five β spectrins are shown. Spectrins are comprised of modular units called spectrin repeats (yellow). Other domains such as the ankyrin binding domain (purple), Src-homology domain 3 (SH3, blue), EF-hand domain (red), and calmodulin-binding domain (green) promote interactions with binding targets important for spectrin function. The pleckstrin homology domain (black) promotes association with the plasma membrane and the actin binding domain (grey) tethers the spectrin-based membrane skeleton to the underlying actin cytoskeleton. (*B*) The spectrin tetramer, the fundamental unit of the spectrin-based membrane skeleton. The spectrin repeat domains of α and β spectrin associate end-to-end to form heterodimers. Heterodimers associate laterally in an antiparallel fashion to form tetramers. The tetramers can then associate end-to-end to form extended macromolecules that link into a geodesic dome shape directly underneath the plasma membrane. (*C*) Molecular domains present in canonical ankyrins. The membrane binding domain of ankyrin isoforms (orange) is comprised of 24 ANK repeats. The spectrin binding domain (green-blue) allows ankyrins to coordinate integral membrane proteins to the membrane skeleton. The death domain (pink) is the most highly conserved domain. The regulatory domain (brown) is the most variable region of ankyrins. The regulatory domain interacts intramolecularly with the membrane binding domain to modulate ankyrin's affinity for other binding partners. All ankyrins and spectrins are subject to alternative splicing, which further increases their functional diversity.

adducin, tropomyosin, tropomodulin, and dematin (Bennett and Baines 2001) (Table 1).

Spectrin is coupled to the inner surface of the erythrocyte membrane primarily through association with ankyrin, which is in turn linked to the cytoplasmic domains of the anion exchanger (Bennett 1978; Bennett and Stenbuck 1979a,b) and Rh/RhAG ammonium transporter (Nicolas et al. 2003). The spectrin-based membrane skeleton and its connections through ankyrin to membrane-spanning proteins are essential for survival of erythrocytes in the circulation, and mutations in these proteins result in hereditary hemolytic anemia

Table 1. Binding partners of spectrin and ankyrins

Spectrin Binding Partners			
Alpha	**Beta**		
Transporters/ion channels	*Membrane anchors*		
EnNaC (sodium)	PI lipids		
NHE2 (ammonium)	Band 4.1		
	Ankyrin		
	EAAT4 (glutamate)		
Membrane receptors	*Signaling*		
NMDA receptor	RACK-1		
Signaling	*Cytoskeleton/cellular transport*		
HsSH3pb1	F-actin		
Calmodulin	Adducin		
	Dynactin		
Ankyrin Binding Partners			
Membrane BD	**Spectrin BD**	**DD**	**REG D**
Ion channels:	Spectrin	FasL	Hsp40
Anion exchanger			Obscurin
Na+/K+ATPase			PP2A
Voltage-gated			
Na+ channels			
Na+/Ca^{2+} Exchanger			
KCNG2/3			
Rh antigen			
IP3 receptor			
Ryanodine receptor			
Cell adhesion molecules:			
L1-CAMs			
CD44			
E-cadherin			
Dystroglycan			
Cellular transport:			
Tubulin			
Clathrin			

(Bennett and Healy 2008). The ankyrin-binding sites of β spectrins 1–4 are located in the 15th spectrin repeat, which is folded identically to other repeats but has distinct surface-exposed residues (Davis et al. 2008; Ipsaro et al. 2009; Stabach et al. 2009) (Figs. 1A, 2A). Mammalian β-5 spectrin and its ortholog β-H spectrin in *Drosophila* and *Caenorhabditis elegans* are the only β spectrins lacking ankyrin-binding activity (Dubreuil et al. 1990; Thomas et al. 1998; McKeown et al. 1998; Stabach and Morrow 2000).

Ankyrin interacts with β spectrins through a ZU5 domain (Mohler et al. 2004a; Kizhatil et al. 2007a; Ipsaro et al. 2009) (Fig. 1B), and with most of its membrane partners through ANK repeats (Bennett and Baines 2001) (Fig. 2C,D). In addition, ankyrins have a highly conserve "death domain" and a carboxy-terminal regulatory domain (see the following discussion). The 24 ANK repeats are stacked in a superhelical array to form a solenoid (Michaely et al. 2002). Interestingly, the ANK repeat stack behaves like a reversible spring when stretched by atomic force microscopy, and may function in mechano-coupling in tissues such as the heart (Lee et al. 2006). ANK repeats are components of many proteins and participate in highly diverse protein interactions (Mosavi et al. 2004) (Fig. 2C).

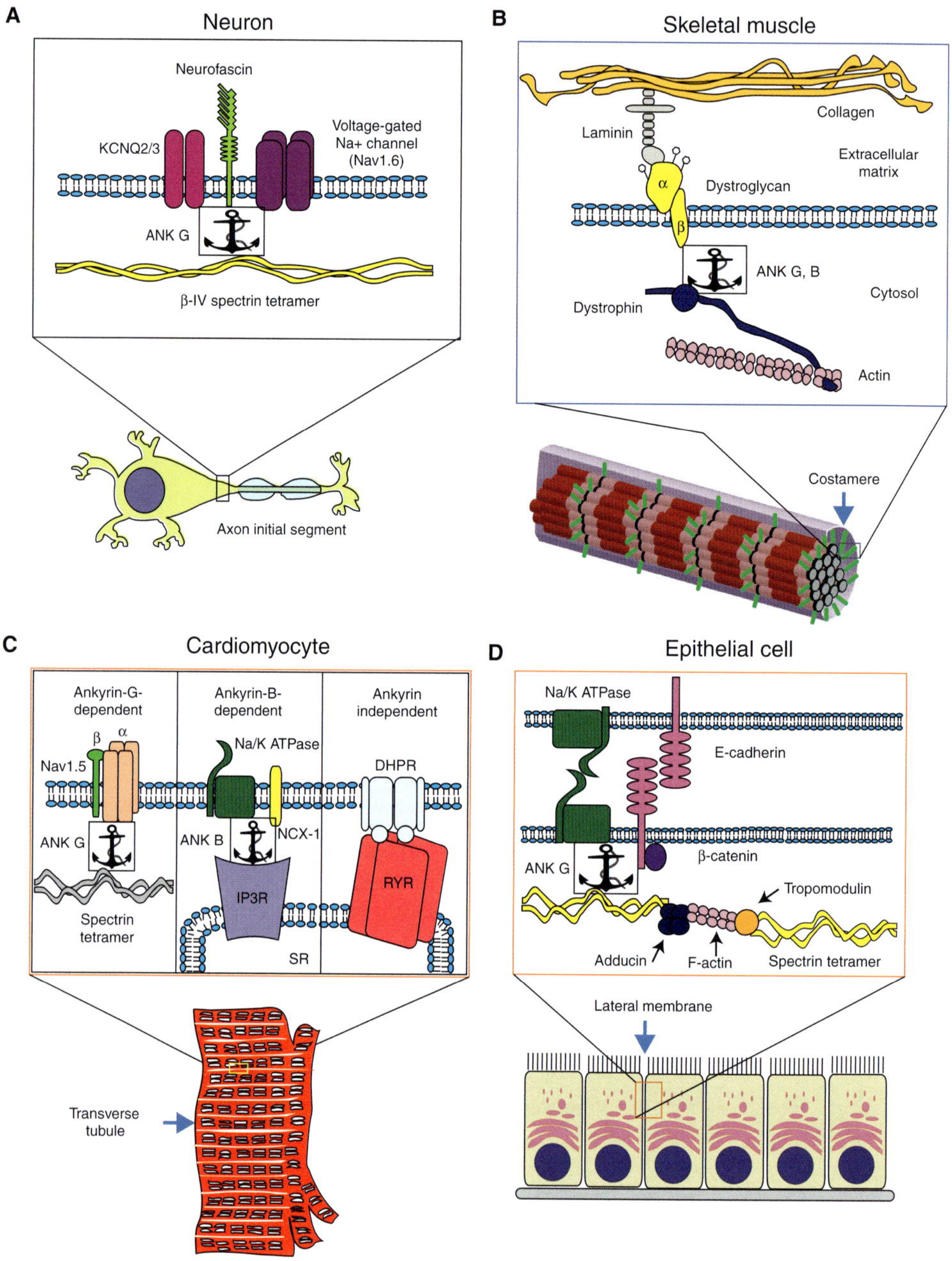

Figure 2. Ankyrins and spectrins organize macromolecular complexes in diverse types of specialized membranes. (*A*) Ankyrin-G forms a complex with β-IV spectrin, neurofascin (a cell adhesion protein), and ion channels (KCNQ2/3 and voltage-gated sodium channel) at axon initial segments in Purkinje neurons. (*B*) In force buffering costameres of skeletal muscle, ankyrins -B and -G cooperate to target and stabilize key components of the dystroglycoprotein complex. At the membrane, ankyrin-G binds to dystrophin and β-dystroglycan. (*C*) In cardiomyocyte transverse tubules, ankyrins -B and -G coordinate separate microdomains. Ankyrin-B binds $Na+/K+$ ATPase, $Na+/Ca^{2+}$ exchanger (NCX-1), and the inositol triphosphate receptor (IP3R). Ankyrin-G forms a complex with Nav1.5 and spectrin. (*D*) Ankyrin-G in epithelial lateral membrane assembly. Ankyrin-G binds to E-cadherin, β-2 spectrin, and the $Na+/K+$ ATPase. Spectrins are connected via F-actin bridges bound to α/γ adducin and tropomodulin.

This versatile motif currently is being exploited using designed ANK repeat proteins (DARPins) engineered to interact with specific ligands that can function as substitutes for antibodies (Stumpp and Amstutz 2007; Steiner et al. 2008).

Spectrin and ankyrin family members are expressed in most, if not all, animal (metazoan) cells, but are not present in bacteria, plants, or fungi. Spectrins are believed to have evolved from an ancestral α-actinin containing calponin homology domains and two spectrin repeats but not other domains (Thomas et al. 1997; Pascual et al. 1997). Ankyrin repeats are expressed in all phyla, presumably because of a combination of evolutionary relationships and in cases of bacteria and viruses by horizontal gene transfer. However, the spectrin-binding domain of ankyrin is present only in metazoans (Fig 1B). It is possible that evolution of ankyrins and spectrins could have been one of the adaptations required for organization of cells into tissues in multicellular animals.

The human spectrin family includes two α subunits and five β subunits, whereas *Drosophila* and *C. elegans* have a single α subunit and two β subunits (Bennett and Baines 2001). Vertebrate ankyrins are encoded by three genes: ankyrin-R (ANK1) (the isoform first characterized in erythrocytes and also present in a restricted distribution in brain and muscle), ankyrin-B (ANK2), and ankyrin-G (ANK3). Vertebrate ankyrins evolved from a single gene in early chordates (Cai and Zhang 2006). *C. elegans* ankyrin is encoded by a single gene termed *unc-44* (Otsuka et al. 1995), whereas the *Drosophila* genome contains two ankyrin genes: *ankyrin* (Dubreuil and Yu 1994) and *ankyrin2* (Bouley et al. 2000).

Mammalian ankyrins -B and -G are coexpressed in most cells, although they have distinct functions (Mohler et al. 2002; Abdi et al. 2006). Ankyrins -B and -G are closely related in their ANK repeats, and spectrin-binding domains, but diverge in their carboxy-terminal regulatory domains. Regulatory domains are natively unstructured and extended (Abdi et al. 2006). These flexible domains engage in intramolecular interactions with the membrane-binding and spectrin-binding domains (Hall and Bennett 1987; Davis et al. 1992; Abdi et al. 2006) that modulate protein associations and provide functional diversity between otherwise conserved ankyrins.

In addition to the standard versions of ankyrins and spectrin subunits depicted in Figure 1, many variants of these proteins are expressed with the addition and/or deletion of functional domains because of alternative splicing of pre-mRNAs. For example, β spectrins can lack PH domains (Hayes et al. 2000), and giant ankyrins have insertions of up to 2000 residues (Kordeli et al. 1995; Chan et al. 1993; Pielage et al. 2008; Koch et al. 2008), whereas other ankyrins lack either the entire membrane-binding domain (Hoock et al. 1997), or both membrane- and spectrin-binding domains (Zhou et al. 1997). The insertions in 440 kDa ankyrin-B and 480 kDa ankyrin-G (Fig. 1B) have an extended conformation that potentially could have specialized roles in connections between the plasma membrane and cytoskeleton of axons where these giant ankyrins reside (Chan et al. 1993; Kordeli et al. 1995) (Fig. 1B). Interestingly, the inserted sequences in *Drosophila* giant ankyrins interact with microtubules at the presynaptic neuromuscular junction (Pielage et al. 2008) (see the following section).

MEMBRANE-SPANNING PROTEIN PARTNERS

Diverse families of membrane-spanning proteins, including ion channels, pumps, and exchangers as well as cell adhesion molecules have independently acquired ankyrin-binding activity multiple times in metazoan evolution (Table 1). These proteins include the anion exchanger (AE1) (Bennett and Stenbuck 1979b; Stefanovic et al. 2007), voltage-gated sodium channels (Srinivasan et al. 1988; Garrido et al. 2003; Lemaillet et al. 2003), Na/K ATPase (Nelson and Veshnock 1987), Na/Ca exchanger (Li et al. 1993; Mohler et al. 2005; Cunha et al. 2007), IP3 receptors (Mohler et al. 2003; Mohler et al. 2005; Kline et al. 2008), Rh ammonium transporter (Nicolas et al. 2003; Lopez et al. 2005; Sohet et al. 2008), KCNQ2/3 channels (Pan et al. 2006; Chung et al. 2006; Rasmussen et al. 2007), Kv3.1 channels (Devaux et al. 2003; Xu et al. 2007), and cell adhesion molecules including L1 CAMs (Davis and Bennett 1994; Zhang et al. 1998), CD44 (Bourguignon et al. 1992), E- and N-cadherin (Kizhatil et al. 2007a), and β-dystroglycan (Ayalon et al. 2008) (Table 1). The ankyrin-binding sites have been defined in many of these proteins as relatively short stretches of 10–20 amino acids that do not contain a single defining motif (Fig. 3A). However, once ankyrin-binding activity appears in a protein family, the binding sites remain highly conserved. For example, the ankyrin-binding sites of human L1CAMs are nearly identical to the binding site of the L1 homolog of *C. elegans* (Chen et al. 2001), whereas sites of voltage-gated sodium channels and KCNQ2/3 channels are absent in *Drosophila* and but are conserved between humans and zebra fish (Pan et al. 2006).

Some membrane proteins such as NMDA receptors (Wechsler and Teichberg 1998), the neuronal glutamate transporter EAAT4 (Jackson et al. 2001), and the epithelial sodium channel (EnaC) (Rotin et al. 1994; Zuckerman et al. 1999) can associate directly with spectrin independently of ankyrin. Spectrin has also been reported to coimmunoprecipitate with the presynaptic voltage-sensitive calcium channel (Sunderland et al. 2000), although a direct interaction between these proteins has not been shown. Together, ankyrins and spectrins display a remarkable capacity for interactions with physiologically important membrane proteins.

AXON INITIAL SEGMENTS AND NODES OF RANVIER

Axon initial segments (Fig. 2A) and nodes of Ranvier are highly enriched in 480/270 kDa alternatively spliced variants of ankyrin-G (Kordeli et al. 1995), β-4 spectrin (Berghs et al. 2000; Lacas-Gervais et al. 2004), as well as ankyrin-binding proteins including voltage-gated Na channels and 186 kDa neurofascin (Davis et al. 1996). Axon initial segments are of special importance as the integrator sites of neurons in which inputs from dendritic synapses (sometimes on the order of several hundred thousand) are transduced into action potentials. These action potentials result in both signaling to other neurons or target cells when propagated down the axon, and to modulation of synaptic function when back-propagated into dendritic shafts (Waters et al. 2005). In myelinated axons, action potentials are propagated at periodic interruptions in the myelin sheathe known as nodes of Ranvier.

A

	Ankyrin Binding Site
NaV	VPIAGESDFE
KCNQ2	PYIAEGESDTDSD
E-cadherin	KEPLLPPEDDTRDNVYY-YDEEGGGEED
Neurofascin/ L1CAMs	QFNEDGSFIGQY
RhBG	KLPFLDSPP
Dystroglycan	GVPIIFAD ELDDSK
AE1	PAVLTRSGDPS

B

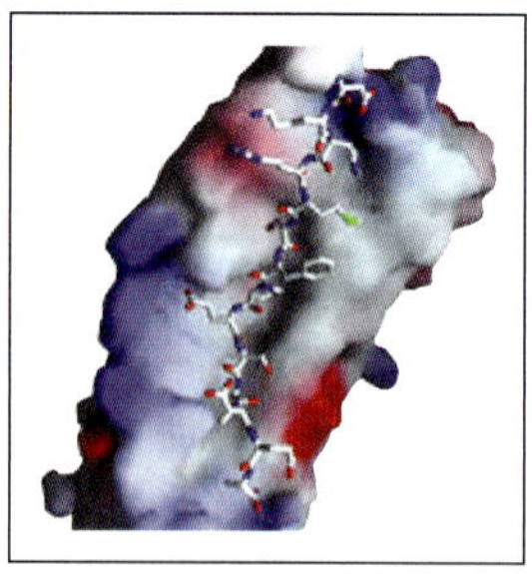

Figure 3. Ankyrins bind to natively unstructured regions of many proteins. (*A*) Known binding sites of ankyrin proteins. All of these regions lie in regions predicted to be intrinsically unstructured (http://iupred.enzim.hu/IUPs.html). Abbreviations: NaV, voltage-gated sodium channels; KCNQ2, voltage-gated potassium channels; RhBG, rhesus blood group antigen; AE1, anion exchanger. (*B*) A theoretical model of how the ankyrin membrane binding domain could bind to unstructured peptides. This pocket is 240 angstroms in length with a variety of surface exposed residues.

Myelin and nodes of Ranvier are adaptations of vertebrates that allow rapid nerve conduction using small caliber axons.

Targeted knockout of ankyrin-G in the postnatal cerebellum in mice results in severe ataxia, and loss of ability to fire action potentials as well as clustering of voltage-gated Na channels (Nav1.6) at axon initial segments of Purkinje neurons (Zhou et al. 1998; Jenkins and Bennett 2001). In addition, β-4 spectrin is absent and neurofascin is no longer restricted to ankyrin-G-deficient initial segments (Jenkins and Bennett 2001). These result in Purkinje neurons that have been recapitulated in cultured hippocampal neurons, where knockdown of ankyrin-G results in mis-localization of β-4 spectrin, voltage-gated sodium channels, as well as neurofascin (Yang et al. 2007; Hedstrom et al. 2007).

β-4 spectrin knockout mice show diminution of the membrane undercoat and increased membrane blebbing at nodes of Ranvier (Lacas-Gervais et al. 2004). Ankyrin-G recruits β-4 spectrin to nodes and initial segments (Jenkins and Bennett 2001; Yang et al. 2007), and is still present in the absence of β-4 spectrin. β-4 spectrin thus plays an important role in stabilizing these excitable membranes, following establishment by ankyrin-G alone at initial segments or by ankyrin-G in collaboration with axonal neurofascin and Schwann cell gliomedin at nodes of Ranvier (Eshed et al. 2005; Dzhashiashvili et al. 2007).

KCNQ2/3(Kv7) channels modulate activity of voltage-gated Na channels, and mutations in these channels results in hyperexcitability phenotypes including epilepsy (Maljevic et al. 2008; Neubauer et al. 2008). KCNQ2/3 channels colocalize with voltage-gated Na channels at nodes of Ranvier and axon initial segments (Devaux et al. 2004; Pan et al. 2006; Rasmussen et al. 2007). KCNQ2/3 channels have ankyrin-binding sites in their cytoplasmic domains, and require ankyrin-G for targeting to axon initial segments (Pan et al. 2006; Chung et al. 2006; Rasmussen et al. 2007).

Strikingly, knockdown of ankyrin-G in cultured hippocampal neurons results in loss of voltage-gated Na channels from initial segments and conversion of these domains into dendrites (Hedstrom et al. 2008). Ankyrin-G thus is required for preservation of the entire axon initial segment. It will be of interest to determine the role of ankyrin-G in the establishment of initial segments as well as whether ankyrin-G operates downstream of axonal polarity pathways.

Axon initial segments receive direct input through axo-axonic synapses of interneurons, which modulate neuronal output and may have roles in diseases such as epilepsy and schizophrenia (Howard et al. 2005). Loss of the neurofascin enrichment at the initial segment in ankyrin-G-deficient Purkinje neurons results in disruption of a class of synapses formed by interneurons that interconnect Purkinje neurons in the cerebellum (Ango et al. 2004). Ankyrin-G thus is responsible for stabilizing transcellular connections as well as organizing the composition of initial segments.

Spectrin and ankyrin are likely to function in other axonal domains. For example, paranodes are specialized zones immediately adjacent to the nodal gap in myelinated axons that contain shaker-type potassium channels and are characterized ultrastructurally by prominent axo-glial junctions. Paranodes contain a membrane skeleton distinct from nodes that includes β-2 spectrin, ankyrin-B, and protein 4.1 B (Ogawa et al. 2006). α-2 Spectrin is present in paranodes where it associates with β-2 spectrin, and also at the nodal gap where it partners with β-4 spectrin. α-2 Spectrin mutations in zebra fish result in abnormal development of nodes of Ranvier, and are responsible for stabilizing initial clusters of voltage-gated Na channels (Voas et al. 2007).

Ankyrin-binding activity of its membrane partners is not "hardwired," but is subject to regulation. Phosphorylation of neurofascin at the FIGQY tyrosine in its ankyrin-binding site abolishes ankyrin-binding (Garver et al. 1997; Whittard et al. 2006) and results in gain of function in binding to doublecortin (Kizhatil et al. 2002). FIGQY-phosphorylated neurofascin is excluded from the node of Ranvier but is concentrated in paranodes (Jenkins et al. 2001). Ankyrin-binding activity of the voltage-gated sodium channel is markedly enhanced by phosphorylation by casein kinase 2 (Bréchet et al. 2008). Interestingly, casein kinase 2 is concentrated at nodes of Ranvier and axon initial segments, and could provide a local activation signal for stabilizing voltage-gated sodium channels at these sites (Bréchet et al. 2008).

NEUROMUSCULAR JUNCTIONS

Spectrin and ankyrin stabilize neuromuscular junctions through both presynaptic and postsynaptic mechanisms. Presynaptic β spectrin is required for normal transmitter release and to stabilize *Drosophila* neuromuscular junctions following initial establishment of these synapses (Featherstone et al. 2001; Pielage et al. 2005). Interestingly, two groups independently identified presynaptic giant ankyrin-2 isoforms through different unbiased forward genetic screens for mutations affecting the *Drosophila* neuromuscular junction (Koch et al. 2008; Pielage et al. 2008). The phenotypes of ankyrin-mutant junctions included retraction of synaptic boutons, loss of axonal microtubules, and misorganization of synaptic cell adhesion molecules. The defects in synaptic stability increased with distance from the neuron cell body and were accompanied by accumulation of synaptic vesicles in axons, suggesting a role of ankyrins in axonal transport in addition to their local function at the synapse (Koch et al. 2008).

Spectrin and ankyrin-G are localized in specialized postsynaptic domains in mammalian neuromuscular junctions that are distinct from the acetylcholine receptor and are enriched in voltage-gated sodium channels (Flucher and Daniels 1989; Wood and Slater 1998; Kordeli et al. 1998; Bailey et al. 2003). Ankyrin-B also is located at the periphery of mammalian neuromuscular junctions (Ayalon et al. 2008). Experiments in flies and mice show that postsynaptic spectrin and ankyrins are required to stabilize neuromuscular junctions. Knockdown of postsynaptic spectrin in the *Drosophila* by siRNA results in misorganization of active zones and abnormal growth of the neuromuscular junction (Pielage et al. 2006). Similarly, knockdown of postsynaptic ankyrin-B in adult mouse muscle results in shrinkage of the neuromuscular junction from its adult form back to the size of neonatal junctions (Ayalon et al. 2008).

COSTAMERES

Costameres (Fig. 2B) are specialized domains formed at the junction of the plasma membrane and Z-discs of peripheral myofibrils in skeletal muscle and cardiomyocytes (Pardo et al. 1983; Rybakova et al. 2000; Ervasti 2003; Bloch et al. 2004). Costameres function as force buffers that transmit force across the plasma membrane from sarcomeres to the extracellular matrix, and protect the muscle plasma membrane from damage during contraction. The dystrophin-glycoprotein complex (DGC) provides a transmembrane connection at costameres through association of dystrophin with γ-actin and dystroglycan, and dystroglycan with extracellular laminin (Ervasti, 2003; Rybakova et al. 2000). The dystrophin-glycoprotein-complex is absent from the plasma membrane in Duchenne muscular dystrophy, which results in membrane damage and contributes to death of muscle cells (Ervasti et al. 1990; Cohn and Campbell 2000; Dalkilic and Kunkel 2003).

Ankyrin-B and ankyrin-G cooperate in localization of dystrophin and β-dystroglycan at costameres and are required to prevent exercise-induced injury (Ayalon et al. 2008). Ankyrin-B is required for transport of β-dystroglycan to the plasma membrane, whereas ankyrin-G retains β-dystroglycan at costameres. Loss of ankyrin-B in skeletal muscle is accompanied by loss of microtubules, both at the neuromuscular junction (see previous discussion) as well as costameres (Ayalon et al. 2008). Ankyrin-B binds directly to dynactin-4/p62 of the dynactin complex, and may capture fast-growing ends of microtubules at costameres and neuromuscular junctions to establish transport routes from the trans-Golgi network for newly synthesized β-dystroglycan (Ayalon et al. 2008).

CARDIOMYOCYTE T-TUBULE MICRODOMAINS

Rhythmic contraction of mammalian hearts requires nearly synchronous waves of calcium release and reuptake throughout the intracellular space of ventricular cardiomyocytes. This is achieved though precise placement of membrane transporters and signaling molecules related to import and export of calcium within microdomains of T-tubules (Fig. 2C). T-tubules are invaginations of the plasma membranes that form a complex three-dimensional network juxtaposed to the sarcoplasmic reticulum (Brette and Orchard 2007). Ventricular myocyte T-tubules contain three microdomains that can be resolved by high resolution light microscopy: A domain containing L-type voltage-gated calcium channels complexed with ryanodine receptors in the sarcoplasmic reticulum, a second domain enriched in the Na/Ca exchanger, and a third domain enriched in voltage-gated sodium channels (Scriven et al. 2000). Voltage-gated sodium channels are responsible for activating voltage-gated calcium channels, which admit a small amount of calcium and activate calcium release from the sarcoplasmic reticulum. The Na/Ca exchanger contributes to calcium homeostasis by export of calcium back across the T-tubule. Evidence will be reviewed below indicating that ankyrin-B is required for the Na/Ca exchanger microdomain, and ankyrin-G is required for the domain enriched in voltage-gated sodium channels.

The Na/Ca exhanger (NCX1) binds to ankyrin-B and colocalizes with ankyrin-B in cardiomyocyte T-tubules (Li et al. 1993; Mohler et al. 2005; Cunha et al. 2007). In addition, NCX1 and ankyrin-B also colocalize with the Na/K ATPase in the T-tubule membrane and with the IP3 receptor in the sarcoplasmic reticulum (Mohler et al. 2005). Haploinsufficiency of ankyrin-B in mice results in selective loss of T-tubule-localized sodium/calcium exchanger, Na/K ATPase as well as IP3 receptor from adult cardiomyocytes (Mohler et al. 2003; 2005). Moreover, ankyrin-B-deficient cardiomyocytes show increased contractility and increased calcium transients (Mohler et al. 2003; 2005). These observations suggest that ankyrin-B-dependent colocalization of NCX1 and Na/K ATPase could result in functional coupling between these transporters, with Na/K ATPase-driven export of sodium ions entering the cell in exchange for calcium ions. The IP3 receptor in the ankyrin-B complex has been proposed to function in coupled calcium export from the sarcoplasmic reticulum directly through NCX1 (Mohler et al. 2005). The physiological importance of the ankyrin-B-microdomain in T-tubules is supported by findings that ankyrin-B-deficient mice and humans heterozygous for loss-of-function mutations of ankyrin-B show stress-induced sudden cardiac death and cardiac arrhythmia symptoms (Mohler et al. 2003; 2004b).

Nav1.5 and ankyrin-G and are both localized to T-tubules and intercalated discs of adult cardiomyoctes (Mohler et al. 2004c). Although high-resolution double-labeling for these proteins has not been reported, several findings strongly indicate that Nav1.5 and ankyrin-G are molecular partners in the heart. E1053K mutation of the ankyrin-binding site of Nav1.5 eliminates ankyrin-G-binding as well as ability of Nav1.5 to accumulate at the cell surface of cardiomyocytes (Mohler et al. 2004c). Interestingly, this Nav1.5 mutation is associated with Brugada syndrome, which is a cardiac arrhythmia associated with loss-of-function mutations of Nav1.5. Nav1.5 requires interaction with ankyrin-G through ank repeats 14 and 15 for expression at the cell surface of neonatal cardiomyocytes (Lowe et al. 2008).

EPITHELIAL LATERAL MEMBRANES

The lateral membrane domain of epithelial cells (Fig. 2D) is of considerable physiological importance because of its roles in salt and water homeostasis and protection of epithelial tissues from mechanical stress. Moreover, loss of this specialized domain is a hallmark of metastatic cancer cells. Early immunofluorescence studies identified ankyrin and spectrin in a polarized pattern localized to lateral membranes of cultured epithelial cells as well as in tissues (Drenckhahn et al. 1985; Nelson and Veshnock 1986; Drenckhahn and Bennett 1987). More recently, studies using siRNA have revealed that ankyrin-G and β-2 spectrin collaborate in formation of the lateral membrane of bronchial epithelial cells (Kizhatil and Bennett 2004; Kizhatil et al. 2007b). Cells depleted of either protein maintain apical–basal polarity, but fail to form new lateral membrane following initiation of cell–cell contact and remain flattened rather than columnar. Moreover, ankyrin-G and β-2 spectrin-depleted cells are incapable of de novo membrane biogenesis during mitosis (Kizhatil et al. 2007b). Ankyrin-G requires β-2 spectrin as a partner because ankyrin-G mutants lacking β-2 spectrin-binding activity are not active in restoring the lateral membrane (Kizhatil et al. 2007b). Ankyrin-G and β-2 spectrin thus work together in bulk delivery of proteins and phospholipids to the lateral membrane.

β spectrins associated with intracellular membranes were initially believed to be distinct from those associated with the plasma membrane. Beck and colleagues reported spectrin immunoreactivity associated with the Golgi (Beck et al. 1994), which was later attributed to β-3 spectrin (Stankewich et al. 1998). However, the cross-reacting protein in Golgi was subsequently determined to be syne-1 (also termed nesprin), which has spectrin-repeats but is otherwise distantly related to β spectrins (Gough et al. 2003). β-3 spectrin is very similar to β-2 spectrin, but has a more restricted pattern of expression primarily in the nervous system and especially the cerebellum (Sakaguchi et al. 1998). β-3 spectrin thus may have roles in intracellular transport similar to β-2 spectrin but is not a specialized component of the Golgi apparatus. Recently, mutations in β-3 spectrin were identified as the cause of a form of spinocerebellar ataxia (SCA5) (Ikeda et al. 2006), indicating an important role in maintenance of certain neurons.

E-cadherin is a key cell adhesion molecule that is required to form the first lateral membrane domains in development and is a ubiquitous component of lateral membranes in epithelial tissues. E-cadherin has recently been reported to bind to ankyrin-G through a highly conserved site in its cytoplasmic domain and to require ankyrin-binding activity for efficient exit from the trans Golgi network (Kizhatil et al. 2007a). Moreover, both ankyrin-G and β-2 spectrin are required for accumulation of E-cadherin at the lateral membrane in both epithelial cells and preimplantation embryos. E-cadherin thus works together with ankyrin-G and β-2 spectrin to coordinate membrane assembly with extracellular interactions of at sites of cell–cell contact. Coupling of E-cadherin to a versatile adaptor protein such as ankyrin-G could promote corecruitment of diverse proteins to sites of cell–cell contact. For example, ankyrin-G associates with other lateral membrane proteins including the Na/K ATPase (Nelson and Veshnock, 1987) and the RhBG ammonium transporter (Lopez et al. 2005). It will be important to determine if ankyrin-binding activity is shared by other proteins residing in the lateral membrane.

Spectrin–actin complexes are stabilized by accessory proteins such as adducin, which recruits spectrin to the fast-growing end of actin filaments (Gardner and Bennett 1987; Li et al. 1998; Kuhlman et al, 1996), and tropomodulin, which caps the slow-growing ends of actin filaments (Littlefield and Fowler 2008). Adducin and tropomodulin are both required to stabilize spectrin–actin networks on the lateral membrane of epithelial cells (Abdi and Bennett 2008; Weber et al. 2007). Depletion of either protein by siRNA results in loss of lateral membrane height. Moreover, depletion of adducin increases long-range mobility of E-cadherin on the lateral membrane (Abdi and Bennett 2008). Adducin is phosphorylated and inactivated by protein kinase C (Matsuoka et al. 1998), suggesting the possibility of signals that modulate the stability of the spectrin-actin network in the lateral membrane. Interestingly in this regard, pleiotrophin is a cytokine that promotes adducin phosphorylation and increased proliferation of epithelial cells (Pariser et al. 2005).

CORE MECHANISMS AND EVOLUTION OF MEMBRANE DOMAINS

Having gone into details of individual membrane domains, it is worthwhile to consider common requirements and the special features of ankyrins and spectrins that satisfy these core needs. All of the domains considered in this review depend on corecruitment of functionally related but structurally distinct membrane partners. For example, axon initial segments are enriched in voltage-gated Na channels, KCNQ2/3 channels, and 186 kDa neurofascin (Fig. 2A), lateral membranes have E-cadherin and the Na/K ATPase (Fig. 2D), and ankyrin-B-based cardiomyocyte T-tubule domains contain the Na/K ATPase together with the Na/Ca exchanger (Fig. 2C). These proteins all have independently evolved ability to bind to ankyrin. A shared feature of many ankyrin-binding sites is that they are either predicted or shown to be extended peptides lacking secondary structure. For example the anion exchanger site is an 11 amino-acid loop based on the crystal structure (Stefanovic et al. 2007), the cytoplasmic domains of E-cadherin and L1 CAMS are established to be natively unstructured by biophysical methods (Huber et al. 2001; Zhang et al. 1998), and sites of Nav channels, KCNQ2/3 channels, RhBG ammonium transporter, and β-dystroglycan are predicted to be unstructured (Fig. 3A).

A possible binding site for unstructured peptides could be the ankyrin groove that runs the 240-angstrom length of the repeat stack (Michaely et al. 2002) (Fig. 3A,B). A groove of this length with variation in surface-exposed residues could potentially accommodate multiple types of partners. Interestingly, ANK repeats can bind to more than one partner at a time and thus are capable of forming homo- and hetero-complexes (Michaely and Bennett, 1995).

Natively unstructured domains of proteins are widely used in protein recognition (Dyson and Wright 2005). One advantage of such a code is that unstructured proteins can multitask and also engage endocytic machinery and other adaptor proteins depending on cellular requirements. Another advantage is that the affinity for ankyrin can vary: The Kd for ankyrin is 10 nM for the anion exchanger, 50 nM for neurofascin, and 500 nM for E-cadherin. This variable affinity allows for flexibility depending on the physiological context. Finally, intrinsically unstructured proteins represent the most rapidly evolving part of the genome (Brown et al. 2002), and have the potential to adjust readily to new physiological demands such as the rapid acquisition of myelination.

Another emerging theme as we learn more about ankyrin-based membrane domains is that they assemble through direct targeting of components along microtubules to specific sites. A similar direct targeting pathway has been proposed for assembly of gap junction subunits at adherens junctions (Shaw et al. 2007). Direct targeting is in contrast to many current models that invoke endocytosis and transcytosis as primary mechanisms for sorting. Ankyrins can bind directly to microtubules (Bennett and Davis 1981), and also can serve as adaptors for the dynactin complex, which can stabilize the fast-growing ends of microtubules, at least in skeletal muscle (Ayalon et al. 2008). Ankyrins also stabilize microtubules at the presynaptic neuromuscular junction, and also may have roles in axonal transport (Pielage et al. 2008; Koch et al. 2008). A current mystery is how newly synthesized membrane proteins are coupled to the appropriate microtubules, especially in epithelial cells and neurons where multiple types of microtubules coexist.

In addition to their roles in stabilizing proteins at the plasma membrane and in directed transport, ankyrin and spectrin may also establish specialized membrane domains. For example, ankyrin-G and β-2 spectrin are required for biogenesis of epithelial lateral membranes (Kizhatil and Bennett 2004; Kizhatil et al, 2007b). Ankyrin-G also is required to form axon initial segments, which lose all initial segment markers and develop dendritic properties in ankyrin-G-knockdown cells (Hedstrom et al. 2008). Clues to how ankyrins and β spectrins could participate in bulk transport of proteins and phospholipids in assembly of membrane domains come from the findings that β-2 spectrin interacts with membrane phospholipids through multiple sites (An et al. 2004; Muresan et al. 2001), and with PI lipids through its PH domain (Hyvönen et al. 1995). Moreover, spectrins also interact with microtubule-based motors, either directly (Takeda et al. 2000), or through dynactin (Muresan et al. 2001; Holleran et al. 2001; Holleran et al. 1996). The combination of ankyrin, with its diversity in protein recognition, and β spectrin, with its capacity to connect membrane compartments with microtubule-based motors, seems well suited for segregation and transport of membrane proteins and lipids to specialized domains. A central unanswered question is the identity of the initial polarity signals that define the site of delivery for ankyrin/spectrin cargo.

SUMMARY AND PERSPECTIVES

Ankyrins and spectrins were first discovered as partners in plasma membrane of erythrocytes, but now are established to be required for specialized membrane domains in many types of cells. Spectrins can both form a two-dimensional actin-based network on the plasma membrane that restricts membrane-spanning proteins, as well as participate in microtubule-dependent transport of membrane lipids and proteins. Ankyrins serve as membrane adaptors that connect spectrin to diverse membrane-spanning proteins through recognition by ANK repeats. Currently characterized ankyrin-binding sites are short 10–20 amino acid stretches that do not have a single shared motif but are intrinsically unstructured or configured as a loop. Ankyrins, in many cases operating with spectrins, are involved in coordinate assembly of a variety of ion transporters and cell adhesion molecules at axon initial segments and nodes of Ranvier in myelinated nerves, the neuromuscular junction, T-tubule microdomains in cardiomyocytes, costameres in striated muscle, and the lateral membrane domain of epithelial cells. Cardiac arrhythmias result from mutation or deficiency of ankyrin-B or mutation of ankyrin-G-binding site of the cardiac voltage-gated sodium channel. It is likely that additional human diseases will turn out to result from defects in organization of membrane proteins. Important questions for future work include elucidation of the polarity signals that define where these domains are localized in cells as well as resolving their assembly mechanisms involving sorting and preferential transport. It is clear that future resolution of the core mechanisms for roles of ankyrins and spectrins in membrane domains will have pervasive implications for physiology as well as clinical medicine.

ACKNOWLEDGMENTS

JH was supported by a predoctoral fellowship from the American Heart Association. VB was supported in part by a grant from the Muscular Dystrophy Association.

REFERENCES

Abdi KM, Bennett V. 2008. Adducin promotes micrometer-scale organization of β2-spectrin in lateral membranes of bronchial epithelial cells. *Mol Biol Cell* **19:** 536–545.

Abdi KM, Mohler PJ, Davis JQ, Bennett V. 2006. Isoform specificity of ankyrin-B: A site in the divergent C-terminal domain is required for intramolecular association. *J Biol Chem* **281:** 5741–5749.

An X, Guo X, Sum H, Morrow J, Gratzer W, Mohandas N. 2004. Phosphatidylserine binding sites in erythroid spectrin: Location and implications for membrane stability. *Biochemistry* **43:** 310–315.

Ango F, di Cristo G, Higashiyama H, Bennett V, Wu P, Huang ZJ. 2004. Ankyrin-based subcellular gradient of neurofascin, an immunoglobulin family protein, directs GABAergic innervation at purkinje axon initial segment. *Cell* **119:** 257–272.

Ayalon G, Davis JQ, Scotland PB, Bennett V. 2008. An ankyrin-based mechanism for functional organization of dystrophin and dystroglycan. *Cell* **135:** 1189–1200.

Bailey SJ, Stocksley MA, Buckel A, Young C, Slater CR. 2003. Voltage-gated sodium channels and ankyrin-G occupy a different postsynaptic domain from acetylcholine receptors from an early stage of neuromuscular junction maturation in rats. *J Neurosci* **23:** 2102–2111.

Bañuelos S, Saraste M, Djinović Carugo K. 1998. Structural comparisons of calponin homology domains: Implications for actin binding. *Structure* **6:** 1419–1431.

Beck KA, Buchanan JA, Malhotra V, Nelson WJ. 1994. Golgi spectrin: Identification of an erythroid β-spectrin homolog associated with the Golgi complex. *J Cell Biol* **127:** 707–723.

Bennett V. 1978. Purification of an active proteolytic fragment of the membrane attachment site for human erythrocyte spectrin. *J Biol Chem* **253:** 2292–2299.

Bennett V, Baines AJ. 2001. Spectrin and ankyrin-based pathways: Metazoan inventions for integrating cells into tissues. *Physiol Rev* **81:** 1353–1392.

Bennett V, Davis J. 1981. Erythrocyte ankyrin: Immunoreactive analogues are associated with mitotic structures in cultured cells and with microtubules in brain. *Proc Natl Acad Sci* **78:** 7550–7554.

Bennett V, Healy J. 2008. Organizing the fluid membrane bilayer: Diseases linked to spectrin and ankyrin. *Trends Mol Med* **14:** 28–36.

Bennett V, Stenbuck PJ. 1979a. Identification and partial purification of ankyrin, the high affinity membrane attachment site for human erythrocyte spectrin. *J Biol Chem* **254:** 2533–2541.

Bennett V, Stenbuck PJ. 1979b. The membrane attachment protein for spectrin is associated with band 3 in human erythrocyte membranes. *Nature* **280:** 468–473.

Bennett V, Davis JQ, Fowler WE. 1982. Brain spectrin, a membrane-associated protein related in structure and function to erythrocyte spectrin. *Nature* **299:** 126–131.

Berghs S, Aggujaro D, Dirkx R Jr, Maksimova E, Stabach P, Hermel JM, Zhang JP, Philbrick W, Slepnev V, Ort T, et al. 2000. βIV spectrin, a new spectrin localized at axon initial segments and nodes of ranvier in the central and peripheral nervous system. *J Cell Biol* **151:** 985–1002.

Bloch RJ, Reed P, O'Neill A, Strong J, Williams M, Porter N, Gonzalez-Serratos H. 2004. Costameres mediate force transduction in healthy skeletal muscle and are altered in muscular dystrophies. *J Muscle Res Cell Motil* **25:** 590–592.

Bouley M, Tian MZ, Paisley K, Shen YC, Malhotra JD, Hortsch M. 2000. The L1-type cell adhesion molecule neuroglian influences the stability of neural ankyrin in the *Drosophila* embryo but not its axonal localization. *J Neurosci* **20:** 4515–4523.

Bourguignon LY, Lokeshwar VB, He J, Chen X, Bourguignon GJ. 1992. A CD44-like endothelial cell transmembrane glycoprotein (GP116) interacts with extracellular matrix and ankyrin. *Mol Cell Biol* **12:** 4464–4471.

Bréchet A, Fache MP, Brachet A, Ferracci G, Baude A, Irondelle M, Pereira S, Leterrier C, Dargent B. 2008. Protein kinase CK2 contributes to the organization of sodium channels in axonal membranes by regulating their interactions with ankyrin G. *J Cell Biol* **183:** 1101–1114.

Brette F, Orchard C. 2007. Resurgence of cardiac t-tubule research. *Physiology (Bethesda).* **22:** 167–173.

Brown CJ, Takayama S, Campen AM, Vise P, Marshall TW, Oldfield CJ, Williams CJ, Dunker AK. 2002. Evolutionary rate heterogeneity in proteins with long disordered regions. *J Mol Evol* **55:** 104–110.

Byers TJ, Branton D. 1985. Visualization of the protein associations in the erythrocyte membrane skeleton. *Proc Natl Acad Sci* **82:** 6153–6157.

Chen L, Ong B, Bennett V. 2001. LAD-1, the *Caenorhabditis elegans* L1CAM homologue, participates in embryonic and gonadal morphogenesis and is a substrate for fibroblast growth factor receptor pathway-dependent phosphotyrosine-based signaling. *J Cell Biol* **154:** 841–855.

Cai X, Zhang Y. 2006. Molecular evolution of the ankyrin gene family. *Mol Biol Evol* **23:** 550–558.

Chan W, Kordeli E, Bennett V. 1993. 440-kD ankyrin-B: Structure of the major developmentally regulated domain and selective localization in unmyelinated axons. *J Cell Biol* **123:** 1463–1473.

Chung HJ, Jan YN, Jan LY. 2006. Polarized axonal surface expression of neuronal KCNQ channels is mediated by multiple signals in the KCNQ2 and KCNQ3 C-terminal domains. *Proc Natl Acad Sci* **103:** 8870–8875.

Clarke M. 1971. Isolation and characterization of a water-soluble protein from bovine erythrocyte membranes. *Biochem Biophys Res Commun* **45:** 1063–1070.

Cohn RD, Campbell KP. 2000. Molecular basis of muscular dystrophies. *Muscle Nerve* **23:** 1456–1471.

Cunha SR, Bhasin N, Mohler PJ. 2007. Targeting and stability of Na/Ca exchanger 1 in cardiomyocytes requires direct interaction with the membrane adaptor ankyrin-B. *J Biol Chem* **282:** 4875–4883.

Dalkilic I, Kunkel LM. 2003. Muscular dystrophies: genes to pathogenesis. *Curr Opin Genet Dev* **13:** 231–238.

Davis J, Bennett V. 1983. Brain spectrin. Isolation of subunits and formation of hybrids with erythrocyte spectrin subunits. *J Biol Chem* **258:** 7757–7766.

Davis JQ, Bennett V. 1994. Ankyrin binding activity shared by the neurofascin/L1/NrCAM family of nervous system cell adhesion molecules. *J Biol Chem* **269:** 27163–27166.

Davis LH, Davis JQ, Bennett V. 1992. Ankyrin regulation: An alternatively spliced segment of the regulatory domain functions as an intramolecular modulator. *J Biol Chem* **267:** 18966–18972.

Davis JQ, Lambert S, Bennett V. 1996. Molecular composition of the node of Ranvier: Identification of ankyrin-binding cell adhesion molecules neurofascin (mucin+/third FNIII domain-) and NrCAM at nodal axon segments. *J Cell Biol* **135:** 1355–13567.

Davis L, Abdi K, Machius M, Brautigam C, Tomchick DR, Bennett V, Michaely P. 2008. Localization and structure of the ankyrin-binding site on β2-spectrin. *J Biol Chem* **284:** 6982–6987.

Devaux J, Alcaraz G, Grinspan J, Bennett V, Joho R, Crest M, Scherer SS. 2003. Kv3.1b is a novel component of CNS nodes. *J Neurosci* **23:** 4509–4518.

Devaux JJ, Kleopa KA, Cooper EC, Scherer SS. 2004. KCNQ2 is a nodal K+ channel. *J Neurosci* **24:** 1236–1244.

Drenckhahn D, Bennett V. 1987. Polarized distribution of Mr 210,000 and 190,000 analogs of erythrocyte ankyrin along the plasma membrane of transporting epithelia, neurons and photoreceptors. *Eur J Cell Biol* **43:** 479–486.

Drenckhahn D, Schlüter K, Allen DP, Bennett V. 1985. Colocalization of band 3 with ankyrin and spectrin at the basal membrane of intercalated cells in the rat kidney. *Science* **230:** 1287–1289.

Dubreuil RR, Yu J. 1994. Ankyrin and β-spectrin accumulate independently of α-spectrin in *Drosophila. Proc Natl Acad Sci* **91:** 10285–10289.

Dubreuil RR, Byers TJ, Stewart CT, Kiehart DP. 1990. A β-spectrin isoform from *Drosophila* (β H) is similar in size to vertebrate dystrophin. *J Cell Biol* **111:** 1849–1858.

Dyson HJ, Wright PE. 2005. Intrinsically unstructured proteins and their functions. *Nat Rev Mol Cell Biol* **6:** 197–208.

Dzhashiashvili Y, Zhang Y, Galinska J, Lam I, Grumet M, Salzer JL. 2007. Nodes of Ranvier and axon initial segments are ankyrin G-dependent domains that assemble by distinct mechanisms. *J Cell Biol* **177:** 857–870.

Ervasti JM. 2003. Costameres: The Achilles' heel of Herculean muscle. *J Biol Chem* **278:** 13591–13594.

Ervasti JM, Ohlendieck K, Kahl SD, Gaver MG, Campbell KP. 1990. Deficiency of a glycoprotein component of the dystrophin complex in dystrophic muscle. *Nature* **345:** 315–319.

Eshed Y, Feinberg K, Poliak S, Sabanay H, Sarig-Nadir O, Spiegel I, Bermingham JR Jr, Peles E. 2005. Gliomedin mediates Schwann cell-axon interaction and the molecular assembly of the nodes of Ranvier. *Neuron* **47:** 215–229.

Featherstone DE, Davis WS, Dubreuil RR, Broadie K. 2001. *Drosophila* α- and β-spectrin mutations disrupt presynaptic neurotransmitter release. *J Neurosci* **21:** 4215–4224.

Flucher BE, Daniels MP. 1989. Distribution of Na+ channels and ankyrin in neuromuscular junctions is complementary to that of acetylcholine receptors and the 43 kd protein. *Neuron* **3:**163–175.

Gardner K, Bennett V. 1987. Modulation of spectrin-actin assembly by erythrocyte adducin. *Nature* **328:** 359–362.

Garrido JJ, Giraud P, Carlier E, Fernandes F, Moussif A, Fache MP, Debanne D, Dargent B. 2003. A targeting motif involved in sodium channel clustering at the axonal initial segment. *Science* **300:** 2091–2094.

Garver TD, Ren Q, Tuvia S, Bennett V. 1997. Tyrosine phosphorylation at a site highly conserved in the L1 family of cell adhesion molecules abolishes ankyrin binding and increases lateral mobility of neurofascin. *J Cell Biol* **137:** 703–714.

Gough LL, Fan J, Chu S, Winnick S, Beck KA. 2003. Golgi localization of Syne-1. *Mol Biol Cell* **14:** 2410–2424.

Hall TG, Bennett V. 1987. Regulatory domains of erythrocyte ankyrin. *J Biol Chem* **262:** 10537–10545.

Hayes NV, Scott C, Heerkens E, Ohanian V, Maggs AM, Pinder JC, Kordeli E, Baines AJ. 2000. Identification of a novel C-terminal variant of β II spectrin: Two isoforms of β II spectrin have distinct intracellular locations and activities. *J Cell Sci* **113:** 2023–2034.

Hedstrom KL, Ogawa Y, Rasband MN. 2008. Ankyrin-G is required for maintenance of the axon initial segment and neuronal polarity. *J Cell Biol* **183:** 635–640.

Hedstrom KL, Xu X, Ogawa Y, Frischknecht R, Seidenbecher CI, Shrager P, Rasband MN. 2007. Neurofascin assembles a specialized extracellular matrix at the axon initial segment. *J Cell Biol* **178:** 875–886.

Hoock TC, Peters LL, Lux SE. 1997. Isoforms of ankyrin-3 that lack the NH2-terminal repeats associate with mouse macrophage lysosomes. *J Cell Biol* **136:** 1059–1070.

Howard A, Tamas G, Soltesz I. 2005. Lighting the chandelier: New vistas for axo-axonic cells. *Trends Neurosci* **28:** 310–316.

Ikeda Y, Dick KA, Weatherspoon MR, Gincel D, Armbrust KR, Dalton JC, Stevanin G, Dürr A, Zühlke C, Bürk K, et al. 2006. Spectrin mutations cause spinocerebellar ataxia type 5. *Nat Genet* **38:** 184–190.

Ipsaro JJ, Huang L, Mondragon A. 2009. Structures of the spectrin-ankyrin interaction binding domains. *Blood* doi: 10.1182/blood-2008-10-184358.

Jackson M, Song W, Liu MY, Jin L, Dykes-Hoberg M, Lin CI, Bowers WJ, Federoff HJ, Sternweis PC, Rothstein JD. 2001. Modulation of the neuronal glutamate transporter EAAT4 by two interacting proteins. *Nature* **410:** 89–93.

Jenkins SM, Bennett V. 2001. Ankyrin-G coordinates assembly of the spectrin-based membrane skeleton, voltage-gated sodium channels, and L1 CAMs at Purkinje neuron initial segments. *J Cell Biol* **155:** 739–746.

Jenkins SM, Kizhatil K, Kramarcy NR, Sen A, Sealock R, Bennett V. 2001. FIGQY phosphorylation defines discrete populations of L1 cell adhesion molecules at sites of cell-cell contact and in migrating neurons. *J Cell Sci* **114:** 3823–3835.

Holleran EA, Ligon LA, Tokito M, Stankewich MC, Morrow JS, Holzbaur EL. 2001. β III spectrin binds to the Arp1 subunit of dynactin. *J Biol Chem* **276:** 36598–36605.

Holleran EA, Tokito MK, Karki S, Holzbaur EL. 1996. Centractin (ARP1) associates with spectrin revealing a potential mechanism to link dynactin to intracellular organelles. *J Cell Biol* **135:** 1815–1829.

Huber AH, Stewart DB, Laurents DV, Nelson WJ, Weis WI. 2001. The cadherin cytoplasmic domain is unstructured in the absence of β-catenin. A possible mechanism for regulating cadherin turnover. *J Biol Chem* **276:** 12301–12309.

Hyvönen M, Macias MJ, Nilges M, Oschkinat H, Saraste M, Wilmanns M. 1995. Structure of the binding site for inositol phosphates in a PH domain. *EMBO J* **14:** 4676–4685.

Kizhatil K, Bennett V. 2004. Lateral membrane biogenesis in human bronchial epithelial cells requires 190-kDa ankyrin-G. *J Biol Chem* **279:** 16706–16714.

Kizhatil K, Davis JQ, Davis L, Hoffman J, Hogan BL, Bennett V. 2007a. Ankyrin-G is a molecular partner of E-cadherin in epithelial cells and early embryos. *J Biol Chem* **282:** 26552–26561.

Kizhatil K, Wu YX, Sen A, Bennett V. 2002. A new activity of doublecortin in recognition of the phospho-FIGQY tyrosine in the cytoplasmic domain of neurofascin. *J Neurosci* **22:** 7948–7958.

Kizhatil K, Yoon W, Mohler PJ, Davis LH, Hoffman JA, Bennett V. 2007b. Ankyrin-G and β2-spectrin collaborate in biogenesis of lateral membrane of human bronchial epithelial cells. *J Biol Chem* **282:** 2029–2037.

Kline CF, Cunha SR, Lowe JS, Hund TJ, Mohler PJ. 2008. Revisiting ankyrin-InsP3 receptor interactions: Ankyrin-B associates with the cytoplasmic N-terminus of the InsP3 receptor. *J Cell Biochem* **104:** 1244–1253.

Koch I, Schwarz H, Beuchle D, Goellner B, Langegger M, Aberle H. 2008. *Drosophila* ankyrin-2 is required for synaptic stability. *Neuron* **58:** 210–222.

Kordeli E, Lambert S, Bennett V. 1995. Ankyrin-G. A new ankyrin gene with neural-specific isoforms localized at the axonal initial segment and node of Ranvier. *J Biol Chem* **270:** 2352–2359.

Kordeli E, Ludosky MA, Deprette C, Frappier T, Cartaud J. 1998. Ankyrin-G is associated with the postsynaptic membrane and the sarcoplasmic reticulum in the skeletal muscle fiber. *J Cell Sci* **111:** 2197–2207.

Kuhlman PA, Hughes CA, Bennett V, Fowler VM. 1996. A new function for adducin. Calcium/calmodulin-regulated capping of the barbed ends of actin filaments. *J Biol Chem* **271:** 7986–7991.

Lacas-Gervais S, Guo J, Strenzke N, Scarfone E, Kolpe M, Jahkel M, De Camilli P, Moser T, Rasband MN, Solimena M. 2004. β IV Sigma1 spectrin stabilizes the nodes of Ranvier and axon initial segments. *J Cell Biol* **166:** 983–990.

Lee G, Abdi K, Jiang Y, Michaely P, Bennett V, Marszalek PE. 2006. Nanospring behaviour of ankyrin repeats. *Nature* **440:** 246–249.

Lemaillet G, Walker B, Lambert S. 2003. Identification of a conserved ankyrin-binding motif in the family of sodium channel α subunits. *J Biol Chem* **278:** 27333–27339.

Li ZP, Burke EP, Frank JS, Bennett V, Philipson KD. 1993. The cardiac Na+-Ca^{2+} exchanger binds to the cytoskeletal protein ankyrin. *J Biol Chem* **268:** 11489–11491.

Li X, Matsuoka Y, Bennett V. 1998. Adducin preferentially recruits spectrin to the fast growing ends of actin filaments in a complex requiring the MARCKS-related domain and a newly defined oligomerization domain. *J Biol Chem* **273:** 19329–19338.

Littlefield RS, Fowler VM. 2008. Thin filament length regulation in striated muscle sarcomeres: Pointed-end dynamics go beyond a nebulin ruler. *Semin Cell Dev Biol* **19:** 511–519.

Lopez C, Métral S, Eladari D, Drevensek S, Gane P, Chambrey R, Bennett V, Cartron JP, Le Van Kim C, Colin Y. 2005. The ammonium transporter RhBG: Requirement of a tyrosine-based signal and ankyrin-G for basolateral targeting and membrane anchorage in polarized kidney epithelial cells. *J Biol Chem* **280:** 8221–8228.

Lowe JS, Palygin O, Bhasin N, Hund TJ, Boyden PA, Shibata E, Anderson ME, Mohler PJ. 2008. Voltage-gated Nav channel targeting in the heart requires an ankyrin-G dependent cellular pathway. *J Cell Biol* **180:** 173–186.

Macias MJ, Musacchio A, Ponstingl H, Nilges M, Saraste M, Oschkinat H. 1994. Structure of the pleckstrin homology domain from β-spectrin. *Nature* **369:** 675–677.

Maljevic S, Wuttke TV, Lerche H. 2008. Nervous system KV7 disorders: Breakdown of a subthreshold brake. *J Physiol* **586:** 1791–1801.

Marchesi VT, Steers E Jr. 1968. Selective solubilization of a protein component of the red cell membrane. *Science* **159:** 203–204.

Matsuoka Y, Li X, Bennett V. 1998. Adducin is an in vivo substrate for protein kinase C: Phosphorylation in the MARCKS-related domain inhibits activity in promoting spectrin-actin complexes and occurs in many cells, including dendritic spines of neurons. *J Cell Biol* **142:** 485–497.

McKeown C, Praitis V, Austin J. 1998. sma-1 encodes a βH-spectrin homolog required for *Caenorhabditis elegans* morphogenesis. *Development* **125:** 2087–2098.

Michaely P, Bennett V. 1995. The ANK repeats of erythrocyte ankyrin form two distinct but cooperative binding sites for the erythrocyte anion exchanger. *J Biol Chem* **270:** 22050–22057.

Michaely P, Tomchick DR, Machius M, Anderson RG. 2002. Crystal structure of a 12 ANK repeat stack from human ankyrin-R. *EMBO J* **21:** 6387–6396.

Mohler PJ, Davis JQ, Bennett V. 2005. Ankyrin-B coordinates the Na/K ATPase, Na/Ca exchanger, and InsP3 receptor in a cardiac T-tubule/SR microdomain. *PLoS Biol* **3:** e423.

Mohler PJ, Gramolini AO, Bennett V. 2002. The ankyrin-B C-terminal domain determines activity of ankyrin-B/G chimeras in rescue of abnormal inositol 1,4,5-trisphosphate and ryanodine receptor distribution in ankyrin-B (-/-) neonatal cardiomyocytes. *J Biol Chem* **277:** 10599–10607.

Mohler PJ, Yoon W, Bennett V. 2004a. Ankyrin-B targets β2-spectrin to an intracellular compartment in neonatal cardiomyocytes. *J Biol Chem* **279:** 40185–40193.

Mohler PJ, Rivolta I, Napolitano C, LeMaillet G, Lambert S, Priori SG, Bennett V. 2004c. Nav1.5 E1053K mutation causing Brugada syndrome blocks binding to ankyrin-G and expression of Nav1.5 on the surface of cardiomyocytes. *Proc Natl Acad Sci* **101:** 17533–17538.

Mohler PJ, Schott JJ, Gramolini AO, Dilly KW, Guatimosim S, duBell WH, Song LS, Haurogné K, Kyndt F, Ali ME, et al. 2003. Ankyrin-B mutation causes type 4 long-QT cardiac arrhythmia and sudden cardiac death. *Nature* **421:** 634–639.

Mohler PJ, Splawski I, Napolitano C, Bottelli G, Sharpe L, Timothy K, Priori SG, Keating MT, Bennett V. 2004b. A cardiac arrhythmia syndrome caused by loss of ankyrin-B function. *Proc Natl Acad Sci* **101:** 9137–9142.

Mosavi LK, Cammett TJ, Desrosiers DC, Peng ZY. 2004. The ankyrin repeat as molecular architecture for protein recognition. *Protein Sci* **13:** 1435–1448.

Muresan V, Stankewich MC, Steffen W, Morrow JS, Holzbaur EL, Schnapp BJ. 2001. Dynactin-dependent, dynein-driven vesicle transport in the absence of membrane proteins: A role for spectrin and acidic phospholipids. *Mol Cell* **7:** 173–183.

Musacchio A, Noble M, Pauptit R, Wierenga R, Saraste M. 1992. Crystal structure of a Src-homology 3 (SH3) domain. *Nature* **359:** 851–855.

Nelson WJ, Veshnock PJ. 1986. Dynamics of membrane-skeleton (fodrin) organization during development of polarity in Madin-Darby canine kidney epithelial cells. *J Cell Biol* **103:** 1751–1765.

Nelson WJ, Veshnock PJ. 1987. Ankyrin binding to (Na++K+)ATPase and implications for the organization of membrane domains in polarized cells. *Nature* **328:** 533–536.

Neubauer BA, Waldegger S, Heinzinger J, Hahn A, Kurlemann G, Fiedler B, Eberhard F, Muhle H, Stephani U, Garkisch S, et al. 2008. KCNQ3 mutations contribute to different idiopathic epilepsy syndromes. *Neurology* **71:** 177–183.

Nicolas V, Le Van Kim C, Gane P, Birkenmeier C, Cartron JP, Colin Y, Mouro-Chanteloup I. 2003. Rh-RhAG/ankyrin-R, a new interaction site between the membrane bilayer and the red cell skeleton, is impaired by Rh(null)-associated mutation. *J Biol Chem* **278:** 25526–25533.

Ogawa Y, Schafer DP, Horresh I, Bar V, Hales K, Yang Y, Susuki K, Peles E, Stankewich MC, Rasband MN. 2006. Spectrins and ankyrin-B constitute a specialized paranodal cytoskeleton. *J Neurosci* **26:** 5230–5239.

Otsuka AJ, Franco R, Yang B, Shim KH, Tang LZ, Zhang YY, Boontrakulpoontawee P, Jeyaprakash A, Hedgecock E, Wheaton VI. 1995. An ankyrin-related gene (unc-44) is necessary for proper axonal guidance in *Caenorhabditis elegans*. *J Cell Biol* **129:** 1081–1092.

Pan Z, Kao T, Horvath Z, Lemos J, Sul JY, Cranstoun SD, Bennett V, Scherer SS, Cooper EC. 2006. A common ankyrin-G-based mechanism retains KCNQ and NaV channels at electrically active domains of the axon. *J Neurosci* **26:** 2599–2613.

Pardo JV, Siliciano JD, Craig SW. 1983. A vinculin-containing cortical lattice in skeletal muscle: Transverse lattice elements ("costameres") mark sites of attachment between myofibrils and sarcolemma. *Pro Natl Acad Sci* **80:** 1008–1012.

Pariser H, Herradon G, Ezquerra L, Perez-Pinera P, Deuel TF. 2005. Pleiotrophin regulates serine phosphorylation and the cellular distribution of β-adducin through activation of protein kinase C. *Pro Natl Acad Sci* **102:** 12407–12412.

Pascual J, Castresana J, Saraste M. 1997. Evolution of the spectrin repeat. *Bioessays* **19:** 811–817.

Pielage J, Fetter RD, Davis GW. 2005. Presynaptic spectrin is essential for synapse stabilization. *Curr Biol* **15:** 918–928.

Pielage J, Fetter RD, Davis GW. 2006. A postsynaptic spectrin scaffold defines active zone size, spacing, and efficacy at the *Drosophila* neuromuscular junction. *J Cell Biol* **175:** 491–503.

Pielage J, Cheng L, Fetter RD, Carlton PM, Sedat JW, Davis GW. 2008. A presynaptic giant ankyrin stabilizes the NMJ through regulation of presynaptic microtubules and transsynaptic cell adhesion. *Neuron* **58:** 195–209.

Rasmussen HB, Frøkjaer-Jensen C, Jensen CS, Jensen HS, Jørgensen NK, Misonou H, Trimmer JS, Olesen SP, Schmitt N. 2007. Requirement of subunit co-assembly and ankyrin-G for M-channel localization at the axon initial segment. *J Cell Sci* **120:** 953–963.

Rybakova IN, Patel JR, Ervasti JM. 2000. The dystrophin complex forms a mechanically strong link between the sarcolemma and costameric actin. *J Cell Biol* **150:** 1209–1214.

Rotin D, Bar-Sagi D, O'Brodovich H, Merilainen J, Lehto VP, Canessa CM, Rossier BC, Downey GP. 1994. An SH3 binding region in the epithelial Na+ channel (α rENaC) mediates its localization at the apical membrane. *EMBO J* **13:** 4440–44450.

Sakaguchi G, Orita S, Naito A, Maeda M, Igarashi H, Sasaki T, Takai Y. 1998. A novel brain-specific isoform of β spectrin: Isolation and its interaction with Munc13. *Biochem Biophys Res Commun* **248:** 846–851.

Scriven DR, Dan P, Moore ED. 2000. Distribution of proteins implicated in excitation-contraction coupling in rat ventricular myocytes. *Biophys J* **79:** 2682–2691.

Shaw RM, Fay AJ, Puthenveedu MA, von Zastrow M, Jan YN, Jan LY. 2007. Microtubule plus-end-tracking proteins target gap junctions directly from the cell interior to adherens junctions. *Cell* **128:** 547–560.

Shotton DM, Burke BE, Branton D. 1979. The molecular structure of human erythrocyte spectrin. Biophysical and electron microscopic studies. *J Mol Biol* **131:** 303–329.

Simonovic M, Zhang Z, Cianci CD, Steitz TA, Morrow JS. 2006. Structure of the calmodulin αII-spectrin complex provides insight into the regulation of cell plasticity. *J Biol Chem* **281:** 34333–34340.

Sohet F, Colin Y, Genetet S, Ripoche P, Métral S, Le Van Kim C, Lopez C. 2008. Phosphorylation and ankyrin-G binding of the C-terminal domain regulate targeting and function of the ammonium transporter RhBG. *J Biol Chem* **283:** 26557–26567.

Speicher DW, Marchesi VT. 1984. Erythrocyte spectrin is comprised of many homologous triple helical segments. *Nature* **311:** 177–180.

Srinivasan Y, Elmer L, Davis J, Bennett V, Angelides K. 1988. Ankyrin and spectrin associate with voltage-dependent sodium channels in brain. *Nature* **333:** 177–180.

Stabach PR, Morrow JS. 2000. Identification and characterization of βV spectrin, a mammalian ortholog of *Drosophila* β H spectrin. *J Biol Chem* **275:** 21385–21395.

Stabach PR, Simonović I, Ranieri MA, Aboodi MS, Steitz TA, Simonović M, Morrow JS. 2009. The structure of the ankyrin-binding site of β-spectrin reveals how tandem spectrin-repeats generate unique ligand-binding properties. *Blood* **113:** 5377–5384.

Stankewich MC, Tse WT, Peters LL, Ch'ng Y, John KM, Stabach PR, Devarajan P, Morrow JS, Lux SE. 1998. A widely expressed βIII spectrin associated with Golgi and cytoplasmic vesicles. *Proc Natl Acad Sci* **95:** 14158–14163.

Stefanovic M, Markham NO, Parry EM, Garrett-Beal LJ, Cline AP, Gallagher PG, Low PS, Bodine DM. 2007. An 11-amino acid β-hairpin loop in the cytoplasmic domain of band 3 is responsible for ankyrin binding in mouse erythrocytes. *Proc Natl Acad Sci* **104:** 13972–13977.

Steiner D, Forrer P, Plückthun A. 2008. Efficient selection of DARPins with sub-nanomolar affinities using SRP phage display. *J Mol Biol* **382:** 1211–1227.

Stumpp MT, Amstutz P. 2007. DARPins: A true alternative to antibodies. *Curr Opin Drug Discov Devel* **10:** 153–159.

Sunderland WJ, Son YJ, Miner JH, Sanes JR, Carlson SS. 2000. The presynaptic calcium channel is part of a transmembrane complex linking a synaptic laminin (α4β2γ1) with non-erythroid spectrin. *J Neurosci* **20:** 1009–1019.

Takeda S, Yamazaki H, Seog DH, Kanai Y, Terada S, Hirokawa N. 2000. Kinesin superfamily protein 3 (KIF3) motor transports fodrin-associating vesicles important for neurite building. *J Cell Biol* **148:** 1255–1265.

Thomas GH, Newbern EC, Korte CC, Bales MA, Muse SV, Clark AG, Kiehart DP. 1997. Intragenic duplication and divergence in the spectrin superfamily of proteins. *Mol Biol Evol* **14:** 1285–1295.

Thomas GH, Zarnescu DC, Juedes AE, Bales MA, Londergan A, Korte CC, Kiehart DP. 1998. *Drosophila* β Heavy-spectrin is essential for development and contributes to specific cell fates in the eye. *Development* **125:** 2125–2134.

Travé G, Pastore A, Hyvönen M, Saraste M. 1995. The C-terminal domain of α-spectrin is structurally related to calmodulin. *Eur J Biochem* **227:** 35–42.

Voas MG, Lyons DA, Naylor SG, Arana N, Rasband MN, Talbot WS. 2007. αII-spectrin is essential for assembly of the nodes of Ranvier in myelinated axons. *Curr Biol* **17:** 562–568.

Waters J, Schaefer A, Sakmann B. 2005. Backpropagating action potentials in neurones: Measurement, mechanisms and potential functions. *Prog Biophys Mol Biol* **87:** 145–170.

Weber KL, Fischer RS, Fowler VM. 2007. Tmod3 regulates polarized epithelial cell morphology. *J Cell Sci* **120:** 3625–3632.

Wechsler A, Teichberg VI. 1998. Brain spectrin binding to the NMDA receptor is regulated by phosphorylation, calcium and calmodulin. *EMBO J* **17:** 3931–3939.

Whittard JD, Sakurai T, Cassella MR, Gazdoiu M, Felsenfeld DP. 2006. MAP kinase pathway-dependent phosphorylation of the L1-CAM ankyrin binding site regulates neuronal growth. *Mol Biol Cell* **17:** 2696–2706.

Wood SJ, Slater CR. 1998. β-Spectrin is colocalized with both voltage-gated sodium channels and ankyrin-G at the adult rat neuromuscular junction. *J Cell Biol* **140:** 675–684.

Xu M, Cao R, Xiao R, Zhu MX, Gu C. 2007. The axon-dendrite targeting of Kv3 (Shaw) channels is determined by a targeting motif that associates with the T1 domain and ankyrin-G. *J Neurosci* **27:** 14158–14170.

Yang Y, Ogawa Y, Hedstrom KL, Rasband MN. 2007. βIV spectrin is recruited to axon initial segments and nodes of Ranvier by ankyrin-G. *J Cell Biol* **176:** 509–519.

Zhang X, Davis JQ, Carpenter S, Bennett V. 1998. Structural requirements for association of neurofascin with ankyrin. *J Biol Chem* **273:** 30785–30794.

Zhang P, Talluri S, Deng H, Branton D, Wagner G. 1995. Solution structure of the pleckstrin homology domain of *Drosophila* β-spectrin. *Structure* **3:** 1185–1195.

Zhou D, Birkenmeier CS, Williams MW, Sharp JJ, Barker JE, Bloch RJ. 1997. Small, membrane-bound, alternatively spliced forms of ankyrin 1 associated with the sarcoplasmic reticulum of mammalian skeletal muscle. *J Cell Biol* **136:** 621–631.

Zhou D, Lambert S, Malen PL, Carpenter S, Boland LM, Bennett V. 1998. Ankyrin-G is required for clustering of voltage-gated Na channels at axon initial segments and for normal action potential firing. *J Cell Biol* **143:** 1295–1304.

Zuckerman JB, Chen X, Jacobs JD, Hu B, Kleyman TR, Smith PR. 1999. Association of the epithelial sodium channel with Apx and α-spectrin in A6 renal epithelial cells. *J Biol Chem* **274:** 23286–23295.

Planar Cell Polarity: Keeping Hairs Straight Is Not So Simple

Helen McNeill

Samuel Lunenfeld Research Institute, Department of Molecular Genetics, University of Toronto, Toronto, Canada

Correspondence: mcneill@mshri.on.ca

The fur on a cat's back, the scales on a fish, or the bristles on a fly are all beautifully organized, with a high degree of polarization in their surface organization. Great progress has been made in understanding how individual cell polarity is established, but our understanding of how cells coordinate their polarity in forming coherent tissues is still fragmentary. The organization of cells in the plane of the epithelium is known as planar cell polarity (PCP), and studies in the past decade have delineated a genetic pathway for the control of PCP. This review will first briefly review data from the *Drosophila* field, where PCP was first identified and genetically characterized, and then explore how vertebrate tissues become polarized during development.

DROSOPHILA PCP GENES, MUTANT PHENOTYPES, AND PATHWAY ANALYSIS

Drosophila PCP genes were first identified because of spontaneous mutations that caused a loss of organization in the surface bristles on the body, hairs on the wing, and photoreceptors in the eye (Fig. 1). Mutations in genes that affect PCP gave rise to a disorganized surface appearance and were named accordingly (for example, *dishevelled* [*dsh*] and *frizzled* [*fz*]) (Fahmy and Fahmy 1959; Gubb and Garcia-Bellido 1982). The distinctive swirls that occur in the wing of flies mutant for other PCP genes reminded their discoverers of the works of Vincent Van Gogh, leading to gene names such as *van gogh* (*vang*) (Taylor et al. 1998) and *starry night* (*stan*) (Chae et al. 1999). Genetic analysis indicated that these genes worked together and were needed for PCP in all tissues, thus they were named core PCP genes.

Core PCP genes include *dsh*, *fz*, *vang*, *stan*, *prickle* (*pk*), and *diego* (*dgo*) (Feiguin et al. 2001). *fz* encodes a seven-transmembrane receptor that functions as a receptor for Wingless (Wg), and *dsh* encodes a cytoplasmic transducer of the Wg pathway, implicating the Wnt pathway in PCP signaling and leading to the PCP pathway being referred to as the non-canonical Wnt pathway (for details of the canonical Wg/Wnt pathway, see Cadigan and Peifer 2009). Loss of Wg does not affect planar polarity, so original models in *Drosophila* PCP speculated that another Wnt might be the instructive cue for PCP, and that a gradient of a Wnt might direct PCP establishment. A great deal of effort went into investigating

 doi: 10.1101/cshperspect.a003376
Cite this article as *Cold Spring Harb Perspect Biol* 2009;1:a003376

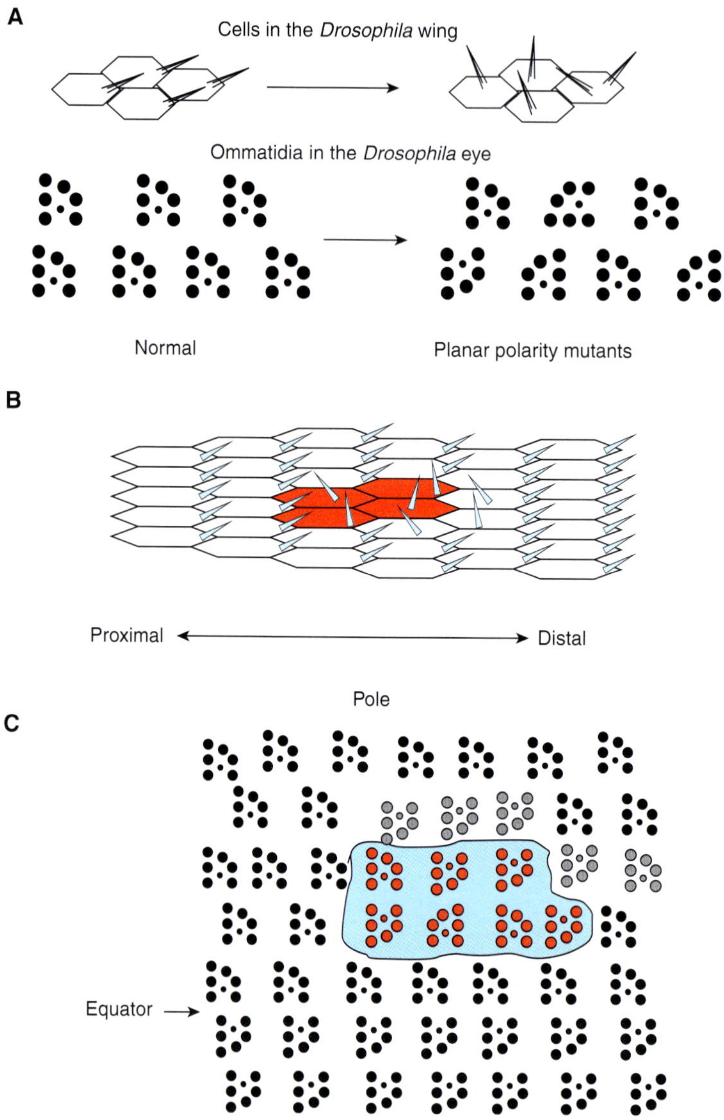

Figure 1. Planar polarity in the fly eye and wing. (*A*) Planar polarity can be clearly seen in the orientation of hairs on the wing of a fly (*top*) and in the orientation of photoreceptor clusters in sections of the fly eye (*bottom*). Mutations in PCP proteins results in loss of the planar organization of these tissues without affecting the cell fate or apical-basal polarity of the affected cells. Directional nonautonomy is a characteristic feature of PCP mutants. (*B*) For example, loss of *fz* in a clone of cells in the wing disrupts the polarity of mutant cells (marked in red) as well as wild-type cells on the distal side of the clones. Tissue on the proximal side is unaffected. (*C*) Similarly, loss of *fat* in a clone of cells in the eye (*fat* mutant tissue is outlined) leads to disruptions of polarity on the polar side of the clone in genotypically wild-type tissue, whereas the equatorial side of the clone is unaffected (disrupted polarity of wild-type cells is shown in gray).

this possibility, with largely negative results. Neither overexpression of different Wnts nor loss of function of multiple Wnts caused defects in planar polarity in the wing. Thus, current models (in *Drosophila*) have discounted the Wnts as instructive factors in PCP, however, vertebrate data suggests Wnts may also play permissive roles in PCP (see the following discussion).

The molecular nature of the other PCP genes reveals a diverse group of players. *pk* and *dgo* encode cytoplasmic proteins (Gubb et al. 1999; Feiguin et al. 2001), whereas *vang* (also known as *strabismus*) encodes a four-pass transmembrane protein (Taylor et al. 1998; Wolff and Rubin 1998) and *stan* (also known as *flamingo* [*fmi*]) encodes a seven-transmembrane protein with a large extracellular domain that contains multiple cadherin repeats (Usui et al. 1999). Fmi can engage in homotypic cell adhesion, and Fmi-based adhesions are essential for the establishment of PCP.

Core PCP proteins become localized to apical junctions early in development (Axelrod 2001; Bastock et al. 2003; Das et al. 2004; Jenny et al. 2003; Strutt et al. 2002; Tree et al. 2002). Strikingly, during pupal development this distribution becomes reorganized into an asymmetric distribution in the cell. This asymmetric localization occurs before any overt morphological evidence of PCP and is thought to be important for PCP signaling (reviewed in McNeill 2002). There are two distinct classes of asymmetric distribution: Fz, Dsh, and Dgo are localized to the distal side of each cell in the wing (where the hair will eventually emerge), whereas Vang and Pk are localized to the proximal side of each cell. Stan is localized to both proximal and distal ends of each hair cell (Fig. 2). There is a similar polarized localization of PCP proteins in the eye; however, the pattern is more complex, consistent with the more complex structure of the eye. Instead of an identical, repetitive pattern of proximal distal accumulation, core PCP proteins only accumulate in a subset of cells, most importantly at the photoreceptor R3/photoreceptor R4 interface.

Loss of any one of the core PCP genes leads to loss of the polarized localization of all the other components, suggesting complex positive and negative interactions. Supporting this is the finding of direct binding between many of the components, at least in in vitro settings: Diego and Pk bind Dsh (Tree et al. 2002; Jenny et al. 2003); Fmi binds Vang (Wu and Mlodzik 2008) and Fz (Chen et al. 2008); and Vang binds Pk and Dsh (Bastock et al. 2003; Park and Moon 2002). Expression of Fz in tissue culture cells causes the recruitment of Dsh to the membrane (Axelrod et al. 1998; Axelrod 2001) and there are quite weak but direct interactions between Fz and Dsh (Wong et al. 2003). Together these findings suggest that multiple direct protein–protein interactions underlie the asymmetric distribution of the core PCP proteins. It is still very unclear, however, how the direction of the asymmetry is established.

Another set of genes are only needed in subsets of tissues to direct planar polarity: These tissue-specific PCP genes include genes such as *multiple wing hair* (*mwh*) (Strutt and Warrington 2008), *fuzzy* (*fy*) (Collier and Gubb 1997), and *inturned* (*in*) (Park et al. 1996), which are only needed for wing PCP, and *mirror*, *fringe*, and *roulette*, which are only necessary for PCP in the eye (Choi and Benzer 1994; McNeill et al. 1997; Cho and Choi 1998; Dominguez and de Celis 1998; Yang et al. 1999; Strutt and Strutt 2003).

A more recently discovered group of PCP genes include the large cadherins Fat and Dachsous (Ds) and the Golgi-associated kinase Four-jointed, and are referred to as the Fat/Ds/Fj PCP cassette (Zeidler et al. 1999; Casal et al. 2002; Rawls et al. 2002; Yang et al. 2002; Matakatsu and Blair 2004; Simon 2004). These genes regulate PCP in all tissues in *Drosophila*; however, they form a distinct class of PCP genes from the core genes. Unlike the core PCP proteins, Fat and Ds are not distributed asymmetrically in the cell—instead the expression of Ds is graded over the tissue. In the wing, Ds is expressed at high levels at the proximal region of the wing and is absent from the distal portions. Conversely, Fj is expressed strongly at the distal edge of the wing and expression fades toward the proximal region. Fat is expressed evenly throughout the

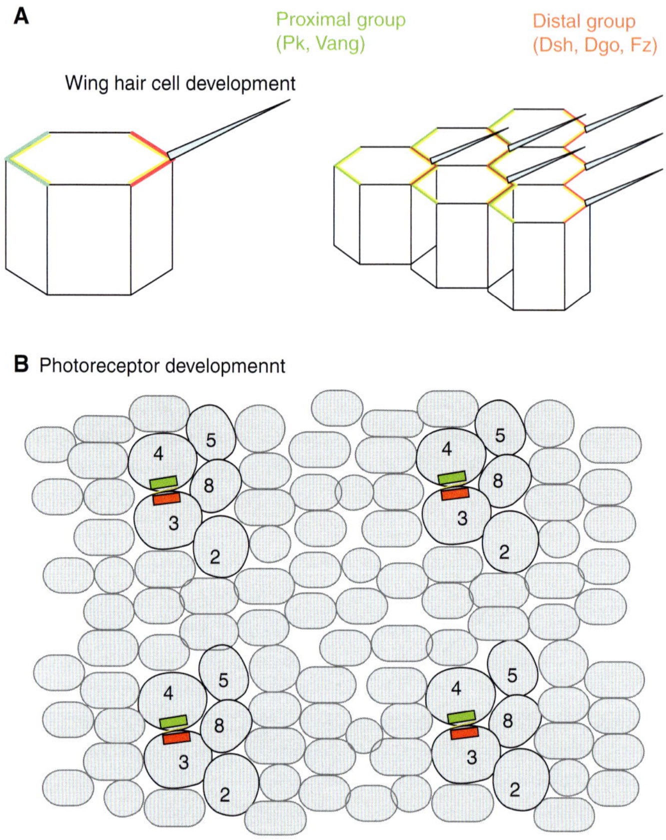

Figure 2. Asymmetry of core PCP proteins. During pupal development, the core PCP proteins become asymmetrically distributed in the plane of the epithelium. Fz, Dsh, and Dgo (shown in green) are enriched on the distal side of each cell, whereas Pk and Stbm (red) are enriched on the proximal side of each cell. Fmi (yellow) is enriched on both proximal and distal sides of each cell.

wing (Fig. 3). Current models propose that Ds binding to Fat inhibits Fat PCP activity (Yang et al. 2002; Ma et al. 2003). Fat regulates the expression of *fj* via direct binding of the transcriptional repressor, Atrophin (Fanto et al. 2003). Original models of PCP suggested that the Fat/Ds/Fj cassette is upstream of the core PCP genes and directs their asymmetric distribution. This has been recently challenged (Casal et al. 2006), and it is currently unclear if the Fat/Ds cassette is upstream or in parallel of the core PCP, and if it is, how the information is propagated from one system to the other. Intriguingly, recent studies have shown that Fat binds to the kinase Dco (Sopko et al. 2009), which has been shown to regulate PCP via the core PCP protein Dsh (Strutt et al. 2006), providing a possible link between core and Fat/Ds PCP regulation. Interestingly, Fat forms *cis*-dimers and Ds binding to Fat promotes Fat dimerization and phosphorylation by Dco (Sopko et al. 2009). Recent data also indicates that the Fat/Ds cassette regulates tissue growth through the Hippo pathway (reviewed in Sopko and McNeill 2009), but the regulation of growth appears to be separable from the control of PCP.

NONAUTONOMOUS PLANAR CELL POLARITY

One of the most striking yet mysterious aspects of planar cell polarity is the phenomena known as domineering nonautonomy. Analysis of patches of mutant tissues (induced by recombination and called mosaic clones) revealed that

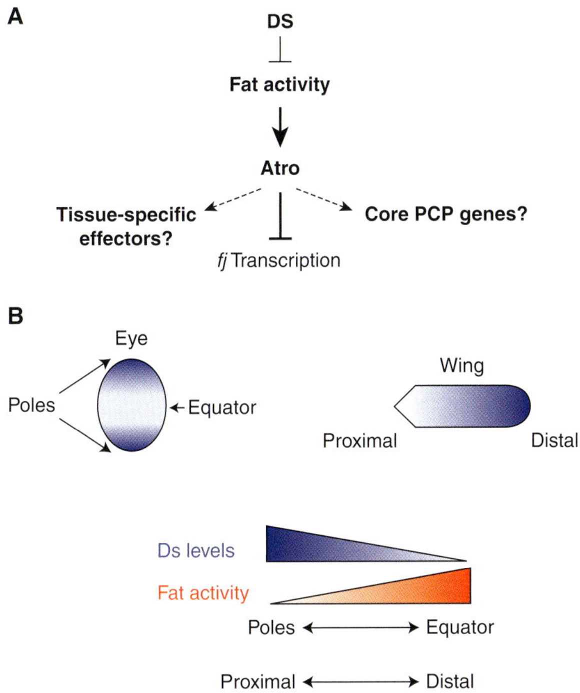

Figure 3. Fat and Ds model and gradients. Fat is a large cadherin that functions as a receptor for another large cadherin, Ds. Ds binding to Fat inhibits Fat activity. Ds is expressed in a gradient in the eye and the wing, resulting in an inverse gradient of Fat activity. Fat represses *fj* transcription, acting through the transcriptional corepressor, Atrophin. The PCP effects of *fj* mutants are much weaker than the PCP effects of loss of *fat* or *atrophin*, suggesting that there are other targets. One potential target is the activity of core PCP proteins such as Fz.

wild-type tissue bordering the mutant tissue lost polarity (Fig. 1). This phenomena is seen in all tissues, but is easiest to describe in the wing, which is planar polarized in the proximal–distal axis. The direction of the nonautonomy is consistent for each genotype: For $fz^{-/-}$ clones, wild-type cells distal to the clone lose planar polarization, but those proximal to the clone are normal (Vinson and Adler 1987). Conversely, clones of $Vang^{-/-}$ cells disrupt the organization of wild-type cells on the proximal side of clones, whereas distal cells are unaffected (Taylor et al. 1998). Nonautonomous disruptions of PCP are also seen with Fat/Ds/Fj clones. In this case, clones of Ds disrupt polarity on the opposite side of the clones from clones of Fat, which is one of the arguments that Ds inhibits Fat activity in PCP. Interestingly, not all elements of the core PCP pathway display nonautonomous polarity defects: Clones of Dsh, Pk, Diego, and Fmi do not affect the polarity of adjacent wild-type cells (Strutt and Strutt 2007). In addition, there are separable mutations that affect Fz autonomous and nonautonomous PCP activity (Vinson and Adler 1987), and different times at which Fz autonomous and nonautonomous signaling function during development (Strutt and Strutt 2002). This has led to the proposal that there are separable cell-autonomous and -nonautonomous routes for PCP signal. The mechanistic basis of nonautonomous polarity is hotly contested at present (Chen et al. 2008; Lawrence et al. 2008; Strutt and Strutt 2008; Wu and Mlodzik 2008).

MAMMALIAN PCP

Mammalian tissues also show clear planar organization, and for many years it was speculated that the coordinate organization of hair cells in the inner ear or the fur on an animal's back represented a similar process as *Drosophila* planar polarity (Eaton 1997) and might be subject to control by homologs of *Drosophila* PCP genes. This has been strongly supported by findings in the past few years. Loss of Fz6, for example, leads to disorganized hair on the back of a mouse with swirls that look strikingly similar to the swirls seen on the wing of a fly with mutations in PCP genes (Guo et al. 2004). Similarly, mutations in Vangl2 disrupt the coordinate organization of hair cells in the inner ear of mice (Montcouquiol et al. 2003). In depth analysis of the processes underlying the planar organization of these tissues has found some striking similarities between *Drosophila* and mammalian PCP establishment, but also some important differences.

PLANAR POLARITY IN THE INNER EAR

The inner ear is a complex structure that requires precise planar organization to carry out its function in hearing and in balance (Fig. 4). The cochlea is essential for hearing, whereas balance is dependent upon the otolith organs and semicircular canals. The organ of Corti, in the cochlea, contains of rows of sensory cells known as hair cells. Hair cells have apical microvilli-derived stereocila and a single primary cilium known as the kinocilium. The stereocilia and kinocilum form a V-shaped structure that is polarized across a field of hair cells. The cochlea has one row of inner hair cells and three rows of outer hair cells, which are interdigitated with support cells. The vertices of the hair bundles all point toward the periphery of the cochlea, thus displaying planar polarity. There is also a planar polarity of the sensory hair cells of the vestibular system, where, intriguingly, the hair cells "point" toward a central line of polarity reversal. This is reminiscent of the reversion of PCP that occurs at the equator of the fly eye.

The planar organization of hair cells in the inner ear is disrupted in a number of core PCP mutants, supporting the proposal that this coordinate organization is a form of vertebrate PCP. Mutation of *Vangl2* (Montcouquiol et al. 2003; Qian et al. 2007; Torban et al. 2008) or *Celsr1* (a homolog of *Fmi/Stan*) (Curtin et al. 2003) disrupts hair cell orientation. Single loss of

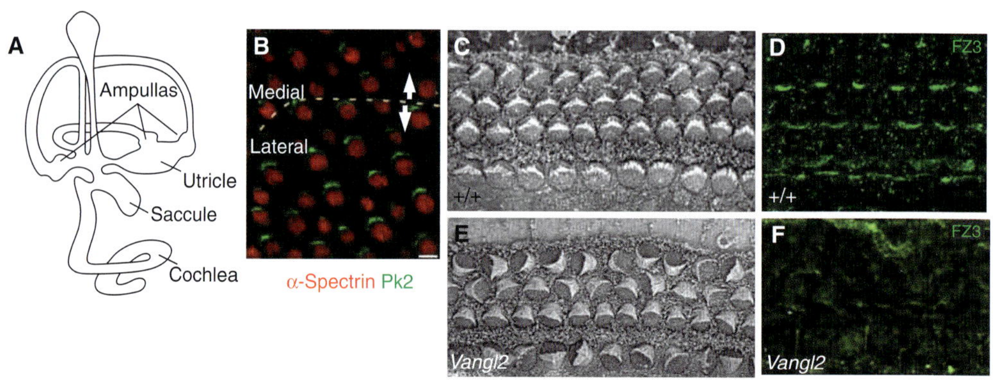

Figure 4. PCP in the vertebrate inner ear. (*A*) Structure of the inner ear. (*B*) Cells in the saccule are planar polarized, with an inversion of polarity near the middle of the tissue. The planar polarity of cells is strikingly seen by staining with antibodies to Prickle 2, which light up one side of all the sensory cells (*A* and *B* reprinted with permission from Deans et al. 2007). (*C*) Hair cells in the cochlea are characterized by apical actin-rich stereocilia and a single sensory cilum, known as a kinocilium. Hair cells are divided into three rows of outer hair cells and a single row of inner hair cells. All the stereocilia are polarized in the plane of the epithelium. This organization is lost in PCP mutants such as *Vangl2* (*E*). Hair cell polarity is also seen with staining with antibodies to Frizzled, which stains one side of the cells, indicating the conserved planar polarity of the tissues (*D*). This staining is lost in *Vangl2* mutants (*F*). (*C–F* reprinted, with permission, from Montcouquiol et al. 2003 [Macmillana], 2006.)

other PCP genes does not cause planar defects in the ear, presumably because of redundancy between related family members. Consistent with this interpretation, *Fz3;Fz6* double mutants (Wang et al. 2006) and *Dvl1;2* double mutants have defective orientation of hair cells (Wang et al. 2005), although single mutants have normal inner ear PCP.

There are often strong genetic interactions between PCP genes in planar polarity of the inner ear. For example, *Dvl3*$^{-/-}$ cochleas have only minor PCP defects and there are no detectable PCP defects in *Vangl2*$^{Lp/+}$ cochlea, yet *Dvl3*$^{-/-}$;*Vangl2*$^{Lp/+}$ hair cells have dramatic misorientations of the hair cells (Etheridge et al. 2008). Loss of Fat4, the homolog of *Drosophila* Fat, also causes defects in orientation of hair cells of the inner ear (Saburi et al. 2008), although these effects are weaker than those found in *Vangl2*$^{Lp/Lp}$ *or Celsr*$^{-/-}$ mutants. Thus the phenotypes of many mutants suggest that vertebrate homologs of *Drosophila* PCP genes control similar planar tissue organization. There may be some dependence on a variety of PCP genes in different portions of the ear.

Excitingly, close analysis of the inner ear has also shown that vertebrate core PCP proteins display asymmetric protein distribution, analogous to *Drosophila* core PCP proteins. Fz3 and Fz6 proteins are localized to the lateral faces of sensory and supporting cells in all sensory epithelia in a pattern that correlates with the axis of planar polarity (Wang et al. 2006). Vangl2, Dvl2, and Pk2 have also been shown to have asymmetric localization in vertebrates (Montcouquiol et al. 2006; Wang et al. 2006; Deans et al. 2007; Devenport and Fuchs 2008). Therefore, a polarized localization of PCP components is a conserved feature of the PCP signaling cascade. This suggests that the mechanism of asymmetric localization of PCP proteins is conserved and likely to play an important role in the establishment of planar polarized tissues. This asymmetry has been seen in other tissues and in other species: Pk, for example, is asymmetric in the fish neural keel during early development (Ciruna et al. 2006).

And, as in flies, loss of components of the core pathway can disrupt the localization of other core PCP proteins: For example, loss of Vangl2 leads to loss of polarized localization of Fz3 and Fz6 (Montcouquiol et al. 2006; Wang et al. 2006), again highlighting conservation of mechanisms between flies and mice. It is not a completely simple story, however, as some of the proteins that show distinct protein localizations in *Drosophila* are found in overlapping patterns in the mouse. For example, in flies, Fz and Dsh colocalize on the distal edges of hair cells, whereas Pk and Vang are localized to the proximal side of cells. It was surprising, therefore, to find that Vangl2, Fz, and Fz6 are found on the same side of inner ear sensory cells (Montcouquiol et al. 2006) in the mouse. Thus, the details of the implementation of the PCP signal may be more complicated in mammals, and one cannot simply extrapolate from the fly data.

PLANAR POLARITY IN VERTEBRATE HAIR

The sleek organization of hairs on the back of a horse or a mouse is clearly analogous to the hairs on a fly's wing, but very different in the underlying biology. Mammalian hairs are made up of groups of cells, whereas the hair on a flies wing is a single actin-rich extension present on each cell. Yet, satisfyingly, recent data has indicated that the planar organization of these very different tissues is controlled by the conserved PCP signaling molecules.

Initial studies showed that *Fz6*$^{-/-}$ mice have disorganized fur and suggested that the strongest disorganization occurred with interspersed *Fz6*$^{-/-}$ patches in chimeras (Guo et al. 2004). Inversin is a homolog of the core PCP protein Dgo, and *inversin* mutants have hair patterning defects similar to those observed in *Fz6*$^{-/-}$ mice (Simons et al. 2005). More recent studies have shown that many core PCP genes are involved in establishing organization of hair in vertebrates (Devenport and Fuchs 2008). Fur is not visible on newborn mice, and loss of many PCP genes causes embryonic lethality. Thus the full contribution of PCP genes to planar organization of hair can only be seen in transplants and chimera experiments.

Hair formation begins in the mouse at E14.5. Inductive signals from the dermis cause placodes to bud from the overlying epithelium. Groups of cells aggregate and invade the dermis. Eventually, the entire structure becomes polarized in the plane of the epithelium with a resulting tilt that causes the hair to lay flat as it emerges. As the hair germ develops, cells on the anterior side constrict basally and more posterior cells become elongated and columnar. Clearly this is a process that is far more complicated than causing a single actin-rich cellular extension to arise from one side of the cell, as in the fly wing. It is perhaps more analogous to the development of polarity in the *Drosophila* eye, which also forms from clusters of cells that rotate 90° to adopt their final morphological polarity. Although every cell in the fly wing shows a proximal–distal distribution of core PCP proteins, in the eye only a subset of cells display asymmetric core PCP protein distribution. In a similar manner, examination of the distribution of core PCP proteins in the developing epidermis revealed that Vangl2, Fz6, and Celsr1 are asymmetrically distributed (Devenport and Fuchs 2008). Vangl2 and Celsr1 are restricted to the lateral sides of the basal epidermal cells and enriched along the anterior–posterior axis. In addition, there also appears to be a nonautonomous component to PCP, as chimeric experiments revealed that *Vangl2*$^{Lp/Lp}$ tissue disrupts the polarity of adjacent wild-type hair follicles.

PLANAR POLARITY GENES CONTROL COORDINATE CELL MOVEMENTS AND PROMOTE LENGTHENING AND NARROWING OF TISSUES

The planar polarity of the inner ear and of the fur of mammals is conceptually similar to the organization of cells in the wing and the eye of the fly. However, there are also novel PCP phenotypes in mammals that do not directly correlate with classic *Drosophila* PCP phenotypes. Most notable is the role of PCP genes in polarized cell movements needed to establish the correct shape of many organs, and even of animals. Loss of many PCP genes leads to problems in neural tube closure, eyelid closure, and cochlear extension. Underlying all of these defects is a loss of polarized cell movements that cause tissues to narrow and lengthen.

The first indications of a role for PCP proteins in this process were found in frogs and fish, where altering Dsh, Fz7, or Wnt11 disrupted the cell movements of convergent extension during gastrulation (Djiane et al. 2000; Heisenberg et al. 2000; Wallingford et al. 2000). This resulted in animals with short, broad tissues. Excitingly, point mutations in Dsh, which specifically altered PCP in flies, led to convergent-extension defects, thereby supporting the proposal that the morphogenetic defects observed were because of loss of PCP signaling. Altering *Vang* also lead to convergence and extension defects (Goto and Keller 2002). Because of these original observations, many other PCP components have been shown to affect convergence and extension movements in fish and frogs (reviewed in Tada and Kai 2009).

Analysis of mice mutants for various core PCP components soon revealed that the neural tube failed to close in many cases and that a lack of convergence and extension movements in the developing tissue underlies the neural tube closure defects. The neural tube closure defects are at times extreme (when they are called craniorachisis), or can be more subtle. Defects in neural tube closure are found in *Vangl2*$^{Lp/Lp}$ mutants (Kibar et al. 2001), Celsr mutants (Curtin et al. 2003), *Fz3/Fz6* double mutants (Wang et al. 2005), and *Dvl1/Dvl2* double mutants. Early studies showed that the midline fails to narrow and extend properly in *Vangl2*$^{Lp/Lp}$ embryos. Although the neural folds form and elevate normally, they are spaced widely apart as a result of intervening cells that have not intercalated.

Many developmental defects caused by loss of convergence and extension movements occur in PCP mutants. One example is lack of eyelid closure. Eyelid closure normally occurs at about E16 in the mouse and, like neural tube closure, involves a medial convergence of a pair of flanking epithelial sheets. Defects in eyelid closure are found in *Fz3,Fz6* mutants and *Celsr* mutants. A narrowing and lengthening analogous to convergent extension also occurs during development of the organ of Corti in the mammalian cochlea and this shape change fails to occur in *Vangl2* mutants, *Dvl1;Dvl2* double-mutants, *Fz3;Fz6* double-mutants, and *Fat4* mutants (Montcouquiol et al. 2003; Wang et al. 2005; Wang et al. 2006; Saburi et al. 2008). Thus shaping elongated tissues is a crucial role of many PCP genes.

CYSTIC KIDNEY DISEASE AND VERTEBRATE PCP

An unexpected defect caused by loss of PCP signaling is cystic kidney disease. One of the first indications of a link between cystic kidney disease and PCP in mammals was the finding that mutations in *inversin* cause cystic kidney disease. Inversin is a homolog of the core PCP protein Dgo, and controls the balance between canonical and noncanonical Wnt (PCP) signaling (Simons et al. 2005). This provided a potential link between PCP and cystic kidney disease. At that time, however, it was unclear how loss of normal PCP signaling might lead to cyst formation.

More recent studies have suggested how PCP signaling may function in normal kidney development and why defective PCP signaling causes cystic dilations of kidney tubules. Using cell lineage tracing and examination of cell in mitosis, it was shown that when tubular epithelial cells divide, there is a high degree of oriented cell division (Fischer et al. 2006): 95% of cells divide within 34° of the axis of the tubule. This suggested that oriented cell division is essential for the normally thin elongated tubes that make up much of the nephron (Fig. 5). Importantly, two different mouse models of polycystic kidney disease display

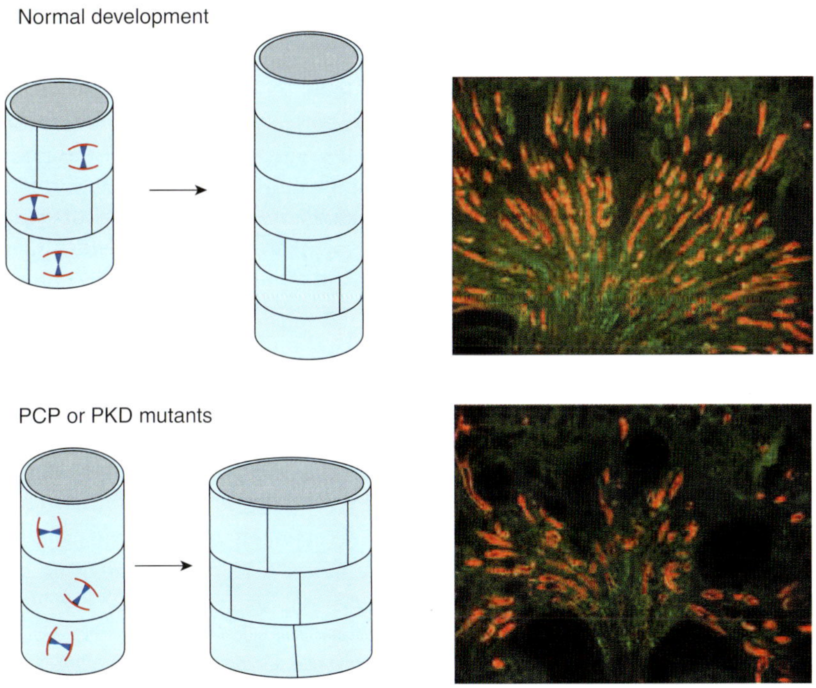

Figure 5. Oriented Cell division is lost in PCP mutants, leading to cyst formation. During normal development, epithelial cells in the kidney tubule have oriented division, resulting in an elongation of the tubule during development (*top*). In PCP mutants such as Fat4, this strict oriented cell division is lost, leading to a widening of the tubule diameter and eventually cyst formation (*bottom*).

loss of oriented cell division. Previous studies had shown that loss of PCP genes leads to defects in oriented cell division in flies and in fish (Gho and Schweisguth 1998; Garoia et al. 2000; Bellaiche et al. 2001; Gong et al. 2004; Baena-Lopez et al. 2005). Given the known role of PCP signaling in oriented cell division, cystic kidney disease might be caused by defective PCP.

This hypothesis was recently strengthened by the observation that mutants in Fat4, an ortholog of *Drosophila* Fat, display many classic PCP phenotypes, including inner ear defects and neural tube defects (Saburi et al. 2008). The Fat/Ds/Fj signaling cassette seems to function in the same way in mice as in flies, because in *Fat4* mutants, *Fjx1*, the *fj* ortholog is transcriptionally up-regulated, as it is in flies in *fat* mutants. Importantly, *Fat4* mutant kidneys have dilated tubules and cysts. Analysis of the oriented cell division revealed that loss of Fat4 leads to randomization of spindle orientation at birth. The cystic phenotype of Fat4 mutant kidneys was enhanced by loss of *Fj* ortholog, *fjx1*, as well as loss of one copy of the core PCP gene *Vangl2*, further supporting the hypothesis that loss of PCP signaling leads to cyst formation. Loss of Drosophila Fat leads to loss of oriented cell division in the wing, suggesting the role of Fat in controlling cell division is conserved (Baena-Lopez et al. 2005). Interestingly, Fat4 is localized to the primary cilium, an organelle already implicated in polycystic kidney disease and in vertebrate PCP signaling.

The finding that the majority of protein products of cystic disease genes are localized on the primary cilia has led to the ciliary hypothesis of cystic disease (Kim and Walz 2007; Yoder 2007; Harris and Torres 2008). A potential role for cilia in the kidney is in flow sensing: Renal cilia project into the tubule lumen and will bend in response to fluid flow. Bending of the cilia stimulates an increase in cytosolic calcium concentration (Praetorius and Spring 2003) leading to the proposal that primary cilia function as mechanosensors of urine flow in the renal tubules, and that loss of flow sensing leads to cyst formation. However, recent papers (Piontek et al. 2007; Patel et al. 2008) have challenged the notion that tubular geometry is maintained by the mechanical bending of cilia in response to flow. Intriguingly, recent data has suggested that PCP signaling may be linked to cilia function. Many ciliary proteins that have been linked to cystic disease, such as Bardel–Biedl syndrome and Oculo-Facial-Digital syndrome, also show PCP defects (Davenport and Yoder 2005; Ferrante et al. 2008; Ross et al. 2005). In addition, cyst formation in kidneys occurs when cilia are ablated, and, intriguingly, the control of oriented cell division is also lost in cells lacking cilia (Patel et al. 2008).

CILIA AND PCP

Many links have been made between PCP and cilia, although the exact nature of the relationship between these is still unclear. Loss of the PCP effector genes *fuzzy* and *inturned* leads to disruption of the cytoskeleton (Park et al. 2006) and defects in cilia formation. Ciliary associated proteins such as Inversin are thought to regulate the balance between canonical Wnt signaling and noncanonical (PCP) signaling, potentially through regulation of Dsh (Simons et al. 2005; Simons and Walz 2006; Benzing et al. 2007). It has also been found that interfering with the core PCP protein Dsh causes loss of cilia in bronchial epithelial cells because of defects in docking of basal bodies (Park et al. 2008). Loss of cilia leads to defects in convergent extension of the cochlea (Jones et al. 2008). And, as mentioned, PCP proteins such as Fat4 and Vangl2 localize to the base of cilia (Ross et al. 2005; Saburi et al. 2008) in cultured cells.

What might be the role of cilia in PCP? Cilia are thought to be chemosensors and may be important in sensing PCP signals (of as yet undefined nature). Another place for cilia might be in response to PCP signals. Recent studies in the cochlea support the placement of cilia downstream of PCP signaling, as loss of cilia does not affect the polarized distribution of core PCP proteins (Jones et al. 2008).

VERTEBRATE-SPECIFIC PCP GENES

The role of cilia in PCP and cystic kidney regulation highlights a function in PCP that is vertebrate specific. Few cells in *Drosophila* are ciliated, and cilia have no role in the control of PCP in the wing or eye of the fly. The role of cilia in vertebrate PCP is further underscored by the observation that many genes identified as susceptibility loci for Bardet–Biedl syndrome (BBS) show characteristic PCP defects, including open eyelids, neural tube defects and inner ear polarity disruptions (Ross et al. 2005). BBS is associated with cystic disease and obesity, and BBS proteins localize to the primary cilia.

Another class of vertebrate-specific PCP genes are related to Wnt signaling. There are 19 mammalian Wnt ligands, the majority of which signal predominantly in canonical signaling via binding to Fz receptors and activation of Dsh homologs (Dvl). Activation of Dvl disrupts the β-catenin disruption complex (composed of Axin, APC, GSK3, and CK1), resulting in stabilization of β-catenin. β-catenin can translocate to the nucleus, where it activates transcription of target genes in conjunction with TCF/LEF transcription factors (reviewed in Huang and He 2008) (see also Cadigan and Peifer 2009). In contrast, noncanonical Wnt signaling is β-catenin independent. A number of Wnts appear to regulate noncanonical signaling. For example, hair cells are misoriented in the cochlea of Wnt5a mutant mice (Qian et al. 2007) and both Wnt5 and Wnt11 are required for convergent extension in fish and frogs (Heisenberg et al. 2000; Tada and Smith 2000; Kilian et al. 2003). Surprisingly, the convergent extension defects of Wnt11 mutants can be rescued by exogenous expression of Wnt11, suggesting that Wnt11 plays a permissive role in PCP (Heisenberg et al. 2000). Recent studies suggested that the ability of a specific Wnt to signal via canonical versus noncanonical pathways may be caused by the presence of specific coreceptors or additional kinases. (Hikasa et al. 2002; Lu et al. 2004; Yamamoto et al. 2008), which may reflect the need for more complex regulation of PCP in vertebrates.

Similarly, the receptor tyrosine kinase PTK7 functions in vertebrate PCP; mutants have neural tube closure defects and inner ear PCP defects (Lu et al. 2004). However, the fly homolog of PTK7, Otk, has been studied, and there are no PCP defects in either the eye or the wing in *otk* mutants (Winberg et al. 2001). Thus perhaps PTK7 has evolved additional functional interactions with the PCP pathway during mammalian evolution.

Of particular interest is the finding that mutations in vertebrate Scribble (Scrb) and Discs Large (Dlg) affects PCP in mammals. Scrb and Dlg are core components of the laterally acting Scrb group, which counteracts the apical-promoting Crumbs complex (reviewed in Dow and Humbert 2007; Assemat et al. 2008; Martin-Belmonte and Mostov 2008; Yamanaka and Ohno 2008). Although *Drosophila* Scrb and Dlg have no role in PCP, current data indicate a role for these genes in PCP control in mammals.

Scrb1 mutants have neural tube closure defects (Murdoch et al. 2003) and hair cell polarity defects in the inner ear (Montcouquiol et al. 2003). Vangl2 and Scrb1 colocalize, and there are very strong genetic interactions between Scrb1 and Vangl2: Scrb1$^{-/+}$; Vangl2$^{-/+}$ mice have clear PCP defects. Interestingly, Scrb1 binds directly to Vangl2 (Montcouquiol et al. 2006), suggesting these genetic interactions are caused by a direct physical interaction between these proteins.

One possible explanation for the role of Scrb in PCP might be that it is needed for the correct localization of Vangl2. Interestingly, Scrb also binds Lgl, at a site distinct from the Vangl2 binding domain (Kallay et al. 2006), and Dsh binds Lgl (Dollar et al. 2005). Although none of these genes affect PCP in *Drosophila*, there are data indicating that loss of apical polarity components, such as aPKC, can weakly disrupt PCP in flies (Djiane et al. 2005). Another possible link could be via JNK, because loss of Scrb complex components in *Drosophila* results in activation of the JNK pathway. This points out another paradoxical aspect of PCP signaling in flies and mice. Original studies in *Drosophila* suggested that

JNK was a key component of PCP. These studies were based primarily on overexpression experiments (Boutros et al. 1998), and later studies showed that there were little if any PCP defects found in JNK mutants (Weber et al. 2000; Strutt et al. 2002). Therefore, in current models of PCP in flies, JNK is no longer considered to play a key role in PCP signaling. The role of JNK in mammalian PCP is generally accepted (Bloor and Kiehart 2002; Kaltschmidt et al. 2002; Kuhl 2002; Schwarz-Romond et al. 2002; Habas et al. 2003; Takeuchi et al. 2003; Cong et al. 2004; Medina et al. 2004; Unterseher et al. 2004; Schambony and Wedlich 2007; Tahinci et al. 2007; Bikkavilli et al. 2008), although some recent studies have suggested that inhibiting JNK either genetically or with pharmacological inhibitors does not affect convergent extension (Ybot-Gonzalez et al. 2007). In any case, the exact contribution of JNK to mammalian PCP is not yet clear and requires further investigation.

SUMMARY AND FUTURE DIRECTIONS

Work from flies, fish, frogs, and mice has shed light on how tissue organization is regulated in development. Yet many questions remain unanswered. How spatial information is communicated to cells in a tissue is a crucial question. Recent studies have revealed a role for Wnt11 in orienting muscle fibers (Gros et al. 2009), providing perhaps the best example of information provided by Wnts in tissue organization. But is this direct or indirect? Work from *Drosophila* has indicated that the Fat/Ds/Fj cassette may provide positional cues for PCP downstream of Wg, and this is a possibility that should be examined in the mouse. In no system is it certain if the Fat/Ds/Fj cassette is upstream of core PCP proteins: Understanding the relationship of the core PCP proteins to the Fat pathway is an important goal. And if it is upstream, how is PCP information conveyed from Fat to the core PCP proteins? Perhaps the biggest challenge of all is to understand how information from the PCP genes is transformed into cell reorganization to generate the nearly flawless planar polarization evident in the animals we see around us every day.

REFERENCES

Assemat E, Bazellieres E, Pallesi-Pocachard E, Le Bivic A, Massey-Harroche D. 2008. Polarity complex proteins. *Biochim Biophys Acta* **1778:** 614–630.

Axelrod JD. 2001. Unipolar membrane association of Dishevelled mediates Frizzled planar cell polarity signaling. *Genes Develop* **15:** 1182–1187.

Axelrod JD, Miller JR, Shulman JM, Moon RT, Perrimon N. 1998. Differential recruitment of Dishevelled provides signaling specificity in the planar cell polarity and Wingless signaling pathways. *Genes Develop* **12:** 2610–2622.

Baena-Lopez LA, Baonza A, Garcia-Bellido A. 2005. The orientation of cell divisions determines the shape of *Drosophila* organs. *Curr Biol* **15:** 1640–1644.

Bastock R, Strutt H, Strutt D. 2003. Strabismus is asymmetrically localised and binds to Prickle and Dishevelled during *Drosophila* planar polarity patterning. *Development* **130:** 3007–3014.

Bellaiche Y, Gho M, Kaltschmidt JA, Brand AH, Schweisguth F. 2001. Frizzled regulates localization of cell-fate determinants and mitotic spindle rotation during asymmetric cell division. *Nat Cell Biol* **3:** 50–57.

Benzing T, Simons M, Walz G. 2007. Wnt signaling in polycystic kidney disease. *J Am Soc Nephrol* **18:** 1389–1398.

Bikkavilli RK, Feigin ME, Malbon CC. 2008. G αo mediates WNT-JNK signaling through dishevelled 1 and 3, RhoA family members, and MEKK 1 and 4 in mammalian cells. *J Cell Sci* **121:** 234–245.

Bloor JW, Kiehart DP. 2002. *Drosophila* RhoA regulates the cytoskeleton and cell-cell adhesion in the developing epidermis. *Development* **129:** 3173–3183.

Boutros M, Paricio N, Strutt DI, Mlodzik M. 1998. Dishevelled activates JNK and discriminates between JNK pathways in planar polarity and wingless signaling. *Cell* **94:** 109–118.

Cadigan KM, Peifer M. 2009. Wnt signaling from development to disease: Insights from model systems. *Cold Spring Harb Perspect Biol* **1:** a002881.

Casal J, Struhl G, Lawrence PA. 2002. Developmental compartments and planar polarity in *Drosophila*. *Curr Biol* **12:** 1189–1198.

Casal J, Lawrence PA, Struhl G. 2006. Two separate molecular systems, Dachsous/Fat and Starry night/Frizzled, act independently to confer planar cell polarity. *Development* **133:** 4561–4572.

Chae J, Kim MJ, Goo JH, Collier S, Gubb D, Charlton J, Adler PN, Park WJ. 1999. The *Drosophila* tissue polarity gene starry night encodes a member of the protocadherin family. *Development* **126:** 5421–5429.

Chen WS, Antic D, Matis M, Logan CY, Povelones M, Anderson GA, Nusse R, Axelrod JD. 2008. Asymmetric homotypic interactions of the atypical cadherin flamingo mediate intercellular polarity signaling. *Cell* **133:** 1093–1105.

Cho KO, Choi KW. 1998. Fringe is essential for mirror symmetry and morphogenesis in the *Drosophila* eye. *Nature* **396:** 272–276.

Choi KW, Benzer S. 1994. Rotation of photoreceptor clusters in the developing *Drosophila* eye requires the nemo gene. *Cell* **78:** 125–136.

Ciruna B, Jenny A, Lee D, Mlodzik M, Schier AF. 2006. Planar cell polarity signalling couples cell division and morphogenesis during neurulation. *Nature* **439:** 220–224.

Collier S, Gubb D. 1997. *Drosophila* tissue polarity requires the cell-autonomous activity of the fuzzy gene, which encodes a novel transmembrane protein. *Development* **124:** 4029–4037.

Cong F, Schweizer L, Varmus H. 2004. Casein kinase Iepsilon modulates the signaling specificities of dishevelled. *Mol Cell Biol* **24:** 2000–2011.

Curtin JA, Quint E, Tsipouri V, Arkell RM, Cattanach B, Copp AJ, Henderson DJ, Spurr N, Stanier P, Fisher EM, et al. 2003. Mutation of Celsr1 disrupts planar polarity of inner ear hair cells and causes severe neural tube defects in the mouse. *Curr Biol* **13:** 1129–1133.

Das G, Jenny A, Klein TJ, Eaton S, Mlodzik M. 2004. Diego interacts with Prickle and Strabismus/Van Gogh to localize planar cell polarity complexes. *Development* **131:** 4467–4476.

Davenport JR, Yoder BK. 2005. An incredible decade for the primary cilium: A look at a once-forgotten organelle. *Am J Physiol* **289:** F1159–1169.

Deans MR, Antic D, Suyama K, Scott MP, Axelrod JD, Goodrich LV. 2007. Asymmetric distribution of prickle-like 2 reveals an early underlying polarization of vestibular sensory epithelia in the inner ear. *J Neurosci* **27:** 3139–3147.

Devenport D, Fuchs E. 2008. Planar polarization in embryonic epidermis orchestrates global asymmetric morphogenesis of hair follicles. *Nat Cell Biol* **10:** 1257.

Djiane A, Riou J, Umbhauer M, Boucaut J, Shi D. 2000. Role of frizzled 7 in the regulation of convergent extension movements during gastrulation in *Xenopus* laevis. *Development* **127:** 3091–3100.

Djiane A, Yogev S, Mlodzik M. 2005. The apical determinants aPKC and dPatj regulate Frizzled-dependent planar cell polarity in the *Drosophila* eye. *Cell* **121:** 621–631.

Dollar GL, Weber U, Mlodzik M, Sokol SY. 2005. Regulation of Lethal giant larvae by Dishevelled. *Nature* **437:** 1376–1380.

Dominguez M, de Celis JF. 1998. A dorsal/ventral boundary established by Notch controls growth and polarity in the *Drosophila* eye. *Nature* **396:** 276–278.

Dow LE, Humbert PO. 2007. Polarity regulators and the control of epithelial architecture, cell migration, and tumorigenesis. *Int Rev Cytol* **262:** 253–302.

Eaton S. 1997. Planar polarization of *Drosophila* and vertebrate epithelia. *Current Opinion in Cell Biol* **9:** 860–866.

Etheridge SL, Ray S, Li S, Hamblet NS, Lijam N, Tsang M, Greer J, Kardos N, Wang J, Sussman DJ, et al. 2008. Murine dishevelled 3 functions in redundant pathways with dishevelled 1 and 2 in normal cardiac outflow tract, cochlea, and neural tube development. *PLoS Genetics* **4:** e1000259.

Fahmy OG, Fahmy M. 1959. New mutants report. *Dros Inf Serv* **33:** 82–94.

Fanto M, Clayton L, Meredith J, Hardiman K, Charroux B, Kerridge S, McNeill H. 2003. The tumor-suppressor and cell adhesion molecule Fat controls planar polarity via physical interactions with Atrophin, a transcriptional co-repressor. *Development* **130:** 763–774.

Feiguin F, Hannus M, Mlodzik M, Eaton S. 2001. The ankyrin repeat protein Diego mediates Frizzled-dependent planar polarization. *Develop Cell* **1:** 93–101.

Ferrante MI, Romio L, Castro S, Collins JE, Goulding DA, Stemple DL, Woolf AS, Wilson SW. 2008. Convergent extension movements and ciliary function are mediated by ofd1, a zebrafish orthologue of the human oral-facial-digital type 1 syndrome gene. *Human Mol Gen* **18:** 289–303

Fischer E, Legue E, Doyen A, Nato F, Nicolas JF, Torres V, Yaniv M, Pontoglio M. 2006. Defective planar cell polarity in polycystic kidney disease. *Nat Gen* **38:** 21–23.

Garoia F, Guerra D, Pezzoli MC, Lopez-Varea A, Cavicchi S, Garcia-Bellido A. 2000. Cell behaviour of *Drosophila* fat cadherin mutations in wing development. *Mech Develop* **94:** 95–109.

Gho M, Schweisguth F. 1998. Frizzled signalling controls orientation of asymmetric sense organ precursor cell divisions in *Drosophila*. *Nature* **393:** 178–181.

Gong Y, Mo C, Fraser SE. 2004. Planar cell polarity signalling controls cell division orientation during zebrafish gastrulation. *Nature* **430:** 689–693.

Goto T, Keller R. 2002. The planar cell polarity gene strabismus regulates convergence and extension and neural fold closure in *Xenopus*. *Develop Biol* **247:** 165–181.

Gros J, Serralbo O, Marcelle C. 2009. WNT11 acts as a directional cue to organize the elongation of early muscle fibres. *Nature* **457:** 589–593.

Gubb D, García-Bellido A. A genetic analysis of the determination of cuticular polarity during development in *Drosophila melanogaster J Embryol Exp Morphol* 1982 Apr; **68:** 37–57.

Gubb D, Green C, Huen D, Coulson D, Johnson G, Tree D, Collier S, Roote J. 1999. The balance between isoforms of the prickle LIM domain protein is critical for planar polarity in *Drosophila* imaginal discs. *Genes Develop* **13:** 2315–2327.

Guo N, Hawkins C, Nathans J. 2004. Frizzled6 controls hair patterning in mice. *Proc Natl Acad Sci* **101:** 9277–9281.

Habas R, Dawid IB, He X. 2003. Coactivation of Rac and Rho by Wnt/Frizzled signaling is required for vertebrate gastrulation. *Genes Develop* **17:** 295–309.

Harris PC, Torres VE. 2008. Polycystic kidney disease. *Ann Rev Med* **60:** 321–337.

Heisenberg CP, Tada M, Rauch GJ, Saude L, Concha ML, Geisler R, Stemple DL, Smith JC, Wilson SW. 2000. Silberblick/Wnt11 mediates convergent extension movements during zebrafish gastrulation. *Nature* **405:** 76–81.

Hikasa H, Shibata M, Hiratani I, Taira M. 2002. The *Xenopus* receptor tyrosine kinase Xror2 modulates morphogenetic movements of the axial mesoderm and neuroectoderm via Wnt signaling. *Development* **129:** 5227–5239.

Huang H, He X. 2008. Wnt/β-catenin signaling: New (and old) players and new insights. *Current Opinion Cell Biol* **20:** 119–125.

Jenny A, Darken RS, Wilson PA, Mlodzik M. 2003. Prickle and Strabismus form a functional complex to generate a correct axis during planar cell polarity signaling. *EMBO J* **22:** 4409–4420.

Jones C, Roper VC, Foucher I, Qian D, Banizs B, Petit C, Yoder BK, Chen P. 2008. Ciliary proteins link basal body polarization to planar cell polarity regulation. *Nat Gen* **40:** 69–77.

Kallay LM, McNickle A, Brennwald PJ, Hubbard AL, Braiterman LT. 2006. Scribble associates with two polarity proteins, Lgl2 and Vangl2, via distinct molecular domains. *J Cell Biochem* **99:** 647–664.

Kaltschmidt JA, Lawrence N, Morel V, Balayo T, Fernandez BG, Pelissier A, Jacinto A, Martinez Arias A. 2002. Planar polarity and actin dynamics in the epidermis of *Drosophila. Nat Cell Biol* **4:** 937–944.

Kilian B, Mansukoski H, Barbosa FC, Ulrich F, Tada M, Heisenberg CP. 2003. The role of Ppt/Wnt5 in regulating cell shape and movement during zebrafish gastrulation. *Mech Develop* **120:** 467–476.

Kibar Z, Vogan KJ, Groulx N, Justice MJ, Underhill DA, Gros P. 2001. Ltap, a mammalian homolog of *Drosophila* Strabismus/Van Gogh, is altered in the mouse neural tube mutant Loop-tail. *Nat Genet* Jul; **28:** 251–255.

Kim E, Walz G. 2007. Sensitive cilia set up the kidney. *Nat Med* **13:** 1409–1411.

Kuhl M. 2002. Non-canonical Wnt signaling in *Xenopus*: Regulation of axis formation and gastrulation. *Sem Cell Develop Biol* **13:** 243–249.

Lawrence PA, Struhl G, Casal J. 2008. Planar cell polarity: A bridge too far? *Curr Biol* **18:** R959–961.

Lu X, Borchers AG, Jolicoeur C, Rayburn H, Baker JC, Tessier-Lavigne M. 2004. PTK7/CCK-4 is a novel regulator of planar cell polarity in vertebrates. *Nature* **430:** 93–98.

Ma D, Yang CH, McNeill H, Simon MA, Axelrod JD. 2003. Fidelity in planar cell polarity signalling. *Nature* **421:** 543–547.

Martin-Belmonte F, Mostov K. 2008. Regulation of cell polarity during epithelial morphogenesis. *Current Opinion Cell Biol* **20:** 227–234.

Matakatsu H, Blair SS. 2004. Interactions between Fat and Dachsous and the regulation of planar cell polarity in the *Drosophila* wing. *Development* **131:** 3785–3794.

McNeill H. 2002. Planar polarity: Location, location, location. *Curr Biol* **12:** R449–451.

McNeill H, Yang CH, Brodsky M, Ungos J, Simon MA. 1997. Mirror encodes a novel PBX-class homeoprotein that functions in the definition of the dorsal-ventral border in the *Drosophila* eye. *Genes Develop* **11:** 1073–1082.

Medina A, Swain RK, Kuerner KM, Steinbeisser H. 2004. *Xenopus* paraxial protocadherin has signaling functions and is involved in tissue separation. *EMBO J* **23:** 3249–3258.

Montcouquiol M, Rachel RA, Lanford PJ, Copeland NG, Jenkins NA, Kelley MW. 2003. Identification of Vangl2 and Scrb1 as planar polarity genes in mammals. *Nature* **423:** 173–177.

Montcouquiol M, Sans N, Huss D, Kach J, Dickman JD, Forge A, Rachel RA, Copeland NG, Jenkins NA, Bogani D, et al. 2006. Asymmetric localization of Vangl2 and Fz3 indicate novel mechanisms for planar cell polarity in mammals. *J Neurosci* **26:** 5265–5275.

Murdoch JN, Henderson DJ, Doudney K, Gaston-Massuet C, Phillips HM, Paternotte C, Arkell R, Stanier P, Copp AJ. 2003. Disruption of scribble (Scrb1) causes severe neural tube defects in the circletail mouse. *Human Mol Gen* **12:** 87–98.

Park M, Moon RT. 2002. The planar cell-polarity gene stbm regulates cell behaviour and cell fate in vertebrate embryos. *Nat Cell Biol* Jan; **4:** 20–5.

Park WJ, Liu J, Sharp EJ, Adler PN. 1996. The *Drosophila* tissue polarity gene inturned acts cell autonomously and encodes a novel protein. *Development* **122:** 961–969.

Park TJ, Haigo SL, Wallingford JB. 2006. Ciliogenesis defects in embryos lacking inturned or fuzzy function are associated with failure of planar cell polarity and Hedgehog signaling. *Nature Genetics* **38:** 303–311.

Park TJ, Mitchell BJ, Abitua PB, Kintner C, Wallingford JB. 2008. Dishevelled controls apical docking and planar polarization of basal bodies in ciliated epithelial cells. *Nat Gen* **40:** 871–879.

Patel V, Li L, Cobo-Stark P, Shao X, Somlo S, Lin F, Igarashi P. 2008. Acute kidney injury and aberrant planar cell polarity induce cyst formation in mice lacking renal cilia. *Human Mol Gen* **17:** 1578–1590.

Piontek K, Menezes LF, Garcia-Gonzalez MA, Huso DL, Germino GG. 2007. A critical developmental switch defines the kinetics of kidney cyst formation after loss of Pkd1. *Nat Med* **13:** 1490–1495.

Praetorius HA, Spring KR. 2003. The renal cell primary cilium functions as a flow sensor. *Current Opinion NephrolHypertension* **12:** 517–520.

Qian D, Jones C, Rzadzinska A, Mark S, Zhang X, Steel KP, Dai X, Chen P. 2007. Wnt5a functions in planar cell polarity regulation in mice. *Develop Biol* **306:** 121–133.

Rawls AS, Guinto JB, Wolff T. 2002. The cadherins fat and dachsous regulate dorsal/ventral signaling in the *Drosophila* eye. *Curr Biol* **12:** 1021–1026.

Ross AJ, May-Simera H, Eichers ER, Kai M, Hill J, Jagger DJ, Leitch CC, Chapple JP, Munro PM, Fisher S, et al. 2005. Disruption of Bardet-Biedl syndrome ciliary proteins perturbs planar cell polarity in vertebrates. *Nat Gen* **37:** 1135–1140.

Saburi S, Hester I, Fischer E, Pontoglio M, Eremina V, Gessler M, Quaggin SE, Harrison R, Mount R, McNeill H. 2008. Loss of Fat4 disrupts PCP signaling and oriented cell division and leads to cystic kidney disease. *Nat Gen* **40:** 1010–1015.

Schambony A, Wedlich D. 2007. Wnt-5A/Ror2 regulate expression of XPAPC through an alternative noncanonical signaling pathway. *Develop Cell* **12:** 779–792.

Schwarz-Romond T, Asbrand C, Bakkers J, Kuhl M, Schaeffer HJ, Huelsken J, Behrens J, Hammerschmidt M, Birchmeier W. 2002. The ankyrin repeat protein Diversin recruits Casein kinase Iepsilon to the β-catenin degradation complex and acts in both canonical Wnt and Wnt/JNK signaling. *Genes Develop* **16:** 2073–2084.

Simon MA. 2004. Planar cell polarity in the *Drosophila* eye is directed by graded Four-jointed and Dachsous expression. *Development* **131:** 6175–6184.

Simons M, Walz G. 2006. Polycystic kidney disease: Cell division without a c(l)ue? *Kidney Int* **70:** 854–864.

Simons M, Gloy J, Ganner A, Bullerkotte A, Bashkurov M, Kronig C, Schermer B, Benzing T, Cabello OA, Jenny A, et al. 2005. Inversin, the gene product mutated in nephronophthisis type II, functions as a molecular switch between Wnt signaling pathways. *Nat Gen* **37:** 537–543.

Sopko R, McNeill H. 2009. The Skinny on Fat: An enormous cadherin regulating cell adhesion, tissue growth and planar cell polarity. *Current Opinions Cell Biol (in press).*

Sopko R, Silva E, Clayton L, Gardano L, Barrios-Rodiles M, Wrana J, Varelas X, Arbouzova NI, Shaw S, Saburi S, et al. 2009. Phosphorylation of the Tumor Suppressor Fat Is Regulated by Its Ligand Dachsous and the Kinase Discs Overgrown. *Curr Biol (in press)* **PMID:** 19540118.

Strutt D, Strutt H. 2007. Differential activities of the core planar polarity proteins during *Drosophila* wing patterning. *Develop Biol* **302:** 181–194.

Strutt H, Strutt D. 2002. Nonautonomous planar polarity patterning in *Drosophila*: Dishevelled-independent functions of frizzled. *Develop Cell* **3:** 851–863.

Strutt H, Strutt D. 2003. EGF signaling and ommatidial rotation in the *Drosophila* eye. *Curr Biol* **13:** 1451–1457.

Strutt H, Strutt D. 2008. Differential stability of flamingo protein complexes underlies the establishment of planar polarity. *Curr Biol* **18:** 1555–1564.

Strutt D, Warrington SJ. 2008. Planar polarity genes in the *Drosophila* wing regulate the localisation of the FH3-domain protein Multiple Wing Hairs to control the site of hair production. *Development* **135:** 3103–3111.

Strutt D, Johnson R, Cooper K, Bray S. 2002. Asymmetric localization of frizzled and the determination of notch-dependent cell fate in the *Drosophila* eye. *Curr Biol* **12:** 813–824.

Strutt H, Price MA, Strutt D. 2006. Planar polarity is positively regulated by casein kinase I epsilon in *Drosophila. Curr Biol* **16:** 1329–36.

Tada M, Kai M. 2009. Noncanonical Wnt/PCP signaling during vertebrate gastrulation. *Zebrafish* **6:** 29–40.

Tada M, Smith JC. 2000. Xwnt11 is a target of *Xenopus* Brachyury: Regulation of gastrulation movements via Dishevelled, but not through the canonical Wnt pathway. *Development* **127:** 2227–2238.

Tahinci E, Thorne CA, Franklin JL, Salic A, Christian KM, Lee LA, Coffey RJ, Lee E. 2007. Lrp6 is required for convergent extension during *Xenopus* gastrulation. *Development* **134:** 4095–4106.

Takeuchi M, Nakabayashi J, Sakaguchi T, Yamamoto TS, Takahashi H, Takeda H, Ueno N. 2003. The prickle-related gene in vertebrates is essential for gastrulation cell movements. *Curr Biol* **13:** 674–679.

Taylor J, Abramova N, Charlton J, Adler PN. 1998. Van Gogh: A new *Drosophila* tissue polarity gene. *Genetics* **150:** 199–210.

Torban E, Patenaude AM, Leclerc S, Rakowiecki S, Gauthier S, Andelfinger G, Epstein DJ, Gros P. 2008. Genetic interaction between members of the Vangl family causes neural tube defects in mice. *Proc Natl Acad Sci* **105:** 3449–3454.

Tree DR, Shulman JM, Rousset R, Scott MP, Gubb D, Axelrod JD. 2002. Prickle mediates feedback amplification to generate asymmetric planar cell polarity signaling. *Cell* **109:** 371–381.

Unterseher F, Hefele JA, Giehl K, De Robertis EM, Wedlich D, Schambony A. 2004. Paraxial protocadherin coordinates cell polarity during convergent extension via Rho A and JNK. *EMBO J* **23:** 3259–3269.

Usui T, Shima Y, Shimada Y, Hirano S, Burgess RW, Schwarz TL, Takeichi M, Uemura T. 1999. Flamingo, a seven-pass transmembrane cadherin, regulates planar cell polarity under the control of Frizzled. *Cell* **98:** 585–595.

Vinson CR, Adler PN. 1987. Directional non-cell autonomy and the transmission of polarity information by the frizzled gene of *Drosophila. Nature* **329:** 549–551.

Wallingford JB, Rowning BA, Vogeli KM, Rothbacher U, Fraser SE, Harland RM. 2000. Dishevelled controls cell polarity during *Xenopus* gastrulation. *Nature* **405:** 81–85.

Wang J, Mark S, Zhang X, Qian D, Yoo SJ, Radde-Gallwitz K, Zhang Y, Lin X, Collazo A, Wynshaw-Boris A, et al. 2005. Regulation of polarized extension and planar cell polarity in the cochlea by the vertebrate PCP pathway. *Nat Gen* **37:** 980–985.

Wang Y, Guo N, Nathans J. 2006. The role of Frizzled3 and Frizzled6 in neural tube closure and in the planar polarity of inner-ear sensory hair cells. *J Neurosci* **26:** 2147–2156.

Weber U, Paricio N, Mlodzik M. 2000. Jun mediates Frizzled-induced R3/R4 cell fate distinction and planar polarity determination in the *Drosophila* eye. *Development* **127:** 3619–3629.

Winberg ML, Tamagnone L, Bai J, Comoglio PM, Montell D, Goodman CS. 2001. The transmembrane protein Off-track associates with Plexins and functions downstream of Semaphorin signaling during axon guidance. *Neuron* **32:** 53–62.

Wolff T, Rubin GM. 1998. Strabismus, a novel gene that regulates tissue polarity and cell fate decisions in *Drosophila. Development* **125:** 1149–1159.

Wong HC, Bourdelas A, Krauss A, Lee HJ, Shao Y, Wu D, Mlodzik M, Shi DL, Zheng J. 2003. Direct binding of the PDZ domain of Dishevelled to a conserved internal sequence in the C-terminal region of Frizzled. *Mol Cell* **12:** 1251–1260.

Wu J, Mlodzik M. 2008. The frizzled extracellular domain is a ligand for Van Gogh/Stbm during nonautonomous planar cell polarity signaling. *Develop Cell* **15:** 462–469.

Yamamoto S, Nishimura O, Misaki K, Nishita M, Minami Y, Yonemura S, Tarui H, Sasaki H. 2008. Cthrc1 selectively activates the planar cell polarity pathway of Wnt signaling by stabilizing the Wnt-receptor complex. *Develop Cell* **15:** 23–36.

Yamanaka T, Ohno S. 2008. Role of Lgl/Dlg/Scribble in the regulation of epithelial junction, polarity and growth. *Front Biosci* **13:** 6693–6707.

Yang CH, Simon MA, McNeill H. 1999. Mirror controls planar polarity and equator formation through repression of fringe expression and through control of cell affinities. *Development* **126:** 5857–5866.

Yang CH, Axelrod JD, Simon MA. 2002. Regulation of Frizzled by fat-like cadherins during planar polarity signaling in the *Drosophila* compound eye. *Cell* **108:** 675–688.

Ybot-Gonzalez P, Savery D, Gerrelli D, Signore M, Mitchell CE, Faux CH, Greene ND, Copp AJ. 2007 Convergent extension, planar-cell-polarity signalling and initiation of mouse neural tube closure. *Development* **134:**789–799.

Yoder BK. 2007. Role of primary cilia in the pathogenesis of polycystic kidney disease. *J Am Soc Nephrol* **18:** 1381–1388.

Zeidler MP, Perrimon N, Strutt DI. 1999. The four-jointed gene is required in the *Drosophila* eye for ommatidial polarity specification. *Curr Biol* **9:** 1363–1372.

Listeria monocytogenes Internalin and E-cadherin: From Bench to Bedside

Matteo Bonazzi[1,2,3], Marc Lecuit[4,5,6], and Pascale Cossart[1,2,3]

[1]Institut Pasteur, Unité des Interactions Bactéries-Cellules, Paris, F-75015 France

[2]Inserm, U604, Paris, F-75015 France

[3]INRA, USC2020, Paris, F-75015 France

[4]Institut Pasteur, Groupe Microorganismes et Barrières de l'Hôte, Paris, F-75015 France

[5]Inserm Avenir U604, Paris, F-75015 France

[6]Université Paris Descartes, Centre d'Infectiologie Necker-Pasteur, Service des Maladies Infectieuses et Tropicales, Hôpital Necker-Enfants malades, Assistance Publique-Hôpitaux de Paris, Paris, F-75015 France

Correspondence: pcossart@pasteur.fr, mlecuit@pasteur.fr

Listeria monocytogenes is a Gram-positive bacterium responsible for a severe infection associated with different clinical features (gastroenteritis, meningoencephalitis, and abortion in pregnant women). These pathologies are caused by the unusual capacity of the bacterium to cross three host barriers during infection and to invade nonphagocytic cells. To invade host cells, *Listeria* uses two proteins, InlA and InlB, which have specific receptors on the host-cell surface, E-cadherin and Met, respectively. Here, we discuss the specificity of the InlA–E-cadherin interaction, the signaling cascade activated on E-cadherin engagement by InlA, and the role of InlA and E-cadherin in the breaching of host barriers and the dissemination of the infection.

Listeriosis is a potentially lethal food-borne infection with a mortality rate up to 30%. It has emerged as a significant human infection in industrialized countries along with the development of large-scale agro-industrial plants and refrigerated food. Opposite to most food-borne infections, listeriosis is rare but potentially very severe, because it remains often under-diagnosed at its early stages (Lecuit 2007). The etiological agent of listeriosis is *Listeria monocytogenes*, a Gram-positive bacterium that contaminates meat, dairy products, and ready to eat food. Upon ingestion of contaminated food, *L. monocytogenes* can colonize the intestine and gives rise to gastroenteritis in case of the absorption of a high inoculum. Strikingly, *L. monocytogenes* has the capacity to cross the intestinal barrier and disseminate to the mesenteric lymph nodes, spleen, and liver. In immunocompromised individuals, *L. monocytogenes* may replicate in the spleen and liver, cause prolonged and sustained bacteremia, cross the blood–brain barrier and the placental barrier, and disseminate to the brain and

 doi: 10.1101/cshperspect.a003087
Cite this article as *Cold Spring Harb Perspect Biol* 2009;1:a003087

placenta, causing meningitis, encephalitis, abortion in pregnant women, and neonatal infections (Hamon et al. 2006; Bonazzi et al. 2009).

The capacity of *L. monocytogenes* to cross multiple host barriers relies on the ability of the bacterium to invade nonphagocytic cells, such as epithelial cells, by interacting with host cell-surface receptors. Adhesion to host cells is a key step underlying bacterial pathogenicity and it is required to counteract the mechanical clearance at tissue surfaces provided by intestinal peristaltism and blood flow. Internalization allows persistence in a shielded niche, away from the soluble effectors of the host immune system (Cossart and Sansonetti 2004; Pizarro-Cerda and Cossart 2006), and access to target organs, as illustrated by *L. monocytogenes* (Bonazzi et al. 2009).

Bacterial surface proteins that engage host receptors are generally called adhesins, although it is now clear that many of them not only mediate adhesion, but also bacterial internalization (Boyle and Finlay 2003; Hauck et al. 2006; Pizarro-Cerda and Cossart 2006). Adhesion to and internalization of *L. monocytogenes* within epithelial cells is mainly mediated by two bacterial surface protein members of the internalin family, namely internalin (InlA) and InlB, that use E-cadherin and Met as receptors, respectively, on the surface of host cells (Hamon et al. 2006; Pizarro-Cerda and Cossart 2006; Bonazzi et al. 2009). Upon receptor-mediated internalization, *L. monocytogenes* is engulfed into the cell and becomes surrounded by a tight phagocytic vacuole that the bacterium can lyse by means of the pore-forming toxin listeriolysin O (LLO). Once free in the cytoplasm of the host cell, *L. monocytogenes* uses the protein ActA to harness the actin polymerization machinery and facilitate its intracellular movement via the formation of so-called actin "comet tails." Actin-based motility is fundamental for *L. monocytogenes* direct cell-to-cell spread, a typical feature that allows the dissemination of the infection to neighboring cells via the formation of plasma membrane protrusions. Once internalized by neighboring cells, *L. monocytogenes* is confined in a double-membrane vacuole from which it escapes to restart its life cycle (Fig. 1) (Hamon et al. 2006).

Its remarkable adaptation to the cellular environment and its capacity to exploit cellular receptor-mediated signaling pathways and the actin polymerization machinery have made *L. monocytogenes* an exceptional tool for the study of a wide array of cellular functions (Cossart and Sansonetti 2004; Bonazzi and Cossart 2006; Hamon et al. 2006; Pizarro-Cerda and Cossart 2006; Veiga and Cossart 2005a; Bonazzi et al. 2009). Here, we review the interaction of InlA with E-cadherin, the signaling pathway initiated by this interaction that results in the internalization of *L. monocytogenes*, and the role of InlA-E-cadherin interaction during listeriosis.

THE *L. MONOCYTOGENES* SURFACE PROTEIN InlA

As introduced above, internalization of *L. monocytogenes* within nonphagocytic cells is mediated by the two bacterial surface proteins InlA and InlB. In vitro, these bacterial proteins are sufficient to mediate adhesion and internalization into host cells, as bacteria expressing either one of these proteins as well as beads coated with either InlA or InlB are efficiently internalized in nonphagocytic cells as long as their respective receptor is expressed (Lecuit et al. 1997; Pizarro-Cerda et al. 2002).

The targets of InlA and InlB on the surface of the host cell are the adhesion molecule E-cadherin and the hepatocyte growth factor (HGF) receptor Met, respectively (Hamon et al. 2006). Met is ubiquitously expressed, allowing InlB to mediate bacterial internalization in a large number of cell types, whereas InlA shows a more stringent cell tropism, as E-cadherin is only expressed by a limited number of cells of epithelial origin.

InlA and InlB are part of a large family of proteins, the internalin family, composed of 25 proteins sharing a common architecture that includes a signal peptide at the amino-terminal and a number of 22 amino acid leucine-rich repeats (LRR) that can vary from three in the case of Lmo2445 to 28 in the case of InlI.

LRR repeats are common to a large number of proteins and are involved in protein–protein interactions (Kobe and Deisenhofer 1994). Downstream of the LRR, internalins display several other regions that are less conserved within the family and can confer specific functions to different members of the family (Bierne et al. 2007). InlA is an 800 amino-acid protein containing 15 LRRs. Downstream of the LRR region, InlA has an inter-repeat (IR) region that has been shown fundamental for the binding of the LRR repeats domain to E-cadherin (Lecuit et al. 1997). An LPXTG motif for the anchoring of InlA at the bacterial cell wall is present at the carboxy-terminal followed by a sorting peptide (Hamon et al. 2006).

The first evidence that indicated that InlA might act as a *L. monocytogenes* virulence factor derived from studies on the *L. monocytogenes* strain LO28, in which it has been observed that it harbors a frameshift mutation in the *inlA* gene that results in a nonsense codon at position 1729. This generates a truncated open reading frame encoding a protein of 63 kDa that lacks the LPXTG motif (Jonquieres et al. 1998). As a consequence, in the LO28 strain, InlA can not be detected on the surface of the bacteria but is found released in the supernatant (Jonquieres et al. 1998), and LO28 invasion of epithelial cell lines is far lower than other strains expressing a functional InlA. Epidemiological studies performed on 300

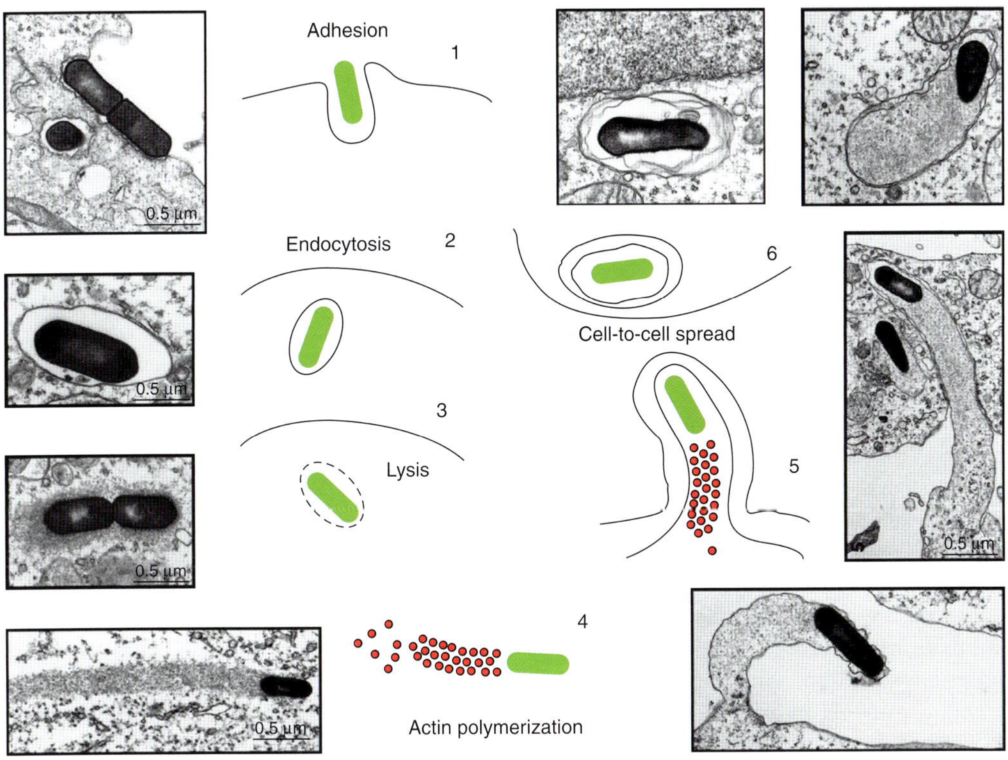

Figure 1. The cell cycle of *L. monocytogenes.* (1) *L. monocytogenes* adheres to the surface of epithelial cells via the interactions of the surface proteins InlA and InlB with E-cadherin and the Met receptor, respectively. (2) On internalization, *L. monocytogenes* is engulfed in a phagocytic vacuole. (3) *L. monocytogenes* lyses vacuolar membranes by means of the toxin LLO. (4) *L. monocytogenes* uses the protein ActA to harness the actin polymerization machinery and facilitate its intracellular movement via the formation of so-called actin "comet tails." (5) *L. monocytogenes* exploits actin-based motility for direct cell-to-cell spread to allow the dissemination of the infection to neighboring cells via the formation of plasma membrane protrusions. (6) Once internalized by neighboring cells, *L. monocytogenes* is confined in a double-membrane vacuole from which it escapes to restart its life cycle.

clinical strains obtained in France and 150 strains from food isolates showed that, at least in food isolates, the truncation in the *InlA* gene is not a rare event (Jonquieres et al. 1998). More recently, a thorough epidemiological study showed that full length internalin was identified in 96% of the clinical strains and in 65% of the strains derived from food isolates (Jacquet et al. 2004). These observations indicated that InlA might serve as a virulence determinant of *L. monocytogenes* in humans, a hypothesis that has now been confirmed in humanized animal models (see the following). The crystal structure of InlA, alone or in complex with the EC1 domain of E-cadherin, has been solved (Schubert et al. 2002). The LRR domain of InlA consists of fifteen and a half 22-residue repeats that form a right-handed curved solenoid structure. Each repeat begins with a β strand of five residues and these strands combine to form a 16-stranded β sheet that is twisted out of a central plane to form a pseudo-helical surface. The cavity thus created forms the binding site for the EC1 domain of E-cadherin (Fig. 2B).

E-CADHERIN, THE RECEPTOR FOR INLA, MEDIATES *L. MONOCYTOGENES* ENTRY INTO HOST CELLS

E-cadherin belongs to the family of classical cadherins that mediate the formation of adherens junctions, Ca^{2+}-dependent intercellular adhesion sites that develop between polarized epithelial cells (Troyanovsky 2005; Meng and Takeichi 2009). Cadherins have been classified into several subfamilies on the basis of shared properties and sequence similarities. Type I cadherins, also called "classical" cadherins, all possess five 110-amino-acid immunoglobulin-like extracellular domains (EC1 to EC5) with a conserved tryptophan in position 2 and interact with actin through their intracellular domain (Shapiro and Weis 2009). This subfamily includes E-cadherin, which is found in epithelial cells, N-cadherin that was first detected in neuronal cells, C-cadherin expressed in the *Xenopus* embryo, and P-cadherin on the placenta and other cell types. T-cadherins are still classified as members of the type I subfamily, although they lack the transmembrane and cytoplasmic domain and are tethered to the plasma membrane via a GPI-anchor (Ranscht and Dours-Zimmermann 1991; Patel et al. 2003). Type II cadherins differ from members of the type I subfamily in that they have two conserved tryptophan residues in position 2 and 4. They also interact with the actin cytoskeleton via specific adaptors and the most common member of this subfamily is VE-cadherin, which is expressed by endothelial cells (Vestweber et al. 2009; Delva et al. 2009). Desmosomal cadherins form another subfamily of proteins: they share with type I cadherins the presence of 5 EC repeats and a conserved tryptophan in position 2. Nevertheless, members of this family differ in the cytoplasmic region that allows the interaction with intermediate filaments instead of actin. Desmoglein and Desmocollin belong to this family and as their names indicate are localized at the level of desmosomes (Green and Gaudry 2000). The Protocadherins subfamily can be further divided in α, β, and γ protocadherins. They all present six EC repeats and they are expressed at the level of synaptic complexes in neurons (Kohmura et al. 1998; Patel et al. 2003).

E-cadherin is a protein of 882 amino acids with a 555-amino-acids-long amino-terminal extracellular domain, a transmembrane domain, and a relatively short cytoplasmic domain of 152 amino acids. The extracellular portion is comprised of five 110-amino-acid repeats (EC1 to EC5), immunoglobulin-like domains involved in intercellular homotypic interactions and in intracellular clustering (Troyanovsky 2005). The analysis of their crystal structure revealed that multiple cadherin domains form Ca^{2+}-dependent rodlike structures with a Ca^{2+}-binding pocket at the interface between two contacting EC domains (Shapiro and Weis 2009). Besides E-cadherin itself, the only known eukaryotic E-cadherin ectodomain interactor identified is αEβ7 integrin (Karecla et al. 1996), which is solely expressed in epithelial lymphocytes, and has E-cadherin as its unique binding partner. E-cadherin is also the receptor for *Listeria*

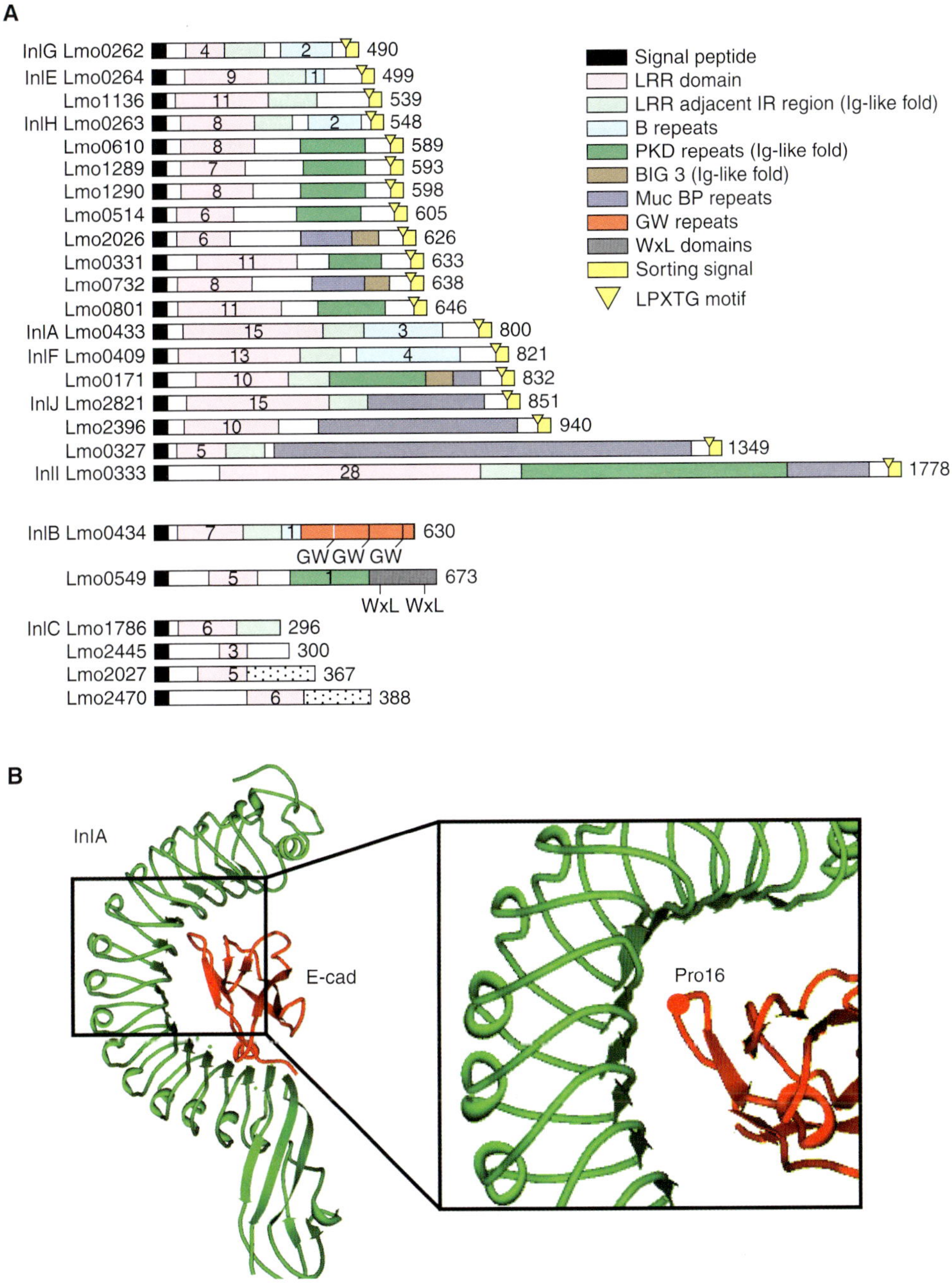

Figure 2. The internalin family and InlA. (*A*) Schematic representation of the internalin family of proteins of *L. monocytogenes*. Homologous regions are color coded as indicated in the legend. Numbers within different domains indicate the number of repeats (Bierne et al. 2007). (*B*) Crystal structure of InlA in complex with the EC1 domain of E-cadherin. The LRR domain of InlA consists of fifteen and a half 22-residue repeats that form a right-handed curved solenoid structure. Each repeat begins with a β strand of five residues: these strands combine in a 16-stranded β sheet that forms a pseudo-helical surface. The presence of a proline in position 16 of E-cadherin allows the terminal loop of E-cadherin to be hydrophobic and uncharged, therefore strengthening the interaction with InlA.

InlA (Mengaud et al. 1996), and the *Candida albicans* invasin Als3 (Phan et al. 2007) and it is the target for the *Bacterioides fragilis* metalloprotease toxin BFT (Wu et al. 1998).

The cytoplasmic domain of E-cadherin is divided into a juxtamembrane domain that spans from aa 734 to aa 770, approximately and a distal domain from aa 770 to 882. Phosphorylation of the cytoplasmic domain regulates the binding of E-cadherin to its interactors (Fig. 3) (McCrea et al. 2009). The nonreceptor tyrosine kinase Src phosphorylates E-cadherin juxtamembrane domain at residues 753 and 754. This post-translational modification triggers the release of the E-cadherin direct interactor p120 catenin and the recruitment of the ubiquitin-ligase Hakai (Fujita et al. 2002; Reynolds 2007). In its E-cadherin bound state, p120 stabilizes E-cadherin at the plasma membrane by sterically impeding the binding of Hakai to E-cadherin. In its cytosolic form, p120 regulates cell motility by activating Rac1 and Cdc42 and inhibiting RhoA (Reynolds 2007). Upon E-cadherin phosphorylation, Hakai binds to and ubiquitinates E-cadherin, triggering its clathrin-mediated internalization (Fujita et al. 2002). Deletion of the last 35 amino acids of the cytoplasmic tail of E-cadherin impairs the binding to β-catenin (Ozawa et al. 1990; Lecuit et al. 2000) that binds α-catenin, a key regulator of actin dynamics (Fig. 3).

Binding of E-cadherin to β-catenin plays a fundamental role in the regulation of adherens junction dynamics and is negatively regulated by phosphorylation of both E-cadherin and β-catenin. Casein-kinase-1-mediated serine phosphorylation of E-cadherin at residue 846 has been reported to impede binding to β-catenin (Dupre-Crochet et al. 2007). Constitutively phosphorylated E-cadherin at serine 846 fails to localize at adherens junctions and to bind β-catenin efficiently (Dupre-Crochet et al. 2007). Similarly, β-catenin can be phosphorylated by Src at tyrosine 654 (Roura et al. 1999) and by Met at tyrosine 142 (Brembeck et al. 2004), which induces the dissociation of β-catenin from both E-cadherin and α-catenin. Phosphorylation of β-catenin at tyrosine 489 has been reported in neuronal cells, it is mediated by the tyrosine kinase Abl

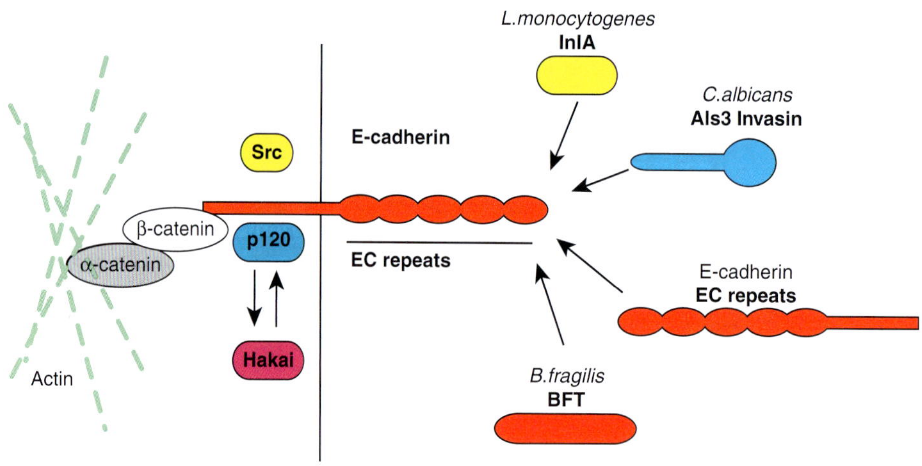

Figure 3. E-cadherin interactors. E-cadherin is a single-pass transmembrane protein with a 152-amino-acid intracellular domain and a 555-amino-acid extracellular domain. The intracellular domain of E-cadherin interacts with proteins of the catenin family. p120 catenin binds to the juxtamembrane domain of E-cadherin and stabilizes E-cadherin at the plasma membrane. On E-cadherin phosphorylation by the tyrosine kinase Src p120 is released and the ubiquitin-ligase Hakai can bind and ubiquitinate E-cadherin. β-catenin binds to the carboxy-terminal domain of E-cadherin and mediates the interaction with the actin cytoskeleton via α-catenin. The extracellular domain of E-cadherin interacts with the EC domains of E-cadherin molecules from neighboring cells, it serves as the receptor for the *L. monocytogenes* surface protein InlA and the *C. albicans* invasin Als3, and is the target of the *B. fragilis* metalloprotease toxin BFT.

on activation of the surface receptor Robo, and, it induces the dissociation of β-catenin from E-cadherin (Rhee et al. 2007). Most cellular β-catenin is associated with E-cadherin at adherens junctions. Cytoplasmic pools of β-catenin can be either phosphorylated at tyrosine 412 or nonphosphorylated and are rapidly translocated to the nucleus. Here, Wnt signaling can induce a conformational change in nonphosphorylated β-catenin, whereby its carboxyl terminus folds into a close conformation that impedes binding to E-cadherin and favors the binding to the T-cell factor (TCF) in a transcription complex (Gottardi and Gumbiner 2004; see Heuberger and Birchmeier 2009). Alternatively, in its phosphorylated form, β-catenin regulates transcription by binding BCL9–2, a homolog of the human oncogene product BCL9 (Brembeck et al. 2004).

A direct interaction between β-catenin and Met has also been documented and the regulation of such interaction depends on tyrosine residues 654 and 670 on β-catenin (Zeng et al. 2006). Of note, InlB, the other major *L. monocytogenes* internalin, is a potent activator of Met, suggesting a possible link between the two signaling pathways exploited by the bacterium for its internalization. Concerning the interactions of β-catenin with the actin cytoskeleton, it has been reported that α-catenin switches from a monomeric form with a high affinity for β-catenin to a dimeric form that has a higher affinity for actin: This equilibrium would regulate actin dynamics at adherens junctions (Drees et al. 2005; Yamada et al. 2005). It has been recently shown that in *Drosophila* embryos, where homophilic E-cadherin interactions generate the so-called spot adherens junctions (SAJ), α-catenin-mediated interaction of E-cadherin with actin is not required at the SAJ itself but rather at its sides to maintain the mosaic distribution of homo-E-cadherin clusters (Cavey et al. 2008).

DOWNSTREAM EFFECTORS OF InlA E CADHERIN INTERACTION

The observation that α- and β-catenin, as well as actin and p120 catenin, are efficiently recruited at the site of InlA-mediated bacterial entry (Lecuit et al. 2000), suggested not only that E-cadherin serves as a receptor for InlA, but also that on ligand/receptor engagement, *L. monocytogenes* is able to exploit the machinery involved in the formation of adherens junctions. A link between E-cadherin and the actin cytoskeleton is also fundamental: InlA-mediated infection is impaired in cells expressing an E-cadherin mutant that lacks either the complete intracellular domain or the β-catenin-binding domain (Lecuit et al. 2000), which is in agreement with the previous observation that inhibiting actin polymerization by cytochalasin D impairs bacterial internalization (Gaillard et al. 1987). Interestingly, the deletion of the juxtamembrane domain of E-cadherin, important for p120 catenin binding, and E-cadherin endocytosis (Fujita et al. 2002), has no effect on the internalization of *L. monocytogenes* (Lecuit et al. 2000), which suggests the presence of an alternative pathway exploited by the bacterium. Indeed, our recent data indicate that in the case of the InlA pathway of *L. monocytogenes,* caveolin also participates in bacteria internalization and this pathway is independent of the presence of the E-cadherin juxtamembrane domain (Bonazzi et al. 2008).

Myosins were previously identified as important players in the stabilization of E-cadherin-mediated cell–cell junction and it has been proposed that their activity as actin-based motor proteins was necessary to generate the force that maintains mature adherens junctions (Vasioukhin and Fuchs 2001). Nonconventional myosin VIIa is also involved in InlA-mediated internalization of *L. monocytogenes* (Sousa et al. 2004). Myosin VIIa and the ubiquitous transmembrane protein vezatin were found to be recruited at the bacterial entry site when *L. monocytogenes* follows the InlA-mediated entry pathway and the inhibition of either protein severely affects bacterial internalization by the InlA pathway (Sousa et al. 2004). A relevant role of α-catenin in the recruitment of vezatin and myosin VIIa was shown, stressing the relevance of catenins during InlA-mediated infections, not only as regulators of actin dynamics but also as scaffolding proteins for the recruitment of other components of the adherens junction machinery (Sousa et al. 2004).

The study of *L. monocytogenes* interactions with eukaryotic cells has also proven to be a powerful tool to identify new proteins involved in both bacterial internalization and adherens junction dynamics. The Rho GAP protein ARHGAP10 is a novel component of the adherens junction complex identified in a yeast two-hybrid screen as an interactor of α-catenin (Sousa et al. 2005). In cells that do not express E-cadherin, ARHGAP10 is predominantly localized at the Golgi complex, but in epithelial cells expressing E-cadherin, a large fraction of the protein colocalizes with E-cadherin, actin, and α-catenin at adherens junctions (Sousa et al. 2005). In these cells, the knockdown of ARHGAP10 results in loss of α-catenin localization at the adherens junctions, whereas E-cadherin localization is unaffected (Sousa et al. 2005). It has been proposed that ARHGAP10 recruits α-catenin to newly formed cell–cell contacts once "passive" transcellular E-cadherin interaction has occurred, to mediate the maturation of adherens junctions (Sousa et al. 2005). In agreement with this hypothesis, ARHGAP10 colocalizes with α-catenin and actin at the bacterial entry site during InlA-mediated infections. Furthermore, the inhibition of ARHGAP10 impairs the InlA-mediated internalization of *L. monocytogenes* (Sousa et al. 2005). ARHGAP10 has been also identified as the binding partner of GTP-bound Arf1 (and Arf6) at the Golgi complex where it plays an important role as a Cdc42-specific GAP protein that regulates actin-mediated Golgi complex architecture (Dubois et al. 2005). Of note, when ARHGAP10 was identified as the binding partner for GTP-bound Arf6, vezatin was also found among the screened preys (Dubois et al. 2005).

LISTERIA TAKES CONTROL OF THE PHAGOCYTIC AND ENDOCYTIC MACHINERIES

As for other bacteria, actin rearrangements are critical for internalization. In the case of *L. monocytogenes*, the signaling that directly orchestrates actin remodeling and bacterial engulfment into host cells involves the Arp2/3 complex, a highly conserved actin nucleator, which is recruited at the bacterial entry site. Arp2/3 recruitment precedes actin cup formation around bacteria, in agreement with the idea that its function is required for actin polymerization (Sousa et al. 2007). Indeed, Arp2/3 inhibition impairs bacterial and InlA-coated bead internalization. A search for putative activators of Arp2/3 suggested that cortactin could be a candidate for the activation of Arp2/3, whereas WAVE and NWASP are not involved (Sousa et al. 2007).

Cortactin modulation of the actin cytoskeleton is regulated by the combined activity of Erk and Src kinases. It has been proposed that an initial phosphorylation by Erk at serines 405 and 418 liberates the SH3 domain of cortactin, otherwise folded in a close conformation on the proline-rich domain of cortactin (Campbell et al. 1999). In this conformation, cortactin is able to bind NWASP. Further phosphorylation of Src at tyrosines 421, 466, and 482 (Huang et al. 1998) impairs cortactin interaction with NWASP (Martinez-Quiles et al. 2004) and dramatically reduces the ability of cortactin to crosslink F-actin (Huang et al. 1997). In its Src-phosphorylated state, cortactin promotes cell migration (Huang et al. 1998). Of note, InlA-mediated *L. monocytogenes* internalization induces the activation of Src early during infection and the pharmacological inhibition of the Src family of tyrosine kinases inhibits entry as well as cortactin recruitment at the bacterial entry site (Sousa et al. 2007). Besides its possible role in the activation of cortactin, Src has been implicated in the initial phosphorylation of E-cadherin that initiates the internalization of the receptor from adherens junctions (Fujita et al. 2002). Src-mediated phosphorylation of E-cadherin is also triggered by InlA during bacterial infections (Bonazzi et al. 2008). This post-translational modification results in the recruitment of the ubiquitin-ligase Hakai at bacterial entry site and in the ubiquitination of E-cadherin, which is necessary for efficient bacterial internalization (Bonazzi et al. 2008).

As mentioned above, Src-mediated phosphorylation of E-cadherin also induces the release of p120 catenin in the cytoplasm, where it activates Rac1 and Cdc42 (Reynolds 2007). The activation of Rac1 has been shown to have a key role in the recruitment of cortactin at the plasma membrane (Weed et al. 1998).

Upon InlB interaction with the Met receptor, *L. monocytogenes* can hijack the host ubiquitination machinery to enter epithelial cells via clathrin-dependent endocytosis (Veiga and Cossart 2005b). More recently, this observation has been extended, revealing that clathrin-dependent endocytosis is an entry pathway shared by a number of zippering bacteria (Veiga et al. 2007). This observation showed that the maximal size of particles that can be internalized via clathrin can exceed 100 nm and, in the case of InlA-mediated internalization, suggested a possible role of the endocytosis machinery in the establishment of adherens junctions (Veiga et al. 2007; Bonazzi et al. 2008). As recently shown, caveolin is required for E-cadherin clustering at the bacterial entry site (Bonazzi et al. 2008). Caveolin and clathrin may then mediate bacterial internalization (Fig. 4). Interestingly, both endocytic proteins are substrates of Src (Wilde et al. 1999; Joshi et al. 2008). Clathrin-mediated bacterial internalization specifically depends on the presence of the juxtamembrane domain of E-cadherin (Bonazzi et al. 2008). Clathrin and caveolin are also required for the internalization of InlA-coated beads and E-cadherin-coated beads (Bonazzi et al. 2008), suggesting that the endocytic machinery is involved in the initial steps of adherens junctions formation.

FROM STRUCTURE TO PATHOGENESIS

InlA interacts in a specific manner with EC1, the amino-terminal EC repeat of E-cadherin also involved in the initial interaction between E-cadherin molecules on the surface of adjacent cells (Lecuit et al. 1999). The region of the EC1 domain involved in InlA binding differs from that involved in the formation of homotypic interactions between E-cadherins, suggesting that E-cadherin molecules engaged in adherens junctions are still able to bind InlA. Despite the large degree of similarities between E-cadherin orthologs, InlA shows a species specificity for some E-cadherins, as its interaction is completely impaired in rat and mouse, whereas it is functional in other mammalian species such as humans, dogs, rabbits, guinea pigs, and gerbils (Lecuit et al. 1999). This species specificity depends on the nature of a single amino acid, a proline in position 16 within the EC1 domain of E-cadherin in permissive species, and a glutamic acid in nonpermissive mouse and rat E-cadherin (Lecuit et al. 1999). The E-cadherin loop that harbors a proline in position 16 is hydrophobic and uncharged, whereas a glutamic acid at this location results in a charged and hydrophilic region, suggesting that InlA/E-cadherin interaction involves hydrophobic interactions. Of note, in humans, a proline in position 16 is present in all members of the classical cadherins. However, a conserved proline in position 16, although necessary, is not the only critical residue for InlA interaction with the EC1 domain of cadherins, as InlA does not bind N-cadherin, although it harbors a proline in position 16 (Mengaud et al. 1996).

These observations have been confirmed when the crystal structure of InlA, alone or in complex with the EC1 domain of E-cadherin, was solved (Fig. 2) (see previous discussion) (Schubert et al. 2002). Replacing proline 16 with glutamic acid in silico in the EC1 domain of E-cadherin, changes the hydrophilic properties of the binding site for InlA, therefore hampering a close interaction between receptor and ligand (Schubert et al. 2002). Overall, the affinity of InlA for E-cadherin is low and bacterial attachment and internalization is thought to reflect the cooperation of multiple interactions. It has been proposed that the lack of a high affinity binding may be of advantage for the bacterium, as it allows the potential dissociation of InlA from E-cadherin once the bacterium has been internalized, but this has not been investigated experimentally (Wollert et al. 2007a).

A key result was obtained when a transgenic mouse expressing human E-cadherin at the enterocyte level was created. Both the

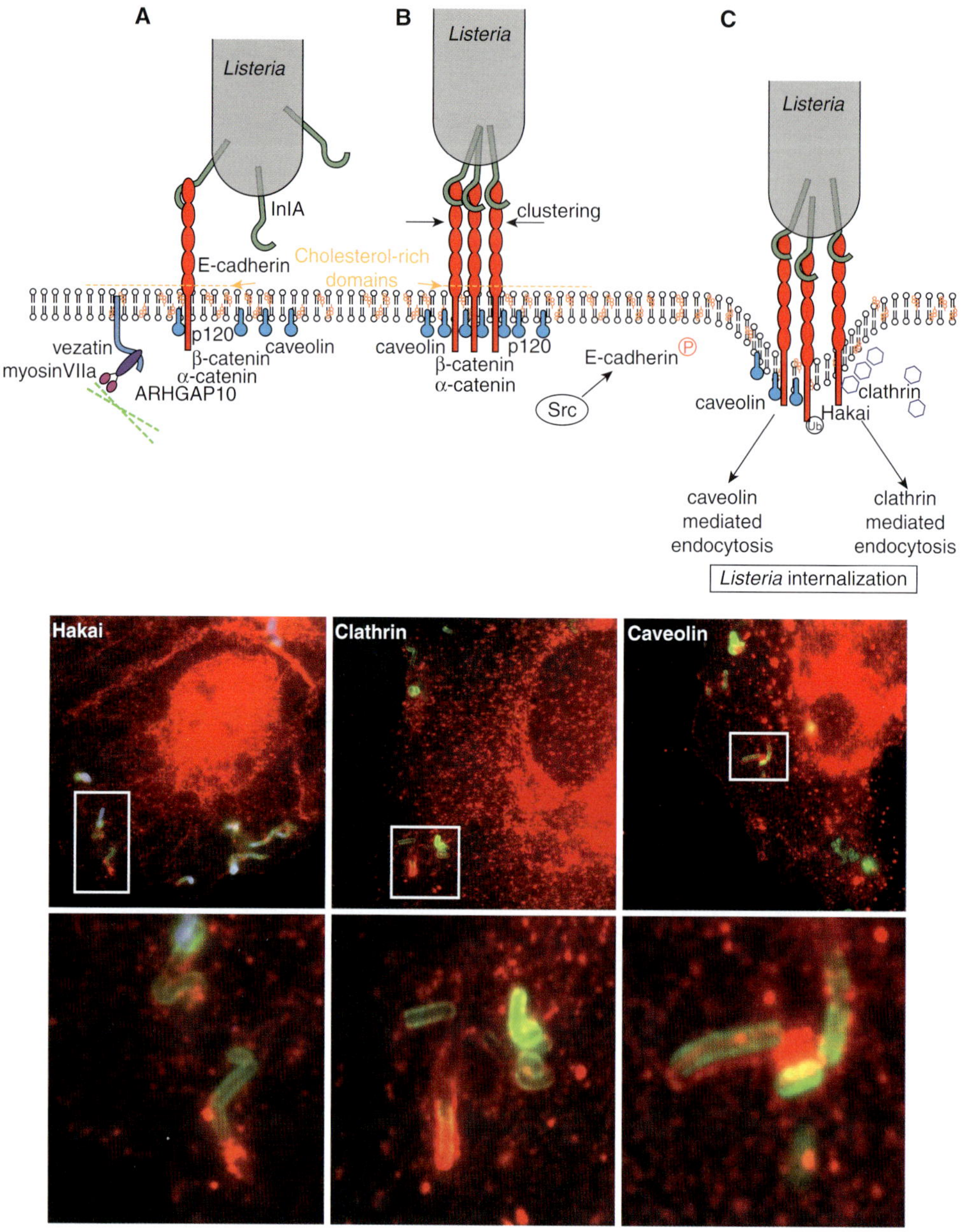

Figure 4. Dynamics of the InlA/E-cadherin interaction (*upper* panel). (*A*) On engagement of E-cadherin by InlA, the adherens junction machinery is activated, inducing the recruitment of the junctional proteins α-catenin, p120 catenin, ARHGAP10, and myosin VIIa. (*B*) InlA interaction with E-cadherin induces the caveolin-dependent clustering of E-cadherin and the activation of the tyrosine kinase Src. (*C*) The Src-dependent phosphorylation of E-cadherin triggers the recruitment of the ubiquitin-ligase Hakai and the ubiquitination of E-cadherin. E-cadherin ubiquitination induces the sorting of E-cadherin within clathrin-coated pits for bacterial internalization. Alternatively, E-cadherin may persist within caveolin-rich domains and bacterial internalization can occur via caveosomes. *Bottom* panels: *Listeria* (green) recruits Hakai (*left* panel), clathrin (*middle* panel), and caveolin (*right* panel) at the bacterial entry site when invading epithelial cells. Lower images are a magnified view of the highlighted areas.

observation that proline 16 is conserved in E-cadherin of guinea pigs and the generation of these transgenic mice, allowed for the first time the study of InlA-mediated *Listeria* infections in a small animal model (Lecuit et al. 2001). These studies showed that InlA (Lecuit et al. 1997; Lecuit et al. 2001), but not InlB (Khelef et al. 2006), plays a critical role in the crossing of the intestinal barrier (Fig. 5). In contrast, when animals are infected intravenously, InlA plays no role, suggesting that InlA is a key virulence factor during the early stage of the infections (Lecuit et al. 2001).

More recently, InlA has been engineered so as to enhance its affinity for human E-cadherin, and this modified InlA also showed an ability to interact with mouse E-cadherin (Wollert et al. 2007b). In particular, replacing serine in position 192 with an asparagine bridges the gap between InlA and E-cadherin and the substitution of the tyrosine in position 369 with a serine allows the formation of a hydrogen bond. Of note, these mutations increase by four orders of magnitude the affinity of this "murinized" InlA ($InlA^m$) for murine E-cadherin, which is in the same range of affinity than that of InlA with human E-cadherin (Wollert et al. 2007b). Under these conditions mice appear to become susceptible to $InlA^m$-mediated infection. Studies on the role of $InlA^m$ confirmed the previously described role of InlA during early oral infections (Lecuit et al. 2001; Wollert et al. 2007b). Surprisingly, whereas murinized InlA has a greatly increased binding affinity for hEcad, this does not translate into a significantly higher level of internalization in in vitro cultured cells expressing hEcad, such as Caco-2 cells (Wollert et al. 2007b).

Interestingly, although epidemiological data strongly argued for a role for InlA in

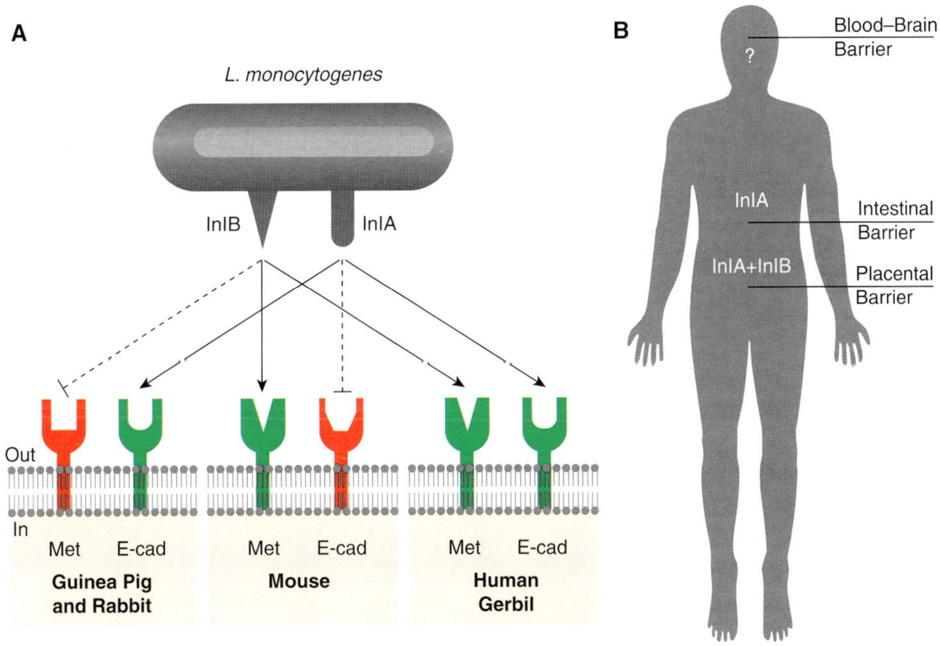

Figure 5. Species specificity of the InlA-E-cadherin interaction (*A*) The species specificity of the InlA/E–cadherin interaction depends on the presence of a proline in position 16 of the amino-terminal E-cadherin repeat. Species that substitute a glutamate to proline in position 16 of E-cadherin are not sensitive to *Listeria* infection, as InlA/E–cadherin interactions cannot occur. Based on the observation that some species are not sensitive to InlB, recent studies revealed that species specificity occurs in the case of InlB. (*B*) InlA/E-cadherin interaction is fundamental to cross the intestinal barrier, whereas InlB does not play any role at this stage. A coordinated action of the two internalins is necessary to cross the placental barrier, whereas the role of InlA and InlB at the level of the blood–brain barrier is still an unresolved issue.

fetoplacental listeriosis (Jacquet et al. 2004), and ex vivo experiments using human placenta also showed a role for InlA in placental invasion (Lecuit et al. 2004), this had not been confirmed in vivo until recently. On the contrary, results obtained in pregnant guinea pigs were in apparent contradiction with these human epidemiological and ex vivo findings (Bakardjiev et al. 2005). The important discovery of the species specificity of InlB, for which guinea pig Met does not act as a receptor (Khelef et al. 2006), led to the hypothesis that for InlA to exert its effect at the placental level, InlB was also required. In a search for a small animal model permissive to both InlA and InlB, a natural host for *L. monocytogenes*, the gerbil, which is naturally susceptible to *L. monocytogenes*, was tested. Results showed that gerbil primary epithelial cells are permissive to both the InlA and InlB pathways, as human epithelial cells (Disson et al. 2008). Gerbils were then infected orally to assess the role of InlA in the crossing of the intestinal barrier, and as in guinea pigs and transgenic mice expressing human E-cadherin, InlA was shown to contribute to the crossing of the intestinal barrier, but not InlB (Disson et al. 2008). Importantly, wt *L. monocytogenes* were able to colonize the placenta of pregnant gerbils infected both orally and intravenously, whereas the *Listeria* mutants that lack either one of the internalins or both (Δ*inlA*, Δ*inlB*, or Δ*inlAB*) were far less efficient in crossing the placental barrier and infecting the fetus. Similar data were also obtained in a novel knock-in mouse model where a humanized E16P mouse E-cadherin is expressed in place of mouse E-cadherin (Disson et al. 2008). Together, these data suggest that for the crossing of the placental barrier, both internalin InlA and InlB have to be able to recognize their respective receptors, E-cadherin and Met, to cooperate and allow efficient placental invasion and fetal infection.

E-CADHERIN IS ALSO A RECEPTOR FOR THE OPPORTUNISTIC FUNGUS *C. ALBICANS*

E-cadherin has been recently identified as the target for another pathogen surface protein, the *C. albicans* invasin Als3 (Phan et al. 2007). It has been previously shown that N-cadherin mediates *C. albicans* endocytosis into endothelial cells, a step that is crucial for the development of systemic infections, but the fungal proteins responsible for cadherin engagement were unknown (Phan et al. 2005). The *C. albicans als* gene family encodes a set of proteins that have a potential role as adhesins and at their amino-termini share a remarkable similarity with *Yersinia pseudotuberculosis* invasin (Phan et al. 2007). In fact, it has been observed that the protein Als3 is capable of binding E-cadherin and N-cadherin on the surface of epithelial and endothelial cells, respectively, and *als3Δ/als3Δ* mutants are no longer internalized within host cells and lose the ability of inducing cellular damage (Phan et al. 2007). Similarly to InlA for *Listeria*, latex beads coated with recombinant Als3 are efficiently internalized within E-cadherin- and N-cadherin-expressing cells. Interestingly, in silico models of the secondary structure of the Als3 amino terminus revealed a striking similarity with the more distal extracellular domain of E-cadherin, suggesting that, differently from InlA, Als3 may engage E-cadherin at the same sites than those required for adherens junctions formation (Phan et al. 2007).

CONCLUSIONS

Pathogen mimicry of host natural ligands and subversion of host cellular functions has been a recurrent theme in the relatively new field of cellular microbiology. It has helped decipher complex signaling pathways and unravel novel cellular functions. Studies on the internalization of bacteria such as *Listeria*, *Staphylococcus*, and *Yersinia*, allowed the identification of unsuspected functions for important cell adhesion molecules (Isberg and Barnes 2001; Agerer et al. 2005; Hamon et al. 2006) and opened new perspectives for well-established cellular models (Veiga and Cossart 2005b; Veiga et al. 2007).

Among bacterial pathogens, *L. monocytogenes* has emerged as a model microorganism for the characterization of key molecular mechanisms, both in bacteria and eukaryotic cells: For example, the study of post-transcriptional regulation in *Listeria* highlighted a novel mechanism of thermoregulation for the promoter PrfA (Johansson et al. 2002); the intracellular actin-based motility of *L. monocytogenes* has helped unravel the mechanism underlying actin dynamics and cell motility (Welch et al. 1998; Gouin et al. 2005). InlA was the first *L. monocytogenes* surface protein identified as a mediator of internalization (Gaillard et al. 1991; Mengaud et al. 1996) and the identification of E-cadherin as the cellular receptor for InlA opened the field of adherens junction biology to microbiologists. Since then, *L. monocytogenes* has been a powerful tool in understanding the complex dynamics of adherens junctions and in building a comprehensive model to describe its functioning.

The development of a humanized small-animal model, a transgenic mouse expressing human E-cadherin, was critical in establishing the importance of a single amino acid in the EC1 domain of E-cadherin in the stringent species-specificity to *L. monocytogenes*. It also helped highlight the unique role of InlA in the crossing of the intestinal barrier (Lecuit et al. 2001). Further advances in the development of new animal models led to the important discovery of an interdependent role for the two major internalins, InlA and InlB, in the crossing of the fetoplacental barrier, a key step of the infection (Disson et al. 2008). A current challenge is to understand the molecular mechanisms of the conjugated action of InlA and InlB at the single cell level. Furthermore, recent insights in the cell biology of InlA-mediated infections have also revealed promising interactions between the adhesion and the endocytic machineries (Bonazzi et al. 2008) that deserve further investigation.

REFERENCES

Agerer F, Lux S, Michel A, Rohde M, Ohlsen K, Hauck CR. 2005. Cellular invasion by Staphylococcus aureus reveals a functional link between focal adhesion kinase and cortactin in integrin-mediated internalisation. *J Cell Sci* **118:** 2189–2200.

Bakardjiev AI, Stacy BA, Portnoy DA. 2005. Growth of *Listeria monocytogenes* in the guinea pig placenta and role of cell-to-cell spread in fetal infection. *J Infect Dis* **191:** 1889–1897.

Bierne H, Sabet C, Personnic N, Cossart P. 2007. Internalins: A complex family of leucine-rich repeat-containing proteins in *Listeria monocytogenes*. *Microbes Infect* **9:** 1156–1166.

Bonazzi M, Cossart P. 2006. Bacterial entry into cells: A role for the endocytic machinery. *FEBS Lett* **580:** 2962–2967.

Bonazzi M, Lecuit M, Cossart P. 2009. Listeria monocytogenes internalin and E-cadherin: From structure to pathogenesis. *Cell Microbiol* **11:** 693–702.

Bonazzi M, Veiga E, Pizarro-Cerda J, Cossart P. 2008. Successive post-translational modifications of E-cadherin are required for InlA-mediated internalization of *Listeria monocytogenes*. *Cell Microbiol* **10:** 2208–2222.

Boyle EC, Finlay BB. 2003. Bacterial pathogenesis: Exploiting cellular adherence. *Curr Opin Cell Biol* **15:** 633–639.

Brembeck FH, Schwarz-Romond T, Bakkers J, Wilhelm S, Hammerschmidt M, Birchmeier W. 2004. Essential role of BCL9–2 in the switch between β-catenin's adhesive and transcriptional functions. *Genes Dev* **18:** 2225–2230.

Campbell DH, Sutherland RL, Daly RJ. 1999. Signaling pathways and structural domains required for phosphorylation of EMS1/cortactin. *Cancer Res* **59:** 5376–5385.

Cavey M, Rauzi M, Lenne PF, Lecuit T. 2008. A two-tiered mechanism for stabilization and immobilization of E-cadherin. *Nature* **453:** 751–756.

Cossart P, Sansonetti PJ. 2004. Bacterial invasion: The paradigms of enteroinvasive pathogens. *Science* **304:** 242–248.

Delva E, Tucker DK, Kowalczyk AP. 2009. The desmosome. *Cold Spring Harb Perspect Biol* **1:** a002543.

Disson O, Grayo S, Huillet E, Nikitas G, Langa-Vives F, Dussurget O, Ragon M, Le Monnier A, Babinet C, Cossart P. 2008. Conjugated action of two species-specific invasion proteins for fetoplacental listeriosis. *Nature* **455:** 1114–1118.

Drees F, Pokutta S, Yamada S, Nelson WJ, Weis WI. 2005. α-catenin is a molecular switch that binds E-cadherin-β-catenin and regulates actin-filament assembly. *Cell* **123:** 903–915.

Dubois T, Paleotti O, Mironov AA, Fraisier V, Stradal TE, De Matteis MA, Franco M, Chavrier P. 2005. Golgi-localized GAP for Cdc42 functions downstream of ARF1 to control Arp2/3 complex and F-actin dynamics. *Nat Cell Biol* **7:** 353–364.

Dupre-Crochet S, Figueroa A, Hogan C, Ferber EC, Bialucha CU, Adams J, Richardson EC, Fujita Y. 2007. Casein kinase 1 is a novel negative regulator of E-cadherin-based cell-cell contacts. *Mol Cell Biol* **27:** 3804–3816.

Fujita Y, Krause G, Scheffner M, Zechner D, Leddy HE, Behrens J, Sommer T, Birchmeier W. 2002. Hakai, a c-Cbl-like protein, ubiquitinates and induces endocytosis of the E-cadherin complex. *Nat Cell Biol* **4:** 222–231.

Gaillard JL, Berche P, Frehel C, Gouin E, Cossart P. 1991. Entry of *L. monocytogenes* into cells is mediated by internalin, a repeat protein reminiscent of surface antigens from gram-positive cocci. *Cell* **65:** 1127–1141.

Gaillard JL, Berche P, Mounier J, Richard S, Sansonetti P. 1987. In vitro model of penetration and intracellular growth of *Listeria monocytogenes* in the human enterocyte-like cell line Caco-2. *Infect Immun* **55:** 2822–2829.

Gottardi CJ, Gumbiner BM. 2004. Distinct molecular forms of β-catenin are targeted to adhesive or transcriptional complexes. *J Cell Biol* **167:** 339–349.

Gouin E, Welch MD, Cossart P. 2005. Actin-based motility of intracellular pathogens. *Curr Opin Microbiol* **8:** 35–45.

Green KJ, Gaudry CA. 2000. Are desmosomes more than tethers for intermediate filaments? *Nat Rev Mol Cell Biol* **1:** 208–216.

Hamon M, Bierne H, Cossart P. 2006. *Listeria monocytogenes*: A multifaceted model. *Nat Rev Microbiol* **4:** 423–434.

Hauck CR, Agerer F, Muenzner P, Schmitter T. 2006. Cellular adhesion molecules as targets for bacterial infection. *Eur J Cell Biol* **85:** 235–242.

Heuberger J, Birchmeier W. 2009. Interplay of cadherin-mediated cell adhesion and canonical Wnt signaling. *Cold Spring Harb Perspect Biol* **2:** a002915.

Huang C, Liu J, Haudenschild CC, Zhan X. 1998. The role of tyrosine phosphorylation of cortactin in the locomotion of endothelial cells. *J Biol Chem* **273:** 25770–25776.

Huang C, Ni Y, Wang T, Gao Y, Haudenschild CC, Zhan X. 1997. Down-regulation of the filamentous actin cross-linking activity of cortactin by Src-mediated tyrosine phosphorylation. *J Biol Chem* **272:** 13911–13915.

Isberg RR, Barnes P. 2001. Subversion of integrins by enteropathogenic Yersinia. *J Cell Sci* **114:** 21–28.

Jacquet C, Doumith M, Gordon JI, Martin PM, Cossart P, Lecuit M. 2004. A molecular marker for evaluating the pathogenic potential of foodborne *Listeria monocytogenes*. *J Infect Dis* **189:** 2094–2100.

Johansson J, Mandin P, Renzoni A, Chiaruttini C, Springer M, Cossart P. 2002. An RNA thermosensor controls expression of virulence genes in *Listeria monocytogenes*. *Cell* **110:** 551–561.

Jonquieres R, Bierne H, Mengaud J, Cossart P. 1998. The inlA gene of *Listeria monocytogenes* LO28 harbors a nonsense mutation resulting in release of internalin. *Infect Immun* **66:** 3420–3422.

Joshi B, Strugnell SS, Goetz JG, Kojic LD, Cox ME, Griffith OL, Chan SK, Jones SJ, Leung SP, Masoudi H, et al. 2008. Phosphorylated caveolin-1 regulates Rho/ROCK-dependent focal adhesion dynamics and tumor cell migration and invasion. *Cancer Res* **68:** 8210–8220.

Karecla PI, Green SJ, Bowden SJ, Coadwell J, Kilshaw PJ. 1996. Identification of a binding site for integrin αEβ7 in the N-terminal domain of E-cadherin. *J Biol Chem* **271:** 30909–30915.

Khelef N, Lecuit M, Bierne H, Cossart P. 2006. Species specificity of the *Listeria monocytogenes* InlB protein. *Cell Microbiol* **8:** 457–470.

Kobe B, Deisenhofer J. 1994. The leucine-rich repeat: A versatile binding motif. *Trends Biochem Sci* **19:** 415–421.

Kohmura N, Senzaki K, Hamada S, Kai N, Yasuda R, Watanabe M, Ishii H, Yasuda M, Mishina M, Yagi T. 1998. Diversity revealed by a novel family of cadherins expressed in neurons at a synaptic complex. *Neuron* **20:** 1137–1151.

Lecuit M. 2007. Human listeriosis and animal models. *Microbes Infect* **9:** 1216–1225.

Lecuit M, Dramsi S, Gottardi C, Fedor-Chaiken M, Gumbiner B, Cossart P. 1999. A single amino acid in E-cadherin responsible for host specificity towards the human pathogen *Listeria monocytogenes*. *Embo J* **18:** 3956–3963.

Lecuit M, Hurme R, Pizarro-Cerda J, Ohayon H, Geiger B, Cossart P. 2000. A role for α-and β-catenins in bacterial uptake. *Proc Natl Acad Sci* **97:** 10008–10013.

Lecuit M, Nelson DM, Smith SD, Khun H, Huerre M, Vacher-Lavenu MC, Gordon JI, Cossart P. 2004. Targeting and crossing of the human maternofetal barrier by *Listeria monocytogenes*: Role of internalin interaction with trophoblast E-cadherin. *Proc Natl Acad Sci* **101:** 6152–6157.

Lecuit M, Ohayon H, Braun L, Mengaud J, Cossart P. 1997. Internalin of *Listeria monocytogenes* with an intact leucine-rich repeat region is sufficient to promote internalization. *Infect Immun* **65:** 5309–5319.

Lecuit M, Vandormael-Pournin S, Lefort J, Huerre M, Gounon P, Dupuy C, Babinet C, Cossart P. 2001. A transgenic model for listeriosis: Role of internalin in crossing the intestinal barrier. *Science* **292:** 1722–1725.

Martinez-Quiles N, Ho HY, Kirschner MW, Ramesh N, Geha RS. 2004. Erk/Src phosphorylation of cortactin acts as a switch on-switch off mechanism that controls its ability to activate N-WASP. *Mol Cell Biol* **24:** 5269–5280.

McCrea PD, Gu D, Balda M. 2009. Junctional music that the nucleus hears. *Cold Spring Harb Perspect Biol* **1:** a002923.

Meng W, Takeichi M. 2009. Adherens junction: Molecular architecture and regulation. *Cold Spring Harb Perspect Biol* **1:** a002899.

Mengaud J, Ohayon H, Gounon P. 1996. E-cadherin is the receptor for internalin, a surface protein required for entry of *L. monocytogenes* into epithelial cells. *Cell* **84:** 923–932.

Ozawa M, Ringwald M, Kemler R. 1990. Uvomorulin-catenin complex formation is regulated by a specific domain in the cytoplasmic region of the cell adhesion molecule. *Proc Natl Acad Sci* **87:** 4246–4250.

Patel SD, Chen CP, Bahna F, Honig B, Shapiro L. 2003. Cadherin-mediated cell-cell adhesion: Sticking together as a family. *Curr Opin Struct Biol* **13:** 690–698.

Phan QT, Fratti RA, Prasadarao NV, Edwards JE Jr, Filler SG. 2005. N-cadherin mediates endocytosis of *Candida albicans* by endothelial cells. *J Biol Chem* **280:** 10455–10461.

Phan QT, Myers CL, Fu Y, Sheppard DC, Yeaman MR, Welch WH, Ibrahim AS, Edwards JE, Filler SG. 2007. Als3 is a *Candida albicans* invasin that binds to cadherins and induces endocytosis by host cells. *PLoS Biol* **5:** e64. doi: 10.1371/journal.pbio.0050064.

Pizarro-Cerda J, Cossart P. 2006. Bacterial adhesion and entry into host cells. *Cell* **124:** 715–727.

Pizarro-Cerda J, Jonquieres R, Gouin E, Vandekerckhove J, Garin J, Cossart P. 2002. Distinct protein patterns associated with *Listeria monocytogenes* InlA- or InlB-phagosomes. *Cell Microbiol* **4:** 101–115.

Ranscht B, Dours-Zimmermann MT. 1991. T-cadherin, a novel cadherin cell adhesion molecule in the nervous system lacks the conserved cytoplasmic region. *Neuron* **7:** 391–402.

Reynolds AB. 2007. p120-catenin: Past and present. *Biochim Biophys Acta* **1773:** 2–7.

Rhee J, Buchan T, Zukerberg L, Lilien J, Balsamo J. 2007. Cables links Robo-bound Abl kinase to N-cadherin-bound β-catenin to mediate Slit-induced modulation of adhesion and transcription. *Nat Cell Biol* **9:** 883–892.

Roura S, Miravet S, Piedra J, Garcia de Herreros A, Dunach M. 1999. Regulation of E-cadherin/Catenin association by tyrosine phosphorylation. *J Biol Chem* **274:** 36734–36740.

Schubert WD, Urbanke C, Ziehm T, Beier V, Machner MP, Domann E, Wehland J, Chakraborty T, Heinz DW. 2002. Structure of internalin, a major invasion protein of *Listeria monocytogenes*, in complex with its human receptor E-cadherin. *Cell* **111:** 825–836.

Shapiro L, Weis WI. 2009. Structure and biochemistry of cadherins and catenins. *Cold Spring Harb Perspect Biol* **1:** a003053.

Sousa S, Cabanes D, Archambaud C, Colland F, Lemichez E, Popoff M, Boisson-Dupuis S, Gouin E, Lecuit M, Legrain P, et al. 2005. ARHGAP10 is necessary for α-catenin recruitment at adherens junctions and for Listeria invasion. *Nat Cell Biol* **7:** 954–960.

Sousa S, Cabanes D, Bougneres L, Lecuit M, Sansonetti P, Tran-Van-Nhieu G, Cossart P. 2007. Src, cortactin and Arp2/3 complex are required for E-cadherin-mediated internalization of Listeria into cells. *Cell Microbiol* **9:** 2629–2643.

Sousa S, Cabanes D, El-Amraoui A, Petit C, Lecuit M, Cossart P. 2004. Unconventional myosin VIIa and vezatin, two proteins crucial for Listeria entry into epithelial cells. *J Cell Sci* **117:** 2121–2130.

Troyanovsky S. 2005. Cadherin dimers in cell-cell adhesion. *Eur J Cell Biol* **84:** 225–233.

Vasioukhin V, Fuchs E. 2001. Actin dynamics and cell-cell adhesion in epithelia. *Curr Opin Cell Biol* **13:** 76–84.

Veiga E, Cossart P. 2005a. Ubiquitination of intracellular bacteria: A new bacteria sensing system? *Trends Cell Biol* **15:** 2–5.

Veiga E, Cossart P. 2005b. Listeria hijacks the clathrin-dependent endocytic machinery to invade mammalian cells. *Nat Cell Biol* **7:** 894–900.

Veiga E, Guttman JA, Bonazzi M, Boucrot E, Toledo-Arana A, Lin AE, Enninga J, Pizarro-Cerda J, Finlay BB, Kirchhausen T, et al. 2007. Invasive and adherent bacterial pathogens co-opt host clathrin for infection. *Cell Host Microbe* **2:** 340–351.

Vestweber D, Winderlich M, Cagna G, Nottebaum AF. 2009. Cell adhesion dynamics at endothelial junctions: VE-cadherin as a major player. *Trends Cell Biol* **19:** 8–15.

Weed SA, Du Y, Parsons JT. 1998. Translocation of cortactin to the cell periphery is mediated by the small GTPase Rac1. *J Cell Sci* **111:** 2433–2443.

Welch MD, Rosenblatt J, Skoble J, Portnoy DA, Mitchison TJ. 1998. Interaction of human Arp2/3 complex and the *Listeria monocytogenes* ActA protein in actin filament nucleation. *Science* **281:** 105–108.

Wilde A, Beattie EC, Lem L, Riethof DA, Liu SH, Mobley WC, Soriano P, Brodsky FM. 1999. EGF receptor signaling stimulates SRC kinase phosphorylation of clathrin, influencing clathrin redistribution and EGF uptake. *Cell* **96:** 677–687.

Wollert T, Heinz DW, Schubert WD. 2007a. Thermodynamically reengineering the listerial invasion complex InlA/E-cadherin. *Proc Natl Acad Sci* **104:** 13960–13965.

Wollert T, Pasche B, Rochon M, Deppenmeier S, van den Heuvel J, Gruber AD, Heinz DW, Lengeling A, Schubert WD. 2007b. Extending the host range of *Listeria monocytogenes* by rational protein design. *Cell* **129:** 891–902.

Wu S, Lim KC, Huang J, Saidi RF, Sears CL. 1998. *Bacteroides fragilis* enterotoxin cleaves the zonula adherens protein, E-cadherin. *Proc Natl Acad Sci* **95:** 14979–14984.

Yamada S, Pokutta S, Drees F, Weis WI, Nelson WJ. 2005. Deconstructing the cadherin-catenin-actin complex. *Cell* **123:** 889–901.

Zeng G, Apte U, Micsenyi A, Bell A, Monga SP. 2006. Tyrosine residues 654 and 670 in β-catenin are crucial in regulation of Met-β-catenin interactions. *Exp Cell Res* **312:** 3620–3630.

Cadherins as Targets for Genetic Diseases

Aziz El-Amraoui[1,2,3] and Christine Petit[1,2,3,4]

[1]Institut Pasteur, Unité de Génétique et Physiologie de l'Audition, 25 rue du Dr Roux, 75015 Paris, France
[2]INSERM UMRS587, 75015 Paris, France
[3]UPMC, F75015 Paris, France
[4]Collège de France, 75005 Paris, France
Correspondence: christine.petit@pasteur.fr, aziz.el-amraoui@pasteur.fr

The 6-billion human population provides a vast reservoir of mutations, which, in addition to the opportunity of detecting very subtle defects, including specific cognitive dysfunctions as well as late appearing disorders, offers a unique background in which to investigate the roles of cell–cell adhesion proteins. Here we focus on inherited human disorders involving members of the cadherin superfamily. Most of the advances concern monogenic disorders. Yet, with the development of single nucleotide polymorphism (SNP) association studies, cadherin genes are emerging as susceptibility genes in multifactorial disorders. Various skin and heart disorders revealed the critical role played by desmosomal cadherins in epidermis, hairs, and myocardium, which experience high mechanical stress. Of particular interest in that respect is the study of Usher syndrome type 1 (USH1), a hereditary syndromic form of deafness. Studies of USH1 brought to light the crucial role of transient fibrous links formed by cadherin 23 and protocadherin 15 in the cohesion of the developing hair bundle, the mechanoreceptive structure of the auditory sensory cells, as well as the involvement of these cadherins in the formation of the tip-link, a key component of the mechano-electrical transduction machinery. Finally, in line with the well-established role of cadherins in synaptic formation, maintenance, strength, and plasticity, a growing number of cadherin family members, especially protocadherins, have been found to be involved in neuropsychiatric disorders.

Cell–cell adhesion maintains the structural and functional integrity of multicellular organisms. It ensures cell recognition, sorting, and signaling in various tissues and organs. In line with the diversity of the structural and physiological properties of the cell–cell junctions (tight junctions, adherens junctions, desmosomes, and gap junctions), related disorders span a highly heterogeneous group of diseases. Abnormal or loss of cell–cell adhesion and/or associated signaling are hallmarks of tumor growth, malignant transformation, and metastases (see Berx and van Roy 2009). In addition, nonneoplastic diseases caused by defects in genes encoding proteins of the tight junctions, adherens junctions, desmosomes, and gap junctions have been identified (Table 1) (see also Lai-Cheong et al. 2007;

 doi: 10.1101/cshperspect.a003095
Cite this article as *Cold Spring Harb Perspect Biol* 2009;1:a003095

Table 1. Integral membrane proteins of cell–cell junctions involved in human diseases.

Junction	Protein	Associated disease(s)
Tight junction	Claudin 1	Neonatal sclerosing cholangitis with ichthyosis (OMIM 603718 and 607626)
Tight junction	Claudin 16	Familial hypercalciuric hypomagnesemia with nephrocalcinosis (OMIM 603959 and 248250)
Tight junction	Claudin 19	Familial hypercalciuric hypomagnesemia with nephrocalcinosis, and severe visual impairment (OMIM 610036 and 248190)
Tight junction	Claudin 14	Sensorineural deafness, DFNB29 (OMIM 605608)
Tight junction	Tricellulin	Sensorineural deafness, DFNB49 (OMIM 610153 and 610572)
Adherens junction	P-cadherin	Hypotrichosis, juvenile macular dystrophy (OMIM 601553) Ectodermal dysplasia, ectrodactyly, macular dystrophy syndrome (OMIM 225280)
Adherens junction	M-cadherin	Intellectualdisability (OMIM 114019)
Adherens junction	Protocadherin 19	X-linked epilepsy and mental retardation limited to females (OMIM 300088) and Dravet syndrome (OMIM 607208)
Adherens junction	Protocadherin 11X	Suceptibility to late-onset Alzheimer's disease, LOAD (OMIM 300246)
Adherens junction	Cadherin 23	Sensorineural deafness, DFNB12 and Usher syndrome, Type 1D (OMIM 605516)
Adherens junction	Protocadherin 15	Sensorineural deafness, DFNB23 and Usher syndrome, Type 1F (OMIM 605514)
Adherens junction	H-cadherin[a]	Suceptibility to blood pressure variations (OMIM 300246)
Desmosome	Desmoglein 1	Striate palmoplantar keratoderma (OMIM 125670 and 148700)
Desmosome	Desmoglein 4	Localized recessive hypotrichosis (OMIM 607892 and 607903) ± Recessive monilethrix (OMIM 158000 and 252200)
Desmosome	Desmoglein 2	Arrhythmogenic right ventricular cardiomyopathy, ARVC10 (OMIM 125671 and 610193)
Desmosome	Desmocollin 2	Arrhythmogenic right ventricular cardiomyopathy, ARVC11 (OMIM 125645 and 610476) ± Palmoplantar keratoderma and woolly hair
Gap junction	Connexin 26 (OMIM 121011)	Sensorineural deafness, DFNB1, DFNA3 ± Plamoplantar keratoderma ± Keratitis-Ichtyosis...
Gap junction	Connexin 30 (OMIM 604418)	Sensorineural deafness, DFNB1, DFNA3 ± Ectodermal dysplasia, hidrotic
Gap junction	Connexin 30.3 (OMIM 605425)	Erythrokerato-derma variabilis (OMIM 133200)
Gap junction	Connexin 31 (OMIM 603324)	Erythrokerato-derma variabilis ± Deafness, DFNA2 ± Peripheral neuropathy (OMIM 133200)
Gap junction	Connexin 32 (OMIM 304040)	X-linked Charcot Marie Toot (OMIM 302800)
Gap junction	Connexin 40 (OMIM 121013)	Atrial fibrillation (OMIM 608583)
Gap junction	Connexin 43 (OMIM 121014)	Oculodento-digital dysplasia (OMIM 164200)
Gap junction	Connexin 46 (OMIM 121015)	Zonular pulverent cataract-3 (OMIM 601885)
Gap junction	Connexin 50 (OMIM 116200)	Zonular pulverent cataract-1 (OMIM 600897)

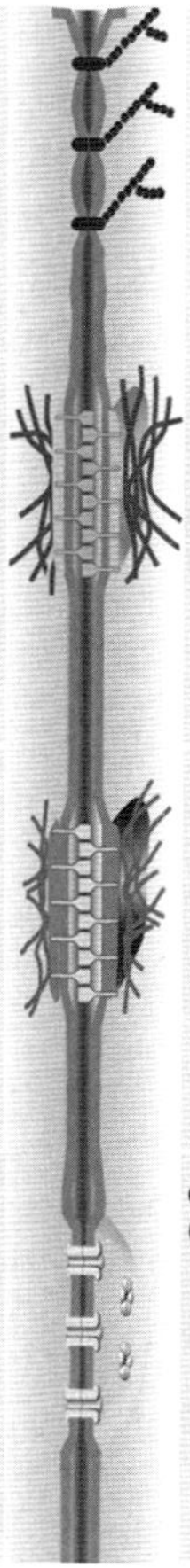

[a] H-cadherin (also referred to as cadherin 13) lacks transmembrane and cytoplasmic domains (see Org et al. 2009). OMIM numbers are listed for references and details on the indicated protein and/or associated human diseases.

Awad et al. 2008; Org et al. 2009). In this article, we summarize the current understanding of human inherited disorders because of defects in members of the cadherin superfamily, i.e., classical cadherins, desmosomal cadherins, protocadherins, and atypical cadherins.

Cadherins are characterized by the presence of 1–34 extracellular cadherin (EC) domains (comprised of ~110 amino acids). Variations in their cytoplasmic domains impart functional specificity by conferring to each molecule the ability to interact with different ligands (see Shapiro and Weis 2009). These cadherins initiate and build up two types of specialized cell–cell contacts: (1) adherens junctions, and (2) desmosomes, where they tether the plasma membrane to actin microfilaments and intermediate filaments, respectively.

Clinical interest in cadherins stems from the discovery, about 25 years ago, of autoantibodies directed against desmosomal cadherins, desmoglein 1 and desmoglein 3, in the autoimmune blistering skin diseases pemphigus foliaceous and pemphigus vulgaris, respectively (Waschke 2008). Pemphigus-affected patients display a loss of intercellular adhesion between keratinocytes (acantholysis) caused by the binding of autoantibodies to desmosomal cadherins (Waschke 2008). Later on, when a dominantly inherited skin disease, striate palmoplantar keratoderma (SPPK; OMIM 148700), was mapped at chromosome 18q12.1, near the cluster of desmosomal cadherin genes (Hennies et al. 1995), the corresponding genes were considered as candidate causative genes for this disorder (Allen et al. 1996). However, it was not until 1999 that the first mutation—an in-frame deletion—was detected in the desmoglein 1 gene in SPPK patients (Rickman et al. 1999).

Genome sequence analysis has revealed an impressive diversity of the cadherin superfamily, with 113 cadherins identified in humans to date (Hulpiau and van Roy 2009). Despite functional redundancy among cell–cell junction proteins, several genetic defects affecting the skin, heart, nervous system, ear, or eye have been reported in mice and humans carrying mutations in some cadherin genes (Table 1). So far, mutations in genes encoding 11 cadherins (two classical, four desmosomal, three protocadherins, and two atypical) have been associated with hereditary human diseases (Table 1). These have highlighted a particular need for dynamic and positional cues for axon guidance during the establishment of neural circuits, or for high adhesive strength in tissues that experience mechanical stress, such as the epidermis and hairs, myocardium, and inner ear sensory epithelia.

CADHERINS IN SKIN AND HEART

Skin, Retinal, and Limb Defects Caused by P-Cadherin Deficiency

Sixteen members of the classical cadherin subfamily have been identified in the human genome (Hulpiau and van Roy 2009). P-cadherin (also referred to as cadherin 3) displays a wide tissue distribution. In 2001, mutations in P-cadherin were identified in a human disease exclusively affecting the hair follicle and the retina (Fig. 1B,C), termed hypotrichosis with juvenile macular dystrophy (HJMD syndrome; OMIM 601553). HJMD patients have short, sparse hair, and progressive blindness (Sprecher et al. 2001). Mutations in P-cadherin were later reported to cause another autosomal recessive disease, ectodermal dysplasia, ectrodactyly, and macular dystrophy (EEM syndrome; OMIM 225280) (Kjaer et al. 2005; Shimomura et al. 2008). The common features of the two diseases are early hair loss followed by progressive degeneration of the central retina, leading to blindness between the first and third decades. EEM affected patients, however, also display hypodontia and limb defects such as a split hand/foot malformation (see Fig. 1C). Genetic and clinical analyses revealed that HJMD patients carrying identical P-cadherin mutations display varying degrees of visual impairment (Kjaer et al. 2005; Shimomura et al. 2008). Whether modifier genes or possible digenic inheritance could explain the phenotype variability remains to be established.

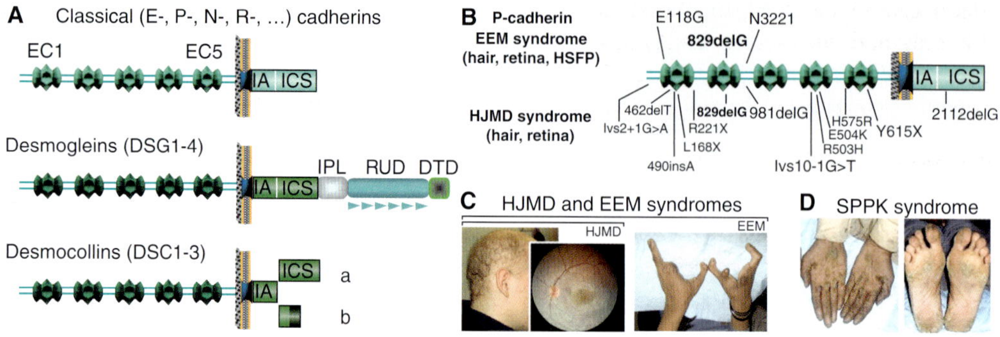

Figure 1. Mutations in P-cadherin, desmogleins (1 and 4), and desmocollin 2 cause hair, retina, and skin disorders. (*A*) Predicted structure of classical and desmosomal cadherins. These cadherins are composed of four highly conserved extracellular domains (EC1–4), a more variable extracellular anchor (EA, also referred to as EC5), a single transmembrane domain (TM), an intracellular anchor (IA), and additional cytoplasmic domains. The four desmogleins are unique in having extended tails beyond the ICS (intracellular catenin-binding site segment), namely an intracellular proline-rich linker domain (IPL), a variable number of repeated-unit domains (RUD; five in Dsg1, six in Dsg2, two in Dsg3, and three in Dsg4) and a desmoglein-specific terminal domain (DTD). Each of the three desmocollins (DSC1-3) have two spliced forms, a and b, with the "b" form lacking the ICS and therefore being unable to bind to plakoglobin. (*B*) P-cadherin mutations identified in HJMD and EEM affected patients are indicated. Note that a similar mutation (bold) can result either in EEM or HJMD disease (see Shimomura et al. 2008). (*C, D*) Clinical features. (*C*) HJMD patients show sparse, light and short hairs over the scalp and macular pigmentary abnormalities. EEM patients in addition show limb defects (adapted, with permission, from Indelman et al. 2007 [© Blackwell], and Shimomura et al. 2008). (*D*) SPPK patients show hyperkeratotic lesions on the palms and soles (adapted, with permission, from Dua-Awereh et al. 2009 [© Elsevier]).

Differences in the phenotypes caused by P-cadherin deficiency in humans and mice have slowed down the deciphering of mechanisms underlying skin, limb, and eye anomalies in HJMD and EEM syndromes. P-cadherin null mice show precocious mammary gland development, but surprisingly hairs, eyes, and limbs develop normally (Radice et al. 1997). Chromatin immunoprecipitation assays revealed that P-cadherin is a direct target for p63, a transcription factor known to be involved in hair follicle and limb development, and whose defects also cause split hand/foot malformations in humans. The p63 protein directly interacts with two distinct regions of the P-cadherin promoter, which could be crucial for hair follicle and limb bud development (Shimomura et al. 2008). Interestingly, one of the p63-target sites of the human P-cadherin gene is located within an intronic region not conserved in the mouse gene (Yang et al. 2006), thereby possibly accounting for the species-related phenotypic differences associated with mutations in the P-cadherin gene.

The strong expression of the P-cadherin gene that is observed in the hair follicle matrix, the apical ridge of limb buds, and the retinal pigment epithelium (RPE) correlate with the affected sites in humans carrying mutations in this gene (Xu et al. 2002). The changes in the spatio-temporal expression pattern of this cadherin during development may explain the restricted spectrum of clinical manifestations resulting from mutations in the P-cadherin gene (Indelman et al. 2007). In the developing epidermis, P-cadherin up-regulation correlated specifically with follicle downgrowth (Jamora et al. 2003). Also, when the outer layer of the optic cup differentiates into RPE cells, an increase in P-cadherin expression is observed concomitantly with a down-regulation of N-cadherin (the most abundant cadherin expressed by RPE cells) (Xu et al. 2002). Details on the subcellular defects caused by P-cadherin dysfunction in RPE cells and the underlying photoreceptor cells in HJMD patients are however still missing.

Desmosomal Cadherins in the Skin and Hair Follicles

Desmosomal cadherins form the core of desmosomes, adhesion structures often present in tissues under high mechanical stress, such as skin epidermis and hair follicles (see Delva et al. 2009). In humans, there are seven desmosomal cadherins: four desmogleins (DSG1–4) and three desmocollins (DSC1–3, each of them having two splice variants) (Fig. 1A). These cadherins resemble the classical cadherins, such as E-cadherin. They contain five tandemly repeated EC domains, a single transmembrane region, and a cytoplasmic domain. The desmogleins, however, differ from classical cadherins by the presence of several additional cytoplasmic domains that interact with the intermediate filament-based cytoskeleton through plakoglobin, plakophilins, and desmoplakin (Figs. 1A and 2).

Desmogleins and desmocollins are both needed for normal adhesive function. They are expressed in a tissue-specific and differentiation step-specific manner (Bazzi et al. 2006). Monolayered epithelia express only the desmoglein 2–desmocollin 2 pair, whereas stratified epithelia, such as the epidermis and the hair follicle, express combinations of the seven desmosomal cadherins (see Delva et al. 2009).

Desmoglein 1: Desmoglein 1 was the first cadherin found as a target of a human hereditary disease, striate palmoplantar keratoderma (Rickman et al. 1999). Sixteen different mutations have been identified in the DSG1 gene, most of which result in premature stop codons (Dua-Awereh et al. 2009; Hershkovitz et al. 2009). SPPK is likely caused by desmoglein 1 haploinsufficiency, even though a dominant-negative effect of mutant alleles cannot be excluded (Dua-Awereh et al. 2009).

SPPK patients display longitudinal hyperkeratotic lesions on the palms and soles, usually evident on the pressure areas that are under continuous mechanical stress and friction (see Fig. 1D). To date, no desmoglein 1 defective mice have been reported. Three Dsg1 genes, *Dsg1α*, *Dsg1β*, and *Dsg1γ*, have been identified in the mouse genome, whereas a single DSG1 gene is present in man. The murine Dsg1 genes display redundant expression patterns, suggesting that a targeted deletion of all three genes (technically difficult because of their genomic proximity) may be required to obtain

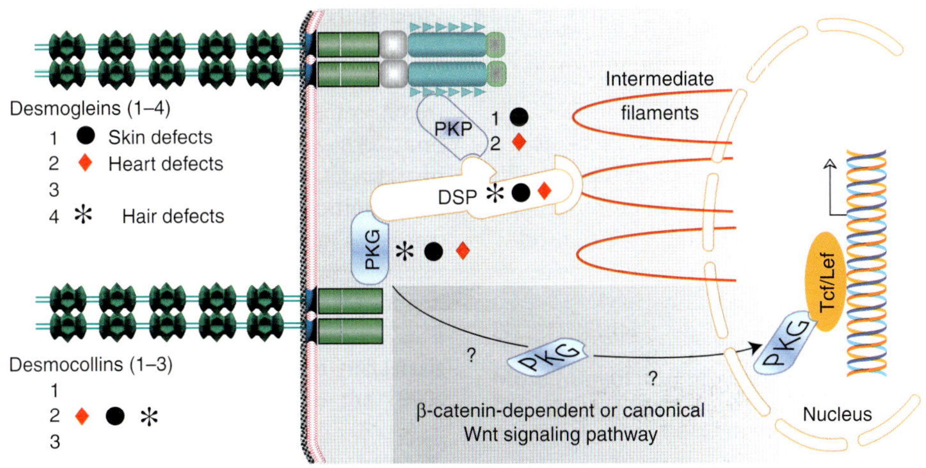

Figure 2. The desmosomal molecular network. The tissues that are affected in human individuals who carry mutations in genes encoding desmosomal proteins are indicated. (PKP) plakophilins 1 and 2; (PKG) plakoglobin; (DSP) desmoplakin.

a murine model of the human disease. Palmar skin analysis by EM in SPPK patients showed a reduced number of desmosomes in suprabasal layers that are abnormally small and poorly formed. At the molecular level, the expression of keratins K5, K14, and K10 was reduced, whereas that of K16 was increased (Wan et al. 2004). The up-regulation of K16, a known marker of the onset of epidermal proliferation and regeneration, may account for the incomplete differentiation of keratinocytes and skin thickening characteristic of the phenotype (Wan et al. 2004).

Desmoglein 4: Mutations in the DSG4 gene have been associated with an inherited human hair disorder, the Localized autosomal recessive hypotrichosis (LAH; OMIM 607903). The first identified mutation was a large, intragenic deletion, which leads to an in-frame deletion of the EC2 domain and part of the EC1 and EC3 domains (Kljuic et al. 2003). The frequent occurrence and high incidence of this mutation in Pakistani individuals from different geographical regions suggest that this is an ancestral mutation (Schweizer 2006). Additional mutations, missense, and null mutations have been reported (Wajid et al. 2007).

LAH affects the scalp, trunk, and extremities, largely sparing the facial, pubic, and axillary hairs. Hairs are fragile and break easily, leaving short sparse scalp hairs with a characteristic appearance (Kljuic et al. 2003). DSG4 is abundant in the hair follicle shaft cortex, where the abnormal phenotype is observed in LAH patients. Skin biopsy analysis from patients showed widening of spaces between keratinocytes throughout most of the epidermis (apart from the basal layer), with ultrastructural evidence of a reduced number of desmosomes that are abnormally small and poorly formed. A clinical overlap was observed between LAH and monilethrix, another human disease that also manifests a beaded hair phenotype (Schweizer 2006). Monilethrix is inherited either as a dominant or a recessive disorder. Homozygous mutations in *DSG4* were identified in monilethrix-affected patients (Schaffer et al. 2006; Shimomura et al. 2006; Zlotogorski et al. 2006). The existence of potential modifying genes or different types/combinations of mutations may explain the interfamilial variations in phenotype among individuals carrying *DSG4* mutations.

Desmoglein 2 and Desmocollin 2 in Heart Function

Myocardium desmosomes are essentially made of desmoglein 2 and desmocollin 2. Several studies have shown that heterozygous mutations in desmoglein 2 and desmocollin 2 in humans cause arrhythmogenic right ventricular cardiomyopathy (ARVC; OMIM 107970). ARVC is estimated to affect approximately one in 5000 individuals. The disease is frequently familial and typically involves autosomal dominant transmission with low penetrance and variable expressivity. Clinical features include right ventricular enlargement and dysfunction, characteristic electrocardiographic abnormalities with ventricular arrhythmia, and fibrofatty replacement of myocytes (see Awad et al. 2008 and references therein).

Desmoglein 2: Complete loss of desmoglein 2 in mice is embryonic lethal, shortly after implantation (Eshkind et al. 2002). Heterozygous mutations in the human DSG2 gene cause isolated ARVC without skin or hair abnormalities (Awad et al. 2006; Pilichou et al. 2006). Compensation by other DSG genes likely takes place in the epidermis but not in the myocardium of *DSG2* mutation carriers. More than 22 mutations have been identified in ARVC patients, most of which are missense mutations located in the extracellular region of DSG2 (Awad et al. 2008). Histopathological analyses in several patients revealed a loss of cardiac cells and replacement with fibrofatty tissue, a decreased desmosome number and intercellular gap widening.

Desmocollin 2: A few families with autosomal dominant ARVC have been described to date (see Awad et al. 2008). Although inactivation of *Dsc2* in the mouse has not yet been reported, knockdown of this gene in zebrafish results in dose-dependent impaired contractility and dilation of cardiac ventricles and bradycardia, associated with consistent loss of the desmosomal extracellular electron-dense midlines. The cardiac phenotype can be rescued with wild-type human *DSC2* mRNA but not with the mutated *DSC2* mRNA that carries the M211fs221 mutation (a frame shift mutation at amino acid position 211, leading to a premature stop codon at position 221) identified in ARVC patients. The markedly decreased levels of mRNA observed with the mutant *DSC2* allele (M211fs221) led the authors to suggest that nonsense-mediated mRNA decay and consequent haploinsufficiency likely contribute to the disease mechanism (Heuser et al. 2006). Recently, sequence analysis of *DSC2* revealed a homozygous single-base deletion in exon 12 (1841delG) in a family affected by autosomal recessive ARVC associated with palmoplantar keratoderma and woolly hair (Simpson et al. 2008), a phenotype similar to Naxos disease (OMIM 601214).

The Desmosomal Molecular Complex in Skin and Heart

Abnormal skin and heart phenotypes are not restricted to mutations of desmosomal cadherins. Twelve ARVC loci have been identified so far (Awad et al. 2008). In approximately 50% of ARVC patients, a mutation in one of the genes encoding the desmosomal cadherins and four major components of the cardiac desmosome have been identified (see Fig. 2). The four genes encode plakoglobin (skin, hair, heart; OMIM 601214), plakophilin 1 (skin; OMIM 604536), plakophilin 2 (heart; OMIM 609040), and desmoplakin (skin, hair, and heart; OMIM 605656) (see Fig. 2) (McGrath 2005; Syrris et al. 2006; Awad et al. 2008).

Two nonmutually exclusive models have been proposed to explain the phenotypes in skin, hair, and heart diseases. The first one involves a structural defect and suggests that the mechanical disruption of cell–cell adhesion underlies the observed anomalies. The second one involves signaling pathways, likely initiated by the translocation of the armadillo proteins, β-catenin and/or plakoglobin (γ-catenin), to the nucleus (Martin et al. 2009). It has been proposed that the translocation of plakoglobin to the nucleus decreases canonical Wnt signaling through Tcf/Lef1, and decreases the transcription of β-catenin target genes, e.g., c-Myc and cyclin D1 (see Fig. 2) (for details of the Wnt signaling pathway, see Cadigan and Peifer 2009). The resulting suppression of the Wnt pathway ultimately enhances the expression of adipogenic genes, thus promoting the transdifferentiation of cardiac myocytes to adipocytes, as observed in ARVC patients (Garcia-Gras et al. 2006).

The existence of crosstalk between desmosomes, adherens junctions, and gap junctions is well established (Fidler et al. 2008; Li et al. 2008). Consistently, abnormal expression of the gap junction protein connexin 43 has been reported in autosomal dominant forms of ARVC, Naxos, and Carvajal diseases, caused by homozygous mutations of desmosomal proteins (Fidler et al. 2008) (for details of gap junction structure and function, see Goodenough and Paul 2009). Connexin 43-containing gap junctions were significantly reduced in the heart both in number and in size, after N-cadherin depletion, which led the authors to suggest that germline mutations in the human N-cadherin gene may predispose patients to increased risk of cardiac arrhythmias (Li et al. 2008). Additional studies are thus necessary to discern how the interplay between the distinct specialized cell–cell junctions relates to the pathogenesis of the previously described diseases. Disentangling the respective contributions of adhesion and signaling functions of the desmosomal cadherins in tissue development and physiology as well as their respective contribution in the associated human diseases is an important issue regarding the development of appropriate therapeutic approaches.

CADHERINS IN THE CENTRAL NERVOUS SYSTEM

Synapses of the central nervous system are specialized asymmetric cell-adhesion junctions (Yamada and Nelson 2007) (for details of the neuronal synapse, see Giagtzoglou et al. 2009). In vitro experiments and analyses of spontaneous as well as engineered mutant mice have shed light on the roles played by members of the cadherin family in both synapse morphogenesis and plasticity (Morishita and Yagi 2007; Takeichi 2007; Arikkath and Reichardt 2008). Given their wide, differential, and combinatorial expression patterns and their critical roles in dendrite spine morphogenesis, synapse formation and remodeling, cadherins have over the past years been considered as possible targets for cognitive disorders (Bourgeron 2007; Arikkath and Reichardt 2008). Accordingly, causative mutations in several adhesion proteins (e.g., neuroligins and neurexins) have been identified in patients that manifest autistic spectrum disorders (Bourgeron 2007; Sudhof 2008). Nonetheless, it is only recently that causative mutations in some cadherin genes have been reported in cognitive disorders.

M-cadherin Deficiency Associated with Mild to Severe Intellectual Disability

In 2008, a balanced translocation in a female patient with severe intellectual disability (ID) was characterized (Bhalla et al. 2008). Intellectual disability, also referred to as mental retardation, is the most frequently reported developmental disability, affecting cognitive function in about 1%–3% of people worldwide. Whereas the origin of at least half of all ID cases is still unknown, chromosome abnormalities, single-gene mutations, and multifactorial interactions may account for approximately 30% of ID cases. The t(11;16) translocation truncates two genes encoding cell-adhesion molecules, M-cadherin (cadherin 15; OMIM 114019) and KIRREL3, Kin of irregular chiasm-like 3 (member of the immunoglobulin superfamily; OMIM 607761) (Bhalla et al. 2008). Subsequent analysis revealed four nonsynonymous variants in the M-cadherin gene, three of which (R60C, R92W, and A122V) were found to alter M-cadherin-mediated adhesiveness (Bhalla et al. 2008).

How defects in M-cadherin lead to ID in humans is still unknown. Expression studies in mice suggested that M-cadherin might play a critical role in skeletal muscle and cerebellum, where this cadherin is mainly expressed. Nevertheless, M-cadherin–deficient mice do not show defects in skeletal muscle development. Because adherens junctions in the cerebellum were normal in the mutant mice, it has been proposed that other cadherins, in particular N-cadherin, substitute for the mutated M-cadherin (Hollnagel et al. 2002). Whether these mice show impaired learning and memory performances remains to be determined.

The second gene affected by the translocation is *KIRREL3*, a mammalian homolog of the *Drosophila* gene *kirre* (kin of irregular chiasm C-roughest). *KIRREL3* expression was detected exclusively in human fetal and adult brain. Interestingly, KIRREL3 interacts with CASK, calcium/calmodulin-dependent serine protein kinase, which is defective in an X chromosome-linked ID (OMIM 300172). CASK interacts with GluR6, which is defective in an autosomal recessive ID (OMIM 138244), and with δ-catenin, mutations of which are associated with the severe mental retardation in cri-du-chat syndrome (OMIM 116806). Thus, a cadherin-catenin-KIRREL3-CASK molecular network has been proposed to operate in cognitive functions.

Mutations in Protocadherin 19 Associated with Epilepsy and Mental Retardation

Protocadherins are predominantly expressed in the nervous system and constitute the largest of the cadherin subfamilies (see Giagtzoglou et al. 2009). Although most of these genes are found in three clusters (Pcdh-α, Pcdh-β, and Pcdh-γ) located at chromosome 5q31, others are found dispersed in the genome (see Morishita and Yagi 2007; Hulpiau and van Roy 2009). Protocadherin genes are characterized by the presence of a large first exon coding for extracellular ectodomain repeats (typically six or seven), a transmembrane domain, and a part of the cytoplasmic domain (Morishita and Yagi 2007).

Mutations in the protocadherin 19 gene, a member of the δ2 subclass of nonclustered protocadherins, cause an X chromosome-linked epilepsy and mental retardation limited to females (EFMR), which is characterized by seizure onset in infancy or early childhood and cognitive impairment (OMIM 300088). EFMR has a unique inheritance pattern. Indeed, disorders caused by defects in genes located on the X chromosome typically affect males and leave carrier females unaffected. In contrast, EFMR spares male carriers of the mutations and only affects females (Dibbens et al. 2008). To explain the sex-limited mode of inheritance of EFMR, Dibbens et al. proposed that males lacking functional PCDH19 are unaffected because a related but nonparalogous protocadherin gene specific to the Y chromosome, *PCDH11Y*, compensates for the lack of PCDH19. By contrast, females, as a result of the X-inactivation process (Heard et al. 2004), display mosaicism with some neurons expressing the mutated PCDH19 gene and others the wild-type gene. The resulting mosaics would scramble neuronal identity and interfere with cell–cell communication, thus causing EFMR (Dibbens et al. 2008). Interestingly, additional point mutations (missense and truncating mutations, all located in the sequence encoding the extracellular region of the protein) were identified in female patients displaying Dravet syndrome, a genetically determined epileptic encephalopathy that mainly affects females (Depienne et al. 2009). In this study, a hemizygous deletion encompassing the PCDH19 gene was, however, found also in one male patient. Analysis of fibroblasts from the affected male revealed cell mosaicism for the *PCDH19* deletion, which further argues in favor of "cellular interference" being the pathogenic mechanism in *PCDH19*-linked diseases (Depienne et al. 2009).

Genetic Variations in Cadherins and the Susceptibility to Psychiatric Disorders

Besides causative mutations, sequence variants, such as SNPs, in cadherin genes (see the following discussion) have recently been associated with an increased risk of developing psychiatric disorders. Given the roles that the adhesion proteins play in the development and the connectivity between individual neurons, corresponding genes have been considered as good candidates for susceptibility to psychiatric disorders.

Over the years, several laboratories have tested and excluded the involvement of sequence variants in *PCDH8*, *PCDH11Y*, or *NrCAM*, in patients with cognitive disorders including autism, attention deficit hyperactivity disorder, obsessive-compulsive disorder, and schizophrenia. More recently, a preliminary report analyzed 32 polymorphisms in 30 genes, including protocadherin α3, β11, cluster α (PCDHα8 and PCDHα10), and protocadherin 12 genes, in schizophrenic patients with abnormal neurodevelopment. These studies revealed an association between a polymorphism in protocadherin 12 (S640 N) and cortical folding (asymmetry coefficient of gyrification index) (Gregorio et al. 2009). Yet, further studies are needed to confirm this finding. Moreover, a recent two-stage genome-wide association study conducted in large cohorts of patients affected by late-onset Alzheimer's disease (a neurodegenerative disease characterized by senile plaques and neurofibrillary tangles in the brain) and control cases identified a significant association between the disease and SNPs located within a haplotype block that encompasses *PCDH11X* at Xq21.3 (Carrasquillo et al. 2009). One of the SNPs (rs2573905) is more likely to alter PCDH11X function, as it resides in a 100-bp region in intron 2 that is conserved (70% identity) between the human and mouse sequences. To date, no mutation affecting protocadherin 11 encoded by the X chromosome has been identified in the patients. Additional studies are therefore needed to determine whether risk for Alzheimer's disease is mediated by specific *PCDH11X* variants.

CADHERINS IN THE INNER EAR AND RETINA

The cell–cell junctions of the mammalian inner ear that house two sensory organs, the cochlea, the auditory sensory organ, the vestibule, the balance organ (see Fig. 3A), have focused a long-lasting interest as they are continuously submitted to tensions evoked by sound or acceleration stimuli. Surprisingly, the presence

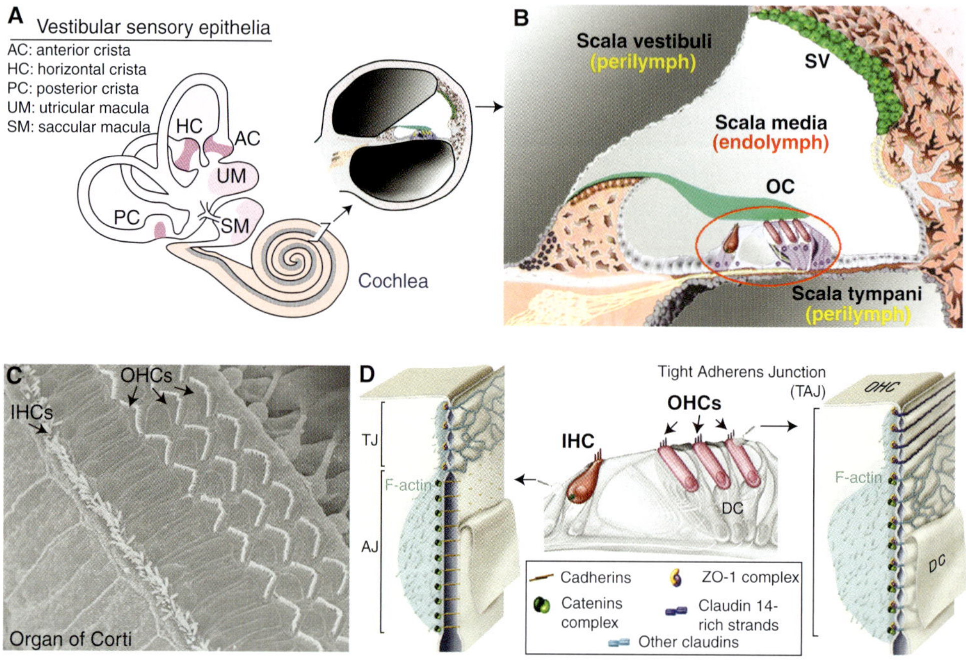

Figure 3. Atypical tight-adherens junctions are found between outer hair cells and their supporting cells in the mammalian auditory organ (cochlea). (*A*) A cross section through the cochlea, which is the auditory organ, shows the three fluid-filled compartments, the scala vestibuli, scala media, and scala tympani. (*B*) Focus on the scala media that houses the auditory sensory epithelium, the organ of Corti (OC), which contains the mechanosensitive inner (IHC) and outer (OHCs) hair cells. sv, stria vascularis. (*C*) Top view of the organ of Corti, showing the highly organized hair bundles of IHCs and OHCs. (*D*) At the TAJ junction between OHCs and Deiters cell (DC), claudins partition into claudin 14- and claudin-9/6 containing subdomains that are distinguishable by their strand morphology and their associated cytoskeletal network. Note the colocalization of catenin complexes with claudin-containing domains (modified, with permission, from Nunes et al. 2006).

of cadherins at the junctions between mature sensory cells and their supporting cells is still under debate (Leonova and Raphael 1997; Nunes et al. 2006). The identification of genes which cause nonsyndromic (isolated) forms of deafness in humans combined with the study of the corresponding mouse mutants (see the following discussion) has highlighted the roles of the auditory neuroepithelium cell–cell junctions in cochlear physiology and the underlying molecular mechanisms. In addition, the study of a human syndromic form of deafness, Usher syndrome type 1 (USH1), has uncovered the key role of two members of the cadherin superfamily, cadherin 23 and protocadherin 15, in the development and function of the hair bundle, the mechanosensitive structure operating the conversion of acoustic pressure signals into electrical signals by the auditory sensory hair cells (Petit 2001; Petit and Richardson 2009).

Typical and Atypical Cell–Cell Junctions in the Cochlea

The cochlea is made up of three fluid-filled compartments of different ionic compositions—the scala vestibuli (perilymph), the scala media (endolymph), and the scala tympani (perilymph) (Fig. 3A,B). The perilymph has an ion composition typical of extracellular fluids (high Na^+ and low K^+ concentrations), whereas the endolymph has high K^+ and low Na^+ concentrations, and thus displays characteristics of an intracellular fluid (see Wangemann 2006). In addition, a transepithelial electric potential difference of ~100 mV exists between the endolymphatic and perilymphatic compartments of the cochlea. This so-called endocochlear potential drives the cationic transduction current through hair cells in response to sound stimulation.

The auditory sensory epithelium, the organ of Corti, which keeps apart the endolymphatic and the perilymphatic compartments, contains sensory hair cells of two types, the inner (IHC) and the outer (OHC) hair cells (Fig. 3C,D). Various typical and atypical junctions ensure both the ion barrier between the endolymph and perilymph, and the integrity of the sensory epithelium despite the continuous mechanical stress the organ of Corti experiences. The hair bundle, at the apical region of hair cells, is exposed to endolymph, whereas hair cell basolateral regions are bathed in perilymph. The junctions between outer hair cells and their supporting cells, the Deiters cells, have a unique organization that does not conform to canonical TJs and AJs (Fig. 3D). It has been first proposed that mosaics of several alternating TJs and AJs make up these junctions (Raphael and Altschuler 2003). Later on, freeze-fracture electron microscopy and immunolabeling studies showed that these junctions are instead made up of a single large (3–5 μm) occluding junction with AJ features (Nunes et al. 2006). Strikingly, at least three canonical AJ proteins, α-, β-, and p120-catenins, were present between OHCs and Deiters cells, and colocalizing with TJ proteins (ZO1, claudin 14, and claudins 6/9) (Fig. 3D). This hybrid junction is now referred to as a tight adherens junction, TAJ (Nunes et al. 2006). Defects in two TJ proteins, tricellulin and claudin 14, cause deafness in humans (Table 1). So far, there is no mutant animal model for the tricellulin defect. In the claudin 14-deficient mice, a disruption of the junctional barrier resulting in an elevated K^+ concentration around the hair cell bodies, and thereby chronic cell depolarization has been proposed to lead to hair cell death (Ben-Yosef et al. 2003).

The sound-induced displacement amplitude of the organ of Corti measured in the apical region of the guinea pig cochlea is roughly 10 nm for a sound intensity of 80 dB SPL. This displacement is estimated to generate a force per unit surface of 30–90 Pascals in the radial direction of the neuroepithelium, with similar forces in the transversal and longitudinal directions (Tomo et al. 2007). How cell–cell junctions of the organ of Corti cope with this mechanical stress is poorly understood. However, the critical role that the junctions between hair cells and their supporting cells play in the resilience of the auditory epithelium to sound trauma has recently been pinpointed. Hair cell conditional inactivation of the gene encoding vezatin, a transmembrane ubiquitous AJ protein previously shown to associate with the E-cadherin-catenin complex (Kussel-Andermann et al. 2000), resulted in a delayed spontaneous hearing loss in the mutant mice (Bahloul et al. 2009). Moreover, a single short exposure (1 minute) of these mice to a loud broadband sound (106 dB SPL) was sufficient to cause irreversible hearing impairment, with TAJ disorganization and hair cell loss (Bahloul et al. 2009).

Dual Function of Cadherin 23 and Protocadherin 15 in Auditory Hair Cell Development and Mechanoelectrical Transduction

Our ability to perceive sound and maintain balance depends on mechanoelectrical transduction (MET), a process achieved in the hair bundle. The hair bundle is made up of 20–300 actin-filled, stiff microvilli—the stereocilia—and a single genuine cilium, the kinocilium (only transient in the cochlea) (Fig. 4A). The stereocilia are arranged in three to four rows of increasing heights toward the kinocilium. Altogether they form a "V"-shaped staircase, with the kinocilium located at the vertex of the V. The vertices of the hair bundles of all OHCs are aligned uniformly along the mediolateral axis of the cochlea, thereby defining the planar cell polarity axis of the organ of Corti (Fig. 4A,B). Two types of links interconnect adjacent stereocilia in the mature hair bundle, namely the tip link and the horizontal top connectors (Fig. 4A). The tip link is a single

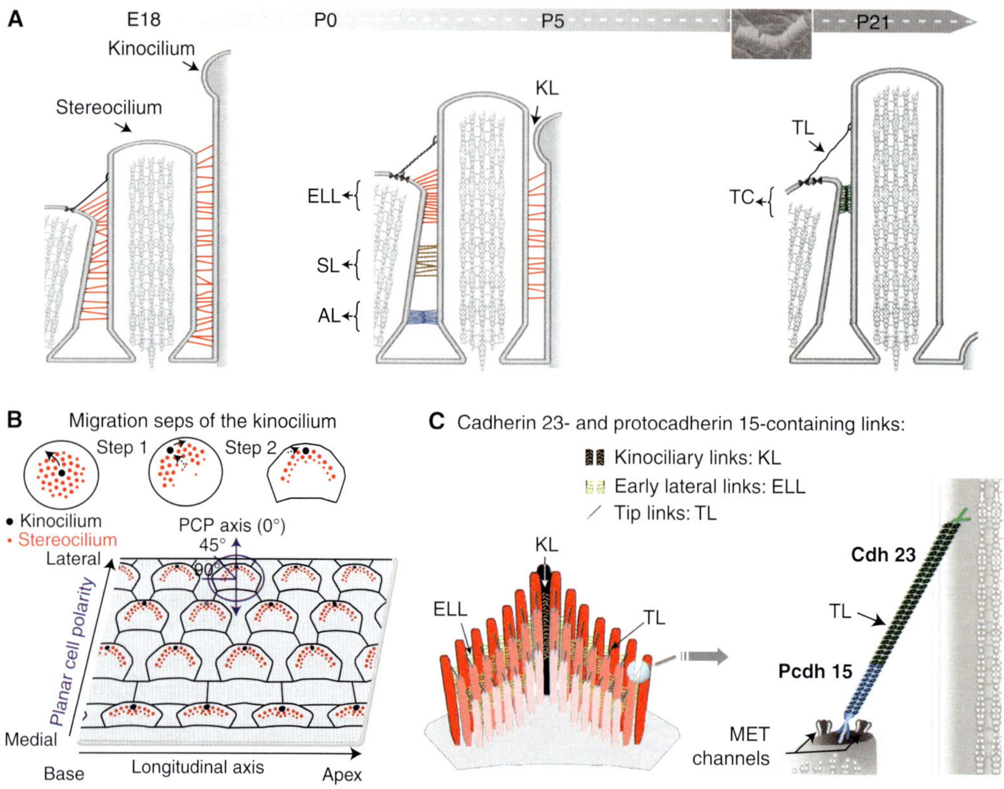

Figure 4. Various links are found between adjacent stereocilia and between stereocilia and the kinocilium in the growing and mature hair bundle of inner ear sensory cells. (*A*) Throughout development and in mature hair cells, the stereocilia are connected by different types of transient and persistent links, respectively. The early transient lateral links (ELL), shaft links (SL), ankle links (AL), and kinociliary links (KL) are progressively replaced by the apical top connectors (TC) found in mature hair cells. The tip links (TL) are already present in the growing hair bundles, and they persist in mature hair cells. The inset shows a scanning electron microscopy image of an OHC hair bundle. (*B*) Kinocilium migration steps and planar cell polarity (PCP) in the organ of Corti. The kinocilium directs the staircase pattern of the hair bundle. (*C*) Cadherin 23 and protocadherin 15 isoforms can form the early transient lateral links, kinociliary links, and the tip link. Note that only protocadherin 15, which forms the lower part of the tip link, may be tightly associated to the MET channels.

oblique apical link that connects the tip of each stereocilium to the side of the nearest taller stereocilium, whereas the top connectors are lateral links that couple adjacent stereocilia both within and between rows. During development, different types of transient lateral links connect the growing stereocilia to each other and the kinocilium to the adjacent stereocilia of the tallest row. These links are the early transient lateral links, shaft links, ankle links, and kinociliary links (Fig. 4A) (see Petit and Richardson 2009). The function of these links was unknown until several deafness genes, including those responsible for the USH1 syndrome, were found to encode components of the different subsets of stereocilia links.

Usher syndrome is the most frequent cause of hereditary deaf-blindness in humans. It is a monogenic disorder, transmitted on an autosomal recessive mode. Of the three clinical forms (USH1–3), USH1 is the most severe. It is characterized by severe to profound congenital hearing impairment, balance deficiency, and prepubertal-onset retinitis pigmentosa leading to blindness. Five USH1 genes have been identified (El-Amraoui and Petit 2005; Leibovici et al.

2008). Two cadherins, cadherin 23 and protocadherin 15, underlie the USH1D (Bolz et al. 2001; Bork et al. 2001) and USH1F (Ahmed et al. 2001; Alagramam et al. 2001b) forms, respectively. The three other USH1 genes encode an actin-based motor, myosin VIIa (USH1B) (Weil et al. 1995), a PDZ-domain-containing protein, harmonin (USH1C) (Bitner-Glindzicz et al. 2000; Verpy et al. 2000), and a putative scaffolding protein with ankyrin repeats, sans (USH1G) (Weil et al. 2003).

Cadherin 23 and Protocadherin 15 Form Transient Links that are Essential for the Cohesion and Proper Orientation of the Growing Hair Bundle

Cadherin 23 and protocadherin 15 consist of several isoforms with distinct extracellular regions and cytoplasmic domains, which result from alternatively spliced *CDH23* and *PCDH15* transcripts. Three classes of cadherin 23 isoforms (a to c) and four classes of protocadherin 15 isoforms (CD1 to CD3, and SI) have been predicted (Lagziel et al. 2005; Michel et al. 2005; Ahmed et al. 2006). The longest cadherin 23 and protocadherin 15 isoforms contain 27 and 11 EC domains, respectively (Fig. 5A,B). Notably, the cytoplasmic domains of both cadherin 23 and protocadherin 15 isoforms show no sequence similarity between each other or with classical cadherins. In particular, they all lack the consensus motif for binding to β-catenin (Cavey et al. 2008; Nelson 2008), raising the question of the existence of a dynamic link between these two cadherins and actin filaments.

Animal models defective for these proteins (Leibovici et al. 2008) have revealed the roles of these two cadherins, i.e., the deaf waltzer mouse (Di Palma et al. 2001) and sputnik zebrafish (Sollner et al. 2004) mutants for cadherin23, and the Ames waltzer mouse (Alagramam et al. 2001a) and orbiter zebrafish (Seiler et al. 2005) mutants for protocadherin 15. Mutations in either of these genes, or in the genes encoding myosin VIIa, harmonin, or sans all lead to similar morphological anomalies in the mouse cochlea, which consist of fragmented and misoriented hair bundles (see Fig. 5C, Lefevre et al. 2008; Petit and Richardson 2009).

When stereocilia emerge at the apical surface of auditory hair cells at E15.5 in the mouse, the kinocilium is located at the center. It then migrates to the cell periphery, thereby initiating the establishment of hair bundle polarity. Then, the kinocilium undergoes a second migratory step that results in a reorientation of the hair bundle, which reaches its final location on the apical cell surface, pointing toward the cochlear lateral wall (Fig. 4B) (Kelly and Chen 2007; Jones and Chen 2008; Petit and Richardson 2009). Analysis of the various USH1 mouse models revealed that although the kinocilium migrated from its central position to the cell periphery, it failed to reach its final correct position and was frequently disconnected from the neighboring stereocilia (Lefevre et al. 2008). In addition, in all Ush1 mutants, the hair cells had undergone early fragmentation of their hair bundle, which was split into a few clumps of stereocilia (Fig. 5C).

How does the absence of any of the USH1 proteins lead to this common hair bundle phenotype? Over the past 10 years, evidence for colocalization in the hair bundle and direct in vitro interactions of these proteins (Fig. 5D) has led to the conclusion that a similar pathogenic mechanism causes deafness in USH1 patients (see Leibovici et al. 2008). This mechanism involves defective early transient stereocilia lateral links and kinociliary links in the growing hair bundles (see Fig. 5C). Myosin VIIa and sans are required for the targeting of harmonin-b into the stereocilia, where it binds to F-actin and thus most likely anchors the cadherin 23- and protocadherin 15-containing links to the actin core of the stereocilium (Boëda et al. 2002; Lefevre et al. 2008). However, because the phenotype of the null harmonin b mutant is less severe than that of the mutant lacking all three harmonin isoforms, additional roles can be predicted for myosin VIIa and sans in the establishment of a cohesive hair bundle (I. Foucher, A. Bahloul, and E. Caberlotto, unpubl.). Cohesive forces applied to the early transient interstereociliary

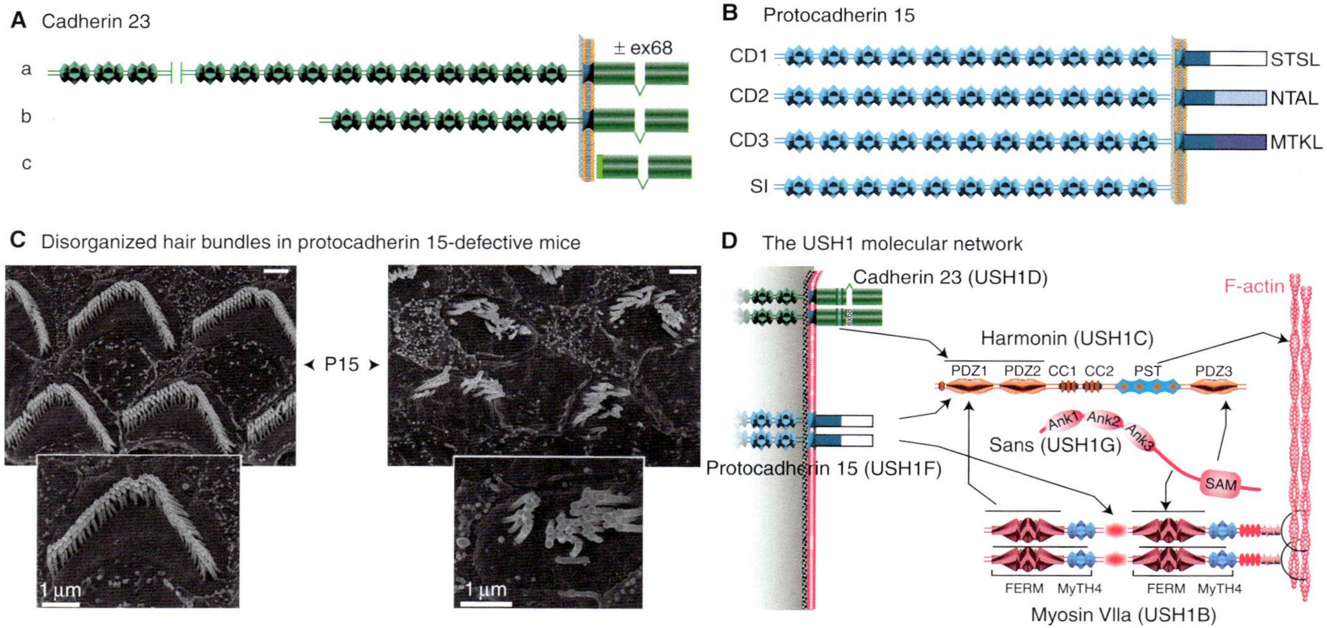

Figure 5. Cadherin 23 and protocadherin 15 defects underlie two genetic forms of Usher type 1 (USH1) syndrome. (*A*, *B*) Three classes of cadherin 23 isoforms (a, b, and c in *A*), and four classes of protocadherin 15 isoforms (CD1, CD2, CD3, and SI in *B*) have been reported. The distinct PDZ domain-binding motifs in the different protocadherin 15 cytodomains (CD) are indicated. SI, secreted isoform (see Ahmed et al. 2008b). (*C*) Scanning electron micrographs of OHC hair bundles in wild-type and protocadherin 15-deficient mice at postnatal day 15 (P15). Note the fragmentation of the hair bundles in the mutant (Source: M. Leibovici, Institut Pasteur, France). (*D*) Summary of the interactions between USH1 proteins. Bar = 1 µm.

and kinociliary links in the forming hair bundle are thus necessary for the formation of a unitary and properly oriented "V"-shaped hair bundles. Conversely, the hair bundle misorientation suggests that the kinocilium final position is controlled by the sum of the forces applied to the entire hair bundle. Whether these forces originate outside the hair cell, for instance from the concomitant convergent-extension process driving cell intercalation and elongation of the cochlear neuroepithelium, or within the hair cell, is still unknown. This calls for both qualitative and quantitative information regarding the movements of the cells and of the kinocilia in the developing auditory neuroepithelium, as well as of actin filaments and microtubules organization over that time period.

Cadherin 23 and Protocadherin 15 Form the Tip Link in Mature Hair Cells

A late onset form of deafness caused by *CDH23* mutations provided the first indication for its role beyond early hair bundle development and possibly also in MET (Noben-Trauth et al. 2003). The prevailing theory for MET states that tension in "gating springs" controls the open probability of MET channels (Corey and Hudspeth 1983). The tip link (Pickles et al. 1984), or a more compliant link in series with the tip link, comprises the gating spring. Sound displacement of the hair bundle toward the tallest stereocilia promotes MET channel opening, leading to an influx of endolymphatic cations. The resulting depolarizing receptor potential ultimately leads to neurotransmitter release by IHCs, and thereby the transfer of the acoustic information to the auditory neurons. Under a 90 dB sound stimulation, a 100 nm displacement of the hair bundle tip occurs in the guinea pig, generating a tension of 3 to 6 pN on the tip link and the molecular complex that bridges it to the stereocilia actin-filaments (Fridberger et al. 2006). This corresponds to a roughly 10% modulation of its resting tension evaluated in the range of 40 pN (Jaramillo and Hudspeth 1993; Prost et al. 2007).

High-resolution ultrastructural studies have shown that tip links consist of two to three helical filaments, approximately 150–200 nm in length (Kachar et al. 2000). Antibodies to specific epitopes of cadherin 23 or protocadherin 15 labeled the tips of stereocilia at the tip link emplacement (Ahmed et al. 2006; Kazmierczak et al. 2007). A detailed subcellular analysis showed that cadherin 23 and protocadherin 15 localize to the upper and lower part of the tip link, respectively (Fig. 4C). Cadherin 23 homodimers have been shown to interact in *trans* in a Ca^{2+}-dependent manner with protocadherin 15 homodimers, forming a filament with structural similarity to the tip link (Kazmierczak et al. 2007; Muller 2008). To date, functional evidence establishing that cadherin 23 and protocadherin 15 form the tip link is a major issue, especially because careful analysis of the hair bundles in mouse mutants with predicted null alleles of cadherin 23 has shown the presence of tip link-like filaments connecting adjacent stereocilia (Rzadzinska and Steel 2009). Yet, a cadherin 23 mouse mutant, salsa, with an E737V substitution within the Ca^{2+} binding motif (LDRE) in the EC7 domain of the a-isoform, the longest one, further supports the involvement of the cadherins in the formation of the tip link because only rare disorganized hair bundles were observed, whereas absence of tension on the stereocilia tips argues for a loss of the tip links (Schwander et al. 2009). Protocadherin 15 isoforms from the CD2 subgroup have been proposed to be components of the early transient lateral links, and isoforms from the CD3 subgroup to form the tip link (Ahmed et al. 2006).

Cadherin 23 and Protocadherin 15 in the Retina

So far, progress in understanding the critical roles of cadherin 23 and protocadherin 15 in the retina has been hampered by the absence of retinitis pigmentosa in Ush1 mutant mice, including cadherin 23 and protocadherin 15 double knockout mutants (Ahmed et al. 2008a, see Leibovici et al. 2008 for a review).

The shorter lifespan of the mouse, interspecies difference in light exposure, existence of a functional redundancy for USH1 proteins in murine but not human retina, or an intrinsic difference between murine and human photoreceptor cell's physiology may account for this phenotypic discrepancy. In regards to the last possibility, the underdevelopment of photoreceptor cells' calycal processes in mouse compared to human is an attractive hypothesis (Petit 2001). Indeed, these microvilli-like extensions of the photoreceptor inner segment of yet unknown functions are bridged to the connecting cilium (Cohen 1963) by fibrous links and some of these links have been shown to share molecular components with a subset of hair bundle transient lateral links, the ankle links (Kremer et al. 2006; Leibovici et al. 2008). To clarify this point, comparative studies determining the subcellular localization of cadherin 23 and protocadherin 15 isoforms in mouse and in human retinas are required.

According to the mutations detected in USH1 patients, the protocadherin 15-CD1 subclass may be the only subclass critical for retinal functions (Ahmed et al. 2008b). Whereas the USH1D mutations that have been identified do affect the a-isoforms of cadherin 23 (see http://webh01.ua.ac.be/hhh/), whether the b- and c-isoforms also play an important role in the retina is still unknown. Results of the comparative analysis will be essential to determine the appropriate animal model to decipher the pathogenesis of retinitis pigmentosa as well as to test future therapeutic approaches. Of note in zebrafish, two orthologues of the mammalian protocadherin 15 gene, *pcdh15a* and *pcdh15b*, encode proteins with highly dissimilar intracellular domains. *Pcdh15a* defects result in the hair bundle disorganization without visual abnormalities, whereas *pcdh15b* defects only cause retinal anomalies, i.e., abnormal alignment of photoreceptor outer disks (Seiler et al. 2005).

CONCLUDING REMARKS

The human genome-wide association studies are likely to extend the list of human disorders involving cadherins in a near future, especially regarding neuropsychiatric disorders. Epigenetic modifications of these genes are likely to broaden their implication in these diseases. Mutation screening will not only allow further deciphering of the cadherin isoforms critically underlying phenotypic features, but should also help identify the role of some of their interacting partners, such as cytoskeletal regulators, protein kinases and phosphatases, and transcription cofactors. The putative correlation between phenotypic features and copy number variants (CNV) of the cadherin genes would be of particular interest regarding the role of adhesion strength. Furthermore, the finding of modifier genes accounting for the failure of phenotype-genotype correlation in cadherin genes should add to our present knowledge of the roles played by the cadherins in various signaling pathways. Generating appropriate animal models remains, however, essential to achieve an in-depth understanding of cadherin functions/dysfunctions and as regards neuropsychiatric disorders, genetically modified animals with more elaborated cognitive functions than those of the mouse are needed.

ACKNOWLEDGMENTS

We are grateful to Jean-Pierre Hardelin for his comments and suggestions, to Michel Leibovici for SEM images. We apologize for not citing all the relevant references because of space limitation. The work in the authors' laboratory is supported by ANR-07-MRARE-009-01, and Fondation pour la Recherche Médicale (FRM) to AE, the R. and G. Strittmatter Foundation, FAUN-Stiftung (Suchert Foundation), Fondation Orange, and the European Commission FP6 Integrated Project EUROHEAR, LSHG-CT-2004-512063 to CP.

REFERENCES

Ahmed ZM, Riazuddin S, Bernstein SL, Ahmed Z, Khan S, Griffith AJ, Morell RJ, Friedman TB, Riazuddin S, Wilcox ER. 2001. Mutations of the protocadherin gene *PCDH15* cause Usher syndrome type 1F. *Am J Hum Genet* **69:** 25–34.

Ahmed ZM, Goodyear R, Riazuddin S, Lagziel A, Legan PK, Behra M, Burgess SM, Lilley KS, Wilcox ER, Griffith AJ, et al. 2006. The tip-link antigen, a protein associated with the transduction complex of sensory hair cells, is protocadherin-15. *J Neurosci* **26:** 7022–7034.

Ahmed ZM, Kjellstrom S, Haywood-Watson RJ, Bush RA, Hampton LL, Battey JF, Riazuddin S, Frolenkov G, Sieving PA, Friedman TB. 2008a. Double homozygous waltzer and Ames waltzer mice provide no evidence of retinal degeneration. *Mol Vis* **14:** 2227–2236.

Ahmed ZM, Riazuddin S, Aye S, Ali RA, Venselaar H, Anwar S, Belyantseva PP, Qasim M, Friedman TB. 2008b. Gene structure and mutant alleles of PCDH15: Nonsyndromic deafness DFNB23 and type 1 Usher syndrome. *Hum Genet* **124:** 215–223.

Alagramam KN, Murcia CL, Kwon HY, Pawlowski KS, Wright CG, Woychik RP. 2001a. The mouse Ames waltzer hearing-loss mutant is caused by mutation of *Pcdh15*, a novel protocadherin gene. *Nature Genet* **27:** 99–102.

Alagramam KN, Yuan H, Kuehn MH, Murcia CL, Wayne S, Srisailpathy CRS, Lowry RB, Knaus R, Van Laer L, Bernier FP, et al. 2001b. Mutations in the novel protocadherin *PCDH15* cause Usher syndrome type 1F. *Hum Mol Genet* **10:** 1709–1718.

Allen E, Yu QC, Fuchs E. 1996. Mice expressing a mutant desmosomal cadherin exhibit abnormalities in desmosomes, proliferation, and epidermal differentiation. *J Cell Biol* **133:** 1367–1382.

Arikkath J, Reichardt LF. 2008. Cadherins and catenins at synapses: Roles in synaptogenesis and synaptic plasticity. *Trends Neurosci* **31:** 487–494.

Awad MM, Dalal D, Cho E, Amat-Alarcon N, James C, Tichnell C, Tucker A, Russell SD, Bluemke DA, Dietz HC, et al. 2006. DSG2 mutations contribute to arrhythmogenic right ventricular dysplasia/cardiomyopathy. *Am J Hum Genet* **79:** 136–142.

Awad MM, Calkins H, Judge DP. 2008. Mechanisms of disease: Molecular genetics of arrhythmogenic right ventricular dysplasia/cardiomyopathy. *Nat Clin Pract Cardiovasc Med* **5:** 258–267.

Bahloul A, Simmler MC, Michel V, Leibovici M, Perfettini I, Roux I, Weil D, Nouaille S, Zuo J, Zadro C, et al. 2009. Vezatin, an integral membrane protein of adherens junctions, is required for the sound-resilience of cochlear hair cells. *EMBO Mol Medicine* **1:** 125–138.

Bazzi H, Getz A, Mahoney MG, Ishida-Yamamoto A, Langbein L, Wahl JK 3rd, Christiano AM. 2006. Desmoglein 4 is expressed in highly differentiated keratinocytes and trichocytes in human epidermis and hair follicle. *Differentiation* **7:** 129–140.

Ben-Yosef T, Belyantseva IA, Saunders TL, Hughes ED, Kawamoto K, Van Itallie CM, Beyer LA, Halsey K, Gardner DJ, Wilcox ER, et al. 2003. Claudin 14 knockout mice, a model for autosomal recessive deafness DFNB29, are deaf due to cochlear hair cell degeneration. *Hum Mol Genet* **12:** 2049–2061.

Berx G, van Roy F. 2009. Involvement of members of the cadherin superfamily in cancer. *Cold Spring Harb Perspect Biol* **1:** a003129.

Bhalla K, Luo Y, Buchan T, Beachem MA, Guzauskas GF, Ladd S, Bratcher SJ, Schroer RJ, Balsamo J, DuPont BR, et al. 2008. Alterations in CDH15 and KIRREL3 in patients with mild to severe intellectual disability. *Am J Hum Genet* **83:** 703–713.

Bitner-Glindzicz M, Lindley KJ, Rutland P, Blaydon D, Smith VV, Milla PJ, Hussain K, Furth-Lavi J, Cosgrove KE, Shepherd RM, et al. 2000. A recessive contiguous gene deletion causing infantile hyperinsulinism, enteropathy and deafness identifies the Usher type 1C gene. *Nature Genet* **26:** 56–60.

Boëda B, El-Amraoui A, Bahloul A, Goodyear R, Daviet L, Blanchard S, Perfettini I, Fath KR, Shorte S, Reiners J, et al. 2002. Myosin VIIa, harmonin and cadherin 23, three Usher I gene products that cooperate to shape the sensory hair cell bundle. *EMBO J* **21:** 6689–6699.

Bolz H, von Brederlow B, Ramirez A, Bryda EC, Kutsche K, Nothwang HG, Seeliger M, del CSCM, Vila MC, Molina OP, et al. 2001. Mutation of CDH23, encoding a new member of the cadherin gene family, causes Usher syndrome type 1D. *Nature Genet* **27:** 108–112.

Bork JM, Peters LM, Riazuddin S, Bernstein SL, Ahmed ZM, Ness SL, Polomeno R, Ramesh A, Schloss M, Srisailpathy CRS, et al. 2001. Usher syndrome 1D and nonsyndromic autosomal recessive deafness DFNB12 are caused by allelic mutations of the novel cadherin-like gene *CDH23*. *Am J Hum Genet* **68:** 26–37.

Bourgeron T. 2007. The possible interplay of synaptic and clock genes in autism spectrum disorders. *Cold Spring Harb Symp Quant Biol* **72:** 645–654.

Cadigan KM, Peifer M. 2009. Wnt signaling from development to disease: Insights from model systems. *Cold Spring Harb Perspect Biol* **1:** a002881.

Carrasquillo MM, Zou F, Pankratz VS, Wilcox SL, Ma L, Walker LP, Younkin SG, Younkin CS, Younkin LH, Bisceglio GD, et al. 2009. Genetic variation in PCDH11X is associated with susceptibility to late-onset Alzheimer's disease. *Nature Genet* **41:** 192–198.

Cavey M, Rauzi M, Lenne PF, Lecuit T. 2008. A two-tiered mechanism for stabilization and immobilization of E-cadherin. *Nature* **453:** 751–756.

Cohen AI. 1963. Vertebrate retinal cells and their organization. *Biol rev Cambridge Philos Soc* **38:** 427–459.

Corey DP, Hudspeth AJ. 1983. Kinetics of the receptor current in bullfrog saccular hair cells. *J Neurosci* **3:** 962–976.

Delva E, Tucker DK, Kowalczyk AP. 2009. The desmosome. *Cold Spring Harb Perspect Biol* **1:** a002543.

Depienne C, Bouteiller D, Keren B, Cheuret E, Poirier K, Trouillard O, Benyahia B, Quelin C, Carpentier W, Julia S, et al. 2009. Sporadic infantile epileptic encephalopathy caused by mutations in PCDH19 resembles Dravet syndrome but mainly affects females. *PLoS Genet* **5:** pe1000381.

Di Palma F, Holme RH, Bryda EC, Belyantseva IA, Pellegrino R, Kachar B, Steel KP, Noben-Trauth K. 2001. Mutations in *Cdh23*, encoding a new type of cadherin, cause stereocilia disorganization in waltzer, the mouse model for Usher syndrome type 1D. *Nature Genet* **27:** 103–107.

Dibbens LM, Tarpey PS, Hynes K, Bayly MA, Scheffer IE, Smith R, Bomar J, Sutton E, Vandeleur L, Shoubridge C, et al. 2008. X-linked protocadherin 19 mutations cause female-limited epilepsy and cognitive impairment. *Nature Genet* **40:** 776–781.

Dua-Awereh MB, Shimomura Y, Kraemer L, Wajid M, Christiano AM. 2009. Mutations in the desmoglein 1 gene in five Pakistani families with striate palmoplantar keratoderma. *J Dermatol Sci* **53:** 192–197.

El-Amraoui A, Petit C. 2005. Usher I syndrome: Unravelling the mechanisms that underlie the cohesion of the growing hair bundle in inner ear sensory cells. *J Cell Sci* **118:** 4593–4603.

Eshkind L, Tian Q, Schmidt A, Franke WW, Windoffer R, Leube RE. 2002. Loss of desmoglein 2 suggests essential functions for early embryonic development and proliferation of embryonal stem cells. *Eur J Cell Biol* **81:** 592–598.

Fidler LM, Wilson GJ, Liu F, Cui X, Scherer SW, Taylor GP, Hamilton RM. 2008. Abnormal connexin43 in arrhythmogenic right ventricular cardiomyopathy caused by plakophilin-2 mutations. *J Cell Mol Med* PMID: 18662195.

Fridberger A, Tomo I, Ulfendahl M, Boutet de Monvel J. 2006. Imaging hair cell transduction at the speed of sound: Dynamic behavior of mammalian stereocilia. *Proc Natl Acad Sci USA* **103:** 1918–1923.

Garcia-Gras E, Lombardi R, Giocondo MJ, Willerson JT, Schneider MD, Khoury DS, Marian AJ. 2006. Suppression of canonical Wnt/beta-catenin signaling by nuclear plakoglobin recapitulates phenotype of arrhythmogenic right ventricular cardiomyopathy. *J Clin Invest* **116:** 2012–2021.

Giagtzoglou N, Ly CV, Bellen HJ. 2009. Cell adhesion, the backbone of the synapse: "vertebrate" and "invertebrate" perspectives. *Cold Spring Harb Perspect Biol* **1:** a003079.

Goodenough DA, Paul DL. 2009. Gap junctions. *Cold Spring Harb Perspect Biol* **1:** a002576.

Gregorio SP, Sallet PC, Do KA, Lin E, Gattaz WF, Dias-Neto E. 2009. Polymorphisms in genes involved in neurodevelopment may be associated with altered brain morphology in schizophrenia: Preliminary evidence. *Psychiatry Res* **165:** 1–9.

Heard E, Chaumeil J, Masui O, Okamoto I. 2004. Mammalian X-chromosome inactivation: An epigenetics paradigm. *Cold Spring Harb Symp Quant Biol* **69:** 89–102.

Hennies HC, Kuster W, Mischke D, Reis A. 1995. Localization of a locus for the striated form of palmoplantar keratoderma to chromosome 18q near the desmosomal cadherin gene cluster. *Hum Mol Genet* **4:** 1015–1020.

Hershkovitz D, Lugassy J, Indelman M, Bergman R, Sprecher E. 2009. Novel mutations in DSG1 causing striate palmoplantar keratoderma. *Clin Exp Dermatol* **34:** 224–228.

Heuser A, Plovie ER, Ellinor PT, Grossmann KS, Shin JT, Wichter T, Basson CT, Lerman BB, Sasse-Klaassen S, Thierfelder L, et al. 2006. Mutant desmocollin-2 causes arrhythmogenic right ventricular cardiomyopathy. *Am J Hum Genet* **79:** 1081–1088.

Hollnagel A, Grund C, Franke WW, Arnold HH. 2002. The cell adhesion molecule M-cadherin is not essential for muscle development and regeneration. *Mol Cell Biol* **22:** 4760–4770.

Hulpiau P, van Roy F. 2009. Molecular evolution of the cadherin superfamily. *Int J Biochem Cell Biol* **41:** 349–369.

Indelman M, Eason J, Hummel M, Loza O, Suri M, Leys MJ, Bayne M, Schwartz FL, Sprecher E. 2007. Novel CDH3 mutations in hypotrichosis with juvenile macular dystrophy. *Clin Exp Dermatol* **32:** 191–196.

Jamora C, DasGupta R, Kocieniewski P, Fuchs E. 2003. Links between signal transduction, transcription and adhesion in epithelial bud development. *Nature* **422:** 317–322.

Jaramillo F, Hudspeth AJ. 1993. Displacement-clamp measurement of the forces exerted by gating springs in the hair bundle. *Proc Natl Acad Sci USA* **90:** 1330–1334.

Jones C, Chen P. 2008. Primary cilia in planar cell polarity regulation of the inner ear. *Curr Top Dev Biol* **85:** 197–224.

Kachar B, Parakkal M, Kurc M, Zhao Y, Gillespie PG. 2000. High-resolution structure of hair-cell tip links. *Proc Natl Acad Sci USA* **97:** 13336–13341.

Kazmierczak P, Sakaguchi H, Tokita J, Wilson-Kubalek EM, Milligan RA, Muller U, Kachar B. 2007. Cadherin 23 and protocadherin 15 interact to form tip-link filaments in sensory hair cells. *Nature* **449:** 87–91.

Kelly M, Chen P. 2007. Shaping the mammalian auditory sensory organ by the planar cell polarity pathway. *Int J Dev Biol* **51:** 535–547.

Kjaer KW, Hansen L, Schwabe GC, Marques-de-Faria AP, Eiberg H, Mundlos S, Tommerup N, Rosenberg T. 2005. Distinct CDH3 mutations cause ectodermal dysplasia, ectrodactyly, macular dystrophy (EEM syndrome). *J Med Genet* **42:** 292–298.

Kljuic A, Bazzi H, Sundberg JP, Martinez-Mir A, O'Shaughnessy R, Mahoney MG, Levy M, Montagutelli X, Ahmad W, Aita VM, et al. 2003. Desmoglein 4 in hair follicle differentiation and epidermal adhesion: Evidence from inherited hypotrichosis and acquired pemphigus vulgaris. *Cell* **113:** 249–260.

Kremer H, van Wijk E, Marker T, Wolfrum U, Roepman R. 2006. Usher syndrome: Molecular links of pathogenesis, proteins and pathways. *Hum Mol Genet* **15:** R262–270.

Kussel-Andermann P, El-Amraoui A, Safieddine S, Nouaille S, Perfettini I, Lecuit M, Cossart P, Wolfrum U, Petit C. 2000. Vezatin, a novel transmembrane protein, bridges myosin VIIA to the cadherin-catenins complex. *EMBO J* **19:** 6020–6029.

Lagziel A, Ahmed ZM, Schultz JM, Morell RJ, Belyantseva IA, Friedman TB. 2005. Spatiotemporal pattern and isoforms of cadherin 23 in wild type and waltzer mice during inner ear hair cell development. *Dev Dyn* **280:** 295–306.

Lai-Cheong JE, Arita K, McGrath JA. 2007. Genetic diseases of junctions. *J Invest Dermatol* **127:** 2713–2725.

Lefevre G, Michel V, Weil D, Lepelletier L, Bizard E, Wolfrum U, Hardelin JP, Petit C. 2008. A core cochlear phenotype in USH1 mouse mutants implicates fibrous links of the hair bundle in its cohesion, orientation and differential growth. *Development* **135:** 1427–1437.

Leibovici M, Safieddine S, Petit C. 2008. Mouse models for human hereditary deafness. *Curr Top Dev Biol* **84:** 385–429.

Leonova EV, Raphael Y. 1997. Organization of cell junctions and cytoskeleton in the reticular lamina in normal and ototoxically damaged organ of Corti. *Hear Res* **113:** 14–28.

Li J, Levin MD, Xiong Y, Petrenko N, Patel VV, Radice GL. 2008. N-cadherin haploinsufficiency affects cardiac gap junctions and arrhythmic susceptibility. *J Mol Cell Cardiol* **44:** 597–606.

Martin ED, Moriarty MA, Byrnes L, Grealy M. 2009. Plakoglobin has both structural and signalling roles in zebrafish development. *Dev Biol* **327:** 83–96.

McGrath JA. 2005. Inherited disorders of desmosomes. *Australas J Dermatol* **46:** 221–229.

Michel V, Goodyear RJ, Weil D, Marcotti W, Perfettini I, Wolfrum U, Kros CJ, Richardson GP, Petit C. 2005. Cadherin 23 is a component of the transient lateral links in the developing hair bundles of cochlear sensory cells. *Dev Dyn* **280:** 281–294.

Morishita H, Yagi T. 2007. Protocadherin family: Diversity, structure, and function. *Curr Opin Cell Biol* **19:** 584–592.

Muller U. 2008. Cadherins and mechanotransduction by hair cells. *Curr Opin Cell Biol* **20:** 557–566.

Nelson WJ. 2008. Regulation of cell-cell adhesion by the cadherin-catenin complex. *Biochem Soc Trans* **36:** 149–155.

Noben-Trauth K, Zheng QY, Johnson KR. 2003. Association of cadherin 23 with polygenic inheritance and genetic modification of sensorineural hearing loss. *Nature Genet* **35:** 21–23.

Nunes FD, Lopez LN, Lin HW, Davies C, Azevedo RB, Gow A, Kachar B. 2006. Distinct subdomain organization and molecular composition of a tight junction with adherens junction features. *J Cell Sci* **119:** 4819–4827.

Org E, Eyheramendy S, Juhanson P, Gieger C, Lichtner P, Klopp N, Veldre G, Doring A, Viigimaa M, Sober S, et al. 2009. Genome-wide scan identifies CDH13 as a novel susceptibility locus contributing to blood pressure determination in two European populations. *Hum Mol Genet* **18:** 2288–2296.

Petit C. 2001. Usher syndrome: From genetics to pathogenesis. *Annu Rev Genomics Hum Genet* **2:** 271–297.

Petit C, Richardson G. 2009. Linking deafness genes to hair-bundle development and function. *Nature Neurosci* **12:** 703–710.

Pickles JO, Comis SD, Osborne MP. 1984. Cross-links between stereocilia in the guinea pig organ of Corti, and their possible relation to sensory transduction. *Hear Res* **15:** 103–112.

Pilichou K, Nava A, Basso C, Beffagna G, Bauce B, Lorenzon A, Frigo G, Vettori A, Valente M, Towbin J, et al. 2006. Mutations in desmoglein-2 gene are associated with arrhythmogenic right ventricular cardiomyopathy. *Circulation* **113:** 1171–1179.

Prost J, Barbetta C, Joanny JF. 2007. Dynamical control of the shape and size of stereocilia and microvilli. *Biophys J* **93:** 1124–1133.

Radice GL, Ferreira-Cornwell MC, Robinson SD, Rayburn H, Chodosh LA, Takeichi M, Hynes RO. 1997. Precocious mammary gland development in P-cadherin-deficient mice. *J Cell Biol* **139:** 1025–1032.

Raphael Y, Altschuler RA. 2003. Structure and innervation of the cochlea. *Brain Res Bull* **60:** 397–422.

Rickman L, Simrak D, Stevens HP, Hunt DM, King IA, Bryant SP, Eady RA, Leigh IM, Arnemann J, Magee AI, et al. 1999. N-terminal deletion in a desmosomal cadherin causes the autosomal dominant skin disease striate palmoplantar keratoderma. *Hum Mol Genet* **8:** 971–976.

Rzadzinska AK, Steel KP. 2009. Presence of interstereocilial links in waltzer mutants suggests Cdh23 is not essential for tip link formation. *Neuroscience* **158:** 365–368.

Schaffer JV, Bazzi H, Vitebsky A, Witkiewicz A, Kovich OI, Kamino H, Shapiro LS, Amin SP, Orlow SJ, Christiano AM. 2006. Mutations in the desmoglein 4 gene underlie localized autosomal recessive hypotrichosis with monilethrix hairs and congenital scalp erosions. *J Invest Dermatol* **126:** 1286–1291.

Schwander M, Xiong W, Tokita J, Lelli A, Elledge HM, Kazmierczak P, Sczaniecka A, Kolatkar A, Wiltshire T, Kuhn P, et al. 2009. A mouse model for nonsyndromic deafness (DFNB12) links hearing loss to defects in tip links of mechanosensory hair cells. *Proc Natl Acad Sci USA* (in press).

Schweizer J. 2006. More than one gene involved in monilethrix: Intracellular but also extracellular players. *J Invest Dermatol* **126:** 1216–1219.

Seiler C, Finger-Baier KC, Rinner O, Makhankov YV, Schwarz H, Neuhauss SC, Nicolson T. 2005. Duplicated genes with split functions: Independent roles of protocadherin15 orthologues in zebrafish hearing and vision. *Development* **132:** 615–623.

Shapiro L, Weis WI. 2009. Structure and biochemistry of cadherins and catenins. *Cold Spring Harb Perspect Biol* **1:** a003053.

Shimomura Y, Sakamoto F, Kariya N, Matsunaga K, Ito M. 2006. Mutations in the desmoglein 4 gene are associated with monilethrix-like congenital hypotrichosis. *J Invest Dermatol* **126:** 1281–1285.

Shimomura Y, Wajid M, Shapiro L, Christiano AM. 2008. P-cadherin is a p63 target gene with a crucial role in the developing human limb bud and hair follicle. *Development* **135:** 743–753.

Simpson MA, Mansour S, Ahnood D, Kalidas K, Patton MA, McKenna WJ, Behr ER, Crosby AH. 2008. Homozygous Mutation of Desmocollin-2 in Arrhythmogenic Right Ventricular Cardiomyopathy with Mild Palmoplantar Keratoderma and Woolly Hair. *Cardiology* **113:** 28–34.

Sollner C, Rauch GJ, Siemens J, Geisler R, Schuster SC, Muller U, Nicolson T. 2004. Mutations in cadherin 23 affect tip links in zebrafish sensory hair cells. *Nature* **428:** 955–959.

Sprecher E, Bergman R, Richard G, Lurie R, Shalev S, Petronius D, Shalata A, Anbinder Y, Leibu R, Perlman I, et al. 2001. Hypotrichosis with juvenile macular dystrophy is caused by a mutation in CDH3, encoding P-cadherin. *Nature Genet* **29:** 134–136.

Sudhof TC. 2008. Neuroligins and neurexins link synaptic function to cognitive disease. *Nature* **455:** 903–911.

Syrris P, Ward D, Evans A, Asimaki A, Gandjbakhch E, Sen-Chowdhry S, McKenna WJ. 2006. Arrhythmogenic right ventricular dysplasia/cardiomyopathy associated with mutations in the desmosomal gene desmocollin-2. *Am J Hum Genet* **79:** 978–984.

Takeichi M. 2007. The cadherin superfamily in neuronal connections and interactions. *Nat Rev Neurosci* **8:** 11–20.

Tomo I, Boutet de Monvel J, Fridberger A. 2007. Sound-evoked radial strain in the hearing organ. *Biophys J* **93:** 3279–3284.

Verpy E, Leibovici M, Zwaenepoel I, Liu X-Z, Gal A, Salem N, Mansour A, Blanchard S, Kobayashi I, Keats BJB, et al. 2000. A defect in harmonin, a PDZ domain-containing protein expressed in the inner ear sensory hair cells, underlies Usher syndrome type 1C. *Nature Genet* **26:** 51–55.

Wajid M, Bazzi H, Rockey J, Lubetkin J, Zlotogorski A, Christiano AM. 2007. Localized autosomal recessive hypotrichosis due to a frameshift mutation in the desmoglein 4 gene exhibits extensive phenotypic variability within a Pakistani family. *J Invest Dermatol* **127:** 1779–1782.

Wan H, Dopping-Hepenstal PJ, Gratian MJ, Stone MG, Zhu G, Purkis PE, South AP, Keane F, Armstrong DK, Buxton RS, et al. 2004. Striate palmoplantar keratoderma arising from desmoplakin and desmoglein 1 mutations is associated with contrasting perturbations of desmosomes and the keratin filament network. *Br J Dermatol* **150:** 878–891.

Wangemann P. 2006. Supporting sensory transduction: Cochlear fluid homeostasis and the endocochlear potential. *J Physiol* **576:** 11–21.

Waschke J. 2008. The desmosome and pemphigus. *Histochem Cell Biol* **130:** 21–54.

Weil D, Blanchard S, Kaplan J, Guilford P, Gibson F, Walsh J, Mburu P, Varela A, Levilliers J, Weston MD, et al. 1995. Defective myosin VIIA gene responsible for Usher syndrome type 1B. *Nature* **374:** 60–61.

Weil D, El-Amraoui A, Masmoudi S, Mustapha M, Kikkawa Y, Laine S, Delmaghani S, Adato A, Nadifi S, Zina ZB, et al. 2003. Usher syndrome type I G (USH1G) is caused by mutations in the gene encoding SANS, a protein that associates with the USH1C protein, harmonin. *Hum Mol Genet* **12:** 463–471.

Xu L, Overbeek PA, Reneker LW. 2002. Systematic analysis of E-, N- and P-cadherin expression in mouse eye development. *Exp Eye Res* **74:** 753–760.

Yamada S, Nelson WJ. 2007. Synapses: Sites of cell recognition, adhesion, and functional specification. *Annu Rev Biochem* **76:** 267–294.

Yang A, Zhu Z, Kapranov P, McKeon F, Church GM, Gingeras TR, Struhl K. 2006. Relationships between p63 binding, DNA sequence, transcription activity, and biological function in human cells. *Mol Cell* **24:** 593–602.

Zlotogorski A, Marek D, Horev L, Abu A, Ben-Amitai D, Gerad L, Ingber A, Frydman M, Reznik-Wolf H, Vardy DA, et al. 2006. An autosomal recessive form of monilethrix is caused by mutations in DSG4: Clinical overlap with localized autosomal recessive hypotrichosis. *J Invest Dermatol* **126:** 1292–1296.

Involvement of Members of the Cadherin Superfamily in Cancer

Geert Berx[1,2] and Frans van Roy[2,3]

[1]Molecular and Cellular Oncology Unit, Department for Molecular Biomedical Research, VIB, Ghent, Belgium
[2]Department of Biomedical Molecular Biology, Ghent University, Ghent, Belgium
[3]Molecular Cell Biology Unit, Department for Molecular Biomedical Research, VIB, Ghent, Belgium
Correspondence: f.vanroy@dmbr.vib-ugent.be

We review the role of cadherins and cadherin-related proteins in human cancer. Cellular and animal models for human cancer are also dealt with whenever appropriate. E-cadherin is the prototype of the large cadherin superfamily and is renowned for its potent malignancy suppressing activity. Different mechanisms for inactivating E-cadherin/CDH1 have been identified in human cancers: inherited and somatic mutations, aberrant protein processing, increased promoter methylation, and induction of transcriptional repressors such as Snail and ZEB family members. The latter induce epithelial mesenchymal transition, which is also associated with induction of "mesenchymal" cadherins, a hallmark of tumor progression. VE-cadherin/CDH5 plays a role in tumor-associated angiogenesis. The atypical T-cadherin/CDH13 is often silenced in cancer cells but up-regulated in tumor vasculature. The review also covers the status of protocadherins and several other cadherin-related molecules in human cancer. Perspectives for emerging cadherin-related anticancer therapies are given.

INTRODUCTION—DYSREGULATION OF CADHERIN FAMILY MEMBERS IN CANCER

Cell–cell adhesion determines cell polarity and participates in cell differentiation and in establishment and maintenance of tissue homeostasis. During oncogenesis, this organized adhesion is disturbed by genetic and epigenetic changes, resulting in changes in signaling, loss of contact inhibition, and altered cell migration and stromal interactions. A major class of cell–cell adhesion molecules is the cadherin superfamily. Its prototypic member, E-cadherin, was characterized as a potent suppressor of invasion and metastasis in seminal studies dating back to the 1990s (reviewed by van Roy and Berx 2008). Since then, many more cadherins and cadherin-related proteins have been identified (Fig. 1) (reviewed by Hulpiau and van Roy 2009), and an increasing number has been implicated in cancer as putative tumor suppressors or as proto-oncogenic proteins. Here we discuss the structural and functional aberrations of cadherin family members in cancer and their roles in cancer initiation and progression.

E-CADHERIN

Most human tumors are carcinomas derived from epithelial tissues, in which E-cadherin is

 doi: 10.1101/cshperspect.a003129
Cite this article as *Cold Spring Harb Perspect Biol* 2009;1:a003129

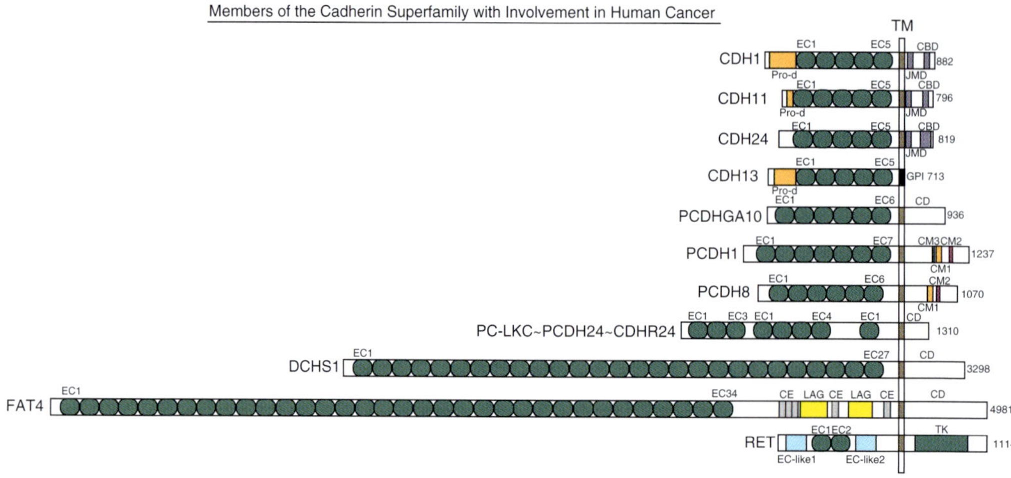

Figure 1. Schematic overview of representative human members of the cadherin superfamily with reported involvement in cancer (modified after Hulpiau and van Roy 2009). All proteins are drawn to scale and aligned at their transmembrane domain (TM). Their total sizes are indicated on the right (number of amino acid residues). The following protein domains are shown: CBD, (conserved cadherin-specific) catenin binding domain; CD, unique cytoplasmic domain; CE, Cysteine-rich EGF repeat-like domain; CM1 to CM3, conserved motifs in the CDs of δ-protocadherins; EC, extracellular cadherin repeat; GPI, glycosylphosphatidylinositol anchor; JMD, (conserved cadherin-specific) juxtamembrane domain; LAG, laminin A globular domain; Pro-d, prodomain; TK, tyrosine kinase domain. On the basis of a phylogenetic analysis (Hulpiau and van Roy 2009), it was proposed that protocadherin LKC (PC-LKC or protocadherin-24) should be renamed (CDHR24).

the prototypic cadherin. Epithelial tumors often lose E-cadherin partially or completely as they progress toward malignancy (reviewed by Birchmeier and Behrens 1994; Christofori and Semb 1999; Strumane et al. 2004). Epithelial ovarian cancers are exceptional because expression of E-cadherin in inclusion cysts, derived from ovarian surface epithelium with little or no E-cadherin, appears to be essential for tumorigenesis in this organ (Sundfeldt 2003; Naora and Montell 2005). Another exception is inflammatory breast cancer, a distinct and aggressive form of breast cancer in which the expression of E-cadherin is consistently elevated regardless of the histologic type or molecular profile of the tumor (Table 1) (Alpaugh et al. 1999; Kleer et al. 2001). However, most studies have shown both strong anti-invasive and antimetastatic roles for E-cadherin (Frixen et al. 1991; Vleminckx et al. 1991; Perl et al. 1998). The possible functional implications, which were reviewed recently (Jeanes et al. 2008), include the sequestering of β-catenin in an E-cadherin-catenin adhesion complex (Fig. 3), leading to inhibition of its function in the canonical Wnt pathway (Fig. 4), besides inhibition of EGF receptor signaling and contribution to epithelial apico-basal polarization. We focus here on the different mechanisms for E-cadherin inactivation in malignant tumors, which include mutations, epigenetic silencing, and increased endocytosis and proteolysis (Fig. 2; Table 1).

Loss of Heterozygosity and Inactivating Mutations in Cancer

Studies on loss of heterozygosity (LOH) of chromosome 16q21-22 hinted at a role for E-cadherin in human cancer (Fig. 2; Table 1). Following the mapping of the human E-cadherin gene *CDH1* to chromosome 16q22.1 (Natt et al. 1989; Berx et al. 1995b), several studies showed frequent LOH of 16q in gastric, prostate, hepatocellular and esophageal carcinomas (reviewed by Strathdee 2002;

Table 1. Representative abnormalities of cadherin superfamily members in cancer

Protein	Human gene	Tumor-associated abnormalities	Tumor type[1]	Clinical correlates[2]	Selected references
E-cadherin	*CDH1*	LOH	Numerous	Malignant progression	(Reviewed in Strumane et al. 2004)
		Promoter methylation	Numerous	Malignant progression	(Reviewed in Strumane et al. 2004)
		Germline mutations	Gastric (DGC) Breast (ILC)	Hereditary diffuse gastric cancer (HDGC) syndrome	(Guilford et al. 1998)
		Somatic mutations	Breast (ILC) Gastric (DGC) Pancreas	Highly invasive growth pattern	(Berx et al. 1998)
		Up-regulated expression	Epithelial ovarian cancer	Tumorigenesis	(Reviewed in Sundfeldt 2003; Naora and Montell 2005)
		Overexpression	Breast (IBC)	Promotes tumor emboli formation	(Kleer et al. 2001)
N-cadherin	*CDH2*	Up-regulation (cadherin switching)	Breast Pancreatic Prostate Melanoma	Enhanced migration and invasion, increased metastasis; poor prognosis	(Reviewed in Hazan et al. 2004; Wheelock et al. 2008)
P-cadherin	*CDH3*	Up-regulation (cadherin switching)	Breast Gastric Pancreatic (PDAC)	Enhanced migration and invasion, poor prognosis	(Paredes et al. 2005; Taniuchi et al. 2005)
		Down-regulation	Melanoma	Increased invasion and metastasis	(Sanders et al. 1999)
R-cadherin	*CDH4*	Promoter methylation	Colorectal Gastric	Early event	(Miotto et al. 2004)
Cadherin-11 = OB-cadherin	*CDH11*	Up-regulation (cadherin switching)	Breast Prostate	High grade cancer; prostate cancer metastasis to bone	(Bussemakers et al. 2000; Tomita et al. 2000; Chu et al. 2008)
VE-cadherin	*CDH5*	Overexpression	Melanoma	Vasculogenic mimicry; malignant progression	(Hendrix et al. 2001)

(*Continued*)

Table 1. *Continued*

Protein	Human gene	Tumor-associated abnormalities	Tumor type[1]	Clinical correlates[2]	Selected references
T-cadherin	*CDH13*	Decreased expression	Breast	Increased tumorigenesis	(Lee 1996)
		Promoter methylation	Breast Lung (NSCLC) Prostate Colon (adenomas and CRC) Pancreas Gastric HCC Esophageal adenocarcinoma Bladder Chronic myeloid leukemia	Increased tumorigenesis; malignant progression; early recurrence; poor prognosis	(Sato et al. 1998; Maruyama et al. 2001; Toyooka et al. 2001; Toyooka et al. 2002; Roman-Gomez et al. 2003; Hibi et al. 2004; Sakai et al. 2004; Kim et al. 2005; Wang et al. 2007; Brock et al. 2008; Jin et al. 2008; Yan et al. 2008)
		LOH and promoter methylation	Ovary Skin (BCC and SCC) Diffuse large B cell lymphoma HCC	Tumor invasion; increased cell proliferation and resistance to apoptosis	(Kawakami et al. 1999; Takeuchi et al. 2002a; Takeuchi et al. 2002b; Ogama et al. 2004; Chan et al. 2008)
		Expression in intratumoral endothelium	Tumor models HCC	Adiponectin-(co)receptor; increased angiogenesis and tumor progression	(Wyder et al. 2000; Adachi et al. 2006; Philippova et al. 2006; Riou et al. 2006; Hebbard et al. 2008)
Cadherin 24	*CDH24*	Nonsense mutation	Acute myeloid leukemia	NA	(Ley et al. 2008)
LKC-protocadherin= cadherin-related protein 24	*PCDH24=PCLKC*	Reduced expression	Colon Liver Kidney	NA	(Okazaki et al. 2002)

Clustered protocadherins	*PCDHBx*, *PCDHGx*	Mutations (mostly missense)	Pancreas	NA	(Jones et al. 2008)
	PCDGGx	Promoter methylation	Astrocytomas Breast	NA	(Waha et al. 2005) (Miyamoto et al. 2005)
Protocadherin-1	*PCDH1*	Decreased expression	Medulloblastoma	Unfavorable survival; independent prognosticator	(Neben et al. 2004)
Protocadherin-8	*PCDH8*	Reduced transcription, somatic missense, LOH, homozygous deletion, promoter methylation	Breast (tumors and cancer cell lines)	Reduced estrogen receptor and progesterone receptor	(Yu et al. 2008)
Protocadherin-9	*PCDH9*	Missense mutation	Pancreas	NA	(Jones et al. 2008)
Protocadherin-10 = OL-protocadherin	*PCDH10*	Frequent promoter methylation	Nasopharyngeal Esophageal Breast Gastric HCC Hematological (multiple)	Decreased overall survival in stage I-III gastric cancer patients	(Miyamoto et al. 2005; Ying et al. 2006; Ying et al. 2007; Yu et al. 2009)
Protocadherin-17	*PCDH17*	Nonsense and missense mutations	Pancreas	NA	(Jones et al. 2008)
Protocadherin-18	*PCDH18*	Missense mutations	Pancreas	NA	(Jones et al. 2008)
Protocadherin-20	*PCDH20*	Promoter methylation	Lung (NSCLC)	Shorter overall survival; stage-independent prognosticator	(Imoto et al. 2006)
Protocadherin-11Y= PC-protocadherin	*PCDH11Y*	Increased expression	Prostate	Androgen-resistance; apoptosis-resistance	(Chen et al. 2002; Terry et al. 2006)
FAT4	*FAT4*	Promoter methylation	Breast	NA	(Qi et al. 2009)
RET	*RET*	Constitutive activation	Thyroid Multiple endocrine neoplasia	Causative	(reviewed by Kondo et al. 2006; Zbuk and Eng 2007)

[1]BCC, basal cell carcinoma; CRC, colorectal carcinoma; DGC, diffuse gastric carcinoma; HCC, hepatocellular carcinoma; IBC, inflammatory breast carcinoma; ILC, invasive lobular carcinoma; NSCLC, nonsmall cell lung carcinoma; PDAC, pancreatic ductal adenocarcinoma; SCC, squamous cell carcinoma.
[2]NA: not analyzed in this study.

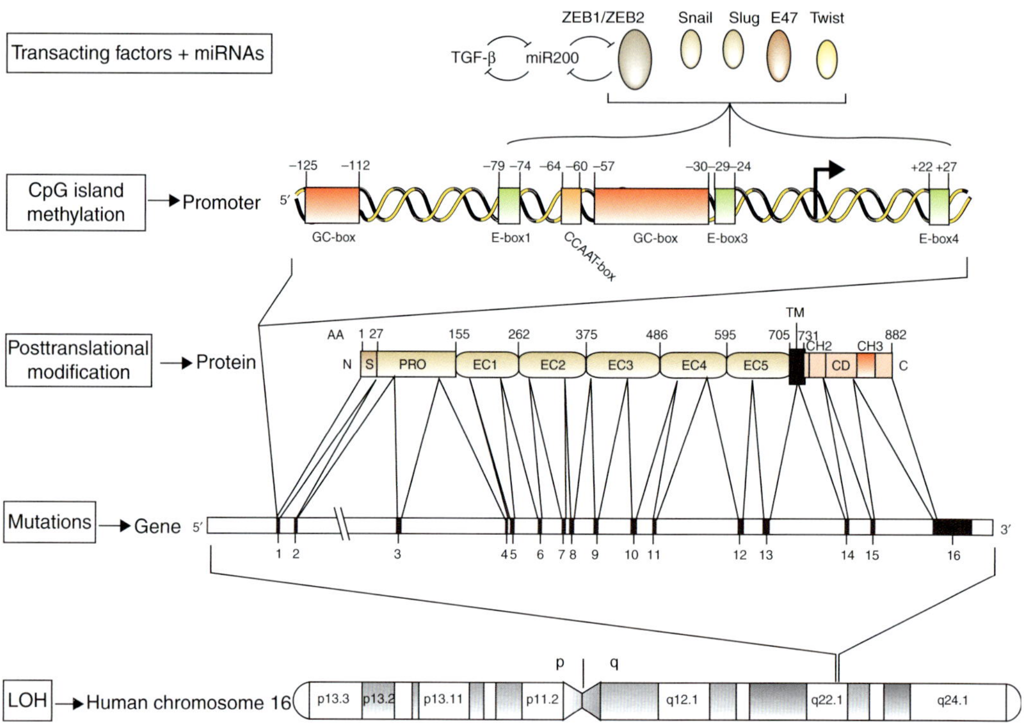

Figure 2. Various levels at which E-cadherin expression is regulated in human tumors (modified, with permission, from van Roy and Berx 2008). The E-cadherin gene *CDH1* is on chromosome 16q22.1 (depicted at the *bottom*). This region frequently shows loss of heterozygosity (LOH) in different human carcinoma types. Specific inactivating mutations are scattered throughout the whole coding region and are particularly abundant in sporadic lobular breast cancer and diffuse gastric cancer. Germline mutations can also occur; they cause the hereditary diffuse gastric cancer syndrome. Furthermore, post-translational modifications, such as phosphorylation and glycosylation, and proteolytic processing can affect E-cadherin protein functionality. Epigenetic silencing has been associated with CpG methylation in the *CDH1* promoter region or with direct binding of specific transcriptional repressors to E-box sequences in this region. The transcriptional repressors ZEB1/δEF1 and ZEB2/SIP1 are repressed in epithelia by the miRNAs of the miR-200 family. In turn, the ZEB transcription factors down-regulate transcription of the miR-200 genes. Thus, a biphasic regulatory system controls the balance between the epithelial and mesenchymal status in response to incoming signals. TGF-β in the tumor microenvironment can induce the expression of ZEB proteins, at least in part by down-regulating the miR-200 family members. This results in a self-enhancing loop that leads to epithelial dedifferentiation and invasion. See text for more details and references. AA, amino acid position; C, carboxy-terminal end; CD, cytoplasmic domain; EC, extracellular cadherin repeat; N, amino-terminal end; PRO, propeptide; S, signal peptide; TM, transmembrane region. The arrows point to the transcriptional initiation start.

Strumane et al. 2004). LOH at 16q is frequent particularly in breast cancer, where it occurs in ~50% of all ductal carcinomas (Cleton-Jansen et al. 2001), and even more frequently in lobular breast cancer (Berx et al. 1996).

E-cadherin-inactivating mutations were first described in diffuse gastric cancer (Becker et al. 1993). In sporadic diffuse gastric cancer, somatic mutations preferentially cause skipping of exon 7 or 9, which results in in-frame deletions. Several truncations have also been reported in this histological subtype of tumors (Becker et al. 1994; Berx et al. 1998). Promoter hypermethylation, rather than LOH, accounts here for biallelic *CDH1* silencing (Machado et al. 2001). In contrast to these mutation

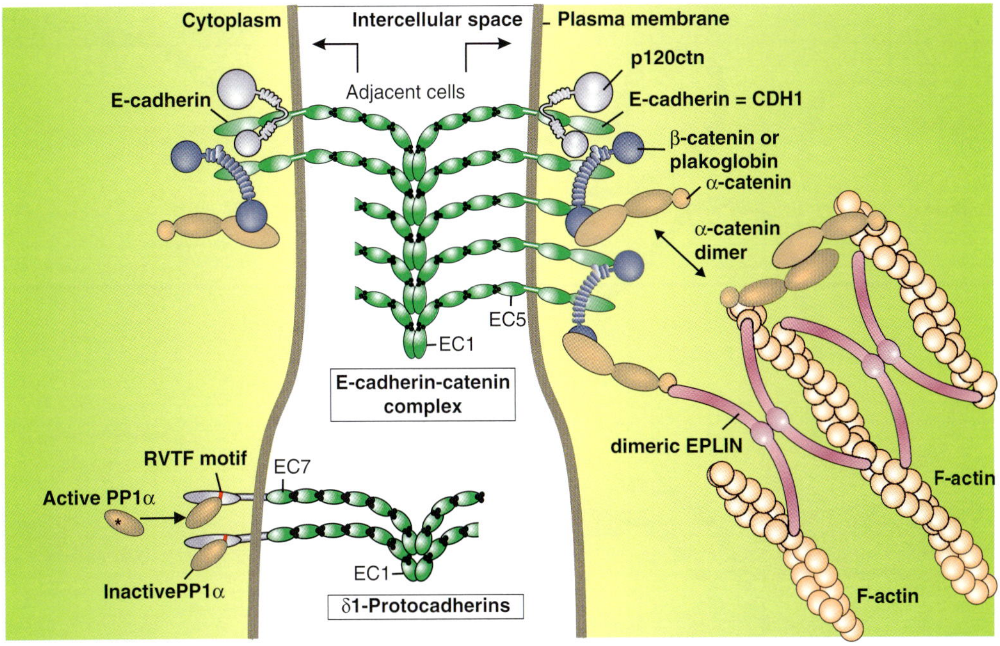

Figure 3. Schematic overview of the E-cadherin-catenin complex and of δ1-protocadherins at the junction between two adjacent cells (modified, with permission, from Redies et al. 2005; van Roy and Berx 2008). *Top*: The armadillo catenins p120ctn and β-catenin/plakoglobin bind to, respectively, membrane-proximal and carboxy-terminal halves of the cytoplasmic domain of E-cadherin. This increases junction strength and stability. As extensively described in the literature, both β-catenin and p120ctn also have cancer-related roles in the cytoplasm and in the nucleus. Monomeric α-catenin binds to the E-cadherin cytoplasmic domain via β-catenin, whereas dimeric α-catenin can bind and cross-link filamentous actin (F-actin). Moreover, dimeric EPLIN forms a link between the E-cadherin-catenin complex and F-actin. See text for more details and references. *Bottom*: The evidence for the depicted δ1-protocadherins structure is circumstantial, but the following features are typical: seven extracellular cadherin repeats (EC) instead of five and a completely different cytoplasmic domain with conserved motifs (CM). The CM3 or RVTF motif has been shown to interact with phosphatase PP1α, probably resulting in its inactivation (reviewed in Redies et al. 2005).

hotspots, E-cadherin inactivating mutations tend to be scattered along the *CDH1* gene in sporadic lobular breast carcinomas (Berx et al. 1995a; Berx et al. 1996). In this highly infiltrating cancer type, most *CDH1* mutations are out-of-frame mutations predicted to yield secreted truncated E-cadherin fragments or no stable protein. E-cadherin expression is silenced because the mutations are accompanied by *CDH1* promoter methylation or LOH (Berx et al. 1996; Berx et al. 1998; Droufakou et al. 2001). Missense mutations are infrequent in these two subtypes of cancer but frequent in monophasic synovial sarcomas (Saito et al. 2001). E-cadherin mutations are rare in carcinomas of bladder, colon, endometrium, lung, esophagus, ovary, and thyroid, and in intrahepatic cholangiocarcinoma (Risinger et al. 1994; Soares et al. 1997; Wijnhoven et al. 1999; Taddei et al. 2000; Endo et al. 2001; Vecsey-Semjen et al. 2002).

Familial aggregation of gastric cancer is well known. These familial cancers can be classified histopathologically into hereditary diffuse gastric cancer (HDGC), familial diffuse gastric cancer (FDGC), and familial intestinal gastric cancer. The International Gastric Cancer Linkage Consortium (IGCLC) established criteria for identifying HDGC families based on the incidence and onset of diffuse gastric cancer (Caldas et al. 1999). Gastric cancers with high incidence and with an index case of

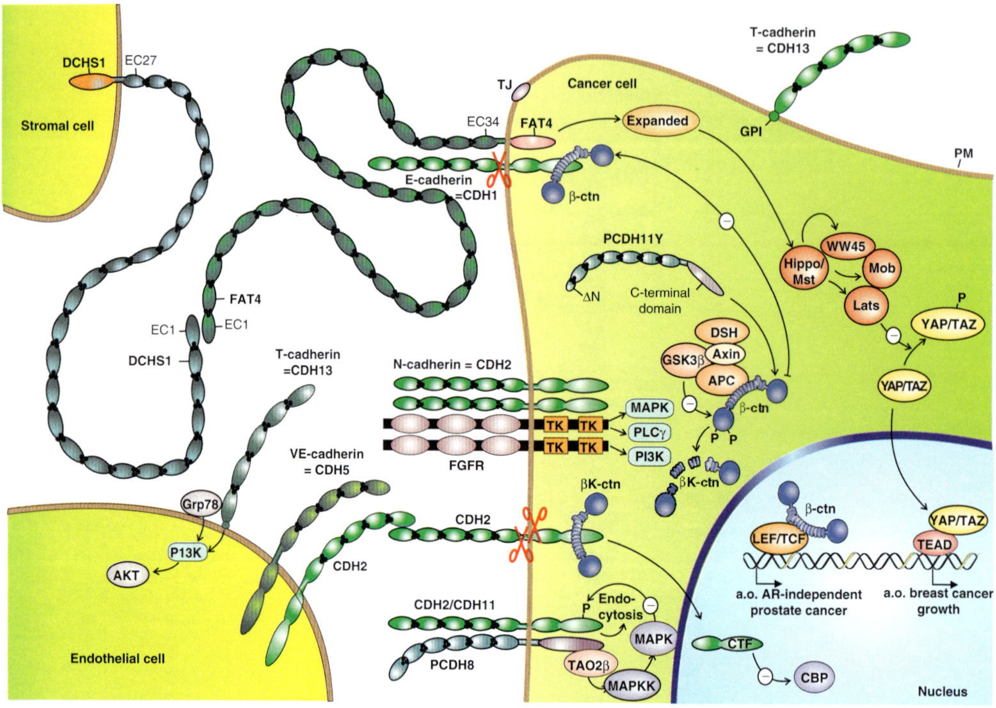

Figure 4. Selection of expression patterns and activities of members of the cadherin superfamily in cancer. Three cell types are partly depicted: a cancer cell, an endothelial cell, and another type of stromal cell. Protein domains in green are strongly homologous to those in the prototypic E-cadherin/CDH1. Domains in other colors deviate substantially in structure and function. Black dots represent Ca^{2+} ions. Arrows with minus sign symbols refer to direct or indirect inhibitory influences. Cadherins shown at the cancer cell surface are expressed at the apical membrane (T-cadherin/CDH13 above the tight junction, TJ), or at lateral membranes (E-cadherin/CDH1, N-cadherin/CDH2, cadherin-11/CDH11). The latter three probably occur as *cis*-homodimers. CDH1 is frequently inactivated in cancer cells, whereas the "mesenchymal" cadherins CDH2 and CDH11 are often up-regulated. CDH2 can interact with fibroblast growth factor receptor (FGFR), potentiating its signaling through the enzymes MAPK, PLCγ, and PI3K. Cadherins are prone to proteolytic processing (scissors symbols), which releases either the ectodomain or a carboxy-terminal fragment (CTF). In the case of CDH2, this CTF has been shown to enter the nucleus and inhibit the CREB binding protein (CBP). On neural activation, CDH2 and CDH11 associate with protocadherin-8 (PCDH8 or Arcadlin), which results in activation of the MAPKKK TAO2β, eventually leading to endocytosis of the cadherins. PCDH8 is often silenced in cancer cells, but it is unclear whether this is causally linked to up-regulation of mesenchymal cadherins. The transcriptional activity of β-catenin (β-ctn) in a nuclear complex with LEF/TCF, leads, amongst other effects, to androgen receptor (AR)-independent prostate cancer growth. This phenomenon is inhibited by sequestration of β-ctn by E-cadherin or by degradation of β-ctn after phosphorylation (-P) by GSK3β. Degradation of β-ctn occurs in a cytoplasmic degradation complex with APC (adenomatous polyposis coli protein), Axin, and Disheveled (DSH). However, nuclear β-ctn-LEF/TCF activity is stimulated by an unknown mechanism by a cytoplasmic variant of protocadherin-11Y (PCDH11Y) lacking a signal peptide because of a truncated aminoterminus (ΔN). Tumor-associated endothelial cells express VE-cadherin/CDH5, CDH2, and T-cadherin/CDH13. The latter is linked to the PM via a glycosylphosphatidylinositol (GPI) anchor and signals via secreted Grp78/BiP to an anti-apoptotic PI3K-AKT pathway. Dachsous-1 (DCHS1) and FAT4 are both huge cadherin-related proteins, interacting with each other in heterophilic (different protein types) and heterotypic (different cell types) ways. Silencing of human FAT4 is seen in breast cancer and its activation is linked in an unresolved way to the Hippo-YAP pathway, which controls organ size in *Drosophila* and is affected in several human cancers. See text for details and references. (a.o.) Amongst others; (EC) extracellular cadherin repeat; (PM) plasma membrane; (TK) tyrosine kinase domain.

diffuse gastric cancer but not fulfilling the IGCLC criteria for HDGC are classified as FDGC (Oliveira et al. 2006).

Germline mutations in the E-cadherin gene were first described in 1998 by Guilford et al. (1998), who identified *CDH1*-inactivating mutations in three Maori families with early-onset diffuse gastric cancer. Since then, 68 families carrying germline *CDH1* mutations have been identified worldwide (Carneiro et al. 2008). *CDH1* mutations were found in 30.5% of HDGC families and 13.8% of FDGC families (reviewed in Oliveira et al. 2006). These mutations resemble sporadic mutations in that most of them are predicted to cause premature stop codons caused by nonsense, splice-site, or frameshift mutations. Only some of them were missense mutations (Pereira et al. 2006). Some of these E-cadherin missense mutants are subjected to ER quality control followed by ER-associated degradation (Simoes-Correia et al. 2008). Germline mutations are scattered along the entire gene.

Multiple cases of lobular breast cancer (including mixed ductal and lobular histology) have been reported in families with HDGC (Caldas et al. 1999; Pharoah et al. 2001; Brooks-Wilson et al. 2004; Suriano et al. 2005). The estimated cumulative lifetime risk of breast cancer in women from HDGC families with germline *CDH1* mutations is 39% (Pharoah et al. 2001). Remarkably, *CDH1* germline mutations can be associated with invasive lobular breast cancer in the absence of diffuse gastric cancer (Masciari et al. 2007). Most of these hereditary tumors are E-cadherin-negative, pointing to a double inactivating mechanism.

Predisposition to diffuse gastric cancer and lobular cancer in patients carrying germline mutations in the E-cadherin gene identify E-cadherin as a tumor suppressor (Dunbier and Guilford 2001). An in vivo role of E-cadherin in cancer progression was supported by demonstrating that expression of a dominant-negative E-cadherin mutant in a mouse pancreatic β-cell tumor model accelerated conversion of adenomas into carcinomas (Perl et al. 1998). In contrast, in mouse models specific loss of E-cadherin in the skin and the mammary gland is not sufficient for tumorigenesis (Boussadia et al. 2002; Tinkle et al. 2004; Tunggal et al. 2005). This ablation caused extensive cell death in the mammary gland, which indicates that the frequent loss of E-cadherin in epithelial cancers, including breast carcinomas, must be accompanied by oncogenic antideath mechanisms. This was shown by combining a floxed *Cdh1* gene with the K14-Cre gene, which is expressed at low, stochastic levels in mammary epithelium (Derksen et al. 2006). No abnormalities were seen in these mice, but combined loss of E-cadherin and p53 resulted in accelerated development of malignant mammary carcinomas resembling human infiltrative lobular carcinomas. Ablating both tumor suppressors induced more metastatic spreading, anoikis resistance, and vascularization than loss of p53 alone.

Epigenetic Silencing of the E-cadherin Locus in Cancer

Promoter hypermethylation is an important epigenetic event associated with loss of E-cadherin gene expression during cancer progression (Fig. 2; Table 1). A large CpG island in the 5′ proximal promoter region of the E-cadherin gene (Berx et al. 1995b) shows aberrant DNA methylation in at least eight human carcinoma types and correlates with reduced E-cadherin protein expression (Graff et al. 1995; Yoshiura et al. 1995; Chang et al. 2002; Kanazawa et al. 2002). In cancer cell lines, promoter methylation of the E-cadherin gene is heterogeneous, dynamic, unstable, and associated with allele to allele variability (Graff et al. 1995; Graff et al. 1998; Graff et al. 2000). This is compatible with heterogeneous loss of E-cadherin protein expression, which might be influenced by the tumor microenvironment. CpG island methylation in the *CDH1* gene seemingly increases during malignant progression of breast and hepatocellular carcinomas (Kanai et al. 2000; Nass et al. 2000).

Causal involvement of hypermethylation in E-cadherin repression is supported by the reactivation of functional E-cadherin in certain cancer cell lines on treatment with the demethylating agent 5-aza-2′-deoxycytidine (5AzaC) (Graff et al. 1995; Yoshiura et al. 1995; Nam et al. 2004). Binding of the methyl-CpG binding proteins MeCP2 and MBP2 to the methylated CpG island of the repressed E-cadherin promoter results in recruitment of HDACs to the promoter area, leading to histone-3 (H3) deacetylation, which is essential for suppressing the methylated E-cadherin gene (Koizume et al. 2002). Interestingly, the methylated *CDH1* promoter status in breast cancer cell lines seems to be part of a general transcriptional program that conforms with EMTand increased invasiveness, but diverges from the specific consequence of E-cadherin mutational inactivation (Lombaerts et al. 2006).

Transcriptional Silencing, EMT, and Cancer

Different repressors of E-cadherin transcription have been associated with the progression of multiple cancer types (Fig. 2). Increased Snail expression, common in ductal breast carcinomas, is strongly associated with reduced E-cadherin gene expression (Cheng et al. 2001). High-grade breast tumors and lymph-node positive tumors consistently show strong Snail expression (Blanco et al. 2002). A new role for Snail in tumor recurrence has been inferred from a reversible HER-2/neu-induced breast cancer mouse model (Moody et al. 2005). Also, abnormal expression of Slug has been associated with disease aggressiveness in metastatic ovarian and breast carcinoma (Elloul et al. 2005). Twist, another EMT-regulating transcription factor, is involved in breast tumor metastasis (Yang et al. 2004), and its expression rises as nodal involvement increases (Martin et al. 2005). Strong expression of SIP1/ZEB2, which is associated with loss of E-cadherin expression, was reported in gastric cancer of the intestinal type, but Snail does not seem to be involved in these tumors (Rosivatz et al. 2002). In contrast, Snail is up-regulated in diffuse gastric cancer, a tumor subtype frequently affected by E-cadherin inactivating mutations (Rosivatz et al. 2002). The transcription factor deltaEF1/ZEB1 seems to be downstream of Snail expression (Guaita et al. 2002). Knock-down of deltaEF1/ZEB1 in dedifferentiated human epithelial colon and breast cancer cell lines results in re-expression of E-cadherin and other epithelial differentiation markers (Eger et al. 2005; Spaderna et al. 2006; Spaderna et al. 2008). Though extensive data show that expression of E-cadherin repressors is inversely correlated with expression of E-cadherin, these results should be interpreted carefully because many data are based on RT-PCR and on the use of antibodies with poorly defined specificity.

Recently, induction of expression of E-cadherin transcriptional repressors has been inversely linked with the expression status of the von Hippel-Lindeau (VHL) tumor suppressor (reviewed in Russell and Ohh 2007). Loss of VHL is an early, requisite step in the pathogenesis of clear-cell renal cell carcinoma (CC-RCC) (Lubensky et al. 1996). Activation of HIFα proteins in cells devoid of VHL, including CC-RCC cells, induces transactivation of several E-cadherin repressors, e.g., SIP1/ZEB2 and Snail, which contributes to the particularly malignant character of this tumor type (Esteban et al. 2006; Krishnamachary et al. 2006; Evans et al. 2007).

Regulatory cross-modulation exists between different E-cadherin-repressing transcription factors and specific microRNAs (Fig. 2). Members of the miR-200 family and miR-205 directly target ZEB1/deltaEF1 and ZEB2/SIP1 mRNAs, thereby repressing ZEB1 and ZEB2 protein translation and hence increasing E-cadherin expression (Christoffersen et al. 2007; Hurteau et al. 2007; Burk et al. 2008; Gregory et al. 2008; Korpal et al. 2008; Park et al. 2008). Expression analysis of miRNAs in normal human tissues showed that the miRNAs targeting ZEB family members are particularly abundant in epithelial tissues (Liang et al. 2007). Experimental up-regulation of miR-200c in dedifferentiated metastatic breast cancer cells suppresses their invasive behavior (Burk et al. 2008). Alternatively, ZEB1/deltaEF1 represses transcription of miR-200 family members by binding to E-box elements in the promoter regions of polycistronic primary miRNAs encoding these miRNAs (Bracken et al. 2008; Burk et al. 2008). These different miRNAs are repressed both by TGF-β and by overexpression of the tyrosine phosphatase Pez, which results in EMT with loss of E-cadherin expression (Gregory et al. 2008). In turn, the EMT activator TGF-β2 is also down-regulated by these miRNAs, indicating that ZEB1/deltaEF1 induces a microRNA-mediated feed-forward loop, which can re-enforce EMT (Lopez and Hanahan 2002; Burk et al. 2008).

Increasing evidence shows that diverse solid tumors are hierarchically organized and sustained by a distinct subpopulation of cancer stem cells (CSCs). For breast cancer, Al-Haij et al. (2003) described a $CD44^{high}/CD24^{low}$ cell population that had greater tumor-initiating capacity. Surprisingly, normal mammary epithelial cells could be induced to adopt the stem-cell like $CD44^{high}/CD24^{low}$ expression profile when exposed to TGF-β or on conditional overexpression of the EMT inducing transcription factor Snail or Twist. The resulting population displayed mesenchymal and stem cell markers and acquired stem-cell like properties, including enhanced growth in mammosphere cultures and increased formation of soft agar colonies (Mani et al. 2008). Moreover, activation of the Ras/MAPK signaling pathway in primary human mammary epithelial cells appeared to be crucial for facilitating the emergence of $CD44^{high}/CD24^{low}$ cells (Morel et al. 2008).

Endocytosis and Proteolytic Processing of E-cadherin in Cancer

Multiple mechanisms other than genetic and epigenetic silencing of E-cadherin could serve as alternative ways for disturbing or inhibiting normal E-cadherin function under pathological conditions. As reviewed (van Roy and Berx 2008), E-cadherin is removed from the plasma membrane by endocytosis and recycled to sites of new cell–cell contacts. Abnormal activation of proto-oncogenes, such as c-Met, Src, and EGFR, results in increased phosphorylation of tyrosine residues in the cytoplasmic domain of E-cadherin, which leads to recruitment of the E3-ubiquitin ligase Hakai and subsequently mediates internalization and ubiquitin-dependent degradation of E-cadherin (Fujita et al. 2002; Shen et al. 2008).

Because p120ctn participates in stabilizing the cadherin–catenin complex, many cancer types are characterized by loss or dislocalization of p120ctn (reviewed in van Hengel and van Roy 2007). We recently showed that p120ctn interacts with hNanos1, the human ortholog of the *Drosophila* zinc-finger protein, Nanos (Strumane et al. 2006). Transcription of hNanos1 mRNA is often suppressed on E-cadherin expression. Conditional expression of hNanos1 in human colon DLD1 cancer cells induces cytoplasmic translocation of p120ctn, up-regulates expression of membrane type 1-matrix metalloproteinase (MT1-MMP) at the mRNA and protein levels, and increases migratory and invasive abilities (Strumane et al. 2006; Bonnomet et al. 2008).

Moreover, matrix metalloproteinases, including stromelysin-1 (MMP3), matrilysin (MMP7), MMP9, and MT1-MMP (MMP14), cleave the E-cadherin ectodomain near the plasma membrane (Fig. 4) (Lochter et al. 1997; Davies et al. 2001; Noë et al. 2001; Covington et al. 2006; Symowicz et al. 2007). Several other proteases, such as the serine protease kallikrein 6 (Klk6), are up-regulated in human squamous skin carcinomas. Ectopic expression of Klk6 in keratinocytes induces E-cadherin ectodomain shedding in parallel with increased levels of mature ADAM10 proteinase (Klucky et al. 2007). Pancreatic adenocarcinomas often overexpress kallikrein-7, which can also generate soluble E-cadherin fragments that may function as pseudoligands to block normal E-cadherin interactions and promote invasion (Johnson et al. 2007). In epithelial ovarian carcinomas, the tumor cells maintain direct contact with ascites, which accumulates high concentrations of the solubilized E-cadherin ectodomain, and thereby promotes disruption of cell–cell junctions and metastatic dissemination (Symowicz et al. 2007). ADAM15, which is associated with progression of breast and prostate cancers, also generates a soluble E-cadherin ectodomain (Najy et al. 2008). This E-cadherin fragment appeared to stabilize heterodimerization of the HER2 (ErbB2) receptor tyrosine kinase with HER3, thus leading to Erk signaling, which stimulates both cell proliferation and migration.

The ectodomain of E-cadherin is also a substrate of secreted cathepsins B, L, and S, but not cathepsin C (Gocheva et al. 2006). This correlates with impaired malignant invasion on ablation of any of these three cathepsins in the mouse pancreatic islet cell carcinogenesis model, RIP1-Tag2, whereas tissue-specific cathepsin C knockout had no effect on either tumor formation or progression. Cathepsins are often secreted by various cells in the tumor microenvironment (Mohamed and Sloane 2006).

MESENCHYMAL CADHERINS AND P-CADHERIN

Loss of expression of E-cadherin (CDH1), P-cadherin (CDH3), or both in invasive tumor cell lines and malignant tissues from breast cancers, prostate cancers, and melanomas is often associated with induced expression of the mesenchymal N-cadherin (CDH2), generally referred to as cadherin switching (reviewed by Tomita et al. 2000; Hazan et al. 2004; Wheelock et al. 2008) (Table 1). E-cadherin down-regulation together with de novo N-cadherin expression is also observed in a Rip1-Tag2 mouse tumor model, in which tumor progression is accelerated by overexpression of IGF1R in the pancreatic β-cells (Lopez and Hanahan 2002). Normally, cadherin-11 (OB-cadherin or CDH11) (Fig. 1) is constitutively expressed in stromal and osteoblastic cells. Its misexpression in carcinoma cell lines and malignant breast and prostate tissues coincides with greater invasiveness and poor prognosis (Table 1) (Pishvaian et al. 1999; Bussemakers et al. 2000; Tomita et al. 2000). Manipulation of cadherin-11 expression in experimental metastasis models for breast and prostate cancers suggests that it promotes homing and migration to bone (Chu et al. 2008; Tamura et al. 2008). In human prostate cancers, cadherin-11 expression increases with progression from primary to metastatic disease in lymph nodes and particularly in bone (Chu et al. 2008).

E-cadherin repressors, such as Snail, ZEB2/SIP1, and Slug, can induce N-cadherin and cadherin-11 expression during EMT, suggesting that this cadherin switch is part of a transcriptional reprogramming of dedifferentiating epithelial cells (Cano et al. 2000; Vandewalle et al. 2005; Sarrio et al. 2008). Overexpression of N-cadherin in epithelial breast tumor cells induces a scattered morphology even when E-cadherin is present, and increases their motility, invasiveness, and metastatic capacity (Nieman et al. 1999; Hazan et al. 2000). This enhanced malignancy can be partly explained: Tumor cells, which express N-cadherin, are more able to interact with N-cadherin-positive tissues, including stroma and endothelium, potentially promoting their access to the vasculature and penetration and survival in secondary organs (Fig. 4) (Hazan et al. 1997; Hazan et al. 2000).

In contrast to the malignant contribution of N-cadherin overexpression in breast tumor cell models, mammary gland tumors arising in a bi-transgenic mouse model overexpressing both N-cadherin and ErbB2/HER-2/neu in a tissue-specific manner are not pathologically different from the tumors in single ErbB2/HER-2/neu transgenic mice (Knudsen et al. 2005). However, coexpression of N-cadherin with polyomavirus middle-T antigen in the mammary epithelium produces breast cancers with greater potential for metastasis to the lung (Hulit et al. 2007). N-cadherin did not enhance tumor onset but affected tumor progression by potentiating oncogenic ERK signaling, leading to MMP-9 up-regulation. These studies suggest that the effects of cadherin switching could occur late in tumor progression and that the impact of abnormal cadherin expression can depend on the cellular context. The concerted action of additional events, such as overexpression of FGF(R), loss of E-cadherin, or up-regulation of MMPs might be required together with N-cadherin up-regulation to promote mammary tumor invasion and metastasis in vivo. In breast cancer cells,

N-cadherin and FGF2 synergistically increase migration, invasion, and secretion of extracellular proteases (Hazan et al. 2004).

Mechanistically, N-cadherin is believed to functionally interact with the FGF receptor, causing sustained downstream signaling by phospholipase Cγ, PI(3)K, and MAPK-ERK to promote cell survival, migration, and invasion (Fig. 4) (Suyama et al. 2002). Shedding of N-cadherin by proteases might stimulate FGFR signaling in neighboring cells. Furthermore, presenilin 1 (PS1)/γ-secretase cleaves N-cadherin in the cytoplasmic part to release a free carboxy-terminal 35-kDa fragment, which translocates to the nucleus and binds the transcriptional coactivator CBP (CREB binding protein) (Fig. 4) (Marambaud et al. 2003). This targets CBP for degradation and represses CBP/CREB-mediated transcription.

P-cadherin (CDH3) is expressed abnormally in basal-like breast carcinomas and in most pancreatic ductal adenocarcinomas (Table 1) (Paredes et al. 2005; Taniuchi et al. 2005). The p120-catenin binding juxtamembrane domain (JMD) of P-cadherin could be crucial for the enhancement of invasion and migration caused by P-cadherin overexpression in these cancers (Paredes et al. 2004). The underlying mechanism is unclear, but small sequence differences between the JMDs of E-cadherin and P-cadherin might result in differential binding of interacting proteins and modulation of p120ctn functionality in case of association with P-cadherin (Paredes et al. 2007). In melanoma cell lines P-cadherin can counteract invasion (Van Marck et al. 2005), which appears to agree with the shift in human melanoma from active, membranous to cytoplasmic P-cadherin (Sanders et al. 1999).

VE-CADHERIN (CADHERIN-5) AND OTHER TYPE-II CADHERINS

The endothelial specific expression of VE-cadherin (cadherin-5) plays a key role in regulating vascular morphology and stability (Fig. 4) (Dejana 2004; reviewed by Cavallaro et al. 2006). Microarray studies revealed that VE-cadherin is overexpressed in aggressive human cutaneous and uveal melanoma cells, but not in nonaggressive or poorly aggressive melanoma (Table 1) (Bittner et al. 2000). On the other hand, formation of patterned networks of matrix-rich tubular structures in three-dimensional culture is characteristic of highly aggressive but not poorly aggressive melanoma cells (Maniotis et al. 1999). This vascular mimicry is clinically significant and increases the risk of metastatic disease (Warso et al. 2001). Aggressive melanoma cells in which VE-cadherin was repressed could not form vasculogenic-like networks (Hendrix et al. 2001), suggesting that tumor-associated misexpression of VE-cadherin (observed in melanoma cells) is instrumental in mimicking endothelial cells to form vasculogenic networks. Recently, VE-cadherin was found to be targeted to proteasome-mediated degradation by a transmembrane ubiquitin ligase of the Kaposi's sarcoma-associated herpesvirus (Mansouri et al. 2008). This resulted in a reduction of the steady-state levels of interacting catenins and in rearrangement of the actin cytoskeleton. This reprogramming and differentiation of endothelial cells accompanies loss of endothelial barrier function and vascular leakage, which could be relevant during Kaposi's sarcoma tumor formation (Mansouri et al. 2008).

Antiangiogenic tumor therapy is lately receiving much attention in the form of anti-VEGF therapy (Loges et al. 2008). Also VE-cadherin has become a target for inhibition of pathological angiogenesis. Monoclonal antibodies directed against VE-cadherin could inhibit angiogenesis and reduce tumor growth of experimental hemangiomas and gliomas (Corada et al. 2002; Cavallaro et al. 2006). Specificity for angiogenic tumor vasculature was further enhanced by targeting a VE-cadherin epitope that is at least partly engaged in trans-dimer formation, and thus is masked on established vessels (May et al. 2005). This therapy is antiangiogenic during neovascularization because de novo adherens junction formation is inhibited. The potential value of anti-VE-cadherin therapy was further shown by using a cyclic peptide directed against VE-cadherin to inhibit oxygen-induced retinal neovascularization (Navaratna et al. 2008). Regarding angiogenesis, it is worth mentioning that endothelial cells also express N-cadherin, R-cadherin (CDH4), and T-cadherin (cadherin-13; see the following discussion). N-cadherin is involved in recruitment (or binding) of pericytes, which stabilizes blood vessel formation (Cavallaro et al. 2006). The functions of T-cadherin and R-cadherin in endothelial cells are unclear. In gastrointestinal cancer, the *CDH4* promoter is often methylated early in tumor progression, indicating that R-cadherin might be a tumor suppressor (Miotto et al. 2004).

Recently, the cytogenetically normal genome of a typical acute myeloid leukemia sample was fully sequenced and compared with the genome of normal skin cells of the same patient (Ley et al. 2008). Only ten genes with acquired mutations were identified, including a nonsense mutation in the ectodomain of cadherin-24 (CDH24). This type-II cadherin (Fig. 1) (Hulpiau and van Roy 2009) has received little attention. It is widely expressed in human tissues, binds p120 and β-catenin, and can mediate strong cell–cell adhesion (Katafiasz et al. 2003).

CADHERIN-13 (T-CADHERIN)

Cadherin-13 (CDH13), also called T(runcated)- or H(eart)-cadherin, is most peculiar. It is the only known cadherin that is membrane-anchored via a glycosylphosphatidylinositol (GPI) anchor instead of transmembrane domains (Figs. 1 and 4). This explains its unusual location at apical instead of basolateral membranes in polarized epithelial cells (Ranscht and Dours-Zimmermann 1991). Nevertheless, the extracellular domain of CDH13 shows significant homology with the ectodomain of classical cadherins (Hulpiau and van Roy 2009), and its binding is indeed homophilic though relatively weak because of structural peculiarities (Vestal and Ranscht 1992; Dames et al. 2008). This binding weakness suggests that CDH13 functions mainly in signaling.

The human *CDH13* gene is located on chromosome 16q24, and because it is frequently silenced in many different cancers, it has long been considered to be a tumor suppressor (Table 1) (Lee 1996; reviewed by Takeuchi and Ohtsuki 2001). Also, *CDH13* LOH is frequently observed in major cancer types. In several of these studies, treatment of tumor cell lines with a demethylating agent or a histone deacetylase inhibitor reactivated *CDH13* expression. Methylation of the *CDH13* promoter was associated with recurrence of nonsmall-cell lung cancers after surgery (Brock et al. 2008). A genome-wide study of association of single-nucleotide polymorphisms (SNPs) with prostate cancer revealed four candidate susceptibility genes, including *CDH13* (Thomas et al. 2008). *Cdh13* expression was up-regulated in castrated mice but was down-regulated following androgen replacement, which is in line with androgen-responsive elements in its promoter region (Wang et al. 2007). Exogenous expression of CDH13 in DU145, a cell line derived from a prostate cancer metastasis, reduced tumorigenicity, whereas knockdown of *CDH13* transcripts in BPH1, derived from benign prostate hyperplasia, facilitated tumorigenesis (Wang et al. 2007).

Other experiments point to several growth inhibitory effects of CDH13 (Lee 1996; Huang et al. 2003; Mukoyama et al. 2005; Yan et al. 2008), but the underlying mechanism is unclear. Overexpression of CDH13 in carcinoma or glioma cells causes G2/M cell cycle arrest (Huang et al. 2003). In the glioma study, arrest depended on expression of the CDK inhibitor $p21^{CIP1}$ but not on expression of p53. Another study reported that CDH13 interferes with c-Jun oncogenic activity (Chan et al. 2008). Moreover, in vitro invasiveness of breast cancer cells was inhibited by exogenous expression of CDH13 (Lee et al. 1998). However, exceptions do occur: One highly invasive hepatoma cell line autonomously produces high levels of CDH13, and knockdown of this expression reduced invasion and migration activity (Riou et al. 2006).

The recently reported *Cdh13* knockout mouse did not show spontaneous tumors or other overt phenotypic abnormality, but crossing it with the MMTV-polyomavirus-middle-T

(MMTV-PyV-MT) transgenic mouse resulted in offspring with CDH13-deficient mammary tumors that were growth restricted and less vascularized (Hebbard et al. 2008). Nevertheless, these CDH3-deficient tumors had greater metastatic potential than wild-type MMTV-PyV-MT tumors, which might be explained by hypoxia-driven EMT (see also previous discussion). The discrepancy of these in vivo findings with the proposed tumor suppressor function of CDH13 could be partly because of the contribution of tumor vasculature.

Indeed, CDH13 was found to be induced in microvessels associated with tumors and metastases, whereas in normal tissues it was expressed only in larger vessels (Wyder et al. 2000). In hepatocellular carcinoma (HCC), CDH13 is silenced in the cancer cells but overexpressed in the intratumoral endothelial cells (Adachi et al. 2006; Riou et al. 2006; Chan et al. 2008). In a two-chamber system, angiogenic factor FGF-2 derived from HCC cells induced CDH13 expression in sinusoidal endothelial cells (Adachi et al. 2006). Overexpression of CDH13 in a human microvascular endothelial cell line stimulated invasion of these nontumoral cells into multicellular spheroids of undifferentiated melanoma cells lacking CDH13 expression (Ghosh et al. 2007). Overexpression and homophilic ligation of CDH13 in endothelial cells stimulates their proliferation and migration, enhances their protection against oxidative stress, and stimulates angiogenesis under pathological conditions, apparently by potentiating factors such as VEGF (Philippova et al. 2006). Growth of CDH13-expressing MMTV-PyV-MT tumors transplanted in mammary fat pads of *Cdh13*-null recipient mice was slower than in wild-type mice, consolidating the nonautonomous effect of CDH13 on tumor growth (Hebbard et al. 2008).

How does CDH13 act as a stromal factor favoring tumor growth whereas the tumor cells themselves tend to silence its expression? On homophilic ligation of CDH13 on surfaces of endothelial cells it becomes linked to different interacting proteins, including Grp78/BiP (Philippova et al. 2008). The latter is normally retained within the ER but becomes surface exposed under pathological conditions. Endothelial cell survival is apparently promoted by surface exposed Grp78 *via* the phosphatidylinositol 3-kinase and Akt kinase pathway (Fig. 4). Another interesting CDH13 partner in endothelial cells is $\alpha_v\beta_3$ integrin. CDH13 colocalizes with this integrin mostly at the leading edge of migrating cells, which agrees with a proangiogenic CDH13 effect (Philippova et al. 2008).

CDH13 was also identified as a "third adiponectin receptor" (reviewed by Takeuchi et al. 2007). Adiponectin, secreted by adipocytes, is a hormone with insulin-sensitizing effects. In serum it exists as trimers, hexamers, or high molecular weight oligomers. CDH13 probably serves as coreceptor with the more classic adiponectin-receptor-1 and -2 for binding hexameric and larger adiponectin forms (Hug et al. 2004). The possible relevance of this coreceptor function of CDH13 to cancer is supported by evidence from the MMTV-PyV-MT mammary tumor system (Hebbard et al. 2008). When *Cdh13* was ablated, adiponectin was no longer associated with the tumor vasculature and its serum levels increased. As adiponectin can sequester various growth factors, its CDH13-dependent accumulation in a tumoral microenvironment may favor both tumor-associated neoangiogenesis and tumor growth.

PROTOCADHERINS AND CANCER

As recently reviewed (Hulpiau and van Roy 2009), protocadherins differ in many aspects from classic cadherins (Figs. 1 and 3). They can be subdivided into clustered and nonclustered protocadherins, reflecting the localization of their genes in vertebrate genomes. Transcripts of clustered protocadherins are composed of one of many variable exons combined with a few constant exons. Three subfamilies of clustered protocadherins are discerned in mammals: α-, β-, and γ-protocadherins. Evidence for the involvement of clustered protocadherins in cancer is limited. In a global genomic analysis of pancreatic cancers, more than 23,000 pancreatic cancer-derived transcripts were sequenced (Jones et al. 2008)

(Table 1). This revealed cancer-associated somatic missense mutations in some members of the clustered protocadherin subfamilies, but their functional relevance remains unclear. Other studies pointed at the involvement of methylation of the *PCDHGA11* or *PCDHGB6* promoter region in the development of astrocytomas or breast cancer (Miyamoto et al. 2005; Waha et al. 2005).

Considerably more information is available about the nonclustered protocadherins, a family comprising 10 members in man and other mammals (Hulpiau and van Roy 2009). In the same large-scale analysis of pancreatic cancers (Jones et al. 2008), missense mutations were found in protocadherin-9, -17, and -18. Expression of protocadherin-1 (PCDH1) in medulloblastoma patients predicted survival (Table 1) (Neben et al. 2004).

In breast cancer, protocadherin-8 (PCDH8) was proposed as a candidate tumor suppressor (Table 1) (Yu et al. 2008), as several somatic missense mutations affecting the ectodomain or a conserved motif in the cytoplasmic domain were found in tumor samples and cell lines (Yu et al. 2008). Proliferation and migration of breast cancer cell lines was inhibited by forced expression of wild type PCDH8 but not by two of the ectodomain missense mutations. Moreover, the E146K mutation, assumed to affect a calcium-coordinating amino acid in PCDH8, conferred a transformed phenotype on "normal" MCF10A cells (Yu et al. 2008). Contrary to common descriptions, PCDH8 is not the homolog but the close paralog of the paraxial protocadherin (PAPC) of *Xenopus laevis* (Hulpiau and van Roy 2009). So one should be careful when interpreting PCDH8 functions based on the numerous reported findings on *Xenopus* PAPC (e.g., Unterseher et al. 2004). Interestingly, neural stimulation induces a short isoform of Arcadlin (= rat Pcdh8) in hippocampal synapses, where it associates with N-cadherin and cadherin-11 (Fig. 4) (Yasuda et al. 2007). Induction of this isoform activated a MAPKKK, TAO2β, triggering a signaling pathway leading to endocytosis of N-cadherin. Whether the putative tumor suppressor activity of PCDH8 is based on a similar phenomenon is unknown.

Within the same protocadherin subfamily, protocadherin-10 (PCDH10 or OL-protocadherin) and protocadherin-20 (PCDH20) were also proposed as tumor suppressors (Table 1). First, strong evidence was reported for frequent epigenetic inactivation of PCDH10 in various human cancers but not in matched normal tissues (Miyamoto et al. 2005; Ying et al. 2006; Ying et al. 2007; Yu et al. 2009). For gastric cancer, *PCDH10* methylation was detected at early stages of carcinogenesis and was associated with poor prognosis (Yu et al. 2009). Ectopic PCDH10 expression in nasopharyngeal and esophageal carcinoma cell lines reduced clonogenicity, anchorage-independent growth, migration potential, and in vitro invasion into Matrigel (Ying et al. 2006). Re-expression of PCDH10 in gastric cancer cells inhibited tumor growth in mice, induced cell apoptosis, and inhibited cell invasion (Yu et al. 2009). Also, the promoter of the *PCDH20* gene is frequently methylated in human tumors, specifically in nonsmall-cell lung cancers (Imoto et al. 2006). Again, ectopic expression in a carcinoma cell line reduced clonogenicity and anchorage-independent growth. The molecular mechanisms underlying these tumor suppressive functions of protocadherins are unknown.

In contrast to the previously mentioned protocadherins, protocadherin-11Y (PCDH11Y, also named protocadherin-PC) is a candidate proto-oncogene. Human *PCDH11X* and *PCDH11Y* are highly homologous and are located in the hominid-specific nonpseudoautosomal homologous region Xq21.3/Yp11.2, which had undergone duplication by transposition from the X to the Y chromosome after the divergence of the hominid and chimpanzee lineages. In other words, *PCDH11Y* is found exclusively in man (Blanco et al. 2000). Transcription occurs mainly in the brain. The encoded PCDH11 proteins belong to the δ1-protocadherin subfamily (Vanhalst et al. 2005).

The start codon in the *PCDH11X* transcript is absent from the *PCDH11Y* transcript because of a short genomic deletion. Use of an AUG further downstream results in a PCDH11Y protein lacking the signal peptide and therefore remaining cytoplasmic (Fig. 4) (Chen et al. 2002). Transcription of *PCDH11Y* is upregulated in apoptosis-resistant cell variants compared with the parental prostate carcinoma cell line LNCaP. PCDH11Y mRNA levels are increased in androgen-refractory prostate cancers, and its ectopic expression in human prostate cancer cell lines induced anchorage-independent growth in hormone-deprived medium and tumorigenicity in castrated nude mice (Terry et al. 2006). Its expression correlated with occurrence of nuclear β-catenin and with increased TCF/Lef transcriptional activity (Yang et al. 2005; Terry et al. 2006). Recently, an interesting link was established between Wnt/β-catenin signaling and androgen independence during prostate cancer progression (Placencio et al. 2008; Schweizer et al. 2008; Wang et al. 2008). Although β-catenin coimmunoprecipitated with PCDH11Y (Chen et al. 2002), the interaction between the two proteins is not direct (our unpublished data).

OTHER CADHERIN-RELATED MOLECULES AND CANCER

Expression of the mRNA and protein of a cadherin-related protein, named protocadherin LKC or PC-LKC and later renamed protocadherin-24 (Fig. 1), was reported to be reduced in carcinomas derived from the colon, liver, and kidney (Table 1) (Okazaki et al. 2002). Ectopic expression of this cadherin-related protein in human colorectal carcinoma cells (HCT116) reduced tumorigenicity in nude mice by an unknown mechanism.

Another interesting cadherin-related protein is Fat4, which has 34 cadherin repeats and is the homologue of *Drosophila* Fat (Fig. 1) (Hulpiau and van Roy 2009). In the fly, Fat modulates cell contact inhibition and organ size by functioning upstream of the (Hippo/Salvador)-(Warts/Mats)-Yorkie-Scalloped signaling pathway (reviewed in Harvey and Tapon 2007; Zeng and Hong 2008; Zhao et al. 2008). Several components of this pathway are protein kinases or adaptors, whereas Yorkie is a transcriptional coactivator that is inactivated by cytoplasmic retention on Hippo activation, and Scalloped belongs to the TEAD transcription factor family. Fat appears to be upstream in this pathway and to signal through Expanded, a Merlin-like protein, or through inhibition of Dachs (not to be confused with Dachsous or Ds), an atypical myosin that inhibits Warts. The mechanism by which Expanded activates the Hippo/Salvador complex is unknown. Anyhow, removal of Fat induces overgrowth of imaginal disc tissues. Also, Fat functions in planar cell polarity processes. There is no evidence for homophilic binding in *trans* between cells expressing Fat (Matakatsu and Blair 2006). In contrast, the ectodomain of Fat can interact with the ectodomain of another huge cadherin-related protein, Dachsous (Ds), on juxtaposed cell surfaces (Figs. 1 and 4). The Fat-Ds interaction can be compared with ligand-receptor binding rather than to cell–cell adhesion, and it might function mainly to activate the intracellular domain of Fat. The intracellular domain of Ds seems to contribute to growth control also independently of the Fat intracellular domain (Matakatsu and Blair 2006).

Each of the Hippo pathway components mentioned above is conserved in mammals, in which the pathway can be abbreviated as (Mst1/Mst2)-(LATS/Mob/WW45)-(YAP/TAZ)-TEAD (Fig. 4). Several of these proteins or their corresponding genes are mutated, silenced, or overexpressed in human tumors and mouse cancer models (reviewed in Harvey and Tapon 2007). A putative role for mammalian Fat4 as tumor suppressor remained unproven until recently, when Qi et al. (2009) reported that biallelic Fat4 inactivation conferred tumorigenicity to a mouse mammary epithelial cell line (Table 1). Re-expression of Fat4 in Fat4-deficient tumor cells suppressed their tumorigenicity. Moreover, Fat4 expression was lost in many human breast tumor cell lines and primary tumors, and this silencing was associated with human Fat4 promoter methylation. Nonetheless, loss of Fat4 by itself is insufficient to generate tumors, as *Fat4* knockout mice show various defects, explainable by defective planar cell polarity signaling, but no tumors

(Saburi et al. 2008). No association with cancer has been reported for mammalian Dachsous homologs.

The human RET protein is a proto-oncogenic receptor tyrosine kinase with cadherin-like domains in the ectodomain. Bioinformatic analysis and molecular modeling revealed the presence of 4 cadherin-like domains in vertebrate Ret proteins (Fig. 1) (Anders et al. 2001). It was suggested that Ret resulted from recombination of a tyrosine kinase receptor gene and an ancestral cadherin gene during early metazoan evolution. Whether RET can be considered a cell adhesion receptor is undetermined, but its involvement in cancer is beyond doubt (reviewed by Kondo et al. 2006; Zbuk and Eng 2007). Glial cell derived neurotrophic factor (GDNF) ligands and GNDF receptor-α bind the extracellular domain of RET, leading to RET dimerization and autophosphorylation. Constitutive activation can result from missense mutations in Cys residues in the ectodomain or from fusion to PTC in a chimeric oncoprotein. The resulting cancers are various sporadic thyroid carcinomas as well as familial cancers such as the multiple endocrine neoplasias MEN2A and MEN2B (Table 1). The role of the cadherin domains in the RET ectodomain is unknown.

CONCLUSIONS AND PERSPECTIVES

The effectiveness of conventional anticancer treatment is often limited by serious toxicity. Even after 30 years of conventional therapies (surgery, irradiation, chemotherapy, and combinations thereof), overall survival from metastatic cancer has not improved much. Among the causes of therapeutic failure are the lack of adequate tumor specificity and the development of drug resistance. Future anticancer drugs should be made more effective and more selective: They should act on or somehow exploit the specific molecular abnormalities driving malignant progression.

Research during the past 20 years has shown that dysregulation of cadherins contributes to different aspects of cancer progression, including drug resistance, angiogenesis, cancer cell invasion, and metastasis. Thus, cadherins and their regulators can become valuable diagnostic and prognostic indicators, as well as potential therapeutic targets.

The most compelling data for the involvement of the cadherin family in cancer progression are available for E-cadherin. The causal relationship between E-cadherin dysfunction and cancer progression has been convincingly shown both in vitro and in vivo. Furthermore, the clinical relevance of E-cadherin deficiency has been confirmed by immunohistochemical evidence of changes in E-cadherin expression and localization in most human cancers (reviewed in Strumane et al. 2004). Consequently, circulating cadherin fragments received as much attention as potential cancer markers (reviewed in De Wever et al. 2007). Indeed, multiple cadherins are targets for ectodomain shedding because of elevated protease activity in the tumoral microenvironment. However, current experimental data support the use only of soluble N-cadherin as a circulation tumor marker for prostate and pancreas cancers. Expression of N-cadherin and cadherin-11 might be used as indicators of unfavorable diagnosis or poor prognosis because they are normally not expressed in epithelia but are frequently up-regulated in invading cancer cells. A cyclic pentapeptide, ADH-1, has been developed as an extracellular N-cadherin antagonist for use as a systemic anticancer agent (reviewed by De Wever et al. 2007; Mariotti et al. 2007). ADH-1 has entered clinical testing, and has been reported to have significant antitumor activity in a mouse model for pancreatic cancer (Shintani et al. 2008) and to strongly potentiate chemotherapy of human melanoma xenografts (Augustine et al. 2008).

Various aberrations that negatively affect E-cadherin can occur during cancer progression, such as mutations, promoter methylation, and transcriptional repression. In sporadic diffuse gastric cancer and possibly in HDGC, *CDH1* mutations are accompanied by epigenetic silencing of the second allele (Grady et al. 2000). This opens the way for future E-cadherin-directed epigenetic therapy by using DNA methylation inhibitors (e.g.,

5AzaC) and histone deacytelase inhibitors (e.g., trichostatin A; Wu et al. 2007), both of which have been shown to reactivate E-cadherin expression in vitro. It is noteworthy that several protocadherins were recently reported to be silenced by promoter methylation in cancers. Nonetheless, one has to consider that reactivating expression of cell–cell adhesion molecules in cancer cells may trigger intrinsic or acquired resistance to anticancer drug-induced apoptosis (St Croix and Kerbel 1997).

Further, transcriptional deregulation of epithelial differentiation has been much explored lately (Peinado et al. 2007). As pointed out previously, different transcriptional repressors of E-cadherin have been identified, such as Twist, Snail, and ZEB family members and their respective miRNA regulators. Because of the central role of these transcription factors in EMT and their possible contribution to tumorigenesis, they can be regarded as candidates for molecular targeting (Berx et al. 2007). Specifically inhibiting them by small interfering RNA (siRNA) or by chemical compounds could improve the management of malignant cancer by enhancing E-cadherin expression. Expression of these EMT-driving transcriptional repressors has also been associated with anti-apoptotic functions, poor response to pharmacological treatments, and chemoresistance (Yauch et al. 2005; Shah and Gallick 2007). Identification of an EMT signature in tumors could therefore be instrumental to developing more effective therapeutics. For this, we should carefully investigate the aberrations in signaling pathways driving EMT in different tumor types to enable use of pathway-specific inhibitors as anti-invasive cancer therapies. For instance, TGF-β is a strong inducer of EMT: It stabilizes and induces the expression of different E-cadherin repressors and can enhance the cancer stem cell phenotype (Mani et al. 2008). Long-term use of TGF-β antagonists has been shown to be effective in reducing metastasis in experimental mouse models (Akhurst 2002) and could therefore be a means to block pathological EMT in cancer.

In conclusion, it is most important to investigate further the pleiotropic effects of different cadherins, protocadherins, and other cadherin-related molecules during cancer progression. This implies unraveling complex regulatory pathways, including molecular inducers, signaling pathways, and molecular effectors, under both physiological and pathological conditions. This is well advanced for classic cadherins, but for numerous other members of the cadherin superfamily, it is still in its infancy. Such scrutiny might eventually lead to development of better tailored anticancer therapies based on cadherin-targeting strategies.

ACKNOWLEDGMENTS

This research was funded by grants from the FWO, the Geconcerteerde Onderzoeksacties of Ghent University, the Belgian Federation against Cancer, the Association for International Cancer Research (Scotland), and FP7 (TUMIC) of the European Union. We acknowledge Dr. Amin Bredan for critical reading of the manuscript, and the members of our research groups for valuable discussions.

REFERENCES

Adachi Y, Takeuchi T, Sonobe H, Ohtsuki Y. 2006. An adiponectin receptor, T-cadherin, was selectively expressed in intratumoral capillary endothelial cells in hepatocellular carcinoma: Possible cross talk between T-cadherin and FGF-2 pathways. *Virchows Archiv* **448:** 311–318.

Akhurst RJ. 2002. TGF-β antagonists: Why suppress a tumor suppressor? *J Clin Invest* **109:** 1533–1536.

Al-Hajj M, Wicha MS, Benito-Hernandez A, Morrison SJ, Clarke MF. 2003. Prospective identification of tumorigenic breast cancer cells. *Proc Natl Acad Sci USA* **100:** 3983–3988.

Alpaugh ML, Tomlinson JS, Shao ZM, Barsky SH. 1999. A novel human xenograft model of inflammatory breast cancer. *Cancer Res* **59:** 5079–5084.

Anders J, Kjaer S, Ibanez CF. 2001. Molecular modeling of the extracellular domain of the RET receptor tyrosine kinase reveals multiple cadherin-like domains and a calcium-binding site. *J Biol Chem* **276:** 35808–35817.

Augustine CK, Yoshimoto Y, Gupta M, Zipfel PA, Selim MA, Febbo P, Pendergast AM, Peters WP, Tyler DS. 2008. Targeting N-cadherin enhances antitumor activity of cytotoxic therapies in melanoma treatment. *Cancer Res* **68:** 3777–3784.

Becker K-F, Atkinson MJ, Reich U, Huang H-H, Nekarda H, Siewert JR, Höfler H. 1993. Exon skipping in the E-cadherin gene transcript in metastatic human gastric carcinomas. *Hum Mol Genet* **2:** 803–804.

Becker KF, Atkinson MJ, Reich U, Becker I, Nekarda H, Siewert JR, Höfler H. 1994. E-cadherin gene mutations provide clues to diffuse type gastric carcinomas. *Cancer Res* **54:** 3845–3852.

Berx G, Cleton-Jansen A-M, Nollet F, de Leeuw WJF, van de Vijver MJ, Cornelisse C, van Roy F. 1995a. E-cadherin is a tumor/invasion suppressor gene mutated in human lobular breast cancers. *EMBO J* **14:** 6107–6115.

Berx G, Staes K, van Hengel J, Molemans F, Bussemakers MJG, van Bokhoven A, van Roy F. 1995b. Cloning and characterization of the human invasion suppressor gene E-cadherin (CDH1). *Genomics* **26:** 281–289.

Berx G, Cleton-Jansen A-M, Strumane K, de Leeuw WJF, Nollet F, van Roy FM, Cornelisse C. 1996. E-cadherin is inactivated in a majority of invasive human lobular breast cancers by truncation mutations throughout its extracellular domain. *Oncogene* **13:** 1919–1925.

Berx G, Becker K-F, Höfler H, van Roy F. 1998. Mutation Update: Mutations of the human E-cadherin (CDH1) gene. *Hum Mutat* **12:** 226–237.

Berx G, Raspe E, Christofori G, Thiery JP, Sleeman JP. 2007. Pre-EMTing metastasis? Recapitulation of morphogenetic processes in cancer. *Clin Exp Metastas* **24:** 587–597.

Birchmeier W, Behrens J. 1994. Cadherin expression in carcinomas: Role in the formation of cell junctions and the prevention of invasiveness. *Biochim Biophys Acta Rev Cancer* **1198:** 11–26.

Bittner M, Meitzer P, Chen Y, Jiang Y, Seftor E, Hendrix M, Radmacher M, Simon R, Yakhini Z, BenDor A, et al. 2000. Molecular classification of cutaneous malignant melanoma by gene expression profiling. *Nature* **406:** 536–540.

Blanco P, Sargent CA, Boucher CA, Mitchell M, Affara NA. 2000. Conservation of PCDHX in mammals; expression of human X/Y genes predominantly in brain. *Mamm Genome* **11:** 906–914.

Blanco MJ, Moreno-Bueno G, Sarrio D, Locascio A, Cano A, Palacios J, Nieto MA. 2002. Correlation of Snail expression with histological grade and lymph node status in breast carcinomas. *Oncogene* **21:** 3241–3246.

Bonnomet A, Nawrocki-Raby B, Strumane K, Gilles C, Kiletzky C, Berx G, van Roy F, Polette M, Birembaut P. 2008. The E-cadherin-repressed hNanos1 gene induces tumor cell invasion by upregulating MT1-MMP expression. *Oncogene* **27:** 3692–3699.

Boussadia O, Kutsch S, Hierholzer A, Delmas V, Kemler R. 2002. E-cadherin is a survival factor for the lactating mouse mammary gland. *Mech Develop* **115:** 53–62.

Bracken CP, Gregory PA, Kolesnikoff N, Bert AG, Wang J, Shannon MF, Goodall GJ. 2008. A double-negative feedback loop between ZEB1-SIP1 and the microRNA-200 family regulates epithelial-mesenchymal transition. *Cancer Res* **68:** 7846–7854.

Brock MV, Hooker CM, Ota-Machida E, Han Y, Guo MZ, Ames S, Glockner S, Piantadosi S, Gabrielson E, Pridham G, et al. 2008. DNA methylation markers and early recurrence in stage I lung cancer. *New Engl J Med* **358:** 1118–1128.

Brooks-Wilson AR, Kaurah P, Suriano G, Leach S, Senz J, Grehan N, Butterfield YSN, Jeyes J, Schinas J, Bacani J, et al. 2004. Germline E-cadherin mutations in hereditary diffuse gastric cancer: Assessment of 42 new families and review of genetic screening criteria. *J Med Genet* **41:** 508–517.

Burk U, Schubert J, Wellner U, Schmalhofer O, Vincan E, Spaderna S, Brabletz T. 2008. A reciprocal repression between ZEB1 and members of the miR-200 family promotes EMT and invasion in cancer cells. *EMBO Rep* **9:** 582–589.

Bussemakers MJG, Van Bokhoven A, Tomita K, Jansen CFJ, Schalken JA. 2000. Complex cadherin expression in human prostate cancer cells. *Int J Cancer* **85:** 446–450.

Caldas C, Carneiro F, Lynch HT, Yokota J, Wiesner GL, Powell SM, Lewis FR, Huntsman DG, Pharoah PDP, Jankowski JA, et al. 1999. Familial gastric cancer: overview and guidelines for management. *J Med Genet* **36:** 873–880.

Caldas C, Carneiro F, Lynch HT, Yokota J, Wiesner GL, Powell SM, Lewis FR, Huntsman DG, Pharoah PDP, Jankowski JA, et al. 2000. The transcription factor Snail controls epithelial-mesenchymal transitions by repressing E-cadherin expression. *Nat Cell Biol* **2:** 76–83.

Cano A, Perez-Moreno MA, Rodrigo I, Locascio A, Blanco MJ, del Barrio MG, Portillo F, Nieto MA. 2000. The transcription factor Snail controls epithelial-mesenchymal transitions by repressing E-cadherin expression. *Nat Cell Biol* **2:** 76–83.

Carneiro F, Oliveira C, Suriano G, Seruca R. 2008. Molecular pathology of familial gastric cancer, with an emphasis on hereditary diffuse gastric cancer. *J Clin Pathol* **61:** 25–30.

Cavallaro U, Liebner S, Dejana E. 2006. Endothelial cadherins and tumor angiogenesis. *Exp Cell Res* **312:** 659–667.

Chan DW, Lee JMF, Chan PCY, Ng IOL. 2008. Genetic and epigenetic inactivation of T-cadherin in human hepatocellular carcinoma cells. *Int J Cancer* **123:** 1043–1052.

Chang HW, Chow V, Lam KY, Wei WI, Yuen APW. 2002. Loss of E-cadherin expression resulting from promoter hypermethylation in oral tongue carcinoma and its prognostic significance. *Cancer* **94:** 386–392.

Chen MW, Vacherot F, De La Taille A, Gil-Diez-De-Medina S, Shen R, Friedman RA, Burchardt M, Chopin DK, Buttyan R. 2002. The emergence of protocadherin-PC expression during the acquisition of apoptosis-resistance by prostate cancer cells. *Oncogene* **21:** 7861–7871.

Cheng CW, Wu PE, Yu JC, Huang CS, Yue CT, Wu CW, Shen CY. 2001. Mechanisms of inactivation of E-cadherin in breast carcinoma: Modification of the two-hit hypothesis of tumor suppressor gene. *Oncogene* **20:** 3814–3823.

Christoffersen NR, Silahtaroglu A, Orom UA, Kauppinen S, Lund AH. 2007. miR-200b mediates post-transcriptional repression of ZFHX1B. *RNA* **13:** 1172–1178.

Christofori G, Semb H. 1999. The role of the cell-adhesion molecule E-cadherin as a tumour-suppressor gene. *Trends Biochem Sci* **24:** 73–76.

Chu K, Cheng CJ, Ye X, Lee YC, Zurita AJ, Chen DT, Yu-Lee LY, Zhang S, Yeh ET, Hu MC, et al. 2008. Cadherin-11 promotes the metastasis of prostate cancer cells to bone. *Mol Cancer Res* **6:** 1259–1267.

Cleton-Jansen AM, Callen DF, Seshadri R, Goldup S, McCallum B, Crawford J, Powell JA, Settasatian C, van Beerendonk H, Moerland EW, et al. 2001. Loss of heterozygosity mapping at chromosome arm 16q in 712 breast tumors reveals factors that influence delineation of candidate regions. *Cancer Res* **61:** 1171–1177.

Corada M, Zanetta L, Orsenigo F, Breviario F, Lampugnani MG, Bernasconi S, Liao F, Hicklin DJ, Bohlen P, Dejana E. 2002. A monoclonal antibody to vascular endothehal-cadherin inhibits tumor angiogenesis without side effects on endothelial permeability. *Blood* **100:** 905–911.

Covington MD, Burghardt RC, Parrish AR. 2006. Ischemia-induced cleavage of cadherins in NRK cells requires MT1-MMP (MMP-14). *Am J Physiol Renal Physiol* **290:** F43-F51.

Dames SA, Bang E, Haussinger D, Ahrens T, Engel J, Grzesiek S. 2008. Insights into the low adhesive capacity of human T-cadherin from the NMR structure of Its N-terminal extracellular domain. *J Biol Chem* **283:** 23485–23495.

Davies G, Jiang WG, Mason MD. 2001. Matrilysin mediates extracellular cleavage of E-cadherin from prostate cancer cells: A key mechanism in hepatocyte growth factor/scatter factor-induced cell-cell dissociation and in vitro invasion. *Clin Cancer Res* **7:** 3289–3297.

De Wever O, Derycke L, Hendrix A, De Meerleer G, Godeau F, Depypere H, Bracke M. 2007. Soluble cadherins as cancer biomarkers. *Clin Exp Metastas* **24:** 685–697.

Dejana E. 2004. Endothelial cell-cell junctions: Happy together. *Nat Rev Mol Cell Biol* **5:** 261–270.

Derksen PWB, Liu XL, Saridin F, vanderGulden H, Zevenhoven J, Evers B, vanBeijnum JR, Griffioen AW, Vink J, Krimpenfort P, et al. 2006. Somatic inactivation of E-cadherin and p53 in mice leads to metastatic lobular mammary carcinoma through induction of anoikis resistance and angiogenesis. *Cancer Cell* **10:** 437–449.

Droufakou S, Deshmane V, Roylance R, Hanby A, Tomlinson I, Hart IR. 2001. Multiple ways of silencing E-cadherin gene expression in lobular carcinoma of the breast. *Int J Cancer* **92:** 404–408.

Dunbier A, Guilford P. 2001. Hereditary diffuse gastric cancer. *Adv Cancer Res* **83:** 55–65.

Eger A, Aigner K, Sonderegger S, Dampier B, Oehler S, Schreiber M, Berx G, Cano A, Beug H, Foisner R. 2005. DeltaEF1 is a transcriptional repressor of E-cadherin and regulates epithelial plasticity in breast cancer cells. *Oncogene* **24:** 2375–2385.

Elloul S, Elstrand MB, Nesland JM, Trope CG, Kvalheim G, Goldberg I, Reich R, Davidson B. 2005. Snail, Slug, and Smad-interacting protein 1 as novel parameters of disease aggressiveness in metastatic ovarian and breast carcinoma. *Cancer* **103:** 1631–1643.

Endo K, Ashida K, Miyake N, Terada T. 2001. E-cadherin gene mutations in human intrahepatic cholangiocarcinoma. *J Pathol* **193:** 310–317.

Esteban MA, Tran MG, Harten SK, Hill P, Castellanos MC, Chandra A, Raval R, O'Brien T, Maxwell PH. 2006. Regulation of E-cadherin expression by VHL and hypoxia-inducible factor. *Cancer Res* **66:** 3567–3575.

Evans AJ, Russell RC, Roche O, Burry TN, Fish JE, Chow VWK, Kim WY, Saravanan A, Maynard MA, Gervais ML, et al. 2007. VHL promotes E2 box-dependent E-cadherin transcription by HIF-mediated regulation of SIP1 and snail. *Mol Cell Biol* **27:** 157–169.

Frixen UH, Behrens J, Sachs M, Eberle G, Voss B, Warda A, Löchner D, Birchmeier W. 1991. E-cadherin-mediated cell-cell adhesion prevents invasiveness of human carcinoma cells. *J Cell Biol* **113:** 173–185.

Fujita Y, Krause G, Scheffner M, Zechner D, Leddy HEM, Behrens J, Sommer T, Birchmeier W. 2002. Hakai, a c-Cbl-like protein, ubiquitinates and induces endocytosis of the E-cadherin complex. *Nat Cell Biol* **4:** 222–231.

Ghosh S, Josh MB, Ivanov D, Feder-Mengus C, Spagnoli GC, Martin I, Erne P, Resink TJ. 2007. Use of multicellular tumor spheroids to dissect endothelial cell-tumor cell interactions: A role for T-cadherin in tumor angiogenesis. *FEBS Lett* **581:** 4523–4528.

Gocheva V, Zeng W, Ke DX, Klimstra D, Reinheckel T, Peters C, Hanahan D, Joyce JA. 2006. Distinct roles for cysteine cathepsin genes in multistage tumorigenesis. *Genes Dev* **20:** 543–556.

Grady WM, Willis J, Guilford PJ, Dunbier AK, Toro TT, Lynch H, Wiesner G, Ferguson K, Eng C, Park JG, et al. 2000. Methylation of the CDH1 promoter as the second genetic hit in hereditary diffuse gastric cancer. *Nat Genet* **26:** 16–17.

Graff JR, Greenberg VE, Herman JG, Westra WH, Boghaert ER, Ain KB, Saji M, Zeiger MA, Zimmer SG, Baylin SB. 1998. Distinct patterns of E-cadherin CpG island methylation in papillary, follicular, Hurthle's cell, and poorly differentiated human thyroid carcinoma. *Cancer Res* **58:** 2063–2066.

Graff JR, Herman JG, Lapidus RG, Chopra H, Xu R, Jarrard DF, Isaacs WB, Pitha PM, Davidson NE, Baylin SB. 1995. E-cadherin expression is silenced by DNA hypermethylation in human breast and prostate carcinomas. *Cancer Res* **55:** 5195–5199.

Graff JR, Gabrielson E, Fujii H, Baylin SB, Herman JG. 2000. Methylation patterns of the E-cadherin 5′CpG island are unstable and reflect the dynamic, heterogeneous loss of E-cadherin expression during metastatic progression. *J Biol Chem* **275:** 2727–2732.

Gregory PA, Bert AG, Paterson EL, Barry SC, Tsykin A, Farshid G, Vadas MA, Khew-Goodall Y, Goodall GJ. 2008. The miR-200 family and miR-205 regulate epithelial to mesenchymal transition by targeting ZEB1 and SIP1. *Nat Cell Biol* **10:** 593–601.

Guaita S, Puig I, Franci C, Garrido M, Dominguez D, Batlle E, Sancho E, Dedhar S, De Herreros AG, Baulida J. 2002. Snail induction of epithelial to mesenchymal transition in tumor cells is accompanied by MUC1 repression and ZEB1 expression. *J Biol Chem* **277:** 39209–39216.

Guilford P, Hopkins J, Harraway J, McLeod M, McLeod N, Harawira P, Taite H, Scoular R, Miller A, Reeve AE. 1998. E-cadherin germline mutations in familial gastric cancer. *Nature* **392:** 402–405.

Harvey K, Tapon N. 2007. The Salvador-Warts-Hippo pathway - an emerging tumour-suppressor network. *Nat Rev Cancer* **7:** 182–191.

Hazan RB, Kang L, Whooley BP, Borgen PI. 1997. N-cadherin promotes adhesion between invasive breast cancer cells and the stroma. *Cell Adhes Commun* **4:** 399–411.

Hazan RB, Phillips GR, Qiao RF, Norton L, Aaronson SA. 2000. Exogenous expression of N-cadherin in breast cancer cells induces cell migration, invasion, and metastasis. *J Cell Biol* **148:** 779–790.

Hazan RB, Qiao R, Keren R, Badano I, Suyama K. 2004. Cadherin switch in tumor progression. *Ann N Y Acad Sci* **1014:** 155–163.

Hebbard LW, Garlatti M, Young LJT, Cardiff RD, Oshima RG, Ranscht B. 2008. T-cadherin supports angiogenesis and adiponectin association with the vasculature in a mouse mammary tumor model. *Cancer Res* **68:** 1407–1416.

Hendrix MJ, Seftor EA, Meltzer PS, Gardner LM, Hess AR, Kirschmann DA, Schatteman GC, Seftor RE. 2001. Expression and functional significance of VE-cadherin in aggressive human melanoma cells: Role in vasculogenic mimicry. *Proc Natl Acad Sci USA* **98:** 8018–8023.

Hibi K, Kodera Y, Ito K, Akiyama S, Nakao A. 2004. Methylation pattern of CDH13 gene in digestive tract cancers. *Br J Cancer* **91:** 1139–1142.

Huang ZY, Wu YL, Hedrick N, Gutmann DH. 2003. T-cadherin-mediated cell growth regulation involves G(2) phase arrest and requires p21 (CIP1/WAF1) expression. *Mol Cell Biol* **23:** 566–578.

Hug C, Wang J, Ahmad NS, Bogan JS, Tsao TS, Lodish HF. 2004. T-cadherin is a receptor for hexameric and high-molecular- weight forms of Acrp30/adiponectin. *Proc Natl Acad Sci USA* **101:** 10308–10313.

Hulit J, Suyama K, Chung S, Keren R, Agiostratidou G, Shan W, Dong X, Williams TM, Lisanti MP, Knudsen K, et al. 2007. N-cadherin signaling potentiates mammary tumor metastasis via enhanced extracellular signal-regulated kinase activation. *Cancer Res* **67:** 3106–3116.

Hulpiau P, van Roy F. 2009. Molecular evolution of the cadherin superfamily. *Int J Biochem Cell Biol* **41:** 343–369.

Hurteau GJ, Carlson JA, Spivack SD, Brock GJ. 2007. Overexpression of the microRNA hsa-miR-200c leads to reduced expression of transcription factor 8 and increased expression of E-cadherin. *Cancer Res* **67:** 7972–7976.

Imoto I, Izumi H, Yokoi S, Hosoda H, Shibata T, Hosoda F, Ohki N, Hirohashi S, Inazawa J. 2006. Frequent silencing of the candidate tumor suppressor PCDH20 by epigenetic mechanism in non-small-cell lung cancers. *Cancer Res* **66:** 4617–4626.

Jeanes A, Gottardi CJ, Yap AS. 2008. Cadherins and cancer: How does cadherin dysfunction promote tumor progression? *Oncogene* **27:** 6920–6929.

Jin Z, Cheng YL, Olaru A, Kan T, Yang J, Paun B, Ito T, Hamilton JP, David S, Agarwal R, et al. 2008. Promoter hypermethylation of CDH13 is a common, early event in human esophageal adenocarcinogenesis and correlates with clinical risk factors. *Int J Cancer* **123:** 2331–2336.

Johnson SK, Ramani VC, Hennings L, Haun RS. 2007. Kallikrein 7 enhances pancreatic cancer cell invasion by shedding E-cadherin. *Cancer* **109:** 1811–1820.

Jones S, Zhang XS, Parsons DW, Lin JCH, Leary RJ, Angenendt P, Mankoo P, Carter H, Kamiyama H, Jimeno A, et al. 2008. Core signaling pathways in human pancreatic cancers revealed by global genomic analyses. *Science* **321:** 1801–1806.

Kanai Y, Ushijima S, Tsuda H, Sakamoto M, Hirohashi S. 2000. Aberrant DNA methylation precedes loss of heterozygosity on chromosome 16 in chronic hepatitis and liver cirrhosis. *Cancer Lett* **148:** 73–80.

Kanazawa T, Watanabe T, Kazama S, Tada T, Koketsu S, Nagawa H. 2002. Poorly differentiated adenocarcinoma and mucinous carcinoma of the colon and rectum show higher rates of loss of heterozygosity and loss of E-cadherin expression due to methylation of promoter region. *Int J Cancer* **102:** 225–229.

Katafiasz BJ, Nieman MT, Wheelock MJ, Johnson KR. 2003. Characterization of cadherin-24, a novel alternatively spliced type II cadherin. *J Biol Chem* **278:** 27513–27519.

Kawakami M, Staub J, Cliby W, Hartmann L, Smith DI, Shridhar V. 1999. Involvement of H-cadherin (CDH13) on 16q in the region of frequent deletion in ovarian cancer. *Int J Oncol* **15:** 715–720.

Kim JS, Han JH, Shim YM, Park J, Kim DH. 2005. Aberrant methylation of H-Cadherin (CDH13) promoter is associated with tumor progression in primary nonsmall cell lung carcinoma. *Cancer* **104:** 1825–1833.

Kleer CG, van Golen KL, Braun T, Merajver SD. 2001. Persistent E-cadherin expression in inflammatory breast cancer. *Modern Pathol* **14:** 458–464.

Klucky B, Mueller R, Vogt I, Teurich S, Hartenstein B, Breuhahn K, Flechtenmacher C, Angel P, Hess J. 2007. Kallikrein 6 induces E-cadherin shedding and promotes cell proliferation, migration, and invasion. *Cancer Res* **67:** 8198–8206.

Knudsen KA, Sauer C, Johnson KR, Wheelock MJ. 2005. Effect of N-cadherin misexpression by the mammary epithelium in mice. *J Cell Biochem* **95:** 1093–1107.

Koizume S, Tachibana K, Sekiya T, Hirohashi S, Shiraishi M. 2002. Heterogeneity in the modification and involvement of chromatin components of the CpG island of the silenced human CDH1 gene in cancer cells. *Nucleic Acids Res* **30:** 4770–4780.

Kondo T, Ezzat S, Asa SL. 2006. Pathogenetic mechanisms in thyroid follicular-cell neoplasia. *Nat Rev Cancer* **6:** 292–306.

Korpal M, Lee ES, Hu G, Kang Y. 2008. The miR-200 family inhibits epithelial-mesenchymal transition and cancer cell migration by direct targeting of E-cadherin transcriptional repressors ZEB1 and ZEB2. *J Biol Chem* **283:** 14910–14914.

Krishnamachary B, Zagzag D, Nagasawa H, Rainey K, Okuyama H, Baek JH, Semenza GL. 2006. Hypoxia-inducible factor-1-dependent repression of E-cadherin in von Hippel-Lindau tumor suppressor-null renal cell carcinoma mediated by TCF3, ZFHX1A, and ZFHX1B. *Cancer Res* **66:** 2725–2731.

Lee SW. 1996. H-cadherin, a novel cadherin with growth inhibitory functions and diminished expression in human breast cancer. *Nat Med* **2:** 776–782.

Lee SW, Reimer CL, Campbell DB, Cheresh P, Duda RB, Kocher O. 1998. H-cadherin expression inhibits in vitro invasiveness and tumor formation in vivo. *Carcinogenesis (London)* **19:** 1157–1159.

Ley TJ, Mardis ER, Ding L, Fulton B, McLellan MD, Chen K, Dooling D, Dunford-Shore BH, McGrath S, Hickenbotham M, et al. 2008. DNA sequencing of a cytogenetically normal acute myeloid leukaemia genome. *Nature* **456:** 66–72.

Liang Y, Ridzon D, Wong L, Chen C. 2007. Characterization of microRNA expression profiles in normal human tissues. *BMC Genomics* **8:** 166.

Lochter A, Galosy S, Muschler J, Freedman N, Werb Z, Bissell MJ. 1997. Matrix metalloproteinase stromelysin-1 triggers a cascade of molecular alterations that leads to stable epithelial-to-mesenchymal conversion and a pre-malignant phenotype in mammary epithelial cells. *J Cell Biol* **139:** 1861–1872.

Loges S, Roncal C, Carmeliet P. 2008. Development of targeted angiogenic medicine. *J Thromb Haemost* **7:** 21–33.

Lombaerts M, van Wezel T, Philippo K, Dierssen JWF, Zimmerman RME, Oosting J, van Eijk R, Eilers PH, van De Water B, Cornelisse CJ, et al. 2006. E-cadherin transcriptional downregulation by promoter methylation but not mutation is related to epithelial-to-mesenchymal transition in breast cancer cell lines. *Br J Cancer* **94:** 661–671.

Lopez T, Hanahan D. 2002. Elevated levels of IGF-1 receptor convey invasive and metastatic capability in a mouse model of pancreatic islet tumorigenesis. *Cancer Cell* **1:** 339–353.

Lubensky IA, Gnarra JR, Bertheau P, Walther MM, Linehan WM, Zhuang Z. 1996. Allelic deletions of the VHL gene detected in multiple microscopic clear cell renal lesions in von Hippel-Lindau disease patients. *Am J Pathol* **149:** 2089–2094.

Machado JC, Oliveira C, Carvalho R, Soares P, Berx G, Caldas C, Seruca R, Carneiro F, Sobrinho-Simões M. 2001. E-cadherin gene (CDH1) promoter methylation as the second hit in sporadic diffuse gastric carcinoma. *Oncogene* **20:** 1525–1528.

Mani SA, Guo W, Liao MJ, Eaton EN, Ayyanan A, Zhou AY, Brooks M, Reinhard F, Zhang CC, Shipitsin M, et al. 2008. The epithelial-mesenchymal transition generates cells with properties of stem cells. *Cell* **133:** 704–715.

Maniotis AJ, Folberg R, Hess A, Seftor EA, Gardner LM, Pe'er J, Trent JM, Meltzer PS, Hendrix MJ. 1999. Vascular channel formation by human melanoma cells in vivo and in vitro: Vasculogenic mimicry. *Am J Pathol* **155:** 739–752.

Mansouri M, Rose PP, Moses AV, Fruh K. 2008. Remodeling of endothelial adherens junctions by Kaposi's sarcoma-associated herpesvirus. *J Virol* **82:** 9615–9628.

Marambaud P, Wen PH, Dutt A, Shioi J, Takashima A, Siman R, Robakis NK. 2003. A CBP binding transcriptional repressor produced by the PS1/epsilon-cleavage of N-cadherin is inhibited by PS1 FAD mutations. *Cell* **114:** 635–645.

Mariotti A, Perotti A, Sessa C, Ruegg C. 2007. N-cadherin as a therapeutic target in cancer. *Expert Opin Investig Drug* **16:** 451–465.

Martin TA, Goyal A, Watkins G, Jiang WG. 2005. Expression of the transcription factors snail, slug, and twist and their clinical significance in human breast cancer. *Ann Surg Oncol* **12:** 488–496.

Maruyama R, Toyooka S, Toyooka KO, Harada K, Virmani AK, Zochbauer Muller S, Farinas AJ, Vakar Lopez F, Minna JD, Sagalowsky A, et al. 2001. Aberrant promoter methylation profile of bladder cancer and its relationship to clinicopathological features. *Cancer Res* **61:** 8659–8663.

Masciari S, Larsson N, Senz J, Boyd N, Kaurah P, Kandel MJ, Harris LN, Pinheiro HC, Troussard A, Miron P, et al. 2007. Germline E-cadherin mutations in familial lobular breast cancer. *J Med Genet* **44:** 726–731.

Matakatsu H, Blair SS. 2006. Separating the adhesive and signaling functions of the Fat and Dachsous protocadherins. *Development* **133:** 2315–2324.

May C, Doody JF, Abdullah R, Balderes P, Xu X, Chen CP, Zhu Z, Shapiro L, Kussie P, Hicklin DJ, et al. 2005. Identification of a transiently exposed VE-cadherin epitope that allows for specific targeting of an antibody to the tumor neovasculature. *Blood* **105:** 4337–4344.

Miotto E, Sabbioni S, Veronese A, Callin GA, Gullini CS, Liboni A, Gramantieri L, Bolondi L, Ferrazzi E, Gafa R, et al. 2004. Frequent aberrant methylation of the CDH4 gene-promoter in human colorectal and gastric cancer. *Cancer Res* **64:** 8156–8159.

Miyamoto K, Fukutomi T, Akashi-Tanaka S, Hasegawa T, Asahara T, Sugimura T, Ushijima T. 2005. Identification of 20 genes aberrantly methylated in human breast cancers. *Int J Cancer* **116:** 407–414.

Mohamed MM, Sloane BF. 2006. Cysteine cathepsins: Multifunctional enzymes in cancer. *Nat Rev Cancer* **6:** 764–775.

Moody SE, Perez D, Pan TC, Sarkisian CJ, Portocarrero CP, Sterner CJ, Notorfrancesco KL, Cardiff RD, Chodosh LA. 2005. The transcriptional repressor snail promotes mammary tumor recurrence. *Cancer Cell* **8:** 197–209.

Morel AP, Lievre M, Thomas C, Hinkal G, Ansieau S, Puisieux A. 2008. Generation of breast cancer stem cells through epithelial-mesenchymal transition. *PLoS ONE* **3:** e2888.

Mukoyama Y, Zhou SX, Miyachi Y, Matsuyoshi N. 2005. T-cadherin negatively regulates the proliferation of cutaneous squamous carcinoma cells. *J Invest Dermatol* **124:** 833–838.

Najy AJ, Day KC, Day ML. 2008. The ectodomain shedding of E-cadherin by ADAM15 supports ErbB receptor activation. *J Biol Chem* **283:** 18393–18401.

Nam JS, Ino Y, Kanai Y, Sakamoto M, Hirohashi S. 2004. 5-Aza-2′-deoxycytidine restores the E-cadherin system in E- cadherin-silenced cancer cells and reduces cancer metastasis. *Clin Exp Metastas* **21:** 49–56.

Naora H, Montell DJ. 2005. Ovarian cancer metastasis: Integrating insights from disparate model organisms. *Nat Rev Cancer* **5:** 355–366.

Nass SJ, Herman JG, Gabrielson E, Iversen PW, Parl FF, Davidson NE, Graff JR. 2000. Aberrant methylation of the estrogen receptor and E-cadherin 5 'CpG islands increases with malignant progression in human breast cancer. *Cancer Res* **60:** 4346–4348.

Natt E, Magenis RE, Zimmer J, Mansouri A, Scherer G. 1989. Regional assignment of the human loci for uvomorulin (UVO) and chymotrypsinogen B (CTRB) with the help of two overlapping deletions on the long arm of chromosome 16. *Cytogenet Cell Genet* **50:** 145–148.

Navaratna D, Maestas J, McGuire PG, Das A. 2008. Suppression of retinal neovascularization with an antagonist to vascular endothelial cadherin. *Arch Ophthalmol* **126:** 1082–1088.

Neben K, Korshunov A, Benner A, Wrobel G, Hahn M, Kokocinski F, Golanov A, Joos S, Lichter P. 2004. Microarray-based screening for molecular markers in medulloblastoma revealed STK15 as independent predictor for survival. *Cancer Res* **64:** 3103–3111.

Nieman MT, Prudoff RS, Johnson KR, Wheelock MJ. 1999. N-cadherin promotes motility in human breast cancer cells regardless of their E-cadherin expression. *J Cell Biol* **147:** 631–643.

Noë V, Fingleton B, Jacobs K, Crawford HC, Vermeulen S, Steelant W, Bruyneel E, Matrisian LM, Mareel M. 2001. Release of an invasion promoter E-cadherin fragment by matrilysin and stromelysin-1. *J Cell Sci* **114:** 111–118.

Ogama Y, Ouchida M, Yoshino T, Ito S, Takimoto H, Shiote Y, Ishimaru F, Harada M, Tanimoto M, Shimizu K. 2004. Prevalent hyper-methylation of the CDH13 gene promoter in malignant B cell lymphomas. *Int J Oncol* **25:** 685–691.

Okazaki N, Takahashi N, Kojima S, Masuho Y, Koga H. 2002. Protocadherin LKC, a new candidate for a tumor suppressor of colon and liver cancers, its association with contact inhibition of cell proliferation. *Carcinogenesis* **23:** 1139–1148.

Oliveira C, Seruca R, Carneiro F. 2006. Genetics, pathology, and clinics of familial gastric cancer. *Int J Surg Pathol* **14:** 21–33.

Paredes J, Stove C, Stove V, Milanezi F, VanMarck V, Derycke L, Mareel M, Bracke M, Schmitt F. 2004. P-cadherin is up-regulated by the antiestrogen ICI 182,780 and promotes invasion of human breast cancer cells. *Cancer Res* **64:** 8309–8317.

Paredes J, Albergaria A, Oliveira JT, Jeronimo C, Milanezi F, Schmitt FC. 2005. P-cadherin overexpression is an indicator of clinical outcome in invasive breast carcinomas and is associated with CDH3 promoter hypomethylation. *Clin Cancer Res* **11:** 5869–5877.

Paredes J, Correia AL, Ribeiro AS, Albergaria A, Milanezi F, Schmitt FC. 2007. P-cadherin expression in breast cancer: A review. *Breast Cancer Res* **9:** 214.

Park SM, Gaur AB, Lengyel E, Peter ME. 2008. The miR-200 family determines the epithelial phenotype of cancer cells by targeting the E-cadherin repressors ZEB1 and ZEB2. *Genes Dev* **22:** 894–907.

Peinado H, Olmeda D, Cano A. 2007. Snail, Zeb and bHLH factors in tumour progression: An alliance against the epithelial phenotype? *Nat Rev Cancer* **7:** 415–428.

Pereira PS, Teixeira A, Ferreira P, Fernandes J, Oliveira C, Seruca R, Suriano G, Casares F. 2006. E-cadherin missense mutations, associated with hereditary diffuse gastric cancer (HDGC) syndrome, display distinct invasive behaviors and genetic interactions with the Wnt and Notch pathways in *Drosophila* epithelia. *Hum Mol Genet* **15:** 1704–1712.

Perl AK, Wilgenbus P, Dahl U, Semb H, Christofori G. 1998. A causal role for E-cadherin in the transition from adenoma to carcinoma. *Nature* **392:** 190–193.

Pharoah PDP, Guilford P, Caldas C. 2001. Incidence of gastric cancer and breast cancer in CDH1 (E- cadherin) mutation carriers from hereditary diffuse gastric cancer families. *Gastroenterology* **121:** 1348–1353.

Philippova M, Banfi A, Ivanov D, Gianni-Barrera R, Allenspach R, Erne P, Resink T. 2006. Atypical GPI-anchored T-cadherin stimulates angiogenesis in vitro and in vivo. *Arterioscler Thromb Vasc Biol* **26:** 2222–2230.

Philippova M, Ivanov D, Joshi MB, Kyriakakis E, Rupp K, Afonyushkin T, Bochkov V, Erne P, Resink TJ. 2008. Identification of proteins associating with glycosylphosphatidylinositol-anchored T-cadherin on the surface of vascular endothelial cells: Role for Grp78/BiP in T-cadherin-dependent cell survival. *Mol Cell Biol* **28:** 4004–4017.

Pishvaian MJ, Feltes CM, Thompson P, Bussemakers MJ, Schalken JA, Byers SW. 1999. Cadherin-11 is expressed in invasive breast cancer cell lines. *Cancer Res* **59:** 947–952.

Placencio VR, Sharif-Afshar AR, Li XH, Huang HX, Uwamariya C, Neilson EG, Shen NM, Matusik RJ, Hayward SW, Bhowmick NA. 2008. Stromal transforming growth factor-β signaling mediates prostatic response to androgen ablation by paracrine Wnt activity. *Cancer Res* **68:** 4709–4718.

Qi C, Zhu YT, Hu L, Zhu YJ. 2009. Identification of Fat4 as a candidate tumor suppressor gene in breast cancers. *Int J Cancer* **124:** 793–798.

Ranscht B, Dours-Zimmermann MT. 1991. T-cadherin, a novel cadherin cell adhesion molecule in the nervous system lacks the conserved cytoplasmic region. *Neuron* **7:** 391–402.

Redies C, Vanhalst K, van Roy F. 2005. Delta-protocadherins: Unique structures and functions. *Cell Mol Life Sci* **62:** 2840–2852.

Riou P, Saffroy R, Chenailler C, Franc B, Gentile C, Rubinstein E, Resink T, Debuire B, PiatierTonneau D, Lemoine A. 2006. Expression of T-cadherin in tumor cells influences invasive potential of human hepatocellular carcinoma. *FASEB J* **20:** 2291–2301.

Risinger JI, Berchuck A, Kohler MF, Boyd J. 1994. Mutations of the E-cadherin gene in human gynecologic cancers. *Nat Genet* **7:** 98–102.

Roman-Gomez J, Castillejo JA, Jimenez A, Cervantes F, Boque C, Hermosin L, Leon A, Granena A, Colomer D, Heiniger A, et al. 2003. Cadherin-13, a mediator of calcium-dependent cell-cell adhesion, is silenced by methylation in chronic myeloid leukemia and correlates with pretreatment risk profile and cytogenetic response to interferon alfa. *J Clin Oncol* **21:** 1472–1479.

Rosivatz E, Becker I, Specht K, Fricke E, Luber B, Busch R, Hofler H, Becker KF. 2002. Differential expression of the epithelial-mesenchymal transition regulators Snail, SIP1, and Twist in gastric cancer. *Am J Pathol* **161:** 1881–1891.

Russell RC, Ohh M. 2007. The role of VHL in the regulation of E-cadherin - A new connection in an old pathway. *Cell Cycle* **6:** 56–59.

Saburi S, Hester I, Fischer E, Pontoglio M, Eremina V, Gessler M, Quaggin SE, Harrison R, Mount R, McNeill H. 2008. Loss of Fat4 disrupts PCP signaling and oriented cell division and leads to cystic kidney disease. *Nat Genet* **40:** 1010–1015.

Saito T, Oda Y, Sugimachi K, Kawaguchi K, Tamiya S, Tanaka K, Matsuda S, Sakamoto A, Iwamoto Y, Tsuneyoshi M. 2001. E-cadherin gene mutations frequently occur in synovial sarcoma as a determinant of histological features. *Am J Pathol* **159:** 2117–2124.

Sakai M, Hibi K, Koshikawa K, Inoue S, Takeda S, Kaneko T, Nakao A. 2004. Frequent promoter methylation and gene silencing of CDH13 in pancreatic cancer. *Cancer Sci* **95:** 588–591.

Sanders DSA, Blessing K, Hassan GAR, Bruton R, Marsden JR, Jankowski J. 1999. Alterations in cadherin and catenin expression during the biological progression of melanocytic tumours. *J Clin Pathol Mol Pathol* **52:** 151–157.

Sarrio D, Rodriguez-Pinilla SM, Hardisson D, Cano A, Moreno-Bueno G, Palacios J. 2008. Epithelial-mesenchymal transition in breast cancer relates to the basal-like phenotype. *Cancer Res* **68:** 989–997.

Sato M, Mori Y, Sakurada A, Fujimura S, Horii A. 1998. The H-cadherin (CDH13) gene is inactivated in human lung cancer. *Hum Genet* **103:** 96–101.

Schweizer L, Rizzo CA, Spires TE, Platero JS, Wu Q, Lin TA, Gottardis MM, Attar RM. 2008. The androgen receptor can signal through Wnt/β-Catenin in prostate cancer cells as an adaptation mechanism to castration levels of androgens. *BMC Cell Biol* **9:** 4.

Shah AN, Gallick GE. 2007. Src, chemoresistance and epithelial to mesenchymal transition: Are they related? *Anti-cancer drugs* **18:** 371–375.

Shen Y, Hirsch DS, Sasiela CA, Wu WJ. 2008. Cdc42 regulates E-cadherin ubiquitination and degradation through an epidermal growth factor receptor to Src-mediated pathway. *J Biol Chem* **283:** 5127–5137.

Shintani Y, Fukumoto Y, Chaika N, Grandgenett PM, Hollingsworth MA, Wheelock MJ, Johnson KR. 2008. ADH-1 suppresses N-cadherin-dependent pancreatic cancer progression. *Int J Cancer* **122:** 71–77.

Simoes-Correia J, Figueiredo J, Oliveira C, van Hengel J, Seruca R, van Roy F, Suriano G. 2008. Endoplasmic reticulum quality control: A new mechanism of E-cadherin regulation and its implication in cancer. *Hum Mol Genet* **17:** 3566–3576.

Soares P, Berx G, van Roy F, Sobrinho-Simões M. 1997. E-cadherin gene alterations are rare events in thyroid tumors. *Int J Cancer* **70:** 32–38.

Spaderna S, Schmalhofer O, Hlubek F, Berx G, Eger A, Merkel S, Jung A, Kirchner T, Brabletz T. 2006. A transient, EMT-linked loss of basement membranes indicates metastasis and poor survival in colorectal cancer. *Gastroenterology* **131:** 830–840.

Spaderna S, Schmalhofer O, Wahlbuhl M, Dimmler A, Bauer K, Sultan A, Hlubek F, Jung A, Strand D, Eger A, et al. 2008. The transcriptional repressor ZEB1 promotes metastasis and loss of cell polarity in cancer. *Cancer Res* **68:** 537–544.

St Croix BS, Kerbel RS. 1997. Cell adhesion and drug resistance in cancer. *Curr Opin Oncol* **9:** 549–556.

Strathdee G. 2002. Epigenetic versus genetic alterations in the inactivation of E-cadherin. *Semin Cancer Biol* **12:** 373–379.

Strumane K, Berx G, van Roy F. 2004. Cadherins in cancer. *In Cell adhesion* (ed. Behrens J., Nelson J.), Handbook of Experimental Pharmacology Vol. 165, pp. 69–103. Springer-Verlag, Heidelberg.

Strumane K, Bonnomet A, Stove C, Vandenbroucke R, Nawrocki-Raby B, Bruyneel E, Mareel M, Birembaut P, Berx G, van Roy F. 2006. E-cadherin regulates human Nanos1, which interacts with p120ctn and induces tumor cell migration and invasion. *Cancer Res* **66:** 10007–10015.

Sundfeldt K. 2003. Cell-cell adhesion in the normal ovary and ovarian tumors of epithelial origin; an exception to the rule. *Mol Cell Endocrinol* **202:** 89–96.

Suriano G, Yew S, Ferreira P, Senz J, Kaurah P, Ford JM, Longacre TA, Norton JA, Chun N, Young S, et al. 2005. Characterization of a recurrent germ line mutation of the E-cadherin gene: Implications for genetic testing and clinical management. *Clin Cancer Res* **11:** 5401–5409.

Suyama K, Shapiro I, Guttman M, Hazan RB. 2002. A signaling pathway leading to metastasis is controlled by N-cadherin and the FGF receptor. *Cancer Cell* **2:** 301–314.

Symowicz J, Adley BP, Gleason KJ, Johnson JJ, Ghosh S, Fishman DA, Hudson LG, Stack MS. 2007. Engagement of collagen-binding integrins promotes matrix metalloproteinase-9-dependent E-cadherin ectodomain shedding in ovarian carcinoma cells. *Cancer Res* **67:** 2030–2039.

Taddei I, Piazzini M, Bartoletti R, Dal Canto M, Sardi I. 2000. Molecular alterations of E-cadherin gene: Possible role in human bladder carcinogenesis. *Int J Mol Med* **6:** 201–208.

Takeuchi T, Ohtsuki Y. 2001. Recent progress in T-cadherin (CDH13, H-cadherin) research. *Histol Histopathol* **16:** 1287–1293.

Takeuchi T, Liang SB, Matsuyoshi N, Zhou SX, Miyachi Y, Sonobe H, Ohtsuki Y. 2002a. Loss of T-cadherin (CDH13, H-cadherin) expression in cutaneous squamous cell carcinoma. *Lab Invest* **82:** 1023–1029.

Takeuchi T, Liang SB, Ohtsuki Y. 2002b. Downregulation of expression of a novel cadherin molecule, T- cadherin, in basal cell carcinoma of the skin. *Mol Carcinogen* **35:** 173–179.

Takeuchi T, Adachi Y, Ohtsuki Y, Furihata M. 2007. Adiponectin receptors, with special focus on the role of the third receptor, T-cadherin, in vascular disease. *Med Mol Morphol* **40:** 115–120.

Tamura D, Hiraga T, Myoui A, Yoshikawa H, Yoneda T. 2008. Cadherin-11-mediated interactions with bone marrow stromal/osteoblastic cells support selective colonization of breast cancer cells in bone. *Int J Oncol* **33:** 17–24.

Taniuchi K, Nakagawa H, Hosokawa M, Nakamura T, Eguchi H, Ohigashi F, Ishikawa O, Katagiri T, Nakamura Y. 2005. Overexpressed P-cadherin/CDH3 promotes motility of pancreatic cancer cells by interacting with p120ctn and activating Rho-family GTPases. *Cancer Res* **65:** 3092–3099.

Terry S, Queires L, Gil-Diez-de-Medina S, Chen MW, de la Taille A, Allory Y, Tran PL, Abbou CC, Buttyan R, Vacherot F. 2006. Protocadherin-PC promotes androgen-independent prostate cancer cell growth. *Prostate* **66:** 1100–1113.

Thomas G, Jacobs KB, Yeager M, Kraft P, Wacholder S, Orr N, Yu K, Chatterjee N, Welch R, Hutchinson A, et al.

2008. Multiple loci identified in a genome-wide association study of prostate cancer. *Nat Genet* **40:** 310–315.

Tinkle CL, Lechler T, Pasolli HA, Fuchs E. 2004. Conditional targeting of E-cadherin in skin: Insights into hyperproliferative and degenerative responses. *Proc Natl Acad Sci USA* **101:** 552–557.

Tomita K, van Bokhoven A, van Leenders G, Ruijter ETG, Jansen CFJ, Bussemakers MJG, Schalken JA. 2000. Cadherin switching in human prostate cancer progression. *Cancer Res* **60:** 3650–3654.

Toyooka KO, Toyooka S, Virmani AK, Sathyanarayana UG, Euhus DM, Gilcrease M, Minna JD, Gazdar AF. 2001. Loss of expression and aberrant methylation of the CDH13 (H-cadherin) gene in breast and lung carcinomas. *Cancer Res* **61:** 4556–4560.

Toyooka S, Toyooka KO, Harada K, Miyajima K, Makarla P, Sathyanarayana UG, Yin J, Sato F, Shivapurkar N, Meltzer SJ, et al. 2002. Aberrant methylation of the CDH13 (H-cadherin) promoter region in colorectal cancers and adenomas. *Cancer Res* **62:** 3382–3386.

Tunggal JA, Helfrich I, Schmitz A, Schwarz H, Gunzel D, Fromm M, Kemler R, Krieg T, Niessen CM. 2005. E-cadherin is essential for in vivo epidermal barrier function by regulating tight junctions. *EMBO J* **24:** 1146–1156.

Unterseher F, Hefele JA, Giehl K, DeRobertis EM, Wedlich D, Schambony A. 2004. Paraxial protocadherin coordinates cell polarity during convergent extension via Rho A and JNK. *EMBO J* **23:** 3259–3269.

van Hengel J, van Roy F. 2007. Diverse functions of p120ctn in tumors. *Biochim Biophys Acta—Mol Cell Res* **1773:** 78–88.

Van Marck V, Stove C, Van Den Bossche K, Stove V, Paredes J, Vander Haeghen Y, Bracke M. 2005. P-cadherin promotes cell-cell adhesion and counteracts invasion in human melanoma. *Cancer Res*: 8774–8783.

van Roy F, Berx G. 2008. The cell-cell adhesion molecule E-cadherin. *Cell Mol Life Sci* **65:** 3756–3788.

Vandewalle C, Comijn J, De Craene B, Vermassen P, Bruyneel E, Andersen H, Tulchinsky E, van Roy F, Berx G. 2005. SIP1/ZEB2 induces EMT by repressing genes of different epithelial cell-cell junctions. *Nucleic Acids Res* **33:** 6566–6578.

Vanhalst K, Kools P, Staes K, van Roy F, Redies C. 2005. Delta-protocadherins: A gene family expressed differentially in the mouse brain. *Cell Mol Life Sci* **62:** 1247–1259.

Vecsey-Semjen B, Becker KF, Sinski A, Blennow E, Vietor I, Zatloukal K, Beug H, Wagner E, Huber LA. 2002. Novel colon cancer cell lines leading to better understanding of the diversity of respective primary cancers. *Oncogene* **21:** 4646–4662.

Vestal DJ, Ranscht B. 1992. Glycosyl phosphatidylinositol-anchored T-cadherin mediates calcium-dependent, homophilic cell adhesion. *J Cell Biol* **119:** 451–461.

Vleminckx K, Vakaet L Jr, Mareel M, Fiers W, van Roy F. 1991. Genetic manipulation of E-cadherin expression by epithelial tumor cells reveals an invasion suppressor role. *Cell* **66:** 107–119.

Waha A, Guntner S, Huang THM, Yan PS, Arslan B, Pietsch T, Wiestler OD. 2005. Epigenetic silencing of the protocadherin family member PCDH-γ-A11 in astrocytomas. *Neoplasia* **7:** 193–199.

Wang XD, Wang BE, Soriano R, Zha JP, Zhang ZM, Modrusan Z, Cunha GR, Gao WQ. 2007. Expression profiling of the mouse prostate after castration and hormone replacement: Implication of H-cadherin in prostate tumorigenesis. *Differentiation* **75:** 219–234.

Wang G, Wang J, Sadar MD. 2008. Crosstalk between the androgen receptor and β-catenin in castrate-resistant prostate cancer. *Cancer Res* **68:** 9918–9927.

Warso MA, Maniotis AJ, Chen X, Majumdar D, Patel MK, Shilkaitis A, Gupta TK, Folberg R. 2001. Prognostic significance of periodic acid-Schiff-positive patterns in primary cutaneous melanoma. *Clin Cancer Res* **7:** 473–477.

Wheelock MJ, Shintani Y, Maeda M, Fukumoto Y, Johnson KR. 2008. Cadherin switching. *J Cell Sci* **121:** 727.

Wijnhoven BPL, de Both NJ, van Dekken H, Tilanus HW, Dinjens WNM. 1999. E-cadherin gene mutations are rare in adenocarcinomas of the oesophagus. *Br J Cancer* **80:** 1652–1657.

Wu Y, Starzinski-Powitz A, Guo SW. 2007. Trichostatin A, a histone deacetylase inhibitor, attenuates invasiveness and reactivates E-cadherin expression in immortalized endometriotic cells. *Reprod Sci* **14:** 374–382.

Wyder L, Vitaliti A, Schneider H, Hebbard LW, Moritz DR, Wittmer M, Ajmo M, Klemenz R. 2000. Increased expression of H/T-cadherin in tumor-penetrating blood vessels. *Cancer Res* **60:** 4682–4688.

Yan Q, Zhanb ZF, Chen XP, Gutmann DH, Xiong M, Xiao ZY, Huang ZY. 2008. Reduced T-cadherin expression and promoter methylation are associated with the development and progression of hepatocellular carcinoma. *Int J Oncol* **32:** 1057–1063.

Yang J, Mani SA, Donaher JL, Ramaswamy S, Itzykson RA, Come C, Savagner P, Gitelman I, Richardson A, Weinberg RA. 2004. Twist, a master regulator of morphogenesis, plays an essential role in tumor metastasis. *Cell* **117:** 927–939.

Yang X, Chen MW, Terry S, Vacherot F, Chopin DK, Bemis DL, Kitajewski J, Benson MC, Guo Y, Buttyan R. 2005. A human- and male-specific protocadherin that acts through the wnt signaling pathway to induce neuroendocrine transdifferentiation of prostate cancer cells. *Cancer Res* **65:** 5263–5271.

Yasuda S, Tanaka H, Sugiura H, Okamura K, Sakaguchi T, Tran U, Takemiya T, Mizoguchi A, Yagita Y, Sakurai T, et al. 2007. Activity-induced protocadherin arcadlin regulates dendritic spine number by triggering N-cadherin endocytosis via TAO2 β and p38 MAP kinases. *Neuron* **56:** 456–471.

Yauch RL, Januario T, Eberhard DA, Cavet G, Zhu W, Fu L, Pham TQ, Soriano R, Stinson J, Seshagiri S, et al. 2005. Epithelial versus mesenchymal phenotype determines in vitro sensitivity and predicts clinical activity of erlotinib in lung cancer patients. *Clin Cancer Res* **11:** 8686–8698.

Ying J, Li H, Seng TJ, Langford C, Srivastava G, Tsao SW, Putti T, Murray P, Chan ATC, Tao Q. 2006. Functional epigenetics identifies a protocadherin PCDH10 as a candidate tumor suppressor for nasopharyngeal, esophageal and multiple other carcinomas with frequent methylation. *Oncogene* **25:** 1070–1080.

Ying JM, Gao ZF, Li HY, Srivastava G, Murray PG, Goh HK, Lim CY, Wang YJ, Marafioti T, Mason DY, et al. 2007. Frequent epigenetic silencing of protocadherin 10 by methylation in multiple haematologic malignancies. *Br J Haematol* **136:** 829–832.

Yoshiura K, Kanai Y, Ochiai A, Shimoyama Y, Sugimura T, Hirohashi S. 1995. Silencing of the E-cadherin invasion-suppressor gene by CpG methylation in human carcinomas. *Proc Natl Acad Sci USA* **92:** 7416–7419.

Yu J, Cheng YY, Tao Q, Cheung KF, Lam CN, Geng H, Tian LW, Wong YP, Tong JH, Ying JM, et al. 2009. Methylation of protocadherin 10, a novel tumor suppressor, is associated with poor prognosis in patients with gastric cancer. *Gastroenterology* **136:** 640–651.e1.

Yu JS, Koujak S, Nagase S, Li CM, Su T, Wang X, Keniry M, Memeo L, Rojtman A, Mansukhani M, et al. 2008. PCDH8, the human homolog of PAPC, is a candidate tumor suppressor of breast cancer. *Oncogene* **27:** 4657–4665.

Zbuk KM, Eng C. 2007. Cancer phenomics: RET and PTEN as illustrative models. *Nat Rev Cancer* **7:** 35–45.

Zeng Q, Hong W. 2008. The emerging role of the Hippo pathway in cell contact inhibition, organ size control, and cancer development in mammals. *Cancer Cell* **13:** 188–192.

Zhao B, Lei QY, Guan KL. 2008. The Hippo-YAP pathway: New connections between regulation of organ size and cancer. *Curr Opin Cell Biol* **20:** 638–646.

Index

L

M

N

O

P

R

S

T

U

V

W

Z